...pidly expanding field that has a part to play ...d range of earth and planetary sciences – from extra-solar system processes to environmental geoscience. This book reviews the field of radiogenic isotope geology in a concise and visual manner to provide an introduction to the subject and its wide variety of applications. The author takes an historical approach and explains the development of the field and the origin of current ideas so that the latest models and theories can be easily understood. The latest ideas and methods, classic ... and case studies all come under scrutiny within this ...th the text gathered around classic diagrams from

Radiogenic Isotope Geology

Radiogenic Isotope Geology

ALAN P. DICKIN

Department of Geology
McMaster University, Hamilton, Ontario

PUBLISHED BY THE PRESS SYNDICATE OF THE UNIVERSITY OF CAMBRIDGE
The Pitt Building, Trumpington Street, Cambridge CB2 1RP, United Kingdom

CAMBRIDGE UNIVERSITY PRESS
The Edinburgh Building, Cambridge CB2 2RU, United Kingdom
40 West 20th Street, New York, NY 10011–4211, USA
10 Stamford Road, Oakleigh, Melbourne 3166, Australia

First published 1995
Reprinted 1997
First paperback edition (with corrections and additional material) 1997

Printed in the United Kingdom at the University Press, Cambridge

Typeset in Monotype Times 9/11 pt

A catalogue record for this book is available from the British Library

Library of Congress cataloguing in publication data

Dickin, Alan P.
Radiogenic isotope geology/Alan P. Dickin.
    p. cm.
Includes bibliographical references and index.
ISBN 0-521-43151-4
1. Isotope geology.   2. Radioactive dating.   I. Title.
QE501.4.N9D53   1995
551.7'01–dc20 93–36765 CIP

ISBN 0 521 43151 4 hardback
ISBN 0 521 59891 5 paperback

DEDICATION

to Margaret

Zmyrna mei Cinnae nonam post denique messem-
quam coepta est nonamque edita post heimem,
milia cum interea quingenta Hortensius uno
                                        Catullus

CONTENTS

PREFACE

The objective of this book is to concisely review the field of Radiogenic Isotope Geology in order to give readers an overview of the subject and to allow them to critically assess the past and future literature.

The approach is historical for three reasons. Firstly, to give an impression of the development of thought in the field so that the reader can understand the origin of present ideas; secondly to explain why past theories have had to be modified; and thirdly to present 'fall back' positions lest current models be refuted at some future date.

The text is focussed on three types of literature. Firstly, it attempts to give accurate attribution of new ideas or methods; secondly, it reviews classic papers, although these may not have historical precedence; and thirdly, it presents case studies that have evoked controversy in the literature, as examples of alternative data interpretations.

The book is organised so that each chapter is a nearly free-standing entity covering one segment of the field of Radiogenic Isotope Geology. However, the reader may benefit from an understanding of the thread, which, in the author's mind, links these chapters together.

Chapter 1 introduces radiogenic isotopes by discussing the synthesis and decay of nuclides within the context of nuclear stability. Decay constants and the radioactive decay law are introduced. Chapter 2 then provides an experimental background to many of the chapters that follow by discussing the details of mass spectrometric analysis and isochron regression fitting.

The next three chapters introduce the three pillars of lithophile isotope geology, comprising the Sr, Nd and Pb isotope methods. The emphasis is on their applications to geochronology and their evolution in terrestrial systems. Chapter 3 covers the Rb–Sr system, since this is one of the simplest and most basic dating methods. Attention is also focussed on seawater Sr evolution. Chapter 4 covers the Sm–Nd system, including the use of Nd model ages to date crustal formation. Chapter 5 examines U–Pb geochronology and introduces the complexities of terrestrial Pb isotope evolution in a straightforward fashion.

Chapters 6 and 7 apply these isotopic methods, as geochemical tracers, to the study of oceanic and continental igneous rocks. This is appropriate, because Sr, Nd and Pb isotopes are some of the basic tools of the isotope geochemist, which together may allow understanding of the complexities of mantle processes and magmatic evolution. These methods are supplemented in Chapters 8 and 9 by insights from the Re–Os, Lu–Hf, La–Ba, La–Ce and K–Ca systems, arising from their distinct chemistry.

Chapter 10 completes the panoply of long-lived isotopic dating systems by introducing the K–Ar and Ar–Ar methods, including their applications to magnetic and thermal histories. This leads us naturally in Chapter 11 to the consideration of rare gases as isotopic tracers, which give unique insights into the degassing history of the earth.

Chapter 12 introduces the short-lived isotopes of the uranium decay series, covering classical and recent developments in the dating of Quaternary-age sedimentary rocks. This prepares us for the complexities of Chapter 13, which examines U-series isotopes as tracers in igneous systems. Short-lived processes in mantle melting and magma evolution are the focus of attention.

Finally, in Chapters 14 to 16, we examine three specialised fields which are on the fringes of radiogenic isotope analysis, but which provide powerful geological tools. The cosmogenic isotopes covered in Chapter 14 are not 'radiogenic' in the strictest sense, but represent a vast and growing field of chronology and isotope chemistry which is especially pertinent to environmental geoscience. The 'extinct' nuclides discussed in Chapter 15 are used to throw light on the early evolution of the solar system. Lastly, Chapter 16 examines the use of (radiogenic) fission track analysis as a specialised dating tool for low-temperature thermal histories.

The text is gathered around a large number of diagrams, many of which are classic figures from the literature. In keeping with this visual emphasis, an attempt is made here to portray the historical scope of each chapter in terms of the publication date of cited figures. Because of the centrality of these figures in the text, they can chart the historical development of each isotopic method. The figure below shows a 'snap-shot' of this data set, in terms of the median source age and 1σ age distribution of figures in each chapter.

Essentially, we can see three generations of method development in the figure. Those such as K–Ar, Rb–Sr, Pb isotopes, Fission Track and (cosmogenic) C-14 reached prominence as dating tools in the 1960s and early 1970s. These methods have remained important, but often in more specialised applications than originally conceived. For example, Ar–Ar and Fission Track for cooling ages, U–Pb for dating plutonism, Sr for seawater evolution. A second group of methods reached prominence in the late 1970s and early 1980s. Examples are the Sm–Nd method, Rare Gas geochemistry and U-series nuclides. These techniques offered new and powerful isotopic tracers for analysing geological systems in the mantle and crust. Finally, the Re–Os method has risen to prominence within the last five years. Together, these methods (and other more specialised techniques) offer an analytical 'tool-box' of isotopic methods. The varied physical and chemical properties of different parent–daughter pairs suits them to a range of different dating and tracing tasks.

Together, they allow constraints to be placed on the behaviour of complex natural systems.

A new appendix, included at the end of the 1997 reprinting, reviews the most significant new work and ideas up to 1996. This appendix follows the same structure as the main text, and is cross-referenced to sections of the main text as appropriate. Numbered figures in the appendix refer to the main text unless an alternative author is named. References are listed at the end of the appendix, except for those denoted 'op. cit.', which were cited in the main text. Cross-references to other parts of the appendix are indicated by the prefix 'A'.

Appendix B, and further updates, can be viewed at the Radiogenic Isotope Geology Web Site using the URL shown below. In case of difficulty with this address, please follow the internal web hierarchy to Alan Dickin at McMaster University, Canada:

http://www.science.mcmaster.ca/Geology/faculty/dickin.html

Alan P. Dickin

McMaster University
June, 1994,
December, 1997

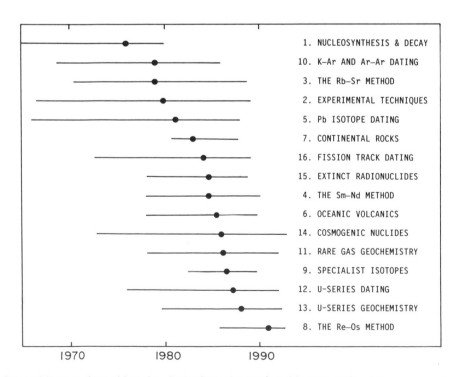

Plot of median publication dates (•) and 1σ limits from the mean publication date of figures reproduced in each chapter of Radiogenic Isotope Geology. Chapters are arranged in order of median publication date.

ACKNOWLEDGEMENTS

I am very grateful to Conrad Guettler for his encouragement to begin this work and continue with it during its lengthy evolution. I also thank many different members of the Geological Community for their helpful comments on the text during this evolution: Louise Corriveau, Mark Epp, Chris Hawkesworth, Scott Hay, Cherylyn Heikoop, Rosemary Lips, Joyce Lundberg, Franco Marcantonio, Chris Martin, Richard McLaughlin, Bob McNutt, Tim Schwartz, Paul Taylor and Derek Vance, along with anonymous reviewers. I thank Brian Watts for his helpful copy editing, and Catharina Jager for help with the references data-base. I am also grateful to many authors and publishers for permission to reproduce their classic figures, and to Jenny Bennet for her excellent re-drafting of these figures. Finally, and especially, I thank my family for their steadfast support and encouragement, without which this book would not have been completed.

I am very grateful to the publishers listed below for permission to redraw figures from journals and books for which they hold the copyright. Author acknowledgement for all figures is given within individual figure captions, and corresponding titles, journal names, volumes and pages are contained in the list of cited references at the end of each chapter.

	No of figures used
Elsevier Science Publishers B.V.	
Earth and Planetary Science Letters:	121
Chemical Geology:	3
Nuclear Instruments and Methods:	7
Lithos:	3
Tectonophysics:	1
Pergamon Press Ltd	
Geochimica et Cosmochimica Acta:	61
Nuclear Tracks:	6
Quaternary International:	1
Macmillan Magazines Ltd	
Nature:	49
American Geophysical Union	
Journal of Geophysical Research:	22
Geophysical Research Letters:	10
Water Resources Research:	1
American Association for the Advancement of Science	
Science:	16
Geological Society of America	
Geology:	9
Geological Society of America Bulletin:	2

No of figures used

Springer Verlag GmbH & Co.
Contributions to Mineralogy and Petrology: 8
Neodymium Isotopes in Geology (DePaolo, 1988):   1

Annual Reviews Inc.
Annual Review of Earth & Planetary Sciences: 4

Blackwell Scientific Publishers
Australian Journal of Earth Sciences: 3

The American Physical Society
Reviews of Modern Physics: 3

Blackie Ltd
Handbook of Silicate Rock Analysis (Potts, 1987):   2

National Research Council of Canada
Canadian Journal of Earth Science: 2

Society for Sedimentary Geology (SEPM)
Journal of Sedimentary Petrology: 2

The University of Chicago Press
Journal of Geology: 1
Radiocarbon Dating (Libby, 1955):   1

Academic Press Inc.
Icarus: 1

American Chemical Society
Environmental Science and Technology: 1

American Geological Institute
Geotimes   1

McGraw-Hill, Inc.
Ages of Rocks, Planets and Stars (Faul, 1966):   1

University of California Press
Nuclear Tracks in Solids (Fleischer *et al.*, 1975):   1

1 Nucleosynthesis and nuclear decay

1.1 The chart of the nuclides

In the field of isotope geology, neutrons, protons and electrons can be regarded as the fundamental building blocks of the atom. The composition of a given type of atom, called a nuclide, is described by specifying the number of protons (atomic number, Z) and the number of neutrons (N) in the nucleus. The sum of these is the mass number (A). By plotting Z against N for all of the nuclides that have been known to exist (at least momentarily), the chart of the nuclides is obtained (Fig. 1.1). In this chart, horizontal rows of nuclides represent the same element (constant Z) with a variable number of neutrons (N). These are isotopes.

Presently 264 stable nuclides are known, which have not been observed to decay (with available detection equipment). These define a central 'path of stability', coloured black in Fig. 1.1. On either side of this path, the zig-zag outline defines the limits of experimentally known unstable nuclides (Hansen, 1987). These tend to undergo increasingly rapid

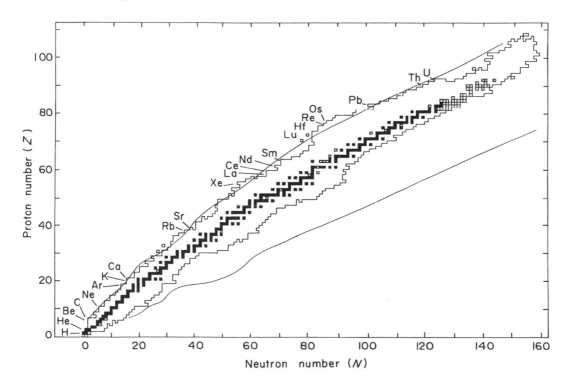

Fig. 1.1. Chart of the nuclides in coordinates of proton number, Z, against neutron number, N. (■) = stable nuclides; (□) = unstable nuclides; (▫) = naturally occurring long-lived unstable nuclides; (▣) = naturally occurring short-lived unstable nuclides. Some geologically useful radionuclides are marked. Smooth envelope = theoretical nuclide stability limits. For a more detailed nuclide chart, see the appendix A.

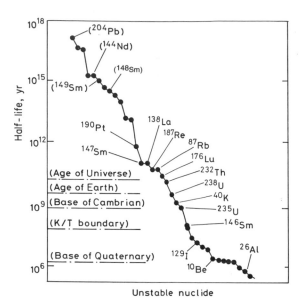

Fig. 1.2. Unstable nuclides with half-lives ($t_{1/2}$) over 0.5 Myr, in order of decreasing stability. Geologically useful parent nuclides are marked. Some very long-lived radionuclides with no geological application are also marked, in brackets.

decay as one moves out on either side of the path of stability. The smooth outer envelopes are the theoretical limits of nuclide stability, beyond which 'prompt' decay occurs. In that case the synthesis and decay of an unstable nuclide occurs in a single particle interaction, giving it a zero effective lifetime. As work progresses, the domain of experimentally known nuclides should approach the theoretical envelope, as has already occurred for nuclides with $Z < 22$ (Hansen, 1987).

A small number of unstable nuclides have sufficiently long half-lives that they have not entirely decayed to extinction since the formation of the solar system. A few other short-lived nuclides are either continuously generated in the decay series of uranium and thorium, or produced by cosmic ray bombardment of stable nuclides. These nuclides, and one or two extinct short-lived isotopes, plus their daughter products, are the realm of radiogenic isotope geology. Those with half-lives over 0.5 Myr are marked in Fig. 1.2. Nuclides with half-lives over 10^{12} yr decay too slowly to be geologically useful. Observation shows that all of the other long-lived isotopes either have been or are being applied in geology.

1.2 Nucleosynthesis

A realistic model for the nucleosynthesis of the elements must be based on empirical data for their 'cosmic abundance'. True cosmic abundances can be derived from stellar spectroscopy or by chemical analysis of galactic cosmic rays. However, such data are difficult to measure at high precision, so cosmic abundances are normally approximated by solar-system abundances. These can be determined by solar spectroscopy or by direct analysis of the most 'primitive' meteorites, carbonaceous chondrites. A comparison of the latter two sources of data (Ross and Aller, 1976) demonstrates good agreement for most elements (Fig. 1.3). Exceptions are the volatile elements, which have been lost from meteorites, and the Li–Be–B group, which are unstable in stars.

It is widely believed (e.g. Weinberg, 1977) that about 30 minutes after the 'hot big bang', the matter of the universe (in the form of protons and neutrons) consisted mostly of ^{1}H and 22–28% by mass of ^{4}He, along with traces of ^{2}H (deuterium) and ^{3}He. Hydrogen is still by far the most abundant element in the universe (88.6% of all nuclei) and with helium, makes up 99% of its mass, but naturally occurring heavy nuclides now exist up to atomic weight 254 or beyond (Fig. 1.1). These heavier nuclei must have been produced by nucleosynthetic processes in stars, and not in the big bang, because stars of different ages have different compositions which can be detected spectroscopically. Furthermore, stars at particular evolutionary stages may have compositional abnormalities, such as the presence of ^{254}Cf in supernovae. If nucleosynthesis of the heavy elements had occurred in the big bang then their distribution would be uniform about the universe.

1.2.1 Stellar evolution

Present-day models of stellar nucleosynthesis are based heavily on a classic review paper by Burbidge et al. (1957), in which eight element-building processes were identified (hydrogen burning, helium burning, α, e, s, r, x and p). Different processes were invoked to explain the abundance patterns of different groups of elements. These processes are, in turn, linked to different stages of stellar evolution. It is therefore appropriate at this point to summarise the life-history of some typical stars (e.g. Iben, 1967). The length of this life-history depends directly on the stellar mass, and can be traced on a plot of absolute magnitude (brightness) against spectral class (colour), referred to as the Hertzsprung–Russell or H–R diagram (Fig. 1.4).

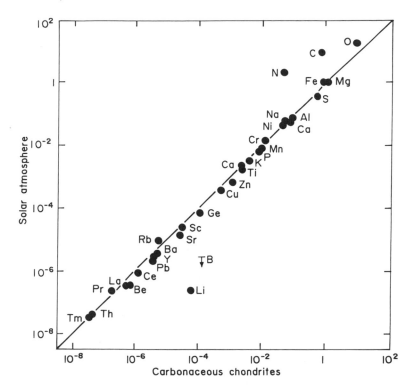

Fig. 1.3. Comparison of solar-system abundances (relative to silicon) determined by solar spectroscopy and by analysis of carbonaceous chondrites. After Ringwood (1979).

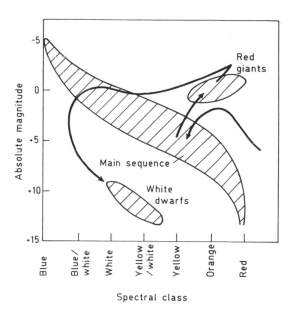

Fig. 1.4. Plot of absolute magnitude against spectral class of stars. Hatched areas show distributions of the three main star groups. The postulated evolutionary path of a star of solar mass is shown.

Gravitational accretion of a star of solar mass from cold primordial hydrogen and helium would probably take about 10^6 yr to raise the core temperature to ca. 10^7 K, when nuclear fusion of hydrogen to helium can begin (Atkinson and Houtermans, 1929). This process is also called 'hydrogen burning'. The star spends most of its life at this stage, as a 'Main Sequence' star, where equilibrium is set up between energy supply by fusion and energy loss in the form of radiation. For the Sun, this stage will probably last ca. 10^{10} yr, but a very large star with 15 times the Sun's mass may remain in the Main Sequence for only 10^7 yr.

When the bulk of hydrogen in a small star has been converted into ^{4}He, inward density-driven forces exceed outward radiation pressure, causing gravitational contraction. However, the resulting rise in core temperature causes expansion of the outer hydrogen-rich layer of the star. This forms a huge low-density envelope whose surface temperature may fall to ca. 4000 K, observed as a 'Red Giant'. This stage lasts only one-tenth as long as the Main Sequence stage. When core temperatures reach 1.5×10^7 K, a more efficient hydrogen-burning reaction becomes possible if the star contains traces of carbon, nitrogen and

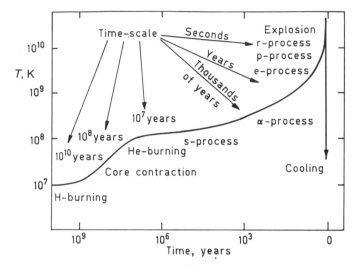

Fig. 1.5. Schematic diagram of the evolution of a large star showing the nucleosynthetic processes that occur along its accelerating life-history in response to increasing temperature (T). Note that time is measured backwards from the end of the star's life on the right. After Burbidge *et al.* (1957).

oxygen inherited from older generations of stars. This form of hydrogen burning is called the CNO cycle (Bethe, 1939).

At some point during the Red Giant stage, core temperatures may reach 10^8 K, when He fusion to carbon is ignited (the 'helium flash'). Further core contraction, yielding a temperature of ca. 10^9 K, follows as helium becomes exhausted. At these temperatures an endothermic process of α-particle emission can occur, allowing the building of heavier nuclides up to mass 40. However, this quickly expends the remaining burnable fuel of the star, which then cools to a White Dwarf.

More massive stars (of several solar masses) have a different life-history. In these stars, the greater gravitationally induced pressure–temperature conditions allow the fusion of helium to begin early in the Red Giant stage. This is followed by further contraction and heating, allowing the fusion of carbon and successively heavier elements. However, as lighter elements become exhausted, gravitationally induced contraction and heating occur at an ever increasing pace (Fig. 1.5), until the implosion is stopped by the attainment of neutron-star density. The resulting shock wave causes a supernova explosion which ends the star's life.

In the minutes before explosion, when temperatures exceed 3×10^9 K, very rapid nuclear interactions occur. Energetic equilibrium is established between nuclei and free protons and neutrons, synthesising elements like Fe by the so-called e-process. The

supernova explosion itself lasts only a few seconds, but is characterised by colossal neutron fluxes. These very rapidly synthesise heavier elements, terminating at ^{254}Cf, which undergoes spontaneous fission. Products of the supernova explosion are distributed through space and later incorporated in a new generation of stars.

1.2.2 Stages in the nucleosynthesis of heavy elements

A schematic diagram of the cosmic abundance chart is given in Fig. 1.6. We will now see how different nucleosynthetic processes are invoked to account for its form.

The element-building process begins with the fusion of four protons to one ^{4}He nucleus, which occurs in three stages:

$$^1\text{H} + {}^1\text{H} \rightarrow {}^2\text{D} + e^+ + \nu$$
$$(Q = +1.44 \text{ MeV}, \ t_{1/2} = 1.4 \times 10^{10} \text{ yr})$$
$$^2\text{D} + {}^1\text{H} \rightarrow {}^3\text{He} + \gamma$$
$$(Q = +5.49 \text{ MeV}, \ t_{1/2} = 0.6 \text{ s})$$
$$^3\text{He} + {}^3\text{He} \rightarrow {}^4\text{He} + 2\,{}^1\text{H} + \gamma$$
$$(Q = +12.86 \text{ MeV}, \ t_{1/2} = 10^6 \text{ yr})$$

where Q is the energy output and $t_{1/2}$ is the reaction time of each stage (the time necessary to consume one-half of the reactants) for the centre of the Sun. The long reaction time for the first step explains the long duration of the hydrogen-burning (Main Sequence) stage for small stars like the Sun. The

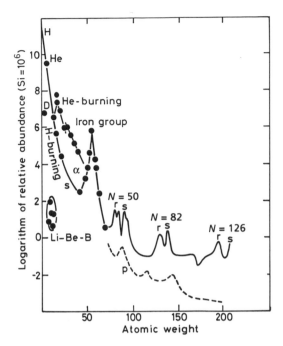

Fig. 1.6. Schematic diagram of the cosmic abundances of the elements, highlighting the nucleosynthetic processes responsible for forming different groups of nuclides. After Burbidge *et al.* (1957).

overall reaction converts four protons into one helium nucleus, two positrons and two neutrinos, plus a large output of energy in the form of high frequency photons. Hence the reaction is very strongly exothermic. Although deuterium and ^{3}He are generated in the first two reactions above, their consumption in the third accounts for their much lower cosmic abundance than ^{4}He.

If heavier elements are present in a star (e.g. carbon and nitrogen) then the catalytic C–N–O sequence of reactions can occur, which also combines four protons to make one helium nucleus:

$$^{12}C + {}^1H \rightarrow {}^{13}N + \gamma$$
$$(Q = +1.95 \text{ MeV}, t_{1/2} = 1.3 \times 10^7 \text{ yr})$$
$$^{13}N \rightarrow {}^{13}C + e^+ + \nu$$
$$(Q = +2.22 \text{ MeV}, t_{1/2} = 7 \text{ min})$$
$$^{13}C + {}^1H \rightarrow {}^{14}N + \gamma$$
$$(Q = +7.54 \text{ MeV}, t_{1/2} = 3 \times 10^6 \text{ yr})$$
$$^{14}N + {}^1H \rightarrow {}^{15}O + \gamma$$
$$(Q = +7.35 \text{ MeV}, t_{1/2} = 3 \times 10^5 \text{ yr})$$
$$^{15}O \rightarrow {}^{15}N + e^+ + \nu$$
$$(Q = +2.70 \text{ MeV}, t_{1/2} = 82 \text{ s})$$
$$^{15}N + {}^1H \rightarrow {}^{12}C + {}^4He$$
$$(Q = +4.96 \text{ MeV}, t_{1/2} = 10^5 \text{ yr})$$

The C–N–O elements have greater potential energy barriers to fusion than hydrogen, so these reactions require higher temperatures to operate than the simple proton–proton (p–p) reaction. However, the reaction times are much shorter than for the p–p reaction. Therefore the C–N–O reaction contributes less than 10% of hydrogen-burning reactions in a small star like the Sun, but is overwhelmingly dominant in large stars. This explains their much shorter lifespan in the Main Sequence.

Helium burning also occurs in stages:

$$^4He + {}^4He \leftrightarrows {}^8Be \qquad (Q = +0.09 \text{ MeV})$$
$$^8Be + {}^4He \leftrightarrows {}^{12}C^* \qquad (Q = -0.37 \text{ MeV})$$
$$^{12}C^* \rightarrow {}^{12}C + \gamma \qquad (Q = +7.65 \text{ MeV})$$

The ^{8}Be nucleus is very unstable ($t_{1/2} < 10^{-15}$ s) and in the core of a Red Giant the Be/He equilibrium ratio is estimated at 10^{-9}. However its life is just long enough to allow the possibility of collision with another helium nucleus. (Instantaneous 3-particle collisions are very rare). The energy yield of the first stage is small, and the second is actually endothermic, but the decay of excited $^{12}C^*$ to the ground state is strongly exothermic, driving the equilibria to the right.

The elements Li, Be and B have low nuclear binding energies, so that they are unstable at the temperatures of 10^7 K and above found at the centres of stars. They are therefore bypassed by stellar nucleosynthetic reactions, leading to low cosmic abundances (Fig. 1.6). The fact that the five stable isotopes ^{6}Li, ^{7}Li, ^{9}Be, ^{10}B and ^{11}B exist at all has been attributed to fragmentation effects (spallation) of heavy cosmic rays (atomic nuclei travelling through the galaxy at relativistic speeds) as they hit interstellar gas atoms (Reeves, 1974). This is termed the x-process.

Following the synthesis of carbon, further helium-burning reactions are possible, to produce heavier nuclei:

$$^{12}C + {}^4He \rightarrow {}^{16}O + \gamma \qquad (Q = +7.15 \text{ MeV})$$
$$^{16}O + {}^4He \rightarrow {}^{20}Ne + \gamma \qquad (Q = +4.75 \text{ MeV})$$
$$^{20}Ne + {}^4He \rightarrow {}^{24}Mg + \gamma \qquad (Q = +9.31 \text{ MeV})$$

Intervening nuclei such as ^{13}N can be produced by adding protons to these species, but are themselves consumed in the process of catalytic hydrogen burning mentioned above.

In old Red Giant stars, carbon-burning reactions can occur:

$$^{12}C + {}^{12}C \rightarrow {}^{24}Mg + \gamma \qquad (Q = +13.85 \text{ MeV})$$
$$\rightarrow {}^{23}Na + {}^1H \qquad (Q = +2.23 \text{ MeV})$$
$$\rightarrow {}^{20}Ne + {}^4He \qquad (Q = +4.62 \text{ MeV})$$

The hydrogen and helium nuclei regenerated in these processes allow further reactions which help to fill in gaps between masses 12 and 24.

When a small star reaches its maximum core temperature of 10^9 K the endothermic α-process can occur:

$$^{20}\text{Ne} + \gamma \rightarrow \, ^{16}\text{O} + \, ^4\text{He} \quad (Q = -4.75 \text{ MeV})$$

The energy consumption of this process is compensated by strongly exothermic reactions such as:

$$^{20}\text{Ne} + \, ^4\text{He} \rightarrow \, ^{24}\text{Mg} + \gamma \, (Q = +9.31 \text{ MeV})$$

so that the overall reaction generates a positive energy budget. The process resembles helium burning, but is distinguished by the different source of ^{4}He. The α-process can build up from ^{24}Mg through the sequence ^{28}Si, ^{32}S, ^{36}Ar and ^{40}Ca, where it terminates, owing to the instability of ^{44}Ti.

The maximum temperatures reached in the core of a small star do not allow substantial heavy-element production. However, in the final stages of the evolution of larger stars, before a supernova explosion, the core temperature exceeds 3×10^9 K. This allows energetic equilibrium to be established by very rapid nuclear reactions between the various nuclei and free protons and neutrons (the e-process). Because ^{56}Fe is at the peak of the nuclear binding-energy curve, this element is most favoured by the e-process (Fig. 1.6). However, the other first-series transition elements V, Cr, Mn, Co and Ni in the mass range 50 to 62 are also attributed to this process.

During the last few million years of a Red Giant's life, a slow process of neutron addition with emission of γ rays (the s-process) can synthesise many additional nuclides up to mass 209 (*see* Fig. 1.7). Two possible neutron sources are:

$$^{13}\text{C} + \, ^4\text{He} \rightarrow \, ^{16}\text{O} + \text{n} + \gamma$$
$$^{21}\text{Ne} + \, ^4\text{He} \rightarrow \, ^{24}\text{Mg} + \text{n} + \gamma$$

The ^{13}C and ^{21}Ne parents can be produced by proton bombardment of the common ^{12}C and ^{20}Ne nuclides.

Because neutron capture in the s-process is relatively slow, unstable neutron-rich nuclides generated in this process have time to decay by β emission before further neutron addition. Hence the nucleosynthetic path of the s-process climbs in many small steps up the path of greatest stability of proton/neutron ratio (Fig. 1.7) and is finally terminated by the α decay of ^{210}Po back to ^{206}Pb and ^{209}Bi back to ^{205}Tl.

The 'neutron capture cross-section' of a nuclide expresses how readily it can absorb incoming thermal neutrons, and therefore determines how likely it is to be converted to a higher atomic mass species by neutron bombardment. Nuclides with certain neutron numbers (e.g. 50, 82 and 126) have unusually small neutron capture cross-sections, making them particularly resistant to further reaction and giving rise to local peaks in abundance at masses 90, 138 and 208.

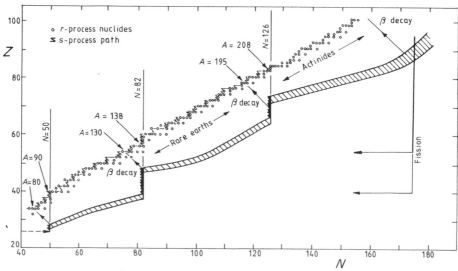

Fig. 1.7. Neutron capture paths of the s-process and r-process shown on the chart of the nuclides. Hatched zone indicates the r-process nucleosynthetic pathway for a plausible neutron flux. Neutron 'magic numbers' are indicated by vertical lines, and mass numbers of nuclide abundance peaks are marked. After Seeger *et al.* (1965).

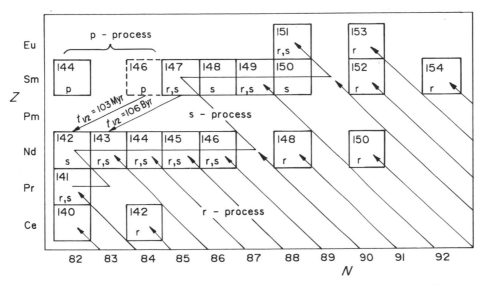

Fig. 1.8. Part of the chart of the nuclides in the area of the light rare earths to show p-, r- and s-process product nuclides. After O'Nions *et al.* (1979).

$N = 50$, 82 and 126 are empirically referred to as neutron 'magic numbers'.

In contrast to the s-process, which may occur over periods of millions of years in Red Giants, r-process neutrons are added in very rapid succession to a nucleus before β decay is possible. The nuclei are therefore rapidly driven to the neutron-rich side of the stability line, until they reach a new equilibrium between neutron addition and β decay, represented by the hatched zone in Fig. 1.7. Nuclides move along this r-process pathway until they reach a configuration with low neutron capture cross-section (a neutron magic number). At these points a 'cascade' of alternating β decays and single neutron additions occurs, indicated by the notched ladders in Fig. 1.7. Nuclides climb these ladders until they reach the next segment of the r-process pathway.

Nuclides with neutron magic numbers build to excess abundances, as with the s-process, but they occur at proton-deficient compositions relative to the s-process stability path. Therefore, when the neutron flux falls off and nuclides on the ladders undergo β decay back to the stability line, the r-process local abundance peaks are displaced about 6–12 mass units below the s-process peaks (Fig. 1.6).

The r-process is terminated by neutron-induced fission at mass 254, and nuclear matter is fed back into the element-building processes at masses of ca. 108 and 146. Thus, cycling of nuclear reactions occurs above mass 108. Because of the extreme neutron flux

postulated for the r-process, its occurrence is probably limited to supernovae.

The effects of r- and s-process synthesis of typical heavy elements may be demonstrated by an examination of the chart of the nuclides in the region of the light rare earths (Fig. 1.8). The step by step building of the s-process contrasts with the 'rain of nuclides' produced by β decay of r-process products. Some nuclides, such as ^{143}Nd to ^{146}Nd are produced by both r- and s-processes. Some, such as ^{142}Nd are s-only nuclides 'shielded' from the decay products of the r-process by intervening nuclides. Others, such as ^{148}Nd and ^{150}Nd are r-only nuclides which lie off the s-process production pathway.

Several heavy nuclides from ^{74}Se to ^{196}Hg lie isolated on the proton-rich side of the s-process growth path (e.g. ^{144}Sm in Fig. 1.8), and are also shielded from r-process production. In order to explain the existence of these nuclides it is necessary to postulate a p-process by which normal r- and s-process nuclei are bombarded by protons at very high temperature ($> 2 \times 10^9$K), probably in the outer envelope of a supernova.

1.3 Radioactive decay

Nuclear stability and decay is best understood in the context of the chart of nuclides. It has already been noted that naturally occurring nuclides define a path in the chart of the nuclides, corresponding to the

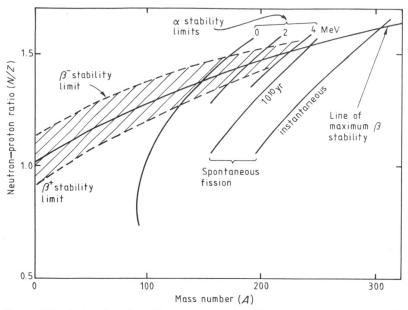

Fig. 1.9. Theoretical stability limits of nuclides illustrated on a plot of N/Z against mass number (A). Lower limits for α emission are shown for α energies of 0, 2 and 4 MeV. Stability limits against spontaneous fission are shown for half-lives of 10^{10} yr and zero (instantaneous fission). After Hanna (1959).

greatest stability of proton/neutron ratio. For nuclides of low atomic mass, the greatest stability is achieved when the number of neutrons and protons are approximately equal ($N = Z$) but as atomic mass increases, the stable neutron/proton ratio increases until $N/Z = 1.5$. Theoretical stability limits are illustrated on a plot of N/Z against mass number (A) in Fig. 1.9 (Hanna, 1959).

The path of stability is in fact an energy 'valley' into which the surrounding unstable nuclides tend to fall, emitting particles and energy. This constitutes the process of radioactive decay. The nature of particles emitted depends on the location of the unstable nuclide relative to the energy valley. Unstable nuclides on either side of the valley usually decay by 'isobaric' processes. That is, a nuclear proton is converted to a neutron, or vice-versa, but the mass of the nuclide does not change significantly (except for the 'mass defect' consumed as nuclear binding energy). In contrast, unstable nuclides at the high end of the energy valley often decay by emission of a heavy particle (e.g. α particle), thus reducing the overall mass of the nuclide.

1.3.1 Isobaric decay

Different decay processes indicated on Fig. 1.9 can best be understood by looking at example sections of

the chart of nuclides. Figure 1.10 shows a part of the chart around the element potassium. The diagonal lines indicate isobars (nuclides of equal mass) which are displayed on energy sections in Fig. 1.11 and Fig 1.12.

Nuclides deficient in protons decay by transformation of a neutron into a proton and an electron. The latter is then expelled from the nucleus as a negative 'β' particle (β^-), along with an anti-neutrino ($\bar{\nu}$). The energy released by the transformation is divided between the β particle and the anti-neutrino as kinetic energy (Fermi, 1934). The observed consequence is that the β particles emitted have a continuous energy distribution from nearly zero to the maximum decay energy. Low-energy β particles are very difficult to separate from background noise in a detector, making the β decay constant of nuclides such as ^{87}Rb very difficult to determine accurately by direct counting (section 3.1).

In many cases the nuclide produced by β decay is left in an excited state which subsequently decays to the ground state nuclide by a release of energy. This may either be lost as a γ ray of discrete energy, or may be transferred from the nucleus to an orbital electron, which is then expelled from the atom. In the latter case, nuclear energy emission in excess of the binding energy of the electron is transferred to the electron as kinetic energy, which is superimposed as a line

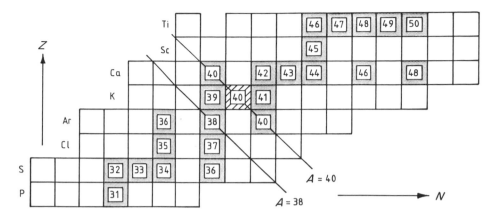

Fig. 1.10. Part of the chart of the nuclides, in coordinates of atomic number (Z) against neutron number (N) in the region of potassium. Stable nuclides are shaded; the long-lived unstable nuclide ^{40}K is hatched. Diagonal lines are isobars (lines of constant mass number, A).

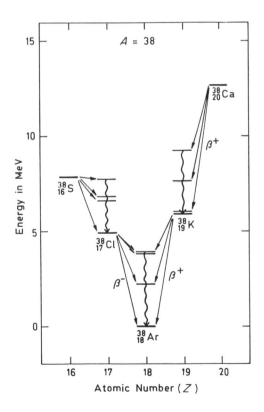

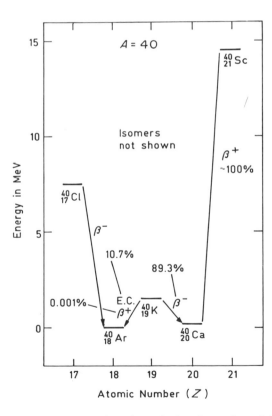

Fig. 1.11. A simple energy section through the chart of nuclides along the isobar A = 38 showing nuclides and isomers. Data from Lederer and Shirley (1978).

Fig. 1.12. Energy section through the chart of nuclides along isobar A = 40. Isomers are omitted for simplicity. For nuclides with more than one decay mechanism the percentage of transitions by different decay routes is indicated. Data from Lederer and Shirley (1978).

spectrum on the continuous spectrum of the β particles. The meta-stable states, or 'isomers' of the product nuclide are denoted by the superfix 'm', and have half-lives from less than a picosecond up to 241 years (in the case of ^{192m}Ir). Many β emitters have complex energy spectra involving a ground state product and more than one short-lived isomer, as shown in Fig. 1.11. The decay of ^{40}Cl can yield 35 different isomers of ^{40}Ar (Lederer and Shirley, 1978), but these are omitted from Fig. 1.12 for the sake of clarity.

Nuclides deficient in neutrons, e.g. ^{38}K (Fig. 1.11), may decay by two different processes: positron emission and electron capture. Both processes yield a product nuclide which is an isobar of the parent, by transformation of a proton to a neutron. In positron emission a positively charged electron (β^+) is emitted from the nucleus along with a neutrino. As with β^- emission, the decay energy is shared between the kinetic energy of the two particles. After having been slowed down by collision with atoms, the positron interacts with an orbital electron, whereby both are annihilated, yielding two 0.511 MeV γ rays. (This forms part of the decay energy of the nuclear transformation.)

In electron capture decay (E.C.) a nuclear proton is transformed into a neutron by capture of an orbital electron, usually from one of the inner shells, but possibly from an outer shell. A neutrino is emitted from the nucleus, and an outer orbital electron falls into the vacancy produced by electron capture, emitting a characteristic x-ray. The product nucleus may be left in an excited state, in which case it decays to the ground state by γ emission.

When the transition energy of a decay route is less than the energy equivalent of the positron mass ($2m_eC^2 = 1.022$ MeV), decay is entirely by electron capture. Thereafter, the ratio β^+/E.C. increases rapidly with increasing transition energy (Fig. 1.12), but a small amount of electron capture always accompanies positron emission even at high transition energies.

It is empirically observed (Mattauch, 1934) that adjacent isobars cannot be stable. Since ^{40}Ar and ^{40}Ca are both stable species (Fig. 1.10), ^{40}K must be unstable, and exhibits a branched decay to the isobars on either side (Fig. 1.12).

1.3.2 Heavy particle decay

Heavy atoms above bismuth in the chart of nuclides often decay by emission of an α particle, consisting of two protons and two neutrons (He^{++}). The daughter product is not an isobar of the parent, and has an atomic mass reduced by four. The product nuclide may be in the ground state, or remain in an excited state and subsequently decay by γ emission. The decay energy is shared between kinetic energy of the α particle and recoil energy of the product nuclide.

The U and Th decay series are shown in Fig. 12.1. Because the energy valley of stable proton/neutron ratios in this part of the chart of the nuclides has a slope of less than unity, α decays tend to drive the products off to the neutron-rich side of the energy valley, where they undergo β decay. In fact β decay may occur before the corresponding α decay.

At intermediate masses in the chart of the nuclides, α decay may occasionally be an alternative to positron or electron capture decay for proton-rich species such as ^{147}Sm. However, α decays do not occur at low atomic numbers because the path of nuclear stability has a Z/N slope close to unity in this region (Fig. 1.1). Any such decays would simply drive unstable species along (parallel to) the energy valley.

A new kind of radioactive decay has recently been discovered in the ^{235}U to ^{207}Pb decay series (Rose and Jones, 1984), whereby ^{223}Ra decays by emission of ^{14}C directly to ^{209}Pb with a decay energy of 13.8 MeV. However this mode of decay occurs with a frequency of less than 10^{-9} of the α decay of ^{223}Ra.

1.3.3. Nuclear fission and the Oklo natural reactor

The nuclide ^{238}U (atomic no. 92) undergoes spontaneous fission into two product nuclei of different atomic number, typically ca. 40 and 55 (Zr and Cs), along with various other particles and a large amount of energy. Because the heavy parent nuclide has a high neutron/proton ratio, the daughter products have an excess of neutrons and undergo isobaric decay by β emission. Although the frequency of spontaneous fission of ^{238}U is less than 2×10^{-6} that of α decay, in heavier transuranium elements spontaneous fission is the principal mode of decay. Other nuclides, such as ^{235}U, may undergo fission if they are struck by a neutron. Furthermore, since fission releases neutrons which promote further fission reactions, a chain reaction may be established. If the concentration of fissile nuclides is high enough, this leads to a thermonuclear explosion, as in a supernova or atomic bomb.

In special cases where an intermediate heavy-element concentration is maintained, a self-sustaining but non-explosive chain reaction may be possible. This depends largely on the presence of a 'moderator'.

Energetic 'fast' neutrons produced by fission undergo multiple elastic collisions with atoms of the moderator. They are decelerated into 'thermal' neutrons, having velocities characteristic of the thermal vibration of the medium, the optimum velocity for promoting fission reactions in the surrounding heavy atoms. One natural case of such an occurrence is known, termed the Oklo natural reactor (Cowan, 1976; Naudet, 1976).

In May 1972, ^{235}U depletions were found in uranium ore entering a French processing plant and traced to an ore deposit at Oklo in the Gabon republic of central Africa. Despite its apparent improbability, there is overwhelming geological evidence that the ^{235}U depletions were caused by the operation of a natural fission reactor ca. 1.8 Byr ago. It appears that in the Early Proterozoic, conditions were such that the series of coincidences needed to create a natural fission reactor were achieved more easily than at the present day.

Uranium dispersed in granitic basement was probably eroded and concentrated in stream-bed placer deposits. It was immobilised in this environment as the insoluble reduced form due to the nature of prevailing atmospheric conditions. With the appearance of blue-green algae, the first organisms capable of photosynthesis, the oxygen content of the atmosphere, and hence river water, probably rose, converting some reduced uranium into more soluble oxidised forms. These were carried down-stream in solution. When the soluble uranium reached a river delta it must have encountered sediments rich in organic ooze, creating an oxygen deficiency which again reduced and immobilised uranium, but now at a much higher concentration (up to 0.5% uranium by weight).

After burial and compaction of the deposit, it was subsequently uplifted, folded and fractured, allowing oxygenated ground-waters to re-mobilise and concentrate the ores into veins over 1 m wide of almost pure uranium oxide. Hence the special oxygen fugacity conditions obtaining in the Proterozoic helped to produce a particularly concentrated deposit. However, its operation as a reactor depended on the greater ^{235}U abundance (3%) at that time, compared with the present day level of 0.72%, reduced by α decay in the intervening time (half-life = 700 Myr).

In the case of Oklo, light water (H_2O), must have acted as a moderator, and the nuclear reaction was controlled by a balance between hot water loss by convective heating or boiling, and replacement by cold ground-water influx. In this way the estimated total energy output (15 000 megawatt years, representing the consumption of six tons of ^{235}U) was probably maintained at an average of only 20 kilowatts for about 0.8 Myr.

Geochemical evidence for the occurrence of fission is derived firstly from the characteristic elemental abundances of fission products. For example, excess concentrations of rare earths and other immobile elements such as Zr are observed. Alkali metal and alkali earths were probably also enriched, but have subsequently been removed by leaching. Secondly, the characteristic isotope abundances of some elements can only be explained by fission (Raffenach et al., 1976).

The Nd isotope composition of the Oklo ore is very distinctive (Fig. 1.13). ^{142}Nd is shielded from isobaric decay of the neutron-rich fission products (Fig. 1.8) so that its abundance indicates the level of normal Nd. After correction for an enhanced abundance of ^{144}Nd and ^{146}Nd due to neutron capture by the large cross-section nuclides ^{143}Nd and ^{145}Nd, Oklo Nd has an isotopic composition closely resembling that of normal reactor fission product waste (Fig. 1.13).

Evidence for a significant neutron flux is also demonstrated by the isotope signatures of actinide elements. For example, the abundant isotope of uranium (^{238}U) readily captures fast neutrons to yield an appreciable amount of ^{239}U, which decays by β emission to ^{239}Np and then ^{239}Pu (Fig. 1.14). The latter decays by α emission with a half-life of 24 400 yr to yield more ^{235}U, contributing an extra 50% to the 'burnable' fuel, as in a 'fast' breeder reactor ('fast' refers to the speed of the neutrons involved). Because the fission products of ^{239}Pu and ^{235}U have distinct isotopic signatures, it is determined that very little ^{239}Pu underwent neutron-induced fission before decaying to ^{235}U. Hence, the low flux and prolonged lifetime of the natural reactor are deduced.

1.4 The law of radioactive decay

The rate of decay of a radioactive parent nuclide to a stable daughter product is proportional to the number of atoms, n, present at any time t (Rutherford and Soddy, 1902):

$$-\frac{dn}{dt} = \lambda n \qquad [1.1]$$

where λ is the constant of proportionality, which is characteristic of the radionuclide in question and is called the decay constant (expressed in units of reciprocal time). The decay constant states the probability that a given atom of the radionuclide

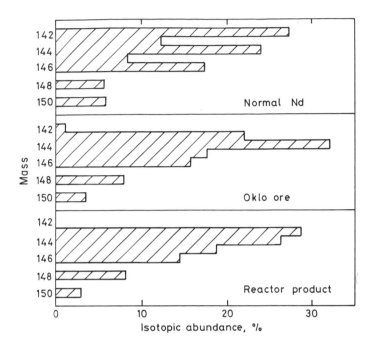

Fig. 1.13. Bar charts of the isotope composition in normal Nd, Oklo ore, and reactor fission product waste. Data from Cowan (1976).

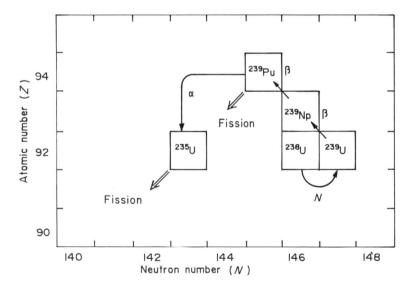

Fig. 1.14. Nuclear reactions leading to 'breeding' of transuranium element fuel in the Oklo natural reactor.

will decay within a stated time. The term dn/dt is the rate of change of the number of parent atoms, and is negative because this rate decreases with time. Rearranging equation [1.1], we obtain:

$$\frac{dn}{dt} = -\lambda n \qquad [1.2]$$

This expression is integrated from $t = 0$ to t, given that the number of atoms present at time $t = 0$ is n_0:

$$\int_{n_0}^{n} \frac{dn}{n} = -\lambda \int_{t=0}^{t} dt \qquad [1.3]$$

Hence:

$$\ln \frac{n}{n_0} = -\lambda t \qquad [1.4]$$

which can also be written as:

$$n = n_0 \, e^{-\lambda t} \qquad [1.5]$$

A useful way of referring to the rate of decay of a radionuclide is the 'half-life', $t_{1/2}$, which is the time required for half of the parent atoms to decay. Substituting $n = n_0/2$ and $t = t_{1/2}$ into equation [1.4], and then rearranging this we obtain:

$$t_{1/2} = \frac{\ln 2}{\lambda} = \frac{0.693}{\lambda} \qquad [1.6]$$

The number of radiogenic daughter atoms formed, D^*, is equal to the number of parent atoms consumed:

$$D^* = n_0 - n \qquad [1.7]$$

but $n_0 = n \, e^{\lambda t}$ (from equation [1.5]); so substituting for n_0 in equation [1.7] yields:

$$D^* = n \, e^{\lambda t} - n \qquad [1.8]$$

i.e.: $\qquad D^* = n \, (e^{\lambda t} - 1) \qquad [1.9]$

If the number of daughter atoms at time $t = 0$ is D_0, then the total number of daughter atoms after time t is given as:

$$D = D_0 + n(e^{\lambda t} - 1) \qquad [1.10]$$

This equation is the fundamental basis of geochronological dating tools.

In the uranium series decay chains, the daughter products of radioactive decay (other than three Pb isotopes) are themselves radioactive. Hence the rate of decay of such a daughter product is given by the difference between its production rate from the parent and its own decay rate:

$$dn_2/dt = n_1 \lambda_1 - n_2 \lambda_2 \qquad [1.11]$$

where n_1 and λ_1 are the abundance and decay

constant of the parent, and n_2 and λ_2 correspond to the daughter.

But equation [1.5] can be substituted for n_1 in equation [1.11] to yield:

$$dn_2/dt = n_{1,\text{initial}} \, e^{-\lambda_1 t} \lambda_1 - n_2 \lambda_2 \qquad [1.12]$$

This equation is integrated for a chosen set of initial conditions, the simplest of which sets $n_2 = 0$ at $t = 0$. Then:

$$n_2 \lambda_2 = \frac{\lambda_1}{\lambda_2 - \lambda_1} n_{1,\text{initial}} \, (e^{-\lambda_1 t} - e^{-\lambda_2 t}) \qquad [1.13]$$

This type of solution was first demonstrated by Bateman (1910) and is named after him. Recently, Catchen (1984) examined more general initial conditions for these equations, leading to more complex solutions.

1.4.1 Uniformitarianism

When using radioactive decay to measure the ages of rocks we must apply the classic principle of uniformitarianism (Hutton, 1788), by assuming that the decay constant of the parent radionuclide has not changed during the history of the Earth. It is therefore important to outline some evidence that this assumption is justified.

The decay constant of a radionuclide depends on nuclear constants, such as a (= elementary charge^2/ Plank's constant/velocity of light). Shlyakhter (1976) argued that the neutron capture cross-section of a nuclide is very sensitively dependent on nuclear constants. Because neutron absorbers (such as ^{143}Nd and ^{145}Nd) in the 1.8 Byr-old Oklo natural reactor give rise to the expected abundance increases in the product isotopes (Fig. 1.13) then this constrains nuclear constants to have remained more or less invariant over the last 2 Byr.

The possibility that physical conditions (e.g. pressure and temperature) could affect radionuclide decay constants must also be examined. Because radioactive decay is a property of the nucleus, which is shielded from outside influence by orbital electrons, it is very unlikely that physical conditions influence α or β decay, but electron capture decay could be affected. Hensley et al. (1973) have demonstrated that the electron capture decay of ^{7}Be to ^{7}Li is increased by 0.59% when BeO is subjected to 270 ± 10 kbars pressure in a diamond anvil. This raises the question of whether the electron capture decay of ^{40}K to ^{40}Ar could be pressure-dependent, affecting K–Ar dates. In fact this is unlikely, because at high pressure–temperature conditions at depth in the Earth, K–Ar systems will be chemically open and unable to yield dates at all, while at crustal depths the pressure-dependence of λ will be negligible compared with experimental errors.

The achievement of concordant K–Ar, Rb–Sr and U–Pb ages on rocks (where chemical systems have remained closed) supports the invariance of decay constants with time, since different radionuclides would be expected to respond differently if decay constants had changed. A final piece of evidence for the invariance of decay constants comes from the agreement between radiometric dates and other time markers such as rates of sedimentation and evolution, sea floor spreading magnetic anomalies (section 10.1.4) and the correspondence of radiocarbon dates with tree ring ages (section 14.1.4) and uranium series dates with coral growth bands (section 12.4.2).

References

Atkinson, R. and Houtermans, F. G. (1929). Zur Frage der Aufbaumoglichkeit der Elemente in Sternen. *Z. Physik* **54**, 656–65.

Bateman, H. (1910). Solution of a system of differential equations occurring in the theory of radio-active transformations. *Proc. Cambridge Phil. Soc.* **15**, 423–7.

Bethe, H. A. (1939). Energy production in stars. *Phys. Rev.* **55**, 434–56.

Burbidge, E. M., Burbidge, G. R., Fowler, W. A. and Hoyle, F. (1957). Synthesis of the elements in stars. *Rev. Mod. Phys.* **29**, 547–647.

Catchen, G. L. (1984). Application of the equations of radioactive growth and decay to geochronological models and explicit solution of the equations by Laplace transformation. *Isot. Geosci.* **2**, 181–95.

Cowan, G. A. (1976). A natural fission reactor. *Sci. Amer.* **235** (1), 36–47.

Fermi, E. (1934). Versuch einer Theorie der β-Strahlen. *Z. Physik* **88**, 161–77.

Hanna, G. C. (1959). Alpha-radioactivity. In Segre, E. (Ed.), *Experimental Nuclear Physics, Vol. 3*, Wiley, pp. 54–257.

Hansen, P. G. (1987). Beyond the neutron drip line. *Nature* **328**, 476–77.

Hensley, W. K., Basset, W. A. and Huizenga, J. R. (1973). Pressure dependence of the radioactive decay constant of beryllium - 7. *Science* **181**, 1164-5.

Hutton, J. (1788). Theory of the Earth; or an investigation of the laws observable in the composition, dissolution, and restoration of land upon the globe. *Trans. Roy. Soc. Edin.* **1**, 209-304.

Iben, I. (1967). Stellar evolution within and off the Main Sequence. *Ann. Rev. Astron. Astrophys.* **5**, 571–626.

Lederer, C. M. and Shirley, V. S. (1978). *Table of Isotopes* (7th Edn), Wiley.

Mattauch, J. (1934). Zur Systematiek der Isotopen. *Z. Physik* **91**, 361–71.

Naudet, R. (1976). The Oklo nuclear reactors: 1800 million years ago. *Interd. Sci.* **1**, 72–84.

O'Nions, R. K., Carter, S. R., Evensen, N. M. and Hamilton P. J. (1979). Geochemical and cosmochemical applications of Nd isotope analysis. *Ann. Rev. Earth Planet. Sci.* **7**, 11–38.

Raffenach, J. C., Menes, J., Devillers, C., Lucas, M. and Hagemann, R. (1976). Etudes chimiques et isotopiques de l'uranium, du plomb et de plusieurs produits de fission dans un echantillon de mineral du reacteur naturel d'Oklo. *Earth Planet. Sci. Lett.* **30**, 94–108.

Reeves, H. (1974). Origin of the light elements. *Ann. Rev. Astron. Astrophys.* **12**, 437–69.

Ringwood, A. E. (1979). Composition and origin of the Earth. In: McElhinny, M. W. (Ed.), *The Earth: its Origin, Structure and Evolution*. Academic Press, pp. 1–58.

Rose, H. J. and Jones, G. A. (1984). A new kind of radioactivity. *Nature* **307**, 245–7.

Ross, J. E. and Aller, L. H. (1976). The chemical composition of the Sun. *Science* **191**, 1223–9.

Rutherford, E. and Soddy, F. (1902). The radioactivity of thorium compounds II. The cause and nature of radioactivity. *J. Chem. Soc. Lond.* **81**, 837–60.

Seeger, P. A., Fowler, W. A. and Clayton, D. D. (1965). Nucleosynthesis of heavy elements by neutron capture. *Astrophys. J. Supp.* **11**, 121–66.

Shlyakhter, A. I. (1976). Direct test of the constancy of fundamental nuclear constants. *Nature* **264**, 340.

Weinberg, S. (1977). *The First Three Minutes*. Andre Deutsch, 190 p.

2 Experimental techniques

In order to use radiogenic isotopes as dating tools or tracers, they must be separated by mass from non-radiogenic isotopes in a 'mass spectrometer'. In a 'magnetic sector' instrument (Fig. 2.1) the nuclides to be separated are ionised under vacuum and accelerated through a high potential (V) before passing between the poles of a magnet. A uniform magnetic field (H) acting on particles in the ion beam bends them into curves of different radius (r) according to the following equation:

$$r^2 = \frac{m}{e} \frac{2V}{H^2} \qquad [2.1]$$

where m/e = mass/charge for the ion in question. Since most of the ions produced are single-charged, nuclides will be separated into a simple spectrum of masses. The relative abundance of each mass is then determined by its corresponding ion current, captured by a Faraday bucket or multiplier detector. Other methods of mass separation are possible (e.g. quadrupole and time-of-flight analysers) but these are less widely used for precise isotope ratio measurement.

Except for the rare gases He, Ne, Ar and Xe, which are analysed in the gas phase, the radiogenic elements of interest to isotope geologists are normally analysed by solid source mass spectrometry. In the spark source mass spectrometer, a solid sample containing a mixture of different elements forms the source. However, using this method, four factors combine to yield poor precision in the measurement of isotope ratios. These are: interferences between atomic and molecular ions with the same masses, dilution of trace elements in a major element matrix, poor ionisation efficiency, and unstable emission. The normal starting point of precise isotopic measurements by mass spectrometry is therefore chemical separation of the element to be analysed. This requires the sample to be converted into a solution.

2.1 Chemical separation

Geological samples, which are commonly silicates, are routinely dissolved in concentrated hydrofluoric acid (HF), although some laboratories use perchloric acid as well. Most rock-forming minerals will dissolve in hot concentrated HF at atmospheric pressure. However, certain resistant minerals such as zircon

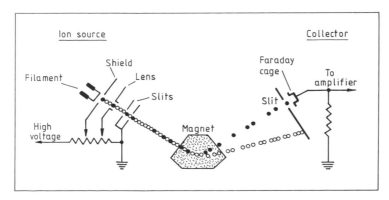

Fig. 2.1. Schematic illustration of the basic features of a magnetic sector mass spectrometer. Solid and open circles represent the light and heavy isotopes of an element, respectively. After Faul (1966).

must be dissolved under pressure in a bomb, in order to achieve the temperatures of up to 220 °C necessary for decomposition. The bomb liners and beakers used for dissolution are almost universally made of poly-fluorinated ethylene (PFE).

The conventional bomb dissolution technique for zircons is described by Krogh (1973). A more recent technique, pioneered by Krogh and described by Parrish (1987), is to place several 'micro-capsules' carrying different samples into one larger bomb. The micro-capsules are open to vapour transfer of HF, but Pb blanks are very low (less than 50 pg), showing that volatile transfer of Pb between samples does not occur.

A major problem which may be encountered after HF dissolution is the formation of fluorides which are insoluble in other 'mineral' acids (e.g. hydrochloric acid, HCl). Refluxing with nitric acid (HNO$_3$) helps to convert these into soluble forms. Experiments by Croudace (1980) suggested that this process was promoted if additional nitric acid was added before complete evaporation of the HF stage. If at some stage complete dissolution is not achieved, then it may be necessary to decant off the solution and return the undissolved fraction to the previous stage of the process for a second acid attack (Patchett and Tatsumoto, 1980). When complete dissolution has been achieved the solution may need to be split into weighed aliquots, so that one fraction can be 'spiked' with an enriched isotope for isotope dilution analysis while another is left 'unspiked' for accurate isotope ratio analysis. Following dissolution, the sample is often converted to a chloride for elemental separation, which is normally performed by ion exchange between a dilute acid (eluent) and a resin (stationary phase) contained in quartz or PFE columns.

2.1.1 Rb–Sr

The separation of Rb and Sr, and the preliminary separation of Sm–Nd, is often performed on a cation exchange column eluted with dilute (e.g. 2.5M) HCl. Columns are normally calibrated in advance of use by means of test solutions. Finally, before separation of the sample, the column resin is cleaned by passing sequential volumes of 50% acid and water.

A small volume of the rock solution is loaded into the column, washed into the resin bed carefully with eluent, and then washed through with more eluent until a fraction is collected when the desired element is released from the resin. This is evaporated to dryness, ready to load onto a metal filament for thermal ionisation in the mass spectrometer. Elements are

eluted from the cation column by HCl in roughly the following order: Fe, Na, Mg, K, Rb, Ca, Sr, Ba, REE (Crock *et al.*, 1984). This series is defined by increasing partition coefficient onto the solid phase (resin), requiring increasing volumes of eluent to release successive elements.

It is very important to remove the 'major elements' of the rock, e.g. Na, K and Ca, from the Sr cut. Nitric acid is not effective for this purpose because it does not separate Sr from Ca (Fig. 2.2). Rb must also be eliminated from Sr because [87]Rb is a direct isobaric interference onto [87]Sr. Small levels of Rb are not a problem in an otherwise clean sample because the Rb burns off before Sr data collection begins. However,

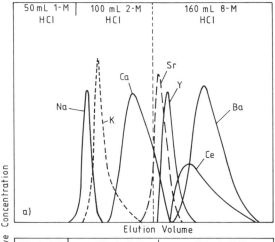

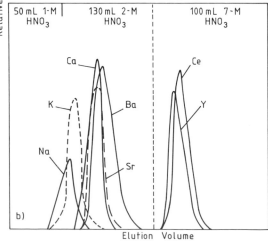

Fig. 2.2. Elution curves for various elements from cation exchange columns: a) with hydrochloric acid; b) with nitric acid. Modified after Crock *et al.* (1984).

the presence of significant Ca in the Sr cut prevents Rb burn-off, causing major interference problems.

Carbonate samples may require two passes through the column to adequately reduce the level of Ca in the Sr cut. However, an alternative separation method for pure carbonates is to precipitate Sr from solution in concentrated nitric acid (Otto et al., 1988). The higher solubility of Ca under such conditions causes it to remain dissolved.

2.1.2 Sm–Nd and Lu–Hf

Rare-earth elements may be separated as a group on a cation exchange column eluted with dilute mineral acid, but because the chemical properties of individual rare earths are so similar, more refined techniques must be used for separations within the REE group. This is necessary because there are several isobaric interferences (e.g. [144]Sm interferes onto [144]Nd). Ba must also be kept to a minimum in the REE cut as it suppresses the ionisation of trivalent REE ions. This can be done by switching to a dilute (e.g. 2M) nitric acid medium after collection of the Sr cut, whereupon Ba is rapidly washed off the column ahead of the rare earths (Crock et al., 1984). The REE can be stripped most quickly as a group in ca. 50% HNO_3 (Fig. 2.2).

Several methods of separation between the rare earths have been used:

1) Hexyl di-ethyl hydrogen phosphate (HDEHP)-coated Teflon powder (stationary phase) with dilute mineral acid eluent, (Richard et al., 1976). In this technique, which may be called the 'reverse phase' method, light REE are eluted first, whereas in the others, heavy REE are eluted first. The reverse phase method yields sharp elution fronts but long tails (Fig. 2.3). It is very effective for removing the Sm interference from [144]Nd, and presently most popular. However, substantial Ce is usually present in the Nd cut, so [142]Nd cannot be measured accurately (section 2.2.5). Similarly, the separation between light REE is not good enough for Ce isotope analysis.

2) Cation exchange resin with hydroxy isobutyric acid (HIBA) eluent, (Eugster et al., 1970; Dosso and Murthy, 1980). This method requires more work in preparation of eluent, whose pH must be carefully controlled. It is therefore less popular than (1) for Nd, but very effective for Ce (Tanaka and Masuda, 1982; Dickin et al., 1987). It has also been recommended by Gruau et al. (1988) for Lu separation in the Lu–Hf method.

3) Anion exchange resin with methanol–dilute acetic acid–dilute nitric acid eluent (Hooker et al., 1975, O'Nions et al., 1977). This method is least popular at present, but may be better than (1) if an analysis of [142]Nd is required.

4) An alternative approach described by Cassidy and Chauvel (1989) is to use high-pressure liquid chromatography (HPLC).

The separation of hafnium (Hf) is difficult because it must be isolated from the very similar element Zr. A method is described in detail by Patchett and Tatsumoto (1980). Much of the procedure was carried out in an HF medium, due to the risk of

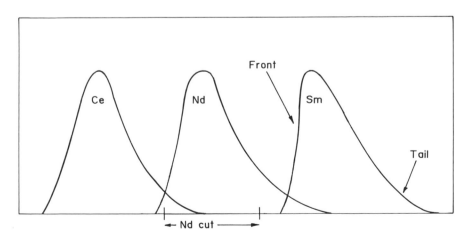

Fig. 2.3. Elution profiles of light REE from a reverse phase HDEHP column showing sharp front but long tail.

deposition of insoluble fluorides. However, if fluorides can be adequately displaced by refluxing with concentrated HNO_3 after dissolution, it may be possible to proceed with the Hf separation on a normal cation column eluted with HCl.

2.1.3 Lead

Lead (Pb) and uranium (U) have normally been separated from zircons by elution with HCl on anion exchange columns (Catanzaro and Kulp, 1964). However, this method is not able to separate Pb from the large quantities of Fe in whole-rock samples, which then causes unstable Pb emission during mass spectrometry. A widely used alternative is to elute all elements except Pb from a miniature anion column with dilute hydrobromic acid (Chen and Wasserburg, 1981). The distribution coefficient for Pb onto the resin has a maximum value at just under 1M HBr, and falls off sharply on either side (Fig. 2.4). Elution

with HBr effectively strips most elements, including Fe, from the column, after which Pb is collected with 6M HCl or water. (Pb also has a distribution coefficient maximum onto the solid phase at 2–3 M HCl which falls off in more dilute and more concentrated acid.) The sample may be passed a second time through a similar ion exchange procedure to further purify it.

Manton (1988) has shown that the distribution coefficient for Pb onto anion resin in dilute HBr is large enough for the absorbtion of small Pb samples onto a single large resin bead. The Pb is subsequently back-extracted into water. Yields from zircon samples were about 50%, after an equilibration period (with stirring) of about 8 hours. The somewhat low efficiency may be outweighed by the high purity of the product.

An alternative method of Pb separation from rocks is to use a two-stage electro-deposition method by which Pb is first deposited on the cathode at an accurately regulated potential, then redissolved and redeposited in a miniature cell on the anode (Arden and Gale, 1974). This method yields a high quality of separation, but is cumbersome to operate. Nevertheless, the anion stage may be useful alone for purifying galena Pb.

Levels of environmental contamination introduced during laboratory procedures are determined by the analysis of 'blanks'. These are measured by taking an imaginary sample through the whole chemical separation procedure, after which the amount of introduced extraneous contamination is measured by isotope dilution (section 2.2.6). Blank levels must be minimised in all of the chemical procedures described above, but particularly strenuous efforts are necessary to limit Pb contamination due to its relatively high concentration in the environment compared to normal rocks.

The minimum laboratory requirements to maintain low blanks would be an overpressure air system, subboiling distillation of all reagents in quartz or PFE stills, and evaporation of all samples under filtered air. For typical terrestrial whole-rock samples, acceptable blanks would not be more than 1 or 2 nanograms (ng $= 10^{-9}$g) total chemical blank for a Pb or Sr sample, and less than 1 ng for Nd (which is easier to control). This is necessary because the samples to be analysed often contain less than 1 microgram (µg $= 10^{-6}$g) of Pb, Sr or Nd. In the analysis of very small samples (e.g. single zircons) blanks of a few picograms (pg $= 10^{-12}$g) are necessary (e.g. Roddick *et al.*, 1987), since the Pb sample itself may weigh less than 1 ng.

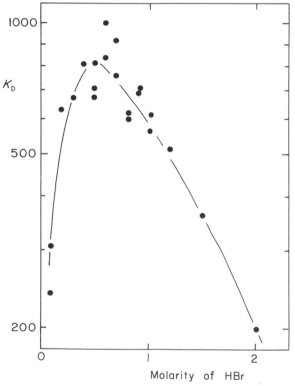

Fig. 2.4. Plot showing the distribution coefficient (K_D) of Pb from dilute HBr onto various types of anion exchange resin, as a function of molarity. The curve represents a best fit to the data points. After Manton (1988).

2.2 **Solid source mass spectrometry**

2.2.1 Sample loading

The most widely used excitation method in solid source mass spectrometry is thermal ionisation (hence TIMS). For some elements, such as Sr, stable emission of metal ions is achieved from a salt deposited directly onto a single metal filament (Fig 2.5a), usually tantalum (Ta). The loading procedure involves evaporating the salt solution onto the filament before insertion into the vacuum system. The sample is often loaded in phosphoric acid, which seems to a) displace all other anion species to yield a uniform salt composition, b) destroy organic residues (such as ion exchange resin) mixed with the sample, and c) glue the sample to the filament. During mass spectrometric analysis, the filament current is raised by means of a stabilised power supply to yield a temperature where effectively simultaneous volatilisation and ionisation of the sample occurs.

However, for many elements, stable volatilisation and ionisation of metal species does not occur at the same temperature. This problem was noted by Ingram and Chupka (1953), who first proposed the use of multiple source filaments (Fig. 2.5b). In this configuration, one or more filaments bearing the sample load can be heated to the optimum tempera-ture for stable volatilisation, while another hotter filament can be used to ionise the atomic cloud by bombarding it with electrons.

This method is particularly effective for REE analysis, where the sample is usually loaded onto one or both of the Ta side filaments of a triple-filament bead. These are held at a moderate temperature (ca. 1400 °C) where REE volatilisation is most stable. The centre filament (usually Re) is held at a much higher temperature (ca. 2000 °C), which promotes ionisation of the metal vapour. To some extent the ratio of metal to oxide species can be controlled by the centre filament temperature, which may help to suppress REE isobaric interferences. The properties of the REE under such conditions vary from light to heavy rare earths. La and Ce tend to form oxides unless extremely high centre filament temperatures are used, while heavier REE tend to form the metal species (Hooker *et al.*, 1975; Thirlwall, 1982).

Uranium and thorium may also be analysed by the triple-filament technique. Again, the temperature of the centre filament controls the metal/oxide ratio of the emitted ions (Li *et al.*, 1989). The triple-filament method was also used in the first successful analysis of Hf (Patchett and Tatsumoto, 1980). However, due to the high ionisation potential of Hf, the centre filament is required to be so hot (ca. 2200 °C) that it must be

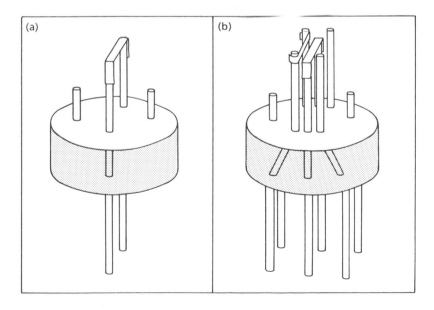

Fig. 2.5. The arrangement of filament ribbons on commonly used single- and triple-filament bead assemblies. Note that only one side filament is shown attached to the 'triple' bead.

made considerably longer to reduce thermal stress and prevent it burning out prematurely.

An alternative approach to using multiple filaments is to use special conditions to control the evaporation–ionisation behaviour from a single filament. For example, in the case of Pb, the sample is usually loaded on a rhenium filament in a suspension of silica gel (Cameron *et al.*, 1969). This is thought to form a blanket over the sample which effectively retards Pb volatilisation so that the filament can be raised to a higher temperature (where Pb fractionation is more reproducible) without burning off the sample uncontrollably.

Early Nd isotope determinations (Lugmair *et al.*, 1975; DePaolo and Wasserburg, 1976) were made using NdO$^+$ ions, whose emission from a single filament source was promoted by relatively high oxygen pressures in the mass spectrometer source housing. Some workers have continued to use this method, rather than the more popular multiple-filament method, since it can yield higher efficiency. Oxygen may be bled into the source in minute amounts to increase oxide emission. Alternatively, loading with silica gel may achieve the same objective without degrading source vacuum (Thirlwall, 1991a). Uranium may also be analysed as the oxide by mixing with TaO_2 powder on a tungsten filament.

A different approach to single filament U and Th analysis is to use graphite to promote the formation of metal ions (Edwards *et al.*, 1987). In this procedure it is critical to prevent oxidation during sample loading by avoiding oxidising acids and maintaining temperatures below the point of visible glowing. Noble *et al.* (1989) applied a similar technique to Nd isotope analysis. The use of platinised graphite was argued to give greater thermal stability to the reducing agent.

A development of this method enabled Hf to also be analysed using a single filament (Corfu and Noble, 1992). In this case, Hf ion production was achieved using very fine-grained metallic iridium (Ir-black). The addition of a little Mo powder (with a higher ionisation potential than Hf) also promotes ionisation, while graphite suppresses REE oxide interferences. It is notable however, that the filament temperature required for this procedure (2100–2300 °C) is similar to the centre filament temperature during triple-filament analysis of Hf.

2.2.2 Fractionation

The process of volatilisation and ionisation on the source filament of a mass spectrometer requires the breaking of chemical bonds, but the strength of these

bonds is mass dependent. This can be explained by approximating the chemical bond between two atoms as a harmonic oscillator.

The energy of a molecule (or part of an ionic lattice) decreases with decreasing temperature, but at absolute zero it has a certain finite value called the zero point energy, equal to $0.5\,h\nu$ (where h is Plank's constant and ν is the vibrational frequency). A bond involving the light isotope of an element has a higher vibration frequency and hence a higher zero point energy than one involving a heavier isotope, as illustrated in Fig. 2.6. The difference in bond energies diminishes as temperature rises, but still persists. Because the potential energy well of the bond involving the lighter isotope is always shallower than for the heavier, the bond with the lighter isotope is more readily broken. Hence it is preferentially released from the hot filament, causing isotopic fractionation.

If the sample on the filament was infinitely large, then mass fractionation (at constant temperature) would produce a constant discrepancy between the isotopic composition of the solid sample and the ion cloud. However, if the sample is finite, the continual process of fractionation starts to 'use up' the lighter isotope on the filament so that the isotopic composition of the sample gets progressively heavier (the 'reservoir effect'). Eberhardt *et al.* (1964) showed that this process follows a Rayleigh fractionation law (Fig. 2.7). The magnitude of this effect could yield totally

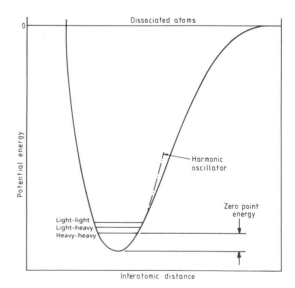

Fig. 2.6. Schematic diagram of potential energy against bond length for a hypothetical molecule made of two isotopes, based on the 'harmonic oscillator' model.

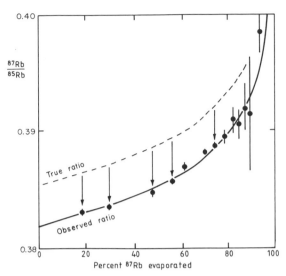

Fig. 2.7. Effect of within-run fractionation, over time, on a sample of natural rubidium undergoing isotopic analysis. Points are observed ratios; dashed line schematically indicates actual composition of Rb on the filament. Data from Eberhardt *et al.* (1964).

unacceptable errors of up to 1% in measured isotope ratio. However, for elements with two or more non-radiogenic isotopes, an internal normalisation for such mass-dependent fractionation can be performed.

In the case of strontium, the fractionation of $^{87}Sr/^{86}Sr$ can be monitored using the $^{88}Sr/^{86}Sr$ ratio, since ^{88}Sr and ^{86}Sr are both non-radiogenic (i.e. produced only by nucleosynthetic processes in stars). The ratio $^{86}Sr/^{88}Sr$ is constant throughout the Earth and is taken to be 0.1194 by international convention.

This value cannot be measured absolutely, but was originally estimated from the average beam composition half-way through very many runs. The deviation of observed $^{86}Sr/^{88}Sr$ from 0.1194 at each point through the run is divided by the difference between the two masses ($\Delta_{mass} = 2.003$) in order to calculate a fractionation factor per mass unit:

$$F = \frac{\dfrac{(^{86}Sr/^{88}Sr)_{obs}}{0.1194} - 1}{\Delta_{mass}} \qquad [2.2]$$

This fractionation factor can then be used to correct the observed (raw) $^{87}Sr/^{86}Sr$ ratio, for which $\Delta_{mass} = 1.003$:

$$\left(\frac{^{87}Sr}{^{86}Sr}\right)_{true} = \left(\frac{^{87}Sr}{^{86}Sr}\right)_{obs} (1 + F\Delta_{mass}) \qquad [2.3]$$

This has the effect of improving the within-run precision of the $^{87}Sr/^{86}Sr$ ratio from ca. 1% to better than 0.01%. $^{145}Nd/^{144}Nd$ metal analyses are similarly normalised for fractionation (Fig. 2.8) using an internationally agreed value of $^{146}Nd/^{144}Nd = 0.7219$ (O'Nions *et al.*, 1979). However, Nd oxide analyses are normalised to different values (Wasserburg *et al.*, 1981) which are incompatible with the Nd metal normalising value.

The fractionation correction described above is usually called the linear law, but the power law (Wasserburg *et al.*, 1981; Thirlwall, 1991b) is effectively identical. Both of these laws assume that fractionation is proportional to mass difference only, and is independent of the absolute masses of the fractionating species. In other words, fractionation

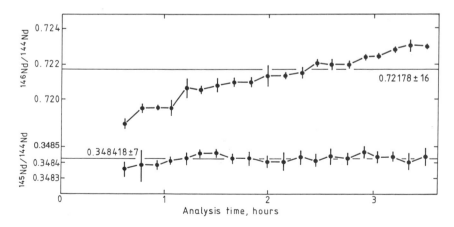

Fig. 2.8 Plot of raw $^{146}Nd/^{144}Nd$ ratios and fractionation-corrected $^{145}Nd/^{144}Nd$ ratios (normalised to $^{146}Nd/^{144}Nd = 0.7219$) for a single mass spectrometer run. Each point is a mean of 10 scans of the mass spectrum, while horizontal lines are grand means. After Noble (pers. comm.).

per mass unit is constant. However, this is an approximation to the real evaporation process, where fractionation per mass unit must vary inversely with the absolute masses of the evaporating species. Russell *et al.* (1978) first observed a breakdown of the linear law in isotopic analysis of the 'light' element Ca. To remedy this, they introduced an 'exponential' law, where the fractionation factor depends also on the mass of the evaporating species. This gave a better fit to Ca isotope data than the linear law (Fig. 2.9).

These problems are much less severe for Sr and Nd isotope analysis because of their heavier masses. However, Thirlwall (1991b) found small deviations from linear law behaviour in a large data set of Sr standard analyses. This is revealed by a correlation between normalised $^{87}Sr/^{86}Sr$ and average observed $^{86}Sr/^{88}Sr$ ratios for complete runs (Fig. 2.10). Thirlwall found that he could eliminate the correlation by retrospectively applying an exponential law correction to the data. This is described as follows:

$$\left[\frac{(^{87}Sr/^{86}Sr)_{norm}}{(^{87}Sr/^{86}Sr)_{corr}}\right]^{\ln(86/88)} = \left[\frac{(^{86}Sr/^{88}Sr)_{obs}}{0.1194}\right]^{\ln(87/86)} \quad [2.4]$$

where 'norm' and 'corr' refer to products of linear normalisation and exponential correction, and where 86, 87 and 88 are the actual masses of evaporating ions. However, this model assumes that strontium evaporates from the filament as the species 'Sr'.

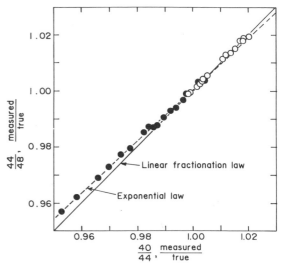

Fig. 2.9. Plot of measured/true $^{44}Ca/^{48}Ca$ *versus* $^{40}Ca/^{44}Ca$ ratios showing fit of linear and exponential fractionation laws to typical data from two runs. After Russell *et al.* (1978).

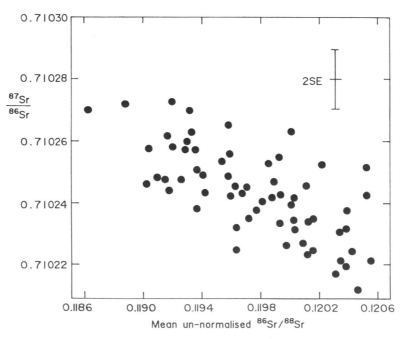

Fig. 2.10. Plot of fractionation-corrected $^{87}Sr/^{86}Sr$ ratios against mean un-corrected $^{86}Sr/^{88}Sr$ ratios for measurements of the SRM987 standard over a period of several months. After Thirlwall (1991b).

Habfast (1983) suggested that a strontium sample on a rhenium filament might evaporate as a species such as $SrReO_4$ rather than atomic Sr. Hence the 'apparent mass' of [88]Sr which should be used in the exponential correction would be ca. 330 rather than 88. Under such conditions the exponential law might actually produce a worse fit to the true fractionation behaviour of the sample than the linear law. However, Thirlwall's data suggest that evaporation from a tantalum filament does occur as the metal species, so the exponential model is an improvement over the linear model.

Internal fractionation correction is only possible where there are two or more isotopes present in a constant ratio. This is not the case for Pb isotope analysis, or in the isotope dilution measurement of Rb, so an external correction must be used. This depends on achieving uniform fractionation behaviour between standards and different samples, so that an across-the-board correction can be made to all runs. In the case of Pb, the use of a silica gel blanket on the filament achieves this objective by reducing the magnitude of fractionation processes drastically, from a previous between-lab. variation of ca. 3% to a present variation of ca. 0.3% . This improvement is partly due to the higher filament temperatures possible using silica gel. Because bond energy levels become closer with increasing temperature, the magnitude of isotope fractionation falls with increasing temperature. In the analysis of uranium, fractionation effects may be reduced by running at high temperature as the oxide. Alternatively, analysis as the metal ion produces larger but relatively consistent degrees of fractionation, which can be corrected by comparison with standard runs.

Where the relevant nuclides exist, double spiking with two artificial isotopes may be used to apply an internal fractionation correction to elements such as Pb with only one natural non-radiogenic isotope (section 2..2.7).

2.2.3 Ion optics

The ion optic properties of an instrument determine how the cloud of ions generated at the source filament is accelerated, focussed into a beam, separated by the magnetic field, and collected for measurement. Correct ion optic alignment is essential to obtain reliable results, because if part of the ion beam hits an obstruction, different masses may be affected to different degrees, leading to a bias in the results.

Most older mass spectrometers follow Nier's (1940) design (Fig. 2.1). From the filament (several kV

positive), the beam traverses a series of focussing source plates ('collimator stack', Fig. 2.11). These plates are at progressively lower potential and bring the beam to a principal focus at the source slit (zero volts). Thereafter, the beam slowly diverges (Fig. 2.12). In the y direction, the magnet brings each nuclide (isotope) beam back into focus at the primary collector slit, which is wide enough to let the whole of one nuclide beam through into the collector.

To bring a heavier nuclide into the collector, the magnetic field may be increased, or the momentum of the ions may be reduced by lowering the total high voltage (HV) potential across the collimator stack. Because the whole of one nuclide beam is focussed into the collector slit in the y direction, an apparently flat topped peak is produced when magnetic field or HV is varied to sweep the mass spectrum across the collector slit (e.g. Fig. 2.13). In practice, the magnetic field rather than the HV is normally switched to bring different nuclide beams into the collector, because the field can be more precisely monitored and controlled, using a 'Hall probe'. This is used to sense the field strength and adjust the magnet power supply in a feed-back loop ('field control'). The magnet (whose pole pieces are perpendicular to the beam) does not focus in the z direction, and in this direction the beam is 'clipped' by baffle plates.

This type of magnet design has a disadvantage, because the process of switching from one mass to another changes 'fringing fields' which are generated by the ends of the magnet poles. The change in these fringing fields may cause slight convergence or

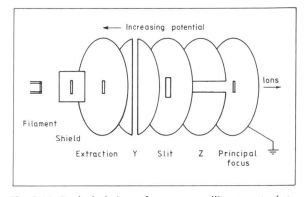

Fig. 2.11. Exploded view of a source collimator stack to show a typical sequence of focussing plates. The principal focus acts as the ion source with respect to the flight tube. However, confusion is caused by the fact that the shield plate over the filaments is often called the source slit. Y plates may also be called D plates.

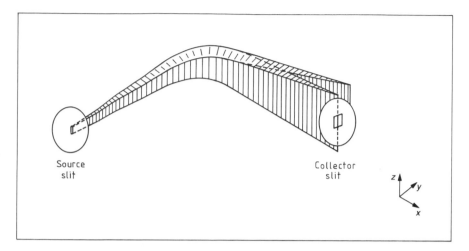

Fig. 2.12. Schematic diagram of the envelope of a double nuclide beam from the source slit to the collector slit in a Nier-type mass spectrometer.

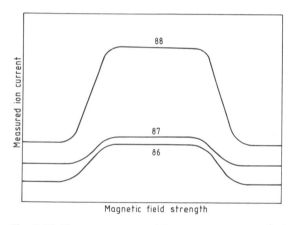

Fig. 2.13. The appearance of flat topped 'peaks' when strontium 88, 87 and 86 nuclide beams are swept across a triple (Faraday) collector system by varying magnetic field strength.

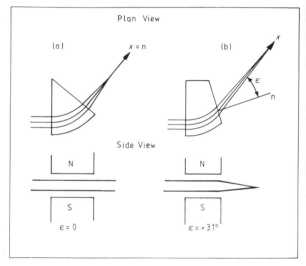

Fig. 2.14. The effect of different shaped pole faces in generating fringing fields which cause focussing of the ion beam in the y and z planes. a) Exit pole piece perpendicular to beam yields short focal length in the x direction but no focussing in z. b) Normal to exit pole face (n) is at an angle ϵ to the beam direction, yielding longer focal length in x, and also z focussing.

divergence of the beam, so that different amounts of different nuclide beams are clipped by the collector slit, and a slight bias is introduced to the beam current reaching the collector according to whether the magnet is switching 'up-mass' or 'down-mass'.

Recently-built machines feature refinements in the design of the magnet pole pieces, based on theoretical work by Cotte (1938). If the pole pieces are set at a slightly oblique angle to the beam, the fringing fields generated by the magnet have the effect of focussing the beam in the z direction (Fig 2.14). At the same time, the distance from the magnet exit pole to the principal focus in the y plane is increased. Therefore

this design is referred to as 'extended geometry'. Ion optics in this type of machine are shown in Fig. 2.15. This configuration has three advantages:

1) Because the whole of the z-focussed beam can pass through the collector slit, the transmission of the machine is improved (defined as the

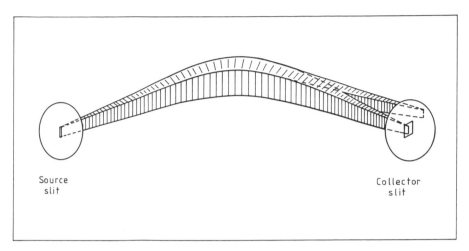

Fig. 2.15. Schematic diagram of the ion optics of an extended geometry machine between the source and collector slits. (Compare with Fig. 2.12.)

number of ions in the source required to yield each ion at the collector).

2) Small variations in fringing field do not cause signal bias, so accuracy is improved.

3) Extended geometry increases the distance between nuclide beams at the collector, so that multiple Faraday buckets can be more easily accommodated. Hence a magnet with 30 cm radius yields a beam separation equivalent to a magnet with 54 cm radius.

If one of the magnet pole faces (e.g. the entry pole) is made slightly convex, then this changes the normally oblique focal plane of nuclide beams in the y direction into a flat plane perpendicular to the beams (Fig. 2.16). This facilitates the installation of multiple collectors, whose spacing can then be adjusted to fit ion beams one atomic mass unit (a.m.u.) apart in any part of the spectrum. However, a more complex adjustable multiple collector configuration can be constructed on the oblique focal plane. At higher mass numbers the spacing of the collectors becomes closer and closer until their outer screens are actually touching during uranium analysis.

A very high vacuum throughout the ion path is essential, otherwise the ion beam becomes scattered, particularly to the low mass side, by inelastic collisions with air molecules. Such beam scattering becomes serious at analyser pressures $> 10^{-8}$ mbar. This causes the formation of a tail from one peak which may interfere with the adjacent nuclide. The problem is particularly severe in the case of a small peak down mass from a very large peak. For example, interference by ^{232}Th onto ^{230}Th may be severe in

silicate rocks with 232/230 ratios approaching 10^6. The magnitude of interference by a peak on a position one a.m.u. lower is called the 'abundance sensitivity' of the instrument (measured in ppm of the peak size). A typical specification for a single-focussing solid source machine with analyser vacuum $< 5 \times 10^{-9}$ mbar is 2 ppm at 1 a.m.u. from ^{238}U.

If a very high abundance sensitivity is essential, it can be obtained by adding a type of ion energy filter between the magnet and collector, thereby creating a double-focussing machine. The ion energy filter has the effect of removing ions with unusually high or low energy (= velocity). Thus, ions which have suffered a collision, and therefore lost energy, should be weeded out. Three types of filter which have been applied to this task are 'electrostatic', quadrupole, and ion retardation types. They typically result in a 10–100 fold improvement in abundance sensitivity.

2.2.4 Detectors

Ion beams in mass spectrometry normally range up to ca. 10^{-10} amps. For beams as small as 10^{-13} amps, the most suitable detector is the Faraday bucket. This is connected to electrical ground (Fig. 2.1) via a large resistance (e.g. 10^{11} ohm). Electrons travel from ground through this output resistor to neutralise the ion beam, and the potential across the resistor is then amplified and converted into a digital signal. A typical ion beam of 10^{-11} amps then generates a potential of 1 volt, converted to, say, 100 000 digital counts. Traditionally, an indefinite life-time has been assumed for Faraday buckets. However, the very

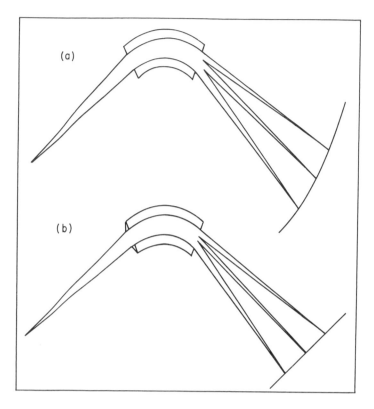

Fig. 2.16. Effect of magnet entry pole face shape. a) Flat pole face yields oblique curved focal plane at collector; b) convex pole face yields flat focal plane perpendicular to flight path.

narrow buckets in modern multi-collector arrays may significantly fill with sample debris if large beams are analysed. Therefore, it is not wise to repeatedly collect beams over 2×10^{-11} amps in size.

For ion beams smaller than ca. 10^{-13} amps, the electrical noise of the Faraday amplifier becomes significant relative to the signal size, so that some form of signal-multiplication is necessary. One of the most useful approaches was pioneered by Daly (1960). In the Daly detector, ions passing into the collector are attracted by a large negative potential (e.g. 20 kV). Collision of each ion with the polished electrode surface (Fig. 2.17) yields a secondary electron shower. When this impinges on a phosphor, the resulting light pulse is amplified by a photo-multiplier (situated behind a glass window, outside the vacuum system). In the analogue mode this system can have a gain (i.e. amplification) about 100 times the Faraday cup. Because ions do not strike the multiplier directly, the detector has a long life-time.

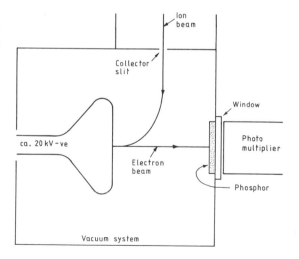

Fig. 2.17. Schematic diagram of a Daly detector showing the means of amplification of an incoming positive ion beam. After Daly (1960).

The Daly detector can be used only with positive ion beams. On the other hand, channel electron multipliers (CEM) can be used to amplify either positive or negative ion beams. These devices are therefore used for Re and Os analysis by negative molecular-ion TIMS (section 8.1). The negative ion enters the orifice of the CEM at a potential near zero, releasing electrons when it strikes the semi-conducting channel wall. These electrons are multiplied during further collisions, as they are attracted to a positive HV collector. Because the collector is at high voltage the signal cannot be amplified directly, but a pulsed ion-counting signal can be transmitted through an isolating capacitor to low voltage pulse-counting electronics (e.g. Kurz, 1979). The drawback of CEM detectors is their tendency to suffer damage when struck by heavy ions. Therefore signal sizes should be minimised to prolong their life.

2.2.5 Data collection

In order to achieve very high precision data, it is necessary to measure the intensity of each ion beam for an hour or more. To achieve this in a single collector machine, the magnetic field is switched to cycle round a sequence of peak positions. On switching to a new peak there is a waiting period of 1–2 s to allow the output resistor and amplifier to reach a steady state in response to the new ion current. Then data is collected for a few seconds. In practice, each peak must normally be corrected for incomplete decay of the signal from previous peaks (termed 'dynamic zero', 'tau' or 'resistor memory' correction).

Lengthy measurement of each peak in the cycle is governed by a law of 'diminishing returns' because while random noise is lowered, the instability of the ion source becomes more apparent between the measurement of successive peaks. The computer therefore cycles round and round a series of peaks, baseline/background(s) and interference monitor position(s), interpolating between successive measurements of the same peak to correct for growth or decay of the beam size. A simple linear time interpolation may be used (Fig. 2.18), but Dodson (1978) has developed a more sophisticated algorithm which can make better allowance for non-linear beam growth or decay.

Before isotope ratios can be determined from the different signals, background electronic noise must be subtracted in order to determine net peak heights. This is done by measuring a baseline position in each collector channel at approximately 0.4 a.m.u. above a

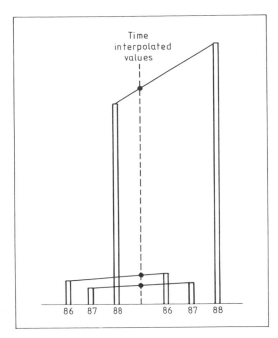

Fig. 2.18. Schematic illustration of the principle of linear time interpolation for a strontium ion beam growing at an (immense) rate of 30% per scan.

whole mass position, usually a few a.m.u. away from the masses of interest. From a single cycle of time-interpolated net peak heights, a set of net peak ratios is extracted. These are often collected in blocks of 10 scans. The cycle time round a set of peaks may be shortened by the measurement of backgrounds and interferences 'between blocks' rather than 'within-scan'. Using a 5 second measurement, 2 second delay cycle and collecting 200 scans in about 3 hours might give a statistical within-run precision of 0.004% (2σ = 2 standard errors on the mean). That is, the scatter of data around the mean suggests that one can be 95% confident that the 'correct' answer lies within 0.004% on either side of the mean. For $^{87}Sr/^{86}Sr$ this is equivalent to 0.71000 ± 0.00003 (2σ).

Occasional signal 'spikes' and other perturbations outside of normal random error are inevitable in a long mass spectrometer run. It is generally regarded as acceptable to run through the data a few times and test for outliers which are more than a certain number of population standard deviations (SD) from the mean (Pierce and Chauvenet, 1868; in Crumpler and Yoe, 1940). The cut-off level should depend on the size of data set, so that only a minimal number of outliers resulting from normal random variation are rejected. In practice the cut-off would normally be between 2 SD and 3 SD.

Ideally, a multiple collector machine can analyse isotope ratios in a 'static' mode without peak jumping. However, this may be limited, firstly by the extent to which each of the Faraday buckets is identical in terms of beam transmission characteristics; and secondly by the extent to which the gain of each bucket's amplifier system can be calibrated. Until recently, these problems did not allow static analysis to achieve the highest levels of analytical precision, for Sr or Nd isotope analysis. However, the quality of static analysis has been steadily improving, and on new machines may now yield data of very high quality. The alternative approach for high-precision analysis is double- or triple-collector peak-jumping (multi-dynamic analysis). The simplest (double-collector) method is given below:

High collector: 87   88   91.4   –

Low collector: 86   87   90.4   85

Place in sequence: 1   2   3   4

After background subtraction, the ^{85}Rb monitor is used to correct both ^{87}Sr peaks. The following algorithm represents an approximation, assuming unit mass differences between the isotopes:

$$\left(\frac{87}{86}\right)_{true} = \left[\left(\frac{87}{86}\right)_1 \cdot \left(\frac{87}{88}\right)_2 \cdot \frac{1}{0.1194}\right]^{0.5} \quad [2.5]$$

where suffixes denote places in the scan sequence. This equation cancels out beam growth or decay and amplifier bias, as well as performing a power law mass fractionation correction, all in a single calculation. To use the exponential law for Sr evaporation as the metal, the function above is raised to the power of 0.5036 (Thirlwall, 1991b). With both of these methods, within-run precision should reach 0.002% (2 sigma) after 3 hours. Triple-collection analysis allows a further improvement in efficiency. In this method, two double-collector determinations on adjacent collectors are averaged to yield a more precise result.

Unfortunately, while within-run precision can be taken to lower and lower levels by more efficient sample ionisation and data collection, between-run precision often reaches a limiting value of ca. 0.004% (2 sigma) which is difficult to break through. Thirlwall (1991b) has attributed this phenomenon to imperfect centring of some peaks during dynamic multiple collection. This occurs because mass separations decrease as absolute masses increase, so that the collectors are actually set slightly different distances apart. Using the above example, if the three collectors

are set up with masses 86–87 perfectly centred (position 1 in the sequence) then at position 2 the peaks will be slightly off-centre. This slight miss-centring may amplify any non-idealities in the optic path of the beam, so that slightly different source configurations (for different bead numbers) yield small systematic variations in isotope ratio.

Thirlwall (1991b) argued that between-bead precision of Nd isotope analyses could be improved to a level comparable with within-run precision by a secondary normalisation using ^{142}Nd/^{144}Nd. This is based on an observed correlation between fractionation-corrected ^{143}Nd/^{144}Nd and ^{142}Nd/^{144}Nd ratios from standard runs over a period of months (Fig. 2.19). The data array follows a mass fractionation line, but is probably not a fractionation effect, because no correlation is observed with ^{146}Nd/^{144}Nd. Thirlwall's approach is only possible if Ce interferences on ^{142}Nd are low, and may not therefore be possible using the popular 'reverse phase' Nd separation (see above). However, it does raise some interesting questions, and may provoke other attempts to improve analytical reproducibility to its theoretical limits.

2.2.6 Isotope dilution

Isotope dilution is generally agreed to be the supreme analytical method for very accurate concentration determinations. In this technique, a sample containing an element of natural isotopic composition is mixed with a 'spike' solution, which contains a known concentration of the element, artificially enriched in one of its isotopes. When known quantities of the two solutions are mixed, the resulting isotopic composition (measured by mass spectrometry) can be used to calculate the concentration of the element in the sample solution. The element in question must normally have two or more naturally occurring isotopes, one of which can be enriched on a mass separator. However, in some cases a long-lived artificial isotope is used.

The isotopic composition of the spike must be accurately measured by mass spectrometry. This measurement cannot be normalized for fractionation, because there is no 'known' ratio to use as a fractionation monitor. Therefore, several long runs are generally made, from which the average midpoint of the run is taken to be the actual spike composition. The concentration of the spike is generally determined by isotope dilution against standard solutions (of natural isotopic composition) whose concentrations are themselves calculated gravimetrically. Metal

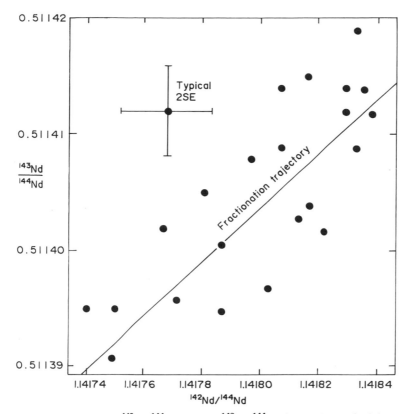

Fig. 2.19. Plot of fractionation-corrected $^{143}Nd/^{144}Nd$ against $^{142}Nd/^{144}Nd$ for analyses of a laboratory standard over a period of months (normalisation is to $^{146}Nd/^{144}Nd = 0.7219$). After Thirlwall (1991b).

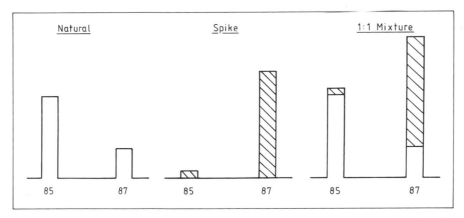

Fig. 2.20. Summation of spike (hatched) and natural ion beams to generate aggregate mixed peaks, as illustrated by rubidium isotope dilution.

oxides are generally weighed out, but if these are hygroscopic (e.g. Nd_2O_3) then accurate weighing requires part of a metal ingot.

A simple example of isotope dilution analysis is the

determination of Rb concentration for the Rb–Sr dating method. Typical mass spectra of natural, spike, and mixed solutions are shown in Fig. 2.20.

In the mixture, each isotope peak is the sum of

spike (S) and natural (N) material. Hence,

$$\frac{^{87}\text{Rb}}{^{85}\text{Rb}} = R = \frac{\text{moles } 87_N + \text{moles } 87_S}{\text{moles } 85_N + \text{moles } 85_S} \quad [2.6]$$

But the number of moles of an isotope is equal to the number of moles of the element as a whole, multiplied by the isotopic abundance. If the total number of moles of natural and spike Rb are represented by M_N and M_S, then:

$$R = \frac{M_N \cdot \%87_N + M_S \cdot \%87_S}{M_N \cdot \%85_N + M_S \cdot \%85_S} \quad [2.7]$$

where percentages indicate the isotopic abundances in the spike and natural solutions. This equation is rearranged in the following steps:

$$R(M_N \cdot \%85_N + M_S \cdot \%85_S)$$
$$= M_N \cdot \%87_N + M_S \cdot \%87_S$$

$$R \cdot M_N \cdot \%85_N + R \cdot M_S \cdot \%85_S$$
$$= M_N \cdot \%87_N + M_S \cdot \%87_S$$

$$R \cdot M_N \cdot \%85_N - M_N \cdot \%87_N$$
$$= M_S \cdot \%87_S - R \cdot M_S \cdot \%85_S$$

$$M_N(R \cdot \%85_N - \%87_N)$$
$$= M_S(\%87_S - R \cdot \%85_S)$$

$$M_N = M_S \cdot \frac{(\%87_S - R \cdot \%85_S)}{(R \cdot \%85_N - \%87_N)} \quad [2.8]$$

If we insert figures for the isotopic abundance of natural Rb, and the isotopic abundance of a typical spike, such that 27.83/72.17 is the natural $^{87}\text{Rb}/^{85}\text{Rb}$ ratio, and 99.4/0.6 is the spike $^{87}\text{Rb}/^{85}\text{Rb}$ ratio, then the number of moles of the natural Rb is given by:

$$M_N = \frac{(99.4 - R \cdot 0.6)}{(R \cdot 72.17 - 27.83)} \cdot M_S \quad [2.9]$$

where R is the measured isotope ratio.

But number of moles = molarity × mass, so the molarity of the natural sample is given by:

$$\text{Molarity}_N = \frac{(99.4 - R \cdot 0.6)}{(R \cdot 72.17 - 27.83)} \cdot \frac{\text{wt}_S}{\text{wt}_N} \cdot \text{Molarity}_S$$
$$[2.10]$$

Molarity is then multiplied by atomic weight to yield concentration:

$$\text{Conc.}_N = \text{At. wt}_N \cdot \frac{(99.40 - R \cdot 0.6)}{(R \cdot 72.17 - 27.83)} \cdot \frac{\text{wt}_S}{\text{wt}_N} \cdot \frac{\text{Conc.}_S}{\text{At.wt}_S}$$
$$[2.11]$$

Because there are only two isotopes of Rb, no internal correction for fractionation is possible in the

measurement of $^{87}\text{Rb}/^{85}\text{Rb}$. However, in the isotope dilution analysis of Sr, fractionation correction based on 88/86 measurement is possible, which allows a much more accurate 84/86 (spike Sr/ natural Sr) measurement to be made.

Given a sufficiently highly enriched ^{84}Sr spike (e.g. 99.9%), the $^{87}\text{Sr}/^{86}\text{Sr}$ ratio can also be determined on the spiked run, rather than by a separate analysis. Accurate correction for the minor ^{86}Sr, ^{87}Sr and ^{88}Sr isotopes of the spike clearly depends on a good determination of the spike isotope composition itself. The error on the determination of these small isotopes attributable to random signal noise is largely independent of their size. However, the fractionation uncertainty in the spike composition analysis is directly proportional to the size of the minor isotopes. This is why precise isotope ratio measurements (e.g. $^{87}\text{Sr}/^{86}\text{Sr}$) on spiked runs are only possible when using highly enriched spikes. Unfortunately, the highly enriched Sr spike is no longer available on the open market.

In the case of Pb, where all naturally occurring isotopes must be measured for the isotope composition, a separate spiked run is essential, unless an unstable artificial isotope is used as spike (e.g. ^{202}Pb or ^{205}Pb). These isotopes are difficult to manufacture, but the procedure for ^{205}Pb was recently described by Parrish and Krogh (1987). Enriched ^{206}Pb was bombarded with protons to yield ^{205}Bi, which is radioactive with a 15 day half-life. After chemical purification, this was allowed to decay to produce a ^{205}Pb spike with low levels of other Pb isotopes. A price of over US $2000 per microgram was quoted on this spike to pay for cyclotron time.

Isotope dilution is potentially a very high-precision method. However, error magnification may occur if the proportions of sample to spike which are mixed are far from unity (Fig. 2.21). Consequently, it is generally believed that the analysed peaks in an isotope dilution mixture should have an abundance ratio of close to unity. In actual fact, the ideal composition of the mixture is half-way between that of the natural and spike compositions. However, the best precision normally required in an isotope dilution analysis is ca. 1 per mil (0.1 %), which is two orders of magnitude worse than the precision normally achieved in Sr or Nd isotope ratio measurements. Hence, significantly non-ideal spike–natural mixtures can be tolerated in normal circumstances.

The only other sources of error in isotope dilution analysis are incomplete homogenisation between sample and spike solution, and weighing errors. The first of these can be overcome by centrifuging the

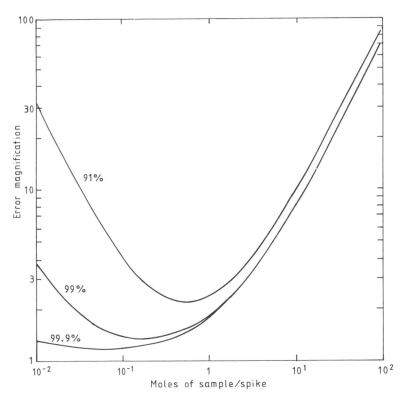

Fig. 2.21 Estimates of error magnification in isotope dilution analysis as a function of the total number of moles of sample/spike. Cases are shown for different percentages of spike isotope enrichment, mixed with a natural sample with a 50–50 isotopic abundance ratio. After DeBievre and Debus (1965).

sample solution to check for any undissolved material, and repeating the dissolution steps as necessary until complete solution is achieved. Given sufficient care, including the use of non-hygroscopic standard material and correct balance calibration, spike solutions can be calibrated to 0.1% accuracy (Wasserburg et al., 1981). The use of mixed spikes (e.g. Sm–Nd, Rb–Sr) then eliminates further weighing errors in the analysis of these ratios in sample material. The result is that isotope dilution accuracy can exceed 1% with ease, and 0.1% if necessary. This compares very favourably with all other analytical methods.

2.2.7 Double spiking

The selection of an arbitrary non-radiogenic ratio for fractionation normalisation (e.g. $^{88}Sr/^{86}Sr = 8.37521$) results in no loss of information for terrestrial samples. However, for meteorites, use of such a procedure means that the true isotopes responsible for

certain anomalous isotope ratios cannot be uniquely identified. An example is provided by inclusions EK 1-4-1 and C1 of the Allende chondritic meteorite (Papanastassiou et al., 1978), for which several mixing models of nucleosynthetic components are possible (e.g. Clayton, 1978).

The double-spike isotope dilution technique can be used to overcome this problem by correcting for fractionation in the mass spectrometer source, thus allowing comparison between all of the isotope ratios in a suite of samples, including those used for fractionation normalisation. The theory of double spiking was first investigated in detail by Dodson (1963). The calculations may be made iteratively (e.g. Compston and Oversby, 1969) or algebraically (e.g. Gale, 1970). However, it is not usually possible to calculate 'absolute' values of isotopic abundance because there is not normally any absolute standard to calibrate the double spike.

Double spiking has been used to study processes of Sr isotope fractionation in the solar nebula, as

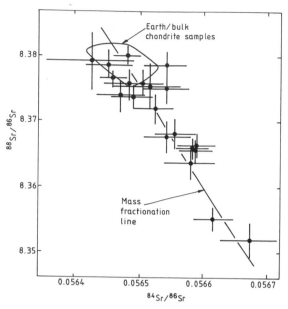

Fig. 2.22. Plot of $^{88}Sr/^{86}Sr$ against $^{84}Sr/^{86}Sr$ for samples from Allende chondrules. Instrumental mass fractionation has been corrected using a double-spike algorithm. After Patchett (1980).

these are thought to be relatively early condensates from the nebula, which should therefore be enriched in heavy isotopes. The simplest explanation for the effect is that the inclusions condensed from an isotopically light region of the nebula.

Several workers have investigated the use of double Pb spikes to allow within-run mass fractionation correction of Pb isotope ratios. Most of these studies utilised double stable isotope spikes such as $^{207}Pb-^{204}Pb$ (Compston and Oversby, 1969; Hamelin *et al.*, 1985), which necessitate two separate mass spectrometer runs, one with spike and one without. This is because the spike interferes with the determination of natural ^{207}Pb and ^{204}Pb. In contrast, the use of a $^{202}Pb/^{205}Pb$ double spike allows both concentration determination and a correction for analytical mass fractionation to be made on a single Pb mass spectrometer run (Todt *et al.*, 1984). This technique has not been widely used for Pb, because between-run fractionation has been controlled by the silica gel loading technique. In addition, the double-spike method may cause error magnification if the spike/sample ratio is not close to optimal, or if the mass spectrometer run is not of high precision (Fig. 2.23).

2.3 Isochron regression line fitting

When $^{87}Sr/^{86}Sr$ compositions for a suite of cogenetic samples of the same age are plotted against their $^{87}Rb/^{86}Sr$ ratios, the points should ideally define a perfect straight line or 'isochron' (section 3.2.2).

sampled by inclusions and chondrules in the Allende meteorite, relative to the composition of all terrestrial samples (Patchett, 1980). $^{88}Sr/^{86}Sr$ *versus* $^{84}Sr/^{86}Sr$ variations were found to fit a mass fractionation line very well (Fig. 2.22). Ca–Al inclusions were enriched in the lighter isotopes of Sr, which is surprising, since

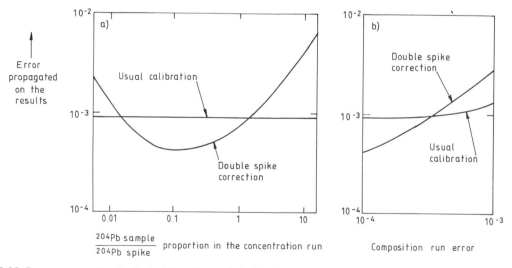

Fig. 2.23. Error propagation in Pb double-spike analysis. (a) Effect of differing proportions of spike to natural Pb; (b) effect of within-run precision. After Hamelin *et al.* (1985).

However, since both quantities involved are measured experimentally, experimental errors are inevitable.

A straight line fitted to an array of points is termed a 'linear regression'. One of the best approaches involves minimising the sum of the squares of the distances that data points lie away from a line drawn through the points, and hence is called a least squares fit. This involves iteration, and is therefore done by calculator or computer.

In simple linear regression programs, one ordinate is defined as 'free of error' and the regression line is calculated to minimise the mis-fit of points in the other ordinate (Fig. 2.24a,b). If the data define a very gentle slope and are somewhat scattered, disastrous fits can be produced by regressing onto the wrong ordinate. Where errors are present in both ordinates, as in the case of the isochron fit, a two-error regression treatment must be used (Fig. 2.24c,d). Several methods of this type have been presented in the literature (e.g. McIntyre et al., 1966; York, 1966; York, 1967; Brooks et al., 1972), sometimes including a ready-to-run computer program.

In cases where the actual deviations of the data points from the regression line are equal to or less than those expected from experimental error, all regression treatments effectively give the same isochron age and initial ratio. In such cases, the only matter of debate is the manner in which experimental errors are assigned.

Ideally, the analysed error in $^{87}Sr/^{86}Sr$ and $^{87}Rb/^{86}Sr$ (for example) would be determined by measuring the reproducibility of an almost infinite number of duplicates (Brooks et al., 1972). Since this is very time-consuming, the best working estimate is probably the long-term reproducibility of standard analyses. Within-run precision of sample analyses is almost certainly an under-estimate of error, since it is typically about 50% of the reproducibility error. For $^{87}Rb/^{86}Sr$, quoted accuracies must include an estimate for sample weighing errors, spike calibration errors etc., as well as mass spectrometry errors (in the case of isotope dilution).

While some of the regression programs in use provide the facility for weighting each data point according to its precision of measurement, this may sometimes be detrimental, as it tends to 'destabilise' the fit. In practice, $^{87}Rb/^{86}Sr$ and $^{87}Sr/^{86}Sr$ errors are probably best assigned as a blanket percentage (e.g. 0.5% and 0.002% 1σ respectively). If one point has a

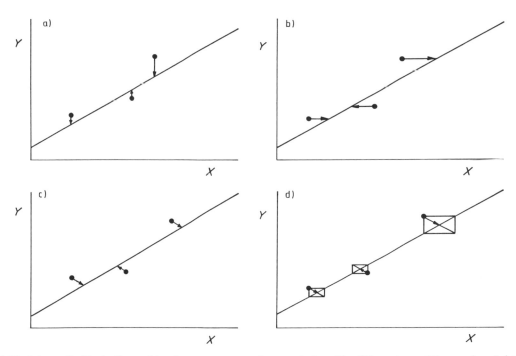

Fig. 2.24. Schematic illustration of least squares regression analysis with different conditions of weighting. a) Infinite weighting of X (all errors in Y); b) infinite weighting of Y; c) fixed weighting of X versus Y; d) individual weighting of each point, inversely proportional to squares of standard deviations. After York (1967).

particularly bad precision, it is better to re-analyse it than give it less weight in the regression.

2.3.1 Regression fitting with correlated errors

In conventional isochron analysis (e.g. Rb–Sr), analytical errors in the two ordinates (isotope ratio and elemental abundance ratio) are effectively uncorrelated (*but see* Zheng, 1989). However, in the lead isotope dating methods, this is far from the case. In common Pb–Pb dating, correlated errors are found between $^{207}Pb/^{204}Pb$ and $^{206}Pb/^{204}Pb$, due to greater analytical uncertainties on the small ^{204}Pb peak (which is common to both ratios) and due to the uncertainty of mass fractionation. These two correlation lines have different slopes, and while the former may be important for very small Pb beam sizes, the latter is normally dominant. Data for the NBS 981 standard shown in Fig. 2.25 yield a correlation coefficient of 0.94 (Ludwig, 1980).

In U–Pb zircon dating, errors may show a much stronger correlation. This is because errors in $^{206}Pb/^{238}U$ and $^{207}Pb/^{235}U$ are mainly attributable to the elemental U/Pb ratio, which may be five or more times less reproducible than the $^{206}Pb/^{207}Pb$ ratio (Davis, 1982). This difference arises from the analytical errors inherent in isotope dilution, and to uncertainties in common Pb correction (section 5.2). Regression treatments for correlated errors using the least squares technique were given by York (1969) and Ludwig (1980). Davis (1982) used the alternative 'maximum likelihood' method, and showed that the two approaches yielded similar estimates of error using test data (*see also* Titterington and Halliday, 1979).

2.3.2 Errorchrons

Brooks *et al.* (1972) argued that 'a line fitted to a set of data that display a scatter about this line in excess of the experimental error is simply not an isochron'. They proposed that Rb–Sr regression fits with excess or 'geological' scatter (McIntyre *et al.*, 1966) should be called 'errorchrons' and treated with a high degree of suspicion. This raises the problem of how to detect the presence of geological scatter, bearing in mind the fact that analytical errors are only probabilities.

The sum of the squares of the mis-fits of each point to the regression line (= squared residuals; York, 1969) or sum of χ^2 (Brooks *et al.*, 1968, 1972), may be divided by the degrees of freedom (number of data points minus two) to yield the Mean Squared Weighted Deviates (MSWD) which is the most

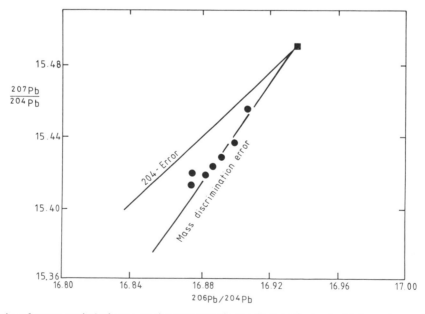

Fig. 2.25. Results of seven analytical runs on the NBS 981 Pb standard performed with large beam sizes at varying filament temperatures. Data cluster near the mass fractionation line. Solid square = 'true value'. After Ludwig (1980).

convenient expression of scatter. If the scatter of data points is, on average, exactly equivalent to that predicted from the analytical errors, then the calculation will yield MSWD = 1. Excess scatter of data points yields MSWD > 1, while less scatter than predicted from experimental errors yields MSWD < 1.

Problems may arise with the interpretation of these MSWD values, since the analytical errors input into the program are only estimates of error. To address this problem, Brooks *et al.* (1972) constructed a table of probabilities (Table 2.1) to distinguish between errorchrons and isochrons from their MSWD values. They established a 'rule of thumb' that on average if MSWD <2.5 then the data define an isochron, and if >2.5 an errorchron. Unfortunately this rule of thumb has been much abused over subsequent years, because the original objectives of Brooks *et al.* (1972) have been misunderstood. They set up the MSWD = 2.5 cut-off in order to reject errorchrons with a 95% certainty of excess scatter over analytical error. This corresponds to only 5% confidence that a fit with MSWD = 2.5 is an isochron (e.g. Wendt and Carl, 1991). However, many workers have wrongly assumed that if MSWD is less than 2.5 then there is a high degree of confidence that the suite is a true isochron (where analytical errors express most or all of the error on the age). In actual fact, MSWD must be near unity in order to have a high degree of confidence that the data represent a true isochron.

Because the number of errorchrons will continually increase as analytical errors decrease, the suggestion that errorchrons be rejected outright is unhelpful. Therefore, other workers have looked for ways of quantifying geological scatter in terms of an error on the age result. McIntyre *et al.* (1966) emphasised that statistical error estimation of errorchrons cannot be properly meaningful unless the geological reasons for the mis-fit are understood. Therefore, they suggested four alternative approaches for error handling. These are as follows:

1) No excess scatter above analytical errors (= true isochron).
2) All excess scatter is attributed to Rb/Sr, equivalent to assuming small differences between the initial ages of the samples.
3) All excess scatter is attributed to $^{87}Sr/^{86}Sr$, equivalent to assuming variation in the initial isotopic ratio of samples.
4) Excess scatter is attributed to some combination of models 2 and 3.

The program of York (1966) allows the analytical errors on X and Y ordinates (e.g. $^{87}Rb/^{86}Sr$ and $^{87}Sr/^{86}Sr$) to be multiplied in equal and uniform proportion by an error factor ($\sqrt{MSWD}$) until the expanded errors equal geological scatter (MSWD = 1). The error on the calculated age will be magnified by this process to give a reasonable estimate of uncertainty which includes both geological and analytical scatter.

Some form of error expansion must *always* be performed if a meaningful geological error estimate is to be given for a data set with MSWD > 1, because this is a definite indicator that excess scatter of some form is present. The only uncertainty is whether the excess scatter is geological or analytical. The York (1966) procedure is the most common method of dealing with excess scatter, but it is an arbitrary procedure which takes no account of geological processes, and their resulting contribution to errors. Where initial ratio variability is suspected, option 3 of McIntyre (above) is preferable (amplification of isotope ratios only). However, this can lead to misinterpretation of a data set if all points are not identical in age.

Misinterpretation of errorchrons is usually due to a failure to properly visualise the distribution of data and attendant errors. This can be avoided by using a graphical presentation. Different methods of graphical assessment using isochron diagrams will be discussed for the Rb–Sr method (section 3.2.2). However, an alternative approach is the so-called 'bootstrap method' of Kalsbeek and Hansen (1989). In this method a set of errorchron data is analysed by computer to see how stable the regression line is to the application of a greater weighting to different points. This test is achieved by successive random selection of a sample of points from the data set, such that this sample is equal in size to the data set. (This is not as

Table 2.1. *MSWD values indicating 95% confidence of an errorchron*

Number of duplicates	Number of samples regressed								
	3	4	5	6	8	10	12	14	26
10	4.96	4.10	3.71	3.48	3.22	3.07	2.98	2.91	2.74
20	4.35	3.49	3.10	2.87	2.60	2.45	2.35	2.28	2.08
30	4.17	3.32	2.92	2.69	2.42	2.27	2.16	2.09	1.89
40	4.08	3.23	2.84	2.61	2.34	2.18	2.08	2.00	1.79
60	4.00	3.15	2.76	2.53	2.25	2.10	1.99	1.92	1.70
120	3.92	3.07	2.68	2.45	2.18	2.02	1.91	1.83	1.61

Numbers underlined just exceed MSWD=2.5 cut-off

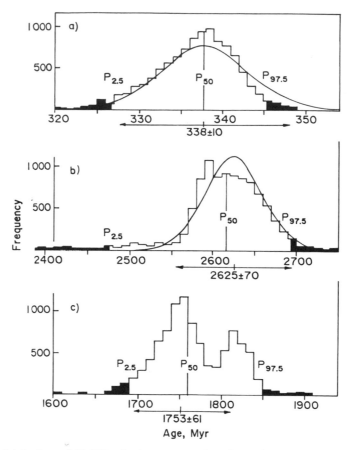

Fig. 2.26. Frequency distribution of 10 000 selection permutations from three sets of errorchron data. 95% (2σ) confidence limits of the 'bootstrap' age determination are indicated ($P_{2.5}$ and $P_{97.5}$). Arrows represent symmetrical 2σ confidence limits resulting from expansion of analytical errors until MSWD=1. After Kalsbeek and Hansen (1989).

strange as it sounds.) What will normally happen is that a few points are selected more than once, while others are omitted. By repeating this process a few thousand times, a probability distribution is set up which portrays the stability of the best-fit line to the influence of certain sub-sets of the data suite (Fig. 2.26).

If geological errors are randomly distributed, the frequency histogram derived from the data set will have a symmetrical (Poisson) distribution. In this case the result is identical to expanding analytical errors by $\sqrt{\text{MSWD}}$. (Of course, a true isochron should always yield a Poisson distribution, because analytical errors are assumed to be random.) However if geological scatter is uneven, the probability histogram of an errorchron may be skewed or even bimodal (Fig. 2.26), and hence highly suspect in terms of age assignment. This diagram therefore represents an

excellent visual test for isochron data quality, and could help to avoid the misinterpretation of problematical data sets.

References

Arden, J. W. and Gale, N. H. (1974). New electrochemical technique for the separation of lead at trace levels from natural silicates. *Anal. Chem.* **46**, 2–9.

Brooks, C., Wendt, I. and Harre, W. (1968). A two-error regression treatment and its application to Rb–Sr and initial Sr^{87}/Sr^{86} ratios of younger Variscan granitic rocks from the Schwarzwald massif, Southwest Germany. *J. Geophys. Res.* **73**, 6071–84.

Brooks, C., Hart, S. R. and Wendt, I. (1972). Realistic use of two-error regression treatments as applied to rubidium–strontium data. *Rev. Geophys. Space Phys.* **10**, 551–77.

Cameron, A. E., Smith, D. H. and Walker, R. L. (1969)

Mass spectrometry of nanogram-size samples of lead. *Anal. Chem.* **41**, 525–6.

Cassidy, R. M. and Chauvel, C. (1989). Modern liquid chromatographic techniques for the separation of Nd and Sr for isotopic analyses. *Chem. Geol.* **74**, 189–200.

Catanzaro, E. J. and Kulp, J. L. (1964). Discordant zircons from the Little Butte (Montana), Beartooth (Montana) and Santa Catalina (Arizona) Mountains. *Geochim. Cosmochim. Acta* **28**, 87–124.

Chen, J. H. and Wasserburg, G. J. (1981). Isotopic determination of uranium in picomole and sub-picomole quantities. *Anal. Chem.* **53**, 2060–7.

Clayton, D. D. (1978). On strontium isotopic anomalies and odd-A p-process abundances. *Astrophys. J.* **224**, L93–5.

Compston, W. and Oversby, V. M. (1969). Lead isotopic analysis using a double spike. *J. Geophys. Res.* **74**, 4338–48.

Corfu, F. and Noble, S. R. (1992). Genesis of the southern Abitibi greenstone belt, Superior Province, Canada: evidence from zircon Hf isotope analyses using a single filament technique. *Geochim. Cosmochim. Acta* **56**, 2081–97.

Cotte, M. (1938). Recherches sur l'optique electronique. *Ann. Physique* **10**, 333–405.

Crock, J. G., Lichte, F. E. and Wildeman, T. R. (1984). The group separation of the rare-earth elements and yttrium from geological materials by cation-exchange chromatography. *Chem. Geol.* **45**, 149–63.

Croudace, I. W. (1980). A possible error source in silicate wet-chemistry caused by insoluble fluorides. *Chem. Geol.* **31**, 153–5.

Crumpler, T. B. and Yoe, J. H. (1940). *Chemical Computations and Errors*, Wiley, pp. 189–90

Daly, N. R. (1960). Scintillation type mass spectrometer ion detector. *Rev. Sci. Instrum.* **31**, 264–7.

Davis, D. W. (1982). Optimum linear regression and error estimation applied to U–Pb data. *Can. J. Earth Sci.* **19**, 2141–9.

DeBievre, P. J. and Debus, G. H. (1965). Precision mass spectrometric isotope dilution analysis. *Nucl. Instrum. Meth.* **32**, 224–8.

DePaolo, D. J. and Wasserburg, G. J. (1976). Nd isotopic variations and petrogenetic models. *Geophys. Res. Lett.* **3**, 249–52.

Dickin, A. P., Jones, N. W., Thirlwall, M. and Thompson, R. N. (1987). A Ce/Nd isotope study of crustal contamination processes affecting Palaeocene magmas in Skye, NW Scotland. *Contrib. Mineral. Petrol.* **96**, 455–64.

Dodson, M. H. (1978). A linear method for second-degree interpolation in cyclical data collection. *J. Phys. E (Sci. Instrum.)* **11**, 296.

Dodson, M. H. (1963). A theoretical study of the use of internal standards for precise isotopic analysis by the surface ionisation technique: Part I - General first-order algebraic solutions. *J. Sci. Instrum.* **40**, 289–95.

Dosso, L. and Murthy, V. R. (1980). A Nd isotope study of the Kerguelen islands: inferences on enriched oceanic mantle sources. *Earth Planet. Sci. Lett.* **48**, 268–76.

Eberhardt, A., Delwiche, R. and Geiss, Z. (1964). Isotopic effects in single filament thermal ion sources. *Z. Natur.* **19a**, 736–40.

Edwards, R. L., Chen, J. H. and Wasserburg, G. J. (1987). ^{238}U–^{234}U–^{230}Th–^{232}Th systematics and the precise measurement of time over the past 500,000 years. *Earth Planet. Sci. Lett.* **81**, 175–92.

Eugster, O., Tera, F., Burnett, D. S. and Wasserburg, G. J. (1970). The isotopic composition of gadolinium and neutron capture effects in some meteorites. *J. Geophys. Res.* **75**, 2753–68.

Faul, H. (1966). *Ages of Rocks, Planets, and Stars*. McGraw-Hill, 109 p.

Gale, N. H. (1970). A solution in closed form for lead isotopic analysis using a double spike. *Chem. Geol.* **6**, 305–10.

Gruau, G., Cornichet, J. and Le Coz-Bouhnik, M. (1988). Improved determination of Lu/Hf ratio by chemical separation of Lu from Yb. *Chem. Geol. (Isot. Geosci. Section)* **72**, 353–6.

Habfast, K. (1983). Fractionation in the thermal ionization source. *Int. J. Mass Spectrom. Ion Phys.* **51**, 165–89.

Hamelin, B., Manhes, G., Albarede, F. and Allegre, C. J. (1985). Precise lead isotope measurements by the double spike technique: a reconsideration. *Geochim. Cosmochim. Acta* **49**, 173–82.

Hooker, P., O'Nions, R. K. and Pankhurst, R. J. (1975). Determination of rare-earth elements in U.S.G.S. standard rocks by mixed-solvent ion exchange and mass spectrometric isotope dilution. *Chem. Geol.* **16**, 189–96.

Ingram, M. G. and Chupka, P. (1953). Surface ionisation source using multiple filaments. *Rev. Sci. Instrum.* **24**, 518–20.

Kalsbeek, F. and Hansen, M. (1989). Statistical analysis of Rb–Sr isotope data by the 'bootstrap' method. *Chem. Geol. (Isot. Geosci. Section)* **73**, 289–97.

Krogh, T. E. (1973). A low contamination method for hydrothermal decomposition of zircon and extraction of U and Pb for isotopic age determination. *Geochim. Cosmochim. Acta* **37**, 485–94.

Kurz, E. A. (1979). Channel electron multipliers. *Amer. Lab.* **11** (3), 67–74.

Li, W. X., Lundberg, J., Dickin, A. P., Ford, D. C., Schwarcz, H. P., McNutt, R. H. and Williams, D. (1989). High-precision mass spectrometric uranium-series dating of cave deposits and implications for paleoclimate studies. *Nature* **339**, 534–6.

Ludwig, K. R. (1980). Calculation of uncertainties of U–Pb isotope data. *Earth Planet. Sci. Lett.* **46**, 212–20.

Lugmair, G. W., Scheinin, N. B. and Marti, K. (1975). Search for extinct ^{146}Sm, 1. The isotopic abundance of ^{142}Nd in the Juvinas meteorite. *Earth Planet. Sci. Lett.* **27**, 79–84.

McIntyre, G. A., Brooks, A. C., Compston, W and Turek, A. (1966). The statistical assessment of Rb–Sr isochrons. *J. Geophys. Res.* **71**, 5459–68.

Manton, W. I. (1988). Separation of Pb from young zircons by single-bead ion exchange. *Chem. Geol. (Isot. Geosci. Section)* **73**, 147–52.

Nier, A. O. (1940). A mass spectrometer for routine isotope abundance measurements. *Rev. Sci. Instrum.* **11**, 212–16.

Noble, S. R., Lightfoot, P. C. and Scharer, U. (1989). A method for single-filament isotopic analysis of Nd using *in situ* reduction. *Chem. Geol. (Isot. Geosci. Section)* **79**, 15–19.

O'Nions, R. K., Hamilton, P. J. and Evensen, N. M. (1977). Variations in $^{143}Nd/^{144}Nd$ and $^{87}Sr/^{86}Sr$ ratios in oceanic basalts. *Earth Planet. Sci. Lett.* **34**, 13–22.

O'Nions, R. K., Carter, S. R., Evensen, N. M. and Hamilton P. J. (1979). Geochemical and cosmochemical applications of Nd isotope analysis. *Ann. Rev. Earth Planet. Sci.* **7**, 11–38.

Otto, J. B., Blank, W. K. and Dahl, D. A. (1988). A nitrate precipitation technique for preparing strontium for isotopic analysis. *Chem. Geol. (Isot. Geosci. Section)* **72**, 173-9.

Papanastassiou, D. A., Huneke, J. C., Esat, T. M. and Wasserburg, G. J. (1978). Pandora's box of the nuclides. In: *Lunar Planet. Sci.* **IX**, Lunar Planet. Sci. Inst., Houston, Texas, pp. 859–61 (abstract).

Parrish, R. R. (1987). An improved micro-capsule for zircon dissolution in U–Pb geochronology. *Chem. Geol. (Isot. Geosci. Section)* **66**, 99–102.

Parrish, R. R. and Krogh, T. E. (1987). Synthesis and purification of ^{205}Pb for U–Pb geochronology. *Chem. Geol. (Isot. Geosci. Section)* **66**, 103–10.

Patchett, P. J. (1980). Sr isotopic fractionation in Ca–Al inclusions from the Allende meteorite. *Nature* **283**, 438–41.

Patchett, P. J. and Tatsumoto, M. (1980). A routine high-precision method for Lu–Hf isotope geochemistry and chronology. *Contrib. Mineral. Petrol.* **75**, 263–7.

Richard, P., Shimizu, N. and Allegre, C. J. (1976). $^{143}Nd/^{146}Nd$, a natural tracer: an application to oceanic basalts. *Earth Planet. Sci. Lett.* **31**, 269–78.

Roddick, J. C., Loveridge, W. D. and Parrish, R. R. (1987). Precise U/Pb dating of zircon at the sub-nanogram Pb level. *Chem. Geol. (Isot. Geosci. Section)* **66**, 11–121.

Russell, W. A., Papanastassiou, D. A. and Tombrello, T. A. (1978). Ca isotope fractionation on the Earth and other solar system materials. *Geochim. Cosmochim. Acta* **42**, 1075–90.

Tanaka, T. and Masuda, A. (1982). The La–Ce geochronometer: a new dating method, *Nature* **300**, 515–18.

Thirlwall, M. F. (1982). A triple-filament method for rapid and precise analysis of rare-earth elements by isotope dilution. *Chem. Geol.* **35**, 155–66.

Thirlwall, M. F. (1991a). High-precision multicollector isotopic analysis of low levels of Nd as oxide. *Chem. Geol. (Isot. Geosci. Section)* **94**, 13–22.

Thirlwall, M. F. (1991b). Long-term reproducibility of multicollector Sr and Nd isotope ratio analyses. *Chem. Geol. (Isot. Geosci. Section)* **94**, 85–104.

Titterington, D. M. and Halliday, A. N. (1979). On the fitting of parallel isochrons and the method of maximum likelihood. *Chem. Geol.* **26**, 183–95.

Todt, W., Cliff, R. A., Hanser, A. and Hofmann, A. W. (1984). $^{202}Pb + ^{205}Pb$ double spike for lead isotopic analysis. *Terra Cognita* **4**, 209 (abstract).

Wasserburg, G. J., Jacobsen, S. B., DePaolo, D. J., McCulloch, M. T. and Wen, T. (1981). Precise determination of Sm/Nd ratios, Sm and Nd isotopic abundances in standard solutions. *Geochim. Cosmochim. Acta* **45**, 2311–23.

Wendt, I. and Carl, C. (1991). The statistical distribution of the mean squared weighted deviation. *Chem. Geol. (Isot. Geosci. Section)* **86**, 275–85.

York, D. (1966). Least-squares fitting of a straight line. *Can. J. Phys.* **44**, 1079–86.

York, D. (1967). The best isochron. *Earth Planet. Sci. Lett.* **2**, 479–82.

York, D. (1969). Least-squares fitting of a straight line with correlated errors. *Earth Planet. Sci. Lett.* **5**, 320–4.

Zheng, Y. F. (1989). Influences of the nature of the initial Rb–Sr system on isochron validity. *Chem. Geol. (Isot. Geosci. Section)* **80**, 1–16.

3 The Rb–Sr method

3.1 Rb decay

Rubidium, a group 1 alkali metal, has two naturally occurring isotopes, ^{85}Rb and ^{87}Rb, whose abundances are 72.17% and 27.83% respectively. These figures yield an atomic abundance ratio of ^{85}Rb/^{87}Rb = 2.593 (Catanzaro *et al.*, 1969), which is a constant throughout the Earth, moon and most meteorites due to isotopic homogenisation in the solar nebula. ^{87}Rb is radioactive, and decays to the stable isotope ^{87}Sr by emission of a β particle and antineutrino ($\bar{\nu}$). The decay energy (Q) is shared as kinetic energy by these two particles.

$$^{87}_{37}\text{Rb} \rightarrow {}^{87}_{38}\text{Sr} + \beta^- + \bar{\nu} + Q$$

The low decay energy for this transformation (275 keV) has always caused problems in the accurate determination of the Rb decay constant. Because the decay energy is divided between the β particle and anti-neutrino, the β particles have a smooth distribution of kinetic energy from the total energy down to zero. When attempting to accurately determine the decay constant by direct counting, the low energy β particles cause great problems because they may be absorbed by surrounding Rb atoms before they ever reach the detector. For example, in a thick (>1 μm deep) solid Rb sample, attenuation is so severe that a false frequency maximum is generated at ca. 10 keV (Fig. 3.1).

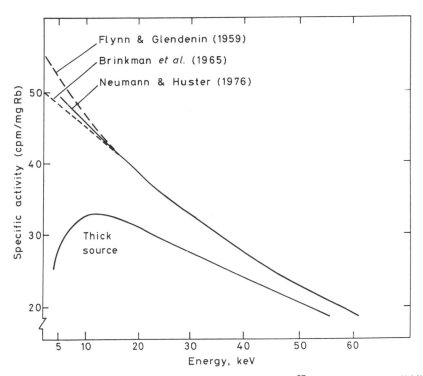

Fig. 3.1. Plot of activity against kinetic energy for β particles generated by ^{87}Rb decay. Lower solid line = thick solid Rb source; upper solid line = extrapolation of solid source to zero thickness. Dashed lines show alternative extrapolations to zero energy for liquid scintillator measurements. After Neumann and Huster (1976).

One way to avoid the attenuation problem is to use a photo-multiplier with a liquid scintillator solution doped with Rb. The β particles will be absorbed by molecules of the scintillator (emitting light flashes) before they can be absorbed by other Rb atoms. The major problem with this method is that a low-energy cut-off at ca. 10 keV must be applied to avoid the high background noise associated with liquid scintillation. The consequent extrapolation of count-rate curves down to zero energy leads to a large uncertainty in the result (Fig. 3.1). Hence this method has given values for the ^{87}Rb half-life from 47.0 ± 1.0 Byr (Flynn and Glendenin, 1959) to 52.1 ± 1.5 Byr (Brinkman et al., 1965).

Another approach to direct counting is to make measurements with progressively thinner solid Rb sources using a proportional counter. The results are then extrapolated to a theoretical source of zero thickness to remove the effect of self-absorption. The proportional counter has a much lower noise level, so the energy cut-off can be set as low as 0.185 keV. Rb films with thicknesses down to 1 μm were measured by Neumann and Huster (1974), and extrapolated to zero thickness by Neumann and Huster (1976) to derive an ^{87}Rb half-life of 48.8 ± 0.8 Byr (equivalent to a decay constant of 1.42×10^{-11} yr^{-1}).

An alternative approach to determine the Rb decay constant is to measure the amount of ^{87}Sr produced by decay of a known quantity of ^{87}Rb in the laboratory over a known period of time. This method was first attempted by McMullen et al. (1966) on a rubidium sample they had purified in 1956, and was repeated on the same sample batch by Davis et al. (1977). Unfortunately, McMullen et al. omitted to measure the small but significant level of residual ^{87}Sr present in their rubidium before they put it away on the shelf. Hence, the accuracy of their determination was compromised. However, this problem contributes less than 1% uncertainty to the later determination of Davis et al. (1977). Their proposed value for the ^{87}Rb half-life (48.9 ± 0.4 Byr, equivalent to a decay constant of $1.42 \pm 0.01 \times 10^{-11}$ yr^{-1}) can therefore be taken to support the value of Neumann and Huster (1976).

A third approach to the determination of the Rb decay constant is to date geological samples whose ages have also been determined by other methods with more reliable decay constants. This method has the disadvantage that it involves geological uncertainties, such as whether all isotopic systems closed at the same time and remained closed. However, it provides a useful check on the direct laboratory determinations. In this respect it is worth noting that

Pinson et al. (1963) proposed a rubidium half-life of 48.8 Byr on the basis of Rb–Sr dating of stony meteorites.

During the last 30 years, values of the decay constant used in geological age calculations have varied between 1.47 and 1.39×10^{-11} yr^{-1} ($t_{1/2} = 46.8$ to 50.0 Byr). The presently used value of $1.42 \ 10^{-11}$ yr^{-1} ($t_{1/2} = 48.8$ Byr) is an international convention (Steiger and Jager, 1977), but may yet have to be revised. For example, very precise U–Pb and Rb–Sr isochrons for chondritic meteorites can only be made to agree if the ^{87}Rb decay constant is reduced to $1.402 \pm 0.008 \times 10^{-11}$ yr^{-1}, equivalent to a half-life of 49.4 ± 0.3 Byr (Minster et al., 1982).

3.2 Dating igneous rocks

The number of ^{87}Sr daughter atoms produced by decay of ^{87}Rb in a rock or mineral since its formation t years ago is given by substituting into the general decay equation [1.10]:

$$^{87}\text{Sr} = {}^{87}\text{Sr}_\text{I} + {}^{87}\text{Rb}(e^{\lambda t} - 1) \qquad [3.1]$$

where ^{87}Sr$_\text{I}$ is the number of ^{87}Sr atoms present initially. However, it is difficult to measure precisely the absolute abundance of a given nuclide. Therefore it is more convenient to convert this number to an isotope ratio by dividing through by ^{86}Sr (which is not produced by radioactive decay and therefore remains constant with time). Hence we obtain:

$$\left(\frac{^{87}\text{Sr}}{^{86}\text{Sr}}\right)_\text{P} = \left(\frac{^{87}\text{Sr}}{^{86}\text{Sr}}\right)_\text{I} + \frac{^{87}\text{Rb}}{^{86}\text{Sr}}(e^{\lambda t} - 1) \qquad [3.2]$$

The present-day Sr isotope ratio (P) is measured by mass spectrometry, and the atomic ratio ^{87}Rb/^{86}Sr is calculated from the weight ratio of Rb/Sr. If the initial ratio $(^{87}\text{Sr}/^{86}\text{Sr})_\text{I}$ is known or can be estimated, then t can be determined, subject to the assumption that the system has been closed to Rb and Sr mobility from time t until the present:

$$t = \frac{1}{\lambda} \ln\left\{ 1 + \frac{^{86}\text{Sr}}{^{87}\text{Rb}} \left[\left(\frac{^{87}\text{Sr}}{^{86}\text{Sr}}\right)_\text{P} - \left(\frac{^{87}\text{Sr}}{^{86}\text{Sr}}\right)_\text{I} \right] \right\} \qquad [3.3]$$

3.2.1 Sr model ages

When the Rb–Sr method was first used in geochronology, the poor precision attainable in mass spectrometry limited the technique to the dating of Rb-rich minerals such as lepidolite. These minerals develop such high ^{87}Sr/^{86}Sr ratios over geological time that a uniform initial ^{87}Sr/^{86}Sr ratio of 0.712 could be assumed in all dating studies without introducing significant errors. Such determinations are called 'model ages' because the initial ratio is predicted by a model rather than measured directly.

Subsequently, the Rb–Sr method was extended to less exotic rock-forming minerals such as biotite, muscovite and K-feldspar, with lower Rb/Sr ratios. However, discordant dates were often generated, by assuming an initial ratio of 0.712 when the real initial ratio was higher. This problem was first recognised by Compston and Jeffery (1959), and overcome by the invention of the isochron diagram (Nicolaysen, 1961). Model ages subsequently re-appeared in more specialised aspects of Rb–Sr dating such as meteorite chronology, and as an important approach in the Sm–Nd method (section 4.2.1).

3.2.2 The isochron diagram

An examination of equation [3.2] shows that it is equivalent to the equation for a straight line:

$$y = c + xm \qquad [3.4]$$

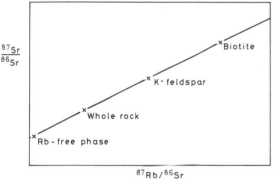

Fig. 3.2. Schematic Rb–Sr isochron diagram for a suite of comagmatic igneous minerals.

This led Nicolaysen (1961) to develop a new way of treating Rb–Sr data, by plotting $^{87}Sr/^{86}Sr$ (y) against $^{87}Rb/^{86}Sr$ (x). The intercept (c) is then the initial $^{87}Sr/^{86}Sr$ ratio of the system. On this diagram, a suite of comagmatic minerals having the same age and initial $^{87}Sr/^{86}Sr$ ratio, and which have since remained as closed systems, define a line termed an 'isochron'. The slope of this line, $m(= e^{\lambda t} - 1)$, yields the age of the minerals. If one of the minerals is very Rb-poor, then this may yield the initial ratio directly. Otherwise, the initial ratio is determined by extrapolating back to the y axis a best-fit line through the available data points (Fig. 3.2). Because λ ^{87}Rb is so small, for geologically young rocks the slope may be quite accurately approximated by λt. Such an approximation does not hold for nuclides with shorter half-lives such as K and U.

The isotopic evolution of a suite of hypothetical minerals in the isochron diagram is illustrated in Fig. 3.3. At the time of crystallisation of the rock, all three minerals have the same $^{87}Sr/^{86}Sr$ ratio, and plot as points on a horizontal line. After each mineral has become a closed system (effectively at the same instant for the minerals in a high-level, fast-cooling intrusion) isotopic evolution begins. On a diagram where the two axes have the same scale (Fig. 3.3), the points move up straight lines with a slope of -1 as each ^{87}Rb decay increases $^{87}Sr/^{86}Sr$ and reduces $^{87}Rb/^{86}Sr$ by the same amount. Each mineral composition remains on the isochron as its slope increases with time. In practice, the y axis is usually very much expanded to display rocks of geological age in a suitable format, and the growth lines are then nearly vertical.

Another development of the Rb–Sr method (Schreiner, 1958), was the analysis of co-genetic

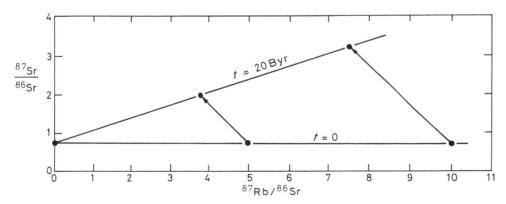

Fig. 3.3. Rb–Sr isochron diagram on axes of equal magnitude showing production of ^{87}Sr as ^{87}Rb is consumed in two hypothetical samples. The third sample has no Rb.

whole-rock sample suites, as an alternative to separate minerals. To be effective, a whole-rock suite must display variation in modal mineral content, such that samples display a range of Rb/Sr ratios, without introducing any variation in initial Sr isotope ratio. In actual fact, perfect initial ratio homogeneity may not be achieved, especially in rocks with a mixed magmatic parentage. However, if the spread in Rb/Sr ratios is sufficient, then any initial ratio variations are swamped, and an accurate age can be determined. Initial ratio heterogeneity is a greater problem in Sm–Nd isochrons, and is therefore discussed under that heading (section 4.1.2). Schreiner's proposal actually preceded the invention of the Rb–Sr isochron diagram, but some of his data are presented on an isochron diagram in Fig. 3.4 to demonstrate the method.

Graphical calculation of isochron ages was super-seded in the 1960s by the application of least squares regression techniques (section 2.3), but the isochron diagram remains a very useful tool to assess the distribution of data points about an isochron. However, Papanastassiou and Wasserburg (1970) found that the scale of axes on an isochron diagram was too compressed to allow clear portrayal of the experimental error bars on their data points. To overcome this problem they developed the ϵ notation, which they defined as the relative deviation of a data

point from the best-fit isochron (in parts per 10^4). This is given by:

$$\epsilon = \left[\frac{(^{87}\mathrm{Sr}/^{86}\mathrm{Sr})_{\mathrm{measured}}}{(^{87}\mathrm{Sr}/^{86}\mathrm{Sr})_{\mathrm{best-fit}}} - 1 \right] \times 10^4 \qquad [3.5]$$

Figure 3.5 shows a combined mineral isochron diagram and ϵ diagram for an Apollo 11 sample from the Sea of Tranquillity. A limitation of the ϵ diagram is that the vertical error bars only describe errors in $^{87}\mathrm{Sr}/^{86}\mathrm{Sr}$, whereas errors in Rb/Sr ratio can also cause points to deviate from the line. In practice, the samples dated by Papanastassiou and Wasserburg (1970) had small Rb/Sr ratios, so errors in this variable were normally subordinate to errors in measured isotope ratio.

Provost (1990) has pointed out that isochrons determined on granitic rocks are dominated by errors in Rb/Sr rather than $^{87}\mathrm{Sr}/^{86}\mathrm{Sr}$ (Fig. 3.6a). He developed a new version of the isochron plot (Fig. 3.6b), with non-linear axes, which attempts to portray both sources of error at once. Uncertainties in isotope ratio and Rb/Sr ratio both generate vertical or sub-vertical error bars, but their meaning changes progressively across the diagram from an error on the initial ratio (left-hand side) to an error on the age (right-hand side). Unfortunately this diagram is conceptually quite difficult to understand, so a more

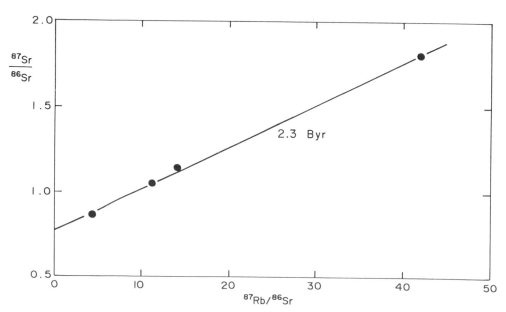

Fig. 3.4. Rb–Sr whole-rock isochron for the 'red granite' of the Bushveld complex, using the data of Schreiner (1958).

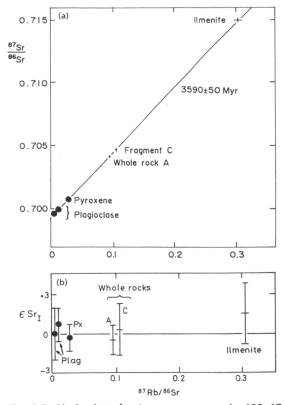

Fig. 3.5. Rb–Sr data for Lunar mare sample 100 17 plotted (a) on a conventional isochron diagram; and (b) on a diagram of epsilon against Rb/Sr. After Papanastassiou and Wasserburg (1970).

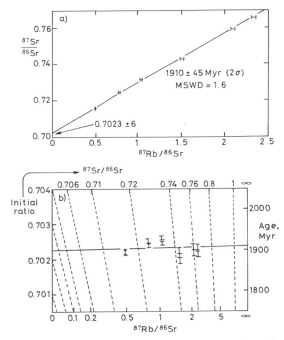

Fig. 3.6. Rb–Sr data for the Agua Branca adamellite, Brazil, plotted (a) on a conventional isochron diagram; and (b) on an 'improved' isochron diagram. After Provost (1990).

practical approach may be to improve the epsilon diagram of Papanastassiou and Wasserburg by adding error bars which represent the effect of Rb/Sr uncertainties on each data point.

3.2.3 Erupted isochrons

A primary basic magma should inherit the isotopic composition of its mantle source, providing that melting occurs in equilibrium conditions. Tatsumoto (1966) first suggested, on the basis of U–Pb data, that primitive basic magmas could also inherit the parent/daughter ratio of their mantle source. If different magma batches were to sample the elemental and isotopic composition of different source domains then this might lead to the eruption of an 'isochron' suite whose slope yields the time over which these sources were isolated. This concept was examined for the Rb–Sr system in alkalic ocean island basalts by Sun and Hansen (1975).

Average compositions for 14 different ocean island basalts were plotted on an isochron diagram (Fig. 3.7). The data are fairly scattered, but form a positive correlation with a slope age of ca. 2 Byr. Individual ocean islands may also define arrays with positive slope, but usually with more scatter. Sun and Hansen attributed the positive correlations between Rb/Sr and isotopic composition to mantle heterogeneity, suggesting that the apparent ages represented the time since mantle domains were isolated from the convecting mantle. Brooks *et al.* (1976a) termed these ages 'mantle isochrons'.

The mantle isochron concept was extended to continental igneous rocks by Brooks *et al.* (1976b). Because these are often ancient (unlike most ocean island basalts), it was necessary to correct measured $^{87}Sr/^{86}Sr$ ratios back to their calculated initial ratios at the time of magmatism, before plotting against Rb/Sr (e.g. Fig. 3.8). Hence Brooks *et al.* termed these plots 'pseudo-isochron' diagrams. They listed 30 examples from both volcanic and plutonic continental igneous rock suites where the data formed a roughly linear array. The controversial aspect of this work was that Brooks *et al.* rejected the possibility that pseudo-isochrons were mixing lines produced by crustal contamination of mantle-derived basic magmas.

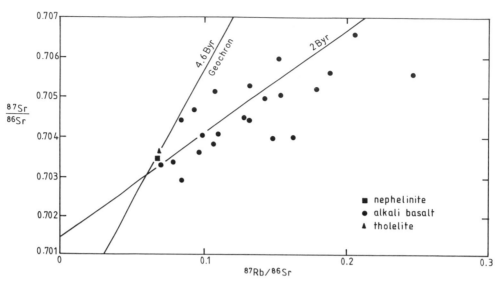

Fig. 3.7. Rb–Sr isochron diagram for young oceanic volcanic rocks (mostly alkali basalts) from ocean islands. After Sun and Hansen (1975).

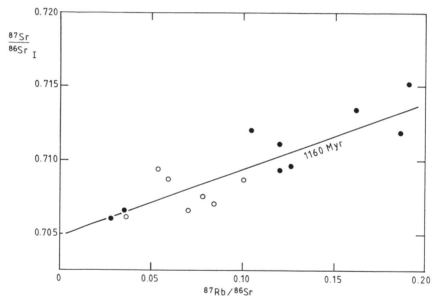

Fig. 3.8. Pseudo-isochron for quartz and olivine norites from Arnage (○) and Haddo House (●) in NE Scotland, yielding an apparent age of 1160 ±420 Myr (2σ) prior to intrusion. After Brooks *et al.* (1976b).

Instead they believed them to date mantle differentiation events which established domains of different Rb/Sr ratio in the subcontinental lithosphere.

It is a fundamental assumption of the mantle isochron model that neither isotope nor elemental ratios are perturbed during magma ascent through the crust. However, it is now generally accepted that this assumption is not upheld with sufficient reliability to attribute age significance to erupted isochrons. For example, the Haddo House norites of N.E. Scotland (Fig. 3.8) are known to contain pelitic xenoliths, so this array must document crustal assimilation. Another pseudo-isochron example from Lower Tertiary lavas of NW Scotland (Beckinsale *et al.*,

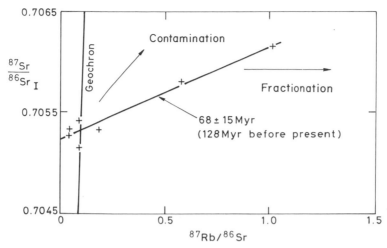

Fig. 3.9. Pseudo-isochron diagram for Tertiary lavas from the Isle of Mull, NW Scotland, showing apparent age of 68 Myr prior to eruption. Vectors illustrate an alternative fractionation–contamination model. Modified after Beckinsale *et al.* (1978).

1978) can be attributed to Sr depletion in the melt by plagioclase fractionation, followed by crustal contamination (Fig. 3.9). Breakdown in the mantle isochron model can also be caused by low degrees of melting in the mantle source, leading to fractionation between Rb, an ultra-incompatible, and Sr, a moderately incompatible element. Hence it is concluded that only isotope–isotope mantle isochrons (such as those provided by the Pb isotope system) can reliably be interpreted as dating the ages of mantle differentiation events.

3.2.4 Meteorite chronology

Meteorites have been the subject of numerous Rb–Sr dating studies, but some of the most important Rb–Sr results on meteorites are initial ratio determinations. These have significance, both as a reference point for terrestrial Sr isotope evolution, and as a model age tool for estimating the relative condensation times of solar system bodies.

The first accurate measurement of meteorite initial ratios was made by Papanastassiou and Wasserburg (1969) on basaltic achondrites. These differ from chondritic meteorites in showing evidence of differentiation after their accretion from the solar nebula. However, they may not have participated in the full planetary differentiation process which generated iron meteorites. Their low Rb/Sr ratios have resulted in only limited radiogenic Sr production since differentiation, so an accurate initial ratio determination is possible.

In order to make this determination, Papanastassiou and Wasserburg analysed whole-rock samples from seven different basaltic achondrites, yielding an isochron (Fig. 3.10) without any excess scatter over analytical error. An age of 4.39 ± 0.26 Byr was calculated using the old decay constant ($\lambda = 1.39 \times 10^{-11}$ yr^{-1}). The initial ratio of 0.69899 ± 5 was referred to by Papanastassiou and Wasserburg as the 'Basaltic Achondrite Best Initial' or BABI. This value represents a bench-mark to which other meteorite initial ratios may be compared. Birck and Allegre (1978) repeated this study with the addition of separated minerals from Juvinas and Ibitira, yielding an identical initial ratio, but an improved age determination of 4.57 ± 0.13 Byr (same decay constant). However, Rb–Sr mineral isochrons are not possible for other achondrites due to a later disturbance.

The determination of precise initial ratios for chondritic meteorites is problematical because of their much higher Rb/Sr ratios than basaltic achondrites. However, by separating out low-Rb/Sr phosphate minerals, Wasserburg *et al.* (1969) and Gray *et al.* (1973) were able to determine good initial ratios for the chondrites Guarena and Peace River. Gray *et al.* also determined accurate initial ratios by analysis of bulk samples from the achondrite Angra dos Reis (ADOR) and Rb-poor inclusions from the carbonaceous chondrite Allende (ALL).

These initial ratios can be translated into a relative chronology for meteorite condensation (Fig. 3.11) by assuming a homogeneous Rb/Sr ratio in the solar

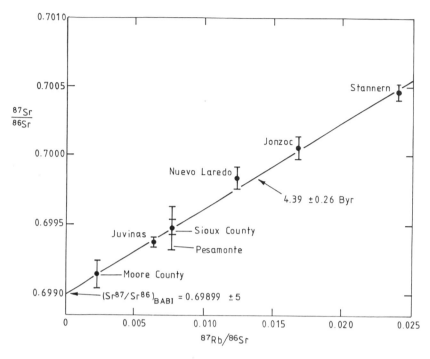

Fig. 3.10. Rb–Sr isochron diagram for whole-rock samples of basaltic achondrites. The initial ratio ('BABI') is 0.69899. After Papanastassiou and Wasserburg (1969).

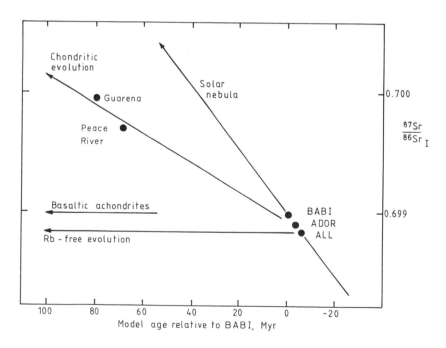

Fig. 3.11. Plot of initial Sr isotope composition for selected meteorites against model ages for condensation or differentiation–metamorphism, based on assumed Rb/Sr ratios in major reservoirs. After Gray *et al.* (1973).

nebula (Papanastassiou and Wasserburg, 1969). The results are only 'model' ages because they depend on an assumed composition for the source reservoir (solar nebula), and they would be rendered invalid if it did not evolve as a homogeneous reservoir. The estimate for Rb/Sr in the solar nebula was based on cited spectroscopic measurements from the Sun, yielding a value of 0.65 which is capable of generating an increase in $^{87}Sr/^{86}Sr$ of ca. 0.0001 in 4 Myr.

If we ignore the possibility of heterogeneous Sr isotope ratios in the solar nebula, the Allende data suggest it to be the oldest known object in the solar system, predating the condensation of basaltic achondrites by ca. 10 Myr (Fig. 3.11). Similarly, Angra dos Reis has a model age ca. 5 Myr older than BABI. Application of the same model to the high initial ratios of Guarena and Peace River would imply unduly late condensation from the solar nebula. Therefore, Gray *et al.* interpreted these as metamorphic ages produced by re-distribution of Rb and Sr between mineral phases within chondritic bodies. Since basaltic achondrites and ADOR are themselves apparently products of planetary differentiation, one might wonder whether the entire model chronology really indicates times of differentiation and metamorphism, rather than condensation.

3.3 Dating metamorphic rocks

3.3.1 Open mineral systems

Mineral and whole-rock Rb–Sr systems may respond differently to metamorphic events. ^{87}Sr generated by Rb decay occupies unstable lattice sites in Rb-rich minerals and tends to migrate out of the crystal if subjected to a thermal pulse, even of a magnitude well below the melting temperature. However, if fluids in the rock remain static, Sr released from Rb-rich minerals such as mica and K-feldspar will tend to be taken up by the nearest Sr sink such as plagioclase or apatite.

The idea of using whole-rock analysis to see back through a metamorphic event which disturbs mineral systems was first conceived by Compston and Jeffery (1959). The model was illustrated graphically by Fairbairn *et al.* (1961) on a plot of isotope ratio against time (Fig. 3.12). Different minerals move along different growth lines after the formation of the rock at time t_0, until they are homogenized by a thermal event at time t_M. Thereafter, isotopic evolution again continues along different growth lines to the present day (t_P). Individual minerals in this model are open systems during the metamorphism. Therefore, a mineral isochron yields the age of

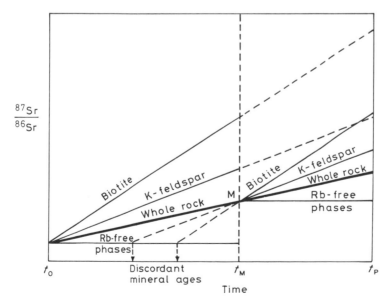

Fig. 3.12. Plot of Sr isotope ratio against time to model the effect of a metamorphic event which opens Rb–Sr mineral systems, but not the whole-rock system. Growth lines of different steepness reflect different Rb/Sr ratios. t_0 = age of rock; t_M = age of metamorphism; t_P = present. After Fairbairn *et al.* (1961).

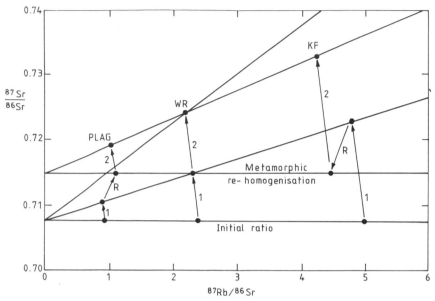

Fig. 3.13. Hypothetical behaviour of a partially disturbed mineral–whole-rock isochron. Evolution lines: 1 = period from igneous crystallisation to metamorphism; R = metamorphic re-homogenisation; 2 = period from metamorphism to present day. Only for the whole-rock system (WR) are stages 1 and 2 directly additive.

cooling from the thermal event, when each mineral again became a closed system. However, a whole-rock domain of a certain minimum size remains as an effectively closed system during the thermal event, and can be used to date the initial crystallisation of the rock.

The effects of metamorphism on mineral and whole-rock systems can also be demonstrated on the isochron diagram, Fig. 3.13 (Lanphere *et al.*, 1964). All systems start on a horizontal cord. Isotopic evolution then occurs along near-vertical parallel paths (due to the extreme amplification of the *y* axis). During the thermal event, isotope ratios are homogenised to the whole-rock value. If this only involved ^{87}Sr then vertical vectors would be produced. However, a possible complication illustrated in Fig. 3.13 involves limited Rb re-mobilisation. Rb-rich minerals tend to suffer some Rb loss, while Rb-poor phases may be contaminated by growth of Rb-rich alteration products, leading to somewhat unpredictable vectors (R). After the event, whole-rock evolution continues undeflected, while mineral systems define an isochron whose slope yields the age of metamorphism.

A practical example of dating plutonism and metamorphism by whole-rock and mineral analysis of the same body was provided by the work of Wetherill *et al.* (1968) on the Baltimore gneiss (Fig.

3.14). Several mineral isochrons all yield ages of ca. 290 Myr, interpreted as the time of closure of mineral systems after isotopic homogenisation associated with the Appalachian orogeny. The good fit of points to the mineral isochrons is evidence that complete isotopic homogenisation on a mineralogical scale was achieved during the metamorphic event. In contrast, whole-rock samples define an isochron whose slope corresponds to an age of 1050 Myr. This was interpreted as the time of crystallisation of igneous precursors of the gneiss. However, more recent studies have shown that even whole-rock Rb–Sr systems may be open during metamorphism. Therefore the 1050 Myr age of the Baltimore gneiss could alternatively represent the time of closure of Rb–Sr whole-rock systems after high-grade metamorphism. More examples of open whole-rock systems will be discussed below.

3.3.2 Blocking temperatures

After Rb–Sr mineral systems have been opened in the thermal pulse of a regional metamorphic event, there must come a time when mineral systems are again closed to element mobility. By dating the closure or 'blocking' of different mineral systems, Rb–Sr ages give information about the cooling history of metamorphic terranes. This was first demonstrated

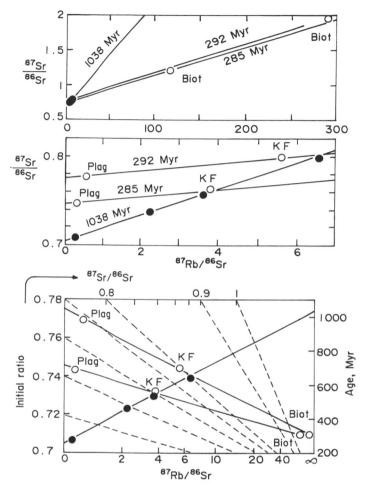

Fig. 3.14. Rb–Sr data for the Baltimore gneiss showing 1038 Myr 'plutonic' and 285–292 Myr metamorphic ages defined by Rb–Sr whole-rock systems (•) and minerals (○). The data are presented in conventional isochron diagrams and in the diagram of Provost (1990) for comparison. After Provost (1990).

by Jager *et al.* (1967) and Jager (1973), working on the central European Alps.

Jager *et al.* found that in rocks of low metamorphic grade round the exterior of the Central Alps, Hercynian Rb–Sr ages (>200 Myr) were preserved in both biotites and muscovites. On moving to a higher metamorphic grade characterised by the appearance of stilpnomelane (which Jager *et al.* believed equivalent to a temperature of 300 ± 50 °C), Rb–Sr biotite ages of 35–40 Myr were measured. Jager *et al.* attributed these younger ages to Rb–Sr biotite systems opened at the peak of Lepontine metamorphism. They argued that the 300 °C temperature at which biotites were just opened at the peak of metamorphism would correspond to the

temperature at which biotites would re-close up to several Myr after suffering a higher peak temperature (e.g. >500 °C within the central staurolite isograd). In other words, Jager *et al.* concluded that biotite had a blocking temperature of 300 ± 50 °C for the Rb–Sr system.

The blocking temperature of white mica (muscovite and phengite) was similarly constrained to 500 ± 50 °C by the first re-setting of the white mica Rb–Sr ages 'somewhat outside the staurolite–chloritoid boundary' (Purdy and Jager, 1976). However, unlike biotite, white micas can undergo primary crystallisation below the Rb–Sr blocking temperature, so that ages as low as 35–40 Myr have been obtained even from the outer zones of low-grade Alpine metamorphism.

These ages are argued to date new mica growth at the peak of metamorphism (Hunziker, 1974). This makes the muscovite Rb–Sr system a more problematical tool than biotite for studying post-orogenic cooling processes.

Jager *et al.* (1967) obtained biotite ages of ca. 12–16 Myr from the Simplon and Gotthard areas of the Central Alps, and results averaging ca. 8 Myr older in coexisting muscovites. Clark and Jager (1969) used these data to make two different estimates of cooling rate for the Central Alps. Firstly, the age difference between muscovite and biotite closure (200 °C) leads to a cooling rate of ca. 25 °C/Myr between 500 and 300 °C. Secondly, the biotite ages yield cooling rates of ca. 20–25 °C/Myr between 300 and 0 °C (average surface temperature at the present day). Division of these results by an estimated geothermal gradient (25–40 °C/km) allows the calculation of uplift rates between 0.5 and 1.0 km/Myr for the Central Alps, which compare well with modern uplift rates of 0.4–0.8 mm/yr from geodetic measurements. More recent calculations of past uplift rate make use of combined Rb–Sr, K–Ar and fission-track cooling ages (section 16.6).

Purdy and Jager (1976) recognised that the 300 ± 50 °C blocking temperature for biotite might need to be revised if new experimental data for stilpnomelane stability was obtained. Most workers continue to use a value of 300 °C; however, experimental work (e.g. Brown, 1971) points to an upper stilpnomelane stability limit of 440–480 °C at ca. 4 kb, implying a biotite Rb–Sr blocking temperature of over 400 °C.

This would be consistent with evidence from SW Norway, where biotites subjected to temperatures of over 400 °C in the Caledonian orogeny nevertheless preserve Sveco-Norwegian (800 Myr) ages (Verschure *et al.*, 1980).

A more direct method of determining blocking temperatures is to measure mineral ages in deep boreholes. Del Moro *et al.* (1982) determined biotite–whole-rock Rb–Sr ages at depths of up to 3.8 km in the Sasso 22 well in the Larderello geothermal field, Italy. All of the biotites show almost complete retention of ^{87}Sr at directly measured in-hole temperatures up to nearly 380 °C (Fig. 3.15), supporting a biotite closure temperature of ca. 400 °C. However, Cliff (1985) has argued that in active geothermal systems, convective heat transport could generate localised thermal pulses whose duration is too short to allow significant diffusional Sr loss, thus implying an anomalously high blocking temperature.

Blocking temperatures can also be determined theoretically, based on calculations of the temperature-dependence of volume diffusion processes (Dodson, 1973; 1979). Ideally, closure of the Rb–Sr system represents an instantaneous transition from a time when Rb and Sr were completely mobile to when they were completely immobile. In a fast-cooling igneous body the moment of crystallisation is a good approximation to this ideal. However, in a slow-cooling regional metamorphic terrane there is a continuous transition from a high-temperature regime, when radiogenic ^{87}Sr escapes from crystal

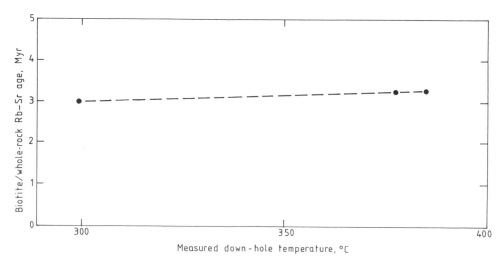

Fig. 3.15. Profile of down-hole temperature against measured Rb–Sr biotite age for the Sasso 22 hole, Larderello, Italy. Data from Del Moro *et al.* (1982)

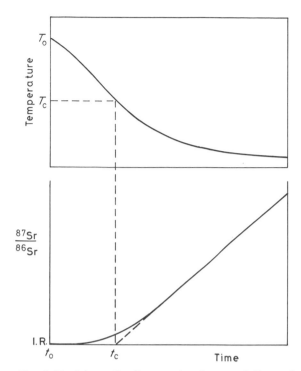

Fig. 3.16. Schematic diagram to show variation of temperature and Sr isotope ratio with time in a mineral cooling from a regional metamorphic event. T_0 = peak metamorphic temperature; T_C = closure or 'blocking' temperature; I.R. = initial isotopic ratio; t_C = apparent closure age. After Dodson (1973).

lattices by diffusion as fast as it is formed by Rb decay, to low-temperature conditions when there is negligible ^{87}Sr escape (Fig 3.16). In such a system, the apparent age of a mineral such as biotite corresponds to a linear extrapolation of the low-temperature ^{87}Sr growth line back into the x axis. The temperature prevailing in the system at the time of the apparent age of the mineral is then defined as the blocking temperature of the mineral in question (Dodson, 1973). This blocking temperature is dependent on cooling rate, since the slower the cooling, the longer will be the time during which partial loss of daughter product may occur, and the lower will be the apparent age (Fig. 3.16).

If a mineral is in contact with a fluid phase which can remove radiogenic Sr from its surface, then the rate of loss of ^{87}Sr depends on the rate of volume diffusion across a certain size of lattice. In the case of biotite, this diffusion will be predominantly parallel to cleavage planes rather than across them. Assuming that the Arrhenius Law is obeyed, Dodson (1979)

calculated blocking temperatures (at a cooling rate of 30 °C/Myr) of 300 °C for the Rb–Sr system in biotites of 0.7 mm diameter. This was based on experimental work for argon diffusion in biotite (Hofmann and Giletti, 1970), because the two elements are thought to have similar diffusional behaviour in crystal lattices.

A problem with the volume diffusional control of blocking temperature is that large (30 cm) fissure-filling biotites in the Central Alps have the same ages, and hence apparent blocking temperatures, as small (< 1 mm) ground-mass biotites in adjacent gneisses. Dodson (1979) suggested three possible explanations:

1) Diffusion geometry is independent of grain size. This could be due to the effects of stress on the crystal lattice.
2) Sr loss is controlled by the rate at which radiogenic atoms leave the site in which they were formed.
3) Blocking temperature is not kinetically controlled, but depends on a change in the biotite lattice at the blocking temperature.

The susceptibility of Sr to mobilisation by fluids increases complexity in the interpretation of Sr blocking temperature. Such problems do not arise for argon, because it is an inert gas. Therefore the latter element is a more reliable tool for studies of 'thermochronology' (section 10.2.7). For example, Harrison et al. (1985) suggested that argon blocking temperature was independent of grain-size because it is controlled by diffusion from small domains with radii of ca. 0.2 mm. They also determined an experimental blocking temperature for argon in biotite which was similar to Dodson's calculation. Therefore, the weight of evidence may now be swinging back in support of the 300 °C blocking temperature for biotite originally proposed by Jager.

3.3.3 Open whole-rock systems

The Rb–Sr whole-rock method was widely used as a dating tool for igneous crystallisation during the 1960s and 1970s, but lost credibility during the 1980s as evidence of whole-rock open-system behaviour mounted. For example, Rb–Sr isochrons in metamorphic terranes can yield good linear arrays whose slope is nevertheless a meaningless average of the protolith and metamorphic ages. This problem is probably caused by the need to sample over a relatively large geographical area, in order to maximise the range of Rb/Sr ratios. A good example is provided by the Arendal charnockites of south Norway (Field and Raheim, 1979a,b).

Eight whole-rock sample suites were collected from individual outcrops of Arendal charnockite over an area of several km². They yielded ages which were dominantly in two groups, of ca. 1540 Myr and 1060 Myr. Field and Raheim (1979a) interpreted the older age as the time of formation of the high-grade charnockite mineralogy, and the younger as dating a subsequent low-grade event. This was manifested as slight mineralogical alteration, probably associated with irregularly spaced narrow fractures which traverse the area. The younger re-setting event also fell within error of the 1063 ± 20 Myr age of undeformed granite sheets in the area.

In order to test the effect of making a regional sample collection from an area of slightly disturbed gneisses, Field and Raheim (1979b) collected a suite of eight samples over an area of 1 km². The data (Fig. 3.17) define a good linear array with an apparent age of 1259 ± 26 Myr. The MSWD value of 1.58 implies that the scatter of data about the line could probably be accounted for by analytical error, but there is no geological evidence for an event at this time. Therefore, Field and Raheim attributed the linear array to a series of closely spaced *en echelon* arrays with slopes corresponding to the age of re-setting, defined by a 1035 Myr mineral isochron. Because the range in Rb/Sr at each locality is small (e.g. 'locality 4', Fig. 3.17), samples lying on each sub-isochron do not deviate much from the fictitious composite 'isochron'. It is therefore concluded that in areas where Rb–Sr systems may have been disturbed, detailed sampling is necessary to measure the mobility of the species before regional geochronological interpretations are made.

Whole-rock open-system behaviour can occur at even lower grades of metamorphism in fine-grained acid volcanic rocks. Such units are attractive for absolute calibration of the stratigraphic column because they are conformable with sedimentary strata. They tend to have large and variable Rb/Sr ratios, thus yielding good isochrons. However, experience has shown that they are particularly susceptible to radiogenic Sr loss. A good example is provided by the Stockdale rhyolite of northern England.

The Stockdale rhyolite is a fine-grained, flow-banded lava, included in the uppermost Ordovician succession, and is argued to have a bio-stratigraphic uncertainty of less than 0.5 Myr. Gale *et al.* (1979) determined a 16 point whole-rock isochron, which yielded an age of 421 ± 3 Myr (2σ) with MSWD =

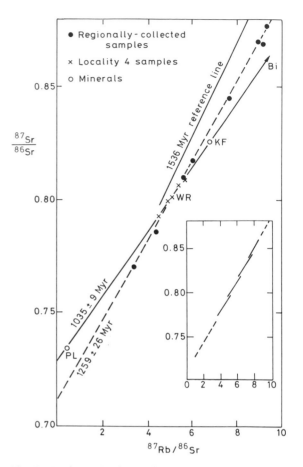

Fig. 3.17. Rb–Sr 'isochron' diagram for Arendal charnockites showing fictitious 1259 Myr regional isochron composed of a series of *en echelon* local isochrons with the same slope as separated minerals. After Field and Raheim (1979b).

1.92. They argued that because of the relatively small number of data points, this MSWD value could be attributed to experimental errors (section 2.3.2), and hence that the 421 Myr age probably represented the time of eruption of the lava. However, if this age was correct, it would require substantial revision of the Ordovician time-scale determined by other methods.

McKerrow *et al.* (1980) argued that because a section of the Stockdale rhyolite which lay inside the Shap granite aureole gave the same age (424 ± 18 Myr) as the rest of the lava (421 Myr), the whole unit was probably disturbed by some kind of hydrothermal event after extrusion and subsequent burial. Compston *et al.* (1982) sought to explain the excess scatter over analytical error by a re-setting event postdating the extrusion of the lava (estimated at 440 Myr

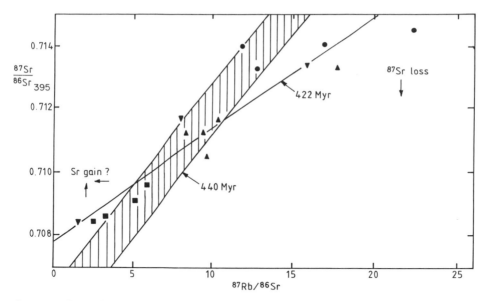

Fig. 3.18. Rb–Sr pseudo-isochron diagram for the Stockdale rhyolite at the time of Shap granite emplacement (395 Myr ago) to show possible open-system behaviour of Sr in samples outside the hatched zone. Four sampling sites are distinguished by different symbols. After Compston *et al.* (1982).

from McKerrow *et al.*, 1980). Consistent with this, re-examination of the probability table of Brooks *et al.* (section 2.3.2) indicates that for 16 data points, an MSWD value of 1.92 indicates up to a 95% probability that the result is not an isochron.

A perfect isochron would imply complete re-setting, but apparently this did not occur. Plotting isotope ratios at 395 Myr ago (the date of intrusion of the Shap granite) on a pseudo-isochron diagram (Fig 3.18) allows an assessment to be made of scatter introduced by an event after extrusion. Compston *et al.* found that if the four samples with highest Rb/Sr ratios were removed, along with one sample (no. 5) with an anomalously high Sr content, then all of the other samples lie close to a 440 Myr reference line. In fact a regression through ten of these points yields a 'minimum' age of 430 ± 7 Myr. They noted that isochrons calculated for each of the four collecting localities yield lower MSWD values than the combined data set. This evidence warns that the combined data set is unsuitable for constructing a single isochron, despite the attractively precise result. Compston *et al.* calculated a weighted mean of 412 ± 7 Myr for the four local isochrons and interpreted this as the time of hydrothermal altera-tion of the rhyolite. In so far as the Rb–Sr evidence 'marks a real event' then 412 Myr may be the date of this event.

The evidence for open-system Rb–Sr systematics in numerous environments has now discredited the Rb–Sr isochron method as a dating tool for igneous crystallisation. This is now performed more effec-tively by other methods such as U–Pb zircon analysis (section 5.2). Therefore we will now examine the effectiveness of the Rb–Sr method in dating sedimen-tary rocks.

3.4 Dating sedimentary rocks

Absolute dating of the time of deposition of sedimentary rocks is an important problem, but one that is very difficult to solve. Accurate dates depend on thorough re-setting of isotopic clocks. Hence Rb–Sr dating of sediments rests on the assumption that Sr isotope systematics in the rock were homo-genized during deposition or early diagenesis, and thereafter remained as a closed system until the present day. However, we will see that these two requirements may be mutually exclusive.

In principle, sedimentary rocks may be divided into two groups according to the nature of the Rb-bearing phase present. Allogenic (detrital) minerals are moderately resistant to open-system behaviour dur-ing burial metamorphism, but problems arise from inherited isotopic signatures. Authigenic minerals are deposited directly from seawater and hence display good initial Sr isotope homogeneity. However, they are highly susceptible to recrystallisation after burial and may not remain closed systems.

In practice, the two distinct dating approaches associated with these sediment types have tended to converge. Analysis of detrital sediments has moved towards the analysis of fine-grained, almost authigenic, minerals such as illite, to escape the effects of the detrital component. In contrast, analysis of authigenic minerals has been focussed on the sub-authigenic mineral glauconite, since the truly authigenic Rb-bearing evaporite minerals are too susceptible to burial metamorphism to be viable geochronometers.

3.4.1 Shales

Detrital Rb-bearing minerals (mica, K-feldspar, clay minerals etc.) can be expected to contain inherited old radiogenic Sr, so that dating of such material should give an average of the provenance ages of the sedimentary constituents. However, if sufficiently fine-grained shales are sampled, it appears that the constituent minerals (mainly illite) often suffer substantial Sr exchange during post-depositional diagenesis. In this case they may develop an almost homogeneous initial Sr isotope composition soon after deposition, thereafter remaining effectively closed systems until the present day.

Compston and Pidgeon (1962) pioneered whole-rock Rb–Sr dating of shales, and found that in some circumstances (e.g. the State Circle shale from S.E. Australia) the above conditions were closely approached, while in other cases (e.g. the Cardup shale of W. Australia), gross inherited $^{87}Sr/^{86}Sr$ variations remained, preventing the calculation of a meaningful age. Compston and Pidgeon attributed this to undecomposed detrital micas, probably sericite. In contrast, the carbonaceous shales of the Cardup unit contained much less detrital mica and, taken alone, gave a tentative depositional age of 660 Myr.

Some recent work on the dating of shales has sought to avoid problems of contamination with detrital micas and feldspars by analysing separated clay-mineral fractions, whose purity is checked by x-ray diffraction (XRD). XRD analysis of illites can also yield information about the nature and origin of clay minerals in a shale which is to be dated.

The 'illite crystallinity index' (Kubler, 1966) is defined as the width of the (001) XRD peak at half its height. A well-crystallised illite, characteristic of a relatively high-temperature history, has sharp peaks, and therefore a low index, while low temperature

illites are more disordered, and have irregular peaks with large indices. In addition to this discriminant, illite has high-temperature (2M) and low-temperature (1M) polymorphs which can be distinguished by XRD (Dunoyer, 1969). '1M' illites with a large crystallinity index are characteristic of low-temperature growth and recrystallisation in the sedimentary–diagenetic regime, while '2M' illites with a small index are indicative of temperatures of zeolite-facies metamorphism or above. The latter reflect a detrital component, or post-diagenetic metamorphism.

A comparison of Rb–Sr whole-rock and clay mineral analysis of a Precambrian shale from Mauritania (W. Africa). is shown in Fig. 3.19 (Clauer, 1979). Four clay fractions were analysed, containing smectite and the 1M illite polymorph with a crystallinity index over 6 (very low-grade metamorphism is characterised by an index below 5.75). These define a linear array which is colinear with associated dolomites, yielding an age of 860 ± 35 Myr, and an initial ratio of 0.7088, characteristic of Precambrian seawater. A whole-rock sample (4) shown (by XRD) to be free of detrital feldspar also lay on the isochron, but two whole rocks (2, 3) with traces of microcline, lay slightly above it, while one with 15% microcline (1) was displaced well above the isochron. It appears from this example that whole-rock Rb–Sr dating of shales is an unreliable geochronometer, but that analysis of separated illite fractions may give meaningful ages of diagenesis or low-grade metamorphism. However, there is always a danger that the detrital component may not be completely eliminated from the illite fraction. An important example of this problem is provided by the dating of the Sinian–Cambrian boundary.

In China, the Sinian (youngest Precambrian)–Cambrian boundary is very well exposed, with an apparently continuous fossil-rich succession of black shales across the base of the Cambrian. Rb–Sr analysis of shales would be a very convenient method of dating this boundary if reliable ages for deposition or early diagenesis could be obtained. Some of these results, summarised by Cowie and Johnson (1985) and Odin et al. (1985), are shown on the left-hand side of Table 3.1. They appear to support an age of ca. 600 Myr for the base of the Cambrian. However, evidence for a much younger age came from the Ercall granophyre of southern Britain. This pluton is cut by an erosion surface which is overlain by quartzites bearing lower Cambrian trilobites (Cope and Gibbons, 1987). Patchett et al. (1980) obtained a statistically perfect (MSWD = 0.97) whole-rock Rb–Sr age of 533 ± 13 Myr (2σ) for the

Table 3.1. *Age data (in Myr) for shales from the Yangtse gorge, based on whole-rock or coarse clay (left column) and fine clay separates (right).*

Coarse fraction or whole-rock		Fine fraction	
573 ± 7	(Rb–Sr > 1.5 μm)		
613 ± 23	(Rb–Sr)		
568 ± 12	(U–Pb whole-rock)		
572 ± 14	(Rb–Sr whole-rock)	435 to 415	(Rb–Sr < 1 μm)
570 ± 4	(Rb–Sr)		
574 ± 20	(Rb–Sr)	565 to 490	(Rb–Sr)
602 ± 15	(Rb–Sr ca. 1.5 μm)	460 ± 9	(Rb–Sr < 1 μm)
Stratigraphic Cambrian–Precambrian boundary			
614 ± 18	(Rb–Sr > 1.5 μm)		
700 ± 5	(Rb–Sr > 1.5 μm)	580 ± 25	(Rb–Sr < 1 μm)
691 ± 29	(Rb–Sr)		
580 to 420	(Rb–Sr > 1.5 μm)		
727 ± 9	(Rb–Sr > 1.5 μm)	460 to 340	(Rb–Sr < 1 μm)
728 ± 27	(Rb–Sr)	500 to 360	(Rb–Sr < 1 μm)
608 ± 15	(Rb–Sr > 1.5 μm)		

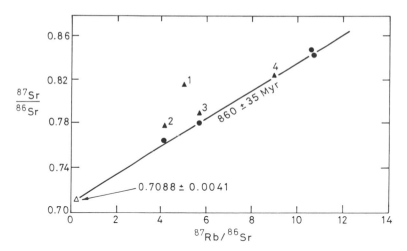

Fig. 3.19. Rb–Sr isochron diagram for whole-rock shales (▲); separated illites (●); and a carbonate sample (△) from Mauritania. Numbered whole-rock samples are discussed in the text. After Clauer (1979).

granophyre, which they interpreted as the age of intrusion.

In view of these conflicting ages for the base of the Cambrian, Rb–Sr shale data must be examined very carefully to judge its reliability. Good agreement between whole-rock Rb–Sr and whole-rock U–Pb isochron dates (^{206}Pb/^{204}Pb–^{238}U/^{204}Pb and ^{207}Pb/^{204}Pb–^{235}U/^{204}Pb) from the east Yangtze gorge area was argued by Zhang *et al.* (1984) to demonstrate the validity of these results for dating

isotopic homogenisation during deposition and early diagenesis. This is because the Rb–Sr system is based on the non-organic fraction and the U–Pb system is largely based on the organic fraction in the shale.

The illite crystallinity index was also used by Zhang *et al.* in an attempt to screen out material with an inherited component. However, the analysis of fine-grained fractions (right-hand column in Table 3.1) almost invariably gave ages significantly lower than whole-rock or coarse clay fractions. This suggests that a diagenetic event affected the rocks some time after deposition, so that the data in the left column of Table 3.1 are probably mixed ages between inherited and diagenetic components, rather than depositional ages.

This interpretation is supported by recent U–Pb zircon dates on tuff and bentonite from near the base of the Cambrian in Morocco, China and Siberia, which confirm a young boundary age near 540 Myr (Compston *et al.*, 1990; Bowring *et al.*, 1993). Therefore, Rb–Sr dating of shales cannot be considered a reliable technique for dating sedimentary deposition.

3.4.2 Glauconite

The mineral glauconite offers an attractive possibility of dating sedimentary rocks directly, due to its high Rb content, easy identification and widespread stratigraphic distribution. Glauconite is a micaceous mineral similar to illite which is best developed in macroscopic pellets (called 'glauconies' by Odin and Dodson, 1982). These are probably formed by the alteration of a very fine-grained clay precursor intermixed with organic matter in a faecal pellet. Glauconies form near the sediment–water interface in the marine environment. However, by studying pellets on the present day ocean floor, Odin and Dodson (1982) have shown that 'glauconitisation' is a slow process which may take hundreds of thousands of years to reach completion. During this process, the potassium content of the pellet increases, and this can therefore be used to monitor the maturation of the pellet.

Rb–Sr analysis of Holocene glauconies (Clauer *et al.*, 1992) shows that Sr isotope equilibrium with seawater is achieved only slowly as the potassium content increases. The Rb–Sr data can be used to calculate a model Sr age for the pellet by making the initial ratio equal to the isotopic composition of seawater Sr at the estimated time of sedimentation (see below). A zero-age pellet starts with a high apparent model age due to a large content of Sr in detrital mineral phases. However, as it matures, the pellet homogenises with seawater so that the model age falls to zero in a fully equilibrated pellet (Fig. 3.20). Analysis of the potassium content of glauconies therefore provides an essential screening procedure, in order to select only fully mature material for dating. Cretaceous and younger glauconies often yield ages concordant with other dating methods (e.g. Harris, 1976), but Paleozoic glauconies commonly give ages that are 10–20% younger than anticipated. Early workers (e.g. Hurley *et al.*, 1960) attributed this

to post-depositional uptake of K and Rb during diagenesis. However, Morton and Long (1980) attributed the young ages to ^{87}Sr loss from the expandable layers of the clay lattice, by some form of ion exchange process with circulating brines.

Morton and Long calculated model ages for a series of glauconite separates, based on the assumed initial ^{87}Sr/^{86}Sr ratio of seawater at the time of deposition (see section 3.5.1). They showed that in some cases erroneous glauconite model ages could be increased to near the stratigraphic age by leaching with ammonium acetate, which is thought to remove excess loosely bound Rb from the expandable layers of the lattice. In contrast, leaching with acetic acid, HCl etc., had unpredictable effects on the glauconite age, probably due to removal of some tightly bound Sr.

Similar experiments were performed on glauconites from the 525 Myr-old Bonneterre Formation (Missouri) by Grant *et al.* (1984). Eight un-leached glauconite pellets gave model ages in the range 413–440 Myr. However, the most radiogenic sample (model age = 426 Myr) converged only slightly on the true age when subjected to ammonium acetate leaching (437 Myr). Therefore, more rigorous criteria are needed to determine whether old glauconites have suffered open-system behaviour, prior to a dating attempt. Until such criteria are developed, glauconite dating in the Paleozoic must be regarded as a monitor of diagenetic processes rather than a viable dating tool for stratigraphic correlation.

3.5 Seawater evolution

Biogenic carbonates fulfil two of the requirements of a sedimentary dating tool: they are fairly resistant to diagenetic alteration, and since they are secreted directly from seawater by the organism, they contain no detrital fraction. Unfortunately, the negligible Rb content of carbonates precludes application of the conventional Rb–Sr dating method. However, calibration of the seawater Sr isotope evolution path would allow the 'initial' ^{87}Sr/^{86}Sr isotope ratios of carbonates to be used as an indirect dating tool. In the following section we will assess the realisation of this concept.

3.5.1 Measurement of the curve

Interest in the strontium isotope composition of seawater dates back to Wickman (1948). He argued that decay of ^{87}Rb to ^{87}Sr in crustal rocks over geological time, and its subsequent release into the

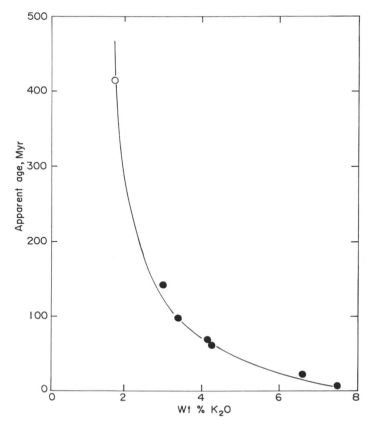

Fig. 3.20. Rb–Sr model ages of Holocene (zero-age) glauconies as a function of potassium content. Open symbol indicates clay fraction. After Clauer *et al.* (1992).

GAST TOO LOW RESOLUTION

hydrosphere by erosion, should lead to a 25% increase in seawater Sr isotope composition over the last 3 Byr. This model was tested by Gast (1955), who analysed carbonates of different ages as a means of characterising the evolution of seawater through geological time. He found that any natural variations were of the same order as the analytical errors of $^{87}Sr/^{86}Sr$ analysis pertaining at that time (ca. 0.004), thus refuting Wickman's model. Evidently the average crustal Rb/Sr ratio assumed by Wickman was an over-estimate.

Resolution of the actual extent of seawater Sr isotope variation with time had to wait 15 years for the advent of more precise mass spectrometry. Peterman *et al.* (1970) measured the $^{87}Sr/^{86}Sr$ composition of macro-fossil shell carbonates with an order of magnitude improvement in precision to 0.0005 (2σ). They found a total isotopic range of 0.0022 (4 × analytical error), which would have been imperceptible using earlier equipment. Peterman *et al.* showed that, contrary to Wickman's prediction,

seawater Sr isotope ratio actually *decreased* during the Paleozoic, reaching a minimum during the Mesozoic before rising quickly to a maximum at the present day.

In order to avoid the effects of post-depositional alteration, Peterman *et al.* rejected any recrystallised shell material, which they claimed to be able to recognise visually. The possibility of Sr exchange between matrix and un-recrystallised shells was rendered unlikely by the good compositional agreement between different shells in a bed. A mixture of mollusc types was used (belemnites, bivalves and brachiopods). Since no variation was seen between such classes at the present day, they were assumed to behave in the same way as fossils.

Additional data were collected by Dasch and Biscaye (1971) and Veizer and Compston (1974) from different types of sample material. Dasch and Biscaye analysed Cretaceous-to-Recent pelagic foraminifera, while Veizer and Compston (1974) studied 'sedimentary carbonate' (in other words not macro-

[handwritten margin notes: PETERMAN BACKED AND VEIZER... DOSCH... WHO GOT THE SAME RESULTS FROM DIFFERENT TESTS]

fossil carbonate) to test its reliability for the determination of seawater Sr isotope ratio. Both studies found general agreement with the data of Peterman *et al.* (1970). This implies global homogenisation of seawater Sr, which can be attributed to the very long residence time of Sr in seawater (ca. 2.5 Myr; Hodell et. al., 1990) compared to the average mixing time of oceanic water (only a few hundred years). However, Veizer and Compston recognised that 'sedimentary carbonate' is more susceptible to post-depositional exchange with pore waters. They argued that since detrital grains would normally have radiogenic Sr isotope signatures, post-depositional exchange would normally be expected to raise $^{87}Sr/^{86}Sr$ ratios. Therefore the minimum Sr isotope ratio found at any given time should be the most reliable guide to contemporaneous seawater composition.

While the analysis of whole-rock carbonate provides fewer constraints on post-depositional processes, it provides more opportunity for sampling, and is essential for Precambrian carbonates. Using the principles outlined above, Veizer and Compston (1976) made a reconnaissance study of the Sr isotope evolution of Precambrian seawater. They found uniformly unradiogenic Sr isotope ratios in Archean carbonates, with values only slightly elevated over contemporaneous upper mantle (Fig.

[handwritten margin notes: OCEAN MIXES 10K TIMES IN THE TIME Sr HAS MIXED THROUGH; PRE CAMBRIAN VIZIER COMPSTON METHOD]

3.21). However, there was a substantial rise in Sr isotope ratio during the Proterozoic, reaching a maximum in the early Cambrian which was similar to the present-day composition.

A major expansion of the seawater Sr data set was achieved by Burke *et al.* (1982). They presented 786 isotopic analyses of marine carbonates, phosphates and evaporites, with good coverage of all of Phanerozoic time except the Lower Cambrian (Fig. 3.22). Subsequent work by Derry *et al.* (1989), Asmerom *et al.* (1991) and Kaufman *et al.* (1993) has extended the curve back to the Late Proterozoic. These studies were made principally on whole-rock carbonates, which are susceptible to contamination by fluid-borne Sr during post-depositional alteration. Burke *et al.* dissolved their samples in hydrochloric or nitric acid, which may be the reason why many of their samples were displaced to radiogenic compositions above the lower bound of the data set (the best estimate for seawater Sr). In the later papers on Precambrian seawater evolution, bulk carbonates were dissolved in dilute acetic acid to reduce the amount of contamination by phases with radiogenic Sr.

Following the wide-ranging study of Burke *et al.* (1982), subsequent work has been devoted to improving precision on small segments of the curve. This requires material to be well-dated stratigraphi-

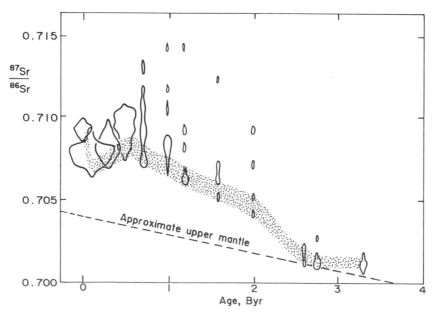

Fig. 3.21. Sr isotope composition of marine carbonates over the last 3.5 Byr, from which the isotopic evolution of seawater is deduced (shaded band). After Veizer and Compston (1976).

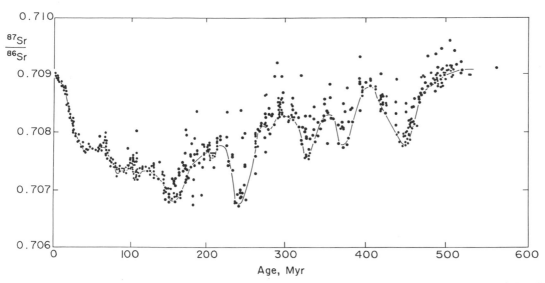

Fig. 3.22. Sr isotope data for Phanerozoic carbonates. Solid line indicates the lower bound of most of the data, which is the most probable seawater Sr composition. After Burke *et al.* (1982).

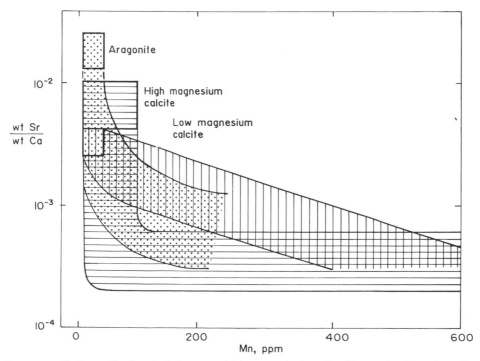

Fig. 3.23. Summary of diagnostic chemical changes which occur during the diagenetic alteration of carbonates. Boxes represent primary fields. After Brand and Veizer (1980).

cally and carefully screened before analysis to exclude post-depositional alteration. In Paleozoic rocks, this screening is best achieved chemically. Brand and Veizer (1980) showed that open-system diagenesis of carbonates was accompanied by a decrease in Sr/Ca ratio and an increase in Mn content (Fig. 3.23). However, Mn-enriched calcite can be detected by cathodoluminescence, so that sections of shell can be

screened for alteration before sample analysis. Popp
et al. (1986) showed that samples of brachiopod shell
prepared in this way gave more reliable results than
whole-rock carbonate (which were usually contaminated by radiogenic Sr) and whole brachiopod shells
(which were sometimes contaminated by unradiogenic Sr).

The availability of Deep Sea Drilling Project
(DSDP) core has allowed the construction of a very
precise seawater Sr isotope evolution curve for the last
100 Myr, based on continuous recovered sections.
Sedimentation rates in these sections are used to
interpolate between biostratigraphic and magneto-
stratigraphic calibration points. This avoids the age
uncertainty involved in correlating short stratigraphic
sections from different localities.

Two different sampling approaches have been
adopted for DSDP core material. DePaolo (1986)
made a study on a single DSDP hole reaching back to
the Early Miocene, but with duplicate analysis of all
samples to improve analytical precision. In his
approach, bulk samples of foram–nano-fossil ooze
were analysed by direct acetic acid leaching of washed
whole-rock samples. This necessitates a correction for
post-depositional exchange in order to determine
original seawater compositions. These corrections

were based on the analysis of pore waters. However,
pore waters displayed relatively small deviations in
$^{87}Sr/^{86}Sr$ from the carbonate fraction (< 0.0001), and
were also found to have Sr contents an order of
magnitude lower (Richter and DePaolo, 1987). Hence
it was argued that corrections for Sr exchange were
smaller than mass spectrometric reproducibilities.

In the other approach (Hess *et al.*, 1986), hand-
picked whole foram tests were analysed. These were
screened for secondary alteration using scanning
electron microscopy (SEM) and elemental chemistry
(e.g. Mn and Sr content). Fig. 3.24 shows data from
eight partially overlapping DSDP sections. Slight
scatter is seen, but much of this can be attributed to
analytical error rather than diagenetic effects. In
selected samples from two sites, pore-waters had very
similar isotope ratios to forams. In one other site,
pore-waters were somewhat more radiogenic, but
there is no evidence that the foram data have been
perturbed. Most subsequent studies have also em-
ployed hand-picked forams. Since less than 50 ng of
Sr is now needed for a precise analysis, this may be
possible on a few or even a single foram. As an
additional precaution, Martin and Macdougall (1991)
were able to break open large Cretaceous forams to
examine them by SEM for internal calcite growth.

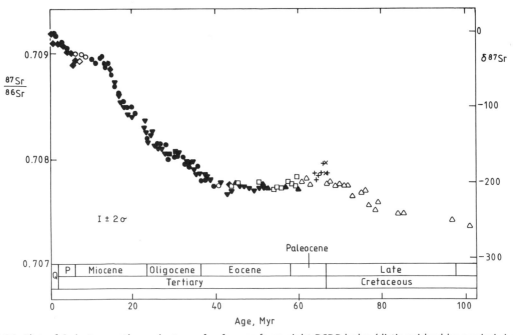

Fig. 3.24. Plot of Sr isotope ratio against age for forams from eight DSDP holes (distinguished by symbol shape).
Solid symbols and crosses indicate most reliable data; open symbols may be slightly disturbed. After Hess *et al.*
(1986).

EVIDENCE FOR BOLIDE IMPACT

The high-precision seawater Sr isotope evolution curve can be used as a stratigraphic dating tool, with a (conservative) precision as good as 0.5 Myr for periods of rapid Sr isotope evolution, but as bad as 2 Myr during periods of slow isotopic evolution. This precision cannot compete with biostratigraphic dating in the Cretaceous and Tertiary periods, but it may be useful for calibration of un-fossiliferous borehole sections (e.g. Rundberg and Smalley, 1989). Another application is in the dating of ferromanganese crusts (Ingram *et al.*, 1990).

An interesting observation by Hess *et al.* (1986) in their data set was a 'spike' in seawater Sr isotope ratio at the Cretaceous–Tertiary (K–T) boundary (Fig. 3.24). They speculated as to whether a meteorite impact could release sufficient Sr, either from the bolide or the terrestrial impact ejecta, to explain this peak. They concluded that this was unlikely, but Macdougall (1988) suggested that a meteorite could have caused massive nitrogen oxide generation by shock-heating of the atmosphere, and that the

resulting acid rain could have released a surge of radiogenic Sr into the hydrosphere by chemical weathering.

If the spike in ^{87}Sr is to be attributed to a meteorite, it is critical to demonstrate that it occurs at exactly the correct stratigraphic level. Martin and Macdougall (1991) collected data from four widely spaced localities around the world which appear to support the model (Fig. 3.25). A fairly rapid increase in seawater Sr below the K–T boundary was also seen by Nelson *et al.* (1991). However, the spike at the boundary itself is very sharp, and appears to be superimposed on the longer-term variations. Therefore the evidence available at present seems to be consistent with a meteorite impact at the K–T boundary.

3.5.2 Modelling of the fluxes

The first model for the Sr isotopic composition of seawater was constructed by Faure *et al.* (1965) to explain the present-day Sr isotope ratio of North Atlantic seawater. They suggested that there was a balance between the supply of unradiogenic Sr by erosion of young volcanics, radiogenic Sr from old crustal rocks, and Sr of intermediate composition from the erosion of carbonates. This model was adopted by Peterman *et al.* (1970) to explain the rise and fall of seawater Sr isotope ratio during the Phanerozoic. Armstrong (1971) supplemented this model, suggesting that peaks in seawater Sr isotope ratio during the Carboniferous and Tertiary periods were due to enhanced glacial erosion of old shields with elevated ^{87}Sr contents (Fig. 3.26). However, in other ways the model remained largely unchallenged.

A major advance in modelling seawater Sr evolution was the proposal of Spooner (1976) that the unradiogenic Sr flux was due to submarine hydrothermal exchange with basaltic crust, rather than sub-aerial erosion of basic rock. Spooner calculated that the hydrothermal flux must be six times the magnitude of the river water Sr flux. However, this was based on high estimates of the isotopic composition of run-off (0.716) and hydro-thermally buffered water (0.708). Subsequent analysis of hydrothermal vent waters from the East Pacific Rise (Albarede *et al.*, 1981) indicated much less-radiogenic compositions. Albarede *et al.* estimated the flux of hydrothermally recycled Sr as less than one-quarter of the flux due to continental run-off. This model predicted an average Sr isotope composition of between 0.710 and 0.711 for run-off, in good

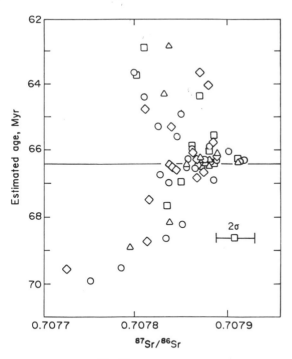

Fig. 3.25. Plot of ^{87}Sr/^{86}Sr ratios for forams from four widely separated DSDP cores of similar age (distinguished by symbol type). Absolute ages are estimated from sedimentation rates relative to the K–T boundary, whose position is determined by a nearly complete turnover of fossil taxa. After Martin and Macdougall (1991).

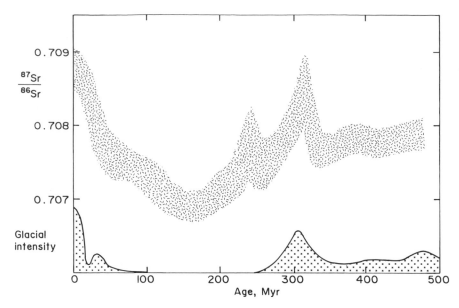

Fig. 3.26. Illustration of a glacial–erosional model to explain the seawater Sr evolution curve of Peterman *et al.* (shaded band). After Armstrong (1971).

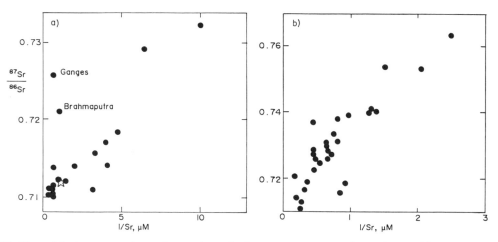

Fig. 3.27. Plot of Sr isotope ratio against reciprocal of Sr concentration. a) For the world's major rivers (star represents global average); b) for the tributaries of the Ganges (also known as Ganga). After Palmer and Edmond (1989).

agreement with major rivers such as the Amazon (Brass, 1976).

These calculations were further refined by Palmer and Edmond (1989), who measured the Sr budget and isotope composition of hydrothermal vent fluids and of most of the world's major rivers. The latter show an inverse relationship between isotope ratio and concentration (Fig. 3.27a), which is attributed to mixing between radiogenic Sr from silicate weathering and less-radiogenic Sr from carbonate weathering. The Ganges and Brahmaputra, which drain the Himalayan uplift, lie off the general trend. However, within the drainage basin of the Ganges, its tributaries themselves display a mixing line, although with steeper slope than other rivers (Fig. 3.27b). Taken together, the complete data set of Palmer and Edmond leads to an estimated global river budget of 3.3×10^{10} mol Sr per year, with $^{87}Sr/^{86}Sr$ of 0.7119, and an ocean ridge hydrothermal Sr flux of about one-half this magnitude, with $^{87}Sr/^{86}Sr$ of 0.7035.

The recognition of competing riverine and hydrothermal fluxes raises the question of how they interact to cause variations in seawater isotope ratio with time. Spooner (1976) assumed that the hydrothermal Sr flux was fairly constant over time. He attributed the increase in $^{87}Sr/^{86}Sr$ since the Cretaceous principally to an increase in continent exposure (and hence Sr run-off) over the last 85 Myr (Fig. 3.28). In contrast, Albarede *et al.* (1981) argued that a drop in the ocean ridge Sr exchange flux from a Mesozoic value nearly four times higher was more important than a rise in the flux of continental run-off. However, these two effects are difficult to separate, since they are bound together as a system. A drop in spreading rate causes ridge collapse and consequent sea-level fall, so that continental exposure should increase as hydrothermal buffering of seawater decreases.

In addition to run-off and hydrothermal exchange, two other fluxes have been proposed to control seawater Sr. One which has been widely accepted, although small in size, represents Sr released from ocean floor carbonates by diagenetic recrystallisation (Elderfield and Gieskes, 1982). This is estimated at

about 10% of the run-off flux and tends to dampen isotopic fluctuations with time because it recycles old seawater Sr. Another proposed flux which has not received widespread acceptance is called 'run-out'. This represents the sub-surface out-flow of continental water from below the water table, and was proposed by Chaudhuri and Clauer (1986) to explain isotopic fluctuations which are not in harmony with changes in sea-level. These authors argued that the present-day run-out Sr flux might be almost as large as the run-off flux (Fig. 3.29). Both fluxes should respond to sea-level changes, but run-out would also be dependent (unlike run-off) on the length of the continental perimeter. Therefore, plate tectonic configurations which form super-continents should be characterised by low run-out, while fragmented continents (such as at the present day) should yield high run-out Sr. This model attributes the rising Sr isotope ratio during early Cretaceous time (despite rising sea-level) to progressive continental break-up at this time.

The best opportunity to study the interaction of competing fluxes in the buffering of seawater Sr is during periods of rapid change in isotope ratio with time. The Tertiary represents one such period, which

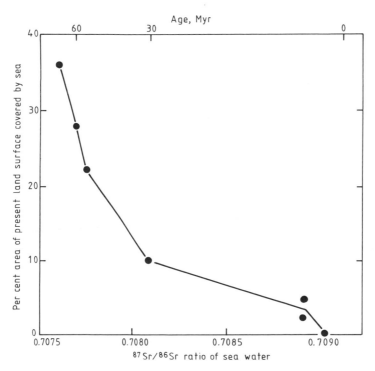

Fig. 3.28. Plot of seawater Sr isotope composition over the last 85 Myr against % continental flooding (relative to the present land area). After Spooner (1976).

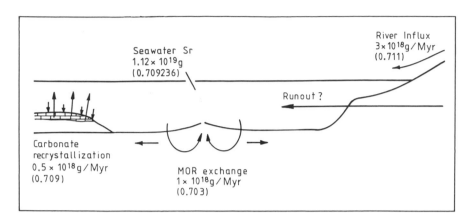

Fig. 3.29. Simplified circulation model for the present-day seawater Sr budget. Modified after DePaolo (1987).

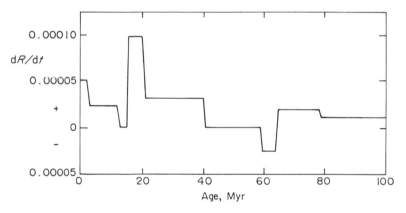

Fig. 3.30. Representation of seawater Sr isotope variation in the last 100 Myr in terms of the rate of change per million years. After Richter *et al.* (1992).

is characterised by an overall trend of increasing Sr isotope ratio, on which several smaller steps are superimposed. These variations can be represented in terms of the rate of change of Sr isotope ratio with time (Fig. 3.30). Raymo *et al.* (1988) attributed the general trend of seawater evolution over the last 40 Myr to increased rates of uplift of the Himalayas, Tibet and Andes. This could have caused a substantial increase in the supply of radiogenic Sr to the oceans, since the rivers which rise in these regions (Ganges-Brahmaputra, Yangtze and Amazon) to-gether supply 20% of the total solid load to the oceans. On the other hand, changes in the hydro-thermal Sr flux are not thought to have occurred, at least in the last 5 Myr, because ocean spreading rates have been nearly uniform during this time.

Additional evidence for the control of Himalayan erosion rates on seawater Sr was provided by Richter *et al.* (1992). Ar–Ar thermochronology was used to date the sudden unroofing of the Quxu granite pluton, corresponding to a period of exceptionally rapid erosion of the Tibetan plateau. The timing of this event, which began 20 Myr ago, matches exactly with the peak rate of change in the seawater Sr isotope record (Fig. 3.30).

Hodell *et al.* (1990) suggested that glaciation is a subordinate mechanism responsible for increased radiogenic Sr supply to rivers in the last few million years. This is possible because glaciation of shield areas generates rock flour, which is then more susceptible to chemical weathering than *in-situ* crystalline basement. On the other hand, Raymo *et*

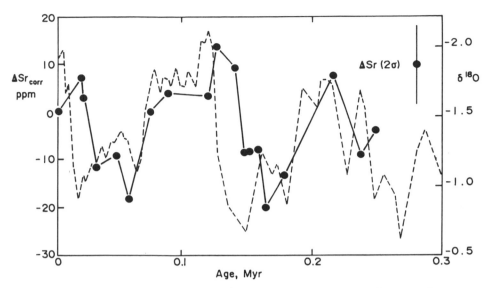

Fig. 3.31. Plot of Sr isotope ratios (•) in planktonic forams over the last 0.3 Myr, relative to the present day. The long-term trend of Sr isotope increase with time has been removed from the data. Dashed line shows oxygen isotope data from the same core. After Dia *et al.* (1992).

al. argued that the climatic cooling which led to glaciation may itself be a product of rapid orogenic erosion. This is because enhanced calcium transport to the oceans causes increased marine carbonate precipitation, which will suck carbon dioxide out of the atmosphere, and thus reduce the greenhouse effect.

Even finer-scale variations in seawater Sr isotope ratio were claimed by Dia *et al.* (1992). By analysing planktonic forams from a deep Pacific Ocean core, these workers found Sr isotope variations with a time-constant of only 0.1 Myr, which matched the oxygen isotope record from the same core (Fig. 3.31). It is generally accepted that the latter variations are a product of cyclic climatic changes (section 12.4.2). Since these variations are much shorter than the 2.5 Myr residence time of Sr in the oceans, they must reflect non-equilibrium effects. Dia *et al.* argued that they could be generated by a 50% change in the magnitude of riverine Sr flux (with constant isotopic composition), or by a change in mean $^{87}Sr/^{86}Sr$ in river water equal to 0.0005 (for a constant Sr flux). They suggested that the latter possibility could be achieved by perturbations of the drainage pattern of the Himalayas in response to climatic changes.

Clemens *et al.* (1993) reported similar short-term Sr isotope variations in forams from an Indian Ocean core. These data also displayed an apparent 0.1 Myr

periodicity which matched the global oxygen isotope record over the past 0.5 Myr (section 12.4.2). However, calculations performed by Clemens *et al.* demanded much larger changes in the riverine Sr flux, if the foram data sets were indicative of global seawater changes (100% in magnitude of riverine flux, or 0.004 in isotope ratio). Similar magnitudes of change were determined by Richter and Turekian (1993), leading them to suggest that the reported isotopic variations could not represent global seawater composition, but probably have a local origin.

References

Albarede, F., Michard, A., Minster, J. F. and Michard, G. (1981). $^{87}Sr/^{86}Sr$ ratios in hydrothermal waters and deposits from the East Pacific Rise at 21 °N. *Earth Planet. Sci. Lett.* **55**, 229–36.

Armstrong, R. L. (1971). Glacial erosion and the variable isotopic composition of strontium in sea water. *Nature Phys. Sci.* **230**, 132–33.

Asmerom, Y., Jacobsen, S. B., Knoll, A. H., Butterfield, N. J. and Swett, K. (1991). Strontium isotopic variations of Neoproterozoic seawater: implications for crustal evolution. *Geochim. Cosmochim. Acta* **55**, 2883–94.

Beckinsale, R. D., Pankhurst, R. J., Skelhorn, R. R. and Walsh, J. N. (1978). Geochemistry and petrogenesis of the early Tertiary lava of the Isle of Mull, Scotland. *Contrib. Mineral. Petrol.* **66**, 415–27.

Birck, J. L. and Allegre, C. J. (1978). Chronology and chemical history of the parent body of basaltic achondrites studied by the [87]Rb–[87]Sr method. *Earth Planet. Sci. Lett.* **39**, 37–51.

Bowring, S. A., Grotzinger, J. P., Isachsen, C. E., Knoll, A. H., Pelechaty, S. M. and Kolosov, P. (1993). Calibrating rates of Early Cambrian evolution. *Science* **261**, 1293–8.

Brand, U. and Veizer, J. (1980). Chemical diagenesis of a multicomponent carbonate system – 1: Trace elements. *J. Sed. Petrol.* **50**, 1219–36.

Brass, G. W. (1976). The variation of the marine [87]Sr/[86]Sr during Phanerozoic time: interpretation using a flux model. *Geochim. Cosmochim. Acta* **40**, 721–30.

Brinkman, G. A., Aten, A. H. W. and Veenboer, J. T. (1965). Natural radioactivity of K-40, Rb-87 and Lu-176. *Physica* **31**, 1305–19.

Brooks, C., Hart, S. R., Hofmann, A. and James, D. E. (1976a). Rb–Sr mantle isochrons from oceanic regions. *Earth Planet. Sci. Lett.* **32**, 51–61.

Brooks, C., James, D. E. and Hart, S. R. (1976b). Ancient lithosphere: its role in young continental volcanism. *Science* **193**, 1086–94.

Brown, E. H. (1971). Phase relations of biotite and stilpnomelane in the green-schist facies. *Contrib. Mineral. Petrol.* **31**, 275–99.

Burke, W. H., Denison, R. E., Hetherington, E. A., Koepnick, R. B., Nelson, H. F. and Otto, J. B. (1982). Variations of seawater [87]Sr/[86]Sr throughout Phanerozoic time. *Geology* **10**, 516–19.

Catanzaro, E. J., Murphy, T. J., Garner, E. L. and Shields, W. R. (1969). Absolute isotopic abundance ratio and atomic weight of terrestrial rubidium. *J. Res. NBS* **73A**, 511–16.

Chaudhuri, S. and Clauer, N. (1986). Fluctuations of isotopic composition of strontium in seawater during the Phanerozoic eon. *Chem. Geol. (Isot. Geosci. Section)* **59**, 293–303.

Clark, S. P. C. and Jager, E. (1969). Denudation rate in the Alps from geochronologic and heat flow data. *Amer. J. Sci.* **267**, 1143–60.

Clauer, N. (1979). A new approach to Rb–Sr dating of sedimentary rocks. In: Jager, E. and Hunziker, J. C. (Eds) *Lectures in Isotope Geology.* Springer, pp. 30–51.

Clauer, N., Keppens, E. and Stille, P. (1992). Sr isotopic constraints on the process of glauconitization. *Geology* **20**, 133–6.

Clemens, S. C., Farrell, J. W. and Gromet, L. P. (1993). Synchronous changes in seawater strontium isotope composition and global climate. *Nature* **363**, 607–10.

Cliff, R. A. (1985). Isotope dating in metamorphic belts. *J. Geol. Soc. Lond.* **142**, 97–110.

Compston, W. and Jeffery, P. M. (1959). Anomalous common strontium in granite. *Nature* **184**, 1792–3.

Compston, W., McDougall, I. and Wyborn, D. (1982). Possible two-stage [87]Sr evolution in the Stockdale rhyolite. *Earth Planet. Sci. Lett.* **61**, 297–302.

Compston, W. and Pidgeon, R. T. (1962). Rubidium–strontium dating of shales by the total-rock method. *J. Geophys. Res.* **67**, 3493–502.

Compston, W., Williams, I. S., Kirschvink, J. and Zhang, Z. (1990). Zircon U–Pb ages relevant to the Cambrian numerical timescale. *Geol. Soc. Australia* **27**, 21 (abstract).

Cope, J. C. W. and Gibbons, W. (1987). New evidence for the relative age of the Ercall Granophyre and its bearing on the Precambrian–Cambrian boundary in southern Britain. *Geol. J.* **22**, 53–60.

Cowie, J. W. and Johnson, M. R. W. (1985). Late Precambrian and Cambrian geological time-scale. In: Snelling, N. J. (Ed.) *The Chronology of the Geological Record. Mem. Geol. Soc. Lond.* **10**, 47–64.

Dasch, E. J. and Biscaye, P. E. (1971). Isotopic composition of strontium in Cretaceous-to-Recent, pelagic foraminifera. *Earth Planet. Sci. Lett.* **11**, 201–4.

Davis, D. W., Gray, J. and Cumming, G. L. (1977). Determination of the [87]Rb decay constant. *Geochim. Cosmochim. Acta* **41**, 1745–9.

Del Moro, A., Puxeddu, M., Radicati do Brocolo, F. and Villa, I. M. (1982). Rb–Sr and K–Ar ages on minerals at temperatures of 300–400 °C from deep wells in the Larderello geothermal field (Italy). *Contrib. Mineral. Petrol.* **81**, 340–9.

DePaolo, D. J. (1986). Detailed record of the Neogene Sr isotopic evolution of seawater from DSDP Site 590B. *Geology* **14**, 103–6.

DePaolo, D. J. (1987). Correlating rocks with strontium isotopes. *Geotimes,* (Dec. 1987), 16–18.

Derry, L. A., Keto, L. S., Jacobsen, S. B., Knoll, A. H. and Swett, K. (1989). Sr isotopic variations in Upper Proterozoic carbonates from Svalbard and East Greenland. *Geochim. Cosmochim. Acta* **53**, 2331–9.

Dia, A. N., Cohen, A. S., O'Nions, R. K. and Shackleton, N. J. (1992). Seawater Sr isotope variation over the past 300 kyr and influence of global climate cycles. *Nature* **356**, 786–8.

Dodson, M. H. (1973). Closure temperature in cooling geochronological and petrological systems. *Contrib. Mineral. Petrol.* **40**, 259–74.

Dodson M. H. (1979). Theory of cooling ages. In: Jager, E. and Hunziker, J. C. (Eds.) *Lectures in Isotope Geology.* Springer, pp. 194–202.

Dunoyer de Segonzac, G. (1969). Les mineraux argileux dans la diagenese. Passage au metamorphisme. *Mem. Serv. Carte Geol. Alsace Lorraine* **29**, 320 p.

Elderfield, H. and Gieskes, J. M. (1982). Sr isotopes in interstitial waters of marine sediments from Deep Sea Drilling Project cores. *Nature* **300**, 493–7.

Fairbairn, H. W., Hurley, P. M. and Pinson, W. H. (1961). The relation of discordant Rb–Sr mineral and rock ages in an igneous rock to its time of subsequent Sr87/Sr86 metamorphism. *Geochim. Cosmochim. Acta* **23**, 135–44.

Faure, G., Hurley, P. M. and Powell, J. L. (1965). The isotopic composition of strontium in surface water from the North Atlantic Ocean. *Geochim. Cosmochim. Acta* **29**, 209–20.

Field D. and Raheim, A. (1979a). Rb–Sr total rock isotope studies on Precambrian charnockitic gneisses from South Norway: evidence for isochron resetting during a low-grade metamorphic-deformational event. *Earth Planet. Sci. Lett.* **45**, 32–44.

Field D. and Raheim, A. (1979b). A geological meaningless Rb–Sr total rock isochron. *Nature* **282**, 497–9.

Flynn, K. F. and Glendenin, L. E. (1959). Half-life and β spectrum of Rb87. *Phys. Rev.* **116**, 744–8.

Gale, N. H., Beckinsale, R. D. and Wadge, A. J. (1979). A Rb–Sr whole rock isochron for the Stockdale Rhyolite of the English Lake District and a revised mid-Paleozoic time-scale. *J. Geol. Soc. Lond.* **136**, 235–42.

Gast, P. W. (1955). Abundance of Sr87 during geologic time. *Bull. Geol. Soc. Amer.* **66**, 1449–64.

Grant, N. K., Laskowski, T. E. and Foland, K. A. (1984). Rb–Sr and K–Ar ages of Paleozoic glauconites from Ohio–Indiana and Missouri, USA. *Isot. Geosci.* **2**, 217–39.

Gray, C. M., Papanastassiou, D. A. and Wasserburg, G. J. (1973). The identification of early condensates from the solar nebula. *Icarus* **20**, 213–39.

Harris, W. B. (1976). Rb–Sr glauconite isochron, Maestrichtian unit of Peedee Formation, North Carolina. *Geology* **4**, 761–2.

Harrison, T. M., Duncan, I. and McDougall, I. (1985). Diffusion of ^{40}Ar in biotite: temperature, pressure and compositional effects. *Geochim. Cosmochim. Acta* **49**, 2461–8.

Hess, J., Bender, M. L. and Schilling, J. G. (1986). Evolution of the ratio of strontium-87 to strontium-86 in seawater from Cretaceous to present. *Science* **231**, 979–84.

Hodell, D. A., Mead, G. A. and Mueller, P. A. (1990). Variation in the strontium isotopic composition of seawater (8 Ma to present): implications for chemical weathering rates and dissolved fluxes to the oceans. *Chem. Geol. (Isot. Geosci. Section)* **80**, 291–307.

Hofmann, A. W. and Giletti, B. J. (1970). Diffusion of geochronologically important nuclides under hydrothermal conditions. *Eclogae Geol. Helv.* **63**, 141–50.

Hunziker, J. C. (1974). Rb–Sr and K–Ar age determination and the Alpine tectonic history of the Western Alps. *Mem. Inst. Geol. Min. Univ. Padova* **31**, 1–54.

Hurley, P. M., Cormier, R. F., Hower, J., Fairbairn, H. W. and Pinson, W. H. (1960). Reliability of glauconite for age measurement by K–Ar and Rb–Sr methods. *Amer. Assoc. Pet. Geol. Bull.* **44**, 1793–808.

Ingram, B. L., Hein, J. R. and Farmer, G. L. (1990). Age determinations and growth rates of Pacific ferromanganese deposits using strontium isotopes. *Geochim. Cosmochim. Acta* **54**, 1709–21.

Jager, E. (1973). Die Alpine Orogenese im Lichte der radiometrischen Altersbestimmung. *Eclogae Geol. Helv.* **66**, 11–21.

Jager, E. Niggli, E. and Wenk, E. (1967). Rb–Sr Altersbestimmungen an Glimmern der Zentralalpen. *Beitr. Geol. Karte Shweiz N. F.* **134**, 1–67.

Kaufman, A. J., Jacobsen, S. B. and Knoll, A. H. (1993). The Vendian record of Sr and C isotopic variations in seawater: implications for tectonics and paleoclimate. *Earth Planet. Sci. Lett.* **120**, 409–30.

Kubler, B. (1966). La cristallinite d'illite et les zones tout a fait superieures du metamorphisme. *Colloque. sur les Etages Tectoniques.* Univ. Neuchatel. A la Baconniere Neuchatel, Suisse, pp. 105–22.

Lanphere, M. A., Wasserburg, G. J., Albee, A. L. and Tilton, G. R. (1964). Redistribution of strontium and rubidium isotopes during metamorphism, World Beater complex, Panamint Range, California. In: Craig, H., Miller, S. L. and Wasserburg, G. J. (Eds.) *Isotopic and Cosmic Chemistry*. North Holland Pub., pp. 269–320.

Macdougall, J. D. (1988). Seawater strontium isotopes, acid rain, and the Cretaceous–Tertiary boundary. *Science* **239**, 485–7.

Martin, E. E. and Macdougall, J. D. (1991). Seawater Sr isotopes at the Cretaceous/Tertiary boundary. *Earth Planet. Sci. Lett.* **104**, 166–80.

McKerrow, W. S., Lambert, R. St J. and Chamberlain V. E. (1980). The Ordovician, Silurian and Devonian time scales. *Earth Planet. Sci. Lett.* **51**, 1–8.

McMullen, C. C., Fritze, K. and Tomlinson, R. H. (1966). The half-life of rubidium-87. *Can. J. Phys.* **44**, 3033–8.

Minster, J-F., Birck, J-L. and Allegre, C. J. (1982). Absolute age of formation of chondrites studied by the ^{87}Rb–^{87}Sr method. *Nature* **300**, 414–9.

Morton, J. P. and Long, L. E. (1980). Rb–Sr dating of Palaeozoic glauconite from the Llano region, central Texas. *Geochim. Cosmochim. Acta* **44**, 663–72.

Nelson, B. K., MacLeod, G. K. and Ward, P. D. (1991). Rapid change in strontium isotopic composition of sea water before the Cretaceous/Tertiary boundary. *Nature* **351**, 644–7.

Neumann, W. and Huster, E. (1974). The half-life of ^{87}Rb measured as a difference between the isotopes of ^{87}Rb and ^{85}Rb. *Z. Physik* **270**, 121–7.

Neumann, W. and Huster, E. (1976). Discussion of the ^{87}Rb half-life determined by absolute counting. *Earth Planet. Sci. Lett.* **33**, 277–88.

Nicolaysen. L. O. (1961). Graphic interpretation of discordant age measurements on metamorphic rocks. *Ann. N. Y. Acad. Sci.* **91**, 198–206

Odin, G. S. and Dodson, M. H. (1982). Zero isotopic age of glauconies. In: Odin, G. S. (Ed.) *Numerical Dating in Stratigraphy*. Wiley, pp. 277–305.

Odin, G. S., Gale, N. H. and Dore, F. (1985). Radiometric dating of Late Precambrian times. In: Snelling, N. J. (Ed.) *The Chronology of the Geological Record. Mem. Geol. Soc. Lond.* **10**, 65–72.

Palmer, M. R. and Edmond, J. M. (1989). The strontium isotope budget of the modern ocean. *Earth Planet. Sci. Lett.* **92**, 11–26.

Papanastassiou, D. A. and Wasserburg, G. J. (1969). Initial strontium isotopic abundances and the resolution of small time differences in the formation of planetary objects. *Earth Planet. Sci. Lett.* **5**, 361–76.

Papanastassiou, D. A. and Wasserburg, G. J. (1970). Rb–Sr ages of lunar rocks from the Sea of Tranquillity. *Earth Planet. Sci. Lett.* **8**, 1–19.

Patchett, P. J., Gale, N. H., Goodwin, R. and Humm, M. J. (1980). Rb–Sr whole-rock isochron ages of late Precambrian to Cambrian igneous rocks from southern Britain. *J. Geol. Soc. Lond.* **137**, 649–56.

Peterman, Z. E., Hedge, C. E. and Tourtelot, H. A. (1970). Isotopic composition of strontium in sea water throughout Phanerozoic time. *Geochim. Cosmochim. Acta* **34**, 105–20.

Pinson, W. H., Schnetzler, C. C., Beiser, E., Fairbairn, H. W. and Hurley, P. M. (1963). Rb–Sr age of stony meteorites. *MIT Geochron. Lab. 11th Ann. Rep.* **NYO-10**, 517.

Popp, B. N., Podosek, F. A., Brannon, J. C., Anderson, T. F. and Pier, J. (1986). $^{87}Sr/^{86}Sr$ ratios in Permo-Carboniferous sea water from the analyses of well-preserved brachiopod shells. *Geochim. Cosmochim. Acta* **50**, 1321–8.

Provost, A. (1990). An improved diagram for isochron data. *Chem. Geol. (Isot. Geosci. Section)* **80**, 85–99.

Purdy, J. W. and Jager, E. (1976). K–Ar ages on rock-forming minerals from the Central Alps. *Mem. Inst. Geol. Mineral. Univ. Padova* **30**, 3–31.

Raymo, M. E., Ruddiman, W. F. and Froelich, P. N. (1988). Influence of late Cenozoic mountain building on ocean geochemical cycles. *Geology* **16**, 649–53.

Richter, F. M. and DePaolo, D. J. (1987). Numerical models for diagenesis and the Neogene Sr isotope evolution of seawater from DSDP Site 590B. *Earth Planet. Sci. Lett.* **83**, 27–38.

Richter, F. M., Rowley, D. B. and DePaolo, D. J. (1992). Sr isotope evolution of seawater: the role of tectonics. *Earth Planet. Sci. Lett.* **109**, 11–23.

Richter, F. M. and Turekian, K. K. (1993). Simple models for the geochemical response of the ocean to climatic and tectonic forcing. *Earth Planet. Sci. Lett.* **119**, 121–31.

Rundberg, Y. and Smalley, P. C. (1989). High-resolution dating of Cenozoic sediments from northern North Sea using $^{87}Sr/^{86}Sr$ stratigraphy. *AAPG Bull.* **73**, 298–308.

Schreiner, G. D. L. (1958). Comparison of the Rb-87/Sr-87 age of the Red granite of the Bushveld complex from measurements on the total rock and separated mineral

fractions. *Proc. Roy. Soc. Lond. A.* **245**, 112–17.

Spooner, E. T. C. (1976). The strontium isotopic composition of seawater, and seawater–oceanic crust interaction. *Earth Planet. Sci. Lett.* **31**, 167–74.

Steiger, R. H. and Jager, E. (1977). Subcommission on geochronology: convention on the use of decay constants in geo- and cosmo-chronology. *Earth Planet. Sci. Lett.* **36**, 359–62.

Sun, S. S. and Hansen, G. N. (1975). Evolution of the mantle: geochemical evidence from alkali basalt. *Geology* **3**, 297–302.

Tatsumoto, M. (1966). Genetic relationships of oceanic basalts as indicated by lead isotopes. *Science* **153**, 1094–101.

Veizer, J. and Compston, W. (1974). $^{87}Sr/^{86}Sr$ composition of seawater during the Phanerozoic. *Geochim. Cosmochim. Acta* **38**, 1461–84.

Veizer, J. and Compston, W. (1976). $^{87}Sr/^{86}Sr$ in Precambrian carbonates as an index of crustal evolution. *Geochim. Cosmochim. Acta* **40**, 905–14.

Verschure, R. H. Andriessen, P. A. M., Boelrijk, N. A. I. M., Hebeda, E. H., Maijer, C. Prien, H. N. A. and Verdurmen, E. A. T. (1980). On the thermal stability of Rb–Sr and K–Ar biotite systems : evidence from co-existing Sveconorwegian (ca. 870 Ma) and Caledonian (ca. 400 Ma) biotites in S. W. Norway. *Contrib. Mineral. Petrol.* **74**, 245–52.

Wasserburg, G. J., Papanastassiou, D. A. and Sanz, H. G. (1969). Initial strontium for a chondrite and the determination of a metamorphism or formation interval. *Earth Planet. Sci. Lett.* **7**, 33–43.

Wetherill, G. W., Davis, G. L. and Lee-Hu, C. (1968). Rb–Sr measurements on whole rocks and separated minerals from the Baltimore Gneiss, Maryland. *Geol. Soc. Amer. Bull.* **79**, 757–62.

Wickman, F. E. (1948). Isotope ratios: a clue to the age of certain marine sediments. *J. Geol.* **56**, 61–6.

Zhang, Z., Ma, G. and Lee, H. (1984). The chronometric age of the Sinian-Cambrian boundary in the Yangtze Platform, China. *Geol. Mag.* **121**, 175–8.

4 The Sm–Nd method

4.1 Sm–Nd isochrons

Sm is a rare earth element with seven naturally occurring isotopes. Of these ^{147}Sm, ^{148}Sm and ^{149}Sm are all radioactive, but the latter two have such long half-lives (ca. 10^{16} yr) that they are not capable of producing measurable variations in the daughter isotopes of ^{144}Nd and ^{145}Nd, even over cosmological intervals (10^{10} yr). However the half-life of ^{147}Sm (106 Byr) is sufficiently short to produce small but measurable differences in ^{143}Nd abundance over periods of several million years, thus providing the basis for the Sm–Nd dating method. This half-life, equivalent to a decay constant of 6.54×10^{-12} yr^{-1}, is the weighted mean of several determinations, and yields ages consistent with U–Pb dating (Lugmair and Marti, 1978).

Another samarium isotope, ^{146}Sm, is not naturally occurring, but has a relatively long half-life of 103 Myr. If Sm/Nd fractionation had occurred in early Earth history, within a few hundred million years of Sm nucleosynthesis, then variations in the abundance of the daughter product, ^{142}Nd, might be generated. Such variations have been detected in the achondritic meteorite Angra dos Reis (Lugmair and Marti, 1977) but their occurrence in terrestrial materials is still a matter of debate. Harper and Jacobsen (1992) have claimed a positive ^{142}Nd anomaly in one sample from Isua, western Greenland, but other workers (Galer and Goldstein, 1992; McCulloch and Bennett, 1993) have observed only normal ^{142}Nd in other Early Archean rocks. In contrast to this situation, the ^{147}Sm–^{143}Nd decay scheme has found widespread geological application.

Considering a given system, such as an igneous rock or mineral, we can write the following equation based on the decay of ^{147}Sm:

$$^{143}\text{Nd} = {}^{143}\text{Nd}_\text{I} + {}^{147}\text{Sm}(e^{\lambda t} - 1) \qquad [4.1]$$

where I signifies the initial abundance and t is the age of the system. In view of the possibility of ^{142}Nd variation (due to ^{146}Sm), it is convenient to divide through by ^{144}Nd, the second-most abundant isotope

of Nd. Thus we obtain:

$$\frac{^{143}\text{Nd}}{^{144}\text{Nd}} = \left(\frac{^{143}\text{Nd}}{^{144}\text{Nd}}\right)_\text{I} + \frac{^{147}\text{Sm}}{^{144}\text{Nd}}\,(e^{\lambda t} - 1) \qquad [4.2]$$

This equation has the same form as that for Rb–Sr (section 3.2) and can be plotted as an isochron diagram. It should be noted however that because of the very similar chemical properties of Sm and Nd (unlike Rb and Sr), large ranges of Sm/Nd in natural rocks are rare; and in particular, low Sm/Nd ratios near the y axis are very rare. Because of the difficulty of obtaining a wide range of Sm/Nd ratio from a single rock body and the greater technical demands of Nd isotope analysis, the Sm–Nd isochron method has generally been applied to problems where the simpler Rb–Sr technique is unsuitable. Most of these instances involve rocks which have been disturbed to some degree by metamorphism. Some case studies will be examined.

4.1.1 Meteorites

Chondritic meteorites have been readily dated by the Rb–Sr method (section 3.1), but achondrites are more problematical. Bulk samples usually have low Rb/Sr ratios, yielding ages of low precision, while separated minerals in many achondrites yield ages below 4.5 Byr, indicative of disturbance. The Sm–Nd system in separated minerals from achondrites is more resistant to re-setting, yielding better age estimates. The first Sm–Nd dating study was performed by Notsu et al. (1973) on the achondrite Juvinas, but with low analytical precision. Lugmair et al. (1975) obtained much more precise results on minerals from the same meteorite (Fig. 4.1) yielding an age of 4560 ± 80 Myr (2σ). Subsequently, numerous other achondrites have been dated, and with the exception of Stannern (Lugmair and Scheinin, 1975), all yield ages in the range 4550–4600 Myr, in good agreement with the U–Pb method.

Sm–Nd dating of chondritic meteorites was not a high priority, due to the success of other methods. However, the isotopic composition of the chondrites

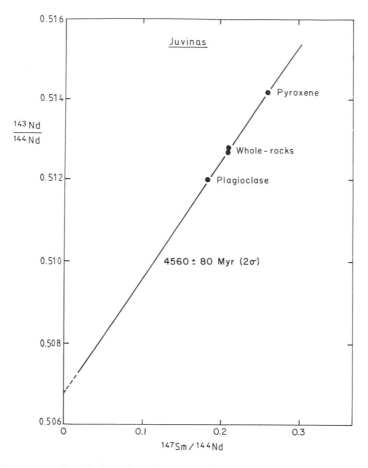

Fig. 4.1. Sm–Nd isochron array for whole-rocks and separated minerals from the basaltic achondrite Juvinas. The initial ratio is affected by the choice of normalising factor for mass fractionation. Data from Lugmair *et al.* (1975).

is a critical benchmark for the evolution of solar system bodies such as the Earth, because chondrites are believed to represent the nearest approach to the primordial solar nebula. DePaolo and Wasserburg (1976a) coined the acronym CHUR (chondritic uniform reservoir) for this benchmark, but in the absence of isotopic data for chondrites had to use Lugmair's (1975) ^{143}Nd/^{144}Nd ratio of 0.511836 from the achondrite Juvinas as an indicator of the present day CHUR value (using a fractionation normal-isation to ^{146}Nd/^{142}Nd = 0.636151 for Nd analysis as the oxide).

This value was tested by direct Sm–Nd analysis of chondrites by Jacobsen and Wasserburg (1980). They obtained a whole-rock isochron with an age of ca. 4.6 Byr and initial ratio of 0.50583 ± 1 (Fig. 4.2). The data points clustered closely around the original

Juvinas value of 0.511836. The intersection of this value with the isochron regression line led to a ^{147}Sm/^{144}Nd ratio of 0.1967 for CHUR. Jacobsen and Wasserburg compared this value to the average of 64 elemental Sm/Nd analyses of chondrites (Fig. 4.3), and demonstrated good agreement between the two values.

In 1981, Wasserburg *et al.* revised the isotopic composition of their oxide correction and modified their recommended ^{143}Nd/^{144}Nd value of CHUR to 0.511847. However, most workers use the alternative normalisation convention (to ^{146}Nd/^{144}Nd = 0.7219) which was proposed by O'Nions *et al.* (1977) for Nd analysis as the metal (section 2.2.2). This leads to corresponding present-day values of ^{143}Nd/^{144}Nd = 0.512638 and ^{147}Sm/^{144}Nd = 0.1966 (Hamilton *et al.*, 1983).

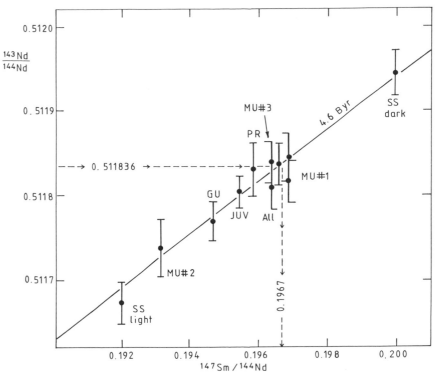

Fig. 4.2. Sm–Nd isochron diagram for whole-rock samples of six different chondrites. SS = St Severin; MU = Murchison; GU = Guarena; PR = Peace River; ALL = Allende. JUV = new analysis of the Juvinas achondrite. The large apparent errors are due to very expanded axis scales. After Jacobsen and Wasserburg (1980).

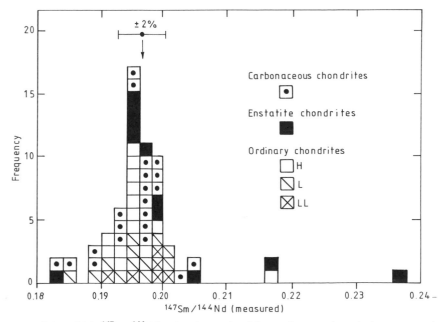

Fig. 4.3. Histogram of chondritic ^{147}Sm/^{144}Nd ratios determined from elemental analysis, compared with the value from Fig. 4.2 (arrow). Ordinary chondrites are sub-divided into compositional classes (H, L, LL). After Jacobsen and Wasserburg (1980).

4.1.2 Low-grade meta-igneous rocks

The long half-life of ^{147}Sm makes it most useful for dating in the Precambrian. Therefore, most early Sm–Nd work was focussed on the determination of crystallisation ages for Archean igneous rocks. In such suites the Rb–Sr or K–Ar methods had often failed to yield an accurate crystallisation age due to remobilisation of the parent or daughter elements during subsequent metamorphism. The Stillwater Complex (DePaolo and Wasserburg, 1979) provides a good example of this application.

Rb–Sr data on three separated minerals from a single adcumulus unit of the Stillwater layered series form a scatter which does not define an isochron, (Fig. 4.4a). However, Sm–Nd data on the same samples yield an excellent linear array (Fig. 4.4b), from which DePaolo and Wasserburg calculated an age of 2701 ± 8 Myr (2σ). In order to test for possible mineralogical re-setting of Sm and Nd, DePaolo and Wasserburg also analysed six whole-rock samples from different layers of the intrusion with a wide range of plagioclase/pyroxene abundances. Sm–Nd data from these samples fell within analytical uncertainty of the mineral isochron (Fig. 4.4c), suggesting that the mineral isochron yields a true crystallisation age for the intrusion, and that the magma had a homogeneous initial Nd isotope composition.

Subsequently, the Sm–Nd mineral age was corroborated by U–Pb dating of zircon from the chilled margin of the intrusion (Nunes, 1981), which gave an age of 2713 ± 3 Myr (2σ). However, Sm–Nd analysis of whole-rock samples from a wider stratigraphic

range in the intrusion demonstrated wider variations of initial ratio (Lambert et al., 1989). This is not surprising, since the initial ratio of DePaolo and Wasserburg falls well away from estimated mantle values at 2.7 Byr, and is best explained by contamination of the magma by old crustal Nd from the Wyoming craton. These findings emphasise the importance of combined mineral and whole-rock isochrons to verify the accuracy of Sm/Nd ages. However, this approach is not possible for fine-grained rocks such as Archean basalts and komatiites. In these situations, whole-rock analysis has often been used alone, but subtle effects on the slope of whole-rock isochrons can be caused by analysing samples with slight variations in crustal contamination. A good example is provided by the Kambalda volcanics of western Australia.

McCulloch and Compston (1981) determined a composite Sm–Nd isochron on a suite of rocks comprising the ore-bearing Kambalda ultramafic unit, the footwall and hanging wall basalts, and an 'associated' sodic-granite and felsic porphyry. Although the whole suite yielded a good isochron age of 2790 ± 30 Myr (Fig. 4.5), the basic and ultra-basic samples alone gave an older best-fit age of 2910 ± 170 Myr.

The danger of constructing a 'composite' Sm–Nd isochron of acid, basic and ultra-basic rocks which might not be co-magmatic was pointed out by Claoue-Long et al. (1984). These workers attempted to date the Kambalda lavas by the Sm–Nd method without utilising acid rocks. However, they were forced to combine analyses from komatiites and basalts in order to achieve a good spread of Sm/Nd

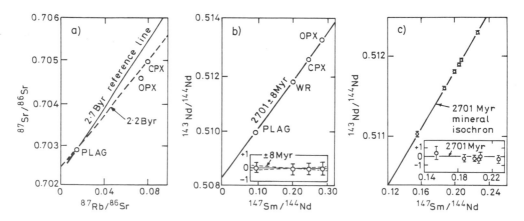

Fig. 4.4. Isochron diagrams for the Stillwater complex. a) Rb—Sr diagram showing scatter of mineral data; b) Sm–Nd mineral isochron; c) whole-rock data with reference isochron from (b). After DePaolo and Wasserburg (1979).

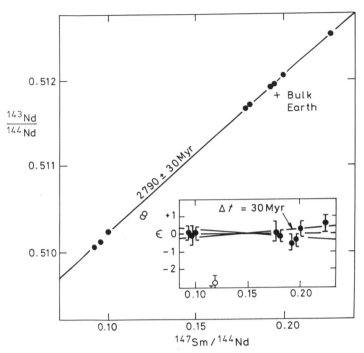

Fig. 4.5. Composite acid-basic Sm–Nd isochron diagram for a suite of Archean rocks from Kambalda, western Australia. Open symbols were omitted from the regression. After McCulloch and Compston (1981).

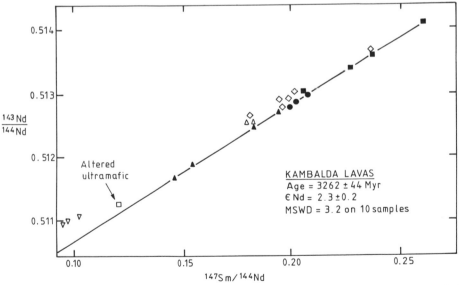

Fig. 4.6. Sm–Nd isochron diagram for whole-rock samples of Kambalda volcanics. (■) = komatiites; (▲) = hanging-wall basalts; (◊) = Bluebush lavas; (△) = 'ocelli' basalts; (▽) = granites. Modified after Claoue-Long *et al.* (1984).

ratios (Fig. 4.6). After the exclusion of one komatiite point from Kambalda and a suite of basalt lavas from Bluebush (40 km south of the main Kambalda sequence), ten data points gave an age of 3262 ± 44

Myr (2σ). Claoue-Long *et al.* interpreted this as the time of eruption.

Chauvel *et al.* (1985) challenged this interpretation on the basis that Pb–Pb dating of the Kambalda

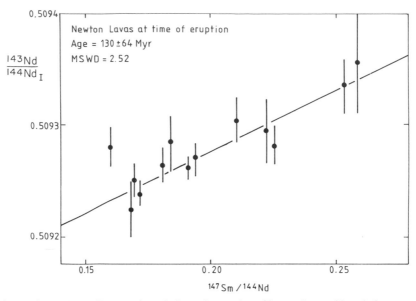

Fig. 4.7. Sm–Nd pseudo-isochron diagram for whole-rock samples of komatiite and basalt from Newton township, Ontario. The large apparent errors and scatter of data are due to a very expanded y-axis scale. After Cattell *et al.* (1984).

volcanics and associated igneous sulphide mineralisation yielded an age of 2726 ± 34 Myr, which they argued to be resistant to re-setting by later events. They attributed the 3.2 Byr apparent Sm–Nd age to either variable crustal contamination of the magma suite by older basement, or possibly a heterogeneous mantle source. U–Pb dating of 3.4 Byr-old zircon xenocrysts in one of the hanging-wall basalts subsequently confirmed the contamination model (Compston *et al.*, 1985).

In retrospect, danger signals can be seen in the whole-rock Sm–Nd data. Taken alone, the komatiites (including the sample rejected by Claoue-Long *et al.*) define a slope of less than 3.2 Byr, as do the Bluebush lavas (Fig. 4.6). Only the hanging-wall basalts define a slope of 3.2 Byr, but these are the samples which have probably suffered most contamination. Hence the data probably consist of a series of sub-parallel isochrons with ca. 2.7 Byr slope.

Similar effects have been demonstrated for komatiitic and basaltic lavas from Newton township in the Abitibi belt of Ontario. Cattell *et al.* (1984) obtained an apparent age of 2.83 Byr from a whole-rock Sm–Nd isochron of basic and ultra-basic lavas. However, a maximum eruption age of 2697 ± 1 Myr was conclusively demonstrated by U–Pb zircon analysis of an underlying dacitic volcaniclastic rock. Cattell *et al.* plotted initial $^{143}Nd/^{144}Nd$ ratios at 2697 Myr against Sm/Nd (Fig. 4.7), demonstrating an

erupted Sm–Nd isochron with an apparent age of 130 ± 64 Myr (MSWD = 2.52). No age significance was attached to this pseudo-isochron, which was attributed to sampling of a variably depleted mantle source. However, contamination by older crustal rock is a strong possibility.

4.1.3 High-grade metamorphic rocks

The Sm–Nd method has frequently been applied to dating the igneous protolith age of high-grade metamorphic basement where other systems are reset. An example is provided by dating work on the Lewisian gneisses of NW Scotland. Whole-rock Rb–Sr, whole-rock Pb–Pb and U–Pb zircon ages on granulite-facies and amphibolite-facies gneisses are concordant at 2630 ± 140, 2680 ± 60 and 2660 ± 20 Myr (2σ) respectively (Moorbath *et al.*, 1975; Chapman and Moorbath, 1977; Pidgeon and Bowes, 1972). However, these gneisses are generally very Rb- and U-depleted, suggesting that even large whole-rock samples were probably open systems for these elements during the depletion event.

A suite of whole-rock samples was dated by the Sm–Nd method (Hamilton *et al.*, 1979) to see whether this system had remained undisturbed during the Badcallian metamorphic event which the other systems are presumed to date. An older age of 2920 ± 50 Myr (2σ) suggested that the gneisses had

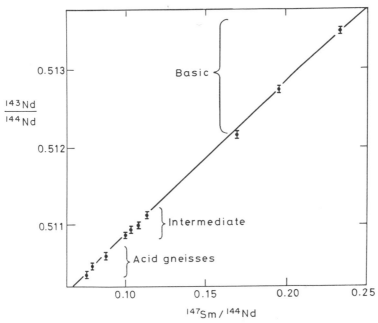

Fig. 4.8. Sm–Nd isochron for a mixed suite of granitic, tonalitic and layered basic gneisses from the Lewisian complex of NW Scotland, yielding an age of 2920 Myr. After Hamilton *et al.* (1979).

remained closed systems for Sm–Nd during granulite-facies metamorphism (Fig. 4.8). Hamilton *et al.* therefore interpreted the age as the time of protolith formation, which occurred on average between 200 and 300 Myr before closing of U–Pb zircon and the whole-rock Rb–Sr and Pb–Pb systems subsequent to the peak of metamorphism.

Two problems with the sampling of the isochron are that it combined amphibolite- and granulite-facies gneisses and also sampled a bimodal suite of tonalitic gneisses and basic rocks from the Drumbeg layered complex. Nevertheless, because the slope ages of the two sub-suites are very similar, the samples as a whole display an MSWD value of only 1.3 (using 1σ errors of 0.1 % for Sm/Nd, and the individual within-run isotopic errors).

More detailed investigation by Whitehouse (1988) showed that the Drumbeg layered basic rocks retain a 2.91 Byr isochron age, but Sm–Nd whole-rock systems in intermediate to acid rocks had been reset to the same age as the U–Pb zircon and other whole-rock systems. Ten samples of the latter suite define an errorchron with MSWD = 5.7, yielding an age (with estimate of geological error) of 2600 ± 155 Myr (2σ), shown in Fig. 4.9. Therefore, the isochron of Hamilton *et al.* (1979) apparently does correctly date the time of protolith formation, but only the basic rocks remained closed systems during the

Badcallian event. However, Sm–Nd model ages for the intermediate gneisses were only slightly upset by the metamorphism, and these do agree with the isochron age for the Drumbeg pluton (section 4.3.2).

4.1.4 High-grade metamorphic minerals

Another area where the Sm–Nd isochron method has been widely applied is to the dating of high-grade metamorphic minerals. Mineral isochrons have the advantage that variations in partition coefficient cause moderately large variations in Sm/Nd ratio, unlike whole-rock systems; thus allowing the determination of precise ages. For example, garnet and clinopyroxene (cpx) have mirror-image distribution coefficients for REE, and can therefore give rise to a large range in Sm/Nd. The classic example of a garnet–cpx rock is eclogite, so this has been a major focus of Sm–Nd mineral dating. However, the relative immobility of the REE, which is such an asset in dating igneous crystallisation, is a problem in using the Sm–Nd method to date metamorphism. Mineral systems may be opened sufficiently to disrupt the original igneous chemistry, but not enough to completely overprint the system. An example is provided by the dating of Caledonian eclogites by Mork and Mearns (1986).

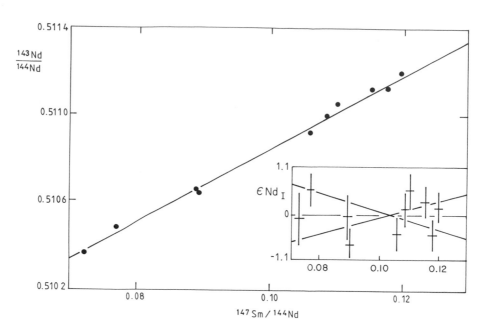

Fig. 4.9. Sm–Nd 'errorchron' for Lewisian tonalitic gneisses, defining an age of 2600 Myr, attributed to granulite-facies metamorphism. After Whitehouse (1988).

Gabbros from western Norway which have been transformed to an eclogite mineralogy (garnet and omphacite), but retain a relict igneous texture, did not reach isotopic equilibrium during Caledonian metamorphism. In contrast, nearby country-rocks which had been transformed to eclogite yield a mineral isochron with very low scatter (MSWD = 0.1), and a typical Caledonian metamorphic age of 400 ± 16 Myr. The contrasting behaviour of the two eclogite types cannot be attributed to variable P,T conditions, since they are from within 1 km of each other. However, the country-rock eclogite had completely lost its pre-existing texture due to penetrative deformation and recrystallisation. Mork and Mearns suggested that such physical disruption might be necessary to achieve complete Nd isotopic equilibrium between mineral phases.

Examination of the metagabbro Sm–Nd data at 400 Myr (Fig. 4.10) suggests that the main obstacle to isotopic homogenisation in this rock was the cpx phase. Because the transformation of augite to omphacite requires relatively minor cation exchange, complete re-setting of the Sm–Nd system in this mineral rarely occurs. In contrast, major chemical exchange and structural reorganisation is required to replace plagioclase with garnet, so complete re-setting is more likely. Hence, garnet–whole-rock isochrons are more reliable than the garnet–cpx pairs used in

early dating work on eclogites (e.g. Griffin and Brueckner, 1980).

Vance and O'Nions (1990) argued that garnet chronology provided a powerful tool for dating prograde metamorphism, in contrast to other methods, such as Ar–Ar and Rb–Sr, which date metamorphic cooling (section 10.2.7). Garnets are widely distributed in meta-pelitic rocks, and develop in response to the changing P,T conditions of prograde metamorphism. Their chemistry (including the Sm–Nd system) is usually preserved during cooling because cation diffusion rates in garnet are very slow. The chemical composition of garnets can be used to calculate the P,T conditions of their growth, which, combined with age data, provide a method of determining progradational P,T -time paths for high-grade metamorphic terranes. An application of this technique was demonstrated by Burton and O'Nions (1991) in a study of Caledonian regional metamorphism of a Proterozoic supracrustal sequence at Sulitjilma, North Norway.

Burton and O'Nions dated garnet growth in adjacent graphite-bearing and graphite-free bands using the Sm–Nd and U–Pb isochron methods. An example is shown in Fig. 4.11 for a case where garnet rims and cores are distinct. The rims yield a slightly younger age, as would be expected. Note that the core is regressed with the whole-rock composition, while

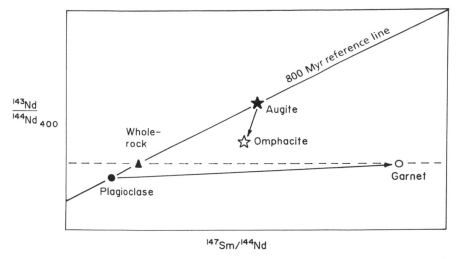

Fig. 4.10. Schematic illustration of the process of Sm–Nd remobilisation during the replacement of gabbro by an eclogite mineralogy. Modified after Mork and Mearns (1986).

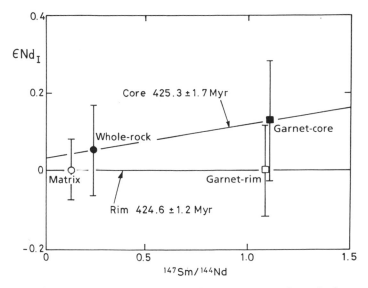

Fig. 4.11. Sm–Nd isochrons for whole-rock–garnet-core, and matrix–garnet-rim pairs from a graphite-free meta-pelite. Error bars indicate within-run precision. After Burton and O'Nions (1991).

the rim is regressed with the matrix of the rock only, since this is the only part of the rock with which the rims were in diffusional contact at the time of their growth.

Concordant results for the Sm–Nd and U–Pb techniques provide strong evidence that ages for garnet–matrix pairs date prograde mineral growth. When this is coupled with temperature data (Fig. 4.12) it indicates that garnet growth occurred first in the graphite-bearing assemblage, and subsequently at higher temperatures in the graphite-free assemblage. Peak metamorphic conditions were registered by the garnet rims of the latter assemblage. Hence, an average heating rate of 9 °C / Myr was calculated. On the other hand, Rb–Sr mineral ages on muscovite and biotite were used to deduce a cooling rate of 4 °C / Myr (Fig. 4.12).

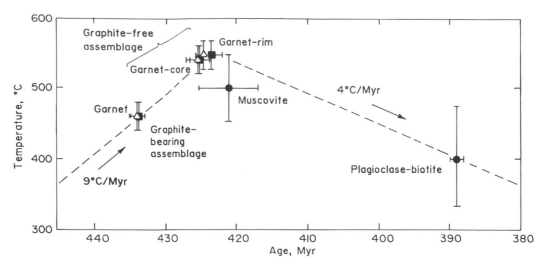

Fig. 4.12. Temperaturetime diagram for Sulitjilma supracrustals, North Norway. Progradational heating rate is from garnet Sm–Nd ages (△) and UPb ages (■). Retrogressive cooling rate is from Rb–Sr ages (•). After Burton and O'Nions (1991).

4.2 Nd Isotope evolution and model ages

DePaolo and Wasserburg (1976a) made the first Nd isotope determinations on terrestrial igneous rocks. When they plotted the ages and initial $^{143}Nd/^{144}Nd$ ratios of these units on a diagram of Nd isotope evolution against time, they found that Archean plutons had initial ratios which were remarkably consistent with the evolution of the Chondritic Uniform Reservoir (CHUR) predicted from meteorites (Fig. 4.13). The CHUR evolution path is normally drawn as a straight line, but in fact it is a very gentle curve, due to the finite half-life of ^{147}Sm (ca. 106 Byr).

Because Sm and Nd are rare earth elements (REE) only two different in atomic number, their chemical properties are very similar and they undergo only slight relative fractionation during crystal–liquid processes. This means that in terrestrial rocks, departures of $^{143}Nd/^{144}Nd$ from the CHUR evolution line are small relative to the steepness of the line (Fig. 4.13). DePaolo and Wasserburg therefore developed a notation whereby initial $^{143}Nd/^{144}Nd$ isotope ratios could be represented in parts per 10^4 deviation from the CHUR evolution line, termed epsilon units (ϵ Nd). Mathematically, this notation is defined as:

$$\epsilon Nd(t) = \left[\frac{(^{143}Nd/^{144}Nd)_{sample}(t)}{(^{143}Nd/^{144}Nd)_{CHUR}(t)} - 1 \right] \times 10^4 \quad [4.3]$$

where t indicates the time at which ϵ Nd is calculated. Another advantage of the ϵ notation is that by normalising all data to CHUR, it removes the effects of the different fractionation corrections which have been applied for Nd analysis as the metal or as the oxide species.

Initial ratios of co-magmatic rock suites are traditionally determined from isochron intercepts. These initial ratios can then be used to calculate ϵ Nd by comparison with the isotopic evolution of the CHUR reservoir. However, the poor spread of Sm/Nd ratios in most rock suites causes error-magnification in the initial ratio determination (Fig. 4.14a). In early Nd isotope studies this problem was combatted by the inclusion of differentiated rocks in the isochron suite, since only by using rocks with high REE-profile slopes (low Sm/Nd) was it possible to calculate precise initial ratios. However, it was shown in section 4.1.2 that this approach may give erroneous results if the differentiation series used to construct the isochron has variable initial ratios due to progressive crustal contamination.

Fletcher and Rosman (1982) argued that these problems could be avoided by calculation of ϵ Nd values directly from the isochron diagram. This is done by translating the y axis of the isochron diagram to the CHUR Sm/Nd ratio, by subtracting 0.1966 from all $^{147}Sm/^{144}Nd$ analyses. The calculated error in the ϵ Nd value will then avoid error magnification (Fig. 4.14b). In this case even poorly constrained isochrons on komatiites can yield precise initial ratios.

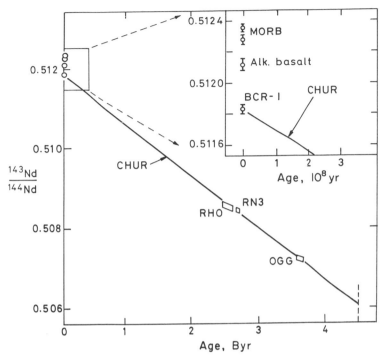

Fig. 4.13. Diagram of ^{143}Nd/^{144}Nd against time showing the close correspondence of early initial Nd isotope ratios of terrestrial rocks to the chondritic meteorite growth line. Data points: OGG = Amitsoq gneiss, W Greenland; RN3 = Preissac-Lacorne batholith, Canada; RHO = Great Dyke, Zimbabwe (Rhodesia). BCR–1 = Columbia River basalt, USA. After DePaolo and Wasserburg (1976a).

However, with the advent of increasing numbers of accurate U–Pb crystallisation ages, these data can be used to calculate initial Nd ratios without the need for Sm–Nd isochron fits.

Using the ϵ notation, DePaolo and Wasserburg (1976b) presented a larger amount of Nd isotope data on a diagram of ϵ Nd against time (Fig. 4.15). They noted that continental igneous rocks through time had ϵ Nd values very close to zero. Indeed, for Archean rocks the error bars overlapped with zero, suggesting that continental igneous rocks were 'derived from a reservoir with a chondritic REE pattern, which may represent primary material remaining since the formation of the Earth.'

4.2.1 Chondritic model ages

DePaolo and Wasserburg (1976b) argued that if the CHUR evolution line defines the initial ratios of continental igneous rocks through time, then measurement of ^{143}Nd/^{144}Nd and ^{147}Sm/^{144}Nd in any crustal rock would yield a model age for the formation of that rock (or its precursor) from the chondritic reservoir. This is true, providing that there was sufficient Nd/Sm fractionation during the process of crustal extraction from the mantle to give a reasonable divergence of crustal and mantle evolution lines (Fig. 4.16), and hence a precise intersection. The model age is then given as:

$$T_{\text{CHUR}} = \frac{1}{\lambda} \cdot \ln\left[1 + \frac{\left(\dfrac{^{143}\text{Nd}}{^{144}\text{Nd}}\right)^0_{\text{sample}} - \left(\dfrac{^{143}\text{Nd}}{^{144}\text{Nd}}\right)^0_{\text{CHUR}}}{\left(\dfrac{^{147}\text{Sm}}{^{144}\text{Nd}}\right)^0_{\text{sample}} - \left(\dfrac{^{147}\text{Sm}}{^{144}\text{Nd}}\right)^0_{\text{CHUR}}}\right]$$

[4.4]

DePaolo and Wasserburg argued that if the Sm/Nd ratio of the sample had not been disturbed since its separation from the chondritic reservoir (taken to be the mantle source), then T_{CHUR} may provide a 'crustal formation' age for a wide variety of rocks. Many elemental investigations have pointed to the relative immobility of REE on a whole-rock scale during the processes of weathering and low-temperature metamorphism associated with sedimentary rock

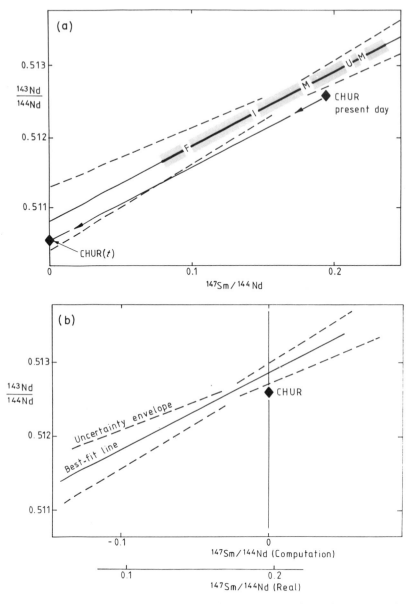

Fig. 4.14. a) Schematic illustration of the conventional determination of initial ratios on the Sm–Nd isochron diagram, showing error-magnification towards the *y*-axis. Shaded band indicates range of Sm/Nd ratios in felsic, intermediate, mafic and ultra-mafic rocks. b) Translation of the *y*-axis of an Sm–Nd isochron to the Bulk Earth value to reduce error-magnification. After Fletcher and Rosman (1982).

formation (e.g. Haskin *et al.*, 1966), and even during high-grade metamorphism (Green *et al.*, 1969). This immobility is schematically illustrated by the lack of deflection in the evolution line of the crustal sample in Fig. 4.16 during metamorphic and sedimentary events.

These premises were applied by McCulloch and

Wasserburg (1978) in a model age study aimed at measuring the crustal formation ages of several cratonic rock samples, mainly from the Canadian Shield. McCulloch and Wasserburg found Nd model ages within the range 2.5–2.7 Byr for composite samples of the Superior, Slave, and Churchill structural provinces. In the first two areas, previously

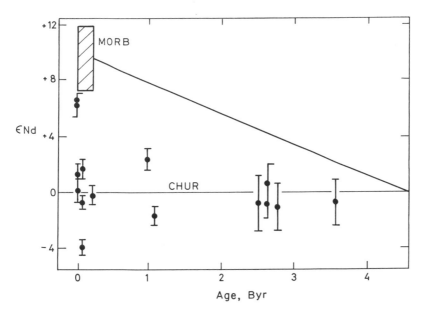

Fig. 4.15. Diagram of Nd isotope evolution against time in the form of deviations from the chondritic evolution line in ϵ units. After DePaolo and Wasserburg (1976b).

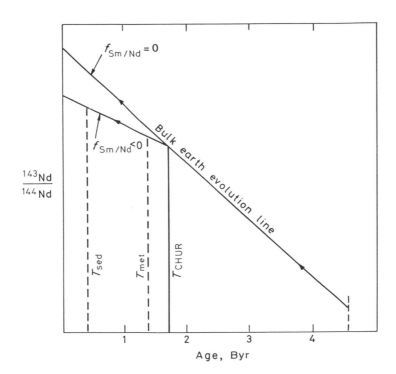

Fig. 4.16. Diagram of Nd isotope evolution against time to show schematically the theory of model ages. T_{met} = age of metamorphic event; T_{sed} = age of erosion–sedimentation event; f = fractionation of sample Sm/Nd relative to Bulk Earth. After McCulloch and Wasserburg (1978).

determined K–Ar and Rb–Sr ages had given the same results, but the 2.7 Byr model age for the Churchill province was 0.8 Byr older than the previously determined K–Ar age, which had presumably been reset by more recent metamorphism. These data supported a model of episodic continental growth, by showing the period 2.5–2.7 Byr ago to be a time of remarkably widespread continental accretion. In contrast, a Grenville Province composite yielded a model age of 0.8 Byr which did not reveal any Archean component, suggesting it to be an addition of more recent crust to the pre-existing shield. However, this sample was by no means representative of the Grenville Province as a whole, which also contains reworked Archean and Early Proterozoic crustal terranes (Dickin and McNutt, 1989).

Although Nd model ages are generally applied to dating the time of crustal separation from the mantle, other more specialised applications have been made. Thus, Richardson *et al.* (1984) investigated the time of diamond formation in the South African mantle lithosphere by dating garnet inclusions in diamonds. Three samples were analysed, each consisting of a composite of several hundred sub-calcic garnet inclusions, and yielding a total of 10–45 ng Nd. The unradiogenic Nd in these samples gave rise to T_{CHUR} ages of 3.19–3.41 Byr (Fig. 4.17). When this evidence is combined with evidence of sub-solidus temperatures for diamond growth (based on equilibrium garnet–olivine inclusions in diamonds), it suggests that sub-continental lithosphere has existed under the African craton since the Early Archean. This material may represent the residue from 3.5 Byr-old komatiite extraction.

4.2.2 Depleted mantle model ages

While observing the good fit of Archean plutons to the CHUR Nd isotope evolution line, DePaolo and Wasserburg (1976b) also noted that young mid ocean

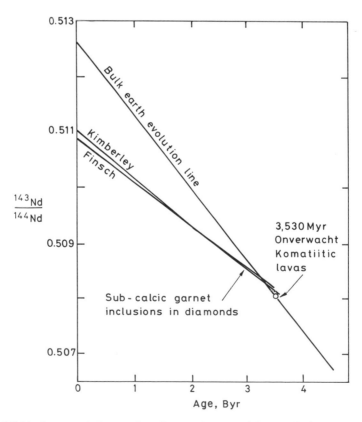

Fig. 4.17. Diagram of Nd isotope evolution against time to show model age calculations for silicate inclusions in South African diamonds. The initial ratio of Onverwacht lavas is shown for comparison. After Richardson *et al.* (1984).

ridge basalts (MORB) lay $+7$ to $+12$ ϵ units above the CHUR evolution line (Fig. 4.15). They recognised that Archean continental igneous rocks which fall within error of the CHUR evolution line could conceivably lie on a depleted mantle evolution line characterised by progressively increasing Sm/Nd and ^{143}Nd/^{144}Nd. However, they rejected this model in favour of a chondritic source for continental igneous rocks on the basis of a comparison with lunar Nd isotope evolution.

Lunar basalts with ages of 3.3–4 Byr show a wide range of initial ^{143}Nd/^{144}Nd ratios, equivalent to a variation from $+7$ to -2 ϵ units relative to CHUR (Fig. 4.18; Lugmair and Marti, 1978). This spread shows that very early Sm/Nd fractionation occurred in the Moon, and that there was no long-lived uniform magma source with a chondritic Sm/Nd ratio. The fact that all the Archean terrestrial rocks analysed by 1976 did not show any dispersion outside error from CHUR led DePaolo and Wasserburg (1976b) to conclude that the Earth did not undergo early differentiation, or that if it did, that this was re-mixed by convection.

The paucity of Nd isotope data for the Proterozoic was a serious weakness in this model, since it left a gap between the Archean CHUR data and the recent MORB depleted source (= elevated Sm/Nd) data, with questions about the relationship between the two. An important stage in filling this gap was a study on Proterozoic metamorphic basement from the

Colorado Front Range (DePaolo, 1981). Four meta-volcanics and two charnockitic granulites from the Idaho Springs Formation were dated by the Sm–Nd isochron method. In addition, Nd isotope and Sm/Nd determinations were made on three plutons previously dated by the Rb–Sr whole-rock method, (the Boulder Creek, Silver Plume and Pikes Peak granitoids). The initial ^{143}Nd/^{144}Nd ratios of all these samples are plotted on an ϵ Nd versus time diagram in Fig. 4.19.

The Idaho Springs meta-igneous rocks cluster at ϵ Nd$(t) = +3.7 \pm 0.3$, showing them to be derived from a depleted mantle reservoir with respect to CHUR at 1.8 Byr. Boulder Creek samples also have positive ϵ Nd ($+1.7$ to $+3.5$), while the Silver Plume and Pikes Peak granites have progressively lower ϵ Nd values which lie on the ^{143}Nd/^{144}Nd evolution line of average Idaho Springs crust, suggesting that they contain a large fraction of re-melted 1.8 Byr-old basement.

DePaolo was able to fit a quadratic curve to Idaho Springs and modern island arc data (Fig. 4.19), representing the Nd isotope evolution of a progressively depleted reservoir which was the source area for calc-alkaline magmatism. This curve is close to the CHUR evolution line in the Early Archean, but diverges progressively to the present day. The composition of the depleted reservoir, relative to CHUR, at time T, is given as:

$$\epsilon \, \mathrm{Nd}(T) = 0.25T^2 - 3T + 8.5 \qquad [4.5]$$

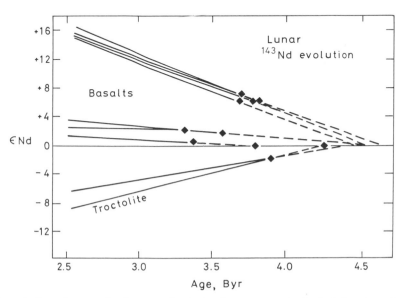

Fig. 4.18. ϵ Nd evolution diagram for lunar rocks indicating very early Sm/Nd fractionation between lunar reservoirs. After Lugmair and Marti (1978).

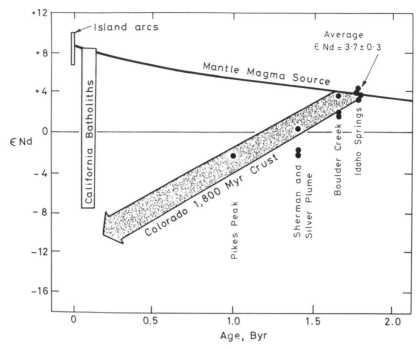

Fig. 4.19. Plot of ϵ Nd against time showing Colorado data relative to a model depleted mantle evolution curve. After DePaolo (1981).

Sm–Nd model ages calculated using this depleted mantle curve are denoted T_{DM}. DePaolo argued that T_{DM} model ages would be a more accurate indication of 'crustal formation age' than T_{CHUR}. For example, an anomalously low T_{CHUR} age of 0.8 Byr for McCulloch and Wasserburg's Grenville composite (section 4.2.1) is revised to a T_{DM} age of 1.3 Byr, consistent with the age of the Llano uplift of Texas, which is interpreted as part of the same province.

Since the discovery of Proterozoic depleted mantle by DePaolo (1981), new analyses have prompted several re-interpretations of the evolution of the depleted mantle reservoir. For example, DePaolo (1983) calculated a more depleted mantle evolution curve with a sinusoidal form, using newly published data for Precambrian basalts. However the depression in the sinusoid in the late Archean was soon filled by analyses of Archean komatiites from Canada and southern Africa (Chauvel et al., 1983). Subsequently, Nelson and DePaolo (1984) proposed using a convex upward depleted mantle curve on the basis of Proterozoic basalt compositions, and the very depleted ϵ Nd value of +3 at 3.2 Byr calculated by Claoue-Long et al. (1984) from Kambalda. Subsequent work showed the Kambalda age to be erroneous (section 4.1.2), but the curve may be re-

instated by the discovery of other high ϵ Nd values in the Archean (section 4.4.3).

An important alternative to DePaolo's (1981) model was proposed by Goldstein et al. in 1984 (Fig. 4.20). This model assumes linear depletion of the mantle from ϵ Nd = 0 at 4560 Myr to ϵ Nd = +10 at 0 Myr (MORB composition), and provides a good fit to Early Proterozoic greenstones from the SW United States and Greenland (Nelson and DePaolo, 1984; Patchett and Arndt, 1986). The most depleted ϵ Nd values in these suites may represent flood basalts erupted in rifting environments that suffered little crustal contamination. However, this is not the most appropriate mantle model for calculating crustal extraction ages of tonalitic crust-forming rocks generated in arc settings, which at the present day have less depleted Nd isotope signatures than spreading ridges.

There has been a tendency for a proliferation of depleted mantle models as new data for different geographical areas becomes available. However, an examination of the literature suggests that the models of DePaolo (1981) and Goldstein et al. (1984) have had the widest application by other workers. This is illustrated in Fig. 4.21 by a comparison of citation rates for these two studies, compared with two control

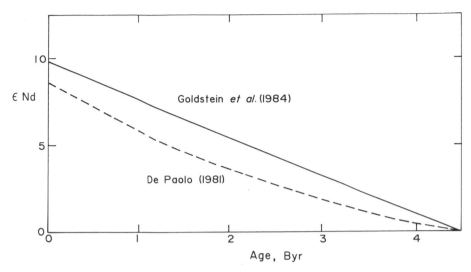

Fig. 4.20. Plot of ϵ Nd against time showing two of the most widely used depleted mantle evolution models. Dashed curve: DePaolo (1981); solid line: Goldstein *et al.* (1984).

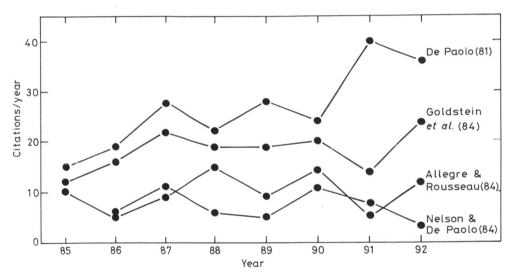

Fig. 4.21. Changing citation rates of papers which introduced new depleted mantle evolution models for Nd. Data from the Science Citation Index.

papers: Nelson and DePaolo (1984), discussed above, and Allegre and Rousseau (1984), who proposed a curved mantle evolution line similar to DePaolo (1981). The durability of citations for DePaolo (1981) indicates the wide usefulness of this mantle model. Hence it is desirable that the T_{DM} and T_{CR} notations should be restricted to the models of DePaolo (1981) and Goldstein *et al.* (1984), while other acronyms can be used to denote different models.

4.3 Model ages and crustal processes

As outlined above, one of the principal uses of the Sm–Nd model age method is to determine what are often called 'crust-formation' or 'crustal-extraction' ages. However, the Sm–Nd method is most often applied when a long or complicated geological history precludes a more direct method of determining crustal age. One of the strengths of the Sm–Nd model age method, as applied to whole-rock systems, is that it provides the opportunity to see back through erosion, sedimentation, high-grade metamorphism and even crustal melting events which may re-set other dating tools. However, these processes may cause complications in the interpretation of model ages. Hence, it is important to examine Sm–Nd systematics in well-constrained conditions in order to predict behaviour in complex environments.

4.3.1 Sedimentation

Behaviour of the Sm–Nd system during erosion can be examined by comparing the calculated model ages of river-borne particulates with the average geological age of sediment sources in the watershed. Goldstein and Jacobsen (1988) performed such a study on particulates in American rivers. They found that rivers draining primary igneous rocks carried sediment which accurately reflected the crustal residence age of the source (Fig. 4.22). Rivers draining sedimentary watersheds were not properly testable, since the crustal residence age of their sources has not been adequately quantified.

Behaviour of the Sm–Nd system during sedimentation can be tested by comparing Nd model ages on different size fractions of sediment. An early study on

bottom sediment from the Amazon River (Goldstein et al., 1984) found that different size fractions yield only a small range of crustal residence ages (1.54–1.64 Byr), despite a large range in total Nd contents (17–47 ppm). Similar agreements in model age were found by Awwiller and Mack (1991) on mud and sand grade sediments from the Rio Grande and Mississippi rivers. However, the bottom sediments of large rivers may be atypical in displaying good chemical homogeneity.

In order to see whether a similar degree of homogeneity is displayed by deep-sea turbidites, McLennan et al. (1989) compared model ages on sand and mud pairs in turbidites from several different tectonic environments (Fig. 4.23). Their findings were rather variable; some pairs demonstrating good agreement in model age, while others were poor. These variations probably reflect the petrological make-up of the sediment. Both a mature passive margin sediment with less than 5% lithic volcanic fragments and a very immature back arc sediment with ca. 90% lithic volcanic fragments displayed good model age agreement between sand and mud fractions (solid symbols). These uniform types of sediment may therefore yield useful model age constraints. In contrast, sediments with intermediate fractions of volcaniclastic material gave inconsistent model ages (open symbols). The latter type was prevalent in continental arcs, and can be attributed to variable

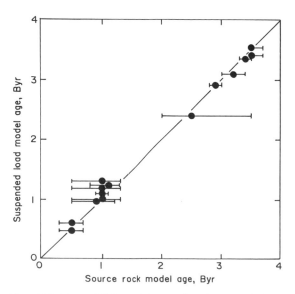

Fig. 4.22. Plot of Nd model ages for river particulates against the area-weighted average crustal residence age of rocks within the watershed. Data are shown for igneous–metamorphic drainage basins only. After Goldstein and Jacobsen (1988).

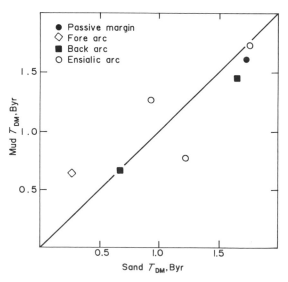

Fig. 4.23. Plot of depleted mantle model ages in mud *versus* sand grade fractions from deep sea turbidites in different tectonic environments. After McLennan *et al.* (1989).

mixing between old continental detritus and young volcanic detritus in different grain-size fractions. Continental arcs therefore provide poor model age constraints on geological events.

Nelson and DePaolo (1988) tested the effects of mixed sediment provenance on Sm–Nd systematics in two small basinal systems. In both cases, the different sediment sources were petrographically and geochemically well-characterised. In order to quantify the mixing process, Nelson and DePaolo plotted ϵ Nd against a petrographic index (percentage of lithic volcanic fragments). The good correlation observed between the end-members and various mixtures (Fig. 4.24) attests to the 'immobile' behaviour of Nd during erosion and sedimentation. This does not *avoid* the problem of mixed provenance, but it shows that coupled isotopic and petrological analysis of a suite of samples can be used to detect and quantify the mixing process.

4.3.2 Metamorphism

Stille and Clauer (1986) and Bros *et al.* (1992) have demonstrated that in carbonaceous (black) shales,

Sm–Nd systematics in the microscopic clay-mineral fraction can be re-set by diagenesis. They showed that in some cases, sub-micron sized particles could yield Sm–Nd isochrons, which they interpreted as dating diagenesis. The accuracy of such ages remains to be proven, given the evidence that Rb–Sr dating of clay minerals can be upset by detrital inheritance (section 3.4.1). However, diagenetic mobilisation of REE on a mineralogical scale does not imply open Sm–Nd systems on a whole-rock scale. A suggestion that such a scenario *could* occur was made by Awwiller and Mack (1991) on the basis of Sm–Nd analysis of borehole samples from Texas. However, the small size of the analysed 'whole-rock' samples (less than 10 g) means that this evidence has little relevance to Sm–Nd studies of crustal residence based on kilogram-sized samples.

In contrast to these suggestions of open Sm–Nd systems during diagenesis, Barovich and Patchett (1992) demonstrated that whole-rock Sm–Nd systems in granitic rocks can remain undisturbed even during severe metamorphic deformation. They studied a 60 m-wide Mesozoic ductile shear zone cutting the Mid

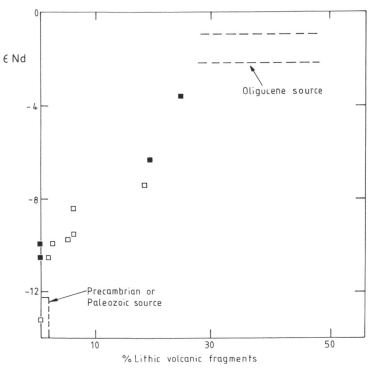

Fig. 4.24. Plot of ϵ Nd against modal % lithic volcanic fragments to show petrographic dependence of the Sm–Nd system in sedimentary basins with mixed provenance. (■) = Hagar basin; (□) = Espanola basin. After Nelson and DePaolo (1988).

Proterozoic Harquahala granite. Samples of increasingly deformed granite were found to yield a narrow range of T_{CHUR} model ages around 1.58 Byr in two different traverses to within 1 m of the thrust plane (Fig. 4.25). Closed-system behaviour was preserved even in samples showing widespread sericitisation of plagioclase and significant epidote growth. Only in ultra-mylonites less than 1m from the main thrust was a reduction in model age of up to 150 Myr observed, possibly due to a high fluid flux which caused calcite veining and intense alteration in the immediate vicinity of the thrust.

The resistance of whole-rock Sm–Nd model ages to significant resetting, even during granulite-facies metamorphism, is demonstrated by the Lewisian granulites from NW Scotland (Whitehouse, 1988). A ten-point Sm–Nd isochron for tonalitic gneisses (section 4.1.3) yields an age of 2.60 Byr, and an initial ratio (ϵ [t]) of −2.4 relative to CHUR (Fig. 4.26a). This isochron is argued to date the metamorphic event. However, T_{DM} model ages of these same gneisses fall in the range 2.84–3.04 Byr, with an average value of 2.93 Byr (Fig. 4.26b). These ages have been slightly scattered by metamorphism, but still yield an average value very close to the undisturbed 2.91 Byr isochron age of the Drumbeg mafic complex.

Nelson and DePaolo (1985) attempted to place upper limits on the disturbance of model ages under conditions of intra-crustal re-working by considering the limiting case of crustal anatexis. From crustal melting models (Hanson, 1978), they estimated that the maximum amount of Sm/Nd fractionation likely to arise by intra-crustal melting processes (Δ) was 20% of the pre-existing fractionation between sample Sm/Nd and CHUR Sm/Nd. This fractionation factor f was defined by DePaolo and Wasserburg (1976):

$$f_{Sm/Nd} = \frac{^{147}Sm/^{144}Nd_{sample}}{0.1967} - 1 \qquad [4.6]$$

Using this notation, the error in depleted mantle model age (T_{DM}) introduced by an intra-crustal fractionation event is given by:

$$\text{Err } T_{DM} = \Delta f_{Sm/Nd} \cdot (T_{CF} - T_m) \qquad [4.7]$$

where T_{CF} is the true crustal formation age and T_m is the age of the partial melting event. This error propagation is illustrated schematically in Fig. 4.27. The problem can be minimised by analysing samples with melting ages fairly close (< 300 Myr ?) to their formation age.

Evidence that intra-crustal melting causes relatively minor perturbations in model age has encouraged the use of granitic plutons to determine crustal formation ages on associated country-rocks (assuming that the granites are the products of anatexis of those country-rocks). The approach has the advantage of allowing basement mapping of large areas with a minimal number of analyses, since each pluton can be expected to have averaged the composition of a large volume of crust. It was used to great effect by Nelson and DePaolo (1985) to map out the crustal extraction ages of huge belts in the central United States (Fig. 4.28).

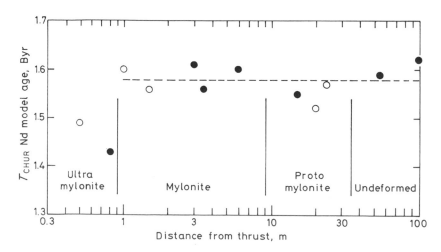

Fig. 4.25. Plot of T_{CHUR} model ages for samples of the Harquahala granite as a function of distance from the Harquahala thrust. Solid and open symbols indicate samples from two different traverses. Approximate boundaries between deformation zones are shown. Data from Barovich and Patchett (1992).

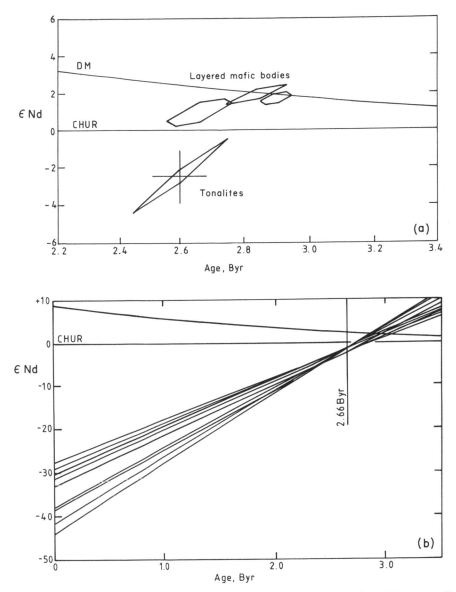

Fig. 4.26. Nd isotope evolution diagrams. a) Showing initial ratios for four suites of Lewisian granulites from NW Scotland (tonalites and layered mafic bodies from Scouriemore, Achiltibuie and Drumbeg); b) showing Sm–Nd evolution lines for individual tonalitic gneisses. After Whitehouse (1988).

The method is appropriate for this application because Phanerozoic cover obscures most of the central US basement, which can only be dated from drill core or drill chips.

Weaknesses in this approach are revealed, however, when model age results do not correspond to known events represented by igneous crystallisation ages. The 2.0–2.3 Byr model ages in Bennet and DePaolo's 'Penokean' and 'Mojavia' terranes exemplify this problem. It is likely that they represent Proterozoic mantle-derived magmas which mixed with large quantities of re-melted Archean crust to generate mixed model ages (Fig. 4.29) which have no meaning as crustal extraction ages (Arndt and Goldstein, 1987).

In a similar situation, McCulloch (1987) demonstrated large areas of Australian basement with Nd model ages of 2.1–2.2 Byr. He rejected a mixing

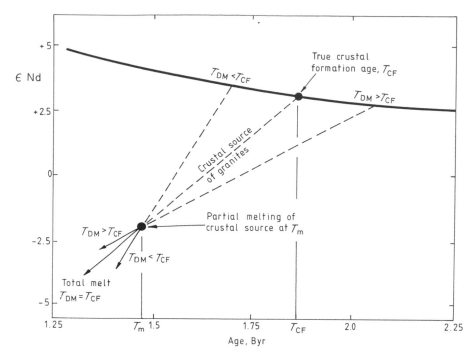

Fig. 4.27. Schematic diagram of Nd isotope systematics to show possible errors in model age arising from Sm/Nd fractionation during intra-crustal melting. After Nelson and DePaolo, (1985).

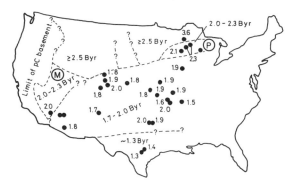

Fig. 4.28. Map of the United States showing Nd model age provinces. Ⓜ = Mojavia terrane; Ⓟ = Penokean terrane. After Bennett and DePaolo (1987)

model between Archean and Proterozoic components for three reasons:

1) The consistency of intermediate ages.
2) The problems of assimilating large fractions of Archean material.
3) The absence of inherited Archean zircons.

Arndt and Goldstein (1987) answered the latter two objections by arguing that the mixing process need

not have involved assimilation of solid crust by a magma, and by pointing out that the zircon argument is invalidated by the lack of 2.1–2.2 Byr-old zircons.

It is concluded that model age mapping of gneiss terranes is a powerful method to delimit the geographical extent of different crustal provinces, but that geochronological confirmation of the resulting model age provinces must be provided by other methods.

4.4 The crustal growth problem

The question of whether the crust has grown over geological time, or maintained an approximately constant volume, is one of the most fundamental in geology, but has proved hard to answer conclusively. A review of the 'crustal growth' model by its most persistent critic (Armstrong, 1991) shows that Nd isotope data provide critical tests for alternative models. Hence, three of the most important lines of Nd isotope evidence will be examined here.

4.4.1 Crustal accretion ages

The ability of the Sm–Nd method to 'see back' through younger thermal events and measure the crustal formation ages of continental rocks makes the

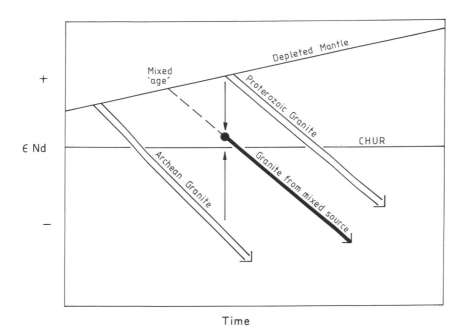

Fig. 4.29. Schematic illustration of magma mixing as a mechanism capable of generating mixed provenance ages which do not date any real geological event. After Arndt and Goldstein (1987).

method ideally suited to chart the present-day age distribution of crustal basement. This yields an apparent profile of continental growth through time which does not take into account crustal recycling into the mantle. Nevertheless, it is an appropriate starting point for this subject.

Attempts to map the age structure of the continents were begun using the Rb–Sr method (Hurley *et al.*, 1962), before the development of Sm–Nd analysis. Hurley *et al.* compared Rb–Sr isochron ages with Sr model ages calculated assuming a mantle $^{87}Sr/^{86}Sr$ ratio of 0.708. On average the two values were correlated, leading them to suppose that the isochron ages dated the time of crustal extraction from a basic source. This was a good approach, although we now know that the Rb–Sr system is too easily re-set to yield reliable crustal extraction ages for old terranes. (Also the mantle growth curve is less radiogenic than 0.708.) Hurley *et al.* applied the method to the North American continent in order to calculate the approximate area of crustal basement attributed to different age provinces (Fig. 4.30).

Hurley and Rand (1969) extended this approach to include two-thirds of the land area of the world (excluding the USSR and China, for which data were not available). K–Ar data were used to geographically extrapolate from the more limited set of Rb–Sr data, bearing in mind the tendency for the former to

be re-set. Rb–Sr model ages were calculated using an improved mantle $^{87}Sr/^{86}Sr$ growth curve, yielding values now somewhat older than apparent crystallisation ages. Hurley and Rand's data are presented on a plot of cumulative crustal age distribution against time (Fig. 4.31, curve 1). From these data it appeared that crustal growth was accelerating somewhat with time. However, more recent studies have yielded different shaped curves.

A study of comparable sweep to Hurley *et al.* (1962) was performed by Nelson and DePaolo (1985), who used Nd model ages to map the age structure of the basement of the United States. Nelson and DePaolo found Nd model ages substantially older than igneous crystallisation ages, leading to a greatly increased estimate of the rate of Lower Proterozoic crustal growth in the mid-continent. These data, along with recently published ages on the Canadian Shield, led to a dramatic increase in the estimated average age of the North American craton, compared with that of Hurley *et al.* (1962). This picture was reinforced by Patchett and Arndt (1986), who further amplified the estimated area of newly accreted Lower Proterozoic (1.9 Byr-old) crustal basement in North America. This has generated a 'sigmoidal' curve of crustal formation against time which suggests that the greatest rates of new crustal accretion occurred in the middle of Earth history (Fig. 4.31, curve 4).

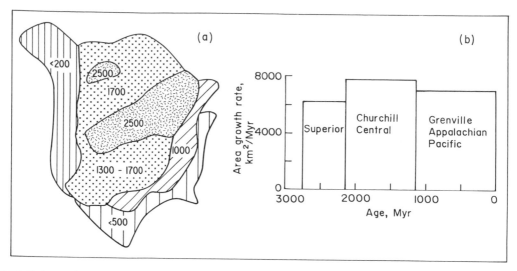

Fig. 4.30. Estimated area of North American crustal basement attributable to different Rb–Sr age provinces. a) Map showing provinces of different ages in Myr; b) histogram of growth rate against time. After Hurley *et al.* (1962).

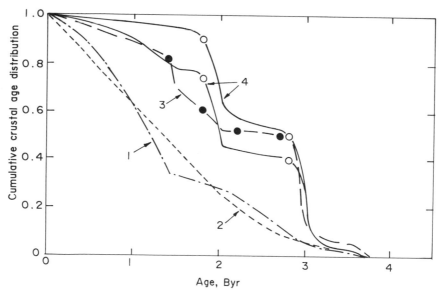

Fig. 4.31. Estimated continental growth rates on a cumulative basis. Curve 1: Hurley and Rand (1969); 2: Tugarinov and Bibikova (1976); 3: Nelson and DePaolo (1985); 4: Patchett and Arndt (1986). After Jacobsen (1988).

As more detailed model age mapping is performed in areas of old crustal basement it is likely that further increases in average crustal extraction age will be found. The Superior and Slave provinces of Canada provide good examples of this. Hurley *et al.* (1962) mapped the Superior craton as 2.7 Byr in age, a concept still widely held today. However, U–Pb geochronology has shown the presence of large areas

of 3 Byr-old crust in the NW Superior Province (reviewed by Thurston *et al.*, 1991). Furthermore, Nd model age mapping in the Winnipeg River sub-province (Dickin *et al.*, 1990) has indicated the presence of heavily reworked Early Archean material, presently exposed in domal structures surrounded by overlying younger gneisses.

Similarly, the Slave province is a largely 2.7 Byr

craton, but a small belt of tonalitic gneisses with Early Archean zircons yield T_{CHUR} model ages of up to 4.1 Byr (Bowring *et al.*, 1989). Therefore, continued mapping of old cratons will probably fill the dip in the Early–Mid Archean segment of the sigmoid (Fig. 4.31), yielding a linear or concave-downwards apparent crustal growth curve with time.

4.4.2 Sediment provenance ages

In response to proponents of the continental growth model (e.g. Moorbath, 1976), Armstrong (1981) argued that the record of continental accretion documented by various methods (as above) did not prove that the continental area had actually grown over geological time. Armstrong argued that a model in which the continental area was approximately the same 4.5 Byr ago as it is today could also generate apparent continental growth with time, provided that crustal recycling into the mantle by sediment subduction equalled that of new crustal formation above subduction zones.

Old crustal terranes may be shortened by orogeny, then flattened again by erosion and sediment subduction. However, some sediment should be expected to escape the recycling process and provide a record of the old, recycled terrane. Therefore, the

search for evidence of constancy or growth in the continental mass turned to the sedimentary record. The ability of the Nd model age method to 'see back' through erosion and sedimentation to an original crustal extraction event made it ideal for these studies.

The data are conveniently portrayed on a diagram of Nd model age (crustal residence age) against stratigraphic age of the sediment in question (Fig. 4.32). Sediments eroded from juvenile mantle-derived sources will have $T_{CR} = T_{STRAT}$ and lie on a 'concordia' line (Allegre and Rousseau, 1984). In contrast, reworking of older sediments without any input of juvenile material will displace compositions to the right along horizontal vectors. A compilation of data from several sources is shown in Fig. 4.32, including clastic sediments (Hamilton *et al.*, 1983; O'Nions *et al.*, 1983; Taylor *et al.*, 1983; Allegre and Rousseau, 1984) and particulates from major river systems at the present day (Goldstein *et al.*, 1984).

Allegre and Rousseau (1984) compared the data with various theoretical models for continental evolution involving different rates of continental growth through time (Fig. 4.32). A 'big bang' model (A), whereby the whole continental mass was extracted at ca. 4 Byr or before, was ruled out. Allegre and Rousseau argued that a model involving uniform growth of the continents from 3.8 Byr to the

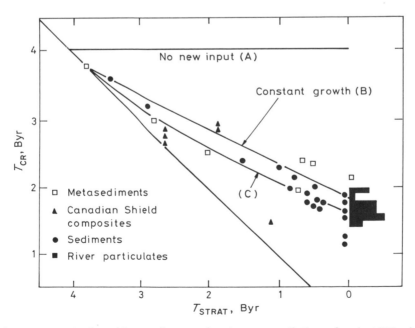

Fig. 4.32. Model age *versus* stratigraphic age diagram showing a compilation of early 1980s data from several clastic sediment studies. Growth lines A, B and C are the products of different crustal evolution models discussed in the text. Data from O'Nions (1984) and Allegre and Rousseau (1984).

present (B) was a better fit to the data, but that the best fit was produced by a curved line (C), representing decreasing growth of the crust through time.

Unfortunately this diagram is not as conclusive as it may appear, due to the great difficulty of determining a global average sediment provenance age at any given time from the very variable provenance ages in individual provinces. This makes the data very susceptible to sampling bias. One source of such bias is preferential recycling of old sediments relative to erosion of more juvenile cratonic material. This will exaggerate the slowing down of continental growth with time, appearing to favour models of type C over type B. Another source of bias is the neglect of young orogenic belts such as the accreted terranes of the Canadian Cordillera (Samson et al., 1989). The inclusion of such data in Fig. 4.32 would favour linear evolution models (type B), suggesting that crustal growth has *not* slowed significantly in the Phanerozoic.

The interpretation of Fig. 4.32. is also heavily influenced by assumptions about the degree of recycling of sediment into the mantle. The so called 'big bang' model shown in Fig. 4.32 involves no recycling of crustal material into the mantle. This does not correspond to Armstrong's model, which involves constant recycling of old crust into the mantle, and replacement by an equal volume of juvenile crust. Armstrong (1991) claimed that his model gave rise to a curve in Fig. 4.33 which looks remarkably like the steady growth model in Fig. 4.32. It is clear then that young sediments provide much too loose a constraint on crustal growth models. Therefore, the argument must focus on the provenance ages of the oldest surviving sediments.

Isua supracrustals from western Greenland, which are the oldest clastic sediments analysed, yield identical stratigraphic and Nd model ages of 3750 Myr, indicating that they did not incorporate a significant amount of older reworked crust. However, the data of Dia et al. (1990) from South Africa show surprisingly old provenance ages for Mid to Late Archean sediments. On balance the sediment data seem to favour a crustal growth model, but ultimately the argument rests on a null hypothesis (no sediments with very old provenance are yet seen, therefore none exist). This is an inherently weak argument upon which to base such an important conclusion.

This weakness comes from the need for representative sampling of old crust using a sediment data set

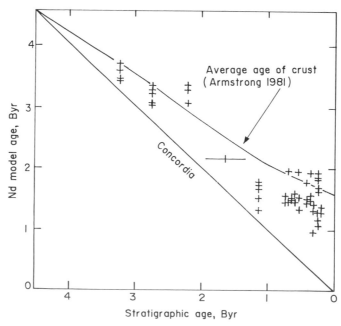

Fig. 4.33. Model age *versus* stratigraphic age diagram showing data of Dia *et al.* (1990) for clastic sedimentary rocks from South Africa, along with the predicted curve for average crustal provenance age from the model of Armstrong (1981). After Armstrong (1991).

which is inherently very noisy. However, an alternative route to assessing the volume of crust at a given time in Earth history is to measure the composition of the depleted reservoir which balances the enriched crustal reservoir, namely upper mantle (section 6.3.1). Because the upper mantle is stirred by convection, we can expect to sample this reservoir (in ancient volcanism) in a much more representative fashion than ancient sediments sample the enriched reservoir. Hence, the problem of crustal growth may be soluble from this direction.

4.4.3 Archean depleted mantle

In the mid 1980s, several studies revealed initial Nd isotope data for Early and Mid Archean rocks which lay well above the chondritic evolution line, and in some cases above the depleted mantle evolution line of Goldstein *et al.* (1984). Smith and Ludden (1989) argued that some of the strongly positive ϵ Nd values calculated for early mafic rocks are in error due to incorrect age assignments. The early Kambalda example has already been mentioned, and doubtless there are problems with some of the other data. However, they concluded that there are enough depleted mantle compositions in the Early Archean for the phenomenon to be real.

Such evidence for very early depletion of the upper

mantle presents a problem for the model in which continental crust grew progressively at the beginning of Earth history. On the other hand, Armstrong (1991) argued that these data supported his model of no crustal growth. In order to examine this claim, the data compilation of Armstrong (1991) is shown in Fig. 4.34, along with an evolution line for the MORB source which he claimed was a product of his 1981 model. However, most of the available Nd data can be satisfied by the less extreme solid evolution line in Fig. 4.34, which is sub-parallel to DePaolo's curve since 4 Byr ago.

The gradual depletion of the upper mantle which is portrayed by the solid line in Fig. 4.34 can only be reconciled with a constant crustal volume model if the average *composition* of the crust changes over geological time. In principle this requirement is met in a model where the Earth begins its evolution with a thick basaltic ('oceanic') crust, which is gradually replaced by continental crust over geological time. This involves a non-plate tectonic model for Archean crustal evolution, (e.g. West, 1980). A similar model was also supported by Galer and Goldstein (1991), who proposed that a thick, long-lived alkali basalt crust was built up in the Archean by small degree melting in the deep mantle. However, as evidence mounts for earlier and earlier operation of plate tectonic processes in Earth history (e.g. Williams *et*

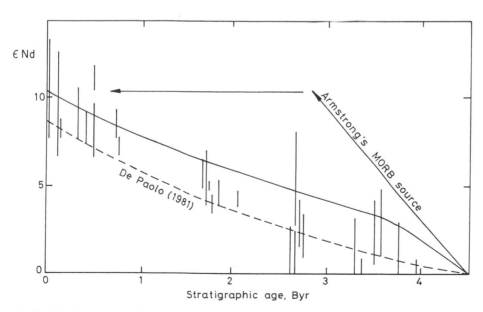

Fig. 4.34. Initial ϵ Nd for terrestrial rocks of different ages, as compiled by Armstrong (1991), along with his proposed 'no growth' MORB evolution line. Solid curve is an alternative MORB depletion line for a crustal growth model. Note that this is not expected to agree with the dashed arc-source model of DePaolo (1981).

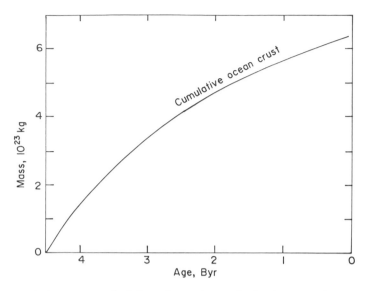

Fig. 4.35. Model for cumulative storage of subducted oceanic crust in the deep mantle over geological time. After Chase and Patchett (1988).

al., 1992), there seems little reason to invoke a prolonged pre-plate tectonic era.

Chase and Patchett (1988) proposed that accelerated early mantle depletion is in fact consistent with plate tectonic processes. They postulated that the storage of subducted oceanic crust in the mantle, before re-homogenisation with the depleted mantle (by convection), would give rise to a hidden enriched reservoir in the deep mantle to balance early depleted mantle. According to this model, the amount of 'stored' subducted oceanic crust has grown over Earth history (Fig. 4.35). Gradual cooling of the earth prevents the system from reaching a steady state by increasing the lifetime of subducted crust over geological time. Taking the cooling process into account, a duration of several hundred Myr to establish Early Archean mantle depletion is consistent with evidence for a 1–2 Byr present-day lifetime of subducted oceanic crust, as deduced from ocean island basalts (section 6.4.1).

McCulloch and Bennett (1993) argued that the effectiveness of the 'slab storage' model would be limited in the Early Archean by a tendency for slab melting, rather than dehydration, during the subduction process. Instead (or in addition), they suggested returning to an old model (section 6.3.1) in which the volume of depleted mantle grows with time. However, rather than smooth growth of this reservoir, they suggested that it grew in one or more jumps, corresponding to the establishment of convective

circulation above successively deeper phase transition boundaries in the upper mantle. For example, a major jump might have been from a 400 km-deep upper mantle to the widely postulated 670 km depth of the present-day MORB source. If this jump occurred between the Early and Mid Archean, then it could have effectively stalled depletion of the upper mantle for nearly 1 Byr.

4.5 Nd in seawater

The abundance of Nd in seawater is about a million times lower than in rocks, at ca. 3 parts per trillion (Goldberg et al., 1963; Piepgras et al., 1979). In contrast, ions such as sodium have similar abundances in rocks and seawater. This led Goldberg et al. to propose that Nd has a very short residence time in seawater, possibly less than 300 yr, and less than the turnover rate of water in the oceans. This can be attributed to effective scavenging of rare earths from seawater by particulate matter.

As with Sr, the isotopic systematics of Nd in seawater are a product of the relative fluxes from different sources. However, the very short oceanic residence time for Nd means that we can expect its isotope systematics in seawater to be quite different from Sr (section 3.5), which has an ocean residence time of over 2 Myr.

The very low Nd concentrations in seawater

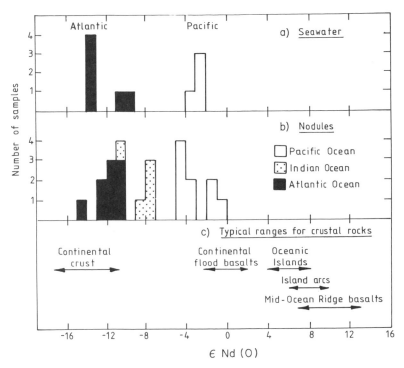

Fig. 4.36. Histograms of Nd isotope analyses of a) seawater and b) manganese nodules from different ocean basins, relative to c) major crustal reservoirs. After Piepgras and Wasserburg (1980).

present some analytical difficulties. In contrast, manganese nodules, which are believed to precipitate directly from seawater, have Nd contents up to hundreds of ppm. Consequently the early studies of O'Nions *et al.* (1978) and Piepgras *et al.* (1979) focussed principally on this material. Significant Nd isotopic variations were found between Mn nodules in different ocean basins (Fig. 4.36b) and attributed by Piepgras *et al.* to real variations in the isotopic composition of seawater.

Piepgras *et al.* justified their interpretation on the grounds that Mn nodules from a wide geographical area within each ocean mass had distinct but reproducible Nd isotope compositions. This was confirmed by the direct analysis of filtered ocean water samples (Fig. 4.36a), which were shown to be consistent with the isotopic composition of sea floor nodules from the same ocean basin. Direct Nd isotope analyses of four water samples from the Pacific (totalling 10 to 20 litres in size) were presented by Piepgras *et al.* (1979), while analyses of Atlantic ocean water were presented by Piepgras and Wasserburg, (1980).

Comparison of seawater isotope compositions with possible source reservoirs (Fig. 4.36c) indicates that Pacific water may be a ca. 50/50 mixture of continental and oceanic crustal Nd, whereas Atlantic ocean water contains up to 80% of continental Nd. This is consistent with a greater river water discharge into the Atlantic. Second-order isotopic variations are also seen within Atlantic ocean water (Fig. 4.37a). Samples from depths of at least 1 km in the Sargasso Sea have very consistent ϵ Nd compositions averaging 13.5, whereas shallower samples from 300 and 50 m depth have more radiogenic ϵ Nd values of −10.9 and −9.6 respectively. This implies isotopic stratification of Atlantic water masses, a conclusion consistent with long-established oceanographic observations of the Atlantic (Wust, 1924).

A more detailed Nd isotope study of the North Atlantic Ocean was performed by Piepgras and Wasserburg (1987) using water samples from five vertical sections. Contoured ϵ Nd values (Fig. 4.37a) are consistent with water masses recognised on the basis of salinity and temperature (Fig. 4.37b). Surface water at mid-latitudes (SW) has ϵ Nd values consistent with the dissolved Nd budget of major rivers such as the Amazon and Mississippi (Piepgras and Wasserburg, 1987; Goldstein and Jacobsen, 1987). Outflow of water from the Mediterranean

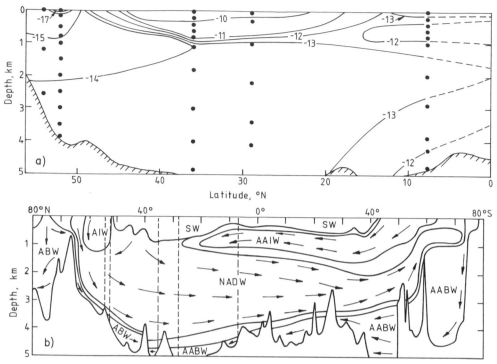

Fig. 4.37. Schematic longitudinal sections through the Atlantic Ocean to show: a) contoured Nd isotope variations in the North Atlantic; b) oceanographically established water masses for the whole Atlantic (with sample locations shown by dashed lines). After Piepgras and Wasserburg (1987).

also has a similar composition (Piepgras and Wasserburg, 1983). In contrast, the major water body of the ocean, North Atlantic Deep Water (NADW) has very uniform unradiogenic Nd ($\epsilon = -13.5$). It is well known that this water largely originates from Arctic Intermediate Water (AIW), shown by Stordal and Wasserburg (1986) to have ϵ Nd as low as -25 in Baffin Bay. Therefore the ϵ Nd composition of NADW must result from mixing of AIW and mid-latitude surface water. Finally, as NADW flows south towards the equator, it becomes sandwiched between two tongues of water with intermediate ϵ Nd, Antarctic Intermediate and Bottom water (AAIW, AABW, Fig. 4.37b).

Following this success in characterising the Nd isotope budget of modern oceans, Shaw and Wasserburg (1985) evaluated different types of material as indicators of the Nd isotope composition of paleo-oceans. They found that carbonate and phosphate in living organisms was very low in Nd (part per billion range), but that fossil carbonate and phosphates had concentrations in the tens to hundreds of ppb and ppm respectively. Shaw and Wasserburg attributed the elevated Nd contents of fossil carbonates largely

to diagenetic remobilisation of detrital Nd, but they attributed the high Nd contents of ancient phosphates (conodonts, fish debris, lingulid brachiopods and inorganic phosphorites) to scavenging directly from seawater (after death). The reliability of conodont and phosphorite for determining ancient seawater Nd was justified by reproducible results from different parts of single ocean basins of a given age.

The demonstration of Nd isotope variations between different present-day oceans implies that such a situation also pertained in the past. Therefore the adoption of a model for the paleo-geography of the oceans is a pre-requisite to establishing the past Nd isotope evolution of seawater. One such model is shown in Fig. 4.38 (Keto and Jacobsen, 1988). It can be seen that the Pacific–Panthalassan ocean is dominant throughout the Phanerozoic, and is all-encompassing in the Late Paleozoic. Early Paleozoic paleo-geography is less well-established, leading to disagreement about the relative importance of the Iapetus ocean.

Keto and Jacobsen (1988) collated conodont and phosphorite data with analyses of fish teeth (Staudigel et al., 1985), ferromanganese coatings on forams

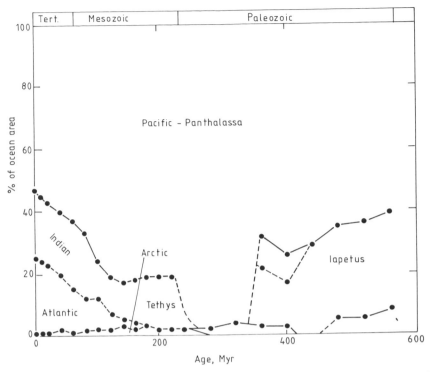

Fig. 4.38. The relative sizes of different ocean basins as a function of time (through the Phanerozoic) based on the paleo-geographic maps of Smith *et al.* (1981). Other interpretations differ for the Paleozoic. After Keto and Jacobsen (1988).

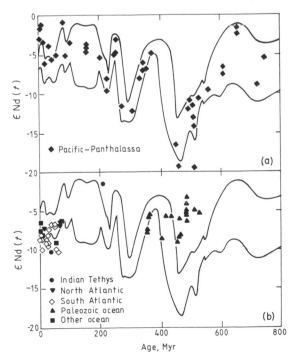

(Palmer and Elderfield, 1986) and conodonts and lingulids (Keto and Jacobsen, 1987; 1988). Because the Pacific–Panthalassan ocean is so dominant, it approximates closely to a global average evolution curve for Nd in seawater (Fig. 4.39a). Other ocean basins with distinct Nd isotope compositions are seen in the Cenozoic and Lower Paleozoic (Fig. 4.39b). These are the Atlantic and the Proto Atlantic (Iapetus) oceans.

Jacobsen and Pimentel-Klose (1988) extended the average seawater Nd isotope evolution curve into the Precambrian by the analysis of Archean and Proterozoic banded iron formations (BIF), which they argued were indicative of the isotopic composition of the Precambrian oceans. Despite the lack of reliable paleogeographic information for this period

Fig. 4.39. ε Nd values for Phanerozoic seawater as a function of age, compared to a calculated global average seawater curve a) Pacific–Panthalassa; b) other ocean and paleo-ocean basins relative to Pacific evolution band. After Keto and Jacobsen (1988).

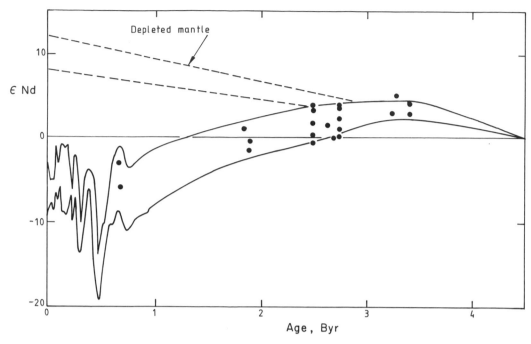

Fig. 4.40. Plot of ϵ Nd values against time for iron formations, used to extend the global average seawater Nd isotope evolution curve into the Precambrian. After Jacobsen and Pimentel-Klose (1988).

of Earth history, it may be justifiable to assume worldwide homogenisation of Nd in seawater, on the grounds that a smaller continental mass during the Precambrian presented less impediment to circulatory mixing of the oceans.

Data from banded iron formations are shown in Fig. 4.40 along with the Phanerozoic average seawater evolution curve of Keto and Jacobsen (1988). This curve is extended into the Precambrian to fit the BIF data, and suggests that unlike the Phanerozoic, where continental run-off is the dominant influence, Archean seawater Nd was controlled by mid ocean ridge hydrothermal circulation. This is consistent with a smaller continental mass and higher global heat flow at this time. The Proterozoic is then a period of transition from the mantle-dominated regime of the Archean to the crust-dominated regime of the Phanerozoic. Similar conclusions were reached about the Sr isotope evolution of Precambrian seawater (section 3.5).

References

Allegre, C. J. and Rousseau, D. (1984). The growth of the continents through geological time studied by Nd isotope analysis of shales. *Earth Planet. Sci. Lett.* **67**, 19–34.

Armstrong, R. L. (1981). Radiogenic isotopes: the case for crustal recycling on a near steady-state no-continental-growth Earth. *Phil. Trans. Roy. Soc. Lond. A***301**, 443–72.

Armstrong, R. L. (1991). The persistent myth of crustal growth. *Aust. J. Earth Sci.* **38**, 613–30.

Arndt, N. T. and Goldstein, S. L. (1987). Use and abuse of crust-formation ages. *Geology* **15**, 893–5.

Awwiller, D. N. and Mack, L. E. (1991). Diagenetic modification of Sm–Nd model ages in Tertiary sandstones and shales, Texas Gulf Coast. *Geology* **19**, 311–14.

Barovich, K. M. and Patchett, P. J. (1992). Behaviour of isotopic systematics during deformation and metamorphism: a Hf, Nd and Sr isotopic study of mylonitized granite. *Contrib. Mineral. Petrol.* **109**, 386–93.

Bennett, V. C. and DePaolo, D. J. (1987). Proterozoic crustal history of the western United States as determined by neodymium isotopic mapping. *Geol. Soc. Amer. Bull.* **99**, 674–85.

Bowring, S. A., King, J. E., Housh, T. B., Isachsen, C. E. and Podosek, F. A. (1989). Neodymium and lead isotope evidence for enriched early Archean crust in North America. *Nature* **340**, 222–5.

Bros, R., Stille, P., Gauthier-Lafaye, F., Weber, F. and Clauer, N. (1992). Sm–Nd isotopic dating of Proterozoic clay material: an example from the Francevillian sedimentary series, Gabon. *Earth Planet. Sci. Lett.* **113**, 207–18.

Burton, K. W. and O'Nions, R. K. (1991). High-resolution garnet chronometry and the rates of metamorphic processes. *Earth Planet. Sci. Lett.* **107**, 649–71.

Cattell, A., Krogh, T. E. and Arndt, N. T. (1984). Conflicting Sm–Nd whole rock and U–Pb zircon ages for Archean lavas from Newton Township, Abitibi Belt, Ontario. *Earth Planet. Sci. Lett.* **70**, 280–90.

Chapman, H. J. and Moorbath, S. (1977). Lead isotope measurements from the oldest recognised Lewisian gneisses of north-west Scotland. *Nature* **268**, 41–2.

Chase, C. G. and Patchett, P. J. (1988). Stored mafic/ ultramafic crust and early Archean mantle depletion. *Earth Planet. Sci. Lett.* **91**, 66–72.

Chauvel, C., Dupre, B. and Jenner, G. A. (1985). The Sm–Nd age of Kambalda volcanics is 500 Ma too old! *Earth Planet. Sci. Lett.* **74**, 315–324.

Chauvel C., Hofmann, A. W. and Arndt, N. T. (1983). New evidence for early mantle depletion from Nd isotopes in greenstones. *Terra Cognita* **3**, 190 (abstract).

Claoue-Long, J. C., Thirlwall, M. F. and Nesbitt, R. W. (1984). Revised Sm-Nd systematics of Kambalda greenstones, Western Australia. *Nature* **307**, 697–701.

Compston, W., Williams, I. S., Campbell, I. H. and Gresham, J. J. (1985). Zircon xenocrysts from the Kambalda volcanics: age constraints and direct evidence for older continental crust below the Kambalda-Norseman greenstones. *Earth Planet. Sci. Lett.* **76**, 299–311.

DePaolo, D. J. (1981). Neodymium isotopes in the Colorado Front Range and implications for crust formation and mantle evolution in the Proterozoic. *Nature* **291**, 193–7.

DePaolo, D. J. (1983). The mean life of continents: estimates of continent recycling rates from Nd and Hf isotopic data and implications for mantle structure. *Geophys. Res. Lett.* **10**, 705–8.

DePaolo, D. J. and Wasserburg, G. J. (1976a). Nd isotopic variations and petrogenetic models. *Geophys. Res. Lett.* **3**, 249–52.

DePaolo, D. J. and Wasserburg, G. J. (1976b). Inferences about magma sources and mantle structure from variations of $^{143}Nd/^{144}Nd$. *Geophys. Res. Lett.* **3**, 743–6.

DePaolo, D. J. and Wasserburg, G. J. (1979). Sm–Nd age of the Stillwater complex and the mantle evolution curve for neodymium. *Geochim. Cosmochim. Acta* **43**, 999–1008.

Dia, A., Allegre, C. J. and Erlank, A. J. (1990). The development of continental crust through geological time: the South African case. *Earth Planet. Sci. Lett.* **98**, 74–89.

Dickin, A. P. and McNutt, R. H. (1989). Nd model age mapping of the southeast margin of the Archean Foreland in the Grenville Province of Ontario: *Geology* **17**, 299–302.

Dickin, A. P., McNutt, R. H., Mueller, E. and Beakhouse, G. (1990). Evidence for an early Archean cratonic nucleus in the Superior province of Ontario. *Geol. Soc. Aust.* **27**, 27 (abstract).

Fletcher, I. R. and Rosman, K. J. R. (1982). Precise determination of initial ϵ Nd from Sm–Nd isochron data. *Geochim. Cosmochim. Acta* **46**, 1983–7.

Galer, S. J. G. and Goldstein, S. L. (1991). Early mantle differentiation and its thermal consequences. *Geochim. Cosmochim. Acta* **55**, 227–39.

Galer, S. J. G. and Goldstein, S. L. (1992). Further ^{142}Nd studies of Archean rocks provide no evidence for early depletion. *EOS* **73**, 622 (abstract).

Goldberg, E. D., Koide, M., Schmidt, R. A. and Smith, R. H. (1963). Rare earth distributions in the marine environment. *J. Geophys. Res.* **68**, 4209–17.

Goldstein, S. L. and Jacobsen, S. B. (1987). The Nd and Sr isotopic systematics of river-water dissolved material: implications for the sources of Nd and Sr in seawater. *Chem. Geol. (Isot. Geosci. Section)* **66**, 245–72.

Goldstein, S. L. and Jacobsen, S. B. (1988). Nd and Sr isotopic systematics of river water suspended material: implications for crustal evolution. *Earth Planet Sci. Lett.* **87**, 249–65.

Goldstein, S. L., O'Nions, R. K. and Hamilton, P. J. (1984). A Sm–Nd isotopic study of atmospheric dusts and particulates from major river systems. *Earth Planet. Sci. Lett.* **70**, 221–36.

Green, T. H., Brunfeldt, A. O. and Heier, K. S. (1969). Rare earth element distribution in anorthosites and associated high grade metamorphic rocks, Lofoten-Vesteraalen, Norway. *Earth Planet. Sci. Lett.* **7**, 93–8.

Griffin, W. L. and Brueckner, H. K. (1980). Caledonian Sm–Nd ages and a crustal origin for Norwegian eclogites. *Nature* **285**, 319–20.

Hamilton, P. J., O'Nions, R. K., Evensen, N. M. and Tarney, J. (1979). Sm–Nd systematics of Lewisian gneisses : Implications for the origin of granulites. *Nature* **277**, 25–8.

Hamilton, P. J., O'Nions, R. K., Bridgwater, D. and Nutman, A. (1983). Sm–Nd studies of Archean metasediments and metavolcanics from West Greenland and their implications for the Earth's early history. *Earth Planet. Sci. Lett.* **62**, 263–72.

Hanson, G. N. (1978). The application of trace elements to the petrogenesis of igneous rocks of granitic composition. *Earth Planet. Sci. Lett.* **38**, 26–43.

Harper, C. L. and Jacobsen, S. B. (1992). Evidence from coupled ^{147}Sm–^{143}Nd and ^{146}Sm–^{142}Nd systematics for very early (4.5-Gyr) differentiation of the Earth's mantle. *Nature* **360**, 728–32.

Haskin, L. A., Frey, F. A., Schmidt, P. A. and Smith, R. H. (1966). Meteoritic, solar and terrestrial rare-earth distributions. *Phys. Chem. Earth* **7**, 167–321.

Hurley, P. M., Hughes, H., Faure, G., Fairbairn, H. W. and Pinson, W. H. (1962). Radiogenic strontium-87 model of continent formation. *J. Geophys. Res.* **67**, 5315–34.

Hurley, P. M. and Rand, J. R. (1969). Pre-drift continental nuclei. *Science* **164**, 1229–42.

Jacobsen, S. B. (1988). Isotopic constraints on crustal growth and recycling. *Earth Planet. Sci. Lett.* **90**, 315–29.

Jacobsen, S. B. and Pimentel-Klose, M. R. (1988). Nd isotopic variations in Precambrian banded iron formations. *Geophys. Res. Lett.* **15**, 393–6.

Jacobsen, S. B. and Wasserburg, G. J. (1980). Sm–Nd isotopic evolution of chondrites. *Earth Planet. Sci. Lett.* **50**, 139–55.

Keto, L. S. and Jacobsen, S. B. (1987). Nd and Sr isotopic variations of Early Paleozoic oceans. *Earth Planet. Sci. Lett.* **84**, 27–41.

Keto, L. S. and Jacobsen, S. B. (1988). Nd isotopic variations of Phanerozoic paleo-oceans. *Earth Planet. Sci. Lett.* **90**, 395–410.

Lambert, D. D., Morgan, J. W., Walker, R. J., Shirey, S. B., Carlson, R. W., Zientek, M. L. and Koski, M. S. (1989). Rhenium–osmium and samarium–neodymium isotopic systematics of the Stillwater Complex. *Science* **244**, 1169–74.

Lugmair, G. W. and Marti, K. (1977). Sm–Nd–Pu timepieces in the Angra dos Reis meteorite. *Earth Planet. Sci. Lett.* **35**, 273–84.

Lugmair, G. W. and Marti, K. (1978). Lunar initial ^{143}Nd/^{144}Nd : differential evolution of the lunar crust and mantle. *Earth Planet. Sci. Lett.* **39**, 349–57.

Lugmair, G. W. and Scheinin, N. B. (1975). Sm–Nd systematics of the Stannern meteorite. *Meteoritics* **10**, 447–8 (abstract).

Lugmair, G. W., Scheinin, N. B., and Marti, K. (1975). Search for extinct ^{146}Sm, I. The isotopic abundance of ^{142}Nd in the Juvinas meteorite. *Earth Planet. Sci. Lett.* **27**, 79–84.

McCulloch, M. T. (1987). Sm–Nd isotopic constraints on the evolution of Precambrian crust in the Australian continent. In: Kroner, A. (Ed.) *Proterozoic Lithospheric Evolution, Amer. Geophys. Union, Geodynamics Series* **17**, 115–30.

McCulloch, M. T. and Bennett, V. C. (1993). Evolution of the early Earth: constraints from ^{143}Nd–^{142}Nd isotopic systematics. *Lithos* **30**, 237–55.

McCulloch, M. T. and Compston, W. (1981). Sm–Nd age of Kambalda and Kanowna greenstones and heterogeneity in the Archean mantle. *Nature* **294**, 322–7.

McCulloch, M. T. and Wasserburg, G. J. (1978). Sm–Nd and Rb–Sr chronology of continental crust formation. *Science* **200**, 1003–11.

McLennan, S. M., McCulloch, M. T., Taylor, S. R. and Maynard, J. B. (1989). Effects of sedimentary sorting on neodymium isotopes in deep-sea turbidites. *Nature* **337**, 547–9.

Moorbath, S. (1976). Age and isotope constraints for the evolution of Archaean crust. In: Windley, B. F. (Ed.), *The Early History of the Earth*, Wiley, 351–60.

Moorbath, S. Powell, J. L. and Taylor, P. N. (1975). Isotopic evidence for the age and origin of the grey gneiss complex of the southern Outer Hebrides, Scotland. *J. Geol. Soc. Lond.* **131**, 213–22.

Mork, M. B. E. and Mearns, E. W. (1986). Sm–Nd isotopic systematics of a gabbro–eclogite transition. *Lithos* **19**, 255–67.

Nelson, B. K. and DePaolo, D. J. (1984). 1,700-Myr greenstone volcanic successions in southwestern North America and isotopic evolution of Proterozoic mantle. *Nature* **312**, 143–6.

Nelson, B. K. and DePaolo, D. J. (1985). Rapid production of continental crust 1.7 to 1.9 b.y. ago: Nd isotopic evidence from the basement of the North American mid-continent. *Geol. Soc. Amer. Bull.* **96**, 746–54.

Nelson, B. K. and DePaolo, D. J. (1988). Application of Sm–Nd and Rb–Sr isotope systematics to studies of provenance and basin analysis *J. Sed. Petrol.* **58**, 348–57.

Notsu, K., Mabuchi, H., Yoshioka, O., Matsuda, J. and Ozima, M. (1973). Evidence of the extinct nuclide ^{146}Sm in

'Juvinas' achondrite. *Earth Planet. Sci. Lett.* **19**, 29–36.

Nunes, P. D. (1981). The age of the Stillwater complex: a comparison of U–Pb zircon and Sm–Nd isochron systematics. *Geochim. Cosmochim. Acta* **45**, 1961–3.

O'Nions, R. K. (1984). Isotopic abundances relevant to the identification of magma sources. *Phil. Trans. Roy. Soc. Lond. A* **310**, 591–603.

O'Nions, R. K., Carter, S. R., Cohen, R. S., Evensen, N. M. and Hamilton, P. J. (1978). Pb, Nd and Sr isotopes in oceanic ferromanganese deposits and ocean floor basalts. *Nature* **273**, 435–8.

O'Nions, R. K., Hamilton, P. J. and Evensen, N. M. (1977). Variations in ^{143}Nd/^{144}Nd and ^{87}Sr/^{86}Sr in oceanic basalts. *Earth Planet. Sci. Lett.* **34**, 13–22.

O'Nions, R. K., Hamilton, P. J. and Hooker, P. J. (1983). A Nd isotope investigation of sediments related to crustal development in the British Isles. *Earth Planet. Sci. Lett.* **63**, 229–40.

Palmer, M. R. and Elderfield, H. (1986). Rare earth elements and neodymium isotopes in ferromanganese oxide coatings of Cenozoic foraminifera from the Atlantic Ocean. *Geochim. Cosmochim. Acta* **50**, 409–17.

Patchett, P. J. and Arndt, N. T. (1986). Nd isotopes and tectonics of 1.9–1.7 Ga crustal genesis. *Earth Planet. Sci. Lett.* **78**, 329–38.

Pidgeon, R. T. and Bowes, D. R. (1972). Zircon U/Pb ages of granulites from the central region of the Lewisian, north western Scotland. *Geol. Mag.* **109**, 247–58.

Piepgras, D. J. and Wasserburg, G. J. (1980). Neodymium isotopic variations in seawater. *Earth Planet. Sci. Lett.* **50**, 128–38.

Piepgras, D. J. and Wasserburg, G. J. (1983). Influence of the Mediterranean Outflow on the isotopic composition of neodymium in waters of the North Atlantic. *J. Geophys. Res.* **88**, 5997–6006.

Piepgras, D. J. and Wasserburg, G. J. (1987). Rare earth element transport in the western North Atlantic inferred from Nd isotopic observations. *Geochim. Cosmochim. Acta* **51**, 1257–71.

Piepgras, D. J., Wasserburg, G. J. and Dasch, E. J. (1979). The isotopic composition of Nd in different ocean masses. *Earth Planet. Sci. Lett.* **45**, 223–36.

Richardson, S. H., Gurney, J. J., Erlank, A. J. and Harris, J. W. (1984). Origin of diamonds in old enriched mantle. *Nature* **310**, 198–202.

Samson, S. D., McClelland, W. C., Patchett, P. J., Gehrels, G. E. and Anderson, G. (1989). Evidence from neodymium isotopes for mantle contributions to Phanerozoic crustal genesis in the Canadian Cordillera. *Nature* **337**, 705–9.

Shaw, H. F. and Wasserburg, G. J. (1985). Sm–Nd in marine carbonates and phosphates: implications for Nd isotopes in seawater and crustal ages. *Geochim. Cosmochim. Acta* **49**, 503–18

Smith, A. D. and Ludden, J. N. (1989). Nd isotopic evolution of the Precambrian mantle. *Earth Planet. Sci. Lett.* **93**, 14–22.

Smith, A. G., Hurley, A. M. and Briden, J. C. (1981). *Phanerozoic Paleocontinental World Maps.* Cambridge University Press. 101 p.

Staudigel, H., Doyle, P. and Zindler, A. (1985). Sr and Nd isotope systematics in fish teeth. *Earth Planet. Sci. Lett.* **76**, 45–56.

Stille, P. and Clauer, N. (1986). Sm–Nd isochron-age and provenance of the argillites of the Gunflint Iron Formation in Ontario, Canada. *Geochim. Cosmochim. Acta* **50**, 1141–6.

Stordal, M. C. and Wasserburg, G. J. (1986). Neodymium isotopic study of Baffin Bay water: sources of REE from very old terranes. *Earth Planet. Sci. Lett.* **77**, 259–72.

Taylor, S. R., McLennan, S. N. and McCulloch, M. T. (1983). Geochemistry of loess, continental crustal composition and crustal model ages. *Geochim. Cosmochim. Acta* **47**, 1897–1905.

Thurston, P. C., Osmani, I. A. and Stone, D. (1991). Northwest Superior province: review and terrane analysis. In: Thurston, P. C., Williams, H. R., Sutcliffe, R. H. and Stott, G. M. (Eds), *Geology of Ontario. Ontario Geol. Surv. Spec. Vol.* **4**, 81–139.

Tugarinov, A. I. and Bibikova, Y. V. (1976). Evolution of the chemical composition of the Earth's crust. *Geokhimiya* **1976**, (8) 1151–9.

Vance, D. and O'Nions, R. K. (1990). Isotopic chronometry of zoned garnets: growth kinetics and metamorphic histories. *Earth Planet. Sci. Lett.* **97**, 227–40.

Wasserburg, G. J., Jacobsen, S. B., DePaolo, D. J., McCulloch, M. T. and Wen, T. (1981). Precise determination of Sm/Nd ratios, Sm and Nd isotopic abundances in standard solutions. *Geochim. Cosmochim. Acta* **45**, 2311–23.

West, G. F. (1980). Formation of continental crust. In: Strangway, D. W. (Ed.), *The Continental Crust and its Mineral Deposits. Geol. Assoc. Canada Spec. Pap.* **8**, 117–48.

Whitehouse, M. J. (1988). Granulite facies Nd-isotopic homogenisation in the Lewisian complex of northwest Scotland. *Nature* **331**, 705–7.

Williams, H. R., Stott, G. M., Thurston, P. C., Sutcliffe, R. H., Bennett, G., Easton, R. M. and Armstrong, D. K. (1992). Tectonic evolution of Ontario: summary and synthesis. In: Thurston, P. C., Williams, H. R., Sutcliffe, R. H. and Stott, G. M. (Eds), *Geology of Ontario. Ontario Geol. Surv. Spec. Vol.* **4**, 1255–1332.

Wust, G. (1924). Florida und Antillenstrom. *Veroffentl. Inst. Meeresh. Univ. Berlin* **12**, 1–48.

5 Lead isotopes

5.1 U–Pb isochrons

Of the four stable isotopes of lead, only ^{204}Pb is non-radiogenic. The other lead isotopes are the final decay products of three complex decay chains from uranium (U) and thorium (Th). However, the intermediate members of each series are relatively short-lived, so they can usually be ignored when geological time-scales of millions of years are involved. Table 5.1 shows the ultimate parent–daughter pairs, of which the highest atomic weight parent (^{238}U) decays to the lowest atomic weight daughter (^{206}Pb) and vice versa. It will be noted that the ^{238}U half-life is comparable with the age of the Earth, whereas that for ^{235}U is much shorter, so that almost all primordial ^{235}U in the Earth has now decayed to ^{207}Pb. The ^{232}Th half-life is comparable with the age of the universe.

If we consider a system of age t, (e.g. a granite intrusion which crystallised from a magma), then we can write an equation for the nuclides involved in each decay scheme, derived from the general equation [1.10]:

$$^{206}\text{Pb}_\text{P} = {}^{206}\text{Pb}_\text{I} + {}^{238}\text{U}(e^{\lambda_{238}t} - 1) \qquad [5.1]$$

$$^{207}\text{Pb}_\text{P} = {}^{207}\text{Pb}_\text{I} + {}^{235}\text{U}(e^{\lambda_{235}t} - 1) \qquad [5.2]$$

$$^{208}\text{Pb}_\text{P} = {}^{208}\text{Pb}_\text{I} + {}^{232}\text{Th}(e^{\lambda_{232}t} - 1) \qquad [5.3]$$

where P indicates the abundance of a given nuclide at the present and I indicates the initial abundance of that nuclide. It is convenient to divide throughout by ^{204}Pb to obtain equations containing isotope ratios

Table 5.1. *Ultimate parent–daughter pairs of uranium and thorium.*

Decay route	$t_{1/2}$, Byr	Decay const λ, yr^{-1}
$^{238}\text{U} \rightarrow {}^{206}\text{Pb}$	4.47	1.55125×10^{-10}
$^{235}\text{U} \rightarrow {}^{207}\text{Pb}$	0.704	9.8485×10^{-10}
$^{232}\text{Th} \rightarrow {}^{208}\text{Pb}$	14.01	0.49475×10^{-10}

Data from Jaffey *et al.* (1971).

rather than absolute nuclide abundances. ^{204}Pb is chosen, as it is the only non-radiogenic isotope. Hence we obtain:

$$\left(\frac{^{206}\text{Pb}}{^{204}\text{Pb}}\right)_\text{P} = \left(\frac{^{206}\text{Pb}}{^{204}\text{Pb}}\right)_\text{I} + \frac{^{238}\text{U}}{^{204}\text{Pb}}(e^{\lambda_{238}t} - 1) \qquad [5.4]$$

$$\left(\frac{^{207}\text{Pb}}{^{204}\text{Pb}}\right)_\text{P} = \left(\frac{^{207}\text{Pb}}{^{204}\text{Pb}}\right)_\text{I} + \frac{^{235}\text{U}}{^{204}\text{Pb}}(e^{\lambda_{235}t} - 1) \qquad [5.5]$$

$$\left(\frac{^{208}\text{Pb}}{^{204}\text{Pb}}\right)_\text{P} = \left(\frac{^{208}\text{Pb}}{^{204}\text{Pb}}\right)_\text{I} + \frac{^{232}\text{Th}}{^{204}\text{Pb}}(e^{\lambda_{232}t} - 1) \qquad [5.6]$$

In principle, the decay equations [5.4] to [5.6] can be used to construct isochron diagrams and hence to date rocks in a manner analogous to the Rb–Sr system (section 3.2.2). Th–Pb and U–Pb isochrons are subject to similar assumptions as for Rb–Sr, the most critical of which is that the samples analysed remained closed to Th, U and Pb during the lifetime of the system being dated. For U–Pb dating it is also necessary to assume that there is no interference by a ^{235}U fission chain reaction. This is very rare in nature, but occurred in the Oklo natural reactor (section 1.3.3). If these assumptions are upheld, then the ^{232}Th/^{208}Pb, ^{235}U/^{207}Pb and ^{238}U/^{206}Pb systems should give concordant ages.

One area where U–Pb isochron dating has been applied with moderate success is the direct dating of marine carbonates (Smith and Farquhar, 1989). Carbonates have proven very difficult to date by other radiometric methods, but because of their low Pb contents (ca. 100–500 ppb), and *relatively* high U (50–100 ppb), rugose corals can yield measurable changes in ^{206}Pb over geological time. Fig. 5.1. shows a U–Pb isochron age for Devonian rugose (Heliophyllum) corals, along with authigenic pyrite from one specimen. The isochron age of 376 ± 10 Myr (2σ, MSWD = 4.7) compares well with a stratigraphic age of ca. 375–385 Myr. However, it can be seen from Fig. 5.1 that one Heliophyllum coral and three out of four Cystiphylloides corals lie off the line. This is probably due to open-system behaviour during

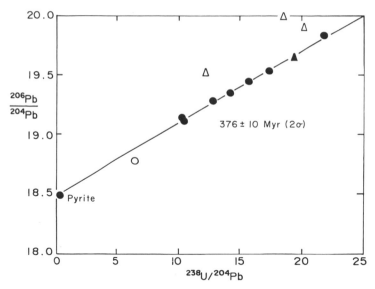

Fig. 5.1. U–Pb isochron diagram for Devonian corals from SW Ontario, Canada. (●, ○) = Heliophyllum; (▲, △) = Cystiphylloides. Open symbols were omitted from the regression. After Smith and Farquhar (1989).

recrystallisation, since the Cystiphylloides corals have a more porous structure which would allow enhanced fluid movement. In subsequent work the U–Pb method has been used to study diagenetic calcite overgrowths, but reliable ages are difficult to determine due to the effects of multiple growth stages (Smith *et al.*, 1991).

Unfortunately, the U–Pb and Th–Pb systems rarely stay closed in silicate rocks, due to the mobility of Pb, Th, and especially U, under conditions of low-grade metamorphism and superficial weathering. For example, Th–Pb analysis of whole-rock samples from the Granite Mountains batholith, Wyoming, yields an approximate age of 2.8 Byr (Fig. 5.2a), but on a U–Pb isochron diagram the system displays disastrous U losses, so that no age can be determined (Fig. 5.2b).

In practice, the mobility of U, Th and Pb renders their use for simple U–Pb isochron dates very limited. This problem would largely invalidate U–Pb dating, but for the existence of a special relationship between the parent and daughter nuclides. Because ^{235}U and ^{238}U on the one hand, and ^{207}Pb and ^{206}Pb on the other, show coherent chemical behaviour, age information can be obtained even from disturbed systems. Three dating techniques exploit this situation, namely the U–Pb 'zircon' method, the common Pb–Pb method, and the galena model age method. The remaining part of this chapter will be devoted to an examination of these methods.

5.2 U–Pb (zircon) dating

If a mineral was available which strongly incorporated uranium at the time of formation but did not incorporate lead, then equation [5.1] above could be simplified by removal of the initial ^{206}Pb term to yield:

$$^{206}Pb = {}^{238}U(e^{\lambda_{238}t} - 1) \qquad [5.7]$$

Taking ^{238}U to the other side yields the equation:

$$\frac{^{206}Pb^*}{^{238}U} = (e^{\lambda_{238}t} - 1) \qquad [5.8]$$

where Pb* represents radiogenic lead only. A similar equation can be derived from [5.2] above:

$$\frac{^{207}Pb^*}{^{235}U} = (e^{\lambda_{235}t} - 1) \qquad [5.9]$$

Minerals which have remained closed systems for U and Pb give concordant values of t when their isotopic compositions are inserted into the left-hand side of equations [5.8] and [5.9]. When compositions yielding such concordant ages are plotted graphically (Fig. 5.3,) they define a curve which was termed the concordia by Wetherill (1956a). The concordia curve can be drawn by substituting decay constants and successive values of t into the right-hand side of equations [5.8] and [5.9], and plotting the results for each value of t.

Uraninite and monazite were the first minerals used

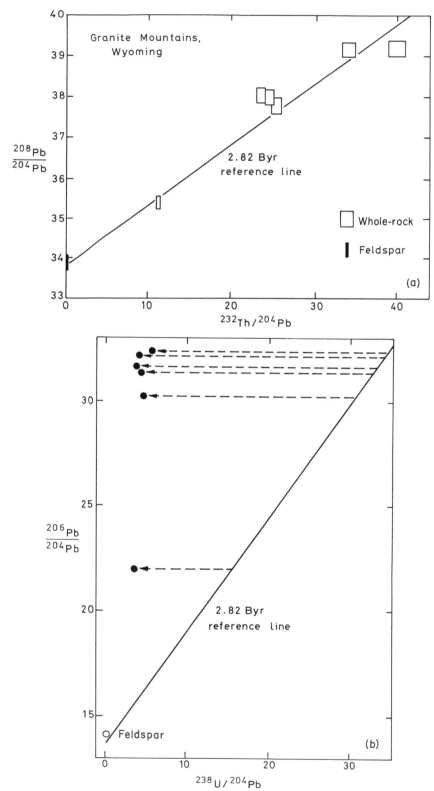

Fig. 5.2. Isochron diagrams for the Granite Mountains batholith. (a) Th–Pb errorchron; (b) U–Pb system showing displacement of whole-rock data points far to the left of the 2.82 Byr reference line, due to disastrous uranium losses. After Rosholt and Bartel (1969).

in U–Pb geochronology, in view of their tendency to incorporate large concentrations of uranium but very little initial (non-radiogenic) lead. However, their limited distribution restricts their usefulness. Zircon is a U-rich mineral with a much wider distribution, present in most intermediate to acid rocks, and has therefore become the principal material used in the U–Pb dating method. A more recent development utilises the zirconium oxide mineral badellyite (Krogh et al., 1987). This is important in dating basic rocks, which often have very low zircon contents.

The small fractions of initial ('common') Pb which are incorporated by zircons are corrected by measuring the amount of (initial)^{204}Pb in the mineral and then using the ^{206}Pb/^{204}Pb and ^{207}Pb/^{204}Pb ratios of the whole-rock to estimate the amount of initial ^{206}Pb and ^{207}Pb incorporated in the zircon. This is subtracted from the present-day ^{206}Pb and ^{207}Pb to yield the radiogenic fraction. For zircons with very low common Pb contents, an adequate correction may be possible by estimating common Pb from a general terrestrial Pb evolution model (e.g. Stacey and Kramers, 1975; section 5.4.3), rather then by direct analysis of the whole-rock sample.

5.2.1 Lead loss models

Early dating work on U-rich minerals soon revealed that most samples yielded discordant ^{206}Pb/^{238}U and ^{207}Pb/^{235}U ages. This discordance was attributed to Pb loss by Holmes (1954). Since that time, much research in U–Pb dating has been devoted to studying the mechanism of lead loss, and to the determination of accurate dates on samples which have suffered lead loss.

Ahrens (1955) found that monazite and uraninite from Zimbabwe (Rhodesia) which yielded discordant U–Pb ages nevertheless defined a linear array on the concordia diagram (Fig. 5.3). Such arrays were later termed discordia. Ahrens argued against lead loss from monazites by leaching, since he (erroneously ?) claimed them not to be metamict due to radiation damage. He thought that lead loss occurred by some kind of continuous diffusional process. This model was elaborated by Russell and Ahrens (1957), who postulated that intermediate members of the uranium decay series were ejected into micro-fissures in the mineral lattice (pitchblende in this case) 'by the recoil energy from α-particle emission'. These nuclides or

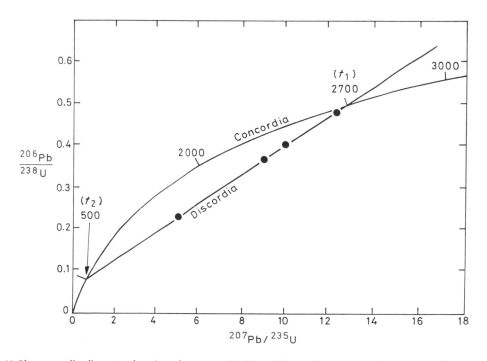

Fig. 5.3. U–Pb concordia diagram showing the concordia line calibrated in Myr, and a discordia line generated by variable Pb loss from 2700 Myr-old U-rich minerals of Zimbabwe (Rhodesia). After Wetherill (1956a).

their decay products could subsequently be removed by diffusion or leaching.

Wetherill (1956a) advanced an alternative interpretation of the data, now called the episodic lead loss model. He agreed that the upper intersection of the discordia with the concordia corresponded to the time of formation of the minerals (t_1). However, Wetherill argued that the lower intersection of the discordia and concordia (t_2) also had age-significance, representing the time of a thermal event which caused lead loss from the minerals. For the 'Rhodesian' minerals these episodes are 2700 and 500 Myr respectively. He supported his model by citing 500 Myr Rb–Sr and K–Ar ages on lepidolite as evidence for a thermal event at that time.

Wetherill (1956b) presented an algebraic proof that the episodic lead loss model could generate the graphically observed results. This can be visualized (Fig. 5.4.) by imagining that the data are plotted at the time of lead loss (500 Myr ago). At that time, points move from the original composition (C) towards the origin (e.g. forming composition B). Subsequent Pb evolution simply rotates the lead loss line in proportion: A → D; B → E; C → F.

Tilton (1960) showed that U-rich minerals with similar formation ages from Archaean shield areas in five continents all lay near a discordia line with a lower intersection at ca. 600 Myr (Fig. 5.5). Under the episodic lead loss model this would imply a worldwide metamorphic event at 600 Myr, but geological evidence for such an event is lacking. Instead, Tilton proposed that the minerals had undergone continuous diffusional lead loss over geological time, yielding a curve on Fig. 5.5 which for much of its length closely resembles a straight line, only curving downwards to an intersection at the origin for relatively recent lead loss.

Goldrich and Mudrey (1972) developed the diffusional lead loss model by arguing that radiation damage of a U-rich mineral was responsible for the formation of a micro-capillary network in the crystal which would become fluid filled. Pb which diffused into these fluids would be lost from the mineral when uplift of basement rock caused the mineral to dilate and expel the capillary-filling fluids. Evidence in support of this 'dilatancy' model was provided by the agreement of various lower intersection ages from North America with times of basement uplift derived from paleo-geographical evidence.

Analytical work on abraded zircons (Krogh, 1982b) has shown that completely unaltered zircons are concordant. Furthermore, Kober (1986) has shown that when Pb is evaporated *in-situ* from zircon grains in the mass spectrometer, discordant Pb can be driven off at low filament temperatures (less than 1350 °C), whereas the concordant Pb fraction is usually emitted between ca. 1400 and 1500 °C. Kober (1987) argued from these high emission temperatures that the concordant radiogenic Pb fraction is substituted into the zircon lattice itself, rather than

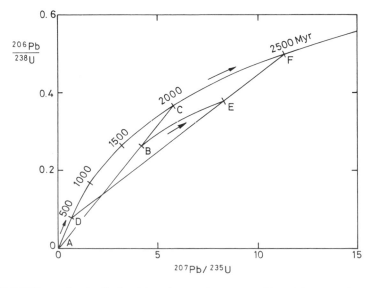

Fig. 5.4. Hypothetical effects of episodic lead loss from a 2500 Myr-old U-rich mineral showing formation of discordia 500 Myr ago and its subsequent rotation due to further uranium decay (see text for discussion).

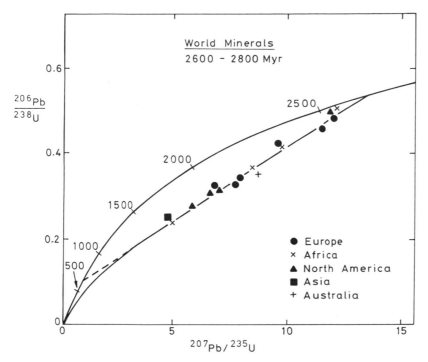

Fig. 5.5. Concordia diagram for Archean U-rich minerals from five continents showing common discordia lower intercept at ca. 600 Myr. Curved line shows hypothetical effects of extreme diffusional lead loss. Dotted line shows extrapolation to apparent episodic lead loss event. After Tilton (1960).

filling defects and voids in the lattice. The ionic radius of Pb^{2+} (1.18–1.29 Å) is much too large to allow substitution for Zr^{4+} (0.72–0.84Å) or Hf^{4+} (0.71–0.83Å). However, Pb^{4+} has an ionic radius of only 0.78–0.94 Å, making it a possible candidate for admission into the lattice. Kober (1987) suggested that the emission of β particles during radioactive decay, and the transformation of emitted He^{2+} (α particles) to neutral He, can effect this oxidation.

This evidence suggests that lead loss from zircons is a fairly 'black and white' process. In other words, unaltered zircon lattices lose very little or no lead, while altered zircon (promoted by metamictisation) loses lead very readily. Any given zircon crystal may contain both kinds of material. For example, Fig. 5.6 shows alteration fronts advancing through the metamict U-rich parts of a zircon.

The exact mechanism for lead loss from altered zircon may in fact be different in different circumstances. Hence it is concluded that the lower intersection of a U–Pb zircon discordia should only be attributed age-significance if this is supported by other geological evidence. However, the interpretation of the upper intersection as the age of formation of the zircons is unaffected.

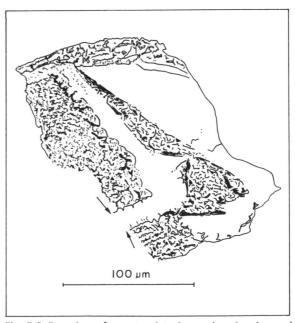

Fig. 5.6. Drawing of a metamict zircon showing inward advance of alteration fronts (arrows). Unaltered material is white. From a photograph by van Breemen *et al.* (1986).

5.2.2 Upper intersection ages

Monazites often lie close to the concordia (the Rhodesian example is exceptional) but zircons are more often quite discordant. In order to obtain the best discordia intersection with the concordia it is desirable to analyse several zircon fractions which have lost different amounts of lead and perform a linear regression on the results. This regression cannot be solved algebraically to yield upper and lower intersection ages; hence these are usually calculated graphically against a calibrated concordia curve, or iteratively by computer.

Silver and Deutsch (1963) made a pioneering case study of lead loss from different zircon fractions in a single rock sample. They found that large zircons lost less lead than small ones (due to the larger surface area/ volume ratio of the latter), and that zircons with low uranium contents lost less lead than high-U

zircons. The latter effect was attributed to the greater radiation damage suffered by U-rich grains. In addition to losing lead, metamict zircons tend to incorporate impurities, including iron. Hence magnetic separation of zircons can yield fractions with variable discordance.

If lead loss processes have operated at different times in the history of a zircon, then the resulting discordia line will probably form a scatter of points which fans out from the upper intercept. This should be reflected in relatively little uncertainty on the upper intercept, but a large uncertainty on the lower intercept. In order to calculate a regression fit to an array of data displaying geological scatter, it is common practice to expand analytical error bars to encompass the scatter (section 2.3.2). However, this would give too much weight to the more discordant data points. Therefore, Davis (1982) suggested that error bars should be magnified in proportion to their discordance (Fig. 5.7).

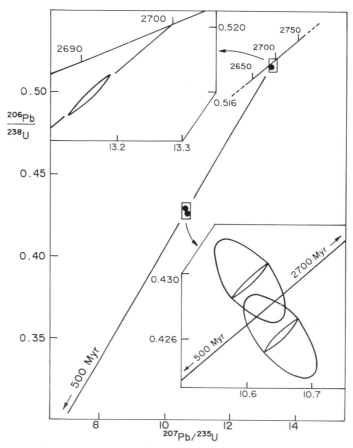

Fig. 5.7. Concordia diagram showing expansion of analytical errors in proportion to discordance to encompass geological scatter on a discordia line. Enlargements are shown for two parts of the discordia. After Davis (1982).

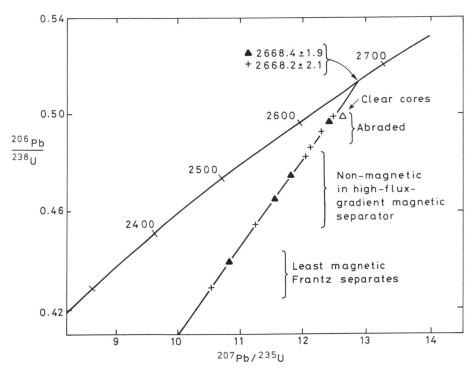

Fig. 5.8. The effect of selecting very non-magnetic zircons, and of abrading off the outer rims, to increase concordance. Symbols (+,▲) indicate two different rock samples. After Krogh (1982b).

Krogh (1982a, b) argued that instead of refining the mathematical treatment of lead loss models to obtain an accurate upper intersection from discordant zircons, it would be better to remove discordant Pb from the sample before analysis. One technique is to use a very-high-flux magnetic separator, which removes all but the least metamict grains (Krogh, 1982a). An alternative approach is to abrade the zircons in a pneumatic mill to remove the outer layers of the crystals, which are usually the most U-rich, and hence metamict (Krogh, 1982b). Spectacular increases in concordance can be obtained in this manner (Fig. 5.8), and the latter technique has become a standard procedure in 'conventional' zircon analysis.

An alternative method of dating zircons with complex geological histories is to break single large zircon crystals into fragments and analyse the individual fragments after a very low blank chemical separation (e.g. Scharer and Allegre, 1982). Corfu (1988) developed this method by breaking the tips off large zircons to analyse the age of new zircon crystallisation over older cores. Aleinikoff et al. (1990) achieved a similar effect by analysing the zircon dust produced by gentle air abrasion of prismatic grains with long terminations. The effect

of the air abrasion process on such a grain is shown in Fig. 5.9. Cracked grains must be excluded to prevent their disintegration during abrasion, which would release core material into the dust.

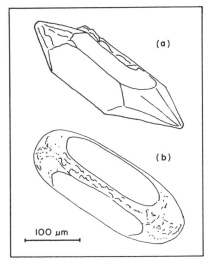

Fig. 5.9. Drawings of zircon grains before and after abrasion to remove tips and interfacial edges for analysis. From photographs by Aleinikoff et al. (1990).

5.2.3 Ion micro-probe analysis

A completely different approach to achieving concordant U–Pb ages is the *in-situ* analysis of Pb isotope
composition (and U/Pb ratio) of zircon grains by ion
microprobe. The general configuration of such an
instrument is shown in Fig. 5.10. A beam of light ions
(e.g. O^-) is used to bombard and sputter a polished
section of the zircon grain to yield a secondary beam
of Pb ions (hence the term secondary ion mass
spectrometry or 'SIMS'). Pb ions are analysed in a
double (magnetic and electrostatic) focussing mass
spectrometer. The electrostatic analyser is necessary
because emitted secondary ions have a range of
energies which would yield bad peak tails in the mass
spectrum if not filtered.

A major problem in SIMS analysis is the
interference of sputtered molecular ions on the
masses of atomic species. In the case of Pb isotope
analysis of zircon, this is caused by species such as
HfO_2^+, which have almost exactly the same mass as
the Pb isotopes, causing isobaric interference (Hinton
and Long, 1979). To overcome this problem the
instrument (including magnet) must have a very large
physical size, in order to generate a large spatial
separation between different masses (equivalent to a
resolution of one mass unit in several thousand). This
allows the separation of Pb from molecular ion
interferences using the 'mass defect' phenomenon
(Fig. 5.11), by which small variations in atomic mass
result from the varying nuclear binding energies of
different atoms. The most successful example of a
SIMS instrument used for U–Pb dating is the
'sensitive high-resolution ion micro-probe' or
'SHRIMP' machine developed at the Australian
National University (Compston *et al.*, 1984).

An important example of the use of the SHRIMP
as a dating tool is provided by the reconnaissance
search for very old rocks (Froude *et al.*, 1983).
Zircons were selected from a formation of Archean
quartzites surrounded by 3.6 Byr-old gneisses (at
Mount Narryer, western Australia) to see whether the
metasediments were derived from a pre-3.6 Byr-old
source. Different areas of single zircon crystals were
analysed using an ion micro-probe, yielding quite
concordant results. Many ages were in the range 3–3.8
Byr, but a few grains gave ages of 4.1–4.2 Byr (Fig.
5.12).

Some of the Mount Narryer zircon spots fell above
the concordia line, displaying what is termed 'reverse
discordance'. This phenomenon is common if whole-
rock compositions are plotted on a concordia
diagram (sometimes done for uranium ore deposits),
and in that case is usually due to uranium loss.

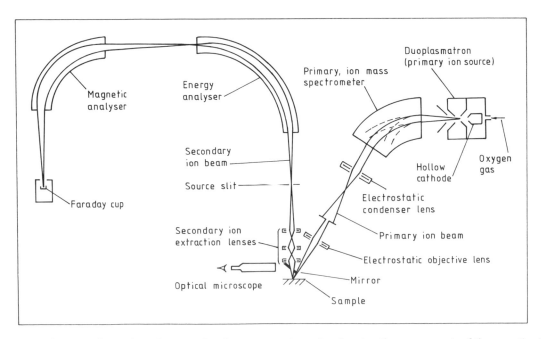

Fig 5.10 Schematic illustration of a secondary ion mass spectrometer showing the components of the negative ion
gun and double focussing analyser. After Potts (1987)

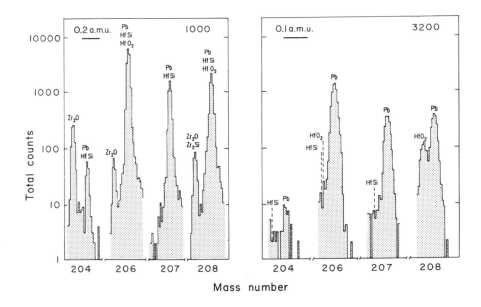

Fig. 5.11. Use of a high-resolution mass spectrometer to separate Pb from interfering molecular ion signals. a) At a resolution of 1000; b) at a resolution of 3200. After Hinton and Long (1979).

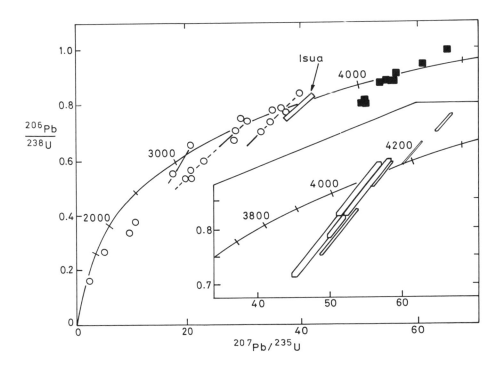

Fig. 5.12. Concordia diagram for ion micro-probe analyses of zircon from Mount Narryer quartzite. (■) = very old zircons (inset shows error ellipses). Isua zircons (box) were analysed as a calibration check. After Froude *et al.* (1983).

Froude *et al.* considered whether such a process could have caused the data to migrate back up the concordia to yield a spuriously old age. In theory, U loss from 3.7 Byr-old zircons during a Late Archean metamorphic episode could have caused points to move to the right and above concordia. This would have to be followed by recent Pb loss (bringing them back down onto the concordia). Froude *et al.* argued that the scatter of data points was too small to be consistent with this model. In addition, Isua zircons were analysed to test the reliability of ion micro-probe analyses for complex metamorphic terranes. These gave the expected age of 3.8 Byr.

An ion micro-probe study on zircons from Mount Sones, Enderby Land, Antarctica, helps to explain the phenomenon of reverse discordance in ion micro-probe analyses (Williams *et al.*, 1984). Uranium concentrations were found to vary quite smoothly as ion sputtering deepened the analysis spots, whereas lead concentrations varied erratically, giving rise to sudden variations in Pb/U ratio (Fig. 5.13). Hence Williams *et al.* suggested that reverse discordance is due to migration of radiogenic Pb between different regions within a zircon crystal, rather than U loss.

5.2.4 207/206 ages

In the dating of Phanerozoic rocks, monazites and zircons may both lie so close to the concordia, sometimes in a clump, that a good discordia line is not generated. In such circumstances it may be necessary to force a discordia line through the origin, assuming that lead loss occurred at the present. The reciprocal of the gradient of this line yields a $^{207}Pb/^{206}Pb$ age, amounting to a simple division of equation [5.9] by equation [5.8] above. 207/206 ages are normally minimum ages, since well-

defined discordia usually have slopes too shallow to go through the origin. However, if the data display reverse discordance (e.g. Fig. 5.12) then 207/206 ages are maximum ages.

Kober (1986, 1987) has demonstrated a new method of zircon dating based on 207/206 ages, in which lead is distilled directly from the zircon crystal in the mass spectrometer. Kober's method is a two-stage process which represents an improvement on techniques previously tried by other workers (e.g. Gentry *et al.*, 1982). A zircon is wrapped in the side filament of a multiple-filament bead, and the temperature of this filament is raised until Pb evaporates directly from the zircon. Some of this lead is re-deposited on the centre filament of the bead assembly, mounted in front of the evaporation filament (Fig. 5.14). After a deposition period of 5–10 minutes, the side filament is turned off and the centre filament is heated to re-emit the deposited lead. It is thought that other species evaporated from the zircon such as ZrO_2 and SiO_2 may form a blanket which holds Pb on the centre filament in a manner similar to the silica gel method for direct Pb analysis (Roddick and Chapman, 1991). When the deposited Pb is exhausted, a new deposition step is made (if possible) at a higher side filament temperature.

Kober's method is based on the premise that discordant lead is contained in less-stable lattice sites than concordant lead. This is driven off at comparatively low temperatures, so that above 1400 °C it can be assumed that all lead is concordant. The results of each run are (ideally) plotted against filament temperature (Fig. 5.15) to demonstrate that the high-temperature Pb emission represents a single phase of lead, rather than mixtures of concordant and discordant lead. The 207/206 ratio will then yield the true crystallisation age of the zircon. Analysis of

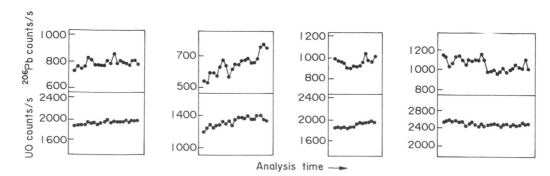

Fig. 5.13. U and Pb concentrations measured as a function of progressive pit deepening during the ion micro-probe analysis of four different zircon spots. Pb emission is seen to be less stable than U. After Williams *et al.* (1984).

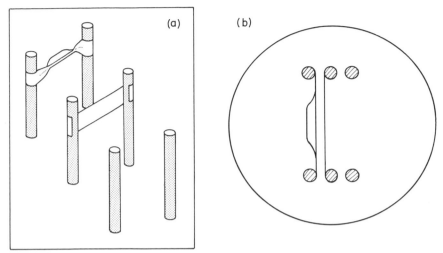

Fig. 5.14. Arrangement of a filament bead for Pb–Pb dating of zircon by the two-stage direct evaporation method. a) exploded view; b) plan view.

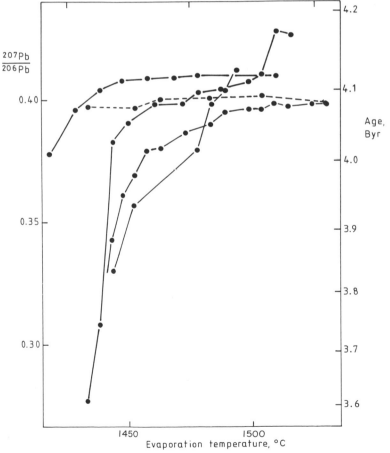

Fig. 5.15. Plot of measured $^{207}Pb/^{206}Pb$ ratios (corresponding to apparent age) against evaporation temperature for five Mount Narryer zircons. Low-temperature domains have suffered Pb loss but high-temperature plateaus are argued to be concordant. After Kober *et al.* (1989).

zircon from the population previously dated by Froude *et al.* (1983) gave similar old ages (Kober *et al.*, 1989).

A development of the direct evaporation method pioneered by Wendt *et al.* (1991) is the determination of the U/Pb ratio of the sample by analysis of ^{210}Pb. U–Pb data would yield direct information on the discordance of the sample, as in the conventional U–Pb zircon method, but cannot be measured directly because of the different evaporation and ionisation behaviour of U and Pb. However, since ^{210}Pb is in the decay chain of ^{238}U (section 12.1), the ^{210}Pb abundance can be used as a monitor of U concentration. Unfortunately, due to the short half-life of ^{210}Pb, the equilibrium ratio of ^{210}Pb/^{238}U is 5×10^{-9}, equivalent to a ^{210}Pb/^{206}Pb ratio of 10^{-7} in a 300 Myr-old sample. This means that a typical zircon of 1.5 µg weight containing 400 ppm U would contain less than 10^4 atoms of ^{210}Pb, which is below detection limit. However, Wendt *et al.* (1991) demonstrated the method on the uranium mineral torbernite, achieving an upper intersection age within error of conventional U–Pb analyses. Subsequently, Wendt *et al.* (1993) were also able to obtain reasonably accurate ^{210}Pb data from bulk (2–20 mg) samples of zircon.

Another recent development in ^{207}Pb/^{206}Pb dating (Feng *et al.*, 1993) utilises a combination of the laser probe (LP) and inductively-coupled plasma mass spectrometry (ICP-MS). The laser probe is used to ablate cylindrical pits from single zircon grains, in a manner analogous to the ion micro-probe. However, laser ablation is performed at atmospheric pressure, yielding a molecular vapour which is carried by argon gas to the plasma torch. Temperatures of several thousand °C in the plasma cause effective atomisation of the sample, destroying potential molecular ion interferences of Pb. The sample then passes into the quadrupole mass spectrometer, where ^{207}Pb/^{206}Pb ratios are analysed. Feng *et al.* were able to obtain ^{207}Pb/^{206}Pb ages from twenty large zircons (> 60 µm grain size) which fell within 1% of conventional U–Pb data. These results suggest that LP-ICPMS has a promising role as a reconnaissance tool in single zircon dating studies, especially in applications such as sediment provenance where conventional U–Pb analysis is too labour intensive.

5.2.5 Alternative U–Pb data presentations

In the classical concordia diagram the variables are strongly correlated, because of the manner in which the data are analysed. ^{207}Pb/^{235}U is calculated from ^{206}Pb/^{238}U, based on the constant value of ^{235}U/^{238}U

and the measured ^{207}Pb/^{206}Pb ratio, which is known much more accurately than U/Pb. The correlation of errors is taken into account when fitting discordia regression lines (section 2.3.1), but it is largely avoided in an alternative presentation of U–Pb data pioneered by Tera and Wasserburg (1973, 1974), where ^{238}U/^{206}Pb is plotted directly against ^{207}Pb/^{206}Pb. This concordia has a different curvature to the conventional presentation, and is preferred by workers dating young rocks (e.g. Scharer *et al.*, 1984) because it displays these discordia lines more clearly than the conventional diagram (Fig. 5.16).

In the case of very young rocks, Pb loss from zircons will yield a discordia which cannot be resolved from the concordia, even on the Tera–Wasserburg plot. In such cases, zircon analysis cannot yield reliable crystallisation ages (Fig. 5.17). Scharer (1984) attempted to overcome this problem in dating the Himalayan Makalu granite by analysis of monazite, which has a greater resistance to Pb loss. However, the high Th contents of monazite cause uptake of a significant content of the U-series isotope ^{230}Th, which subsequently decays to ^{206}Pb (section 12.1), causing an excess abundance of this isotope (Ludwig, 1977). Scharer demonstrated that a correction for this excess production caused apparently discordant analyses to fall properly on the concordia (Fig. 5.17), yielding a precise age of 24 ± 1 Myr for crystallisation of the Makalu pegmatitic granite.

Wendt (1984) has further developed the Tera–Wasserburg plot into a three-dimensional U–Pb diagram by the addition of an axis in ^{204}Pb/^{206}Pb, representing the level of common Pb present. In this construction the discordia is a plane, and ages can be calculated without independent knowledge of the isotopic composition of the common lead component, subject to the assumption that only one such component is present. An example of the application of this method is the dating of Mesozoic uranium minerals from Germany by Carl and Dill (1985) and Wendt *et al.* (1991). The three-dimensional diagram may also allow dating of partially open whole-rock U–Pb systems because it allows more accurate correction of large common Pb fractions (e.g. Carl *et al.*, 1989).

5.2.6 Inherited zircon

If a magma is derived by partial melting of the crust, or assimilates crustal material, old zircons may be entrained into the magma. These 'inherited' zircons may dissolve in per-alkaline magmas, which have a high Zr saturation level. However, they may survive

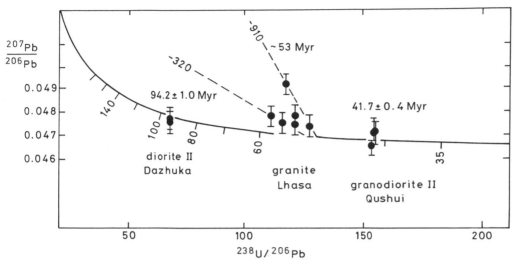

Fig. 5.16. 'Tera–Wasserburg' concordia diagram on axes of $^{207}Pb/^{206}Pb$ against $^{238}U/^{206}Pb$ showing data for Himalayan granites. After Scharer *et al.* (1984).

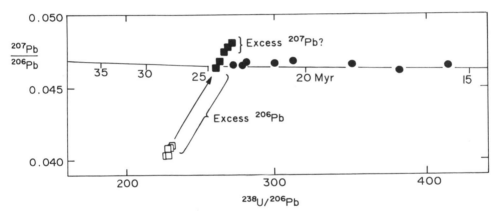

Fig. 5.17. Tera–Wasserburg concordia diagram for the Makalu leucogranite, Himalayas, showing zircon (●) which has lost Pb, and monazite (□, ■) before and after correction for inherited U/Th disequilibrium. After Scharer (1984).

in per-aluminous melts, especially if cool and dry, due to the low Zr saturation levels of such magmas (Watson and Harrison, 1983). Inherited zircon xenocrysts tend to lose much of their old Pb, and may be overgrown by a new zircon crystal. Fig. 5.18 shows an example from the Ben Vuirich granite of Scotland (Pankhurst and Pidgeon, 1976). The lower intersection (514 ± 7 Myr) was interpreted as the age of intrusion and the upper intersection (1316 ± 26 Myr) as the approximate age of assimilated old crustal material.

This study was extended by Pidgeon and Aftalion (1978) to include U–Pb analysis of 24 Caledonian granites from Scotland and northern England. Of this suite, 17 plutons with inherited zircon lie north of the Highland Boundary fault in Scotland and only one (Eskdale granite) lies to the south. In contrast, all 6 granites without inherited zircon lie south of the fault. Pidgeon and Aftalion discussed various possible explanations for these observations.

Because the granites north and south of the fault have similar chemistry, Pidgeon and Aftalion ruled out the dissolution of inherited zircons during magma evolution, or their removal during emplacement. They also rejected contamination of granite magmas by sedimentary zircons north of the fault, since these granites do not have the S-type (per-aluminous) chemistry characteristic of assimilation. (In contrast,

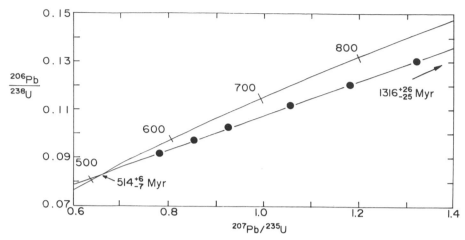

Fig. 5.18. Concordia diagram for Ben Vuirich granite (Scotland) showing mixing between new Caledonian Pb and inherited Mid Proterozoic Pb. After Pankhurst and Pidgeon (1976).

inherited zircon in the S-type Eskdale granite was probably derived by assimilation of sediments containing old zircon.) It was concluded that there is a fundamental difference in granite source rocks between the Scottish Highlands, and crust to the south; a conclusion supported by whole-rock Nd isotope analysis (Halliday, 1984).

The study of Pankhurst and Pidgeon (1976) made use of bulk zircon separates (total quantity of zircon separated was 8 g!). In an attempt to refine and test the old determination, Rogers *et al.* (1989) re-dated the pluton using modern techniques of miniature sample analysis and zircon abrasion. The results (Fig. 5.19) were startlingly different. The lower intercept was increased by 76 Myr to 590 ± 2 Myr, while the upper intercept was increased by 132 Myr to 1448 ± 7 Myr. The lower ages determined from the earlier study can be attributed to the effects of secondary lead loss after intrusion, from a system which already represented a two-component mixing line. This caused rotation of the apparent discordia, yielding erroneously young ages for both upper and lower intercepts.

The occurrence of secondary lead loss from Ben Vuirich zircons is demonstrated by a comparison of abraded and unabraded needle-shaped grains (representing new 590 Myr magmatic zircons). In contrast, abraded stubby grains provided a closer approach to the inherited zircon composition than the bulk fraction of large non-magnetic grains analysed by Pankhurst and Pidgeon. The study of Rogers *et al.* is typical of much recent work showing the dangers of bulk zircon analysis in rocks with complex geological histories. Such samples can yield discordia of high statistical quality which nevertheless yield erroneous ages. Consequently the *painstaking* selection and abrasion of crack-free and inclusion-free grains is essential to ensure the reliability of U–Pb data.

5.3 Common (whole-rock) Pb–Pb dating

The U–Pb decay equations [5.5] and [5.4] can be rearranged to bring the Pb/Pb terms to the left-hand side:

$$\left(\frac{^{207}Pb}{^{204}Pb}\right)_P - \left(\frac{^{207}Pb}{^{204}Pb}\right)_I = \frac{^{235}U}{^{204}Pb}(e^{\lambda_{235}t} - 1) \qquad [5.10]$$

$$\left(\frac{^{206}Pb}{^{204}Pb}\right)_P - \left(\frac{^{206}Pb}{^{204}Pb}\right)_I = \frac{^{238}U}{^{204}Pb}(e^{\lambda_{238}t} - 1) \qquad [5.11]$$

Nier *et al.* (1941) showed that if these two equations refer to the same system, equation [5.10] can be divided by [5.11], and the ^{204}Pb terms in the right-hand side of the equations cancelled, leaving the term $^{235}U/^{238}U$ (which has a constant value of 1/137.88 throughout the solar system). This yields the simplified equation:

$$\frac{\left(\frac{^{207}Pb}{^{204}Pb}\right)_P - \left(\frac{^{207}Pb}{^{204}Pb}\right)_I}{\left(\frac{^{206}Pb}{^{204}Pb}\right)_P - \left(\frac{^{206}Pb}{^{204}Pb}\right)_I} = \frac{1}{137.88}\frac{(e^{\lambda_{235}t} - 1)}{(e^{\lambda_{238}t} - 1)} \qquad [5.12]$$

If we consider a number of systems which have the same age and initial isotopic composition (e.g. whole-rock samples of a granite) then it can be seen from

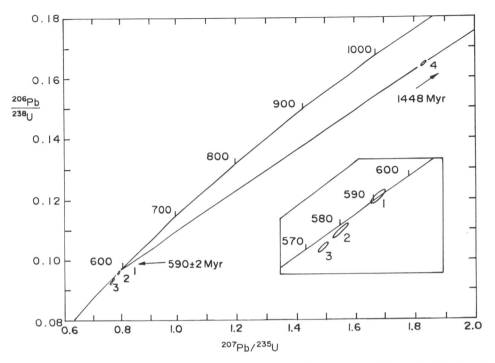

Fig. 5.19. Concordia diagram for Ben Vuirich granite showing mixing between new Caledonian Pb (needle-shaped grains) and inherited Mid Proterozoic Pb (stubby grains). Inset shows Pb-loss line from 590 Myr defined by needle-shaped grains. 1 & 4 = strongly abraded; 2 = slightly abraded; 3 = unabraded, to control Pb-loss line. After Rogers *et al.* (1989).

equations [5.10] and [5.11] that they will develop different Pb isotope compositions, according to their U/Pb ratios, at the present day. Therefore, if the present-day Pb isotope compositions of this suite are plotted (left-hand side of equation [5.12]), they should form a straight-line array, provided that they have remained closed systems. The slope of this array, first termed an 'isochrone' by Houtermans (1947), depends only on t, and does not require any knowledge of the U and Pb concentrations in the samples. It should be noted that the isochron equation [5.12] is 'transcendental'. In other words the term on the right-hand side (equal to the slope), cannot be solved algebraically to yield the age, t, but must therefore be solved iteratively, usually by computer.

Since the closed U–Pb system requirement remains, it might be wondered what advantage this method offers over the discredited whole-rock U–Pb isochron method (section 5.1), in view of the known high mobility of uranium. This question can be answered empirically. Fig. 5.20 shows a whole-rock Pb–Pb isochron diagram for the Granite Mountains, Wyoming (Rosholt and Bartel, 1969) which gives a

geologically correct age of 2.82 Byr. However, it was shown in Fig. 5.2b that these samples had suffered disastrous uranium losses. This can be explained by the fact that U–Pb whole-rock systems were effectively closed from the time of formation of the intrusion until very near the present day, when uranium was lost in recent weathering processes. This invalidates the U–Pb isochron method, but since the Pb isotope ratios in the rock reflect the pre-weathering U concentrations, they are not upset by the recent alteration event.

5.3.1 The geochron

The first application of the common Pb–Pb dating technique was actually on meteorites rather than terrestrial rocks. Patterson (1956) calculated a Pb–Pb age of 4.55 ± 0.07 Byr on a suite of three stony meteorites and two iron meteorites, the least radiogenic sample analysed being troilite (FeS) from the Canyon Diablo iron meteorite (Fig. 5.21). The U/Pb ratio measured on this sample (0.025) was so low that Patterson concluded that 'no observable change in the

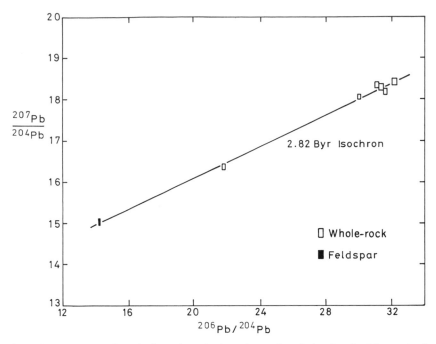

Fig. 5.20. Pb–Pb isochron diagram for whole-rock and mineral samples of the Granite Mountains batholith. After Rosholt and Bartel (1969).

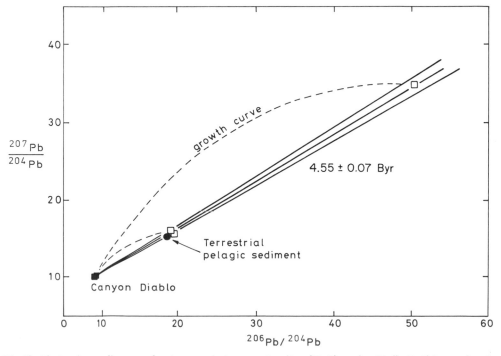

Fig. 5.21. Pb–Pb isochron diagram for iron and stony meteorites (■, ▫) and a 'Bulk Earth' sample of oceanic sediment (●), showing that the Earth lies on the meteorite isochron, therefore also called the 'geochron'. After Patterson (1956).

isotopic composition of lead could have resulted from radioactive decay after the meteorite was formed'. Hence Canyon Diablo troilite represents the primordial Pb isotope composition of the solar system. This is an important benchmark for terrestrial Pb isotope evolution, just as the chondritic reservoir constrains terrestrial Nd isotope evolution.

Patterson also solved a problem that had occupied geochronologists for decades, namely the age of the Earth. A sample of recent oceanic sediment, regarded as the best estimate of the Bulk Earth Pb isotope composition, lay on the meteorite isochron, and furthermore had the appropriate U and Pb concentrations to be generated by radiogenic Pb growth from the Canyon Diablo composition in 4.55 Byr. This finding provided good evidence that the Earth has both the same age and ultimate origin as meteorites. The meteorite isochron was therefore termed the geochron.

Subsequent work has shown that the interpretation of pelagic sediment as a Bulk Earth composition is only a rough approximation to the complexities of terrestrial Pb isotope evolution. Therefore the apparent Pb–Pb age of the Earth must now be revised downwards slightly (section 5.4.3). However, the new value is almost within experimental error of the determination of Patterson.

5.4 Model (galena) ages

5.4.1 The Holmes–Houtermans model

As discussed above, different Pb isotope dating methods address the problem of uranium mobility in different ways. In the U–Pb zircon dating method, a mineral was chosen which held U well and in which Pb loss could be modelled and corrected. In common Pb–Pb dating the recent loss of U can be permitted, provided the system was closed for most of its life. In the galena method discussed here, a phase is analysed which contains no U, so there is no problem of U loss. Since there is no decay in the galena, we are not measuring its age directly back from the present day, but are measuring the age of the galena source from the formation of the Earth until the isolation of the galena. This approach was independently conceived by Holmes (1946) and Houtermans (1946). They divided the isotopic evolution of galena Pb into two parts. The first was regarded as a rock system, which must have remained closed to U and Pb from the formation of the Earth until galena separation. The second was in the galena itself, which must contain no significant amounts of uranium. This model for

terrestrial Pb isotope evolution may be summarised as follows:

$$
\begin{array}{ccc}
 & \text{U decay} & \text{no U decay} \\
 & \text{in rock} & \text{in galena} \\
T & \xrightarrow{\hspace{2cm}} t & \xrightarrow{\hspace{2cm}} P \\
\text{age of Earth} & \text{age of galena} & \text{present}
\end{array}
$$

Given this model, the basic decay eqation for ^{207}Pb is:

$$^{207}\text{Pb}_t = {}^{207}\text{Pb}_T + {}^{235}\text{U}\,(e^{\lambda_{235}T} - e^{\lambda_{235}t}) \qquad [5.13]$$

This decay equation is more complex than [5.1] because 't' is not zero. Each term is next divided through by ^{204}Pb and rearranged. The same procedure is applied to the corresponding equation for ^{206}Pb to yield:

$$\left(\frac{^{207}\text{Pb}}{^{204}\text{Pb}}\right)_t - \left(\frac{^{207}\text{Pb}}{^{204}\text{Pb}}\right)_T = \frac{^{235}\text{U}}{^{204}\text{Pb}}\,(e^{\lambda_{235}T} - e^{\lambda_{235}t}) \qquad [5.14]$$

$$\left(\frac{^{206}\text{Pb}}{^{204}\text{Pb}}\right)_t - \left(\frac{^{206}\text{Pb}}{^{204}\text{Pb}}\right)_T = \frac{^{238}\text{U}}{^{204}\text{Pb}}\,(e^{\lambda_{238}T} - e^{\lambda_{238}t}) \qquad [5.15]$$

Equation [5.14] is divided through by equation [5.15] and the result is simplified as follows:

1) ^{204}Pb terms are cancelled on the right-hand side of the equation. This leaves a factor for the U isotope ratio at the present day, which is a constant, 1/137.88.
2) $(^{207}\text{Pb}/^{204}\text{Pb})_t$ and $(^{206}\text{Pb}/^{204}\text{Pb})_t$ represent the present-day compositions, since galena incorporates no U.
3) The Pb isotope compositions at time 'T' represent the composition of the solar nebula; that is the primordial composition of the Earth, now represented by Canyon Diablo troilite (C.D.).

The equation can then be written as:

$$\frac{\left(\dfrac{^{207}\text{Pb}}{^{204}\text{Pb}}\right)_P - \text{C.D.}}{\left(\dfrac{^{206}\text{Pb}}{^{204}\text{Pb}}\right)_P - \text{C.D.}} = \frac{1}{137.88}\frac{(e^{\lambda_{235}T} - e^{\lambda_{235}t})}{(e^{\lambda_{238}T} - e^{\lambda_{238}t})} \qquad [5.16]$$

If the isotope ratios on the left-hand side of the equation represent a sample extracted from the mantle at time t, then the term on the right-hand side corresponds to the slope of an 'isochron' line joining it to the solar nebula composition (Fig. 5.22).

To apply the Holmes–Houtermans model, the galena source rock is assumed to be a closed system with a 'single-stage' Pb isotope history. A growth curve is then constructed for the galena source, which

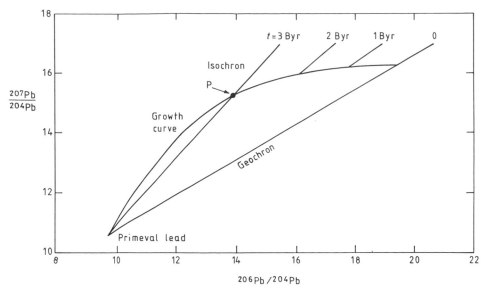

Fig. 5.22. Pb–Pb isochron diagram showing present-day composition (P) of galena extracted from a Bulk Earth reservoir 3 Byr ago. After Russell and Farquhar (1960).

runs from the primordial Pb composition to that of the analysed galena, and is calibrated for different values of t. (Since this is a transcendental curve, t cannot be solved by direct algebra starting with a composition on the left hand side of equation [5.16].) The shape of the growth curve is determined by the decay constants, and its trajectory by the $^{238}U/^{204}Pb$ or 'μ' value of the closed-system galena source. For the single stage model described it is called the μ_1 value and would normally be between 7 and 9. According to the Holmes–Houtermans model, not every galena source rock need have the same growth curve, defined by the same μ value. Galena ore bodies were expected to have concentrated the metal from local continental basement in the vicinity. However, this presupposes that the basement in question has been in existence since near the formation of the Earth, which is now regarded as very unlikely (e.g. see section 4.3).

The major problem encountered with the Holmes–Houtermans model was that as more galenas were analysed they were found to scatter more and more widely on the Pb–Pb isochron diagram (Fig. 5.23). Some of the ages determined were clearly erroneous, since they were in the future. Others, which were outliers to the main trend, often gave ages which could be shown to be geologically impossible. Since galenas of these two types contradicted the Holmes–Houtermans model, they were called 'anomalous leads'. The crucial problem with this situation was

the lack of an *a priori* test which could be performed to predict whether a galena would be anomalous, in the absence of other evidence of its age.

5.4.2 Conformable leads

Given the complexity of Earth evolution, it was realised even in the 1950s that the country-rock to a given galena ore was unlikely to have been a closed system since the formation of the Earth. Alpher and Herman (1951) attempted to overcome this problem by attributing Pb isotope evolution in the galena source rock to a single world-wide homogeneous reservoir, regarded by Russell (1956) as the Earth's mantle. As an explanation of the observed galena Pb isotope variation this model is quite obviously inadequate, but it was the basis of a more geologically realistic model proposed by Stanton and Russell (1959).

A certain class of Pb was found by Stanton and Russell which did lie on a single closed-system growth curve. These were sulphides associated with sediments and volcanics in greenstone belts and island arcs, which were structurally conformable with the host rocks (in contrast to cross-cutting veins). Stanton and Russell regarded these ores as being formed by syngenetic deposition in sedimentary basins associated with volcanic centres, and therefore as representing galena derived directly from the upper mantle without crustal contamination.

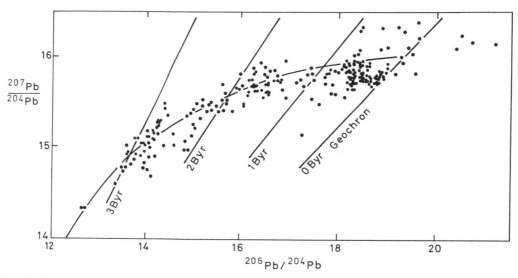

Fig. 5.23. Pb–Pb isochron diagram showing a compilation of many analysed galenas from different environments. After Stanton and Russell (1959).

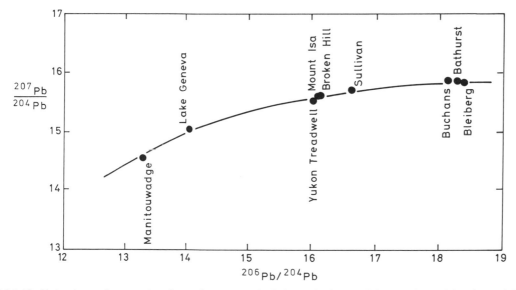

Fig. 5.24. Pb–Pb isochron diagram showing galena ores which form the basis of the 'conformable' Pb model. After Stanton and Russell (1959).

Stanton and Russell selected nine deposits of different ages which met these criteria, and fitted a single stage (upper mantle) growth curve with a μ_1 value of 9.0 (Fig. 5.24). These ores were termed 'conformable' leads because of their structural occurrence, and all galenas that didn't fit this curve were by inference 'anomalous'. Anomalous leads were divided into groups such as 'J-type leads' after a deposit at Joplin, Missouri which gave ages in the future, and some other types such as the 'B' type which gave ages in the past. Unfortunately, due to the mobility of Pb during crustal processes, it is difficult to develop *a priori* criteria to recognise galenas which will fit the conformable Pb evolution model. Therefore, the galena method is largely discredited as a dating tool. Nevertheless, it may provide powerful constraints to Earth evolution, which will be examined below.

5.4.3 Open-system Pb evolution

As early as 1956, Russell considered the possibility that the mantle might not have been a closed system to U and Pb, such that it might have had a variable μ value over time, due to some kind of differentiation mechanism. However, the success of the conformable lead model to explain uncontaminated galena compositions with a closed-system mantle mitigated against such complications.

In the early 1970s, new measurements of the uranium decay constants (Jaffey et al., 1971) and a better estimate of primordial Pb from Canyon Diablo necessitated a re-examination of the conformable Pb model. For example, using the new values, and a revised age of 4.57 Byr for the Earth based on Pb–Pb dating of meteorites (Tatsumoto et al., 1973), single-stage model lead ages gave results up to 1 Byr in error for Phanerozoic rocks. Alternatively, a curve calculated to yield a reasonable fit to conformable galenas gave a low apparent age for the Earth of 4.43 Byr (Doe and Stacey, 1974).

To rectify these problems, Oversby (1974) proposed a model for an evolving (mantle) source of galena Pb

with a progressive increase in μ value with time (approximated by a series of small increments in μ). This model was elaborated upon by Cumming and Richards (1975), who modelled a galena source with a linear increase in μ value. Surprisingly perhaps, Cumming and Richards regarded the galena source as a regional average of the crust. However, this may not be as strange as it sounds, since later work would show that mantle and crustal Pb evolution are in fact coupled together, and that upper mantle Pb is largely buffered by the crust (section 6.4.2). The model of Cumming and Richards yields a good fit to the ages of selected galena data, but still implies a young apparent age for the Earth of 4.50 Byr.

Stacey and Kramers (1975) used Canyon Diablo Pb and average modern Pb (from a mixture of manganese nodules, ocean sediments and island arc rocks) to define the ends of a composite growth curve. This was produced by two closed systems (1 and 2) with different μ values (μ_1 and μ_2), separated in time by a world-wide differentiation event. The closed systems consisted of a combination of the upper mantle and upper crust (lower crust, lower mantle and core isolated). The model gave the best fit to a selection of conformable galenas (dated by the enclosing sediments) when $\mu_1 = 7.2$, $\mu_2 = 9.7$, and the event was at 3.7 Byr (Fig. 5.25). This time was

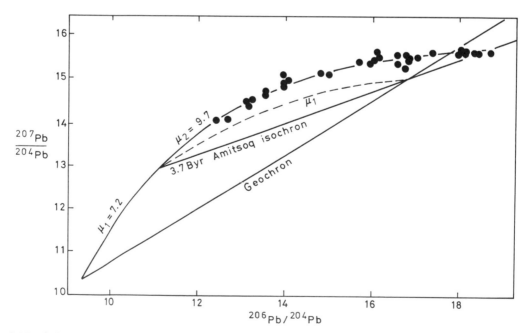

Fig. 5.25. Pb isotope diagram showing a two-stage lead isotope evolution model proposed for the source of galenas (•). After Stacey and Kramers (1975).

interpreted as a peak of crust forming events, made particularly attractive by the 3.7 Byr age determined for the Amitsoq gneisses of western Greenland (section 5.5). However Stacey and Kramers noted that their model was only an approximation of Pb isotope evolution in the real Earth. For example, the discrete 3.7 Byr event in the model might actually represent a slow change in the Earth's evolution during the Early Archean.

The two isotopic systems of uranogenic Pb clearly form a powerful constraint in separating conformable Pb from anomalous Pb with a complex crustal history. However, the curved arrays inherent in the two-isotope system make it hard to evaluate the goodness of fit of conformable Pb data to terrestrial Pb isotope evolution models, as well as making conceptual understanding of the system difficult. Manhes et al. (1979) developed a linear presentation which overcame the first problem, but their formulation was rather complex, which detracted from its usefulness. To simplify the data presentation, Albarede and Juteau (1984) analysed each of the U–Pb systems (and Th–Pb) on a separate diagram of Pb isotope ratio against time, as is done for Nd (section

4.2). However, because of the effectively finite half-lives of U and Th relative to the age of the Earth (unlike Nd), the time dimension must be presented as the exponent (Fig. 5.26) in order to achieve linear evolution lines.

Albarede and Juteau utilised the combined galena data sets from Stacey and Kramers (1975) and Cumming and Richards (1975), with the addition of the oldest terrestrial galena data from Isua (Appel et al., 1978). The latter value considerably strengthens the data base, since it improves the constraints on early terrestrial Pb evolution. Fitting a linear growth trend to the data (equivalent to a constant source μ value), causes this trend to intersect the Canyon Diablo Pb composition at an apparent age of 4.4 Byr (in close agreement to Doe and Stacey, 1974 and Manhes et al., 1979). This is seen most clearly for [206]Pb (Fig. 5.26) but also less strongly for [207]Pb. However, terrestrial accretion at such a late date as 4.4 Byr is inconsistent with the 4.55 Byr age of differentiated meteorites (section 3.2.4). Therefore, as an alternative, an early reservoir with lower μ value is postulated. Hence the model of Stacey and Kramers is supported, but due to noise in the data set, the

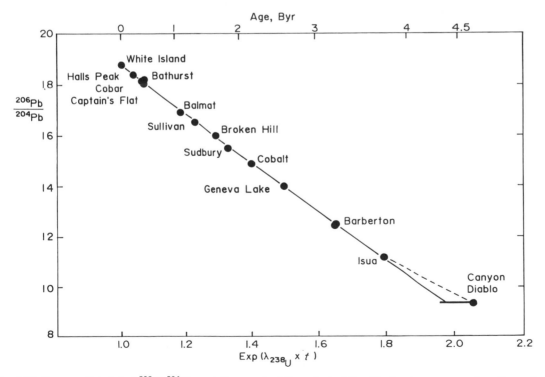

Fig. 5.26. Exponential plot of [206]Pb/[204]Pb evolution against time, to test the fit of galena sources to a linear isotope evolution trend. After Albarede and Juteau (1984).

model of Cumming and Richards also cannot be ruled out. The real situation may be intermediate between these two models, involving rapid change of μ in the Archean, but relatively constant μ thereafter.

The above observations suggest that the galena source evolved for the last 3.8 Byr along a higher-μ growth curve than the geochron, but the reason for this behaviour is not clear. Armstrong (1968) and Russell (1972) argued that these observations could be explained by recycling (bi-directional transport) of Pb between the crust and mantle. This concept was developed further by Doe and Zartman (1979) and presented as a computer program which modelled the Pb isotope evolution of the Earth.

Doe and Zartman defined three reservoirs: upper crust, lower crust and upper mantle (< 500 km depth). Based on evidence that continental accretion began ca. 4 Byr ago, and that frequent orogenies mixed mantle and crustal sources to yield differentiated crustal blocks, they modelled orogenies at 400 Myr intervals, with a decreasing mantle contribution through time. Crustal contributions represented erosion and continental foundering. Orogenies instantaneously extracted U, Th and Pb from the three sources, mixed them, and redistributed them back to

the sources (Fig. 5.27). U fractionation into the upper crust represented granulite-facies metamorphism.

The orogene composition generated by the 'plumbotectonics' model was constrained empirically to fit galena ores, and consequent growth curves generated for the other components are shown in Fig. 5.28. The calculated upper mantle μ value is similar to that for the total crust, but recycling of radiogenic upper crustal Pb into the mantle yields an *apparent* increase in mantle μ value with time. This process is balanced by the development of an unradiogenic lower crustal reservoir, due to preferential retention of Pb relative to U during granulite-facies metamorphism of the lower crust.

While the plumbotectonics model of Doe and Zartman (1979) was a substantial advance in thinking about Pb isotope evolution in the Earth, it represents an over-simplification of galena Pb, in that it provides generalised sources for ore bodies which undoubtedly had different evolutionary histories in the crust. Subsequent work has therefore shifted in attention from the concept of a hypothetical galena source to specific models for mantle and crustal Pb isotope evolution. In order to refine these models (e.g. Zartman and Doe, 1981; Zartman and Haines, 1988)

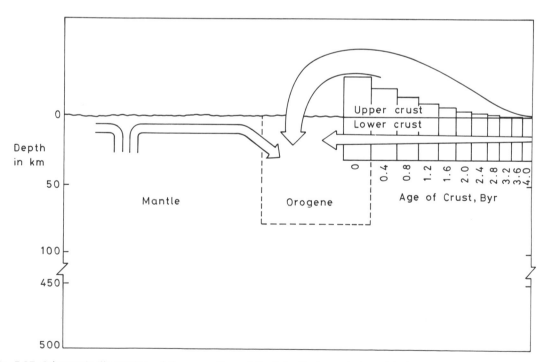

Fig. 5.27. Schematic illustration of the operation of the 'plumbotectonics' model, showing mixing of crustal and mantle reservoirs into the orogene (galena source) reservoir. After Doe and Zartman (1979).

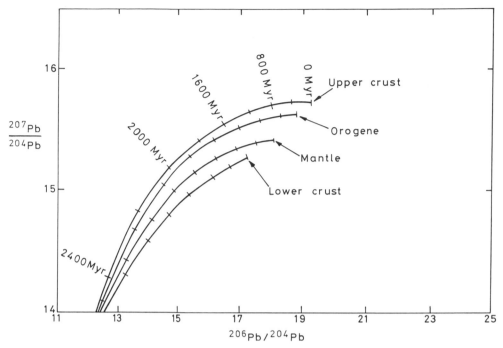

Fig. 5.28. Pb–Pb isochron diagram showing isotopic evolution of the four reservoirs computed by the plumbotectonics model. After Doe and Zartman (1979).

the principal focus of attention moved to Pb analysis of oceanic volcanics (section 6.4). However, it is worth noting here that Doe and Zartman's estimate for the composition of upper crustal Pb was recently confirmed (Asmerom and Jacobsen, 1993) by isotopic analysis of suspended sediment loads from a variety of major rivers.

5.5 Pb–Pb dating and crustal evolution

Because the Pb–Pb whole-rock method depends only on isotopic compositions, it is comparatively resistant to metamorphic re-setting, and can also yield age information for some mixed sources. Good examples of these uses are provided by Pb–Pb dating studies on the gneisses of western Greenland (Taylor *et al.*, 1980; Moorbath and Taylor, 1981).

Nuk gneisses from Fiskanaesset, Nordland and Sukkertopen in western Greenland (filled circles in Fig. 5.29) fall on a reference line with slope age of 2900 Myr. If a single-stage mantle growth curve is calculated to fit this isochron it yields a μ_1 value of 7.5, which is a typical value for the mantle source of juvenile Archean gneiss terranes (Moorbath and Taylor, 1981). This single-stage model mantle

composition is not expected to represent the real Earth, since this was shown above to be an over-simplification; however, the value provides a convenient yard-stick for comparison between different crust-forming events.

The Nuk gneisses approximate a two-stage Pb isotope evolution model, in which the first stage is in the mantle and the second is in each analysed whole-rock system. However, there may in fact be two short extra stages in the middle. The first of these short stages may represent basalt extraction from the mantle before subsequent re-melting to form tonalitic magmas. The second short stage may occur between tonalite emplacement in the crust and its high-grade metamorphism. The two short stages together were termed a crustal accretion–differentiation super-event or CADS by Moorbath and Taylor (1981).

Nuk gneisses from near Godthaab (open circles in Fig. 5.29) fall in a scatter below the 2900 Myr isochron. However, Taylor *et al.* (1980) argued that if these were age-corrected back to 2900 Myr ago, then they would lie on a mixing line between 2900 Myr-old mantle (M), and the local crust, represented by average 3700 Myr-old Amitsoq gneiss (A). Due to their low U contents, the Amitsoq gneisses barely changed in Pb isotope ratio between 3700 and 2900

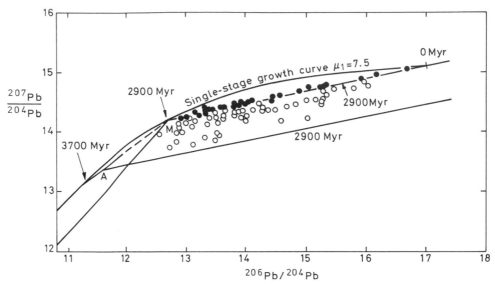

Fig. 5.29. Pb–Pb isochron diagram for basement gneisses from western Greenland. (•) = Nuk gneisses from Fiskanaesset, Nordland and Sukkertopen; (○) = contaminated Nuk gneisses from near Godthaab. M = mantle at 2900 Myr; A = range of Amitsoq gneiss compositions at 2900 Myr. After Taylor *et al.* (1980).

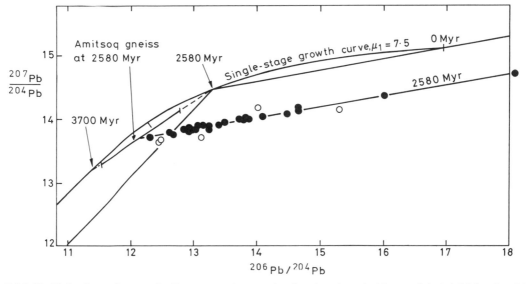

Fig. 5.30. Pb–Pb isochron diagram for Qorqut granite samples showing the coincidence of their initial ratio with the average Amitsoq gneiss composition at 2580 Myr. (open symbols omitted from regression). After Moorbath and Taylor (1981).

Myr ago. Given these end-members, the distance down the mixing line from 'M' to 'A' indicates the fraction of crustal Pb incorporated into the magma. The variable extent of crustal Pb contamination suffered by Late Archean Nuk magmas is consistent with the known extent of Early Archean crust. Thus, while Godthaab is known to lie on Amitsoq gneiss

basement, it is not exposed near Fiskanaesset, Nordland or Sukkertopen.

The Qorqut granite is also exposed within the Amitsoq gneiss terrane near Godthaab (Moorbath *et al.*, 1981). Whole-rock samples define a linear array whose slope corresponds to an age of 2580 Myr (Fig. 5.30). If an attempt is made to fit a single-stage mantle growth curve to this data, an impossibly low μ_1 value of 6.23 is obtained, showing that the Qorqut granite cannot be a mantle-derived melt. In fact, the initial Pb isotope ratio of the Qorqut granite coincides closely with the average composition of analysed Amitsoq crust at 2580 Myr, indicating that the Qorqut is probably a partial melt of Amitsoq gneiss. It therefore approximates to three-stage Pb isotope evolution:– stage 1: mantle; stage 2: Amitsoq crust; stage 3: Qorqut granite. The initial $^{87}Sr/^{86}Sr$ ratio of the Qorqut granite (0.7083 ± 4) supports this model (section 7.3.4).

It was argued above that comparatively short periods (less than 200 Myr) between the crustal accretion and metamorphic differentiation of a gneiss complex do not necessarily upset the dating of the differentiation event using Pb–Pb systematics. However, if the period between the two events is substantial then spurious ages may be obtained. A good example is provided by the Vikan gneiss complex from Lofoten Vesteralen in NW Norway. If we assume that these rocks behaved as closed systems after their generation from an isotopically homogeneous (mantle?) source, we determine a slope age of 3410 ± 70 Myr (Taylor, 1975). However, model Nd dating yields ages of ca. 2.4–2.7 Byr (Jacobsen and Wasserburg, 1978).

Subsequent examination of present-day U/Pb ratios in the gneisses (Griffin *et al.*, 1978) revealed that they were uniformly far too low to 'support' the observed range of Pb isotope compositions. It is now argued that the Pb isotope ratios were generated in a 2680 Myr-old igneous protolith which suffered high-grade metamorphism at ca. 1760 Myr. Pb isotope compositions in the protolith are shown as a palaeo-isochron at the time of metamorphism in Fig. 5.31. If the rocks were depleted in U to a nearly uniform level at 1760 Myr, then subsequent U decay would yield a 'transposed palaeo-isochron' (Griffin *et al.*, 1978; Moorbath and Taylor, 1981) which is almost parallel to the original palaeo-isochron.

The slope of the transposed palaeo-isochron approximates Pb evolution from time T (protolith age) to t (metamorphic age). This is described by an equation which is analogous to [5.16] for galena evolution:

$$\frac{\left(\frac{^{207}Pb}{^{204}Pb}\right)_P - \left(\frac{^{207}Pb}{^{204}Pb}\right)_I}{\left(\frac{^{206}Pb}{^{204}Pb}\right)_P - \left(\frac{^{206}Pb}{^{204}Pb}\right)_I} = \frac{1}{137.88} \frac{(e^{\lambda_{235}T} - e^{\lambda_{235}t})}{(e^{\lambda_{238}T} - e^{\lambda_{238}t})} \quad [5.17]$$

In contrast, the simple Pb/Pb isochron equation [5.12], describing evolution from t to the present,

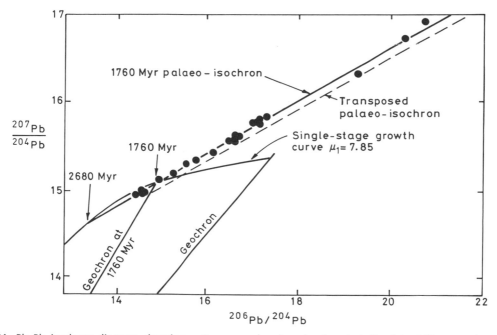

Fig. 5.31. Pb–Pb isochron diagram showing a 'transposed palaeo-isochron' defined by Vikan gneisses of NW Norway. These rocks were formed from 2680 Myr-old precursors which were subjected to a granulite-facies uranium-depletion event at 1760 Myr. After Moorbath and Taylor (1981).

yields too large an age because it is based on the lower $^{235}U/^{238}U$ prevailing at the present day compared to 1760 Myr ago.

Transposed palaeo-isochrons can be detected by checking concordancy of Pb with Sr or Nd dates and by checking that observed Pb isotope compositions are adequately supported by the U/Pb ratio in the samples. Another example was found in upper amphibolite-facies gneisses of the Outer Hebrides, NW Scotland, by Whitehouse (1990). By substituting the 2660 Myr (Badcallian) Pb homogenisation event as T in equation [5.17], and assuming uniform U/Pb ratios after the second event, he was able to estimate the timing (t) of this second event. The calculated age of 1880 ± 270 Myr was consistent with the time-span of the Laxfordian metamorphic event.

References

Ahrens, L. H. (1955). Implications of the Rhodesia age pattern. *Geochim. Cosmochim. Acta* **8**, 1–15.

Albarede, F. and Juteau, M. (1984). Unscrambling the lead model ages. *Geochim. Cosmochim. Acta* **48**, 207–12.

Aleinikoff, J. N., Winegarden, D. L. and Walter, M. (1990). U–Pb ages of zircon rims: a new analytical method using the air-abrasion technique. *Chem. Geol. (Isot. Geosci. Section)* **80**, 351–63.

Alpher, R. A. and Herman, R. C. (1951). The primeval lead isotopic abundances and the age of the Earth's crust. *Phys. Rev.* **84**, 1111–14.

Appel, P. W. U., Moorbath, S. and Taylor, P. N. (1978). Least radiogenic terrestrial lead from Isua, west Greenland. *Nature* **272**, 524-6.

Armstrong, R. L. (1968). A model for Sr and Pb isotope evolution in a dynamic Earth. *Rev. Geophys.* **6**, 175–99.

Asmerom, Y. and Jacobsen, S.B. (1993). The Pb isotopic evolution of the Earth: inferences from river water suspended loads. *Earth Planet. Sci. Lett.* **115**, 245–56.

Carl, C. and Dill, H. (1985). Age of secondary uranium mineralization in the basement rocks of the north eastern Bavaria F. R. G. *Chem. Geol. (Isot. Geosci. Section)* **52**, 295–316.

Carl, C., Wendt, I. and Wendt, J. I. (1989). U/Pb whole-rock and mineral dating of the Falkenburg granite in northeast Bavaria. *Earth Planet. Sci. Lett.* **94**, 236–44.

Compston, W., Williams, I. S. and Meyer, C. (1984). U–Pb geochronology of zircons from lunar breccia 73217 using a sensitive high mass-resolution ion microprobe. *Proc. 14th Lunar and Planet. Sci. Conf., J. Geophys. Res.* **89** *Supp.*, B525–34.

Corfu, F. (1988). Differential response of U–Pb systems in coexisting accessory minerals, Winnipeg River Sub-province, Canadian Shield: Implications for Archean crustal growth and stabilization. *Contrib. Mineral. Petrol.* **98**, 312–25.

Cumming, G. L. and Richards, J. R. (1975). Ore lead isotope ratios in a continuously changing earth. *Earth Planet. Sci. Lett.* **28**, 155–71.

Davis, D. W. (1982). Optimum linear regression and error estimation applied to U-Pb data. *Can. J. Earth Sci.* **19**, 2141–9.

Doe, B. R. and Stacey, J. S. (1974). The application of lead isotopes to the problems of ore genesis and ore prospect evaluation: a review. *Econ. Geol.* **69**, 757–76.

Doe, B. R. and Zartman, R. E. (1979). Plumbotectonics. In: Barnes, H. L. (Ed.) *Geochemistry of Hydrothermal Ore Deposits.* Wiley, pp. 22–70.

Feng, R., Machado, N. and Ludden, J. (1993). Lead geochronology of zircon by LaserProbe Inductively Coupled Plasma Mass Spectrometry (LP–ICPMS). *Geochim. Cosmochim. Acta* **57**, 3479–86.

Froude, D. O., Ireland, T. R., Kinny, I. S., Williams, I. S. and Compston, W. (1983). Ion microprobe identification of 4,100–4,200 Myr-old terrestrial zircons. *Nature* **304**, 616–18.

Gentry, R. V., Sworski, T. J., McKown, H. S., Smith, D. H., Eby, R. E. and Christie, W. H. (1982). Differential lead retention in zircons: implications for nuclear waste containment. *Science* **216**, 296–7.

Goldrich, S. S. and Mudrey, M. G. (1972). Dilatancy model for discordant U-Pb zircon ages. In: Tugarinov (Ed.), *Contributions to Recent Geochemistry and Analytical Chemistry.* Moscow Nauka Publ. Office, 415–18.

Griffin, W. L., Taylor, P. N., Hakkinea, J. W., Heier, K. S., Idea, I. K., Krogh, E. J., Malm, O., Olsen, K. I., Ormaasen, D. E. and Treten, E. (1978). Archaean and Proterozoic crustal evolution in Lofoten-Vesteralen, Norway. *J. Geol. Soc. Lond.* **135**, 629–47.

Halliday, A. N. (1984). Coupled Sm–Nd and U–Pb systematics in Late Caledonian granites and the basement under northern Britain. *Nature* **307**, 229–33.

Hinton, R. W. and Long, J. V. P. (1979). High-resolution ion-microprobe measurement of lead isotopes: variations within single zircons from Lac Seul, Northwest Ontario. *Earth Planet. Sci. Lett.* **45**, 309–25.

Holmes, A. (1946). An estimate of the age of the Earth. *Nature* **157**, 680–4.

Holmes, A. (1954). The oldest dated minerals of the Rhodesian Shield. *Nature* **173**, 612–17.

Houtermans, F. G. (1946). Die Isotopen-Haufigkeiten im naturlichen Blci und dasAalter des Urans. *Naturwissenschaften* **33**, 185–7.

Houtermans, F. G. (1947). Das Alter des Urans. *Z. Naturforsch.* **29**, 322–8.

Jacobsen, S. B. and Wasserburg, G. J. (1978). Interpretation of Nd, Sr and Pb isotope data from Archaean migmatites in Lofoten-Vesteralen, Norway. *Earth Planet. Sci. Lett.* **41**, 245–53.

Jaffey, A. H., Flynn, K. F., Glendenin, L. E., Bentley, W. C. and Essling, A. M. (1971). Precision measurement of the half-lives and specific activities of U235 and U238. *Phys. Rev. C* **4**, 1889–1907.

Kober, B. (1986). Whole-grain evaporation for $^{207}Pb/^{206}Pb$-age investigations on single zircons using a double-filament ion source. *Contrib. Mineral. Petrol.* **93**, 482–90.

Kober, B. (1987). Single-zircon evaporation combined with Pb+ emitter bedding for $^{207}Pb/^{206}Pb$-age investigations using thermal ion mass spectrometry, and implications to zirconology. *Contrib. Mineral. Petrol.* **96**, 63–71.

Kober, B., Pidgeon, R. T. and Lippolt, H. J. (1989). Single-zircon dating by stepwise Pb-evaporation constrains the Archean history of detrital zircons from the Jack Hills, Western Australia. *Earth Planet. Sci. Lett.* **91**, 286–96.

Krogh, T. E. (1982a). Improved accuracy of U–Pb zircon dating by selection of more concordant fractions using a high gradient magnetic separation technique. *Geochim. Cosmochim. Acta* **46**, 631–5.

Krogh, T. E. (1982b). Improved accuracy of U–Pb zircon ages by the creation of more concordant systems using the air abrasion technique. *Geochim. Cosmochim. Acta* **46**, 637–49.

Krogh, T. E., Corfu, F., Davis, D. W., Dunning, G. R., Heaman, L. M., Kamo, S. L. and Machado, N. (1987). Precise U–Pb isotopic ages of diabase dykes and mafic to ultramafic rocks using trace amounts of baddeleyite and zircon. In: Halls, H. C. and Fahrig, W. F. (Eds.) *Mafic Dyke Swarms. Geol. Assoc. Canada Spec. Pap.* **34**, 147–52.

Ludwig, K. R. (1977). Effect of initial radioactive daughter disequilibrium on U–Pb isotope apparent ages of young minerals. *J. Res. U. S. Geol. Surv.* **5**, 663–7.

Manhes, G., Allegre, C. J., Dupre, B. and Hamelin, B. (1979). Lead–lead systematics, the 'age of the Earth' and the chemical evolution of our planet in a new representation space. *Earth Planet. Sci. Lett.* **44**, 91–104.

Moorbath, S., Taylor, P. N. and Goodwin, R. (1981). Origin of granite magma by crustal remobilisation: Rb–Sr and Pb/Pb geochronology and isotope geochemistry of the late Archaean Qorqut Granite complex of southern West Greenland. *Geochim. Cosmochim. Acta* **45**, 1051–60.

Moorbath, S. and Taylor, P. N. (1981). Isotopic evidence for continental growth in the Precambrian. In: Kroner (Ed.), *Precambrian Plate Tectonics.* Elsevier, pp.491–525.

Nier, A. O., Thompson, R. W. and Murphy, B. F. (1941). The isotopic constitution of lead and the measurement of geological time III. *Phys. Rev.* **60**, 112–17.

Oversby, V. M. (1974). A new look at the lead isotope growth curve. *Nature* **248**, 132–3.

Pankhurst, R. J. and Pidgeon, R. T. (1976). Inherited isotope systems and the source region pre-history of early Caledonian granites in the Dalradian series of Scotland. *Earth Planet. Sci. Lett.* **31**, 55–68.

Patterson, C. C. (1956). Age of meteorites and the Earth. *Geochim. Cosmochim. Acta* **10**, 230–7.

Pidgeon, R. T. and Aftalion, M. (1978). Cogenetic and inherited zircon U–Pb systems in granites: Palaeozoic granites of Scotland and England. In: Bowes, D. R. and Leake, B. E. (Eds.), *Crustal Evolution in Northwestern Britain and Adjacent Regions. Geol. Soc. Spec. Issue* **10**, 183–220.

Potts, P. J. (1987). *Handbook of Silicate Rock Analysis.* Blackie, 622 p.

Roddick, J. C. and Chapman, H. J. (1991). $^{207}Pb^*/^{206}Pb^*$ dating by zircon evaporation: mechanism of Pb loss. *EOS* **72**, 531 (abstract).

Rogers, G., Dempster, T. J., Bluck, B. J. and Tanner, P. W. G. (1989). A high precision U–Pb age for the Ben Vuirich granite: implications for the evolution of the Scottish Dalradian Supergroup. *J. Geol. Soc. Lond.* **146**, 789–98.

Rosholt, J. N. and Bartel, A. J. (1969). Uranium, thorium and lead systematics in Granite Mountains, Wyoming. *Earth Planet. Sci. Lett.* **7**, 14–17.

Russell, R. D. (1956). Lead isotopes as a key to the radioactivity of the Earth's mantle. *Ann. N. Y. Acad. Sci.* **62**, 435–48.

Russell, R. D. (1972). Evolutionary model for lead isotopes in conformable ores and in ocean volcanics. *Rev. Geophys. Space Phys.* **10**, 529–49.

Russell, R. D. and Ahrens, L. H. (1957). Additional regularities among discordant lead-uranium ages. *Geochim. Cosmochim. Acta* **11**, 213–18.

Russell, R. D. and Farquhar, R. M. (1960). *Lead Isotopes in Geology.* Interscience Pub., 243 p.

Scharer, U. (1984). The effect of initial ^{230}Th disequilibrium on young U–Pb ages: the Makalu case, Himalaya. *Earth Planet. Sci. Lett.* **67**, 191–204.

Scharer, U. and Allegre, C. J. (1982). Uranium–lead system in fragments of a single zircon grain. *Nature* **295**, 585–7.

Scharer, U., Xu, R. H. and Allegre, C. J. (1984). U–Pb geochronology of Gangdese (Transhimalaya) plutonism in the Zhasa-Xigaze region, Tibet. *Earth Planet. Sci. Lett.* **69**, 311–20.

Silver, L. T. and Deutsch, S. (1963). Uranium–lead isotopic variations in zircons: a case study. *J. Geol.* **71**, 721-58.

Smith, P. E. and Farquhar, R. M. (1989). Direct dating of Phanerozoic sediments by the $^{238}U-^{206}Pb$ method. *Nature* **341**, 518–21.

Smith, P. E., Farquhar, R. M. and Hancock, R. G. (1991). Direct radiometric age determination of carbonate diagenesis using U–Pb in secondary calcite. *Earth Planet. Sci. Lett.* **105**, 474–91.

Stacey, J. S. and Kramers, J. D. (1975). Approximation of terrestrial lead isotope evolution by a two-stage model. *Earth Planet. Sci. Lett.* **26**, 207–21.

Stanton, R. L. and Russell, R. D. (1959). Anomalous leads and the emplacement of lead sulphide ores. *Econ. Geol.* **54**, 588–607.

Tatsumoto, M., Knight, R. J. and Allegre, C. J. (1973). Time differences in the formation of meteorites as determined from the ratio of lead-207 to lead-206. *Science* **180**, 1279–83.

Taylor, P. N. (1975). An early Precambrian age for migmatitic gneisses from Vikan i Bo, Vesteralen, North Norway. *Earth Planet. Sci. Lett.* **27**, 35–42.

Taylor, P. N., Moorbath, S., Goodwin, R. and Petrykowski, A. C. (1980). Crustal contamination as an indicator of the extent of early Archaean continental crust: Pb isotopic evidence from the late Archaean gneisses of West Greenland. *Geochim. Cosmochim. Acta* **44**, 1437–53.

Tera, F. and Wasserburg, G. J. (1973). A response to a comment on U–Pb systematics in lunar basalts. *Earth Planet. Sci. Lett.* **19**, 213–17.

Tera, F. and Wasserburg, G. J. (1974). U–Th–Pb systematics

on lunar rocks and inferences about lunar evolution and the age of the Moon. *Proc. 5th Lunar Sci. Conf.* (Supp. 5), *Geochim. Cosmochim. Acta* **2**, 1571–99.

Tilton, G. R. (1960). Volume diffusion as a mechanism for discordant lead ages. *J. Geophys. Res.* **65**, 2933–45.

van Breemen, O., Davidson, A., Loveridge, W. D. and Sullivan, R. W., (1986). U–Pb zircon geochronology of Grenville tectonites, granulites and igneous precursors, Parry Sound, Ontario. In: Moore, J. M., Davidson, A. and Baer, A. J. (Eds.), *The Grenville Province. Geol. Assoc. Canada Spec. Pap.* **31**, 191–207.

Watson, E. B. and Harrison, T. M. (1983). Zircon saturation revisited: temperature and composition effects in a variety of crustal magma types. *Earth Planet. Sci. Lett.* **64**, 295–304.

Wendt, I. (1984). A three-dimensional U–Pb discordia plane to evaluate samples with common lead of unknown isotopic composition. *Isot. Geosci.* **2**, 1–12.

Wendt, I., Carl, C., Habfast, K., Tuttas, D. and Wendt, J. I. (1991). Complete Pb/U analysis of unspiked samples by measuring Pb isotopes only. *Earth Planet. Sci. Lett.* **107**, 618–24.

Wendt, J. I., Wendt, I. and Tuttas, D. (1993). Determination of U–Pb ages of zircons by direct measurement of the $^{210}Pb/^{206}Pb$ ratio. *Chem. Geol. (Isot. Geosci. Section)* **106**, 467–74.

Wetherill, G. W. (1956a). An interpretation of the Rhodesia and Witwatersrand age patterns. *Geochim. Cosmochim. Acta* **9**, 290–2.

Wetherill, G. W. (1956b). Discordant uranium-lead ages. *Trans. Amer. Geophys. Union* **37**, 320–7.

Whitehouse, M. (1990). Isotopic evolution of the southern Outer Hebridean Lewisian gneiss complex: constraints on Late Archean source regions and the generation of transposed Pb–Pb palaeoisochrons. *Chem. Geol. (Isot. Geosci. Section)* **86**, 1–20.

Williams, I. S., Compston, W., Black, L. P., Ireland, T. R. and Foster, J. J. (1984). Unsupported radiogenic Pb in zircon: a cause of anomalously high Pb–Pb, U–Pb and Th–Pb ages. *Contrib. Mineral. Petrol.* **88**, 322–7.

Zartman, R. E. and Doe, B. R. (1981). Plumbotectonics – the model. *Tectonophys.* **75**, 135–62.

Zartman, R. E. and Haines, S. M. (1988). The plumbotectonic model for Pb isotopic systematics among major terrestrial reservoirs – a case for bi-directional transport. *Geochim. Cosmochim. Acta* **52**, 1327–39.

6 Isotope geochemistry of oceanic volcanics

6.1 Mantle heterogeneity

Some of the most important questions in geology concern the processes which operate in the Earth's mantle. Mantle convection is clearly the driving force behind plate tectonics (Turcotte and Oxburgh, 1967), but the details of its operation are still unclear (e.g. Olson *et al.*, 1990). The depth of mantle convection cells, the fate of subducted lithosphere, and the source of upwelling mantle plumes are all unresolved problems. Geochemical sampling may help to answer these questions by revealing the progress of mantle differentiation into different reservoirs, and the extent to which these reservoirs are re-mixed by convective stirring.

The inaccessibility of the mantle presents a severe problem for geochemical sampling. However, mantle-derived basic magmas provide a prime source of evidence about the chemical structure of the mantle. Isotopic tracers represent a particularly powerful tool for such studies, because unlike elemental concentrations, isotope ratios are insignificantly affected by crystal fractionation. However, isotope ratios are susceptible to contamination in the continental lithosphere. Therefore the simplest approach to studying mantle chemistry through basic magmas is to analyse oceanic volcanics, which are expected to have suffered minimal contamination in the thin oceanic lithosphere.

Isotopic analysis of ocean island basalts (OIB) was first used to demonstrate the existence of mantle heterogeneity (Faure and Hurley, 1963; Gast *et al.*, 1964). Subsequently, variations were found between the isotopic compositions of mid ocean ridge basalts (MORB) and OIB (Tatsumoto, 1966). Isotopic analysis of oceanic basalts can be used both to probe the structure of the mantle and to model its evolution over time. The approach taken here will be to examine the constraints on mantle structure from single isotopic systems (mainly Sr and Pb), then to examine the constraints on mantle evolution from multiple isotopic systems (Sr–Nd), (U–Th–Pb), (Sr–Nd–Pb). Additional evidence will be examined in later chapters.

6.2 Isotopic tracing of mantle structure

6.2.1 Contamination and alteration

Before oceanic volcanics can be used to deduce mantle composition, we must examine and quantify the amounts of alteration and contamination which could occur during magma transport and eruption on ocean islands or the ocean floor.

In their early work on Ascension and Gough islands, Gast *et al.* (1964) considered the possibility of contamination of the analysed lavas by a crustal micro-plate. They tested this possibility by analysing a range of lavas at variable degrees of magmatic differentiation (Fig. 6.1). The lack of any correlation in all but the most evolved rocks was argued to rule out crustal contamination. High Sr isotope ratios in the highly evolved rocks were attributed to radio-active growth after eruption, since these rocks have very high Rb/Sr ratios. No age corrections could be applied to these lavas since their ages were unknown. Similar problems have been encountered in more recent studies of Ascension lavas (Harris *et al.*, 1983). However, most oceanic basalts require no age correction since they have very low Rb/Sr ratios.

Some workers, most notably O'Hara, suggested that the isotopic variations in MORB and OIB could be explained by fractionation or contamination processes affecting magmas during their ascent through oceanic crust. In his early papers on the subject, O'Hara (1973, 1975) suggested that $^{87}Sr/^{86}Sr$ variations could be generated by physical fractionation of the isotopes during magmatic differentiation. This is a misconception, since $^{87}Sr/^{86}Sr$ ratios are always fractionation-corrected to the standard $^{88}Sr/^{86}Sr$ ratio of 8.37521 (section 2.2.2) to eliminate both natural and analytical mass-dependent fractionation. Subsequently, O'Hara and Mathews (1981)

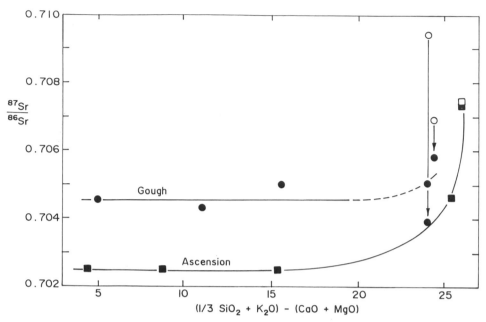

Fig. 6.1. Sr isotope ratios in lavas from Gough and Ascension islands plotted against an index of magmatic differentiation. Radiogenic Sr in highly evolved lavas (open symbols) is attributed to radioactive growth since eruption. Arrows show estimated age corrections. After Gast *et al.* (1964).

argued that large ion lithophile (LIL) elements (including strontium) could be perturbed by contamination with altered oceanic crust in a 'periodically tapped, periodically re-filled, long-lived magma chamber'. This model is now ruled out by thorium isotopic evidence, which severely limits the time between generation and eruption of ocean floor basalt, and hence the ability of an open-system magma chamber model to overprint the source isotopic signature in the lavas (section 13.3.2).

Sub-solidus alteration of analysed samples could result from hydrothermal interaction with seawater, in the case of submarine basalts, or sub-aerial weathering, in the case of ocean island basalts. For example, Dasch *et al.* (1973) found a positive correlation between $^{87}Sr/^{86}Sr$ and water content in dredged oceanic basalts of various ages (Fig. 6.2). Samples with over 1% H_2O had almost invariably suffered Sr contamination from seawater, but those with less than 1% alteration were generally uncontaminated.

Sub-solidus alteration can be reliably avoided in submarine samples by analysing 100% fresh MORB glasses (Cohen *et al.*, 1980). Where crystalline rock must be analysed (e.g. White *et al.*, 1976), alteration can be avoided by analysing fresh material dredged from the median valley of the ocean ridges, where

very young, unmetamorphosed basalts outcrop. Alternatively, leaching of crystalline samples before analysis may remove contaminated alteration minerals, also yielding results which are consistent with glasses (Dupre and Allegre, 1980). Unaltered ocean island basalts are easily obtained by sampling only fresh lavas.

6.2.2 Disequilibrium melting

Various workers in the past (e.g. Harris *et al.*, 1972; O'Nions and Pankhurst, 1973; Flower *et al.*, 1975) have suggested that if mantle temperatures were not high enough to ensure diffusional homogenisation of Sr isotope ratios between different mantle minerals, then grains with higher Rb/Sr ratios could develop more radiogenic $^{87}Sr/^{86}Sr$ compositions over geological time. An example of such a mineral is the magnesian mica phlogopite. 'Disequilibrium' melting of such phases could bias the isotopic composition of a melt towards higher $^{87}Sr/^{86}Sr$ compositions. Small degree partial melts would tend to be enriched in Rb/Sr and $^{87}Sr/^{86}Sr$ relative to large degree partial melts due to the tendency of high Rb/Sr phases such as phlogopite to enter the melt first.

Harris *et al.* (1972) argued in favour of disequilibrium melting during basalt genesis, based on

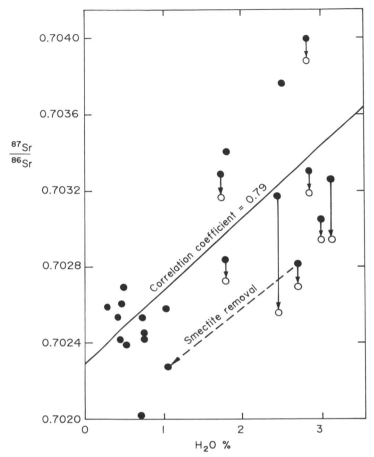

Fig. 6.2. Plot of strontium isotope ratio against water content in ocean-floor basalts. Vertical arrows show the effect of leaching before analysis. Dashed arrow shows the effect of smectite removal from an altered sample. After Dasch *et al.* (1973).

evidence of isotopic disequilibrium in mantle xenoliths carried to the surface in basic magmas. Isotopic disequilibrium in ultramafic xenoliths is very widespread (section 7.1), but such cases represent samples of the solid lithosphere. It is questionable whether these observations can be extrapolated to the higher-temperature environment of basaltic magma genesis in the convecting asthenosphere.

Hofmann and Hart (1978) examined data for the diffusion of Sr in mantle silicates in order to determine the rates at which isotopic disequilibrium could be eradicated at different temperatures. In Fig. 6.3, values of diffusivity (*D*) are used to calculate times for effective equilibration of a species between a 1 cm diameter sphere and an infinite reservoir such as a slowly moving melt. These times are roughly those taken for diffusion over a 'characteristic transport distance' of 0.25 cm, using the equation $X = (Dt)^{1/2}$.

Using the lower of the measured diffusivities, it would take millions of years to eradicate Sr isotope heterogeneity between large grains of phlogopite and clinopyroxene in solid lithospheric mantle at, say, 600 °C. Even in a solid mantle at 1000 °C, equilibration could take millions of years if the phlogopite and clinopyroxene grains were separated by intervening olivine or orthopyroxene, which effectively contain no Sr but lengthen the diffusion pathways between phlogopite and cpx. However, as soon as a melt is present, the surface of each crystal is in diffusional contact with nearby (ca. 2 cm distant) grains over a period of a few years. Therefore isotopic disequilibrium between phlogopite and cpx could be eradicated in a few thousand years at temperatures above the basalt solidus (ca. 1000–1200 °C). Nevertheless, diffusion over long distances, even in a partially molten mantle, is still slow.

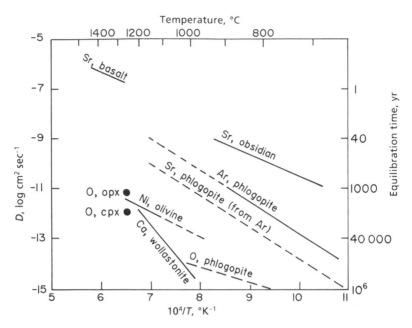

Fig. 6.3. Plot of diffusivity against 1/temperature, showing experimental results for the diffusion of Sr, Ar, Ni, Ca and oxygen in different types of material. Times for effective equilibration are based on 1 cm grain size. Modified after Hofmann and Hart (1978).

Hofmann and Hart (1978) concluded that the evidence favoured 'local equilibrium in a partially molten mantle, local disequilibrium in a completely crystalline mantle, and regional disequilibrium in any mantle that convects only slowly in large convection cells'.

6.2.3 Mantle plumes

Following the acceptance of the plate tectonic model, it was realised that the tectonic setting of basic volcanism was a crucial factor in determining the nature of the mantle source being tapped, and consequent magma chemistry.

Morgan (1971) proposed that the different chemistry of MORB and OIB could be explained if the former were derived directly from the asthenospheric upper mantle, while the latter were generated by upwelling plumes from the lower mantle. Evidence in support of this model was provided by elemental analysis of Iceland basalts (Schilling, 1973). These data suggested a region of mixing between plume (OIB source) and depleted upper mantle (MORB source) on the Reykjanes Ridge south of Iceland. Sr isotope data for the Reykjanes Ridge (Hart et al., 1973) were slightly more equivocal, since they showed a step-like feature in the data (Fig. 6.4). Hart et al. interpreted the data as a mixing phenomenon, but

some workers (e.g. Flower et al., 1975) interpreted this step as resulting from disequilibrium melting of a mantle with variable phlogopite contents.

White et al. (1976, 1979) extended the Sr isotope data set by analysing dredged samples from the axial valley of the Mid Atlantic Ridge (MAR) between 29 and 63 °N, and by sampling across the Azores platform. Isotopic data are plotted against latitude down the MAR in Fig. 6.4, and longitude across the Azores Plateau in Fig. 6.5. There are large variations in the strontium isotope ratio of MORB samples along the Mid Atlantic Ridge, but where MORB and OIB are erupted alongside each other (the Azores Plateau), they have very similar isotope ratios (with the exception of Sao Miguel). Because tholeiitic (MORB) and alkaline (OIB) magmas are attributed to different degrees of mantle melting, the overlap of their compositions across the Azores Plateau is evidence against sampling of isotopic heterogeneities on a mineralogical scale.

The plume–asthenosphere mixing model for the Reykjanes Ridge was strongly confirmed by Pb isotope analysis (Sun et al., 1975), which showed a smooth compositional variation down the ridge (Fig. 6.6a). In contrast, Pb isotope analysis of basalts from the Kolbeinsey Ridge, north of Iceland, did not reveal any contamination of this ridge segment with plume

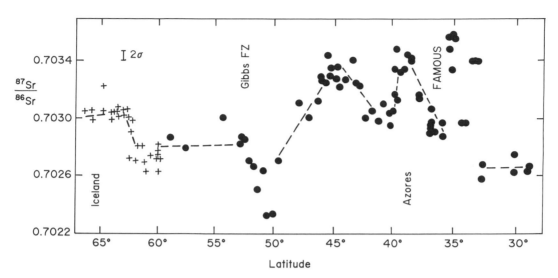

Fig. 6.4. Plot of strontium isotope ratio against latitude for basalts from the Mid Atlantic Ridge. (+) = Iceland–Reykjanes Ridge. Age correction of Sr isotope ratios is unnecessary, due to the low Rb/Sr ratios and young ages of analysed material. After White *et al.* (1976).

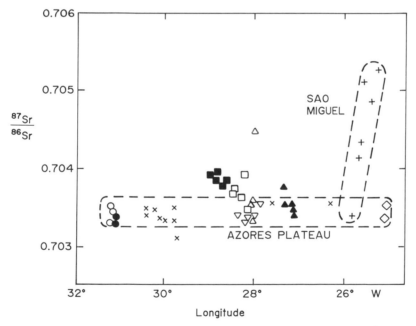

Fig. 6.5. Plot of strontium isotope ratio against longitude for basalt samples from the Azores Plateau. (×) = dredged basalts. Other symbols represent individual islands. After White *et al.* (1979).

material (Mertz *et al.*, 1991). These differences in mixing style north and south of Iceland can be attributed to asymmetrical distortion of the plume by a regional southerly flow of asthenospheric mantle (Fig. 6.6b). Non-uniform Pb isotope contamination of ridge segments has also been observed in the South Atlantic, caused by the off-axis St Helena plume (Hanan *et al.*, 1986).

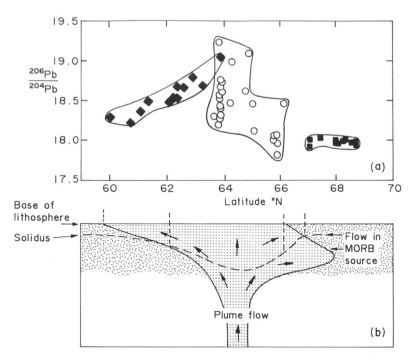

Fig. 6.6. Interpretation of isotopic data for the Iceland plume: a) compilation of Pb isotope data for Iceland (○), the Reykjanes ridge (◆) and the Kolbeinsey ridge (■); b) model cross-section of the upper mantle. After Mertz *et al.* (1991).

6.2.4 Plum pudding mantle

Many workers have questioned whether there might be an intermediate scale of mantle heterogeneity between rare large plumes and mineralogical disequilibrium. Even in their early elemental studies of the Faeroes 'plume', Schilling and Noe-Nygaard (1974) recognised that this structure need not be a continuous column, but could have the form of a train of 'blobs'. Later workers (e.g. Allegre *et al.*, 1980) developed the idea that trains of blobs need not simply pass in streams from a (hypothetical) lower mantle reservoir through the asthenosphere, but could be part of the convecting asthenosphere itself. Allegre identified three alternative models for 'blob heterogeneity' of the asthenosphere (Fig. 6.7).

In an analysis of basaltic glasses from the ocean basins, Cohen and O'Nions (1982) showed that the (comparatively) very large ranges of Pb isotope variation seen in Atlantic MORB were not equalled on the East Pacific Rise. Rather than attributing these differences to a smaller degree of mantle heterogeneity beneath the Pacific, Cohen and O'Nions argued that approximately equal degrees of heterogeneity in

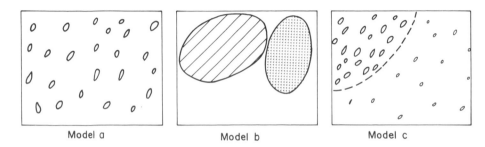

Fig. 6.7. Hypothetical scales of mantle heterogeneity. a) Small scale; b) large scale; c) large and small scale. After Allegre *et al.* (1980).

the Atlantic and Pacific upper mantle were homogenised in the large magma chamber associated with
its fast-spreading ridge.

Batiza (1984) confirmed the inverse effect of ridge
spreading rate on isotopic heterogeneity by plotting
total ranges (Δ) of $^{87}Sr/^{86}Sr$ for different mid ocean
ridges against their spreading rate (Fig. 6.8). He
attributed the small range of compositions on the fast-
spreading ridges to homogenisation, during the
melting process, of a mantle 'ubiquitously heterogeneous on a small scale'. Low isotopic variation on
some slow-spreading ridges was attributed to either
their short length or limited sampling. Batiza adopted
the more gastronomically elegant term of 'plum
pudding' mantle to describe this blob-bearing asthenosphere. Allegre *et al.* (1984) also found an inverse
correlation between ridge spreading rate and isotopic
variation, but argued that homogenisation must be
primarily by (solid state) mantle convection rather
than magma mixing.

Zindler *et al.* (1984) provided support for the plum
pudding model by demonstrating that seamounts near
the East Pacific Rise showed much more variation

than the adjacent ridge. This shows that heterogeneities are widespread in the Pacific mantle, but are
eradicated by the intense diapirism under the ridge.
These observations led Zindler *et al.* to classify mantle
plumes into two types. 'Thermal plumes' were
attributed to preferential fusion of low-melting (?)
blobs (or plums) in a zone of mantle diapirism
promoted by an underlying heat source (Fig. 6.9). In
this case, only heat (Q) is contributed from the
underlying lower mantle or core. Iceland was cited as
a possible example. In contrast, 'chemical plumes'
were recognised as having an exotic composition
distinct from the asthenosphere, due to a net transport
of material from the lower mantle. Hawaii was cited
as the prime example of this type.

Sun (1985) coined the term 'plume pudding' mantle
(*sic*) to express the idea that plume and plum pudding
models should not be thought of as mutually
exclusive, but rather as a continuum of phenomena.
The question of the origin of plums or plumes cannot
effectively be answered by the application of single
isotopic systems, and will be discussed on the basis of
co-variations in multiple isotopic systems.

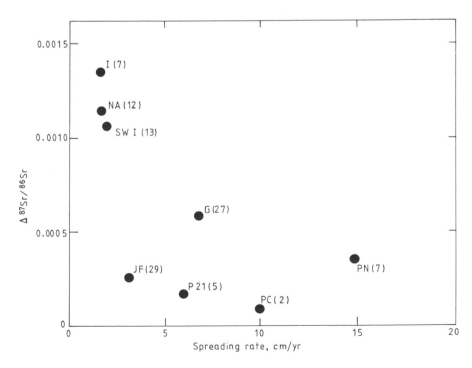

Fig. 6.8. Total ranges of Sr isotope ratio (Δ) for MORB glasses or leached whole-rocks from a given ridge, plotted
against spreading rate on that ridge. Figures in brackets indicate number of analyses. Ridges: I = Indian; NA = North
Atlantic; SWI = SW Indian; G = Galapagos; JF = Juan de Fuca; P21 = East Pacific Rise (EPR) 21 °N; PC = EPR Cocos; PN =
EPR Nasca. After Batiza (1984).

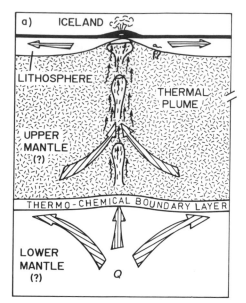
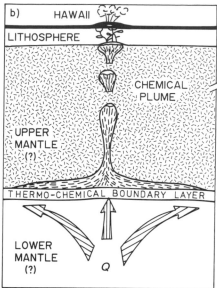

Fig. 6.9. Schematic illustration of a) thermal, and b) chemical plumes. The boundary layer could be located at the 700 km discontinuity or the core–mantle boundary. After Zindler *et al.* (1984).

6.2.5 Marble cake mantle

Fluid dynamic modelling of the convecting astheno-sphere (e.g. Richter and Ribe, 1979; McKenzie, 1979) has suggested that discrete structures in the mantle (e.g. blobs, plums, etc.) cannot remain undeformed for long periods in the convecting asthenosphere. They will tend to be elongated and sheared until they are eventually physically homogenised with the depleted reservoir. Polve and Allegre (1980) argued that evidence of this process was provided in orogenic lherzolites (Fig. 6.10), which contain alternating bands of (depleted) lherzolite and (enriched) pyrox-enite. They suggested that this banding might have been generated by convective 'stirring' and stretching of a two-part sandwich of oceanic crust and under-lying residual lherzolite, which is recycled back into the mantle by subduction. Allegre and Turcotte (1986) coined the term 'marble cake' mantle to describe this concept, and argued that it is represen-tative of the structure of much of the upper mantle.

Prinzhofer *et al.* (1989) argued that random mixing between partial melts of pyroxenite and peridotite in a marble cake mantle could generate the large ranges of incompatible element concentrations and the moder-ate range of radiogenic isotope ratios seen in lavas from a small (40 × 10 km) area of the East Pacific Rise. However, mixing in the magma chamber is not

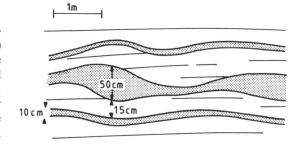

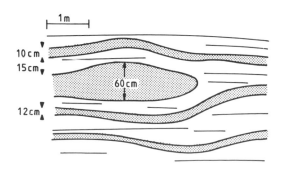

Fig. 6.10. Schematic illustration of 'marble cake' structure of pyroxenite (shaded) and lherzolite layers in the Beni Bousera high-temperature peridotite of Morocco. After Allegre and Turcotte (1986).

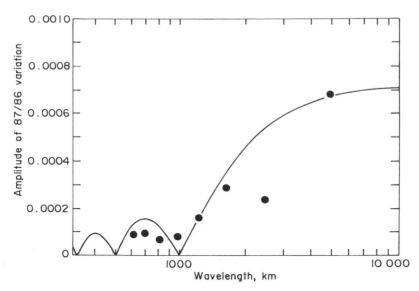

Fig. 6.11. Curve-fit for mixing of isotopic heterogeneity, compared with empirical data for amplitude *versus* wavelength of Sr isotope variation on the South Atlantic Ridge. After Kenyon (1990).

capable of explaining the length-dependence of large-scale isotopic anomalies on ridges (Kenyon, 1990). For example, the isotopic 'texture' of the South Atlantic Ridge requires convective homogenisation over distances up to 1000 km (Fig. 6.11). This is too large for a magma chamber, since it is more than the length of ridge segments between transform faults. Hence it follows that homogenisation must be at a deeper level, by solid-state convection of the marble cake mantle.

6.3 The Nd–Sr isotope diagram

In the mid 1970s, studies of the origins of mantle heterogeneity were revolutionised by the application of Nd isotope analysis to young volcanic rocks (DePaolo and Wasserburg, 1976; Richard *et al.*, 1976). DePaolo and Wasserburg plotted $^{143}Nd/^{144}Nd$ isotope ratios, in the form of ϵ Nd (section 4.2) against $^{87}Sr/^{86}Sr$, and found a negative correlation between them in oceanic and some continental igneous rocks (Fig. 6.12). Based on this evidence, they suggested that the formation of magma *sources* in the mantle involved the coupled fractionation of Sm–Nd and Rb–Sr, while some continental samples (which lay to the right of the main correlation line) could have been contaminated by radiogenic Sr in the crust.

On the basis that the 'Bulk Earth' has a chondritic Sm/Nd ratio (section 4.2), DePaolo and Wasserburg used the intersection of the chondritic (zero) ϵ Nd line with the mantle Nd–Sr correlation line to calculate an unfractionated mantle (= Bulk Earth) $^{87}Sr/^{86}Sr$ ratio of 0.7045 (Fig. 6.12). By using the initial $^{87}Sr/^{86}Sr$ ratio of the solar nebula (0.699), calculated from the 'Basaltic Achondrite Best Initial' (=BABI, section 3.2.4), and the present-day value from Fig. 6.12, they deduced an Rb/Sr ratio for the unfractionated mantle (now referred to as the Bulk Earth) of 0.029.

O'Nions *et al.* (1977) extended the $^{143}Nd/^{144}Nd$ *versus* $^{87}Sr/^{86}Sr$ correlation line in oceanic volcanics by analysing a larger suite of ocean island basalts. This included two samples from Tristan da Cunha with $^{143}Nd/^{144}Nd$ ratios lower than the Bulk Earth, indicative of a mantle source which is slightly enriched in light rare earths relative to Bulk Earth. O'Nions *et al.* argued that enrichment of some mantle sources in Nd/Sm and Rb/Sr and depletion of others (such as MORB) could be explained by trace element partition between solid and liquid silicate phases. In view of the long half-lives of Rb and Sm, they concluded that these heterogeneities had existed for long periods of time.

The observed depleted nature of the MORB source relative to Bulk Earth has very important implications for the evolution of the mantle, and can be explained by the extraction of the continental crust. This was modelled by Jacobsen and Wasserburg (1979) and O'Nions *et al.* (1979), using calculations commonly termed 'box models'.

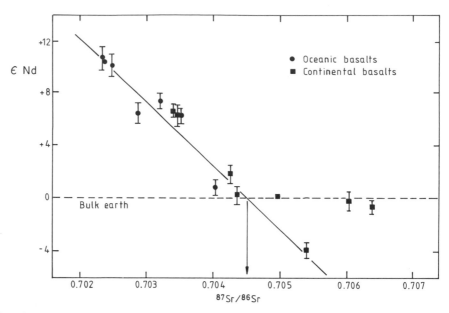

Fig. 6.12. Plot of ϵ Nd against Sr isotope ratio for ocean floor, ocean island, and continental basalts analysed before 1976. Arrow shows calculation of Bulk Earth Sr isotope ratio. After DePaolo and Wasserburg (1976).

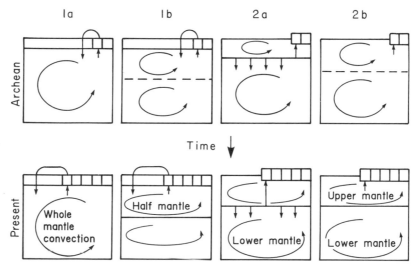

Fig. 6.13. Box models for the geochemical evolution of the mantle. Models 1a and 1b correspond to whole- or half-mantle depletion due to the extraction of continental crust. Models 2a and 2b represent progressive growth of a constantly depleted mantle and progressive depletion of a constant-volume mantle.

6.3.1 Box models for MORB sources

In a box model, the Earth is divided into chemical reservoirs which may exchange matter, grow, shrink, etc., and whose evolution is modelled over the Earth's 4.5 Byr history. Typical reservoirs or 'boxes' are the crust, mantle and core, although these may be subdivided, e.g. into upper and lower mantle. The Earth's evolution is portrayed in some alternative box models in Fig. 6.13, which will be briefly discussed.

O'Nions *et al.* (1979) examined two models of mantle differentiation and crustal growth (1a and 1b in Fig. 6.13). These were based on the numerical solution of upward and downward transport coefficients for several elements in 90 steps, each corresponding to 50 Myr of Earth history. The model was constrained by boundary conditions in the form of the composition of the primitive chondritic mantle 4.55 Byr ago and the estimated composition of the outermost 50 km of the Earth (including the continental and oceanic crust) at the present day. In Fig. 6.14 the results for $^{87}Sr/^{86}Sr$ evolution are shown for cases where (a) the whole mantle is depleted by the extraction of the upper 50 km layer; and (b) only the upper half of the mantle is depleted. (These scenarios correspond to models 1a and 1b in Fig. 6.13). Model (b) is found to yield a much better approximation to the present $^{87}Sr/^{86}Sr$ ratio of the depleted (MORB) source.

Jacobsen and Wasserburg (1979) used box models to examine another aspect of global differentiation (2a and 2b in Fig. 6.13). They simplified their treatment by considering only unidirectional transport of species from the mantle to generate the crust continuously over geological time, and solved the transport equations algebraically. In model 2a (Fig. 6.13) melts are extracted from the primitive mantle, and generate the continental crust and a depleted mantle, both of whose *volumes* grow over geological time. However, the elemental *composition* of the depleted mantle remains constant through time. Mass balance calculations based on Sm–Nd data led Jacobsen and Wasserburg to calculate that only 33% of the mantle need be depleted to generate the continental crust, corresponding to the formation of a depleted MORB reservoir occupying approximately the upper 650 km of the mantle.

In Jacobsen and Wasserburg's second model (2b in Fig. 6.13), the crust is extracted from a *fixed volume* of mantle which therefore becomes more and more depleted through geological time. The mass of this depleted mantle needed to generate the crust was calculated as only 25% of the total. In this model the isotopic composition of new continental crust will reflect a derivation from depleted mantle, whereas in model 2a new continental crust will have a chondritic (primitive mantle) isotopic signature. On the basis of Nd isotope data available to them at the time (section 4.2), Jacobsen and Wasserburg preferred model 2a.

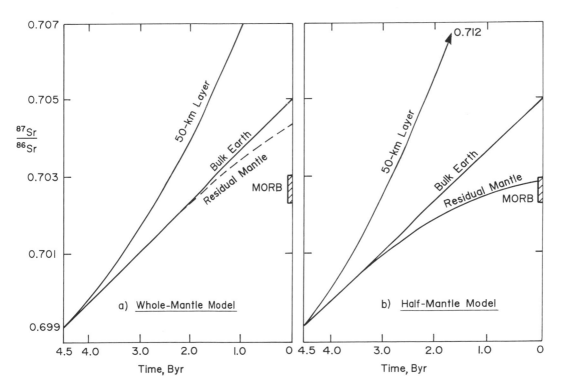

Fig. 6.14. Plots of Sr isotope evolution against time to compare the effects of a) whole-mantle or b) half-mantle convection on the degree of depletion predicted for the residual (MORB) reservoir. Hatched area is the observed composition of MORB at the present day. After O'Nions *et al.* (1979).

However, more recent Nd isotope evidence (section 4.2.2) strongly favours model 2b. The different estimates of O'Nions *et al.* (1979) and Jacobsen and Wasserburg (1979) for the volume of the depleted mantle reflect the uncertainties involved in estimating the trace element and isotopic composition of the crust.

DePaolo (1980) studied a model similar to 2b, but with the possibility of crustal recycling into the mantle, and again concluded that only 25–50% of the mantle need be depleted to generate the continental crust. It appears that modest amounts of crustal recycling have relatively little effect on mantle Nd–Sr isotope systematics, but a large effect on Pb (section 5.4.3).

The box model approach has been used in numerous more recent papers, e.g. Allegre *et al.* (1983). These authors use the so-called 'total inversion method' to attempt to choose between different models, but the uncertainties in the data do not allow significant extra information to be gained over the previous studies.

6.3.2 The mantle array and OIB sources

The Nd–Sr isotope correlation in oceanic rocks was first referred to as the 'mantle array' by DePaolo and Wasserburg (1979). They attributed the OIB which form most of this array to a chondritic lower mantle source contaminated by mixing with melts from the depleted MORB source during ascent. Indeed, much of the early discussion about the Nd isotope systematics of ocean island basalts attempted to explain their composition in terms of the two major reservoirs discussed above, namely Bulk Earth and depleted mantle. Little attention was given to the problem of generating enriched oceanic mantle, since Tristan da Cunha was regarded as more-or-less representing a primitive mantle composition similar to the Bulk Earth (Allegre *et al.*, 1979; O'Nions *et al.*, 1980).

The extension of the mantle array into the 'enriched' lower right quadrant of the Nd–Sr isotope diagram was first convincingly demonstrated in a study of the Kerguelen Islands (Dosso and Murthey, 1980), shown in Fig. 6.15. In contrast, Hawkesworth *et al.* (1979a) discovered that alkali basalts from Sao Miguel in the Azores trended to enriched $^{87}Sr/^{86}Sr$ compositions off to the right of the mantle array. The Sao Miguel trend is extended by more recent data for Samoa and the Society Islands (White and Hofmann, 1982), breaking the simple Nd–Sr isotope correlation in OIB into a 'mantle disarray' (White, 1981). Further disarray is caused by the existence of other ocean islands with compositions to the left of the mantle array, such as St Helena.

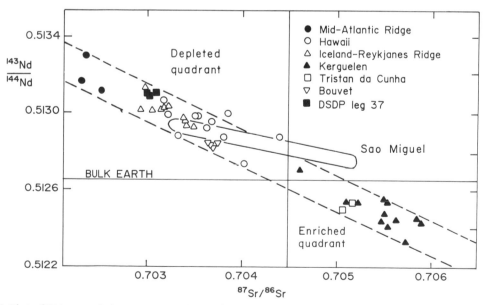

Fig. 6.15. Plot of Nd *versus* Sr isotope compositions for oceanic volcanics showing extension of the 'mantle array' into the 'enriched' quadrant relative to Bulk Earth, based on Kerguelen data (▲), and the extension of Sao Miguel data into the upper right quadrant. Modified from Dosso and Murthey (1980).

Hawkesworth *et al.* examined several possible ways of generating the isotopic features of the Azores data. They ruled out interaction with seawater because this would displace points horizontally to the right on the Nd–Sr isotope diagram (section 6.7). They recognised that ocean floor sediments have suitable Nd and Sr isotope compositions to generate the Sao Miguel array by contamination, but they ruled out this model for two reasons. Firstly, a 1 km hole drilled into the pillow lavas which build the island did not encounter any sediments. Secondly, the primitive chemistry of the basalts shows no sign of crustal contamination. Hence, by a process of elimination, Hawkesworth *et al.* attributed the isotopic signatures to the mantle sources of the basalts.

Hawkesworth *et al.* (1979a) discovered an additional problem when they compared isotopic and trace element data in OIB samples. Iceland, Hawaii and Sao Miguel basalts plot in the upper left quadrant of the Nd–Sr isotope diagram, along with MORB (Fig. 6.15), indicating derivation from a source with a time-integrated depletion in light REE relative to Bulk Earth for all of these samples, and depletion in Rb/Sr for most. However, when trace element abundance ratios in these samples are examined (Fig. 6.16), only MORB samples plot completely in

the depleted quadrant relative to Bulk Earth. Iceland reaches into the LREE enriched (= low Sm/Nd) quadrant, while some Hawaiian samples also reach into the Rb/Sr enriched quadrant, and all Sao Miguel samples plot in the LREE and Rb/Sr enriched quadrant.

One model that can potentially explain both the isotopic and trace element characteristics of the Azores data is mantle metasomatism. Hawkesworth *et al.* suggested that this caused LIL-element enrichment of the source a few tens of millions of years before generation of the Azores magmas. This can explain how a mantle source with a long-term depletion in light REE relative to Bulk Earth (as indicated by Nd isotope compositions, Fig. 6.15) can nevertheless be enriched in LIL trace elements (Fig. 6.16). Because the source was only recently enriched in LREE, there was insufficient time to affect the Nd isotope signature of the source. On the other hand, if the metasomatising fluids came from a region with long-term Rb/Sr enrichment relative to Bulk Earth, they would carry a radiogenic Sr isotope signature which they could impart to the melting region of the Sao Miguel basalts.

In more recent work, Hawkesworth *et al.* (1984) have argued strongly for mantle metasomatism, both

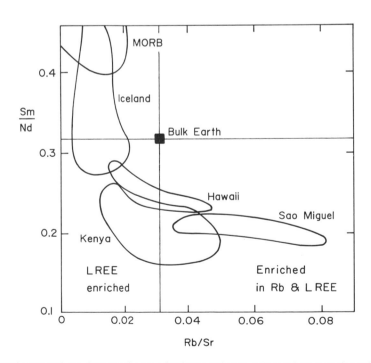

Fig. 6.16. Plot of Sm/Nd *versus* Rb/Sr elemental ratios for basic volcanic suites relative to the calculated ratios for the Bulk Earth. After Norry and Fitton (1983).

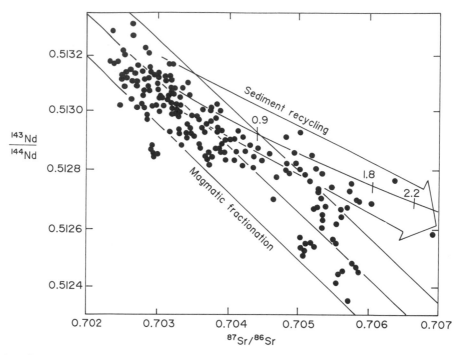

Fig. 6.17. Plot of Nd *versus* Sr isotope composition for oceanic volcanics showing two arrays. The main mantle array was attributed to recycling of magmatically fractionated material such as oceanic crust. The shallow mixing line was attributed to sediment recycling. After Hofmann and White (1982).

for the sources of continental and ocean island basalts. However, the paradox of derivation of LIL-enriched magmas from a source with long-term LIL depletion has been solved by new views of the nature of mantle melting. It is now proposed that very small degree (LIL-enriched) melts can be extracted from the mantle under conditions of shear stress (section 13.3.2). This avoids the need to invoke widespread metasomatic enrichment of mantle sources prior to magmatism.

As an alternative to the metasomatism model, Hofmann and White (1980, 1982) proposed that recycling of ancient oceanic crust into the OIB source could explain relatively enriched trace-element and isotopic compositions within the mantle array. Similarly, the deviation of Azores, Samoa and Society Islands basalts to the right of the mantle array could be explained by the addition of subducted sediment to the recycling of oceanic crust (Fig. 6.17).

Subducted ocean crust was believed to descend to a density compensation level where it was stored and reheated for 1–2 Byr before returning to the surface in a plume. Hofmann and White suggested that this was the core–mantle boundary, but Ringwood (1982) advocated the 670 km phase transition as the

compensation depth where oceanic crust resides. Seismic evidence for depression of the 670 km mantle phase boundary under subduction zones supports this model, by suggesting that the descending slab may be deflected horizontally at this level (Shearer and Masters, 1992). However, other evidence suggests that the density contrast is too small to impede convective transport across this boundary and prevent slab penetration into the lower mantle (Morgan and Shearer, 1993).

6.4 Pb isotope geochemistry

Pb isotopes are a powerful tool in studies of mantle and crustal evolution, because the three different radiogenic isotopes are generated from parents with a wide span of half-lives, two of which are isotopes of the same element. By using the different isotopes in conjunction, it is not only possible to identify the nature of differentiation events, but also to place constraints on their timing.

Early inferences about Pb isotopic evolution of the mantle were based on the analysis of galenas. However, these were plagued by the complex evolutionary history implied by the formation of galena ores, involving both mantle and crustal residence times. As analytical methods improved, it was possible to analyse mantle-derived samples such as basic magmas. These have much lower lead contents, but usually a much simpler history, allowing inferences about the mantle source with greater confidence.

6.4.1 Pb–Pb isochrons and the lead paradox

Ocean Island Basalt (OIB) leads were found by several workers (e.g. Gast *et al.*, 1964; Tatsumoto, 1966; 1978; Sun and co-workers, 1975) to define a series of arrays to the right of the geochron on the Pb–Pb 'isochron' diagram (Fig. 6.18). The slopes of these OIB arrays correspond to apparent ages of between 1 and 2.5 Byr, and can be interpreted in three principal ways: as resulting from discrete mantle differentiation events; as the products of two-component mixing processes; or resulting from continuous evolution of reservoirs with changing μ values. Each of these models may be applicable to different magmatic suites.

The differentiation model was described by Chase (1981), who evaluated OIB data in terms of a two-stage Pb isotope evolution model. This allowed $^{238}U/^{204}Pb$ values to be calculated for an 'original' mantle reservoir (μ_1) and for the secondary sources

(μ_2) which yield OIB Pb–Pb arrays. Chase found that values for μ_2 are variable within each island group and between groups, but the calculated μ_1 value is remarkably constant (7.84–7.96) for all of the data (Fig. 6.19). He therefore concluded that ocean islands were derived from separate OIB sources of variable age, but that these in turn were derived from a single long-lived primary reservoir.

The alternative mixing model was championed by Sun *et al.* (1975), who showed that the array of Pb isotope compositions in Reykjanes Ridge basalts was best explained by two-component mixing of 'plume' and 'low-velocity-zone' (upper mantle) components under Iceland. They suggested that the linear Pb isotope arrays generated by several other ocean islands might be explained by the same mechanism. However, since these arrays have different slopes, a mixing model can only work if each array is attributed to mixing of the MORB reservoir with a different enriched source (Sun, 1980). Therefore, the problem of explaining the origin of these radiogenic sources still remains.

The model of continuous mantle evolution with a changing μ value was adopted by Dupre and Allegre (1980) to explain the Pb isotope composition of leached basalt samples dredged from the Mid Atlantic Ridge. The data define a linear array to the right of the geochron, whose slope yields an apparent Pb–Pb isochron of 1.7 Byr age. However, this result was

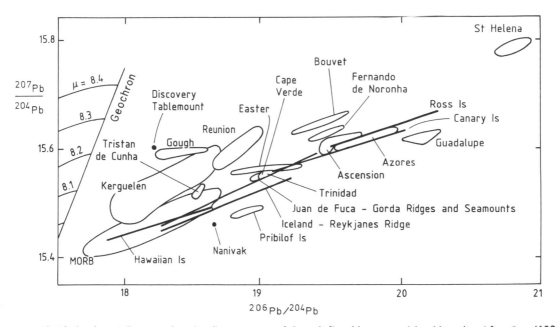

Fig. 6.18. Pb–Pb 'isochron' diagram showing linear arrays of data defined by ocean island basalts. After Sun (1980).

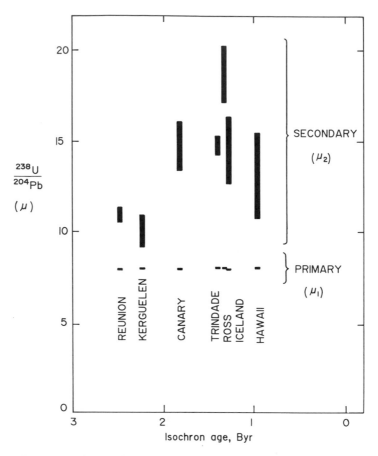

Fig. 6.19. Range of μ values required to explain OIB sources using a two-stage Pb evolution model. Parental mantle (μ_1) undergoes differentiation events at different times to yield discrete OIB source domains (μ_2). After Chase (1981).

interpreted, not as a worldwide mantle differentiation event, but as an average age for continuous differentiation from ca. 3.8 Byr ago till the present. This can occur by the mixing of enriched components with depleted mantle in numerous small events.

The distribution of OIB Pb–Pb arrays to the right of the geochron presents a problem in understanding Pb evolution in the Earth as a whole, since it implies that the depleted mantle has an average composition more radiogenic than the geochron (Bulk Earth). This is the opposite of the expected behaviour, since experimental evidence (e.g. Tatsumoto, 1988) suggests that U is more incompatible than Pb, and should generate low U/Pb ratios in the depleted mantle. This problem has been termed the 'lead paradox'. A complimentary reservoir with unradiogenic Pb must exist to balance the radiogenic depleted mantle, but this other reservoir has proved hard to locate.

Vidal and Dosso (1978) and Allegre (1982) suggested that Pb fractionation from a lower mantle OIB reservoir into the core could increase the μ value of OIB sources so as to generate the Pb–Pb arrays to the right of the geochron. These authors have proposed that while core segregation progressed very rapidly after the Earth's accretion, and was probably almost complete after 100 Myr, it nevertheless continued at a slow rate up to ca. 1.5 Byr ago, preferentially incorporating Pb. However, Newsome et al. (1986) pointed out that other elements such as Mo and W have much higher distribution coefficients from a lithophile (mantle) to siderophile (core) phase than Pb. Hence if Pb partitioning into the core is invoked to explain very radiogenic Pb sources (e.g. St Helena) then these sources should be very depleted in Mo and W.

This model can be tested by comparing Pb isotope

data in OIB sources with Mo elemental data. However, allowance must first be made for the behaviour of Mo during solid–liquid partitioning in OIB magma genesis. This is done by comparing Mo with another element with similar bulk partition coefficients for an upper mantle mineralogy. Experimental evidence suggests that light REEs such as Pr have this behaviour. After removing upper mantle effects by normalising against Pr, Mo abundance shows no correlation with radiogenic Pb isotope ratios in OIB (Fig. 6.20). Hence, recent Pb partition into the core is ruled out. However, early Pb partition into the core could displace the Bulk Silicate Earth to the right of the Geochron.

In the plumbotectonics model (section 5.4.3) Doe and Zartman (1979) proposed that the Earth's unradiogenic Pb reservoir was located in the lower crust. Because U is more mobile than Pb in high-grade metamorphism, it is 'sweated' out of the lower continental crust, causing this reservoir to develop low U/Pb ratios, and hence, over time, unradiogenic Pb. In contrast, uranium-enriched upper crust develops radiogenic Pb signatures, which are recycled into the mantle when uplifted orogens are eroded. A similar process may operate to a lesser extent in the Rb–Sr system, causing it to be somewhat decoupled from Sm–Nd systematics (Goldstein, 1988).

The plumbotectonics model can explain the *general* distribution of oceanic volcanic Pb to the right of the geochron. However, the U/Pb ratios of typical ocean floor sediment (White *et al.*, 1985) are not great

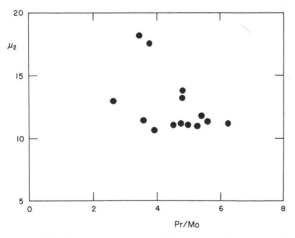

Fig. 6.20. Pb isotope data for OIB, expressed in terms of the μ_2 value of a two-stage evolution model. These show no correlation with a trace element index which measures possible fractionation of siderophile elements into the Earth's core. After Newsome *et al.* (1986).

enough to explain the extremely radiogenic Pb signatures of some OIB. On the other hand, Chase (1981) proposed the radiogenic OIB reservoirs could be generated by subduction of oceanic crust. This model is discussed further in section 6.6.

In order to translate U/Pb enrichments in the modelled OIB source into Pb isotope enrichments, the reservoir must be isolated for 1–2.5 Byr. However, the process of mantle convection naturally tends to streak out any heterogeneities into narrow schlieren (Olson, 1984). These would then be too small to source large volumes of enriched OIB, such as in Hawaii. Ringwood (1982) has attempted to overcome this problem by postulating that subducted oceanic crust and continental sediment collects in a large blob or 'megalith' at the 670 km seismic discontinuity, which is also proposed by some workers as a boundary layer between upper and lower mantle convection.

An alternative model to avoid homogenisation of the OIB source into the MORB source before it has time to develop old isotope signatures is to involve the lithosphere. By definition this material is solid and non-convecting, and Nd model age dating (section 4.2.1) points to its potentially long life-time. Since the lithosphere is thought to be generally depleted, it must undergo secondary enrichment (after consolidation) by the emplacement of LIL-rich metasomatic fluids. McKenzie and O'Nions (1983) suggested that the sub-continental lithosphere has a greater density than the underlying Fe-depleted asthenosphere, so that over-thickening during continental collision might cause some of the lithosphere to constrict off and fall into the upper mantle convection system. If this material was sampled within a few hundred million years then it might yield OIB magmas, before being homogenised into the MORB source by convection. However, Archean lithosphere may be stabilised against this process by its own Fe-depleted signature (section 7.1). Further discussion of lithospheric recycling is found in section 6.6.

6.4.2 The terrestrial Th/U ratio

Over many years, the prime focus of Pb isotope analysis has been on the U–Pb system. However, the combination of ^{208}Pb and ^{206}Pb isotopes also allows constraints to be placed on the atomic Th/U ratio or 'κ' value of Earth reservoirs.

In order to use Pb isotope ratios to determine the Th/U ratio of a reservoir, it is necessary to know the age of the reservoir and its Pb isotope composition at the start and end of its evolution. For the 'Bulk Earth system' the age is defined by the geochron, and the initial ratio at time T (age of the Earth) is given by the Canyon Diablo composition (Tatsumoto et al., 1973). The Pb isotope ratio of a mantle reservoir at time t (end of the period of mantle evolution considered) is determined from the initial Pb isotope composition of a mantle-derived magma at that time. Hence (following Allegre et al., 1986), we can define the radiogenic $^{208}Pb/^{206}Pb$ ratio of a mantle reservoir as:

$$\frac{^{208}Pb^*}{^{206}Pb^*} = \frac{\left(\frac{^{208}Pb}{^{204}Pb}\right)_t - \left(\frac{^{208}Pb}{^{204}Pb}\right)_T}{\left(\frac{^{206}Pb}{^{204}Pb}\right)_t - \left(\frac{^{206}Pb}{^{204}Pb}\right)_T} \qquad [6.1]$$

Given a closed system from time T to t, the Th/U ratio (κ value) of the reservoir can be calculated from $^{208}Pb^*/^{206}Pb^*$ by solving U–Th and U–Pb decay equations for values T and t (Chapter 5). However, we can also calculate the average or 'time-integrated' Th/U ratio of an open system from time T to t.

In the conformable Pb model (section 5.4.2), the closed-system assumption for the mantle implied a constant κ value against time, equal to the meteorite

value of 3.9 ± 0.1 (Tatsumoto et al., 1973). In contrast, the model of Cumming and Richards (1975) proposed a decrease in terrestrial Th/U ratio from an initial value of 4.13 to a present-day value of 3.84. This model was largely overlooked in constraining mantle Th/U evolution because of the near coincidence of κ values in Phanerozoic galenas with the meteorite value. However, an examination of high quality $^{208}Pb/^{204}Pb$ data for conformable galenas (Albarede and Juteau, 1984) confirmed that the closed-system model could not explain terrestrial Th/U systematics, since it yields a young apparent age for the Earth (section 5.4.3).

In a major new examination of this problem, Allegre et al. (1986) used initial Pb isotope ratios for Phanerozoic ophiolite complexes and Archean komatiites to calculate time-integrated Th/U ratios for the upper mantle from T (4.57 Byr ago) until the age of eruption (Fig. 6.21). They concluded that the upper mantle had higher Th/U ratios in the Archean than at the present, and that this reservoir was progressively depleted in Th/U over time, in a similar way to its depletion in Rb/Sr and Nd relative to Sm (section

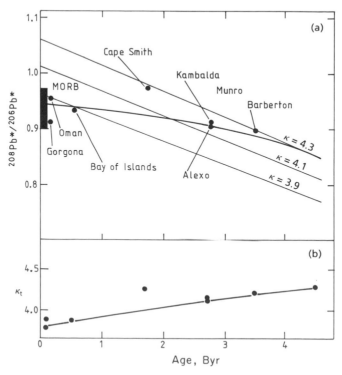

Fig. 6.21. Pb isotope evidence for time-integrated mantle Th/U ratio. a) Plot of radiogenic $^{208}Pb/^{206}Pb$ ratios for mantle-derived Pb, showing best-fit open-system curve relative to closed-system evolution lines. b) Calculated variations of time-integrated mantle Th/U ratio (κ_t) over Earth history. After Allegre et al. (1986).

4.2.2). It follows that the κ value for the oldest rocks (ca. 4.3) might be expected to approximate the Bulk Earth value (i.e. before significant depletion of the upper mantle reservoir). A κ value averaging 4.25 may be calculated from the Isua galena analysed by Appel *et al.* (1978). This is the oldest analysed terrestrial galena, and strongly supports the model of Allegre *et al.*, although not cited by them.

The Bulk Earth κ value derived above was tested by an independent determination from recent oceanic volcanics (Allegre *et al.*, 1986). This is analogous to the determination of the Bulk Earth Sr isotope ratio from the Sr–Nd isotope 'mantle array'. Radiogenic $^{208}Pb^*/^{206}Pb^*$ ratios in oceanic volcanics define fairly good linear arrays when plotted against $^{143}Nd/^{144}Nd$ and $^{87}Sr/^{86}Sr$ (Fig. 6.22). The intersection of Bulk Earth Sr and Nd with these correlation lines yields Bulk Earth $^{208}Pb^*/^{206}Pb^*$, and hence time-integrated

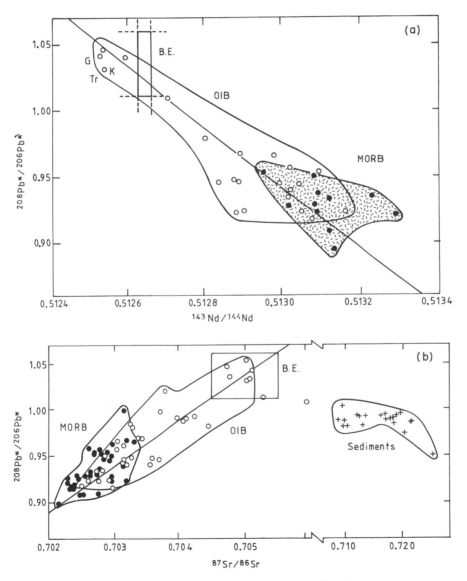

Fig. 6.22. Plot of radiogenic $^{208}Pb/^{206}Pb$ against a) $^{143}Nd/^{144}Nd$; and b) $^{87}Sr/^{86}Sr$ for modern oceanic volcanics, allowing a calculation of the time-integrated Th/U ratio of the Bulk Earth. ($\bullet$) = MORB; ($\circ$) = OIB (G = Gough; K = Kerguelen; Tr = Tristan). Box marked B.E. indicates the uncertainty of the Bulk Earth composition. After Allegre *et al.* (1986).

Bulk Earth κ values of 4.1–4.2. Thus, the two different approaches (old and modern leads) yield Bulk Earth κ values in good agreement, averaging 4.2. Allegre *et al.* argued that this value would also be consistent with the lower Th/U value of meteorites if early U partition into the core is taken into account. The similarity between the time-integrated κ values of MORB and chondrites therefore appears to be a coincidence.

In addition to the time-integrated Th/U ratio for the upper mantle, determined from Pb isotope data, we can also determine an 'instantaneous' present-day Th/U ratio for the mantle from oceanic volcanics. Tatsumoto (1978) estimated this value as about 2.5 in the MORB source, based on elemental Th/U ratios in lavas. This value has been confirmed more recently by ^{232}Th/^{230}Th activity ratios in oceanic volcanics, which can be used to calculate a more accurate instantaneous κ value for the mantle source (section 13.4). Compared with a Bulk Earth κ value near 4, these data indicate strong mantle depletion, which can be attributed to crustal extraction. However, this presents a problem, since the time-integrated κ value

of ca. 3.75 in MORB is much higher than the instantaneous value, and only slightly less than the Bulk Earth value.

Galer and O'Nions (1985) explained this problem by proposing that the MORB reservoir was buffered over geological time by a less depleted reservoir. In other words, Pb recently extracted from the MORB source only had a brief residence time in the depleted reservoir, and spent most of Earth history in a reservoir with a κ value near Bulk Earth. They calculated that a residence time of 600 Myr in a Th-depleted MORB reservoir with $\kappa = 2.5$ and a 4 Byr residence in a reservoir with $\kappa = 3.9$ would give the time-integrated κ value of 3.75 needed to explain MORB lead isotope compositions (Fig. 6.23). Galer and O'Nions examined three possible locations for their proposed Bulk Earth κ lead source: upper continental crust, sub-continental lithosphere and lower mantle. However, the short upper mantle residence time for Pb calculated using their model was a severe test of the ability of all of these sources to account for MORB Pb.

The proposal to buffer MORB by upper crustal Pb

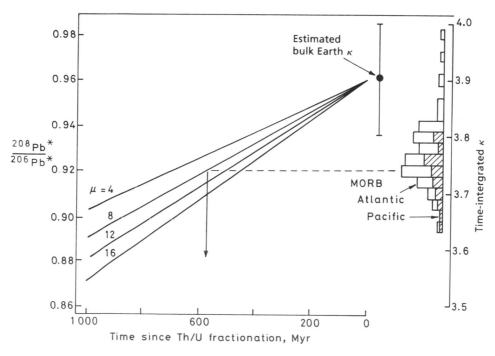

Fig. 6.23. Evolution of time-integrated κ values as a function of residence time in the MORB reservoir, starting at a value of 3.9. Histogram at right indicates present-day composition of Pacific (hatched) and Atlantic MORB. After Galer and O'Nions (1985).

causes problems because the continental crust is characterized by higher $^{207}Pb/^{204}Pb$ than MORB and, due to the almost complete extinction of the ^{235}U parent of ^{207}Pb at the present day, these differences must be long-lived. The low Pb concentration and relatively small volume of sub-continental lithospheric mantle require an unreasonably rapid rate of exchange (complete exchange within 1 Byr) to buffer upper mantle Pb isotope compositions. This seems to be precluded by the old Sm–Nd model ages of inclusions in diamonds (section 4.2.1). Buffering of MORB Pb by the lower mantle necessitates exchange with anywhere between one-quarter and one-half of its mass over geological time. However, fluid dynamic considerations suggest an order of magnitude less exchange between these mantle reservoirs (Olson, 1984).

These problems can be overcome by using Allegre *et al.*'s (1986) Bulk Earth κ value of 4.2, which allows a much longer Pb residence period in the MORB reservoir, of up to 1.8 Byr. This is in good agreement with the 1.7 Byr apparent Pb–Pb age generated by Mid Atlantic Ridge basalts, and suggests the operation of unified mantle depletion processes. A residence time approaching 1.8 Byr permits buffering of the MORB reservoir by crustal Pb without

generating excessive ^{207}Pb levels in MORB. Also, the longer residence period makes lithosphere and lower mantle buffering more viable. Recycling of κ-enriched components into the depleted upper mantle can be achieved via the OIB source. This is demonstrated on a plot of instantaneous κ value against time-integrated κ value for oceanic volcanics (Fig. 6.24). The OIB samples define a reasonable array linking MORB compositions (solid symbols) and Allegre *et al.*'s Bulk Earth point on the geochron.

6.4.3 The upper mantle μ value re-examined

Much discussion has occurred about the time-integrated μ value of the mantle, as revealed by Pb isotope analysis. It is now generally agreed that the *apparent* increase in μ value of the upper mantle over time may be an illusion caused by recycling of radiogenic Pb from the oceanic and continental lithosphere (sections 5.4.3 and 6.4.1). However, in order to properly quantify this process, it is necessary to determine the instantaneous (present-day) μ value of the upper mantle. Unfortunately, there is no isotopic route to this quantity, as was possible for the κ value. Neither can it be determined directly

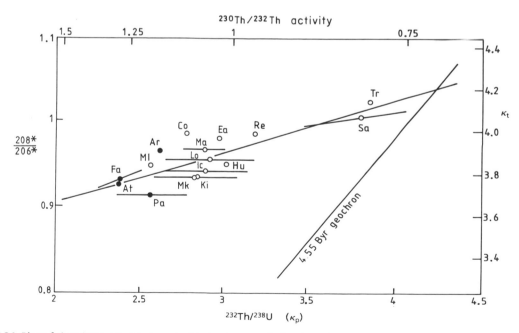

Fig. 6.24. Plot of time-integrated (κ_t) against instantaneous (κ_p) values for oceanic volcanics. (•) = MORB; (○) = OIB. After Allegre *et al.* (1986).

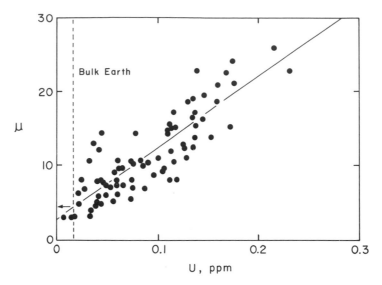

Fig. 6.25. Plot of $^{238}U/^{204}Pb$ (μ) in MORB glasses against uranium content, showing a positive correlation, from which a maximum upper-mantle μ value can be estimated. After White (1993).

from the U/Pb ratios of MORB glasses, since U/Pb fractionation during partial melting is poorly constrained. However, White (1993) has developed an indirect approach to the upper mantle μ value from the relationship between μ and uranium content in MORB glasses.

Analysis of U and Pb in 82 glasses from the Atlantic, Pacific and Indian oceans revealed a strong positive correlation between $^{238}U/^{204}Pb$ and U content (Fig. 6.25). White (1993) attributed this correlation to fractionation (of uranium) during partial melting, and argued that it could be used to estimate the depleted mantle μ value. Since U is incompatible, the U content of the Bulk Silicate Earth estimated from chondrites (0.018 ppm) must be an upper limit for U in the depleted mantle (MORB source). Applying this value to the μ versus U correlation line leads to a maximum instantaneous μ value of 4.5 in the MORB source (compared with a time-integrated μ value of ca. 8.5).

This discrepancy is exactly analogous to the one discussed above for mantle κ and, again, is due to the relatively short residence time of Pb in the upper mantle. White proposed that upper mantle Pb is buffered by the entrainment of radiogenic Pb from plumes. Therefore, the Pb isotope composition of the upper mantle reflects dynamic equilibrium between Pb fluxes into and out of this reservoir. In contrast, radiogenic Sr and Nd in the upper mantle are largely generated by *in-situ* decay of Rb and Sm.

6.5 Mantle reservoirs in isotopic multispace

6.5.1 The mantle plane

The unification of radiogenic $^{208}Pb^*/^{206}Pb^*$ with other isotope systematics, apparent above, breaks down when Pb isotope ratios are plotted against other isotope systems on bivariate diagrams (e.g. Fig. 6.26). This indicates that the isotope systematics of the mantle cannot be explained by a two-component mixing model. An exception to this general observation is provided by Pb–Sr isotope systematics on the North Atlantic, which *do* define a coherent positive correlation (Dupre and Allegre, 1980). However, this can be attributed to coincidental contamination of the MORB reservoir in this area with a single compositional type of enriched plume material.

Zindler *et al.* (1982) argued that the Pb–Sr–Nd isotope compositions of oceanic volcanics could be explained by (solid state?) mixing of three mantle components. The proposed end-members were a pristine chondritic mantle with a Pb composition on the geochron, a MORB source depleted by continental crustal extraction, and a reservoir containing recycled MORB. This made Kerguelen the best candidate for a primitive mantle source, while St Helena was regarded as having the greatest amount of recycled MORB material in its source. Zindler *et al.* argued that average isotopic compositions of ocean

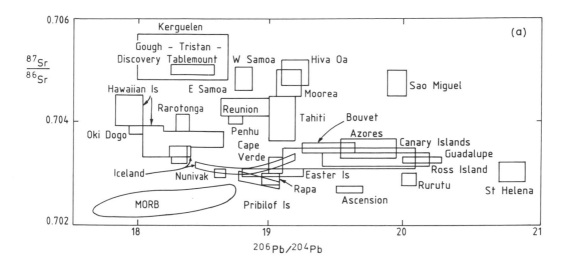

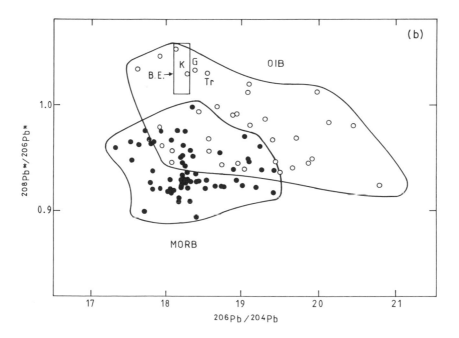

Fig. 6.26. Diagrams to show the decoupling of $^{206}Pb/^{204}Pb$ from other isotopic systems in oceanic volcanics: a) Sr isotope data, after Sun (1980); b) radiogenic $^{208}Pb*/^{206}Pb*$ data, after Allegre *et al.* (1986).

ridges and ocean islands displayed very little scatter away from a plane containing the three end-member components (Fig. 6.27).

Zindler *et al.* justified their three-component model by the high correlation coefficient of 0.98 calculated for their data set. However, such a limited scatter was achieved by excluding some ocean islands. For example, Sao Miguel was not included in the Azores average; but this signature was argued by White (1985) to be part of a much wider compositional field, including data from the Society Islands, Samoa and Marquesas which extend 'above' the 'mantle plane' of Zindler *et al.* (1982). Furthermore, the averaging process also obscured data which lay 'below' the

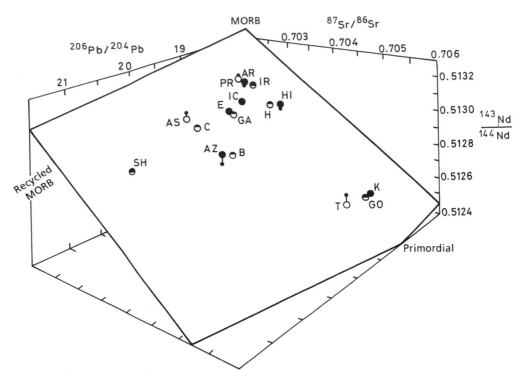

Fig. 6.27. Three-component mantle mixing model for MORB and OIB sources. Solid and open symbols indicate points respectively above and below the mantle plane. AR, IR, PR = Atlantic, Indian and Pacific ridges. HI= Hiva Oa, H= Hawaii, IC= Iceland, E= Easter, GA= Guadeloupe, AS= Ascension, C= Canaries, AZ= Azores, B= Bouvet, SH= St Helena, T= Tristan, G= Gough, K= Kerguelen. After Zindler *et al.* (1982).

mantle plane, such as the Walvis Ridge. Therefore, at least one additional component must be invoked to explain the data.

6.5.2 The mantle tetrahedron

Hart *et al.* (1986) considered that the mantle plane of Zindler *et al.* (1982) might really be a 'co-incidence of similar mixing proportions' of end-members with more extreme compositions, rather than a discrete entity in its own right. This is illustrated in Fig. 6.28, where samples are plotted in terms of part per 10^5 deviation in Nd isotope ratio from the mantle plane, against Pb isotope composition.

Hart *et al.* proposed that the lower bound of individual $^{143}Nd/^{144}Nd$ sample compositions on the Nd–Sr isotope diagram (Fig. 6.29a) might be a more fundamental topological structure, which they termed the 'LoNd' array. The same samples which define this array on the Nd–Sr isotope diagram also fall in a line on the Sr–Pb isotope plot (Fig. 6.29b), despite the fact that this cuts across the middle of the OIB field in this

diagram. $^{208}Pb/^{204}Pb$ ratios in these samples are also coherent with the three other systems.

The LoNd array was itself interpreted as a mixing line between 'HIMU' (high U/Pb) and 'EMI' (enriched mantle I) end-members (Zindler and Hart, 1986). Other important end-members were defined by the most extreme composition of the MORB field (DMM) and the Societies (EMII). In addition, Zindler and Hart (1986) suggested that three other components might be located inside the tetrahedral mixing space in Fig. 6.29. These are a 'primordial helium isotope reservoir', exemplified by Loihi seamount (section 11.1.3); a 'Bulk Earth' U–Th–Pb isotope reservoir exemplified by Gough–Tristan (section 6.4.2); and a 'PREvalent MAntle' or PREMA component, justified on the grounds that the mixing of discrete components may have reached such a stage of completeness that this mixture itself becomes a recognisable entity. These possible components will be discussed further below.

One of the characteristics of the LoNd array (Hart, 1988) is that island groups are not generally elongated

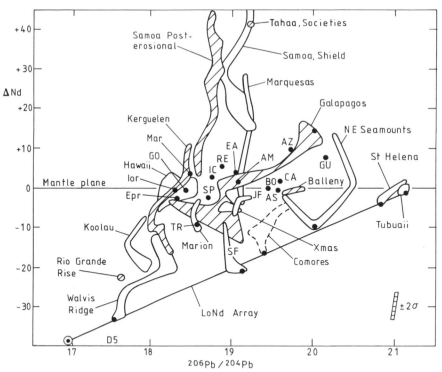

Fig. 6.28. Plot of Δ Nd (part per 10⁵ deviation in $^{143}Nd/^{144}Nd$ ratio from the mantle plane of Zindler *et al.*, 1982) against Pb isotope ratio. OIB compositions are plotted both as fields and discrete points. After Hart *et al.* (1986).

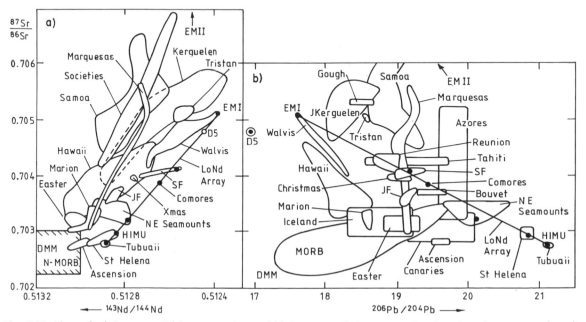

Fig. 6.29. Plots of: a) Sr *versus* Nd isotope ratio, and b) Sr *versus* Pb isotope ratio, showing the proposed end-members of a four-component mixing system: DMM, HIMU, EMI (= EM1) and EMII (= EM2). Dots are compositions argued to lie on an array between the HIMU and EMI end-members, termed the LoNd array. After Hart *et al.* (1986).

along the proposed mixing line, but often trend obliquely off the line. This was used as evidence that the LoNd mixing processes occurred a long time ago. In addition, Hart *et al.* (1986) argued that the straightness of the proposed LoNd mixing line places tight constraints on the nature of the two end-members by requiring them to have similar Nd–Sr–Pb ratios, and an intimately related environment of formation. Since they believed that such conditions would not be expected between recycled crustal and mantle components, Hart *et al.* argued that the two end-members must have resulted by different metasomatic enrichment processes in the sub-continental lithosphere.

Hart (1988) identified another two-component mixing line within the OIB data set, by using an upper $^{87}Sr/^{86}Sr$ cut-off of 0.703 to exclude all samples with an enriched mantle component. On a diagram of $^{143}Nd/^{144}Nd$ against $^{206}Pb/^{204}Pb$ (Fig. 6.30), island groups define a so-called 'no EM' array between the HIMU and DMM end-members. The straightness of this array again suggests that the end-members had similar Nd/Pb ratios, and hence that DMM, HIMU and EMI all have similar Nd/Pb ratios. However, the

geochemical relationship between DMM and HIMU cannot easily be attributed to spatial proximity, as was the EMI–HIMU relationship, because the depleted mantle is a distinct reservoir. This therefore weakens Zindler and Hart's argument for an intimate genetic relationship between the end-members of the LoNd array. Instead, a more general relationship is possible, whereby the three components are generated by similar mantle melting processes, but in different locations.

In contrast to the linear mixing lines described above, mixing with the EMII component tends to generate elongate curved arrays within island groups, as shown in Figs. 6.28 and 6.29. This suggests that elemental ratios between EMII and the other mantle domains are far from unity, consistent with a model in which DMM, HIMU and EMI are generated by mantle differentiation processes, but EMII represents recycled continental crust with a very different trace element signature. Hart (1988) went further in his distancing of EMII from the other components, suggesting that mixing with this end-member was a late phenomenon which occurred after other mixing processes. However, Staudigel *et al.* (1991) found

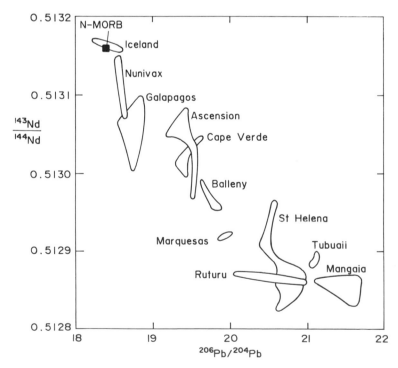

Fig. 6.30. Nd *versus* Pb isotope diagram showing the linear array of OIB samples with $^{87}Sr/^{86}Sr$ below 0.703, attributed to the 'No-EM' mixing line. After Hart (1988).

strong evidence for mixing between HIMU and EMII in the South Pacific isotopic and thermal anomaly (SOPITA), particularly on the Sr–Pb isotope diagram (not shown here). In view of the intimate geographical association of HIMU and EMII in the SOPITA case, it is likely that this array was formed prior to mixing with MORB, and it may constitute one of a family of curved 'HiNd' mixing lines analogous to the LoNd array.

There is considerable danger in looking at isotope variations in a number of two-component systems, since arrays are projected onto these surfaces from a multi-dimensional mixing polygon, and in this process the true trends of the arrays may be misunderstood. In order to analyse the data in a more objective fashion, Allegre et al. (1987) ran a principal component analysis on a large set of $^{87}Sr/^{86}Sr$, $^{143}Nd/^{144}Nd$, $^{206}Pb/^{204}Pb$, $^{207}Pb/^{204}Pb$ and $^{208}Pb/^{204}Pb$ data for MORB and OIB samples. This was also performed on an updated sample set by Hart et al. (1992).

Principal component analysis resolves the oceanic data set into five eigenvectors, representing directions in multi-component space which show the greatest percentage of variance in the data. The magnitudes of these vectors (in the calculation by Hart et al.) are approximately 56%, 37%, 4%, 2% and 1%. The pre-eminence of the first two vectors demonstrates the largely planar form of the data set, as emphasised by Zindler et al. (1982). However, there is enough

residual scatter in the data that a third vector is necessary to properly represent the mixing process. The sum of these three vectors is 97.5% in Hart's analysis, and 99.2% in Allegre's analysis. Hence Hart et al. argued that a three-dimensional analysis is appropriate to analyse the data in more detail than is possible in two dimensions. However, the eigenvectors are so divorced from the familiar isotope ratios that it becomes difficult to understand the data. Therefore, Hart et al. presented the data in the from of a three-dimensional isotope plot (of $^{143}Nd/^{144}Nd$, $^{87}Sr/^{86}Sr$ and $^{206}Pb/^{204}Pb$), but projected in such a way as to approximate the eigenvector directions (Fig. 6.31).

The focussing of data points at the lower corners of the mantle tetrahedron provides evidence that some of the proposed components are real entities, rather than merely theoretical end-members. This is exemplified by the intersection of the LoNd and No-EM arrays, which provide a relatively strong constraint on the composition of HIMU, suggesting that the 'pure end-member' has a composition very similar to the most radiogenic Pb already analysed, from the island of Mangaia. This conclusion is supported by the close agreement between the composition of the widely separated HIMU islands from the South Atlantic and South Pacific. In contrast, the density of samples lying at the EMI and EMII end-member compositions is much lower, and may suggest that these are the most extreme products yet sampled of enrichment

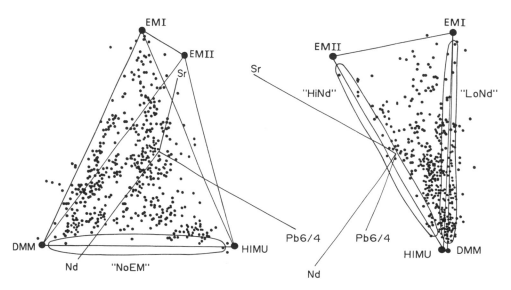

Fig. 6.31. Two views of a three-dimensional mantle tetrahedron representing the mixing relationships of four isotopically proposed mantle components seen in oceanic volcanics. Modified after Hart et al. (1992).

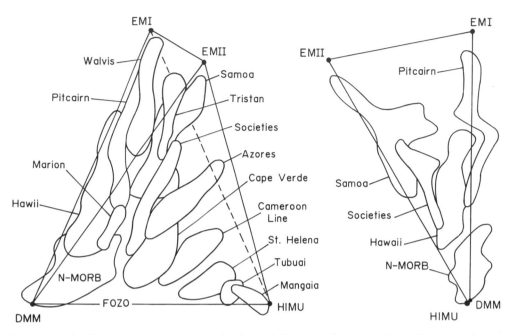

Fig. 6.32. Ocean island arrays shown in two projections of the three-dimensional mantle tetrahedron, to show radiation from a focus zone at the bottom left of the diagram, corresponding to either the MORB field, or the unknown component 'FOZO'. Modified after Hart *et al.* (1992).

processes, rather than significant mantle reservoirs in their own right (Barling and Goldstein, 1990).

When individual ocean island arrays are shown in the projected three-dimensional diagram of Hart *et al.* (1992), they are seen to fan out from near the DMM component (Fig. 6.32). However, Hart *et al.* argued that the arrays are not in fact focussed on DMM, but on a previously unrecognised 'FOcus ZOne' (FOZO) between DMM and HIMU which might be the lower mantle. Unfortunately, it is hard to be certain about the trend lines at the bottom left of the island arrays, because these are 'non-extreme' compositions which are easily confused with enriched (E-) MORB. Nevertheless, there are some difficulties with the concept of a universal depleted component which is distinct from the MORB reservoir.

One of these difficulties is the detailed spatial and temporal control on isotopic variation within individual island groups. We can take as an example the Cameroon line, which may represent the clearest trend in Fig. 6.32 towards the mid point of the DMM–HIMU edge. Much of the isotopic variation in this data set is contributed by the island of Principe. However, Halliday *et al.* (1988) showed that Pb isotope variation in Principe lavas was related to age and to silica saturation (Fig. 6.33), due to mixing processes which occur in the uppermost mantle

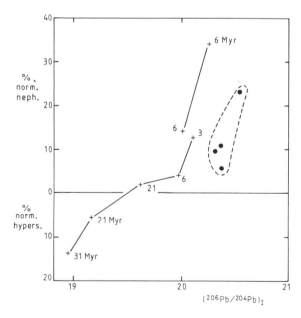

Fig. 6.33. Data for basalts from Principe, Cameroon Line, showing the correlation of Pb isotope ratio with silica saturation (% of normative nepheline or hypersthene), and eruption age (in Myr). Radiogenic Pb from the ocean–continent boundary is shown for reference (●). After Halliday *et al.* (1988).

(Halliday *et al.*, 1990; 1992). Such a process would probably generate a hyperbolic isotope mixing line in which the observed array does not trend directly towards the depleted end-member. This is consistent with typical mixing behaviour in magmatic systems (section 7.2.2). Similar hyperbolic mixing lines between mantle components have been observed in Hawaiian volcanics by Chen and Frey (1983).

Another difficulty with the 'FOZO' model concerns the manner of its incorporation during binary mixing with enriched reservoirs. In order to explain the universality of this mixing, it is best explained by entrainment during plume ascent from the lower mantle. However, the most essential isotopic properties of FOZO are radiogenic Nd and unradiogenic Sr. Therefore, if we reject the MORB reservoir in this role, we are faced with the prospect of a large incompatible-element depleted lower mantle reservoir. But this runs against the conclusions of box models, which suggest that only a fraction of the mantle is depleted (section 6.3.1).

One of the strongest arguments presented by Hart *et al.* was based on correlated lithophile and rare gas isotope systematics in OIB. They argued that in some island chains (e.g. Hawaii, Azores and Samoa), helium isotope compositions approach more primordial compositions as the focus zone is approached. Since primordial helium originates from the lower mantle or core, they suggested that the lithophile signature of FOZO also originates from the lower mantle. However, this does not necessarily follow, since rare gases may be decoupled from lithophile isotope systems. Mantle reservoirs which are generated by lithospheric recycling (EMI, EMII, HIMU) are outgassed during shallow mantle processes and then subsequently enriched in radiogenic helium. This deep mantle helium rises within a plume of enriched mantle, but a separate flux of primordial helium may also escape up the same plume conduit. For lithophile elements in such a plume, we see two-component mixing between recycled and upper mantle components. However, because the MORB reservoir is severely degassed, the helium signatures may instead reflect two-component mixing of primordial and enriched components.

6.6 Identification of enriched mantle components

Since the study of Hart *et al.* (1986), major efforts have been devoted to identifying the proposed enriched mantle components in geological terms, and explaining how they have interacted to generate

OIB sources. To a large extent the debate has been polarised between those who invoke metasomatic enrichment models (e.g. Hart *et al.*, 1986) and those who invoke crustal recycling models (e.g. Weaver, 1991) to explain the enriched components. Some of the arguments will be briefly examined for the different end-members.

6.6.1 HIMU

Several authors have proposed that HIMU represents subducted oceanic crust (e.g. Chase, 1981; Palacz and Saunders, 1986; Staudigel *et al.*, 1991; Chauvel *et al.*, 1992; Hauri *et al.*, 1993). The great advantage of this model is that it attributes HIMU to a known major subducted component, but the mechanism for U/Pb enrichment remains unclear.

Seawater alteration has been invoked as a possible mechanism to elevate U/Pb ratios in oceanic crust (Michard and Albarede, 1985), but this might also elevate Rb/Sr ratios, generating more radiogenic strontium than is seen in the HIMU component. A better model (Weaver, 1991) is to invoke preferential extraction of Pb, relative to U, from the down-going slab in subduction zones. Weaver argued that the characteristic trace element signature necessary to generate HIMU could indeed be produced in the dehydration residue of subducted ocean crust, if fluids are enriched in Pb, but depleted in U. This requires that uranium be held in a U^{4+} state, limiting the formation of soluble U^{6+} complexes. The model is supported by U/Pb ratios nearly an order of magnitude lower in island arc tholeiites than in MORB (Sun, 1980). Mobilisation of Pb from the slab in a fluid phase could also explain the surprising degree of Pb isotope homogeneity in arc-related 'conformable' galena deposits (section 5.4.2).

An additional site for possible U/Pb enrichment of the subducted slab is the sub-oceanic lithosphere. For example, Halliday *et al.* (1990; 1992) argued that shallow Pb–Pb isotope arrays in the Cameroon Line volcanics and other Atlantic islands were best explained by recent strong U/Pb enrichment of the oceanic lithosphere (section 7.3.2). This process cannot directly explain the much steeper correlation between ^{206}Pb and ^{207}Pb in HIMU islands. However, after storage for ca. 1 Byr, and mixing with less radiogenic Pb from other parts of the subducted slab, this represents an additional mechanism to generate the HIMU component.

Class *et al.* (1993) suggested that some plumes (e.g. Ninetyeast–Kerguelen and Tristan–Walvis) could undergo *in-situ* growth of radiogenic Pb, following the delamination of sub-continental lithosphere into the plume source. This model was based on an observed correlation of Pb isotope ratio with age along the Ninetyeast Ridge. However, the relatively radiogenic Nd and unradiogenic Sr on the Ninetyeast Ridge are quite different from the generally recognised composition of the Kerguelen plume composition (e.g. Weis *et al.*, 1993).

Comparison with the Cameroon Line suggests an alternative explanation for the Ninetyeast data. Both suites began with the impingement of a plume on a rifted continental margin, and subsequently witnessed extreme *in-situ* radiogenic Pb production over the following tens of Myr. Therefore, both suites may result from a streaked-out residual blob of young U-enriched lithosphere, generated when the plume was trapped under the continental margin. Evidence that this blob is now nearly exhausted (even as it continues to evolve) comes from the tendency of Kerguelen Archipelago and Heard Island magmas to revert to the original Pb isotope composition seen in the Kerguelen Plateau.

6.6.2 EM II

The case for EMII as subducted continental material is almost universally agreed, since this end-member is squarely located on mixing lines between depleted mantle and marine sediments. This model is further strengthened by evidence from peridotite xenoliths in Samoan lavas (Hauri *et al.*, 1993). Trace element data for these xenoliths point to an origin from carbonate-rich melts within the Samoan plume, and the isotopic compositions of the xenoliths are therefore taken as indicative of the EMII mantle component. These xenoliths extend the EMII array directly into the field of marine sediments, and thus provide a compelling case for this material as the source of the EM II component.

6.6.3 EM I

Some workers have suggested that EMI can also be attributed to sediment recycling. For example, Weaver (1991) suggested that EMI represents recycled pelagic ocean floor sediment (in contrast to EMII which represents terrigenous sediment). However, Hf isotope evidence argues strongly against this model (section 9.1.4), and suggests that sub-continental lithosphere is a more attractive component for EMI.

On the other hand, Woodhead *et al.* (1993) recently interpreted oxygen isotope evidence from submarine

glasses as favouring a sedimentary origin for EMI. Nd, Sr and Pb isotope data for basaltic glasses from the Pitcairn seamounts trend strongly towards the EMI end-member, but oxygen isotope ratios vary far outside the normal range for submarine glasses (e.g. Fig. 6.34.). Unfortunately, there are many ways in which oxygen isotope ratios can be perturbed by sea-floor processes to yield spurious signals. However, Woodhead *et al.* argued that none of these processes could explain their data, which could best be explained by isotopic variations in the plume source itself. Their preferred explanation for this effect was the recycling of marine sediment into the EMI source by subduction.

Because oxygen is more-or-less equally abundant in all rocks, fractions of up to 10% of sediment must be incorporated into the EMI reservoir to explain the Pitcairn data (Fig. 6.34). This amount of assimilation

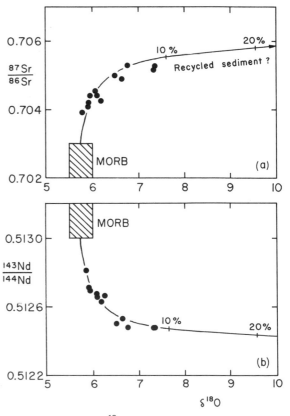

Fig. 6.34. Plot of $\delta^{18}O$ against radiogenic isotopes for Pitcairn submarine glasses. a) Sr isotope ratio; b) Nd isotope ratio. A possible mixing line (for solid-state mixing in the mantle) is drawn to fit the data. After Woodhead *et al.* (1993).

is sufficient to completely overprint the mantle isotope signature for lithophile isotope systems such as Pb, Nd and Sr. However, the Sr isotope signature in Pitcairn basalts (ca. 0.706) is not consistent with the composition of subducted ocean-floor sediment (>0.71). Carbonates represent a better fit to the data (White, 1993), but are not a realistic component for subduction into the mantle, due to their dissolution at abyssal depths. Therefore, it is concluded here that if the oxygen data represent a real mantle signature, then this is best explained by hydrothermal fluxing of the mantle wedge above a subduction zone.

An origin for EMI in metasomatised lithosphere is the preferred geochemical model of many workers, but a good kinematic mechanism for recycling this material has been lacking until recently. However, Tatsumoto and Nakamura (1991) suggested that subduction of sub-continental lithosphere from

craton *margins* could explain this process. Therefore this mechanism will be incorporated below into an integrated model to explain the formation of enriched mantle components.

6.6.4 Kinematic model for mantle recycling

If EMI, EMII and HIMU are attributed to sub-continental lithosphere, continental sediment, and oceanic lithosphere respectively, a simple plate tectonic model can explain recycling of these components into the deeper mantle in two conjugate pairs: EMI–HIMU and EMII–HIMU. This is because there are two recognised styles of subduction zone setting, which were identified by Uyeda and Kanamori (1979) and described in detail by Uyeda (1982). The Peruvian-type setting (Fig. 6.35a) is characterised by a compressional stress regime across the arc–trench gap. This will cause ocean-floor sediments to be scraped off the upper side of the subducting oceanic plate and stacked in a fore-arc wedge, while the underside of the arc lithosphere suffers tectonic erosion. In contrast, the Mariana-type (Fig. 6.35b) is characterised by a tensional stress regime across the arc–trench gap. This causes subsidence of the trench bottom so that ocean-floor sediments are efficiently subducted, but tectonic erosion of sub-arc lithospheric mantle does not occur.

These two arc types can have very different consequences for mantle recycling. The compressional type should subduct a composite sheet consisting of oceanic crust overlain by lithospheric mantle. In continental arcs, this lithosphere may be enriched by metasomatism, thus generating the conjugate pair HIMU–EMI in the OIB source. On the other hand, for oceanic arcs, the lithosphere will be young (but possibly U-enriched). When combined with subducted crust, this can form a nearly-pure HIMU component.

Arcs displaying net extension must also be divided into two types. Trenches such as the north end of the Lesser Antilles (with a low sediment supply) will subduct barren oceanic crust, forming a HIMU component. Trenches such as the south end of the Lesser Antilles (with a large sediment supply) will subduct a composite sheet of oceanic crust and continental sediment. This can generate the conjugate pair HIMU–EMII in the OIB source. Isotopic evidence for these subduction processes will be discussed below in a discussion of the chemistry of arc environments.

The subduction of oceanic crust, along with marine

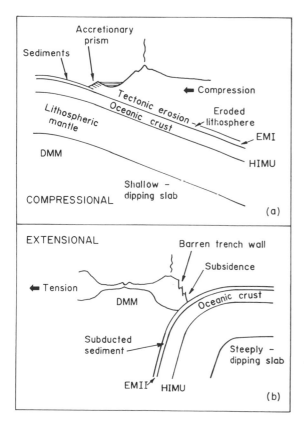

Fig. 6.35. Schematic illustrations of two different tectonic styles at subduction zones which may generate conjugate pairs of enriched mantle signatures: (a) HIMU–EMI; and (b) HIMU–EMII. Modified after Uyeda (1982).

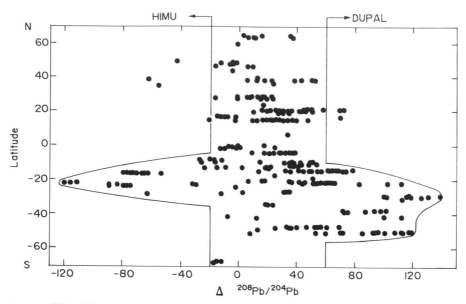

Fig. 6.36. Plot of Δ ^{208}Pb/^{204}Pb (see text) against latitude to show the geographical distribution of the Dupal and associated HIMU components. After Hart (1988).

sediment or eroded sub-continental lithosphere, may give rise to large-scale isotopic structure in the mantle. Hart (1984) argued that recycling into the astheno-sphere was responsible for generating a Pb and Sr isotope anomaly of global scale which he observed to form a small circle of approximately constant latitude encircling the southern hemisphere. He named it the 'Dupal' anomaly because its characteristic signature was first described in Indian ocean volcanics by Dupre and Allegre (1983). The spatial distribution of the Dupal anomaly is shown in Fig. 6.36, in the form of the variation of Δ 208/204 (^{208}Pb/^{204}Pb deviations from the North Atlantic correlation line) against latitude. Not only are the southern tropics charac-terised by the Dupal anomaly with a positive Δ 208/204 value, but a HIMU source with negative Δ 208/204 is also seen in the same area.

Hart (1988) and Castillo (1988) argued that the configuration of these anomalies was an indicator of the convective structure of the deep mantle. Staudigel *et al.* (1991) further suggested that large-scale regional isotope signatures such as the Dupal anomaly could be explained by 'focussed subduction' from a group of destructive plate margins, such as are presently seen in SE Asia.

6.7 Island arcs and mantle evolution

Island arcs are central to the understanding of mantle evolution because they represent the site where crustal

material of various types may be returned to the deep mantle. Island-arc magmatism may allow us to sample this material which is in the process of being recycled. Dewey (1980) showed that the volcanic front is always established about 100 km above the descending slab, whatever the angle of subduction. This shows that de-watering of the slab, triggered by pressure, is central to the operation of island-arc magmatism. However, the petrology of island-arc basalts (IAB) precludes their genesis by fusion of subducted oceanic crust (since this would require nearly 100% melting). Therefore, they must be dominantly produced by melting of the 'mantle wedge' overlying the subduction zone (e.g. Wyllie, 1984). Hence, the central problem in interpreting island-arc basalts is to identify which signatures are derived from the slab (and subducted sediment) and which are derived from the overlying wedge.

Island-arc basalts have enhanced levels of ^{87}Sr/^{86}Sr relative to MORB. However, the origin of these differences is only discernible in the context of other isotope evidence. The first study using combined Sr and Nd isotope data was made by Hawkesworth *et al.* (1977) on island-arc and back-arc tholeiites from the Scotia Sea (South Sandwich Islands). Analysis of back-arc material provides a control condition because it samples a mantle segment which may be similar to the wedge, but without any slab compo-nent.

Hawkesworth *et al.* found that both island-arc and back-arc samples had identical $^{143}Nd/^{144}Nd$, overlapping with MORB. However, the island-arc samples had significantly higher $^{87}Sr/^{86}Sr$ ratios (Fig. 6.37), which could not be explained by subaerial weathering. Hawkesworth *et al.* suggested that the enhanced Sr isotope ratios of the island-arc basalts were a product of subducted ^{87}Sr from seawater or (alternatively) oceanic sediments. Possible processes proposed were the direct partial melting of altered and subducted oceanic crust, or alternatively, metasomatic contamination of the mantle wedge with elements derived from the ocean crust. It is now generally accepted that the latter model is correct for the Scotia arc (e.g. Pearce, 1983).

The Scotia arc provides an example of the role of the slab-derived component in an arc with depleted chemistry. However, in arcs with less-depleted chemistry, material contributions from the slab and wedge are more difficult to resolve. The Lesser

Antilles (Caribbean) arc provides a good test case for the behaviour of arcs with more enriched signatures, since the chemistry of the arc changes along its length. This may help in resolving the origin of enriched components.

Grenada volcanics display variations in both $^{87}Sr/^{86}Sr$ and $^{143}Nd/^{144}Nd$, defining a range similar to those observed by Hawkesworth *et al.* (1979a) from Sao Miguel in the Azores, but further to the right of the main mantle correlation line (Fig. 6.38). In view of these parallels, Hawkesworth *et al.* (1979b) attributed the Grenada data to Sr contamination (from the slab) of a heterogeneous mantle wedge with a range of Nd and Sr isotope compositions along the mantle array. However it is unclear why this particular piece of sub-oceanic lithosphere should display such large heterogeneities, since mantle plumes cannot penetrate above a subduction zone. It would be necessary to propose that the mantle wedge in this location happened to contain an

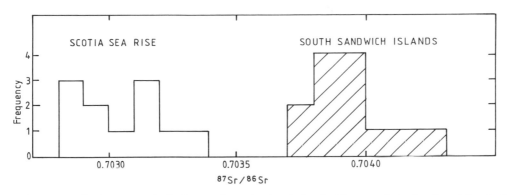

Fig. 6.37. Histograms of Sr isotope ratio for basalts from the South Sandwich arc and the (back-arc) Scotia Sea Rise. After Hawkesworth *et al.* (1977).

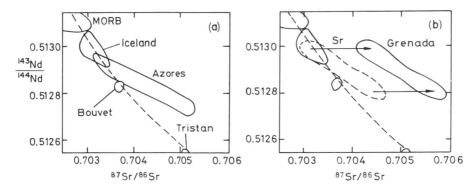

Fig. 6.38. Comparison of Sr–Nd isotope systematics in a) the Azores, and b) Grenada, showing possible derivation from enriched mantle sources. After Hawkesworth *et al.* (1979b).

enriched 'plum', as suggested for the Aleutian arc by Morris and Hart (1983).

Isotopic investigation of other islands in the Lesser Antilles (Davidson, 1983) revealed that St Kitts, situated at the northern end of the arc, has a very small range of isotopic composition close to MORB, while Martinique from the centre of the arc has an extremely large range of isotope composition (Fig. 6.39). If such variations were inherited from the mantle wedge, then 'gross heterogeneity on a scale of kilometres is implied'. Davidson initially ascribed the variations to contamination of the mantle source with subducted sediment. However, more detailed geochemical studies (Davidson, 1987) revealed positive correlations between Sr isotope ratio, oxygen isotope ratio and silica content in Martinique lavas. These are indicative of crustal contamination of ascending magma in the arc crust, which is thickest in the central region of the arc near Martinique. Such processes will not be detailed here, since they will be covered in the next chapter. However, they serve to exclude Martinique data from considerations of magma petrogenesis in the mantle.

White and Dupre (1986) presented Pb isotope data for representative samples from the whole length of the Lesser Antilles arc, showing that they were generally intermediate between MORB and sediment compositions. There is no evidence that these signatures are derived from magma contamination in the arc crust. For example, sedimentary xenoliths in Grenada lavas actually have unradiogenic Pb, inherited from an earlier location of the arc to the west, above the subducting Farallon plate. In contrast, Atlantic ocean floor sediments in front of the present-day arc have radiogenic Pb signatures.

White and Dupre found a general increase in the Pb isotope ratio of Atlantic floor sediment when going southwards in front of the Lesser Antilles subduction zone, probably reflecting sediment carried onto the sea floor at the south end of the arc by the Orinoco river. This trend was matched by the composition of Lesser Antilles volcanics, suggesting the presence of a subducted sediment component in the arc magmas. This model is supported by the covariation of Pb and Nd isotope data in the volcanics (Fig. 6.40). Two-component mixing between a MORB source and average Atlantic sediment can therefore explain the observed Pb–Nd isotope systematics of Lesser

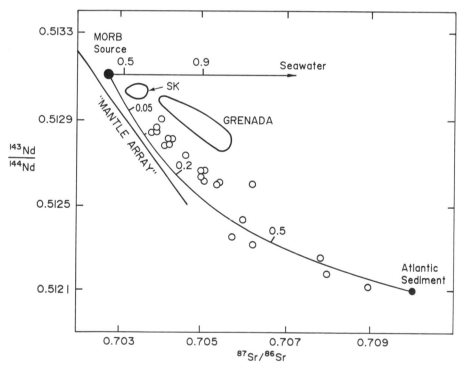

Fig. 6.39 Sr–Nd isotope diagram showing extreme isotopic variation in Martinique lavas (○) compared with St Kitts (SK). Mixing lines model the effects of contamination by sediments or seawater. After Davidson (1983).

Antilles magmas, avoiding the need to invoke an enriched mantle wedge (Ellam and Hawkesworth, 1988).

Rare earth concentration data may present a problem for this model, since light REE enrichment

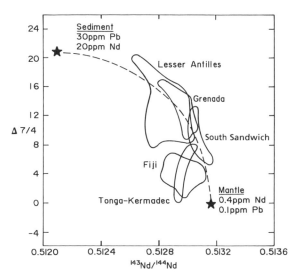

Fig. 6.40. Assessment of a sediment–asthenosphere mixing model for Lesser Antilles volcanics, in terms of Pb and Nd isotope systematics. Δ 7/4 indicates the $^{207}Pb/^{204}Pb$ deviation above the 'Northern Hemisphere Reference Line' of Hart (1984). After Ellam and Hawkesworth (1988).

in some arc volcanics may be too great to be explained by simple mixing between a MORB source and subducted sediment (Hawkesworth *et al.* 1991). This problem is illustrated in Fig. 6.41 on a plot of Ce/Yb (= REE profile slope) against Sr isotope ratio. LREE-enriched basalts and andesites from Grenada, the Sunda arc, and the Aeolian arc of southern Italy fall off the mixing line between depleted arcs and a typical sediment represented by 'post Archean average shale' (=PAAS). However, White and Dupre (1986) argued that the Pb isotope evidence for sediment involvement in arc magma genesis was so conclusive that it over-rides these trace element problems. Given this constraint, the very steep REE profiles must be due to some feature of the melting process. For example, *partial* melting of sediment in the presence of residual garnet could elevate light REE abundances in the melt while depressing the heavy REE.

Most workers now accept the supremacy of Pb isotope evidence for sediment involvement in IAB genesis. For example, Ben Othman *et al.* (1989) observed perfect matching of Pb isotope systematics between the West Sunda arc and ocean-floor sediment in front of the arc (Fig. 6.42). Since the Pb contents of arc volcanics are nearly an order of magnitude lower than typical sediments, it is unlikely that the sediment signature is itself controlled by erosion of arc volcanics. Therefore, it is most likely that the reverse relationship applies: arc volcanic Pb is controlled by

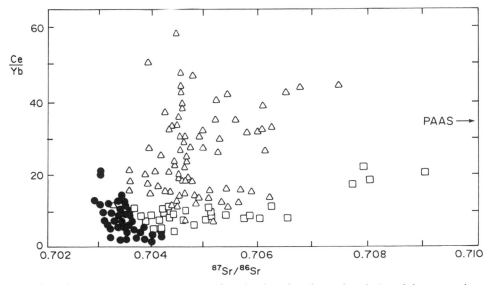

Fig. 6.41. Plot of Ce/Yb ratio against Sr isotope ratio for island-arc basalts and andesites. (●) = normal arc volcanics; (△) = LREE enriched; (▢) = Martinique lavas, contaminated during magma ascent. PAAS (post-Archean average shale) is a typical sediment composition. After Hawkesworth *et al.* (1991).

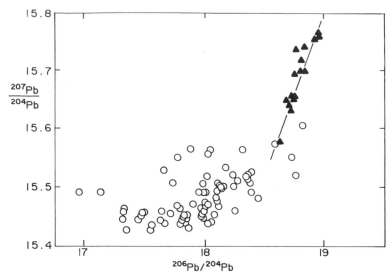

Fig. 6.42. Pb–Pb isotope plot showing collinearity of the West Sunda arc (▲) with ocean-floor sediment in front of the trench (solid line). Indian MORB is shown by open symbols. After Ben Othman *et al.* (1989).

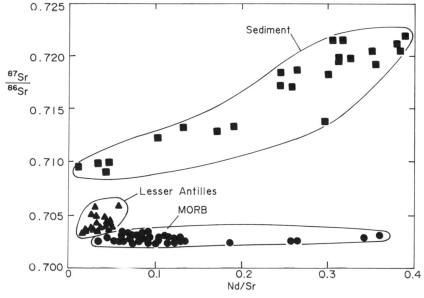

Fig. 6.43. Plot of Sr isotope ratio against Nd/Sr ratio to show the inadequacy of two-component mantle–sediment mixing to explain the high Sr contents (low Nd/Sr ratios) of Lesser Antilles lavas. After White and Dupre (1986).

subducted sediment. Further evidence was provided by McDermott *et al.* (1993), who observed Pb isotopic variations along the North Luzon (Philippine) arc which were correlated with the composition of sediment cores from the South China Sea, in front of the trench.

The two-component model for arc magmatism breaks down when we also consider incompatible trace element abundances in Lesser Antilles volcanics. Abundances of low-field-strength large-ion lithophile elements (LILE) such as Sr, are enriched relative to high-field-strength elements (HFSE) such as REE, when we compare the arc volcanics to MORB (Fig. 6.43). However, it is possible to explain the data by expanding the model to include mixing of three components in arc magma genesis (White and Dupre, 1986; Ellam and Hawkesworth, 1988). This involves contamination of the depleted-mantle source of IAB with partial melts of subducted sediment *and* LILE-enriched slab-derived fluids. Conclusive evidence for this fluid component is provided by beryllium–boron evidence (section 14.3.4).

References

Albarède, F. and Juteau, M. (1984). Unscrambling the lead model ages. *Geochim. Cosmochim. Acta* **48**, 207–12.

Allegre, C. J. (1982). Chemical geodynamics. *Tectonophys.* **81**, 109–32.

Allegre, C. J., Ben Othman, D., Polve, M. and Richard, P. (1979). The Nd–Sr isotopic correlation in mantle materials and geodynamic consequences. *Phys. Earth Planet. Inter.* **19**, 293–306.

Allegre, C. J., Brevart, O., Dupre, B. and Minster, J. F. (1980). Isotopic and chemical effects produced by a continuously differentiating convecting Earth mantle. *Phil. Trans. Roy. Soc. Lond. A* **297**, 447–77.

Allegre, C. J., Dupre, B. and Lewin, E. (1986). Thorium/uranium ratio of the Earth. *Chem. Geol.* **56**, 219–27.

Allegre, C. J., Hamelin, B. and Dupre, B. (1984). Statistical analysis of isotopic ratios in MORB: the mantle blob cluster model and the convective regime of the mantle. *Earth Planet. Sci. Lett.* **71**, 71–84.

Allegre, C. J., Hamelin, B., Provost, A. and Dupre, B. (1987). Topology in isotopic multispace and origin of mantle chemical heterogeneities. *Earth Planet. Sci. Lett.* **81**, 319–37.

Allegre, C. J., Hart, S. R. and Minster, J. F. (1983). Chemical structure and evolution of the mantle and continents determined by inversion of Nd and Sr isotopic data. I. Theoretical methods. *Earth Planet. Sci. Lett.* **66**, 177–90.

Allegre, C. J. and Turcotte, D. L. (1986). Implications of a two-component marble-cake mantle. *Nature* **323**, 123–7.

Appel, P. W. U., Moorbath, S. and Taylor, P. N. (1978). Least radiogenic terrestrial lead from Isua, west Greenland. *Nature* **272**, 524–6.

Barling, J. and Goldstein, S. L. (1990). Extreme isotopic variations in Heard Island lavas and the nature of mantle reservoirs. *Nature* **348**, 59–62.

Batiza, R. (1984). Inverse relationship between Sr isotope diversity and rate of oceanic volcanism has implications for mantle heterogeneity. *Nature* **309**, 440–1.

Ben Othman, D., White, W. M. and Patchett, J. (1989). The geochemistry of marine sediments, island arc magma genesis, and crust-mantle recycling. *Earth Planet. Sci. Lett.* **94**, 1–21.

Castillo, P. (1988). The Dupal anomaly as a trace of the upwelling lower mantle. *Nature* **336**, 667–70.

Chase, C. G. (1981). Oceanic island Pb: Two-stage histories and mantle evolution. *Earth Planet. Sci. Lett.* **52**, 277–84.

Chauvel, C., Hofmann, A. W. and Vidal, P. (1992). HIMU–EM: the French Polynesian connection. *Earth Planet. Sci. Lett.* **110**, 99–119.

Chen, C. Y. and Frey, F. A. (1983). Origin of Hawaiian tholeiite and alkalic basalt. *Nature* **302**, 785–9.

Class, C., Goldstein, S. L., Galer, S. J. G. and Weis, D. (1993). Young formation age of a mantle plume source. *Nature* **362**, 715–21.

Cohen, R. S., Evensen, N. M., Hamilton, P. J. and O'Nions, R. K. (1980). U–Pb, Sm–Nd and Rb–Sr systematics of ocean ridge basalt glasses. *Nature* **283**, 149–53.

Cohen, R. S. and O'Nions, R. K. (1982). Identification of recycled continental material in the mantle from Sr, Nd and Pb isotope investigations. *Earth Planet. Sci. Lett.* **61**, 73–84.

Cumming, G. L. and Richards, J. R. (1975). Ore lead isotope ratios in a continuously changing Earth. *Earth Planet. Sci. Lett.* **28**, 155–71.

Dasch, E. J., Hedge, C. E. and Dymond, J. (1973). Effect of seawater alteration on strontium isotope composition of deep-sea basalts. *Earth Planet. Sci. Lett.* **19**, 177–83.

Davidson, J. P. (1983). Lesser Antilles isotopic evidence of the role of subducted sediment in island arc magma genesis. *Nature* **306**, 253–6.

Davidson, J. P. (1987). Crustal contamination *versus* subduction zone enrichment: examples from the Lesser Antilles and implications for mantle source compositions of island arc volcanic rocks. *Geochim. Cosmochim. Acta* **51**, 2185–98.

DePaolo, D. J. (1980). Crustal growth and mantle evolution: inferences from models of element transport and Nd and Sr isotopes. *Geochim. Cosmochim. Acta* **44**, 1185–96.

DePaolo, D. J. and Wasserburg, G. J. (1976). Inferences about magma sources and mantle structure from variations of $^{143}Nd/^{144}Nd$. *Geophys. Res. Lett.* **3**, 743–6.

DePaolo, D. J. and Wasserburg, G. J. (1979). Petrogenetic mixing models and Nd–Sr isotopic patterns. *Geochim. Cosmochim. Acta* **43**, 615–27.

Dewey, J. (1980). Episodicity, sequence and style at convergent plate boundaries. In: Strangway, D. W. (Ed.), *The Continental Crust and its Mineral Deposits. Geol. Assoc. Canada Spec. Pap.* **8**, pp. 553–73.

Dosso, L. and Murthey, V. R. (1980) A Nd isotope study of the Kerguelen islands: inferences on enriched oceanic mantle sources. *Earth Planet. Sci. Lett.* **48**, 268–76.

Dupre, B. and Allegre, C. J. (1980). Pb–Sr–Nd isotopic correlation and the chemistry of the North Atlantic mantle. *Nature* **286**, 17–22.

Dupre, B. and Allegre, C. J. (1983). Pb–Sr isotope variation in Indian Ocean basalts and mixing phenomena. *Nature* **303**, 142–6.

Ellam, R. M. and Hawkesworth, C. J. (1988). Elemental and isotopic variations in subduction related basalts: evidence for a three component model. *Contrib. Mineral. Petrol.* **98**, 72–80.

Faure, G. and Hurley, P. M. (1963). The isotopic composition of strontium in oceanic and continental basalt. *J. Petrol.* **4**, 31–50.

Flower, M. F. J., Schmincke, H. U. and Thompson, R. N.

(1975). Phlogopite stability and the $^{87}Sr/^{86}Sr$ step in basalts along the Reykjanes Ridge. *Nature* **254**, 404–6.

Galer, S. J. G. and O'Nions, R. K. (1985). Residence time of thorium, uranium and lead in the mantle with implications for mantle convection. *Nature* **316**, 778–82.

Gast, P. W., Tilton, G. R. and Hedge, C. (1964). Isotopic composition of lead and strontium from Ascension and Gough Islands. Science **145**, 1181–5.

Goldstein, S. L. (1988). Decoupled evolution of Nd and Sr isotopes in the continental crust and the mantle. *Nature* **336**, 733–8.

Halliday, A. N., Davidson, J. P., Holden, P., DeWolf, C., Lee, D-C. and Fitton, J. G. (1990). Trace-element fractionation in plumes and the origin of HIMU mantle beneath the Cameroon Line. *Nature* **347**, 523–8.

Halliday, A. N., Davies, G. R., Lee, D-C., Tommasini, S., Paslick, C. R., Fitton, J. G. and James, D. E. (1992). Lead isotope evidence for young trace element enrichment in the oceanic upper mantle. *Nature* **359**, 623–7.

Halliday, A. N., Dickin, A. P., Fallick, A. E. and Fitton, J. G. (1988). Mantle dynamics: A Nd, Sr, Pb and O isotope study of the Cameroon Line volcanic chain. *J. Petrol.* **29**, 181–211.

Hanan, B. B., Kingsley, R. H. and Schilling, J-G. (1986). Pb isotope evidence in the South Atlantic for migrating ridge-hotspot interactions. *Nature* **322**, 137–44.

Harris, C., Bell, J. D. and Atkins, F. B. (1983). Isotopic composition of lead and strontium in lavas and coarse-grained blocks from Ascension Island, South Atlantic–an addendum. *Earth Planet. Sci. Lett.* **63**, 139–41.

Harris, P. G., Hutchison, R. and Paul, D. K. (1972). Plutonic xenoliths and their relation to the upper mantle. *Phil. Trans. Roy. Soc. Lond. A* **271**, 313–23.

Hart, S. R. (1984). A large-scale isotope anomaly in the Southern Hemisphere mantle. *Nature* **309**, 753–7.

Hart, S. R. (1988). Heterogeneous mantle domains: signatures, genesis and mixing chronologies. *Earth Planet. Sci. Lett.* **90**, 273–96.

Hart, S. R., Hauri, E. H., Oschmann, L. A. and Whitehead, J. A. (1992). Mantle plumes and entrainment: isotopic evidence. *Science* **256**, 517–20.

Hart, S. R., Gerlach, D. C. and White, W. M. (1986). A possible new Sr–Nd–Pb mantle array and consequences for mantle mixing. *Geochim. Cosmochim. Acta* **50**, 1551–7.

Hart, S. R., Schilling, J-G. and Powell, J. I.. (1973). Basalts from Iceland and along the Reykjanes Ridge: Sr isotope geochemistry. *Nature Phys. Sci.* **246**, 104–7.

Hauri, E. H., Shimizu, N., Dieu, J. J. and Hart, S. R. (1993). Evidence for hotspot-related carbonatite metasomatism in the oceanic upper mantle. *Nature* **365**, 221–7.

Hawkesworth, C. J., Hergt, J. M., McDermott, F. and Ellam, R. M. (1991). Destructive margin magmatism and the contributions from the mantle wedge and subducted crust. *Aust. J. Earth Sci.* **38**, 577–94.

Hawkesworth, C. J., Norry, M. J., Roddick, J. C. and Vollmer, R. (1979a). $^{143}Nd/^{144}Nd$ and $^{87}Sr/^{86}Sr$ ratios from the Azores and their significance in LIL element enriched mantle. *Nature* **280**, 28–31.

Hawkesworth, C. J., O'Nions, R. K. and Arculus, R. J.

(1979b). Nd and Sr isotope geochemistry of island arc volcanics, Grenada, Lesser Antilles. *Earth Planet. Sci. Lett.* **45**, 237–48.

Hawkesworth, C. J., O'Nions, R. K., Pankhurst, R. J., Hamilton, P. J. and Evensen, N. M. (1977). A geochemical study of island-arc and back-arc tholeiites from the Scotia Sea. *Earth Planet. Sci. Lett.* **36**, 253–62.

Hawkesworth, C. J., Rogers, N. W., van Calsteren, P. W. C. and Menzies, M. A. (1984). Mantle enrichment processes. *Nature* **311**, 331–3.

Hofmann, A. W. and Hart, S. R. (1978). An assessment of local and regional isotopic equilibrium in the mantle. *Earth Planet. Sci. Lett.* **38**, 44–62.

Hofmann, A. W. and White, W. M. (1980). The role of subducted oceanic crust in mantle evolution. *Carnegie Inst. Washington Yearbook* **79**, 477–83.

Hofmann, A. W. and White, W. M. (1982). Mantle plumes from ancient oceanic crust. *Earth Planet. Sci. Lett.* **57**, 421–36.

Jacobsen, S. B. and Wasserburg, G. J. (1979). The mean age of mantle and crustal reservoirs. *J. Geophys. Res.* **84**, 7411–27.

Kenyon, P. M. (1990). Trace element and isotopic effects arising from magma migration beneath mid-ocean ridges. *Earth Planet. Sci. Lett.* **101**, 367–78.

McDermott, F., Defant, M. J., Hawkesworth, C. J., Maury, R. C. and Joron, J. L. (1993). Isotope and trace element evidence for three component mixing in the genesis of the North Luzon arc lavas (Philippines). *Contrib. Mineral. Petrol.* **113**, 9–23.

McKenzie, D. (1979). Finite deformation during fluid flow. *Geophys. J. Roy. Astr. Soc.* **58**, 689–715.

McKenzie, D. P. and O'Nions, R. K. (1983). Mantle reservoirs and ocean island basalts. *Nature* **301**, 229–31.

Mertz, D. F., Devey, C. W., Todt, W., Stoffers, P. and Hofmann, A. W. (1991). Sr–Nd–Pb isotope evidence against plume–asthenosphere mixing north of Iceland. *Earth Planet. Sci. Lett.* **107**, 243–55.

Michard, A. and Albarede, F. (1985). Hydrothermal uranium uptake at ridge crests. *Nature* **317**, 244–6.

Morgan, J. P. and Shearer, P. M. (1993). Seismic constraints on mantle flow and topography of the 660-km discontinuity: evidence for whole-mantle convection. *Nature* **365**, 506–11.

Morgan, W. J. (1971) Convection plumes in the lower mantle. *Nature* **230**, 42–3.

Morris, J. D. and Hart, S. R. (1983). Isotopic and incompatible element constraints on the genesis of island arc volcanics: Cold Bay and Amak Islands, Aleutians. *Geochim. Cosmochim. Acta* **47**, 2015–30.

Newsome, H. E., White, W. M., Jochum, K. P. and Hofmann, A. W. (1986). Siderophile element abundances in oceanic basalts, Pb isotope evolution and growth of the Earth's core. *Earth Planet. Sci. Lett.* **80**, 299–313.

Norry, M. J. and Fitton, J. G. (1983). Compositional differences between oceanic and continental basic lavas and their significance. In: Hawkesworth, C. J. and Norry, M. J. (Eds), *Continental Basalts and Mantle Xenoliths*. Shiva, pp. 5–19.

O'Hara, M. J. (1973). Non-primary magmas and dubious mantle plume beneath Iceland. *Nature* 243, 507–8.

O'Hara, M. J. (1975). Is there an Icelandic mantle plume? *Nature* 253, 708–10.

O'Hara, M. J. and Mathews, R. E. (1981). Geochemical evolution in an advancing, periodically replenished, periodically tapped, continuously fractionated magma chamber. *J. Geol. Soc. Lond.* 138, 237–77.

Olson, P. (1984). Mixing of passive heterogeneities by mantle convection. *J. Geophys. Res.* 89, B425–36.

Olson, P., Silver, P. G. and Carlson, R. W. (1990). The large-scale structure of convection in the Earth's mantle. *Nature* 344, 209–15.

O'Nions, R. K., Evensen, N. M. and Hamilton, P. J. (1979). Geochemical modelling of mantle differentiation and crustal growth. *J. Geophys. Res.* 84 6091–101.

O'Nions, R. K., Evensen, N. M. and Hamilton, P. J. (1980). Differentiation and evolution of the mantle. *Phil. Trans. Roy. Soc. Lond. A* 297, 479–93.

O'Nions, R. K., Hamilton, P. J. and Evensen, N. M. (1977). Variations in ^{143}Nd/^{144}Nd and ^{87}Sr/^{86}Sr ratios in oceanic basalts. *Earth Planet. Sci. Lett.* 34, 13–22.

O'Nions, R. K. and Pankhurst, R. J. (1973). Secular variation in the Sr–isotope composition of Icelandic volcanic rocks. *Earth Planet. Sci. Lett.* 21, 12–21.

Palacz, Z. A. and Saunders, A. D. (1986). Coupled trace element and isotope enrichment in the Cook–Austral–Samoa islands, southwest Pacific. *Earth Planet. Sci. Lett.* 79, 270–80.

Pearce, J. (1983). The role of sub-continental lithosphere in magma genesis at destructive plate margins. In: Hawkesworth, C. J. and Norry, M. J. (Eds), *Continental Basalts and Mantle Xenoliths*. Shiva, pp. 230–49.

Polve, M. and Allegre, C. J. (1980). Orogenic lherzolite complexes studied by ^{87}Rb–^{87}Sr: a clue to understanding the mantle convection process? *Earth Planet. Sci. Lett.* 51, 71–93.

Prinzhofer, A., Lewin, E. and Allegre, C. J. (1989). Stochastic melting of the marble cake mantle: evidence from local study of the East Pacific Rise at 12° 50′ N. *Earth Planet. Sci. Lett.* 92, 189–206.

Richard, P., Shimizu, N. and Allegre, C. J. (1976) ^{143}Nd/^{144}Nd, a natural tracer: an application to oceanic basalts. *Earth Planet. Sci. Lett.* 31, 269–78.

Richter, F. M. and Ribe, N. M. (1979). On the importance of advection in determining the local isotopic composition of the mantle. *Earth Planet. Sci. Lett.* 43, 212–22.

Ringwood, A. E. (1982). Phase transformations and differentiation in subducted lithosphere: implications for mantle dynamics, basalt petrogenesis, and crustal evolution. *J. Geol.* 90, 611–43.

Schilling, J-G. (1973). Iceland mantle plume: geochemical study of Reykjanes Ridge. *Nature* 242, 565–71.

Schilling, J-G. and Noe Nygaard, A. (1974). Faeroe–Iceland plume; rare-earth evidence. *Earth Planet. Sci. Lett.* 24, 1–14.

Shearer, P. M. and Masters, T. G. (1992). Global mapping of topography on the 660-km discontinuity. *Nature* 355, 791–6.

Staudigel, H., Park, K-H., Pringle, M., Rubenstone, J. L., Smith, W. H. F. and Zindler, A. (1991). The longevity of the South Pacific isotopic and thermal anomaly. *Earth Planet. Sci. Lett.* 102, 24–44.

Sun, S. S. (1985). Ocean islands–plums or plumes? *Nature* 316, 103–4.

Sun, S. S. (1980). Lead isotopic study of young volcanic rocks from mid-ocean ridges, ocean islands and island arcs. *Phil. Trans. Roy. Soc. Lond. A* 297, 409–45.

Sun, S. S. and Hanson, G. N. (1975). Evolution of the mantle: geochemical evidence from alkali basalt. *Geology* 3, 297–302.

Sun, S. S., Tatsumoto, M. and Schilling, J-G. (1975). Mantle plume mixing along the Reykjanes ridge axis: lead isotopic evidence. *Science* 190, 143–7.

Tatsumoto, M. (1966). Genetic relations of oceanic basalts as indicated by lead isotopes. *Science* 153, 1094–101.

Tatsumoto, M. (1978). Isotopic composition of lead in oceanic basalt and its implication to mantle evolution. *Earth Planet. Sci. Lett.* 38, 63–87.

Tatsumoto, M. (1988). U, Th and Pb abundances in Hawaiian xenoliths. *Conf. Origin of the Earth. Lunar Planet. Inst.* pp. 89-90.

Tatsumoto, M., Knight, R. J. and Allegre, C. J. (1973). Time differences in the formation of meteorites as determined from the ratio of lead-207 to lead-206. *Science* 180, 1279–83.

Tatsumoto, M. and Nakamura, Y. (1991). Dupal anomaly in the Sea of Japan: Pb, Nd, and Sr isotopic variations at the eastern Eurasian continental margin. *Geochim. Cosmochim. Acta* 55, 3697–708.

Turcotte, D. L. and Oxburgh, E. R. (1967). Finite amplitude convective cells and continental drift. *J. Fluid. Mech.* 28, 29–42.

Uyeda, S. (1982). Subduction zones: an introduction to comparative subductology. *Tectonophys.* 81, 133–59.

Uyeda, S. and Kanamori, H. (1979). Back-arc opening and the mode of subduction. *J. Geophys. Res.* 84, 1049–61.

Vidal, P. and Dosso, L. (1978). Core formation: catastrophic or continuous? Sr and Pb isotope geochemistry constraints. *Geophys. Res. Lett.* 5, 169–72.

Weaver, B. L. (1991). The origin of ocean island basalt end-member compositions: trace element and isotopic constraints. *Earth Planet. Sci. Lett.* 104, 381–97.

Weis, D., Frey, F. A., Leyrit, H. and Gautier, I. (1993). Kerguelen Archipelago revisited: geochemical and isotopic study of the Southeast Province lavas. *Earth Planet. Sci. Lett.* 118, 101–19

White, W. M. (1981). In: *European Colloquium of Geochronology, Cosmochronology and Isotope Geology VII*, meeting abstract.

White, W. M. (1985). Sources of oceanic basalts: radiogenic isotopic evidence. *Geology* 13, 115–18.

White, W. M. (1993a). ^{238}U/^{204}Pb in MORB and open system evolution of the depleted mantle. *Earth Planet. Sci. Lett.* 115, 211–26.

White, W. M. (1993b). Isotopes and a smoking gun. *Nature* 362, 791–2.

White, W. M. and Dupre, B. (1986). Sediment subduction

and magma genesis in the Lesser Antilles: isotopic and trace element constraints. *J. Geophys. Res.* **91**, 5927–41.

White, W. M., Dupre, B. and Vidal, P. (1985). Isotope and trace element geochemistry of sediments from the Barbados Ridge–Demerara Plain region, Atlantic Ocean. *Geochim. Cosmochim. Acta* **49**, 1875–86.

White, W. M. and Hofmann, A. W. (1982). Sr and Nd isotope geochemistry of oceanic basalts and mantle evolution. *Nature* **296**, 821–5.

White, W. M., Schilling, J-G. and Hart, S. R. (1976). Evidence for the Azores mantle plume from strontium isotope geochemistry of the Central North Atlantic. *Nature* **263**, 659–63.

White, W. M., Tapia, M. D. M. and Schilling, J-G. (1979). The petrology and geochemistry of the Azores islands. *Contrib. Mineral. Petrol.* **69**, 201–13.

Woodhead, J. D., Greenwood, P., Harmon, R. S. and Stoffers, P. (1993). Oxygen isotope evidence for recycled crust in the source of EM-type ocean island basalts. *Nature* **362**, 809–13.

Wyllie, P. J. (1984). Constraints imposed by experimental petrology on possible and impossible magma sources and products. *Phil. Trans. Roy. Soc. Lond. A* **310**, 439–56.

Zindler, A. and Hart, S. R. (1986). Chemical geodynamics. *Ann. Rev. Earth Planet. Sci.* **14**, 493–571.

Zindler, A., Jagoutz, E. and Goldstein, S. (1982). Nd, Sr and Pb isotopic systematics in a three-component mantle: a new perspective. *Nature* **298**, 519–23.

Zindler, A., Staudigel, H. and Batiza, R. (1984). Isotope and trace element geochemistry of young Pacific seamounts: Implications for the scale of upper mantle heterogeneity. *Earth Planet. Sci. Lett.* **70**, 175–95.

7 Isotope geochemistry of continental rocks

Oceanic volcanics, erupted through thin, young lithosphere, represent a window on the asthenosphere and deep mantle. In contrast, continental basalts and mantle xenoliths, emplaced through thick, old lithosphere, may tell us about shallow enrichment processes in the Earth. Isotopic data represent a powerful tool for such studies because of their ability to date geological events and their usefulness as tracers of mixing processes.

Unfortunately, continental igneous rocks are difficult to interpret. This is because they can derive an enriched elemental and isotopic signature from three possible sources: mantle plumes, sub-continental lithosphere, and the crust. Resolving these components from one another in continental volcanics and plutons has been a major subject of discussion in geochemistry for three decades. Much progress has been made, but the large number of variables tends to make each case a unique example; or as Read (1948) put it, there are 'granites and granites.' This makes a generalised approach to continental magmas difficult, and forces us to adopt a case study approach as an attempt to illustrate underlying principles.

Mantle xenoliths provide a more direct means of sampling the sub-continental lithosphere. Their texture provides evidence of a solid source, while the peridotite (= lherzolite) petrology of the commonest types is readily distinguished from crustal xenoliths (which will not be dealt with here). Therefore, our approach in this chapter will be firstly to study the lithospheric mantle by means of xenoliths, secondly to examine crustal contamination processes, and lastly to look at some classic case studies in the genesis and evolution of continental igneous rocks.

7.1 Mantle xenoliths

The sub-continental lithosphere is distinguished from the underlying asthenosphere by its non-convecting, rigid state. Hence it was termed the 'tectosphere' by Jordan (1975, 1978). Jordan argued from seismic and heat-flow evidence that this tectosphere was 200–300 km thick under shield areas. Evidence from diamond inclusions in garnets (section 4.2.1) suggests similar thickness of continental lithosphere in the Archean.

Alkaline magmas, kimberlites and carbonatites in many continental areas bring up peridotite xenoliths (also called nodules). On the basis of their mineral chemistry, these must be samples of the mantle rather than the crust. Maaloe and Aoki (1977) analysed the major-element composition of numerous such xenoliths in an attempt to estimate the bulk upper mantle composition. They recognised compositional differences between spinel lherzolite xenoliths, derived from Proterozoic and younger lithosphere, and garnet lherzolites, derived from Archean cratons. Both xenolith types had overlapping ranges of MgO content, but the (Archean) garnet peridotites had distinctly lower FeO contents. In view of their more exotic history, we will direct our main attention to this group.

The world's classic mantle xenolith suites come from South Africa, where they are by-products of diamond mining. Within this collection, two main textural types are observed: granular and sheared. Harte (1983) proposed that the former were samples of the lithosphere, while the latter, which are more often found around the margins of the Kaapvaal craton, were derived from the convecting asthenosphere. Of the granular types, Hart further sub-divided samples showing obvious or 'modal' metasomatism (indicated by hydrous or other exotic minerals) from the more normal garnet peridotites. The latter samples came from the centre of the Kaapvaal craton, at Northern Lesotho and Bultfontein (hence NLB-type), and were regarded as typical samples of the mantle lithosphere.

Various explanations have been proposed to account for the differing FeO contents of spinel and garnet peridotites. However, the most satisfactory was developed by Richter (1988). He proposed that garnet peridotites were residues of komatiite extraction in the Archean, and that the large degrees of melting associated with this process caused FeO depletion. This in turn lowered the density of the residuum, relative to fertile mantle, and allowed its stabilisation as sub-continental lithosphere. This material reached sufficient thicknesses (> 150 km) for diamond crystallisation to occur at its base. In contrast, Proterozoic lithosphere was stabilised only by conductive cooling of the upper mantle (a mechanism which would not have been possible in the hotter Archean mantle). This upper mantle may be residual from basalt extraction, or may not be depleted by melt extraction at all; hence it has higher levels of FeO and other fertile components. The thickness of lithosphere formed in this way is insufficient for diamond stability, while its high density makes it susceptible to delamination from the base of the crust during orogenic shortening of the lithosphere.

The major-element compositional differences between garnet and spinel peridotite xenoliths, described above, are paralleled by isotopic differences. Fig. 7.1 shows a compilation of Sr and Nd isotope data for the two groups (Hawkesworth *et al.*, 1990a), which define fairly distinct fields. Spinel peridotite data are derived mainly from separated clinopyroxene (cpx), but garnet peridotite data are based on a combination of separated mineral and whole-rock analyses. The latter are less reliable because they are susceptible to

contamination by the host magma (usually kimberlite in the case of garnet peridotite).

Menzies (1989) adopted a terminology for interpreting mantle xenoliths (Fig. 7.2) which was based on the DMM, EMI and EMII end-members proposed for OIB sources by Zindler and Hart (1986), (section 6.5.2). He did not propose that the processes which formed these types of lithospheric 'domains' were necessarily the same as those which formed OIB end-members, but the use of such a terminology may imply a genetic relationship. Zindler and Hart did in fact propose (section 6.5.2) that the HIMU and EMI components (forming the LoNd array) were derived from recycled mantle lithosphere. However, the EMI component does not necessarily bear a direct relationship to any given segment of lithosphere. Furthermore, such a model does not fit well to the EMII component of the OIB source, which is widely attributed to sediment subduction. Hence, the present author prefers the use of different names for domain types in plume sources and the lithosphere.

7.1.1 Mantle metasomatism

Spinel peridotite data in Fig. 7.1 are generally depleted relative to Bulk Earth. Therefore, they may represent fairly normal samples of the upper mantle. However, garnet peridotites generally fall in the

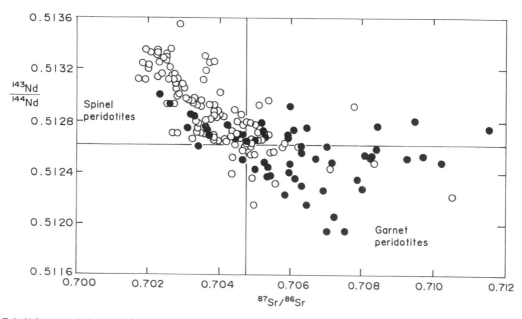

Fig. 7.1. Nd *versus* Sr isotope diagram showing the largely distinct compositional fields of spinel peridotite (o), and garnet peridotite (•). After Hawkesworth *et al.* (1990a).

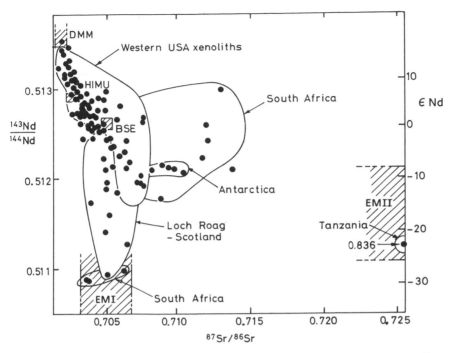

Fig. 7.2. Nd *versus* Sr isotope diagram showing compositional fields for xenolith suites from different provinces, relative to enriched mantle components identified in OIB sources (hatched fields). After Menzies (1989).

enriched quadrant relative to Bulk Earth, despite the fact that they are interpreted as residues of komatiite extraction. This demands a secondary enrichment process, which cannot be due to interaction of the xenoliths with kimberlite liquid, since kimberlites have more depleted isotopic compositions than the xenoliths.

Mantle enrichment may be caused by either silicate melts, or by hydrous or carbonaceous fluids. Only the latter are examples of metasomatism in the strict sense (Menzies and Hawkesworth, 1987), but typically, mantle enrichment is regarded as more-or-less synonymous with mantle metasomatism. Some garnet peridotite xenoliths show chemical evidence of an enrichment process, but no petrographic evidence for this process. This phenomenon was termed 'cryptic' metasomatism by Dawson (1984) to distinguish it from 'patent' metasomatism, which is petrographically recognisable, due to the development of replacement textures or new hydrous phases. The latter is equivalent to the 'modal' metasomatism of Harte (1983).

Having established the importance of mantle metasomatism in generating the observed isotope signatures in peridotite xenoliths, another important question is the timing of this process. We will examine this question for the classic xenolith suites of the Kaapvaal craton in South Africa.

Kramers (1979) analysed the Pb isotope composition of sulphide inclusions in diamonds (and also cpx from eclogite and peridotite xenoliths) in several Cretaceous kimberlite pipes. Both inclusion and cpx data lay close to a 2.5 Byr isochron line (Fig. 7.3), implying that diamonds and xenoliths are co-genetic, and that mineralogical heterogeneity was preserved in the South African sub-continental lithosphere since the Archean. In particular, the very unradiogenic composition of the diamonds, which yield Pb model ages of over 2 Byr, would be very difficult to explain by any recent metasomatic event. In contrast, Pb isotope compositions in 'fertile peridotites' and cpx megacrysts were interpreted as evidence of fairly recent disturbance.

Menzies and Murthy (1980) analysed the Sr and Nd isotope compositions of diopsides in micaceous garnet lherzolite nodules from South African kimberlite pipes. The diopsides showed a strong inverse correlation on the Sr–Nd isotope diagram (Fig. 7.4). This was attributed by Menzies and Murthy to gross mantle heterogeneity, randomly sampled by kimberlite magmas. They suggested that these signatures were generated by an ancient metasomatic event,

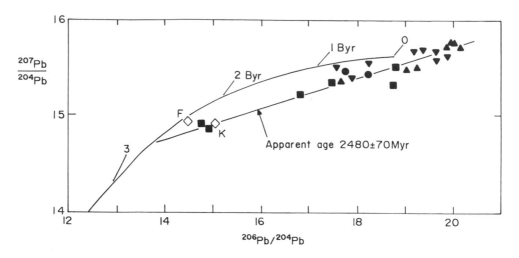

Fig. 7.3. Pb–Pb isochron diagram for nodules from South African kimberlites. Open symbols: sulphide inclusions in diamonds (F = Finsch mine, K = Kimberley); solid symbols: cpx from peridotite and cpx megacrysts (different symbols signify different mines). After Kramers (1979).

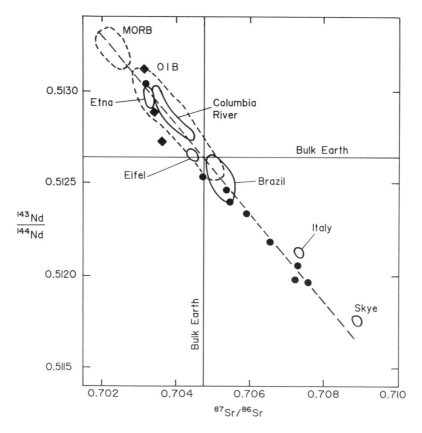

Fig. 7.4. Plot of Nd *versus* Sr isotope ratios for diopsides from South African kimberlite nodules (●), relative to the mantle array of oceanic basalts. (◆) = whole-rock peridotites. After Menzies and Murthy (1980).

probably related to an upwelling mantle plume, which caused LIL element enrichment of the mantle lithosphere.

Hawkesworth *et al.* (1983) estimated from Nd isotope data that the ancient enrichment event postulated by Menzies and Murthy probably occurred ca. 1–4 Byr ago. However, the Rb/Sr ratios of the analysed diopsides (and indeed any mantle diopsides) are much too low to 'support' their observed $^{87}Sr/^{86}Sr$ compositions (i.e., generate the required extra amount of ^{87}Sr by *in-situ* ^{87}Rb decay in the required time). This is demonstrated in Fig. 7.5. The enhanced Sr isotope ratios also cannot be generated by contamination with kimberlite magma, because the latter has unradiogenic ^{87}Sr. Therefore, Hawkesworth *et al.* argued that the diopsides must have crystallised in a recent event, presumably during secondary metasomatism of the enriched mantle which was generated by the ancient metasomatic event.

Hawkesworth *et al.* (1983; 1990b) observed a linear array of whole-rock data for (garnet-free) phlogopite-bearing and richterite-bearing nodules on an Rb–Sr isochron diagram (Fig. 7.5). They attributed this array to a metasomatic event about 150 Myr ago, possibly associated with the Karoo event. This age was supported by whole-rock Pb–Pb data on peridotite lithologies from Kimberley, which yield an age of 200 ± 150 Myr (Hawkesworth *et al.*, 1990b). Unfortunately, the time of metasomatism cannot be pin-pointed any more exactly than this, due to the open-system nature of the whole-rock samples during metasomatism. Similarly, mineral isochrons cannot define the time of metasomatism because they were opened during the thermal event associated with kimberlite emplacement. For example, a suite of

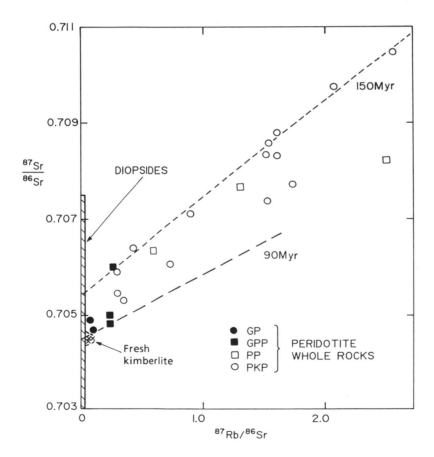

Fig. 7.5. Rb–Sr isochron diagram for South African kimberlite nodules, showing diopside field (hatched) relative to kimberlite host (×) and nodules of different lithologies (GP = garnet peridotite; GPP = garnet–pargasite peridotite; PP = phlogopite peridotite; PKP = phlogopite–K-richterite peridotite). After Hawkesworth *et al.* (1983).

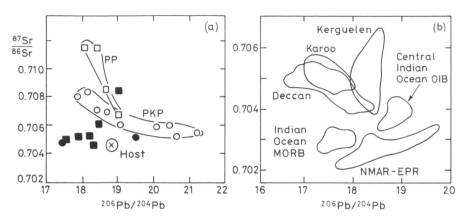

Fig. 7.6. Plots of Sr against Pb isotope data. a) Whole-rock samples of peridotite xenoliths from Kimberley; symbols as in Fig. 7.5. b) Fields for continental flood basalt provinces and oceanic volcanics. After Hawkesworth *et al.* (1990b).

phlogopites from a Bultfontein peridotite yielded a well-fitted Rb–Sr mineral isochron with an age of 84 Myr (Kramers *et al.*, 1983), which is close to the emplacement age of 90 Myr determined from U–Pb data.

When whole-rock Pb isotope data for these nodules are plotted against Sr data (Fig. 7.6a), they define an inverse correlation which cannot be explained by mixing with the kimberlite host. Therefore, Hawkesworth *et al.* attributed the data to phlogopite metasomatism in the postulated 150 Myr event. This is expected to generate sources with elevated Rb/Sr but depressed U/Pb, and was argued to be a distinctive geochemical fingerprint which may identify shallow mantle enrichment processes. Hawkesworth *et al.* (1990b) argued that the trends in Fig. 7.6a match the fields defined by some continental flood basalts (Fig. 7.6b), but were unlike most oceanic basalts (with the exception of the Walvis Ridge). Therefore, these flood basalts may have a lithospheric source. However, such trends could potentially be generated by contamination of magmas by U- and Rb-poor lower continental crust. It therefore seems appropriate at this stage to examine crustal contamination mechanisms.

7.2 Crustal contamination

Many continental igneous rocks have enriched chemical and isotopic signatures. A critical question is whether these signatures were inherited from the mantle or the crust. In principle, isotopic methods represent an ideal tool to solve this problem, since

they are not upset by the crystal fractionation processes which affect most magmas during ascent and emplacement. However, the high degrees of enrichment which can occur in plume or lithospheric mantle sources may generate isotopic signatures similar to the crust. Hence it has been argued (e.g. Thirlwall and Jones, 1983; Hawkesworth *et al.*, 1984) that mantle and crustal sources cannot be distinguished simply on the basis of 'isotopic discriminant diagrams' in which each component has a unique field. Instead, crustal or mantle contributions to magmatism must be recognised by observing the products of *processes* such as magma mixing or crustal assimilation.

Philosophically, one can examine contamination processes in two ways: a predictive model (e.g. DePaolo, 1981a) or an inversion technique (e.g. Mantovani and Hawkesworth, 1990). In the former, we set conditions and then examine consequences. In the latter, we examine products and attempt to reconstruct the original conditions. The predictive model is well-suited to two-component mixing processes, such as progressive contamination of a single magma batch by wall-rock assimilation. Some examples of such models will be examined below. In contrast, volcanic lava piles may involve multi-component mixing. These processes are more difficult to examine using predictive models, because of the plethora of possible mixing scenarios. Therefore, it is more effective to model such suites using the inversion approach, bearing in mind the predictive models already developed for single magma batches. The inversion approach is best illustrated using case studies, which may involve a loss of objectivity, but this, in the writer's opinion, does not invalidate the approach.

7.2.1 Two-component mixing models

In its simplest form, contamination of mantle-derived magma by the continental crust can be regarded as a process of two-component mixing. However, magma–crust mixing processes usually have more than one degree of freedom (such as the compositions and proportions of mixed components). Therefore, to adequately evaluate mixing relations it is usually necessary to apply two or more measured variables to the problem. These variables are usually isotope ratios, elemental ratios, and elemental abundances. In the context of isotope geology, it is logical to begin by examining the behaviour of isotopic tracers as a function of the elemental concentration of the same element. Therefore, we will begin by studying initial $^{87}Sr/^{86}Sr$ ratios as a function of Sr concentration.

Mixing of components with different isotopic and elemental compositions yields a hyperbolic curve on a diagram of initial $^{87}Sr/^{86}Sr$ against Sr concentration (Fig 7.7a). Ideally, initial Sr isotope ratios should be plotted against ^{86}Sr abundance, since the total concentration of strontium is slightly perturbed by variations in ^{87}Sr. This is particularly true for old rocks, where there is a large age correction to obtain the initial ratio. However, the decay constant of Rb is so low that ^{88}Sr makes up the bulk of strontium in most rocks. Therefore ^{86}Sr abundance can be approximated by total Sr without introducing significant errors. (This is not so for Pb in old rocks, where radiogenic Pb can easily swamp the non-radiogenic component).

A bivariate diagram for two ratios with common denominators must yield linear mixing lines. Therefore the hyperbolic mixing curve of Fig. 7.7a can be transformed into a straight line (Fig. 7.7b) by plotting initial $^{87}Sr/^{86}Sr$ against 1/Sr (approximating $1/^{86}Sr$). Briquet and Lancelot (1979) used this format to examine contamination and fractionation processes in a 'selective contamination' model (Fig. 7.8), which envisages two-component mixing between a primary basic magma and a hypothetical Sr-rich extract from the crust. Following the contamination process, plagioclase fractionation may cause the Sr content of the magma to fall as it evolves to dacitic and then rhyolitic compositions (Fig. 7.8a). If these contamination and fractionation steps were repeated sequentially, then they would create the effect seen in Fig. 7.8b. If the steps become very small, the result is simultaneous fractionation and contamination (Fig. 7.8c). However, Briquet and Lancelot's 'selective' model is probably not the most realistic for magma contamination, since Nd isotope evidence suggests that most contamination is by crustal melts (e.g. Thirlwall and Jones, 1983).

Crustal melting and assimilation is an endothermic process. If the magma is on or below the liquidus then it can only obtain heat to power melting by itself undergoing fractional crystallisation. Hence, we may expect these two processes to be coupled into a mechanism which DePaolo (1981a) termed 'assimilation fractional crystallisation' (AFC). In this model, the effect of fractionation on the mixing trajectory

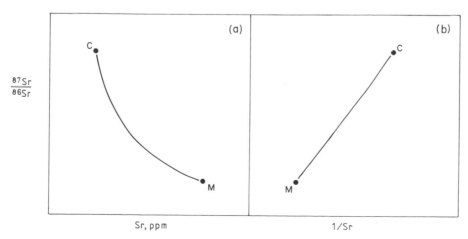

Fig. 7.7. Schematic illustration of two-component mixing on plots of Sr isotope ratio against (a) Sr concentration, and (b) 1/Sr. C = crustal end-member; M = mantle-derived end-member.

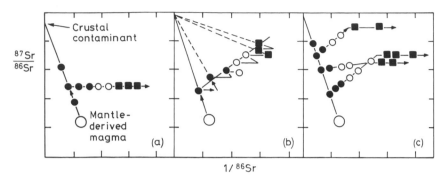

Fig. 7.8. Schematic modelling of selective Sr contamination and fractionation of magmas on plots of Sr isotope ratio against 1/Sr. a) contamination followed by fractionation; b) sequential contamination and fractionation events; c) simultaneous contamination and fractionation, followed by pure fractionation. After Briquet and Lancelot (1979).

will depend on the relative importance of assimilation and fractional crystallisation, and also on the crystal–liquid distribution coefficient (D) pertaining at the time. Fig. 7.9 shows calculated mixing lines for different D^{Sr} values at increasing fractions of assimilate (M_a) relative to initial magma (M_m), at a fixed ratio of assimilation to crystallisation (M_c). (A smaller amount of fractionation relative to assimilation will cause less deviation from the simple mixing line, and a larger relative amount of fractionation will cause more deviation.) As plagioclase joins the crystallising assemblage, this will have a very dramatic effect on D^{Sr} values, changing strontium from an incompatible element ($D^{Sr} \ll 1$) to a compatible element ($D^{Sr} > 1$) in the crystallising material. This may cause a magma to follow the bold dashed curve in Fig. 7.9 during its evolution.

The Sr *versus* Nd isotope diagram provides a useful tool for assessing crustal contamination models. DePaolo and Wasserburg (1979a) showed that simple two-component mixing on this diagram gives rise to hyperbolae whose trajectories depend on the relative Sr/Nd concentration ratio in the two end-members (Fig. 7.10). For the special case were the Sr/Nd ratio is the same in both end-members, the mixing line is straight. When the mantle-derived component has a higher Sr/Nd ratio, Nd compositions are more readily affected by contamination than Sr, yielding a concave upwards curve. This is the normal situation when the mantle-derived component is more basic than the crustal end-member (whose Sr content has been lowered by plagioclase fractionation in its previous history). However, contamination by very plagioclase-rich crust could yield a convex-upward curve (Fig. 7.10).

7.2.2 Inversion modelling of magma suites

The above modelling has considered the evolution of single magma batches during assimilation and/or fractionation processes. However, a suite of analysed lavas may represent magma batches that reached different degrees of differentiation (and hence had different trace element contents) before contamination. Just as different bivariate plots can be used to model progressive contamination of a single magma, the same variety of plots can be used to examine the evolution of magma suites. The Tertiary volcanic province of NW Scotland represents a good natural 'laboratory' in which to examine some of these processes for two main reasons. Firstly, magma–crust interaction was relatively intense, due to the volatile-poor nature of the magmas. This prevented them from punching through the crust quickly. Secondly, isotopic contrasts between mantle and crustal end-members are well developed, because old lithospheric mantle had been melted away from under the Tertiary volcanic centres by earlier magmatism.

An example of the co-variation of Sr isotope ratio and Sr concentration is provided by Tertiary basic-to-intermediate lavas from the Isle of Skye, NW Scotland (termed the Skye Main Lava Series). Moorbath and Thompson (1980) found a weak negative correlation between Sr isotope ratio and concentration in this suite, forming a hyperbolic trend (Fig. 7.11). However, any individual mixing line between a hypothetical mantle-derived precursor and the estimated crustal component has a slope perpendicular to the observed trend. Such a trajectory is displayed by a small suite of low-potassium (low-K) basalts in Fig. 7.11.

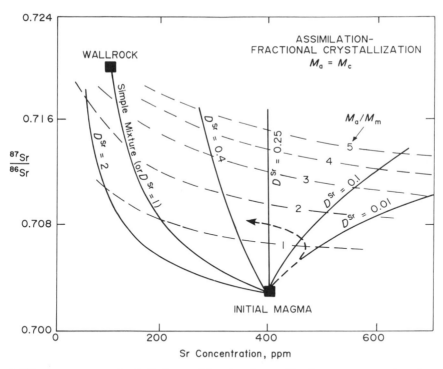

Fig. 7.9. Effect of different amounts of assimilation and fractional crystallisation on magma chemistry, shown on a plot of Sr isotope ratio against concentration. For explanation, see text. After DePaolo (1981a).

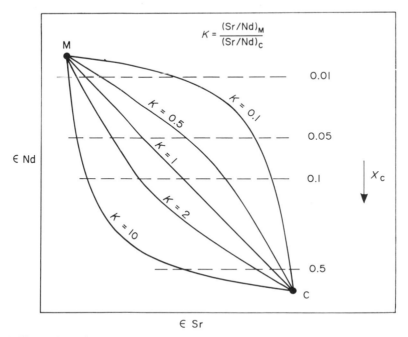

Fig. 7.10. Schematic illustration of two-component mixing on a plot of Nd *versus* Sr isotope ratio (as ϵ, section 4.2.1). M and C are mantle-derived and crustal end-members; X_C = fraction of crustal component in product magma. K = Sr/Nd ratio in mantle-derived relative to crustal end-member. Normally K is between 2 and 10. After DePaolo and Wasserburg (1979a).

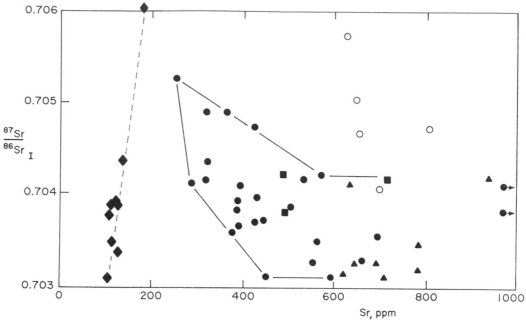

Fig. 7.11. Diagram of initial Sr isotope ratio against concentration for Tertiary lavas from Skye, NW Scotland. Skye Main Lava Series: (●) = basalt; (▲) = hawaiite; (■) = mugearite–benmoreite. Other lavas: (○) = silica-oversaturated intermediates; (◆) = low-K basalts. After Moorbath and Thompson (1980).

To explain the main data set, Moorbath and Thompson proposed that crystal fractionation had occurred in the upper mantle to yield a series of magmas with variable Sr contents. These were then subjected to similar degrees of contamination with radiogenic crustal Sr, so that those with high Sr contents were less affected than those with low Sr contents; yielding a hyperbolic pattern for the suite as a whole. The scatter in the data probably results from somewhat variable degrees of contamination in different magma batches.

Thirlwall and Jones (1983) made Nd isotope determinations on the same suite of Skye lavas. The data are shown (Fig. 7.12) on a plot of Nd isotope ratio against 1/concentration. Most of the basalts define an approximate linear array (equivalent to a hyperbola on a plot of $^{143}Nd/^{144}Nd$ against Nd concentration). However, this linear array does not have the trajectory expected for two-component mixing (steep vector in Fig. 7.12). Instead, it is attributed to contamination of a magma series with variable Nd contents, in which the most 'primitive' magmas, with lowest Nd contents, show the greatest effects of contamination. In contrast, a few basalts, along with silica-rich intermediate lavas, show the effects of an AFC process, in which Nd contents rise rapidly as contamination progresses (Fig. 7.12).

Thirlwall and Jones confirmed this interpretation (Fig. 7.13) using a plot of Nd isotope ratio against the major element differentiation index (FeO/FeO+MgO). They showed that the 'F/M' ratios of the Skye lavas must have been generated by fractionation at the base of the crust, since they were too high in most of the rocks to have been in equilibrium with mantle olivines. It follows that the strong correlation of ϵ Nd with F/M must be the result of a subsequent process, i.e. contamination in the crust. The most primitive basalts (lowest F/M) were the most contaminated, consistent with their lower Nd contents, which rendered them more sensitive to contamination. Again, the linear array in this diagram does not correspond to a two-component mixing line. Crustal contaminants have low Fe and Mg concentrations, so that they do not affect the F/M ratio of the contaminated magma. Hence, sub-vertical mixing vectors are generated in Fig. 7.13. The formation of the array of lava compositions at an oblique angle to these vectors can be ascribed to a regular and predictable contamination mechanism affecting a suite of related differentiates.

Isochron diagrams are a particular example of a bivariate plot involving isotope ratios and trace element ratios, and may therefore be useful for

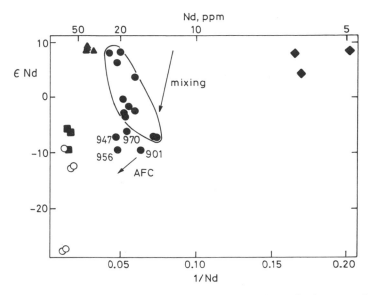

Fig. 7.12. Plot of initial Nd isotope ratio (in the form of ϵ Nd), against reciprocal Nd content in Skye lavas. Symbols as in Fig. 7.11. Numbered lavas display evidence of AFC. After Thirlwall and Jones (1983).

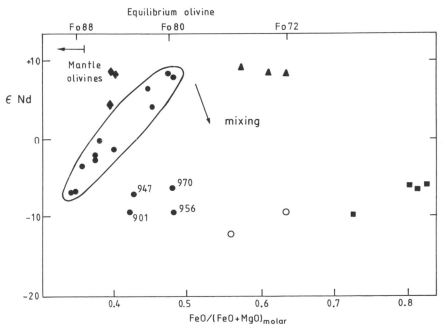

Fig. 7.13. Plot of initial Nd isotope ratios against 'F/M' ratio for Tertiary lavas from Skye, NW Scotland. Symbols as in Fig. 7.11. After Thirlwall and Jones (1983).

studying crustal contamination processes. For old rock suites, initial isotope ratios are plotted on a pseudo-isochron diagram. Because the denominator on both axes is the same, two-component mixing must give rise to products which lie on a straight line between the end-members. However, a magma suite may again generate a data array which does not project to the mixing end-members. The Tertiary lavas from Skye provide a good example of this problem also.

Thirlwall and Jones (1983) found a linear array of ϵ Nd *versus* Sm/Nd ratios in Skye basalts (Fig. 7.14). They interpreted this array as a mixing line between a mantle-derived magma with constant Sm/Nd ratio and a partial melt of intermediate (tonalitic) Lewisian gneiss. The projection of the mixing line onto the Lewisian isochron then indicates an ϵ Nd value (at 60 Myr) of ca. -15. However, Dickin *et al.* (1984) argued that the basalt array was not a single mixing line, but was generated by a series of obliquely angled mixing lines involving mantle-derived magmas with different Sm/Nd ratios. These trajectories point to a crustal end-member with ϵ Nd of ca. -40, corresponding to Lewisian granitic (acid) gneiss. This controversy may be resolvable by Ce isotope evidence (section 9.2), but it serves to reiterate the importance of distinguishing between individual mixing lines and magma evolution trends on all plots where contamination models are considered.

Other trace element ratios may also help in the interpretation of isotopic data. For example, the K/Zr ratio (Fig. 7.15), which compares a highly incompatible element with a less incompatible element, can be used to differentiate between mantle and crustal sources. Correlation between this variable and an isotope ratio is indicative of a contamination process, while variation in isotope ratio which is not correlated with K/Zr may indicate derivation from a heterogeneous mantle source. Thirlwall and Jones (1983) interpreted the hyperbolic array of data for Skye lavas in Fig. 7.15a. as a mixing line. However, Huppert and Sparks (1985) showed that the hyperbolic array could result from a family of separate hyperbolic mixing lines (Fig. 7.15b). They attributed this contamination process to thermal erosion of wall rocks by turbulently flowing magma during its actual ascent through the crust. However, a more probable site for such wall-rock assimilation may be sill complexes in the crust, where the longer magma residence time would allow more opportunity for contamination.

Since the Sr and Nd isotope compositions of a contaminated lava suite may be a complex function of Sr and Nd concentrations, depending on the differentiation history of the suite, these factors must be borne in mind when interpreting the Nd *versus* Sr isotope diagram for a magma suite. In the case of the

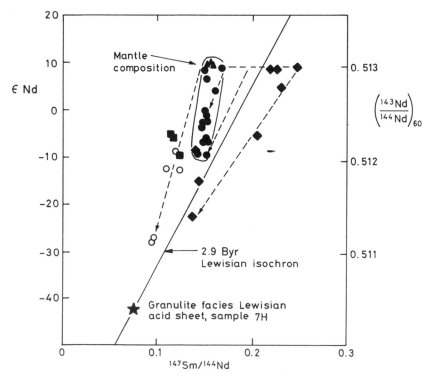

Fig. 7.14. Sm–Nd pseudo-isochron diagram for Tertiary lavas from Skye, NW Scotland. Symbols as in Fig. 7.11. For explanation see text. After Dickin *et al.* (1984).

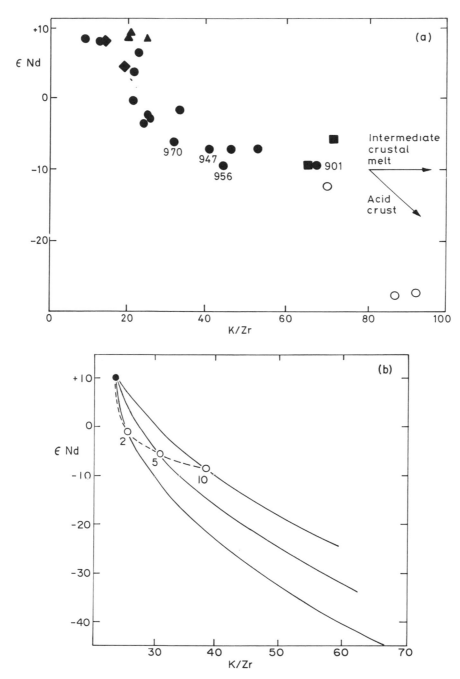

Fig. 7.15. Plots of ϵ Nd against K/Zr ratio. a) Skye lava data, with symbols as in Fig. 7.11; b) hypothetical model involving a family of mixing hyperbolae, where numbers indicate % crustal contamination. After Thirlwall and Jones (1983) and Huppert and Sparks (1985).

Skye data (Fig. 7.16), most samples define an array with negative slope, implying coupled behaviour of Sr and Nd in the mixing process. Since it has been shown that both *separate processes* are controlled by the trace element content of the magma, it follows that both Sr and Nd were behaving as incompatible elements in the Skye magmas before contamination (i.e., evolved magmas were richer in *both* Sr and Nd).

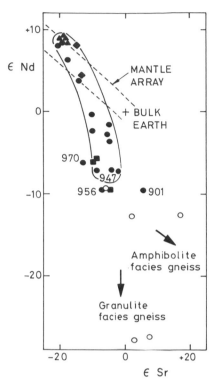

Fig. 7.16. Plot of Nd *versus* Sr isotope ratio for Skye lavas showing a steep negative trend between the two variables. Symbols as in Fig. 7.11. After Thirlwall and Jones (1983).

Hence, little plagioclase fractionation occurred during differentiation of these magmas, consistent with the occurrence of this process at upper mantle depths (e.g. Thompson, 1982).

Pb isotopes are a powerful tool in studies of mantle and crustal evolution, because the three different radiogenic isotopes are generated from parents with a wide span of half-lives, two of which are a common element. By using $^{206}Pb/^{204}Pb$ and $^{207}Pb/^{204}Pb$ ratios in conjunction, it is not only possible to measure the importance of crustal contamination, but also the age of the crustal component. On the other hand, by using $^{206}Pb/^{204}Pb$ and $^{208}Pb/^{204}Pb$ ratios in conjunction, it may be possible to locate the depth of the crustal contaminant, since the crust may develop a stratified signature of these isotopes in response to high-grade metamorphism. Both of these possibilities are illustrated by the Tertiary magmatism of Skye.

Moorbath and Welke (1969) found that both acid and basic Tertiary igneous rocks from Skye lay on a strong linear array on the $^{207}Pb/^{204}Pb$ *versus* $^{206}Pb/^{204}Pb$ diagram, with a slope age of ca. 3 Byr.

They interpreted the linear array as a mixing line between radiogenic mantle-derived Pb and very unradiogenic Archean (Lewisian) crustal Pb. Dickin (1981) repeated this study with more modern techniques and found a mixing line with a slope-age of 2920 ± 70 Myr (Fig. 7.17a), the same as the Sm–Nd age of the Lewisian complex (*see* section 4.1.3). By plotting $^{208}Pb/^{204}Pb$ *versus* $^{206}Pb/^{204}Pb$ ratios (Fig. 7.17b), it is possible to resolve three components in the Skye Tertiary igneous rocks. The lavas are interpreted as mantle-derived magmas that had suffered strong contamination in the granulite-facies lower crust, while the granites are attributed to similar precursors which underwent further differentiation and contamination in shallower amphibolite-facies crust under Skye.

In this model, the crustal end-members were based on average compositions of gneisses from NW Scotland, supported by evidence from crustal xenoliths carried up in a Tertiary intrusion from Skye. The lower crustal rocks were depleted in both U and Th relative to Pb during the 2.7 Byr-old Scourian granulite-facies metamorphism, while the present-day upper crust represents rocks depleted in U but not Th (relative to Pb) in the Archean middle crust. The original upper crust, enriched in U and Th relative to Pb, has largely been removed by erosion.

7.3 Petrogenesis of continental magmas

It is impossible here to attempt a comprehensive review of continental magma suites. Instead, a few case studies will be examined for different magma types which illustrate a variety of approaches to problems of petrogenetic interpretation.

7.3.1 Kimberlites, carbonatites and lamproites

Kimberlites, carbonatites and lamproites are highly incompatible-element-enriched magmas which may be genetically related. Experimental evidence suggests that they are all products of very small-degree partial melting in the deep mantle, and that CO_2 plays an important role in their genesis (e.g. Wendlandt and Mysen, 1980). The volatile-rich nature of these magmas causes rapid ascent through the crust. This, coupled with their high incompatible-element concentrations, renders these magmas very resistant to isotopic modification by crustal contamination.

South African kimberlites are divided into two petrological types, basaltic and micaceous (phlogo-

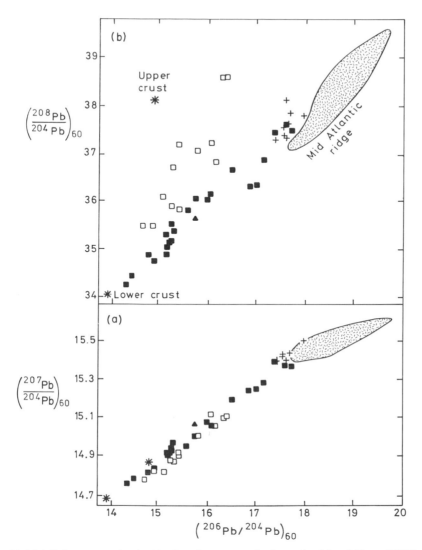

Fig. 7.17. Plot of initial Pb isotope ratios for Tertiary igneous rocks from the Isle of Skye, NW Scotland, to show evidence for three-component mixing in their genesis. Symbols: (∗) = crustal end-members; (■) = lava suite; (▢) = granites; (+) = low-K tholeiites. Mid Atlantic Ridge approximates Hebridean mantle composition. After Thompson (1982).

pitic), according to their groundmass mineralogy (Dawson, 1967). These two groups, referred to as types I and II respectively by Smith (1983), have distinct Sr and Nd isotope compositions. Basaltic (Group I) kimberlites have isotopic compositions which cluster just within the depleted quadrant relative to Bulk Earth, while micaceous (Group II) kimberlites fall well inside the enriched quadrant (Fig. 7.18).

Kimberlites of western Australia graduate from phlogopite kimberlite to lamproite, and were found by McCulloch *et al.* (1983) to extend the composi-

tional range of micaceous kimberlites even further into the enriched quadrant in Fig. 7.18. McCulloch *et al.* calculated T_{DM} model ages for the source region of these rocks, on the assumption that REE fractionation had not occurred during magma genesis. These were in the range 0.9–1.3 Byr. However, they recognised that the magmas might have been fractionated to more light REE enriched compositions than the source, making these minimum ages which may substantially under-estimate the age of the enrichment event.

The resistance of carbonatites to crustal contam-

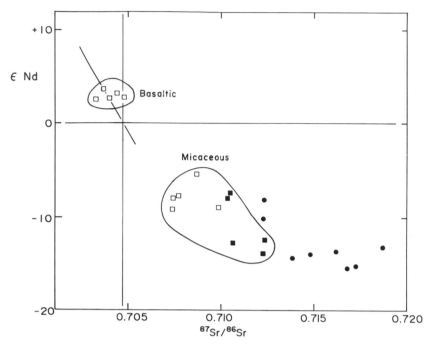

Fig. 7.18. Plot of ϵ Nd against Sr isotopic ratios of basaltic and micaceous kimberlites from South Africa (□) and western Australia (■), along with Australian lamproites (•). After DePaolo (1988).

ination makes them a potential source of data on the composition of the upper mantle under continents. Bell and Blenkinsop (1987) explored this application by Sr and Nd isotope analysis of carbonatites from Ontario and Quebec, ranging in age from 110 to 2700 Myr. Sr isotope ratios lay along a depleted mantle evolution line, which Bell and Blenkinsop attributed to the sub-continental lithosphere of the Superior Province. However, Nd isotope data for the same samples were more scattered. They comprise part of a larger set of Nd data from different continents (Nelson *et al.*, 1988) which scatter between the depleted mantle and reservoirs at least as enriched as Bulk Earth (Fig. 7.19).

Nelson *et al.* argued that the world-wide occurrence of these scattered data mitigated against an origin in the sub-continental lithosphere, and instead favoured a plume origin similar to ocean island basalts. However, such a model can hardly explain the extreme signatures of the Australian lamproite data (Fig. 7.18). In contrast, the mixing model which Nixon *et al.* (1981) originally proposed for kimberlite genesis can serve as a unifying petrogenetic model for kimberlites, carbonatites and lamproites. In this model, very small-degree partial melts originate in comparatively 'fertile' asthenospheric sources (Group

1 signature); they are subsequently contaminated to different degrees in the LIL-enriched but refractory sub-continental lithosphere (Group 2 signature). If this mixing model holds then carbonatite compositions cannot be used to accurately model the composition of source reservoirs.

7.3.2 Alkali basalts

An interesting location in which to study alkali basalt sources is provided by the Cameroon Line of western Africa. This volcanic chain, composed dominantly of alkali basalts with subordinate tholeiites, stretches from the Atlantic island of Pagalu (700 km SW of the Niger delta) to the Biu Plateau (800 km inland). Despite the fact that the volcanic chain is situated half on young oceanic crust and half on ancient continental crust, trace element contents and Sr isotope ratios are identical in the two sections of the line (Fitton and Dunlop, 1985). Given that oceanic and continental lithosphere would be expected to have different signatures, Fitton and Dunlop argued that the magma source must lie below the lithosphere.

The Cameroon line shows no evidence of age progression along its length, and must therefore represent a 'hot zone' rather than a hot-spot trail

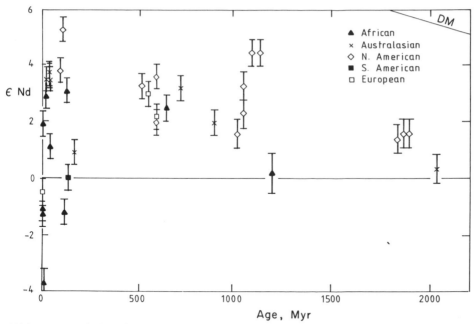

Fig. 7.19. Nd isotope evolution diagram in terms of ϵ Nd against time, showing carbonatite derivation from variably depleted (or mixed) mantle sources. DM = depleted mantle line seen in flood basalts (section 4.2.2). After Nelson *et al.* (1988).

generated by plate motion over a small plume. Fitton and Dunlop argued that because there is no evidence of migration of the area of volcanism over its 65-Myr history (despite movement of the African plate) then 'the mantle source must be coupled to the lithosphere'. Hence they suggested that the source of the Cameroon line magmas must be the convecting upper mantle rather than a deep mantle plume. This was a surprising result, since it appeared to support the disequilibrium melting model rather than the plume model for mantle heterogeneity (section 6.2).

Halliday *et al.* (1988) performed a more detailed isotopic study on the Cameroon Line, including Pb and Nd isotope measurements (Fig. 7.20). A few samples show evidence of contamination by continental basement. After exclusion of these, the most distinctive feature of the data was the very radiogenic Pb isotope compositions displayed by basic lavas from the continent–ocean boundary, about half-way along the volcanic chain. These compositions approach those of the St Helena hot spot, but volcanics on either side (within the oceanic and continental segments) are less radiogenic.

Halliday *et al.* (1988) attributed these features to the 'impregnation' of the upper mantle under the Cameroon Line by material from the St Helena plume. This plume played a major role in promoting

the initial opening of the South Atlantic, 120 Myr ago. It was probably responsible for the actual location of rifting, which subsequently became the continental edge. With time, the African plate moved away from the St Helena plume, but a 'blob' of hot plume material became incorporated into the lithospheric mantle under Cameroon as the continental margin cooled after the rifting event. As the plume component was gradually dispersed laterally along the volcanic chain, its compositional effect was seen at volcanic centres progressively further from the continental edge.

Halliday *et al.* (1990) revised this model as a result of new observations on the Pb isotope data. These revealed that the radiogenic $^{206}Pb/^{204}Pb$ signatures at the continental edge were not accompanied by high enough $^{207}Pb/^{204}Pb$ ratios to represent direct mixing with the St Helena plume. Instead, a positive correlation is observed between $^{206}Pb/^{204}Pb$ and U/ Pb ratio (Fig. 7.21), which Halliday *et al.* interpreted as a 200 Myr-old erupted isochron. However, it was shown in Fig. 7.20 above that $^{206}Pb/^{204}Pb$ correlates with Nd isotope ratio, which cannot develop large variations by Sm decay over periods of only 200 Myr. Therefore the arrays of Pb isotope ratio are probably mixing lines. The radiogenic end-member must represent young lithosphere whose high $^{206}Pb/^{204}Pb$

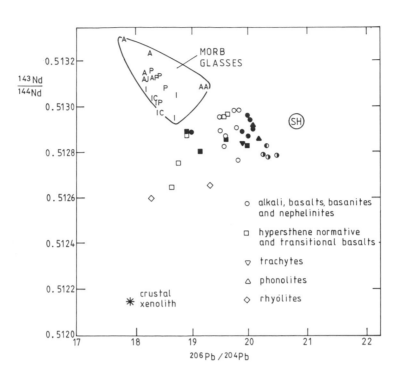

Fig. 7.20. Plot of Nd *versus* Pb isotope compositions for Cameroon Line volcanics. Solid symbols: continental lavas; open symbols: oceanic; half-filled: continental edge; SH= St Helena hot spot; A, I and P = Atlantic, Indian and Pacific MORB. After Halliday *et al.* (1988).

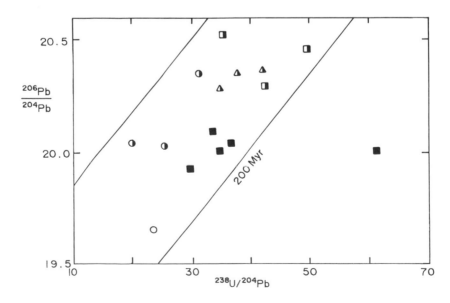

Fig. 7.21. U–Pb isochron diagram for young Cameroon line lavas with over 4% MgO, yielding an apparent age of ca. 200 Myr. Half-open symbols denote individual volcanos from the continental edge. Other symbols indicate continental (○) and oceanic segments (■). After Halliday *et al.* (1990).

signature was generated by magmas with high U/Pb ratios (μ) at the time of continental rifting. Mixing between this component and local asthenospheric upper mantle can explain the isotopic mixing process.

This work has two important conclusions. Firstly, continental rifting episodes can replace old sub-continental lithosphere with young lithosphere which may have an exotic composition. Secondly, U is more incompatible than Pb in magmatic processes, so that preferential extraction of Pb from the mantle cannot be invoked to explain the lead paradox (section 6.4.1).

7.3.3 Flood basalts

The northwestern United States displays one of the world's major flood basalt provinces, and as a case study, illustrates the complexities of their petrogenetic modelling. Early Nd isotope data on the Columbia River basalts clustered near ϵ Nd = 0, leading DePaolo and Wasserburg (1976) to propose an undepleted (primordial type) source for these magmas, in contrast to the depleted mantle source of MORB (ϵ Nd = +10). This model was supported by DePaolo (1983) on the basis of a volume-weighted

histogram (Fig. 7.22) of initial ϵ Nd values measured on the Columbia River basalts. He argued that the marked concentration of data at slightly positive ϵ values, and the sharp cut off at ϵ Nd = 0 was evidence for a chondritic source for the most voluminous Grand Ronde group of lavas, merging into a depleted mantle source for the Imnaha and Picture Gorge basalts. However, the abundance peak at ϵ Nd = .0 in Fig. 7.22 is a product of adding the data of DePaolo and Wasserburg (1976, 1979b) to that of Carlson *et al.* (1981), and is largely a sampling effect.

A compilation of Sr and Nd isotope data for several basaltic suites from the northwestern US appears to present a rather different picture. The samples display a very strong, almost continuous, curved trend, which fans out somewhat in the enriched quadrant (Fig. 7.23). These data imply a relatively simple mixing process, such as crustal contamination, in the genesis of the lavas (Carlson *et al.*, 1981). However, it has become apparent over the last few years that radiogenic isotopes alone may not be able to distinguish between enriched mantle and crustal sources. Incompatible element ratios and stable isotope data may be needed to assist in this distinction.

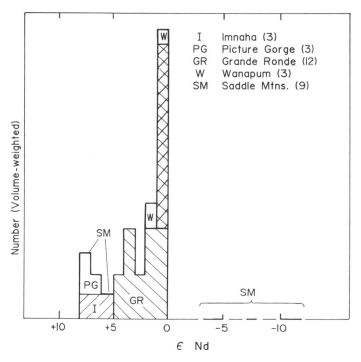

Fig. 7.22. Histogram of ϵ Nd compositions for Columbia River basalts, weighted according to eruptive volume. Double-hatched data are from DePaolo and Wasserburg (1976, 1979b). Modified after DePaolo (1983).

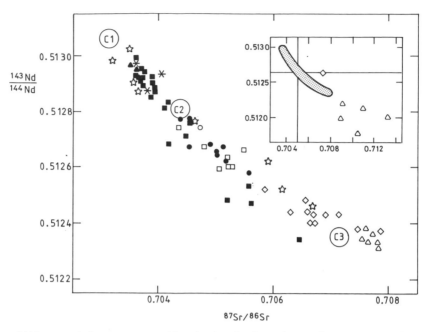

Fig. 7.23. Plot of Nd *versus* Sr isotope composition for basalts from the northwestern United States. Key: (■) = Grand Ronde; (✳) = Imnaha; (●) = Picture Gorge; (▢) = Wanapum; (▲) = Steens Mtn; (△) = Saddle Mountains; (☆) = HAOT; (◇) = SROT. C1 to C3 are possible sources discussed in the text. Data from the main diagram define the shaded field on the inset. After Carlson and Hart (1988).

Carlson and Hart (1988) argued that the ratio of a highly incompatible element against a high-field-strength element (e.g. K_2O/P_2O_5) can be used as an index of (specifically) crustal contamination. This index is plotted against Sr isotope ratio in Fig. 7.24. Some Picture Gorge and Grand Ronde basalts of the Columbia River Basalt Group (CRBG), along with Steens Mountain basalts from the Oregon Plateau, have quite elevated K_2O/P_2O_5 ratios, despite low to intermediate Sr isotope ratios. Carlson and Hart attributed this pattern to contamination of magmas from a 'C1' mantle source by crustal units with a variety of ages. The C1 source was identified as typical asthenospheric upper mantle, whose melting was probably caused by mantle convection behind the Cascades arc. In contrast to the above lavas, some Saddle Mountains CRBG flows, high-Al olivine tholeiites (HAOT) from the Oregon Plateau, and Snake River olivine tholeiites (SROT) have $^{87}Sr/^{86}Sr$ ratios up to 0.708, but low K_2O/P_2O_5. Carlson and Hart attributed these signatures to a lithospheric mantle source ('C3').

A plot of $\delta^{18}O$ against Sr isotope ratio supports this model (Fig. 7.25). Steep vectors result from contamination of basaltic magmas by typical crustal

units. In contrast, sub-horizontal vectors would be produced by mixing with old ^{87}Sr-enriched mantle, or possibly by recent contamination of the Sr-poor mantle source by subducted sediment. This distinction between source and magma contamination vectors on the oxygen–strontium isotope diagram was explored in detail by Taylor (1980) for granitic rocks (section 7.3.5).

Pb isotope data introduce one more complexity into this picture. Basalts with $^{87}Sr/^{86}Sr$ below 0.708 display a triangular distribution on plots of Sr or Nd isotope ratio against $^{206}Pb/^{204}Pb$ (Fig. 7.26). On this diagram, many Grand Ronde lavas trend towards an end-member (C2) with radiogenic Pb which is distinct from the C1 and C3 mantle end-members recognised from other evidence. Carlson and Hart argued that the C2 source may have been derived by contamination of C1 depleted mantle by subducted sediment. In other words, this mantle source is a vestige of the time when the western US continental margin was a convergent plate margin. Such a component is not typical of flood basalts, but its occurrence emphasises the need to use many geochemical indices to interpret petrogenetic processes with several degrees of freedom.

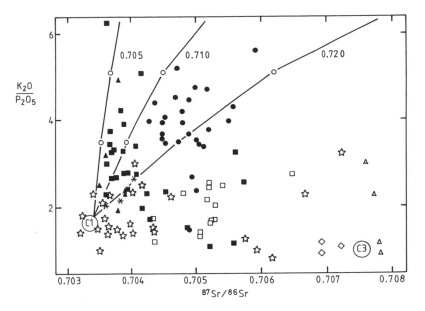

Fig. 7.24. Plot of K_2O/P_2O_5 against Sr isotope ratio for basalts from the northwestern US showing mixing models between C1 magmas and three crustal contaminants with different Sr isotope ratios. Symbols as in Fig. 7.23. After Carlson and Hart (1988).

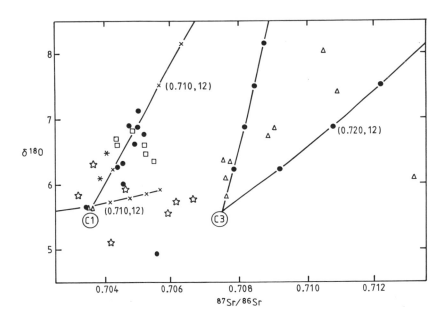

Fig. 7.25. Plot of $\delta^{18}O$ against Sr isotope ratio for basalts from the northwestern US. Mixing models show effects of contamination with material of a given Sr and oxygen isotope composition. Steep mixing lines show contamination of magmas; shallow mixing line shows contamination of MORB-type source with subducted sediment. 10% increments are marked. Symbols as in Fig. 7.23. After Carlson and Hart (1988).

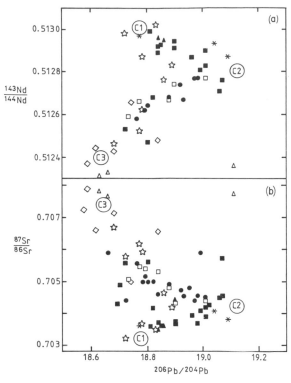

Fig. 7.26. Isotope compositions of basalts from the northwestern US: a) Nd *versus* Pb isotope plot; b) Sr *versus* Pb isotope plot. Three distinct mantle sources are resolved (C1 to C3). Symbols as in Fig. 7.23. After Carlson and Hart (1988).

7.3.4 Precambrian granitoids

One of the most fundamental questions about the continental crust is the extent to which any given block of sialic basement is the product of juvenile separation from the mantle or re-working of older cratonic material. Sr isotope data were originally applied to this problem on the grounds that crustal reservoirs, which have high Rb/Sr ratios, develop higher $^{87}Sr/^{86}Sr$ ratios over geological time than the low-Rb/Sr mantle. Calculation of the initial Sr isotope composition of a plutonic crustal segment should then indicate whether it has a mantle or crustal source. The evolution line for Sr in the depleted mantle is constructed by drawing a linear growth curve from the 'basaltic achondrite best initial' (BABI) value of 0.69899 ± 5 (section 3.2.4) to the $^{87}Sr/^{86}Sr$ composition of recent ocean ridge basalts in the range 0.702–0.704. Data for specific crustal provinces can then be compared with this evolution line to assess their petrogenesis.

A classic example of the application of the Sr isotope evolution diagram to the provenance of crustal basement is provided by studies of the Archean and Proterozoic gneisses of West Greenland by Moorbath and Pankhurst (1976). Average growth lines are drawn in Fig. 7.27 for 3.7 Byr-old Amitsoq gneisses from four localities, 2.8–2.9 Byr-old Nuk gneisses from five localities, 1.8 Byr-old Ketilidian gneisses from two localities in South Greenland, and the 2.52 Byr-old Qorqut granite. The initial ratios of these terranes are compared on Fig. 7.27 with a hypothetical linear upper mantle growth line drawn between BABI and MORB.

Moorbath and Pankhurst argued that the Nuk (and Ketilidian) gneisses could not be derived by re-working of older (e.g. Amitsoq) gneiss, since the growth lines of the Amitsoq samples are much too steep to generate products with initial ratios of only 0.702–0.703. Instead they concluded that the igneous precursors of the Nuk gneisses represented a massive addition of juvenile calc-alkaline crust to the Archean basement of West Greenland. The slight elevation of the calculated initial ratios above the upper mantle evolution line was attributed to a period of crustal Sr isotope evolution, lasting perhaps 100–200 Myr, between the separation of the igneous precursors from the mantle and their subjection to granulite-facies metamorphism (*see* section 5.5). In contrast, Moorbath and Pankhurst recognised the Qorqut granite as a good candidate for a pluton derived by re-working of older crust, since its initial ratio of 0.709 ± 0.007 is well within error of the compositions of Amitsoq gneisses at that time.

Pb isotope analysis of the Nuk gneisses revealed a more complex picture than the Sr isotope data alone, by demonstrating that Nuk magmas emplaced into areas of Amitsoq crust suffered significant contamination with old crustal Pb (section 5.5). In view of the lack of obvious crustal Sr contamination, selective contamination by Pb was invoked to explain these observations (Taylor *et al.*, 1980). In this situation, the application of Nd isotope analysis provides an ideal tool to test petrogenetic models for the Nuk gneisses.

Taylor *et al.* (1984) analysed a selection of both Pb-contaminated and Pb-uncontaminated gneisses for Nd isotope composition (Fig. 7.28). The data indicate a good correlation between ϵ Nd and Rb, an incompatible trace element expected to be enriched in the Amitsoq gneisses. Taylor *et al.* attributed these results to contamination of mantle-derived Nuk magmas by partial melts of Amitsoq gneiss in the lower crust. However, ϵ Nd does not correlate well

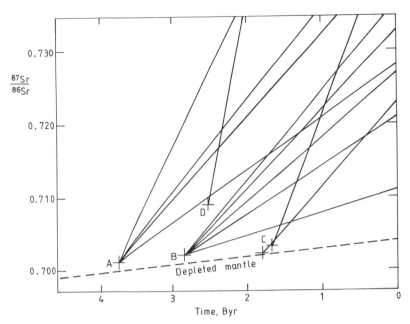

Fig. 7.27. Diagram of Sr isotope evolution with time showing development of four crustal suites relative to depleted mantle. A: Amitsoq gneiss; B: Nuk gneiss: C: Ketilidian gneiss; D: Qorqut granite. After Moorbath and Taylor (1981).

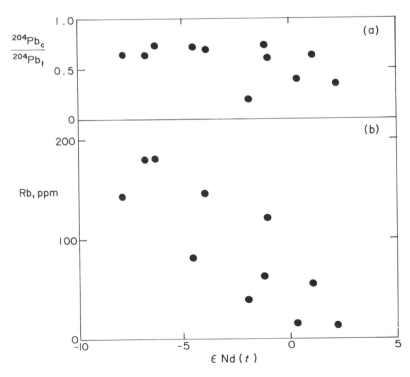

Fig. 7.28. Variation of ϵ Nd in Nuk gneisses, compared with (a) the fraction of isotopic contamination by Amitsoq Pb (from section 5.5); and (b) Rb content. After Taylor *et al.* (1984).

with the degree of Pb isotopic contamination (represented by the index ^{204}Pb contaminant/^{204}Pb total). This suggests additional selective Pb contamination due to Pb-enriched fluids generated by crustal dehydration.

The paradox whereby substantial Pb and Nd contamination of the Nuk magmas was not accompanied by observable Sr isotope disturbance must be attributed to the stratified nature of the Amitsoq crust. Taylor *et al.* argued that the deep crust responsible for contamination must have had lower Rb/Sr ratios than those analysed from the surface outcrops, presumably due to the flushing out of Rb from the lower crust during granulite-facies metamorphism. Hence it did not develop elevated Sr isotope ratios over geological time. It is concluded from this evidence and other studies that Sr isotope data often cannot readily distinguish between mantle and lower crustal source regions. In this situation, Nd isotopes are a particularly powerful petrogenetic tool because Sm/Nd is fractionated during crustal extraction from the mantle but not by intra-crustal processes (section 4.3.2).

7.3.5 Phanerozoic batholiths

Hurley *et al.* (1965) made a strontium isotope study of the Sierra Nevada batholith and concluded that most of the intrusive bodies making up the batholith had initial ^{87}Sr/^{86}Sr ratios of ca. 0.7073 ± 0.001. They recognised that this value was intermediate between expected upper mantle and Precambrian crustal values of ca. 0.703–0.705 and 0.71–0.73, respectively. However they were unable to determine on the basis of the Sr isotope evidence whether the Sierra

Nevada batholith represented mantle-derived magmas subsequently contaminated by the crust, or simply partial melting of geosynclinal sediments and volcanics.

DePaolo (1981b) made a combined Sr and Nd isotope study of both the Sierra Nevada and Peninsular Ranges batholiths in a further attempt to resolve the genesis of these bodies. The data define hyperbolic arrays on ϵ Sr *versus* ϵ Nd diagrams (Fig. 7.29), running from the island-arc basalt field towards the composition of a nearby Precambrian schist. The latter was regarded as representative of the source area which yielded the Paleozoic–Mesozoic geosynclinal sediments into which the batholiths are intruded. DePaolo recognised that the western Peninsular Ranges samples closely conformed to the Sr–Nd mantle array and that they could therefore be products of a heterogeneous mantle without crustal contamination. However, in the context of the Sierra Nevada data, crustal contamination of magmas within the island-arc field seems much more likely.

This interpretation is supported by a comparison of strontium and oxygen isotope data (Taylor and Silver, 1978; DePaolo, 1981b) which together form another powerful tool for studies of granite petrogenesis. (For background to stable isotope geology, see Hoefs, 1987.) Sierra Nevada and Peninsular Range granitoids form a hyperbolic array on the ϵ Sr *versus* δ^{18}O diagram (Fig. 7.30), between mantle-derived and Paleozoic sediment end-members. The shape of the hyperbola is determined by the relative strontium/oxygen concentrations in the two end-members, and is consistent with a simple mixture of high δ^{18}O sedimentary crustal melts with basic magmas.

Three alternative models can all be ruled out

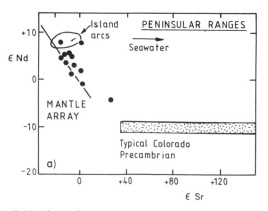
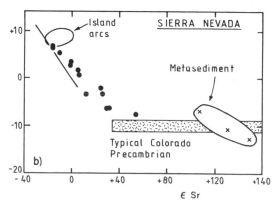

Fig. 7.29. Plots of ϵ Nd against ϵ Sr for a) Peninsular Ranges and b) Sierra Nevada granitoids (•). Compositions of crustal reservoirs and the effect of seawater alteration are also shown. After DePaolo (1981b).

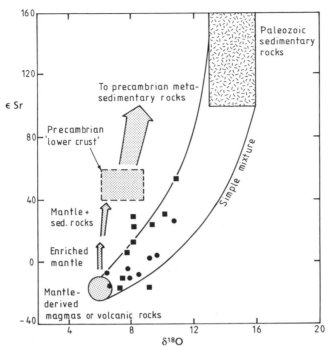

Fig. 7.30. Plot of ϵ Sr against δ^{18}O showing data for the Peninsular Ranges (●) and Sierra Nevada (■) batholiths, relative to various possible magma sources and models. After DePaolo (1981b).

because they would cause vertical vectors in Fig. 7.27, in which Sr isotope increases would not be accompanied by appreciable change in δ^{18}O. These models are:

1) Sr (and Nd) isotopic enrichment of a mantle source along the mantle array in Fig. 7.30.
2) Contamination of the mantle source by sediment subduction. The much lower strontium content of mantle, relative to basic magmas would make it much more susceptible to contamination by subducted sedimentary or seawater Sr, whereas the oxygen content of the mantle and of basic magmas is the same. In other words, the mantle has a lower Sr/O ratio than basic magma, which would yield a mixing hyperbola of steeper slope in Fig. 7.30.
3) Contamination with a hypothetical lower crust of Precambrian basement which would have low δ^{18}O.

Neither Figs. 7.29 or 7.30 can distinguish between genesis of the Peninsular Ranges batholith as a direct mantle-derived differentiate or a re-melt of young basic igneous rock at the base of the crustal geosyncline. However, the San Marcos gabbro unit of the Peninsular Ranges must be a direct mantle melt

because of its basic major element composition. Additional mantle-derived melts must have been present at depth to cause crustal melting. Therefore, the simplest, but not exclusive model, is that these same melts contributed fractionated magmas to the rest of the batholith.

As important products of Phanerozoic crustal evolution, the California batholiths are paralleled by the Berridale and Kosciusko batholiths of the Lachlan fold belt of SE Australia. Chappell and White (1974) distinguished two major granite types there on the basis of chemical and mineralogical criteria. 'S-type' granites with low Ca contents and a tendency to per-aluminous character (Al_2O_3/ [$Na_2O + K_2O + CaO$] > 1.05) were regarded as partial melts of sedimentary rocks; 'I-type' granites with high Ca contents and Al_2O_3/ [$Na_2O + K_2O + CaO$] < 1.05 were regarded as partial melts of young igneous crustal rocks.

McCulloch and Chappell (1982) tested this model by analysing a suite of samples from the Berridale and Kosciusko batholiths for Sr and Nd isotope compositions. The data formed two overlapping fields which together define a hyperbolic array in the lower right quadrant of the ϵ Nd *versus* ϵ Sr diagram (Fig. 7.31). McCulloch and Chappell interpreted these data in

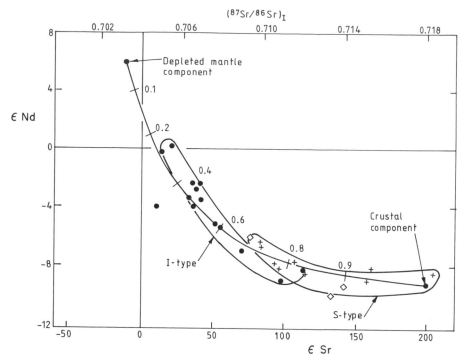

Fig. 7.31. Diagram of ϵ Nd against ϵ Sr for 'I-type' (•) and 'S-type' (+) granites and crustal xenoliths (◇) from SE Australia. A best-fit mixing line is shown between hypothetical crustal and mantle-derived end-members. After McCulloch and Chappell (1982).

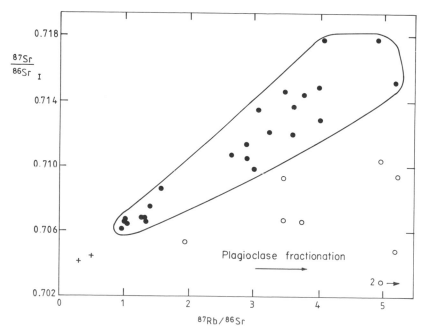

Fig. 7.32. Rb–Sr pseudo-isochron diagram for granites of SE Australia (•) showing possible 'mixing fan'. (+) = gabbros; (o) = granites argued to have fractionated plagioclase after contamination. After Gray (1984).

support of the crustal melting model of Chappell and White (1974). However, Gray (1984, 1990) attributed them to contamination of mantle-derived basic magmas, by mixing with a sedimentary crustal component. Possible end-members are represented by young basic rocks with a mantle-like signature, and Ordovician flysch with a model Nd age of ca. 1400 Myr. The left end of the array projects back to a depleted mantle-like end-member with ϵ Nd of +6. The existence of 'rare gabbros' found in the vicinity of the batholiths demonstrates that such magmas were available in the crust; their rarity at the surface can be attributed to the 'density problem' of raising basic magma through a felsic crust. The crustal end-member is well represented by the Cooma granodiorite, which displays strong structural evidence of being an *in-situ* melt of Ordovician flysch.

Gray supported his model with major-element variation diagrams and by examining Sr isotope compositions on a Rb–Sr isochron diagram (Fig. 7.32). Average initial $^{87}Sr/^{86}Sr$ and Rb/Sr ratios are plotted for two gabbros and for several plutons from the major 'S-type' and I-type' batholiths. Most of the data form a cone-shaped array which Gray argued to represent mixing between a low Rb/Sr basaltic or andesitic end-member ($^{87}Sr/^{86}Sr$ = ca. 0.703–0.704), and a somewhat heterogeneous crustal end-member, typified by the crustally-derived Cooma granodiorite. Compositions to the right of this array were attributed to plagioclase fractionation subsequent to mixing, which would yield horizontal displacements.

McCulloch and Chappell (1982) and Chappell and White (1992) have acknowledged that the isotopic data could be explained by two-component mixing between mafic and sedimentary end-members. However, they rejected this model on the grounds that mixing of basic igneous and greywacke components could not explain the major-element signatures of the rocks (White and Chappell, 1988). This question is disputed by Gray (1990), but is outside the scope of the present discussion. However, the controversy serves to illustrate the limits of isotopic data in petrogenetic studies, and therefore provides an appropriate conclusion to this discussion.

References

Bell, K. and Blenkinsop, J. (1987). Archean depleted mantle: evidence from Nd and Sr initial isotopic ratios of carbonatites. *Geochim. Cosmochim. Acta* **51**, 291–8.

Briquet, L. and Lancelot, J. R. (1979). Rb–Sr systematics and crustal contamination models for calc-alkaline igneous rocks. *Earth Planet. Sci. Lett.* **43**, 385–96.

Carlson, R. W. and Hart, W. K. (1988). Flood basalt volcanism in the northwestern United States. In: Macdougall, J. D. (Ed.), *Continental Flood Basalts.* Kluwer, pp. 35–62.

Carlson, R. W., Lugmair, G. W. and MacDougall, J. D. (1981). Columbia River volcanism: the question of mantle heterogeneity or crustal contamination. *Geochim. Cosmochim. Acta* **45**, 2483–99.

Chappell, B. W. and White, A. J. R. (1974). Two contrasting granite types. *Pacific Geol.* **8**, 173–4.

Chappell, B. W. and White, A. J. R. (1992). I- and S-type granites in the Lachlan Fold Belt. *Trans. Roy. Soc. Edin. : Earth Sci.* **83**, 1–26.

Dawson, J. B. (1967). A review of the geology of kimberlite. In: Wyllie, P. J. (Ed.), *Ultramafic and Related Rocks.* Wiley, pp. 241–51.

Dawson, J. B. (1984). Contrasting types of upper mantle metasomatism? In: Kornprobst, J. (Ed.), *Kimberlites II.* Elsevier, pp. 289–329.

DePaolo, D. J. (1981a). Trace elements and isotopic effects of combined wallrock assimilation and fractional crystallisation. *Earth Planet. Sci. Lett.* **53**, 189–202.

DePaolo, D. J. (1981b). A neodymium and strontium isotopic study of the Mesozoic calc-alkaline granitic batholiths of the Sierra Nevada and Peninsular Ranges, California. *J. Geophys. Res.* **86**, 10470–88.

DePaolo, D. J. (1983). Comment on 'Columbia River volcanism: the question of mantle heterogeneity or crustal contamination' by R. W. Carlson, G. W. Lugmair and J. D. Macdougall. *Geochim. Cosmochim. Acta* **47**, 841–4.

DePaolo, D. J. (1988). *Neodymium Isotopes in Geology.* Springer-Verlag, 187 p.

DePaolo, D. J. and Wasserburg, G. J. (1976). Nd isotopic variations and petrogenetic models. *Geophys. Res. Lett.* **3**, 249–52.

DePaolo, D. J. and Wasserburg, G. J. (1979a). Petrogenetic mixing models and Nd–Sr isotopic patterns. *Geochim. Cosmochim. Acta* **43**, 615–27.

DePaolo, D. J. and Wasserburg, G. J. (1979b). Neodymium isotopes in flood basalts from the Siberian Platform and inferences about their mantle sources. *Proc. Nat. Acad. Sci. USA* **76**, 3056–60.

Dickin, A. P. (1981). Isotope geochemistry of Tertiary igneous rocks from the Isle of Skye, N. W. Scotland. *J. Petrol.* **22**, 155–89.

Dickin, A. P., Brown, J. L., Thompson, R. N., Halliday, A. N. and Morrison, M. A. (1984). Crustal contamination and the granite problem in the British Tertiary Volcanic Province. *Phil. Trans. Roy. Soc. Lond. A* **310**, 755–80.

Fitton, J. G. and Dunlop, H. M. (1985). The Cameroon line, West Africa, and its bearing on the origin of oceanic and continental alkali basalt. *Earth Planet. Sci. Lett.* **72**, 23–38.

Gray, C. M. (1984). An isotopic mixing model for the origin of granitic rocks in southeastern Australia. *Earth Planet. Sci. Lett.* **70**, 47–60.

Gray, C. M. (1990). A strontium isotopic traverse across the granitic rocks of southeastern Australia: petrogenetic and tectonic implications. *Aust. J. Earth Sci.* **37**, 331–49.

Halliday, A. N., Davidson, J. P., Holden, P., DeWolf, C., Lee, D-C, and Fitton, J. G. (1990). Trace-element fractionation in plumes and the origin of HIMU mantle beneath the Cameroon line. *Nature* **347**, 523–8.

Halliday, A. N., Dickin, A. P., Fallick, A. E. and Fitton, J. G. (1988). Mantle dynamics: a Nd, Sr, Pb and O isotopic study of the Cameroon line volcanic chain. *J. Petrol.* **29**, 181–211.

Harte, B. (1983). Mantle peridotites and processes – the kimberlite sample. In: Hawkesworth, C. J. and Norry, M. J. (Eds.), *Continental Basalts and Mantle Xenoliths.* Shiva, pp. 46–91.

Hawkesworth, C. J., Erlank, A. J., Kempton, P. D. and Waters, F. G. (1990b). Mantle metasomatism: isotope and trace element trends in xenoliths from Kimberley, South Africa. *Chem. Geol.* **85**, 19–34.

Hawkesworth, C. J., Erlank, A. J., Marsh, J. S., Menzies, M. A. and van Calsteren, P. W. C. (1983). Evolution of the continental lithosphere: evidence from volcanics and xenoliths in Southern Africa. In: Hawkesworth, C. J. and Norry, M. J. (Eds.), *Continental Basalts and Mantle Xenoliths.* Shiva, pp. 111–38.

Hawkesworth, C. J., Kempton, P. D., Rogers, N. W., Ellam, R. M. and van Calsteren, P. W. C. (1990a). Continental mantle lithosphere, and shallow level enrichment processes in the Earth's mantle. *Earth Planet. Sci. Lett.* **96**, 256–68.

Hawkesworth, C. J., Rogers, N. W., van Calsteren, P. W. C. and Menzies, M. A. (1984). Mantle enrichment processes. *Nature* **311**, 331–3.

Hoefs, J. (1987). *Stable Isotope Geology. 3rd Edn,* Springer-Verlag. 241 p.

Huppert, H. E. and Sparks, R. S. J. (1985). Cooling and contamination of mafic and ultramafic magmas during ascent through continental crust. *Earth Planet. Sci. Lett.* **74**, 371–86.

Hurley, P. M., Bateman, P. C., Fairbairn, H. W. and Pinson, W. H. (1965). Investigation of initial Sr87/Sr86 ratios in the Sierra Nevada plutonic province. *Bull. Geol. Soc. Amer.* **76**, 165–74.

Jordan, T. H. (1975). The continental tectosphere. *Rev. Geophys. Space Phys.* **13** (3), 1–12.

Jordan, T. H. (1978). Composition and development of the continental tectosphere. *Nature* **274**, 544–8.

Kramers, J. D. (1979). Lead, uranium, strontium, potassium and rubidium in inclusion-bearing diamonds and mantle-derived xenoliths from southern Africa. *Earth Planet. Sci. Lett.* **42**, 58–70.

Kramers, J. D., Roddick, J. C. M. and Dawson, J. B. (1983). Trace element and isotopic studies on veined, metasomatic and 'MARID' xenoliths from Bultfontein, South Africa. *Earth Planet. Sci. Lett.* **65**, 90-106.

Maaloe, S. and Aoki, K. (1977). The major element composition of the upper mantle estimated from the composition of lherzolites. *Contrib. Mineral. Petrol.* **63**, 161–73.

Mantovani, M. S. M. and Hawkesworth, C. J. (1990). An inversion approach to assimilation and fractional crystallisation processes. *Contrib. Mineral. Petrol.* **105**, 289–302.

McCulloch, M. T. and Chappell, B. W. (1982). Nd isotopic characteristics of S- and I-type granites. *Earth Planet. Sci. Lett.* **58**, 51–64.

McCulloch, M. T., Jaques, A. L., Nelson, D. R. and Lewis, J. D. (1983). Nd and Sr isotopes in kimberlites and lamproites from Western Australia: an enriched mantle origin. *Nature* **302**, 400–3.

Menzies, M. A. (1989). Cratonic, circumcratonic and oceanic mantle domains beneath the Western United States. *J. Geophys. Res.* **94**, 7899–915.

Menzies, M. A. and Hawkesworth, C. J. (1987). *Mantle Metasomatism.* Academic Press. 472 p.

Menzies, M. A. and Murthy, V. R. (1980). Enriched mantle: Nd and Sr isotopes in diopsides from kimberlite nodules. *Nature* **283**, 634–6.

Moorbath, S. and Pankhurst, R. J. (1976). Further rubidium–strontium age and isotope evidence for the nature of the late Archean plutonic event in West Greenland. *Nature* **262**, 124–6.

Moorbath, S. and Taylor, P. N. (1981). Isotopic evidence for continental growth in the Precambrian. In: Kroner, A. (Ed.), *Precambrian Plate Tectonics.* Elsevier, pp. 491–525.

Moorbath, S. and Thompson, R. N. (1980). Strontium isotope geochemistry and petrogenesis of the early Tertiary lava pile of the Isle of Skye, Scotland and other basic rocks of the British Tertiary Province: an example of magma crust interaction. *J. Petrol.* **21**, 217–231.

Moorbath, S. and Welke, H. (1969). Lead isotope studies on igneous rocks from the Isle of Skye, Northwest Scotland. *Earth Planet. Sci. Lett.* **5**, 217–30.

Nelson, D. R., Chivas, A. R., Chappell, B. W. and McCulloch, M. T. (1988). Geochemical and isotopic systematics in carbonatites and implications for the evolution of ocean-island sources. *Geochim. Cosmochim. Acta* **52**, 1–17.

Nixon, P. H., Rogers, N. W., Gibson, I. L. and Grey, A. (1981). Depleted and fertile mantle xenoliths from southern African kimberlites. *Ann. Rev. Earth Planet. Sci.* **9**, 285–309.

Read, H. H. (1948). Granites and granites. In: Gilluly, J. (Ed.), *Origin of Granite. Geol. Soc. Amer. Mem.* **28**, 1–19.

Richter, F. M. (1988). A major change in the thermal state of the Earth at the Archean–Proterozoic boundary: consequences for the nature and preservation of continental lithosphere. *J. Petrol. Spec. Vol.*, 39–52.

Smith, C. B. (1983). Pb, Sr and Nd isotopic evidence for sources of southern African kimberlites. *Nature* **304**, 51–4.

Taylor, H. P. (1980). The effects of assimilation of country rocks by magmas on $^{18}O/^{16}O$ and $^{87}Sr/^{86}Sr$ systematics in igneous rocks. *Earth Planet. Sci. Lett.* **47**, 243–54.

Taylor, H. P. and Silver, L. T. (1978). Oxygen isotope relationships in plutonic igneous rocks of the Peninsula Ranges Batholith, southern and Baja California. *US Geol. Surv. Open File Rep.* **79–701**, 423–6.

Taylor, P. N., Jones, N. W. and Moorbath, S. (1984). Isotopic assessment of relative contributions from crust and mantle sources to the magma genesis of Precambrian granitoid rocks. *Phil. Trans. Roy. Soc. Lond. A* **310**, 605–25.

Taylor, P. N., Moorbath, S., Goodwin, R. and Petrykowski, A. C. (1980). Crustal contamination as an indicator of the extent of Early Archean continental crust: Pb isotopic evidence from the Late Archean gneisses of West Greenland. *Geochim. Cosmochim. Acta* **44**, 1437–53.

Thirlwall, M. F. and Jones, N. W. (1983). Isotope geochemistry and contamination mechanisms of Tertiary lavas from Skye, northwest Scotland. In: Hawkesworth, C. J. and Norry, M. J. (Eds.), *Continental Basalts and Mantle Xenoliths*. Shiva, pp. 186–208.

Thompson, R. N. (1982). Magmatism of the British Tertiary Volcanic Province. *Scott. J. Geol.* **18**, 49–107.

Wendlandt, R. F. and Mysen, B. O. (1980). Melting phase relations of natural peridotite + CO_2 as a function of degree of partial melting at 15 and 30 kbar. *Amer. Mineral.* **65**, 37–44.

White, A. J. R. and Chappell, B. W. (1988). Some supracrustal (S-type) granites of the Lachlan Fold Belt. *Trans. Roy. Soc. Edin.: Earth Sci.* **79**, 169–81.

8 The Re–Os system

Rhenium is a refractory metal with two isotopes, ^{185}Re and ^{187}Re. The latter, which makes up 62% of total rhenium, undergoes β decay to ^{187}Os, a platinum group element with seven naturally occurring isotopes. Like Pb, osmium is an element with siderophile–chalcophile affinities, but unlike the U–Pb system, Re and Os also display marked fractionation between mantle and crustal systems. Rhenium is a moderately incompatible element which is partitioned fairly readily from the mantle into magmatic liquids, whereas osmium is a highly compatible element which is held strongly in the mantle. These unique geochemical properties mean that the Re–Os method has great potential in geochronological and geochemical studies of ore formation, magma genesis, and mantle evolution.

8.1 Analytical methods

Despite its great potential as a geological tool, analytical difficulties have, until recently, limited the application of the osmium isotope method. The chief of these difficulties is the high ionisation potential of Os (ca. 9 eV) which prevents the formation of positive osmium ions at temperatures attainable in conventional thermal ionisation mass spectrometry. Alternative methods of excitation therefore had to be sought.

Hirt *et al.* (1963) analysed osmium isotopes as the gaseous species OsO_4, but precision was low (10% on a 200 ng sample of pure radiogenic osmium). This was probably due to dissociation of OsO_4 during thermal ionisation of the molecule. Consequently, this method was not pursued for over 25 years. Instead, subsequent work focussed on the enhanced production of atomic osmium ions using more energetic ion sources. One of the most successful of these methods is secondary ion mass spectrometry (SIMS). In this technique, a beam of light negative ions (e.g. O^-) is used to bombard and sputter a purified solid concentrate of osmium metal to yield a positive Os ion beam which is analysed in a double focussing mass spectrometer (section 5.2.3).

In view of the specialized equipment involved, few labs were able to perform this work. However, an osmium isotope research program was developed by Luck, Allegre and co-workers in the early 1980s. Luck *et al.* (1980) established a chemical extraction method in which samples were oxidised after dissolution, allowing distillation of volatile osmium tetroxide. The resulting concentrate was further purified on an anion exchange column. Rhenium was extracted from the acidic residue of distillation by an organic solvent before back-extraction into aqueous solution with ammonia, and purification by anion exchange.

An alternative ion bombardment method which has been applied to Os isotope analysis by Fehn *et al.* (1986) is accelerator mass spectrometry. In this method, purified osmium is bombarded with positive caesium ions to generate negative osmium ions in a 'caesium sputter source'. While this is an inefficient process, the very high energy of the bombardment assures reasonable Os⁻ ion emission. The negative ions are converted to normal positive ions by charge exchange in the tandem accelerator (section 14.2). The subsequent magnetic analyser and multiplier detector are comparatively normal pieces of equipment.

Another excitation method is by high-energy (short-wavelength) laser. The use of the laser to promote ionisation leads to the method of 'resonance ionisation' mass spectrometry (RIMS). The energy of the laser can be tuned so accurately to the ionising reaction that it can be used to ionise ^{187}Os without exciting its isobaric interference ^{187}Re (Walker and Fassett, 1986). However, it is not possible to raise a ground-state atom directly to the ionised state because of the size of the energy jump. Therefore the process must be achieved in two or three steps. Because the second and third photon interactions must occur before the decay of intermediate species back to the ground state, extreme energy densities (in the mega watt range per mm³) must be achieved. This demands the use of pulsed lasers, which have a very poor duty cycle (i.e. the duration of the pulse relative to the time between pulses is ca. 10^{-6}). Therefore pulsed evaporation must also be used. This could take the form of pulsed thermal volatilisation (Walker and Fassett, 1986); laser ablation mass analysis (LAMMA), as used by Lindner *et al.* (1986); or ion sputtering (e.g. Blum *et al.*, 1990).

A totally different approach to osmium excitation is to use an inductively coupled plasma (ICP) source, which was first adapted to mass spectrometry by Houk *et al.* (1980). Temperatures of over 5000 °K in an argon plasma torch assure efficient osmium ionisation (Houk, 1986). The plasma feeds ions through a two-stage orifice system into a quadrupole mass analyser (Dawson, 1976). By constant evacuation of the zone between the two orifice plates, the interface can maintain a pressure differential of over seven orders of magnitude between the source and analyser.

The normal form of sample introduction for ICP–MS is to nebulise an analyte solution into a fine mist, which is carried to the torch in a flow of argon. This method has been used for osmium isotope analysis (e.g. Lichte *et al.*, 1986). However, solution nebulisation is generally less than 10% efficient, and limits precision on small samples. A more effective approach pioneered by Bazan (1987) and Russ *et al.* (1987) is to distil osmium directly into the plasma, using the argon gas flow to 'sparge' it from an oxidising solution as osmium tetroxide. This process has the added advantage of providing an excellent separation of osmium from matrix elements and the isobaric interference ^{187}Re.

All of these excitation methods for atomic osmium ions were rendered largely obsolete by the discovery that a solid osmium sample could yield negative Os molecular ions by conventional thermal ionisation (Volkening *et al.*, 1991). This N–TIMS method allows levels of precision over an order of magnitude better than the positive ion techniques described above (Fig. 8.1).

In the N–TIMS method Os is measured as the species OsO_3^-, using platinum filaments coated with a barium salt to lower the work function of the filament. This enhances the emission of negative ions relative to electrons. The formation of the oxide species may be enhanced by loading with barium nitrate (e.g. Creaser *et al.*, 1991) and by bleeding oxygen into the source (Walczyk *et al.*, 1991). The same method may be used to perform isotope dilution analysis of other platinum group elements (PGE) as well as rhenium, which forms the ReO_4^- species (Fig. 8.2). This method can generate beams large enough for analysis by Faraday detector from a few ng of osmium, while multiplier analysis allows picogram size samples to be analysed (Creaser *et al.*, 1991). It is anticipated that this technical advance will shortly bring the Re–Os method to the same wide range of applications as the Sr, Nd and Pb isotope methods.

A further development in Re–Os analysis has

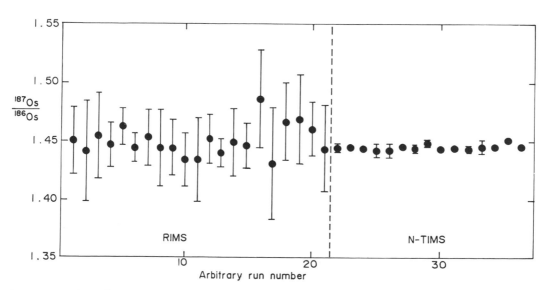

Fig. 8.1. A comparison of analytical data and uncertainties for an osmium standard analysed by RIMS and N–TIMS at the Carnegie Institute. After Shirey and Carlson (1991).

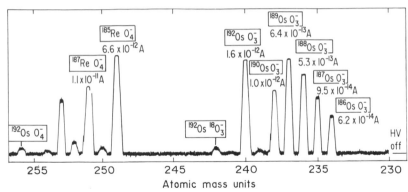

Fig. 8.2. Mass spectrum of Re and Os molecular ions produced from a Ba-doped Pt filament at 770 °C (ca. 2 A), loaded with 5 ng Os and 3 ng Re. After Creaser *et al.* (1991).

recently been suggested by Yin *et al.* (1993). These workers showed that neutron activation (with subsequent β decay) could be used to convert a fraction of sample ^{185}Re and ^{187}Re into ^{186}Os and ^{188}Os. If the activation process is adequately calibrated it allows the Re/Os ratio determination to be incorporated into the mass spectrometric Os isotope analysis. This approach is analogous to the ^{40}Ar^{39}–Ar method in K–Ar dating (section 10.2). Yin *et al.* demonstrated the method by dating molybdenites from a mineralised granite in Australia. The Os–Os ages which they determined were in good agreement with the emplacement ages of associated leucogranite intrusions. This technique may prove to be very useful for Re–Os analysis, because the homogenisation of sample and spike Re and Os has been a recurring problem in the determination of Re/Os ratios by isotope dilution (see below).

The main disadvantage of the activation technique is that it induces interfering reactions from other PGEs (especially osmium itself). These interferences are minor when dating rhenium-rich material such as molybdenite, so that Yin *et al.* were able to determine an accurate ^{187}Os abundance from the activated sample. However, if the method was applied to PGE-rich minerals with low Re/Os ratios, the interferences would be very large. The method might still be viable under these conditions, but ^{187}Os would need to be measured separately. This would make the analysis sensitive to heterogeneous Os distribution in rocks and minerals, known as the 'nugget effect'.

8.2 Determination of the Re decay constant

The maximum ^{187}Re β decay energy of 2.65 keV is extremely low, even compared with ^{87}Rb (275 keV).

This makes measurement of the decay constant by direct counting very difficult. Accurate counting of solid samples is almost impossible, due to absorption of β particles by surrounding Re atoms. An alternative technique is to use either a gaseous Re compound to replace the gas filling of a proportional counter, or to use a liquid Re compound in a scintillation detector. In both cases it is difficult to find compounds with suitable properties, but Brodzinski and Conway (1965) obtained a ^{187}Re half-life of 66 ± 13 Byr by the former method, while Naldrett (1984) obtained a value of 35 ± 4 Byr by the latter.

The difficulty with counting determinations has encouraged geological measurements of the half-life. However, these have also encountered technical difficulties. In view of the low concentrations of Re in normal rocks, early attempts at age determination (e.g. Hirt *et al.*, 1963) were made on molybdenite (MoS_2), which strongly concentrates Re at contents of ca. 10–50 ppm. Since molybdenites effectively incorporate no initial osmium, only the total abundance of the daughter need be measured, but this still involves mass spectrometry, since isotope dilution is the only method sufficiently sensitive. Samples were plotted on a diagram of Os growth against time. Although the data scattered significantly, they yielded an improved estimate for the half-life of 43 ± 5 Byr.

Luck and Allegre (1982) made further studies of the potential for Re–Os dating with molybdenite. They selected samples of known age over a wide range of geological time and analysed Re and Os concentrations by isotope dilution. Since insignificant amounts of common osmium were found, no isotope ratio determination was necessary. High-precision results were obtained, sometimes in good agreement with published ages, but often giving older ages. It was

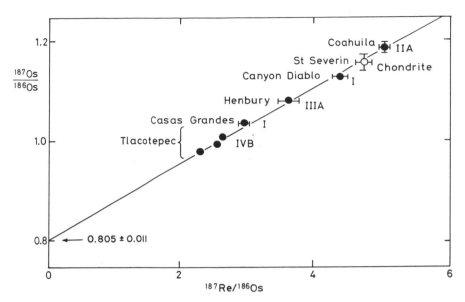

Fig. 8.3. Re–Os isochron diagram showing best-fit regression for five iron meteorites (notation indicates sub-group) and one chondrite. For $T = 4.55$ Byr, $\lambda = 1.62 \pm 0.08 \times 10^{-11}$ yr^{-1} ($t_{1/2} = 42.8$ Byr). After Luck *et al.* (1980).

concluded from this study that Re–Os dating of molybdenite is an unreliable geochronometer, probably due to preferential Re leaching during alteration.

More recently, there has been disagreement over the susceptibility of molybdenite to open-system behaviour. Suzuki *et al.* (1993) claimed that recent Re–Os dates were usually concordant with other methods, and suggested that errors in the earlier work might be due to poor sample spike homogenisation rather than geological disturbance. However, other workers such as McCandless *et al.* (1993) have maintained that open-system behaviour is a major problem, at least in old molybdenites. They suggested that a combination of microprobe analysis, electron back-scatter imaging and x-ray diffraction should be used to screen samples for alteration prior to analysis.

In view of the problems associated with molybdenites, other attempts at geological half-life determination were focussed principally on iron meteorites. These have moderately large Re and Os contents, commonly in the high ppb (parts-per-billion) to low ppm range, and also display the good range of Re/Os ratios necessary for a precise age. Since these samples contain initial Os, the age must be calculated on an isochron diagram. Luck *et al.* (1980) ratioed radiogenic ^{187}Os against ^{186}Os, to obtain the following isochron equation (after Hirt *et al.*, 1963):

$$\left(\frac{^{187}Os}{^{186}Os}\right)_P = \left(\frac{^{187}Os}{^{186}Os}\right)_I + \frac{^{187}Re}{^{186}Os}(e^{\lambda t} - 1) \qquad [8.1]$$

Luck *et al.* (1980) determined a good isochron fit for 'whole-rock' (bulk) samples of five iron meteorites, suggesting that iron meteorites of different types were all formed during a narrow time interval. Good isochron fits were also obtained in subsequent work on iron meteorites by Luck and Allegre (1983), Walker and Morgan (1989) and Morgan *et al.* (1992). These isochrons can be used to calculate values for the Re half-life by substituting a value of t into equation [8.1]. Because type II and III iron meteorites represent the cores of differentiated planetessimals, their age is constrained to be slightly younger than chondrites. However, iron meteorites cannot be more than about 50 Myr younger than chondrites, since they incorporated the short-lived nuclide ^{107}Pd with a half-life of 6.5 Myr (section 15.2.5). Hence t must be near 4.55 Byr.

Unfortunately, these meteoritic half-life determinations have also been dogged by analytical problems, principally involving spike calibration. This is probably due to the existence of Re and Os in variable oxidation states. For example, Luck *et al.* (1980) determined a half-life of 42.8 ± 2.4 Byr (Fig. 8.3), in close agreement with the estimate of Hirt *et al.* (1963). However, Luck and Allegre (1983) retracted this value on the grounds that isotope dilution analysis of Os in their 1980 data set was upset by a change of osmium species in the spike solution subsequent to its calibration. After re-calibration of the spike, Os concentrations were 6% lower than

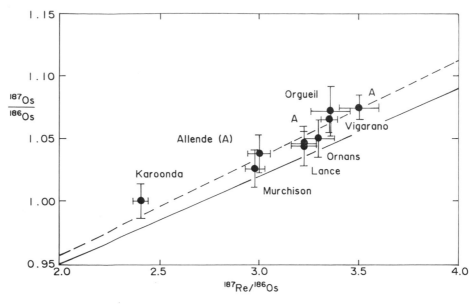

Fig. 8.4. Re–Os isochron diagram for chondrite whole-rock analyses relative to the iron meteorite isochron (solid line) of Luck and Allegre (1983). Three analyses of Allende are included. After Walker and Morgan (1989).

previously determined. By revising the older data and analysing further samples, Luck and Allegre (1983) calculated a new half-life of 45.6 ± 1.2 Byr. However, in the light of new results described below, one wonders in retrospect whether the revision in spike concentration was justified.

Further Re–Os studies of meteorites were made by Walker and Morgan (1989). They obtained two analyses of iron meteorites which were consistent with Luck and Allegre (1983). However, whole-rock samples of seven chondrites all gave results lying above the irons' isochron (Fig. 8.4). Morgan *et al.* (1992) re-examined the spike calibration used by Walker and Morgan (1989), and again revealed errors in concentration; this time due to an over-estimate of the Re content of the spike by 6%. After this correction, a composite isochron of group II and III iron meteorites gave a half-life of 41.4 Byr, assuming a meteorite age of 4.55 Byr. This was confirmed by a more precise isochron of group II irons (Horan *et al.*, 1992). Using the same assumptions, the isochron slope yields a (maximum) half-life of 41.9 ± 0.3 Byr (2σ) which is very close to a laboratory half-life determination described below. Hopefully, the Re–Os systematics of chondritic meteorites will likewise soon be clarified, but until that time these data must remain under suspicion.

In contrast to the above attempts at geological determination of the [187]Re half-life, Lindner *et al.*

(1986) used the 'laboratory shelf' technique to make an independent half-life determination. A 1 kg sample of purified perrhenic acid ($HReO_4$) was spiked with two different non-radiogenic Os isotopes (190 and 192), set aside for two years to allow radiogenic Os growth, and then sampled for Os isotope composition over a further two-year interval. Although this is a laboratory determination, it yields results equivalent to the geological determinations rather than β counting, because it measures the production of [187]Os rather than the decay of [187]Re. Os isotope measurements by LAMMA and ICP–MS gave results of comparable precision which were in good agreement, although the two spikes gave results differing by 2%. Results for [187]Os/[190]Os by ICP–MS are shown in Fig. 8.5.

Unfortunately, the starting material used by Lindner *et al.* had a non-zero level of initial radiogenic Os, as indicated by the positive intercept in Fig. 8.5. Hence the first two years of storage were effectively wasted. Despite this setback, a half-life of 43.5 ± 1.3 Byr was determined, within error of Luck and Allegre (1983). However, further work by Lindner *et al.* (1989), using five times as much data as the earlier work, led to a lower value of 42.3 ± 1.3 Byr, equivalent to a decay constant of 1.64×10^{-11} yr^{-1}. This value is outside error of Luck and Allegre (1983), but is within error of the value of Morgan *et al.* (1992) and Horan *et al.* (1992). It is

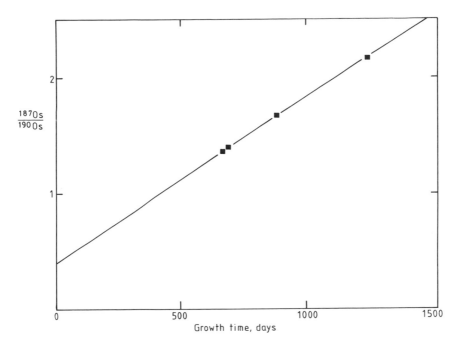

Fig. 8.5. Least squares growth line of ^{187}Os/^{190}Os as a function of time 'on the shelf' for a Re stock solution. Note non-zero initial ratio. After Lindner *et al.* (1986).

therefore the most appropriate for use in geological age studies.

8.3 Os normalisation and the Pt–Os decay scheme

Hirt *et al.* (1963) established the convention of ratioing ^{187}Os data against ^{186}Os, and was followed in this practice by Luck and Allegre (1983), who normalised Os data for within-run fractionation to a ^{192}Os/^{188}Os value of 3.0827, which has been followed by most other workers. However, ^{186}Os is itself the α decay product of the rare long-lived unstable isotope ^{190}Pt. This is not a significant problem in meteorite dating, since ^{190}Pt makes up only 0.0122 % of total platinum (Walker *et al.*, 1991). However, some of the more important terrestrial applications of the Re–Os method involve PGE-bearing ores, where platinum/osmium ratios may be very high. In these conditions, a significant fraction of ^{186}Os could be radiogenic.

Walker *et al.* (1991) observed significant quantities of radiogenic ^{186}Os in ore from the Strathcona deep copper zone at Sudbury (see below). They presented an analysis of this material, which can be used to construct a Pt–Os isochron (Fig. 8.6) by substituting into the general decay equation:

$$\left(\frac{^{186}\text{Os}}{^{188}\text{Os}}\right)_{\text{P}} = \left(\frac{^{186}\text{Os}}{^{188}\text{Os}}\right)_{\text{I}} + \frac{^{190}\text{Re}}{^{188}\text{Os}}\left(e^{\lambda t} - 1\right) \qquad [8.2]$$

Assuming that the system has been closed since formation of the Sudbury Complex at 1.85 Byr, Walker *et al.* calculated a ^{190}Pt decay constant of $7.9 \pm 0.6 \times 10^{-13}$ yr^{-1}, which is in quite good agreement with a counting determination of 1×10^{-12} yr^{-1} (Macfarlane and Kohman, 1961).

To avoid perturbation by radiogenic ^{186}Os, Walker *et al.* determined ^{187}Os abundances of Sudbury ores relative to ^{188}Os, then re-normalised their data to ^{186}Os by multiplying by 8.340 (equivalent to division by 0.1199). Some other workers avoided this problem by quoting osmium data directly against ^{188}Os (e.g. Fehn *et al.*, 1986; Dickin *et al.*, 1992; Ellam *et al.*, 1992). However, this option will also be excluded if the Re–Os method of Yin *et al.* (1993) is used, since ^{188}Os is the activation decay-product of ^{187}Re (section 8.1). Hence, there may be no ideal isotope for normalisation of Os data, but in view of the widespread use of ^{186}Os in the published literature, this isotope will be used here.

An alternative notation proposed by Walker *et al.* (1989a) is to present Os isotope ratios in the form of

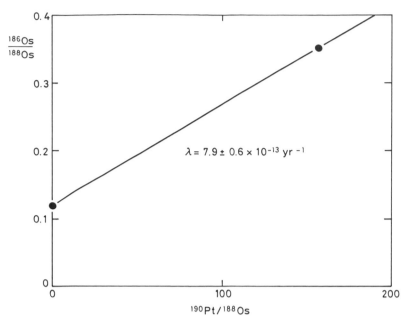

Fig. 8.6. Two-point Pt–Os isochron between deep copper ore from Strathcona and non-radiogenic osmium, yielding a ^{190}Pt decay constant of $7.9 \pm 0.6 \times 10^{-13}$ yr^{-1} ($t_{1/2} = 880$ Byr). Data from Walker *et al.* (1991).

percentage deviations (γ) from a chondritic reference point. Unfortunately, the Bulk Silicate Earth may not have a perfect chondritic Os signature (see below), so the CHUR reference point is less powerful for Os than in the Sm–Nd system (section 4.2). Nevertheless, if we wish to compare Os isotope data of different ages, we may need to use this notation, so it will be used here to a limited extent. The present-day average chondrite reference values chosen by Walker *et al.* were ^{187}Os/^{186}Os $= 1.06$ and ^{187}Re/^{186}Os $= 3.3$ (see also Fig. 8.4).

8.4 Mantle osmium

8.4.1 Asthenospheric evolution

Allegre and Luck (1980) used the initial ratio from the meteorite isochron as the starting point of an attempt to define the osmium isotope evolution of the Earth's mantle. They also analysed placer samples of the platinoid alloy osmiridium, whose crystallisation age could be estimated. Not only can osmiridium be analysed directly on the ion probe, but since its Re content is zero, no age correction is necessary to obtain initial ratios. The data define a good linear array for all samples except the Urals (Fig. 8.7), suggesting that the mantle system was effectively

closed for Re–Os over Earth history (in other words it does not display evidence of progressive depletion similar to Nd).

Allegre and Luck showed from concentration data that Re was strongly partitioned into the crust relative to Os, since the former is an incompatible element in the mantle while the latter is not. Therefore they attributed the apparent closed-system osmium isotope evolution of the mantle to the small total osmium budget of the crust relative to the mantle. The Re–Os analysis of whole-rock chondrites by Walker and Morgan (1989) allows a meaningful comparison between chondritic osmium and the mantle growth curve of Allegre and Luck. This shows that the range of chondrite Os isotope compositions (Fig. 8.4) strongly overlaps the mantle growth curve (Fig. 8.7). This is surprising, in view of the possibility of fractionation of mantle Re–Os by partition into the core.

Based on the siderophile chemistry of Re and Os, most of the inventory of these elements originally accreted to the Earth must have been partitioned into the core. Since it is very unlikely that Re and Os should have identical partition coefficients between the mineralogy of the mantle and core, we should expect to see Re/Os fractionation during core formation. Therefore, if Re and Os in the mantle

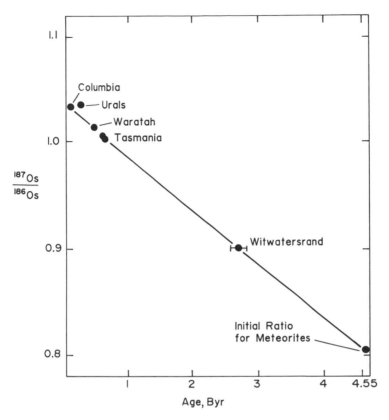

Fig. 8.7. Diagram of Os isotope evolution against time showing the first determination of the mantle growth curve. After Allegre and Luck (1980).

represent the residue from core formation, we would not expect them to display a Bulk Earth (= chondritic) ratio. To explain this co-incidence, Morgan (1985) suggested that the Re–Os budget of the mantle was produced by late accretion of chondritic material to the Earth, after core formation.

Given a mantle growth curve for osmium, platinoids of unknown age could be dated by analysis of their Os isotope composition. However, the rarity of such material limits the application of the method. In an attempt to widen the usefulness of the Re–Os method as a geochronometer, Luck and Allegre (1984) applied it to the dating of a Ni–Cu sulphide ore from the Cape Smith komatiite of NE Quebec. This material has a moderate concentration of both Re and Os at ca. 0.4 ppm, and can be used to calculate a model Re–Os isotope age in a manner analogous to model Nd ages.

Using a mantle ^{187}Re/^{186}Os ratio of 3.2 ± 0.14, Luck and Allegre calculated a model Os age of 1740 ± 60 Myr for the sulphide ore (Fig. 8.8). This result just overlapped the error limits of the Sm–Nd isochron age of 1871 ± 75 Myr determined on the host komatiite by Zindler (1982). However, using the 1.64×10^{-11} yr^{-1} decay constant of Lindner et al. (1989), the Re–Os model age is reduced to 1615 Myr, 250 Myr younger than the Sm–Nd age. This suggests that Ni–Cu sulphides may be susceptible to re-setting in the same way as Mo sulphide. Similar evidence of open-system behaviour was found by Walker et al. (1989b) in Archean schists from India.

The data set of Allegre and Luck (1980) for mantle Os evolution was very rudimentary, and more recently has been supplemented by the analysis of a wider variety of mantle-derived materials. For example, Walker et al. (1988) determined a Re–Os isochron on whole-rock samples of an Archean komatiite from Munro Township in Ontario, Canada (Fig. 8.9). This yields an age of 2.65 ± 0.09 Byr (using the decay constant of Lindner et al., 1989), in good agreement with the 2.70 Byr U–Pb age of nearby units from Newton township (section 4.1.2).

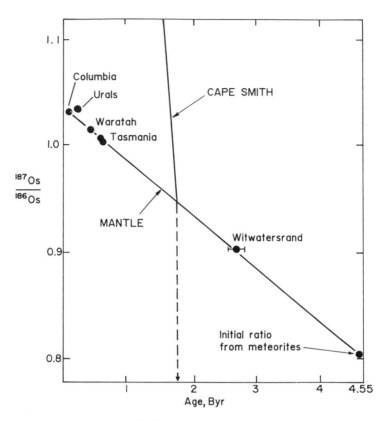

Fig. 8.8. Osmium isotope evolution diagram displaying model age determination on a sulphide from the Cape Smith komatiite. After Luck and Allegre (1984).

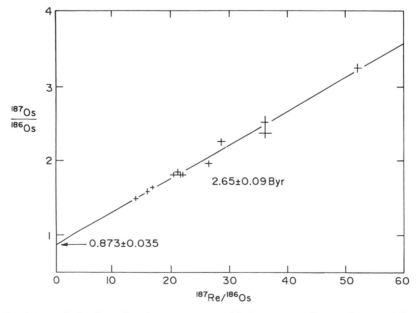

Fig. 8.9. Re–Os isochron for the Munro Township komatiite. After Walker *et al.* (1988).

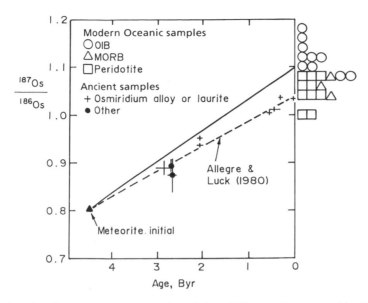

Fig. 8.10. Histogram of osmium isotope ratios in oceanic and chondritic samples analysed by 1991, along with two proposed mantle growth lines. Dashed = Allegre and Luck (1980); solid = Martin (1991). Modified after Martin *et al.* (1991).

In contrast, Sm–Nd data formed a scatter of points consistent with contamination by older crustal material, while whole-rock Rb–Sr data were scattered due to open-system behaviour during subsequent greenschist facies metamorphism.

The initial ratio for the Munro Township komatiite isochron was 0.873 ± 0.035, which falls slightly below, but within error of the mantle evolution line of Allegre and Luck. Unfortunately, the elevated Re/Os ratios of the analysed rocks precluded the calculation of a more precise initial ratio. However, this point represented (until recently) the only reliable constraint on osmium isotope evolution in the Archean, since further analysis of the Witwatersrand ores by Hart and Kinloch (1989) showed the original mantle-like signature determined by Allegre and Luck to be a coincidence.

Analysis of relatively young mantle-derived products avoids the uncertainties of age correction which are encountered in ancient rocks. Therefore the analysis of oceanic volcanics and xenoliths has been a major focus of attention. Martin and co-workers (1991) presented a compilation of new and published osmium analyses of mid ocean ridge basalt (MORB) and oceanic peridotites from various localities (Fig. 8.10). These data were supplemented by Luck and Allegre (1991), who analysed whole-rock samples of Phanerozoic ophiolites, as well as performing repli-

cate analyses of samples analysed in their 1980 paper. All of these data continue to support a relatively homogeneous upper mantle composition with a $^{187}Os/^{186}Os$ ratio close to 1.05, strongly overlapping with the range in chondritic meteorites.

In another attempt to constrain the Os isotope ratio of the oceanic mantle, Hattori and Hart (1991) analysed platinum group minerals in placer deposits derived from zoned and un-zoned ultramafic bodies. The former are attributed to ultramafic magmas, while the latter are interpreted as tectonically uplifted asthenospheric mantle. Since magmas may be susceptible to minor contamination by crustal Os, the tectonically emplaced bodies are more reliable indicators of mantle compositions. Variable ranges of Os isotope ratios were determined from different localities (Fig. 8.11), and were attributed to mantle processes. If the average value is then taken from each locality, we obtain an average upper mantle composition in close agreement to the other estimates above. However, if the lower limit from each locality is accorded greatest significance (diamond symbol in Fig. 8.11), then a more depleted upper mantle composition is determined. Whether such an approach would be justified remains an open question.

Luck and Allegre (1991) interpreted the empirical upper mantle evolution line as representative of the Bulk Silicate Earth. However, Martin (1991) argued

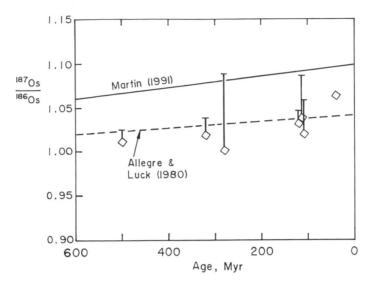

Fig. 8.11. Ranges of initial Os isotope ratio in Phanerozoic placer deposits derived from massive ultramafic bodies, relative to proposed mantle evolution lines. After Hattori and Hart (1991).

for a different Bulk Silicate Earth model on the basis of Os data from ocean island basalts (OIB). These display a range of $^{187}Os/^{186}Os$ ratios from 1.08 to 1.18 which is significantly higher than the upper mantle value. Martin plotted these data against $^3He/^4He$ (Fig. 8.12) in an attempt to discern the osmium isotope signature of primordial mantle, as distinct from recycled lithospheric components. High $^3He/^4He$ ratios (as seen at Loihi) are indicative of a primordial (relatively non-outgassed) helium reservoir, which might also represent primordial mantle osmium. This argument leads to a much higher estimate of the Bulk Silicate Earth $^{187}Os/^{186}Os$ ratio around 1.10. However, it is argued in section 6.5.2 that the lithophile isotope signatures of plumes may be decoupled from their rare gas signatures. Therefore it is dangerous to use Loihi to infer a Bulk Earth osmium isotope composition.

The choice between the above values for Bulk Silicate Earth has profound implications for terrestrial Re–Os evolution. For example, Martin et al. (1991) estimated that extraction of the continental crust from the mantle was insufficient or barely sufficient to balance Re/Os depletion in the sub-continental lithosphere (see below), let alone to balance a Re-depleted asthenospheric upper mantle. They argued that depletion of the upper mantle from a $^{187}Os/^{186}Os$ ratio of 1.10 to 1.05 could only be achieved by storage (somewhere in the mantle) of oceanic crust equal to ten times the volume of the

continental crust. Since this is unlikely, it casts additional doubt on the model of Martin et al.

New osmium isotope data for OIB (Hauri and Hart, 1993) help to explain the Loihi data in the context of recycling of radiogenic osmium into the mantle. Initial $^{187}Os/^{186}Os$ ratios as high as 1.25 were found in lavas from the HIMU island of Mangaia, in addition to less-elevated values in numerous other islands (Fig. 8.13). These radiogenic compositions are best explained by recycling of enriched crustal material back into the upper mantle. Hauri and Hart (1993) calculated that 16% of recycled 2.1 Byr-old oceanic crust can explain the osmium signature of the HIMU component. On the other hand, 1–2% of recycled continental crust in the EMII source does not markedly affect its osmium isotope signature.

Reisberg et al. (1993) found even more radiogenic osmium isotope ratios in the HIMU islands of the Comores and St Helena, with initial $^{187}Os/^{186}Os$ ratios as high as 1.7. However, they cautioned that, because these signatures were carried by osmium-poor lavas (0.01–0.03 ppb), their elevated ^{187}Os signatures could be introduced by processes in the oceanic lithosphere. This could involve contamination of OIB magmas with ocean floor sediments included in the volcanic edifice, or accumulated lithospheric olivine xenocrysts. Until these effects have been quantified, it is dangerous to attribute elevated osmium isotope signatures in osmium-poor OIB lavas to deep mantle reservoirs.

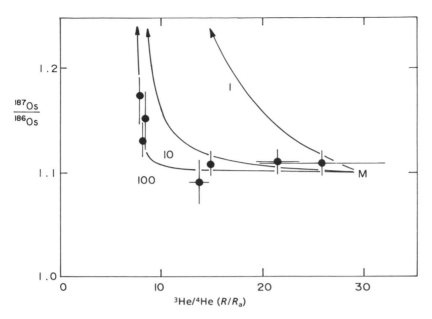

Fig. 8.12. Plot of Os isotope ratio against He isotope ratio (relative to air) for OIB. Curves represent mixing lines between plume and upper mantle components, assuming different Os/He ratios between these components. Coupled behaviour of Os and He might imply that $^{187}Os/^{186}Os = 1.10$ in the Bulk Silicate Earth (= undepleted mantle, M). After Martin (1991).

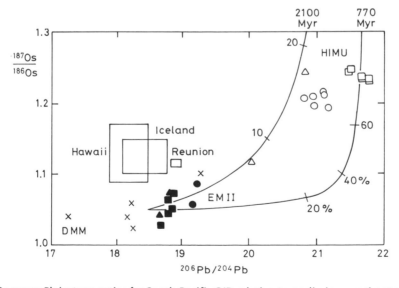

Fig. 8.13. Plot of Os *versus* Pb isotope ratios for South Pacific OIB relative to preliminary estimates for the osmium isotope composition of mantle components. Symbols: (●, ■, ▲) = Tahaa, Samoa, Rarotonga; (○, ☐, △) = Tubuai, Mangaia, Rurutu; (×) = peridotite xenoliths. Curves show the effect of re-cycling basaltic crust of different ages into the mantle. After Hauri and Hart (1993).

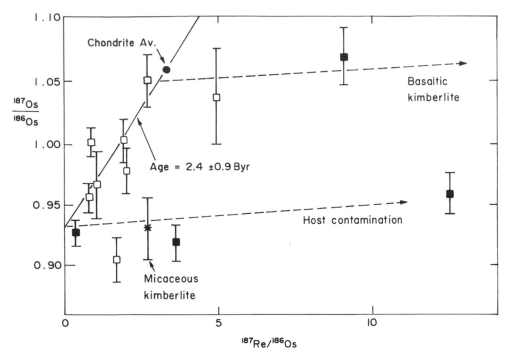

Fig. 8.14. Re–Os isochron diagram for South African peridotite xenoliths. (□) = High-temperature peridotite; (■) = Low-temperature peridotite; (✱) = Group II (micaceous) kimberlite. After Walker *et al.* (1989a).

8.4.2 Lithospheric evolution

The first osmium isotope data on ancient sub-continental lithosphere was provided by Walker *et al.* (1989a), who measured osmium isotope compositions in peridotite xenoliths from South African kimberlites. The Re–Os isochron diagram (Fig. 8.14) allows the effects of contamination by the host (Group I type) kimberlite to be assessed. Walker *et al.* argued that such a process causes samples to move along a sub-horizontal vector towards high Re/Os ratios characteristic of the host magma. Samples argued not to have suffered this effect define a steep array in Fig. 8.14, corresponding to an age of 2.4 ± 0.9 Byr.

A micaceous kimberlite (Group II) analysed by Walker *et al.* had an unradiogenic composition (Fig. 8.14), attributed to a magma source in the sub-continental lithosphere. In contrast, the initial ratio of the Re-rich basaltic kimberlite (Group I) was similar to average chondrites. Hence, the Group I kimberlite was probably derived from an un-depleted astheno-spheric source. These deductions are consistent with Sr and Nd isotope evidence for kimberlite genesis (section 7.3.1).

Walker *et al.* attributed the relatively coherent Re–Os isotope systematics of the peridotite xenoliths to separation of South African lithosphere from the convecting asthenosphere in the Archean. Their unradiogenic compositions, relative to Allegre and Luck's upper mantle $^{187}Os/^{186}Os$ ratio of 1.04, suggest that they represent residues from partial melting, consistent with gross incompatible-element depletion of the sub-continental lithosphere (Jordan, 1978). Hence, Walker *et al.* suggested that the Re–Os system could provide different insights from LIL tracers such as Sr and Nd, which commonly chart secondary *enrichment* of the sub-continental lithosphere by metasomatism.

In contrast to these findings, Re–Os analysis of the Ronda ultramafic complex (Reisberg *et al.*, 1991) provided an example of enriched osmium isotope systematics in an orogenic lherzolite massif which is widely interpreted as mantle peridotite. The massif contains both mafic and ultramafic layers, which Reisberg *et al.* interpreted as relics of old recycled oceanic lithosphere (the 'marble cake mantle' model, section 6.2.5), or as the products of an ancient melting event. Whole-rock samples of mafic and ultramafic

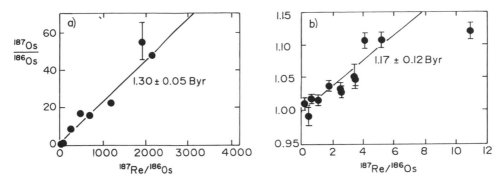

Fig. 8.15. Re–Os isochron diagrams for the Ronda ultramafic complex. a) Mafic units; b) ultramafic units. After Reisberg *et al.* (1991).

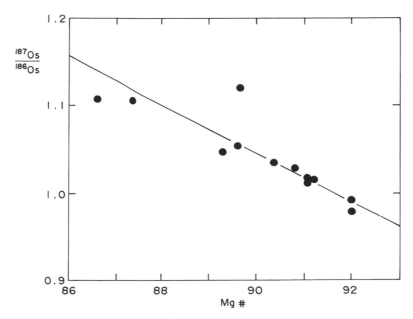

Fig. 8.16. Plot of Os isotope ratio against Mg number (100 × Mg/[Mg + Fe]), where abundances are atomic. After Reisberg *et al.* (1991).

layers define Re–Os errorchrons with similar apparent ages (Fig. 8.15), but the ranges of isotopic composition are very different. Ultramafic units display both isotopic enrichment and depletion relative to the estimated Bulk Earth composition. In contrast, the mafic units display extreme isotopic enrichment similar to crustal rocks.

Reisberg *et al.* found a negative correlation between Os isotope ratio and Mg number in the ultramafic Ronda samples (Fig. 8.16), but observed no correlation between Os and Nd isotope data. They suggested that the isochron arrays and the inverse correlation with Mg number could be explained either by a single ancient differentiation event, or by continuous mixing over a long period of time, as proposed in the marble cake model. However, in view of the structural context of the massif, Reisberg *et al.* concluded that it may best be interpreted as ancient

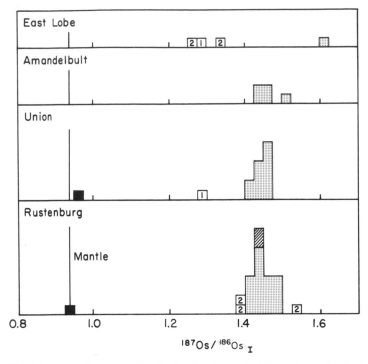

Fig. 8.17. Histogram of initial osmium isotope ratios for laurites from the Merensky reef (stipple) and the UG1 and UG2 chromitites (1 and 2), grouped by mining area. Solid box = erlichmanite; hatch = other PGM. Vertical line = mantle composition. Data from Hart and Kinloch (1989) and McCandless and Ruiz (1991).

asthenospheric mantle which was incorporated into the lithosphere ca. 1.2 Byr ago as the residue of a partial melting event.

8.5 Osmium as a petrogenetic tracer

As one of the platinum group elements (PGE), osmium is uniquely suited as a tracer of petrogenetic and ore-forming processes of noble metal deposits. PGE deposits are generally associated with major mafic complexes, and have been attributed to mixing process between mantle-derived and crustal components (e.g. Naldrett, 1989). The strong partitioning between Re and Os in crust-forming processes, which generates very radiogenic osmium in the crust relative to the mantle, makes osmium a powerful tracer for such studies. Two of the world's largest basic layered intrusions have been subjected to Re–Os analysis, but the complexities of their chemistry, relative to the density of isotopic sampling, has impeded an understanding of their genesis.

The Bushveld Complex is the world's largest layered mafic intrusion and principal PGE producer.

Most of these PGEs come from the Merensky reef, but the UG1 and UG2 chromitites are also major sources. Hart and Kinloch (1989) made an ion probe study of PGE sulphides from the Merensky reef on the western lobe of the intrusion. They found consistent initial $^{187}Os/^{186}Os$ ratios around 1.45 for grains of laurite (RuS_2) from the Rustenburg, Union and Amandelbult mining areas (Fig. 8.17). These values are far above the mantle growth line for osmium at the time of intrusion of the Bushveld (2.05 Byr ago), indicating a large crustal component in the ore. Two other types of platinum group mineral (PGM) from Rustenburg gave similar results, but two grains of erlichmanite (OsS_2) from Rustenburg and Union gave low $^{187}Os/^{186}Os$ ratios on the mantle growth curve.

The Bushveld laurites might be interpreted as the products of crustally contaminated magmas, as has been proposed to explain Sr isotope data for the intrusion (Sharpe, 1985). However, the erlichmanite results pose a major problem for this interpretation. They cannot be attributed to open-system perturbation of age corrections, since they have no rhenium. If

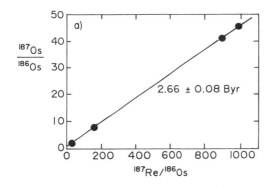
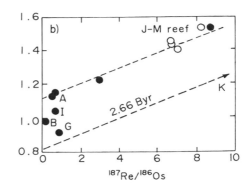

Fig. 8.18. Re–Os isochron diagrams for whole-rock samples from the Stillwater Complex. a) Re-rich samples; b) Re-poor samples, including the J–M reef, chromitite bands A to K, and un-named chromitites. After Lambert *et al.* (1989).

the osmium isotope variations in the laurites are attributed to magmatic processes then the erlichmanites must represent mantle-derived PGE xenocrysts which were somehow carried into the intrusion (which seems unlikely). Alternatively osmium isotope variations in the laurites must be hydrothermal in origin, and some component of the Merensky reef mineralisation must therefore be attributed to hydrothermal PGE introduction. McCandless and Ruiz (1991) also analysed a small suite of UG1, UG2 and Merensky reef laurites from the E and W lobes. These data displayed more scatter than those of Hart and Kinloch (Fig. 8.17), but did not resolve the question of magmatic or hydrothermal origins.

The Stillwater Complex in Montana has similarities to the Bushveld, although on a smaller scale. Two distinct magmas have been proposed to explain the petrology and chemistry of the pluton. An ultramafic liquid gives rise to the lower UltraMafic Series (UMS), while a magma similar to high-Al basalt forms most of the overlying Banded Series. The PGE-bearing J–M reef is located near the stratigraphic boundary between these two liquids, whose mixing may have promoted the segregation of a PGE-bearing sulphide liquid to form the reef. However, chromite layers scattered through the ultramafic series may also be due to small influxes of Al-rich magma into a magma chamber crystallising an ultramafic liquid.

Lambert *et al.* (1989) determined an excellent Re–Os isochron for the complex from four Re-rich whole-rock samples (Fig. 8.18a). These comprise two sulphide-rich cumulates, a bronzite pegmatite and the K-seam chromitite band. Although these units are widely separated through the intrusion, any initial

ratio heterogeneity is swamped by their high Re/Os ratios. Using the decay constant of Lindner *et al.* (1989), the isochron yields an age of 2.66 ± 0.08 Byr (MSWD = 0.03), in good agreement with U–Pb and Sm–Nd ages for the intrusion (section 4.1.2). This shows that on a whole-rock scale, the Re–Os system remained closed during a thermal event which reset Rb–Sr *mineral* systems in the complex.

Analysis of whole-rock samples with low Re/Os ratios by Lambert *et al.* revealed a degree of initial ratio heterogeneity which was similar to the Bushveld (Fig. 8.18b). Chromitite bands from the ultramafic series had a range of initial $^{187}Os/^{186}Os$ compositions from the mantle value of 0.9 to a ratio of 1.15 in the J–M reef. Martin (1989) obtained a similar range of initial ratios in whole-rock samples from the G chromite (0.90 to 1.00), but slightly higher ratios from the J–M reef. However, using the decay constant of Lindner *et al.* (1989), Martin's age-corrected J–M reef data come down almost within error of Lambert *et al.* (1989). Lambert *et al.* attributed the chromitites to crustal contamination of mantle-derived magmas by an enriched crustal component, while Martin attributed isotopic variation in the reefs to variable mixing between chromite cumulates and contaminated intercumulus liquid.

Further study of fresh Stillwater chromites by Marcantonio *et al.* (1993) did not yield such large variations in initial ratio. Chromite separates and chromitite whole-rocks from four horizons had initial $^{187}Os/^{186}Os$ ratios within error of each other, with a total range (0.91–0.95) which is very close to the 'chondritic' mantle ratio of 0.92 at the time of intrusion. This implies minimal contamination of

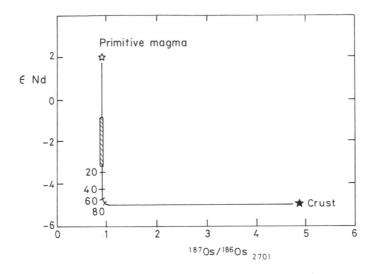

Fig. 8.19. Plot of ε Nd against initial Os isotope ratio for the Stillwater Complex (hatched zone) showing extreme hyperbolic mixing line between ultramafic magma (Nd/Os =400) and the crust (Nd/Os = 400 000). Modified after Lambert *et al.* (1989).

the cumulus-forming magma by crustal melts. On the other hand, Marcantonio *et al.* measured an initial $^{187}Os/^{186}Os$ ratio of 6.55 in a molybdenite-bearing whole-rock from the G-chromitite. They attributed the elevated initial ratios in this and other chromitite samples to hydrothermal introduction of radiogenic Os after crystallisation. A Re–Os age of 2.74 ± 0.08 Byr from the above-mentioned molybdenite shows that hydrothermal mineralisation occurred soon after emplacement of the complex.

The minimal degree of crustal Os contamination permitted by the unaltered chromites creates problems in the interpretation of Sm–Nd data for the Stillwater Complex. These data imply substantial variation in (initial) ε Nd, which was attributed by Lambert *et al.* to crustal assimilation at the magmatic stage (section 4.1.2). Different responses of Os and Nd to crustal contamination can be explained, however, by the extreme difference in Nd/Os ratios between ultramafic magmas and the crust. These components may have ratios of ca. 400 and 400 000 respectively, so that crustal contamination can generate extremely hyperbolic mixing lines, as shown in Fig. 8.19.

A combination of Os and Nd isotope data has been used to study mixing between source components of the picritic Karoo flood basalts from southern Africa (Ellam *et al.*, 1992). Initial Os–Nd isotope data for these samples (at 190 Myr) are shown relative to

possible source reservoirs in Fig. 8.20. These data reveal some of the potential advantages of Os isotope data, relative to other tracers such as Nd, in discriminating between components involved in basalt genesis. Nd isotope compositions may be similar in the continental crust and sub-continental lithosphere, making these components hard to resolve. In addition, the greater incompatible-element inventories of lithospheric reservoirs, relative to asthenospheric melts, makes it hard to quantify contamination processes which may occur during magma ascent through the lithosphere. On the other hand, Os isotope data can resolve these components because the sub-continental lithosphere has an unradiogenic signature relative to plume sources and crustal units (Walker *et al.*, 1989a). The model mixing lines in Fig. 8.20 suggest variable contributions of material from asthenospheric and lithospheric mantle to magma genesis.

8.6 Osmium in impact sites

The radiogenic Os isotope ratios characteristic of continental crust are readily resolvable from the unradiogenic signature of most meteorites. Therefore, Os isotopes are potentially a powerful tracer in distinguishing terrestrial and extra-terrestrial components. The first application of osmium isotopes to this

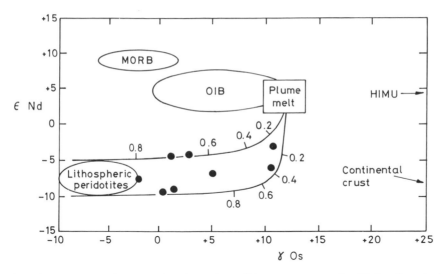

Fig. 8.20. Plot of ϵ Nd against γ Os for Karoo basalts and possible source reservoirs. Mixing lines between plume and lithospheric components are shown. After Ellam *et al.* (1992).

problem, by Luck and Turekian (1983), was to test the proposed extra-terrestrial source of the iridium anomaly at the Cretaceous–Tertiary (K–T) boundary (Alvarez *et al.*, 1980). A meteoritic source should yield ^{187}Os/^{186}Os abundances of ca. 1.1 at the present day. In contrast, if the PGE were concentrated from seawater they should have a signature largely reflecting continental run-off. The strong fractionation of Re/Os in the crust yields a ^{187}Re/^{186}Os ratio of ca. 400, which over an average continental age of 2 Byr should yield sediments with ^{187}Os/^{186}Os ratios of about 10.

Luck and Turekian tested the composition of oceanic chemical sediments by analysis of manganese nodules from several oceans. ^{187}Os/^{186}Os ratios of between 6 and 9 attested to a major continental component, with a subsidiary fraction from juvenile mantle sources (e.g. via ocean floor hydrothermal alteration), or from the cosmic dust flux. In contrast, analysis of K–T boundary sediments gave ^{187}Os/^{186}Os ratios of 1.3–1.65. The data set was enlarged by the analysis of a New Zealand sample by Lichte *et al.* (1986), yielding a ^{187}Os/^{186}Os ratio of 1.12. Taken together, the data suggest that a chondritic-type source was an important component in the sediments. Hence, these data show that the iridium anomaly is an external feature superimposed on the sedimentary sequence, and not the result of chemical PGE remobilisation. However, additional evidence is necessary to exclude an origin from mafic volcanism.

A similar resolution of terrestrial and meteoritic components was achieved by Fehn *et al.* (1986) for the East Clearwater crater in Quebec, Canada, and the Ries crater in Germany. A sub-surface melt layer from the East Clearwater crater yielded ^{187}Os/^{186}Os ratios of 0.9–1.0, within error of iron meteorites, suggesting an extra-terrestrial origin. In contrast, melt rocks from the Ries crater yielded radiogenic ^{187}Os/^{186}Os ratios between 8 and 9, bracketed within the range of shocked gneiss samples. Hence, in this case the melt rocks appear to be entirely of crustal origin.

A possible example of an ancient impact site is the Sudbury structure in Ontario, Canada. The Sudbury Intrusive Complex (SIC) is the world's largest nickel source, and consists of a stratified tholeiitic pluton which is overlain by the Onaping tuff and underlain by a 'Sublayer' containing Fe–Ni–Cu–PGE sulphide ore. Various features of Sudbury geology, such as the widespread distribution of brecciated rock outside the complex, the presence of shatter cones in the footwall rocks, and the lopolith form of the SIC were argued by Dietz (1964) to support an origin by meteorite impact.

Nd isotope data for the silicate rocks of the SIC were presented by Faggart *et al.* (1985) and Naldrett *et al.* (1986). Both groups showed that the silicate rocks had a remarkably strong crustal signature, with ϵ Nd at 1.85 Byr averaging about −7.5. However, Naldrett *et al.* found a larger range of values (from

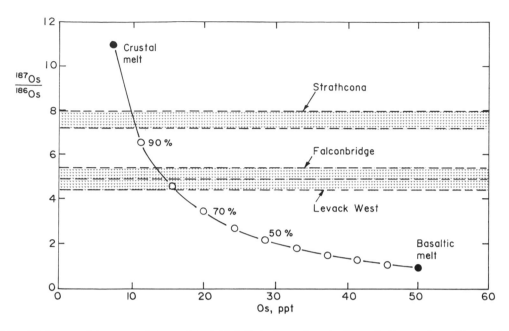

Fig. 8.21. Mixing model proposed to explain initial Os isotope data in three Sudbury mines. The crustal end-member is based on Superior Province Archean gneisses, which outcrop to the north of the complex. 10 % increments of crustal component are shown. After Walker *et al.* (1991).

−5 to −9). Faggart *et al.* argued that their data could be explained by an exclusively crustal origin for the SIC, while Naldrett *et al.* preferred a model involving gross crustal contamination of a mantle-derived magma. However, because Nd is a lithophile element, this evidence cannot reliably be extrapolated to deduce such an origin for the nickel-bearing sulphide ores of the complex. Furthermore, the enrichment of Nd in crustal relative to mantle-derived melts makes it an insensitive tracer for a small mantle-derived source component, which would tend to be swamped by crustal Nd. In this situation, Os data may be more diagnostic.

Walker *et al.* (1991) demonstrated approximate agreement between Re–Os isochron ages for sulphide ores and the 1.85 Byr U–Pb age of the silicate rocks (Krogh *et al.*, 1984). This substantiated previous geochemical evidence that the sulphide and silicate melts were co-genetic. Walker *et al.* interpreted initial Os isotope data from the Levack West, Falconbridge and Strathcona mines (Fig. 8.21) as indicative of substantial isotopic heterogeneity in the intrusion. This was attributed to variable mixing between mantle-derived osmium and radiogenic crustal osmium, the latter probably derived from underlying gneisses of the Superior Province.

In contrast, Dickin *et al.* (1992) noted strong initial ratio homogeneity between the Creighton, Falconbridge and Levack West mines (Fig. 8.22). Tails to lower initial ratios in two of these mines, and the large scatter of initial ratios from the Strathcona mine, were attributed to post-intrusive open-system behaviour of the Re–Os system, possibly in response to the Grenville orogeny. The consensus of initial ratios for Sudbury mines falls within the range of estimated crustal compositions at 1.85 Byr, and was attributed by Dickin *et al.* to an entirely crustal source for osmium in the Sudbury ores. This is consistent with an origin of the SIC as an impact melt sheet (Dietz, 1964). However, no evidence of material contribution from the meteorite itself is seen.

Since Lower Proterozoic sediments of the Huronian Supergroup composed the upper crust at the time of the Sudbury impact, we should expect these rocks, in addition to the underlying Archean gneisses, to have contributed substantially to the melt. This is consistent with the data in Fig. 8.22, since the ores fall within the range of estimated Huronian crustal compositions at 1.85 Byr. The precise origin of the sulphide melt at Sudbury remains unclear, but this melt must have been in intimate contact with the pool of fused crustal material for considerable time.

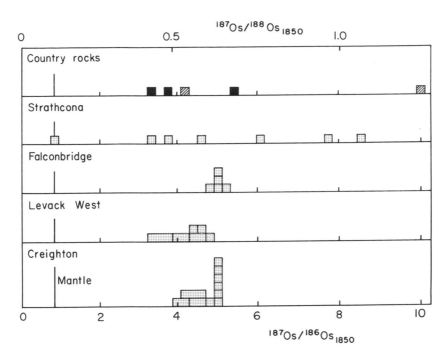

Fig. 8.22. Histogram of calculated Os isotopic data for Sudbury ores and country rocks 1.85 Byr ago. Boxes represent single samples and ranges of duplicates. Stipple = ores; hatch = Archean gneiss; solid = Huronian Supergroup; vertical line = mantle composition. After Dickin *et al.* (1992).

During this time, PGE were probably partitioned from the bulk crustal melt into the sulphide phase in a process analogous to the nickel sulphide fire assay method (e.g. Hofmann *et al.*, 1978). Hence, Sudbury may represent 'nature's largest fire assay'.

8.7 Seawater osmium evolution

A comparison with Sr isotope systematics suggests that the wide range in osmium isotope composition between crustal and mantle reservoirs should generate large variations in the Os isotope composition of seawater. The very low osmium content of seawater presents a major obstacle to such analysis, but organic-rich pelagic sediments act to pre-concentrate osmium by scavenging it and other metals from seawater. Ravizza and Turekian (1992) showed that this osmium can be preferentially released from the substrate by leaching in order to perform an isotopic analysis. By this means they demonstrated radiogenic $^{187}Os/^{186}Os$ ratios around 8.5 in modern seawater. In contrast, residues from leaching have significantly

lower Os isotope ratios, due to the presence of a micro-meteorite (cosmic dust) component which is constantly raining down upon the Earth.

Pegram *et al.* (1992) applied this method to a study of seawater osmium isotope evolution through the Cenozoic. Osmium was extracted by acidic hydrogen peroxide leaching of pelagic black shales from a large piston core recovered from the North Pacific. An overall correlation was observed between the Os and Sr isotope composition of samples of the same age. Pegram *et al.* interpreted the osmium isotope ratios as primary signatures of carbonaceous sediments, and not the product of secondary mixing between re-mobilised terrestrial and meteoritic osmium. However, if this is correct then the data show a surprisingly rapid increase in isotope ratio through the Cenozoic (Fig. 8.23). This was tentatively attributed to weathering of uplifted black shale strata in the Himalayas. The magnitude of the observed isotopic variations suggest that osmium may be a very sensitive tracer of the competing fluxes which control seawater geochemistry, and therefore that this method will become a powerful tool for paleo-oceanographic studies.

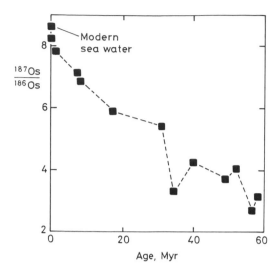

Fig. 8.23. Plot of Os isotope ratio of sediment leachates against age, attributed to changing seawater Os isotope ratio. After Pegram *et al.* (1992).

References

Allegre, C. J. and Luck, J. M. (1980). Osmium isotopes as petrogenetic and geological tracers. *Earth Planet. Sci. Lett.* **48**, 148–54.

Alvarez, L. W., Alvarez, W., Asaro, F. and Michel, H. V. (1980). Extraterrestrial cause for the Cretaceous–Tertiary extinction. *Science* **208**, 1095–108.

Bazan, J. M. (1987). Enhancement of osmium detection in inductively coupled plasma atomic emission spectrometry. *Anal. Chem.* **59**, 1066–9.

Blum, J. D., Pellin, M. J., Calaway, W. F., Young, C. E., Gruen, D. M., Hutcheon, I. D. and Wasserburg, G. J. (1990). *In-situ* measurement of osmium concentrations in iron meteorites by resonance ionization of sputtered ions. *Geochim. Cosmochim. Acta* **54**, 875–81.

Brodzinski, R. L. and Conway, D. C. (1965). Decay of rhenium187. *Phys. Rev.* **138**, B1368–71.

Creaser, R. A., Papanastassiou, D. A. and Wasserburg, G. J. (1991). Negative thermal ion mass spectrometry of osmium, rhenium, and iridium. *Geochim. Cosmochim. Acta* **55**, 397–401.

Dawson, P. H. (1976) (Ed.), *Quadrupole Mass Spectrometry and its Applications.* Elsevier, 349 p.

Dickin, A. P., Richardson, J. M., Crocket, J. H., McNutt, R. H. and Peredery, W. V. (1992). Osmium isotope evidence for a crustal origin of platinum group elements in the Sudbury nickel ore. *Geochim. Cosmochim. Acta* **56**, 3531–7.

Dietz, R. S. (1964). Sudbury structure as an astrobleme. *J. Geol.* **72**, 412–34.

Ellam, R. M., Carlson, R. W. and Shirey, S. B. (1992). Evidence from Re–Os isotopes for plume–lithospheric mixing in Karoo flood basalt genesis. *Nature* **359**, 718–21.

Faggart, B. E., Basu, A. R. and Tatsumoto, M. (1985). Origin of the Sudbury Complex by meteoritic impact: neodymium isotope evidence. *Science* **230**, 436–9.

Fehn, U., Teng, R., Elmore, D. and Kubik, P. (1986). Isotopic composition of osmium in terrestrial samples determined by accelerator mass spectrometry. *Nature* **323**, 707–10.

Hart, S. R. and Kinloch, E. D. (1989). Osmium isotope systematics in Witwatersrand and Bushveld ore deposits. *Econ. Geol.* **84**, 1651–5.

Hattori, K. and Hart, S. R. (1991). Osmium-isotope ratios of platinum-group minerals associated with ultramafic intrusions: Os-isotopic evolution of the oceanic mantle. *Earth Planet. Sci. Lett.* **107**, 499–514.

Hauri, E. H. and Hart, S. R. (1993). Re–Os isotope systematics of HIMU and EMII oceanic island basalts from the south Pacific Ocean. *Earth Planet. Sci. Lett.* **114**, 353–71.

Hirt, B., Tilton, G. R., Herr, W. and Hoffmeister, W. (1963). The half life of [187]Re. In: Geiss, J. and Goldberg, E. (Eds.), *Earth Science Meteoritics*. North Holland Pub., pp. 273–80.

Hofmann, E. L., Naldrett, A. J., van Loon, J. C., Hancock, R. G. V. and Manson, A. (1978). The determination of all the platinum group elements and gold in rocks and ore by neutron activation analysis after preconcentration by a nickel sulfide fire-assay technique on large samples. *Anal. Chim. Acta* **102**, 157–66.

Horan, M. F., Morgan, J. W., Walker, R. J. and Grossman, J. N. (1992). Rhenium–osmium isotope constraints on the age of iron meteorites. *Science* **255**, 1118–21.

Houk, R. S. (1986). Mass spectrometry of inductively coupled plasmas. *Anal. Chem.* **58**, 97A–105A.

Houk, R. S., Fassel, V. A., Flesch, G. D., Svec, H. J., Gray, A. L. and Taylor, C. E. (1980). Inductively coupled argon plasma as an ion source for mass spectrometric determination of trace elements. *Anal. Chem.* **52**, 2283–9.

Jordan, T. H. (1978). Composition and development of the continental tectosphere. *Nature* **274**, 544–8.

Krogh, T. E., Davis, D. W. and Corfu, F. (1984). Precise U–Pb zircon and badeleyite ages from the Sudbury area. In: Pye, E. G., Naldrett, A. J. and Giblin, P. E. (Eds) *The Geology and Ore Deposits of the Sudbury Structure. Ont. Geol. Surv. Spec. Pub. Vol. 1*, pp. 431–47.

Lambert, D. D., Morgan, J. W., Walker, R. J., Shirey, S. B., Carlson, R. W., Zientek, M. L. and Koski, M. S. (1989). Rhenium–osmium and samarium–neodymium isotopic systematics of the Stillwater Complex. *Science* **244**, 1169–74.

Lichte, F. E., Wilson, S. M., Brooks, R. R., Reeves, R. D., Holzbecher, J. and Douglas E. R. (1986). New method for the measurement of osmium isotopes applied to a New Zealand boundary shale. *Nature* **322**, 816–17.

Lindner, M., Leich, D. A., Borg, R. J., Russ, G. P., Bazan, J.

M., Simons, D. S. and Date, A. R. (1986). Direct laboratory determination of the ^{187}Re half-life. *Nature* **320**, 246–8.

Lindner, M., Leich, D. A., Russ, G. P., Bazan, J. M. and Borg, R. J. (1989). Direct determination of the half-life of ^{187}Re. *Geochim. Cosmochim. Acta* **53**, 1597–606.

Luck, J. M. and Allegre, C. J. (1982). The study of molybdenites through the ^{187}Re–^{187}Os chronometer. *Earth Planet. Sci. Lett.* **61**, 291–6.

Luck, J. M. and Allegre, C. J. (1983). ^{187}Re–^{187}Os systematics in meteorites and cosmochemical consequences. *Nature* **302**, 130–2.

Luck, J. M. and Allegre, C. J. (1984). ^{187}Re–^{187}Os investigation in sulphide from Cape Smith komatiite. *Earth Planet. Sci. Lett.* **68**, 205–8.

Luck, J. M. and Allegre, C. J. (1991). Osmium isotopes in ophiolites. *Earth Planet. Sci. Lett.* **107**, 406–15.

Luck, J. M., Birck, J. L. and Allegre, C. J. (1980). ^{187}Re–^{187}Os systematics in meteorites: early chronology of the solar system and the age of the galaxy. *Nature* **283**, 256–9.

Luck, J. M. and Turekian, K. K. (1983). Osmium-187/Osmium-186 in manganese nodules and the Cretaceous–Tertiary boundary. *Science* **222**, 613–15.

Macfarlane, R. D. and Kohman, T. P. (1961). Natural α radioactivity in medium-heavy elements. *Phys. Rev.* **121**, 1758–69.

Marcantonio, F., Zindler, A., Reisberg, L. and Mathez, E. A. (1993). Re–Os isotopic systematics in chromitites from the Stillwater Complex, Montana, USA. *Geochim. Cosmochim. Acta* **57**, 4029–37.

Martin, C. E. (1989). Re–Os isotopic investigation of the Stillwater Complex, Montana. *Earth Planet. Sci. Lett.* **93**, 336–44.

Martin, C. E. (1991). Osmium isotopic characteristics of mantle-derived rocks. *Geochim. Cosmochim. Acta* **55**, 1421–34.

Martin, C. E., Esser, B. K. and Turekian, K. K. (1991). Re–Os isotopic constraints on the formation of mantle and crustal reservoirs. *Aust. J. Earth Sci.* **38**, 569–76.

McCandless, T. E. and Ruiz, J. (1991). Osmium isotopes and crustal sources for platinum-group mineralization in the Bushveld Complex, South Africa. *Geology* **19**, 1225–8.

McCandless, T. E., Ruiz, J. and Campbell, A. R. (1993). Rhenium behaviour in molybdenite in hypogene and near-surface environments: implications for Re–Os geochronometry. *Geochim. Cosmochim. Acta* **57**, 889–905.

Morgan, J. W. (1985). Osmium isotope constraints on Earth's accretionary history. *Nature* **317**, 703–5.

Morgan, J. W., Walker, R. J. and Grossman, J. N. (1992). Rhenium–osmium isotope systematics in meteorites I: magmatic iron meteorite groups IIAB and IIIAB. *Earth Planet. Sci. Lett.* **108**, 191–202.

Naldrett, A. J. (1989). *Magmatic Sulphide Deposits.* Oxford Univ. Press, 186 p.

Naldrett, A. J., Rao, B. V. and Evensen, N. M. (1986). Contamination at Sudbury and its role in ore formation, In: Gallagher, M. J., Ixer, R. A., Neary, C. R. and Pritchard, H. M. (Eds.), *Metallogeny of Basic and Ultrabasic Rocks.* Spec. Pub. Inst. Mining & Metall., pp. 75–92.

Naldrett, S. N. (1984). Half-life of rhenium: geologic and cosmologic ages. *Can. J. Phys.* **62**, 15–20.

Pegram, W. J., Krishnaswami, S., Ravizza, G. E. and Turekian, K. K. (1992). The record of seawater ^{187}Os/^{186}Os variation through the Cenozoic. *Earth Planet. Sci. Lett.* **113**, 569–76.

Ravizza, G. E. and Turekian, K. K. (1992). The osmium isotopic composition of organic-rich marine sediments. *Earth Planet. Sci. Lett.* **110**, 1–6.

Reisberg, L. C., Allegre, C. J. and Luck, J. M. (1991). The Re–Os systematics of the Ronda Ultramafic Complex of southern Spain. *Earth Planet. Sci. Lett.* **105**, 196–213.

Reisberg, L. C., Zindler, A., Marcantonio, F., White, W., Wyman, D. and Weaver, B. (1993). Os isotope systematics in ocean island basalts. *Earth Planet. Sci. Lett.* **120**, 149–67.

Russ, G. P., Bazan, J. M. and Date, A. R. (1987). Osmium isotopic ratio measurements by inductively coupled plasma source mass spectrometry. *Anal. Chem.* **59**, 984–9.

Sharpe, M. R. (1985). Strontium isotope evidence for preserved density stratification in the main zone of the Bushveld Complex, South Africa. *Nature* **316**, 119–26.

Shirey, S. B. and Carlson, R. W. (1991). The Re–Os isotopic system: new applications in geochemistry at DTM. *Carneg. Inst. Wash. Yearbook* **90**, 58–71.

Suzuki, K., Lu, Q., Shimizu, H. and Masuda, A. (1993). Reliable Re–Os age for molybdenite. *Geochim. Cosmochim. Acta* **57**, 1625–8.

Volkening, J., Walczyk, T. and Heumann, K. G. (1991). Osmium isotope ratio determinations by negative thermal ionization mass spectrometry. *Int. J. Mass Spectrom. Proc.* **105**, 147–59.

Walczyk, T., Hebeda, E. H. and Heumann, K. G. (1991). Osmium isotope ratio measurements by negative thermal ionization mass spectrometry (NTI-MS). *Fres. J. Anal. Chem.* **341**, 537–41.

Walker, R. J., Carlson, R. W., Shirey, S. B. and Boyd, F. R. (1989a). Os, Sr, Nd, and Pb isotope systematics of southern African peridotite xenoliths: implications for the chemical evolution of subcontinental mantle. *Geochim. Cosmochim. Acta* **53**, 1583–95.

Walker, R. J. and Fassett, J. D. (1986). Isotopic measurement of sub-nanogram quantities of rhenium and osmium by resonance ionization mass spectrometry. *Anal. Chem.* **58**, 2923–7.

Walker, R. J. and Morgan, J. W. (1989). Rhenium–osmium isotope systematics of carbonaceous chondrites. *Science* **243**, 519–22.

Walker, R. J., Morgan, J. W., Naldrett, A. J. and Li, C. (1991). Re–Os isotopic systematics of Ni–Cu sulfide ores, Sudbury Igneous Complex, Ontario: evidence for a major crustal component. *Earth Planet. Sci. Lett.* **105**, 416–29.

Walker, R. J., Shirey, S. B., Hanson, G. N., Rajamani, V. and Horan, M. F. (1989b). Re–Os, Rb–Sr, and O isotopic systematics of the Archean Kolar schist belt, Karnataka, India. *Geochim. Cosmochim. Acta* **53**, 3005–13.

Walker, R. J., Shirey S. B. and Stecher O. (1988). Comparative Re–Os, Sm–Nd and Rb–Sr isotope and trace element systematics for Archean komatiite flows from Munro Township, Abitibi Belt, Ontario. *Earth Planet. Sci. Lett.* **87**, 1–12.

Yin, Q. Z., Jagoutz, E., Verhovskiy, A. B. and Wanke, H.

(1993). ^{187}Os–^{186}Os and ^{187}Os–^{188}Os method of dating: an introduction. *Geochim. Cosmochim. Acta* **57**, 4119–28.

Zindler, A. (1982). Nd and Sr isotopic studies of komatiites and related rocks. In: Arndt, N. T. and Nisbet, E. G. (Eds.), *Komatiites.* George Allen and Unwin, pp. 399–420.

9 Specialist isotopic schemes

9.1 The Lu–Hf system

9.1.1 Geochronology

Lutetium lies at the end of the lanthanide series as the 'heaviest' of the rare earth elements (REE). It has two isotopes ^{175}Lu and ^{176}Lu whose respective abundances are 97.4 and 2.6%. ^{176}Lu displays a branched isobaric decay, by β^- to ^{176}Hf and electron capture to ^{176}Yb. However the latter makes up only a few percent at most of the total activity and can be more-or-less ignored (Dixon et al., 1954). ^{176}Hf is left in an excited state after β emission, and decays to the ground state by γ emission. It is one of six isotopes and makes up 5.2% of total hafnium, an element which is not a rare earth but resembles Zr very closely in its crystal chemical behaviour.

The decay scheme:

$$^{176}_{71}\text{Lu} \rightarrow {}^{176}_{72}\text{Hf} + \beta^- + \bar{\nu} + Q$$

yields a decay equation:

$$^{176}\text{Hf} = {}^{176}\text{Hf}_I + {}^{176}\text{Lu}\,(e^{\lambda t} - 1) \qquad [9.1]$$

This is conveniently divided through by ^{177}Hf:

$$\frac{^{176}\text{Hf}}{^{177}\text{Hf}} = \left(\frac{^{176}\text{Hf}}{^{177}\text{Hf}}\right)_I + \frac{^{176}\text{Lu}}{^{177}\text{Hf}}(e^{\lambda t} - 1) \qquad [9.2]$$

The first Lu–Hf geochronological measurement was made by Herr et al. (1958), who attempted to determine the half-life of ^{176}Lu by analysing the isotopic composition of Hf in the heavy-REE-rich mineral gadolinite (containing several thousand ppm Lu). However, application of the Lu–Hf method to geological problems was prevented until recently by the difficulty of low blank chemical isolation of Hf, and by the difficulty of accurate isotopic analysis of Hf. These problems were overcome by Patchett and Tatsumoto (1980a) whose experimental methods are outlined in section 2.1.2.

Patchett and Tatsumoto (1980b) presented the first Lu–Hf isochron, on a suite of eucrite meteorites with a known age of 4550 Myr. From this they were able to calculate the decay constant of ^{176}Lu and the initial ^{176}Hf/^{177}Hf ratio of the inner solar system at the time of its accretion. Their original ten-point whole-rock isochron was improved by the addition of three extra points (Tatsumoto et al., 1981) to yield a half-life of 35.7 ± 1.2 Byr (equivalent to a decay constant of $1.94 \pm 0.07 \times 10^{-11}$ yr^{-1}) and an initial ^{176}Hf/^{177}Hf ratio of 0.27978 ± 9 (2σ), Fig. 9.1.

This half-life compares very well with the value of 35.4 ± 1.1 Byr which Patchett and Tatsumoto (1980b) calculated as a weighted mean of five physical half-life determinations made since 1960 (Faure, 1977). The agreement is better than might have been expected, since the physical determinations are based on the activity of the β decay branch to Hf only, whereas the geochronological measurement is a total half-life incorporating also the electron capture branch. The good agreement implies that the latter branch is very minor, but if the geochronologically determined half-life is used in future then the minor branch can be completely ignored since 'the percentage effect on age of an unaccounted-for minor branching decay is essentially linear with time when the decay is slow, as for ^{176}Lu' (Patchett and Tatsumoto, 1980b). This does not apply to the shorter half-life branched decay of ^{40}K (section 10.1).

9.1.2 Mantle evolution

Patchett et al. (1981) explored the usefulness of initial Hf isotope ratios as a tracer of mantle Hf through time, using crustal rocks with a clear mantle derivation. In order to calculate accurate initial ^{176}Hf/^{177}Hf ratios for many rock bodies without making numerous isochron determinations, Patchett et al. principally analysed zircon separates. These are an ideal subject for Hf isotope analysis for several reasons:

1) Hf forms an integral part of the zircon lattice,

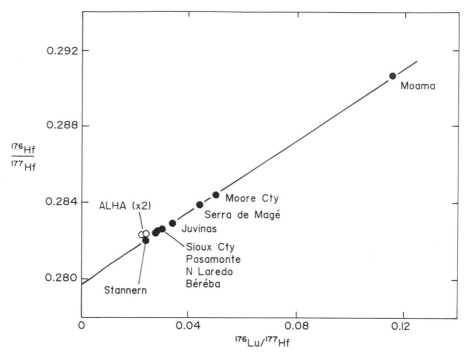

Fig. 9.1. Lu–Hf isochron for eucrite meteorites. The antarctic meteorite, ALHA (open symbols) was omitted from the regression. After Patchett and Tatsumoto (1980b).

which is therefore very resistant to Hf mobility and contamination;

2) The very high Hf concentrations in zircon (ca. 10 000 ppm) yield very low Lu/Hf ratios and consequently minute age corrections;

3) There are large quantities of zircon separates previously prepared for Pb isotope analysis;

4) Accurate U–Pb dates are available on the same material;

5) Metamorphic overprinting or zircon inherited from a previous crustal history are clearly revealed by the Pb data.

Pettingill and Patchett (1981) confirmed the usefulness of zircon for Hf isotope studies in a dating study on the Amitsoq gneisses of west Greenland. Despite metamorphic events at 2.9 and 1.7 Byr, zircon separates and whole-rock gneiss samples define a coherent array on a Lu–Hf isochron diagram (Fig. 9.2). The regression yields an age of 3.58 ± 0.22 Byr (2σ) using the ^{176}Lu decay constant of 1.94×10^{-11} yr^{-1}.

Initial ^{176}Hf/^{177}Hf ratios of presumed mantle-derived igneous rocks of different ages are plotted on a hafnium isotope evolution diagram in Fig. 9.3 (Patchett et al., 1981). The chondritic evolution line is

based on the initial ratio of the eucrite meteorite isochron described above, and a ^{176}Lu/^{177}Hf ratio of 0.0334 derived from the carbonaceous chondrites Murchison and Allende (Patchett and Tatsumoto, 1981). The present-day chondritic ^{176}Hf/^{177}Hf ratio derived from this evolution line (0.28286) compares well with the Bulk Earth Hf isotope ratio of 0.28295 determined from the intersection of Bulk Earth (= chondritic) ^{143}Nd/^{144}Nd with the Nd–Hf isotope array defined by ocean island basalts (see below). Hf isotope ratios can also be expressed using the epsilon notation developed for Nd (parts per 10 000 deviation from the chondritic evolution line).

All of the igneous rocks with a mantle-derived signature analysed by Patchett et al. (1981) lie within error of, or above, the chondrite evolution line (Fig. 9.3). Only one sample, a meta-tholeiite dyke cutting the SuomussalmiKuhmo greenstone belt in eastern Finland, has an initial ^{176}Hf/^{177}Hf above a linear depleted mantle evolution line drawn from the primordial solar system value to the most radiogenic MORB analysis (Fig. 9.3). In contrast, many compositions lie scattered between the chondritic and the depleted-mantle evolution lines. One explanation for these data is their derivation from a

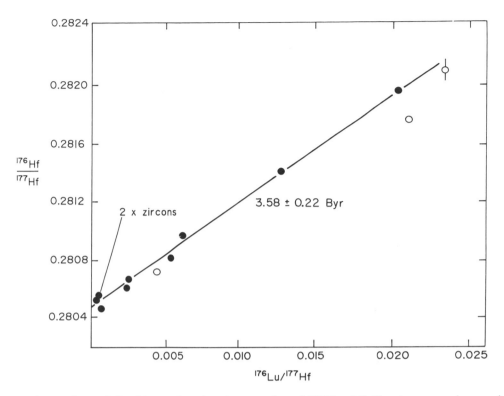

Fig. 9.2. Lu–Hf 'errorchron' defined by a suite of Amitsoq gneisses (MSWD = 11). The zircon samples are shown to be colinear with the whole-rock array. Open symbols were omitted from the regression. After Pettingill and Patchett (1981).

heterogeneous mantle showing variable trace element depletion of Hf relative to Lu through space and time.

An alternative to variably depleted mantle sources would be to derive magmas from a more depleted homogeneous source (such as defined by the dashed evolution line in Fig. 9.3) and subject them to contamination by older crustal basement. Patchett *et al.* (1981) preferred this model for the 1.4 Byr-old Silver Plume and 1.0 Byr-old Pikes Peak batholiths of Colorado, which have Nd *and* Hf initial ratios near the chondritic evolution line, and were argued to contain large fractions of 1.7 Byr-old crust by DePaolo (1981). Two 1.8 Byr-old post-tectonic granites intruded into the Archean craton of North Finland had spectacularly low initial ratios, corresponding to ϵ Hf values of −10 and −12. These two samples are clearly of crustal derivation on the basis of Pb and Sr isotope data, and were selected to demonstrate the effects of crustal re-working on Hf isotope systematics.

It is important to remember that at the time that this paper was published (1981) the only conclusive

Nd isotope evidence for depleted mantle in the Proterozoic was provided by the work of DePaolo (1981) on the Front Ranges of Colorado (see section 4.2.2). Nd isotope evidence from komatiites for Archean depleted mantle (section 4.4.3) was not yet available. Hence the major significance of this paper at the time.

Because Hf involves greater analytical difficulties, relative to Nd, the latter has been the primary tool for studying continental growth through geological time (section 4.4). However, the resistance of zircon to chemical weathering means that this mineral should demonstrate enhanced preservation of ancient sediment provenance. Hence Hf isotope analysis of zircon represents a good test for Nd isotope data on ancient sediments, which have been used to constrain crustal growth–recycling models (section 4.4.2).

Stevenson and Patchett (1990) performed Lu–Hf analyses on Archean and Proterozoic sedimentary zircons from the Canadian, Wyoming, North Atlantic and South African cratons. These data were summarised on a Lu–Hf model age *versus* stratigraphic

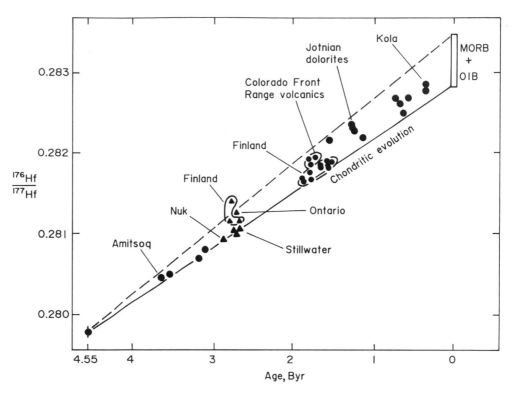

Fig. 9.3. Diagram of Hf isotope evolution over geological time. Initial ratios of uncontaminated mantle-derived magmas show them to be derived from a slightly depleted source relative to chondrites. Data from Patchett *et al.* (1981).

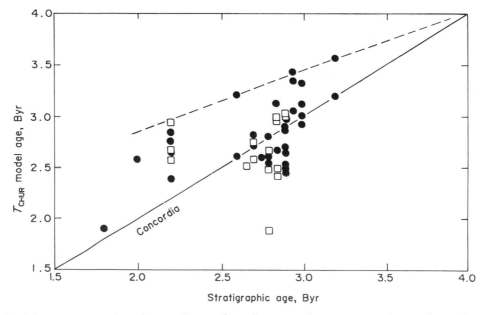

Fig. 9.4. Model age *versus* stratigraphic age diagram for sedimentary zircons to constrain crustal growth models for the Archean. (●) = Hf data; (□) = Nd data. Dashed line accents the upper envelope of the data. After Stevenson and Patchett (1990).

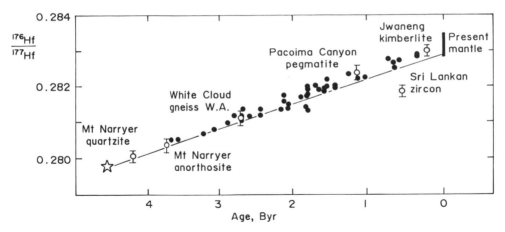

Fig. 9.5. Comparison of zircon Hf isotope data by SIMS (○) and TIMS (•), on a hafnium isotope evolution diagram. After Kinny *et al.* (1991).

age diagram (Fig. 9.4), using a chondritic mantle model. A significant number of samples show model ages in excess of stratigraphic age (i.e. above the 'concordia' line). The excess of model age over stratigraphic age decreased in older sediments, leading Stevenson and Patchett to argue that there were 'no great quantities of continental crust in the early Earth providing ancient zircons...'. However, a significant number of points lie below the concordia line in Fig. 9.4, indicating that the chondritic mantle model is an under-estimate of the provenance ages of these samples. The use of a depleted mantle model then implies provenance ages ca. 200 Myr older, suggesting a significant volume of Early Archean crust. Nevertheless, there is no evidence for a 'big bang' model in which the volume of Early Archean crust was as great as that at the present day. Consequently, the data provide some support for a model involving progressive crustal growth with time.

Hafnium isotope measurements in zircons have also been performed by ion probe (Kinny *et al.*, 1991). Unfortunately the precision attained was about an order of magnitude lower than by conventional TIMS analysis (Fig. 9.5). Consequently these data do not provide a strong constraint on crust and mantle evolution.

9.1.3 Hf–Nd systematics of mantle depletion

Patchett and Tatsumoto (1980c) made Lu/Hf concentration determinations and Hf isotope measurements on selected MORB and OIB samples

previously analysed for $^{143}Nd/^{144}Nd$ and $^{87}Sr/^{86}Sr$. These data (augmented by Patchett, 1983) show that $^{176}Hf/^{177}Hf$ very closely parallels $^{143}Nd/^{144}Nd$ in ocean island basalts (Fig. 9.6). However, MORB samples display a proportionally greater degree of spread in $^{176}Hf/^{177}Hf$ (60% of the total range for oceanic basalts) than $^{143}Nd/^{144}Nd$ (only 30% of the total range). Hence the MORB arrays in Fig. 9.6 are nearly three times steeper than the OIB arrays. Patchett and Tatsumoto attributed these differences to stronger fractionation of Lu/Hf than Sm/Nd and Sr/Rb in very trace-element depleted source regions such as MORB, due to the greater incompatibility displayed by Hf relative to Lu than Nd/Sm or Rb/Sr.

Compared to their radiogenic Hf isotope signatures (indicative of a depleted source), many MORB samples are enriched in Hf/Lu ratio (corresponding to low Lu/Hf). This same phenomenon has also been observed for other incompatible-element systems, and is generally attributed to sequential extraction of very small degree melts (sections 6.3.2, 13.3.2); however, it is more marked in the Lu–Hf system than Sm–Nd. Salters and Hart (1989) attributed this 'hafnium paradox' to residual garnet in the MORB source. Because Hf is not a true rare earth element its chemistry is not coherent with the heavy rare earths. This is illustrated by the mineral–magma partition coefficients shown in Fig. 9.7. Hence there is an opportunity for more extreme fractionation in Lu/Hf than Sm/Nd when low-degree melting occurs at depths greater than 80 km, within the garnet stability zone.

Salters and Hart (1991) argued that variable garnet

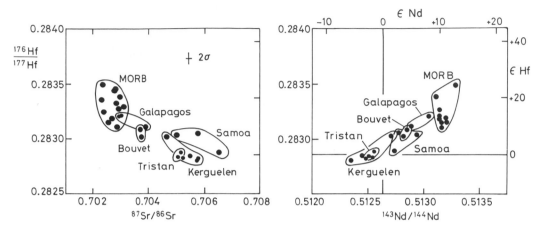

Fig. 9.6. Hf *versus* Sr and Nd isotope diagrams showing good 'mantle arrays' of OIB compositions, but some decoupling of Hf systematics in MORB. After Patchett (1983).

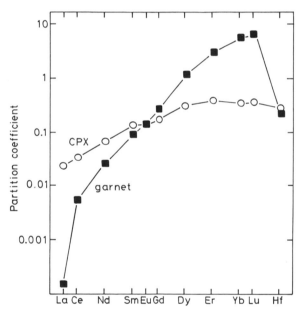

Fig. 9.7. Partition coefficients for REE and Hf between minerals (cpx, garnet) and kimberlite magma. After Fujimaki *et al.* (1984).

free) spinel peridotites will have lower Lu/Hf ratios, resulting in less radiogenic Hf isotope signatures over geological time. Johnson and Beard (1993) termed these two types of depleted mantle signature DMMI and DMMII respectively, and showed that the latter type was also represented in the source of Tertiary basalts in the southern Rio Grande rift (USA). They argued that the DMMI and DMMII components were generated by ancient depletion events, and that intermediate MORB compositions were generated by mixing between them (Beard and Johnson, 1993).

Gruau *et al.* (1990) analysed 3.5 Byr-old komatiites from the Onverwacht Group of South Africa to see whether Archean depleted mantle displayed extreme ranges of Lu/Hf depletion due to garnet fraction. They compared Al-depleted and Al-enriched komatiites whose chemistry is attributed to garnet fractionation in the mantle source. However, calculated initial Hf and Nd isotope ratios were indicative of similar degrees of mantle depletion (ϵ from 0 to +2) in both Al-enriched and depleted types. Hence there is no evidence for widespread garnet fractionation in the early Earth, such as might have resulted from a magma ocean. Consequently the major-element composition of the magma was attributed to garnet fractionation in the source during the komatiite melting event itself.

9.1.4 Hf–Nd systematics of mantle enrichment

Another geological environment where Lu/Hf can undergo strong fractionation relative to Sm/Nd is the

contents in ancient melting events could explain the partial decoupling of Hf–Nd isotope systematics in MORB. When garnet is present in the residue from melting, this residue develops high Lu/Hf ratios, and hence, over time, a radiogenic Hf isotope signature. On the other hand, the melting residues of (garnet-

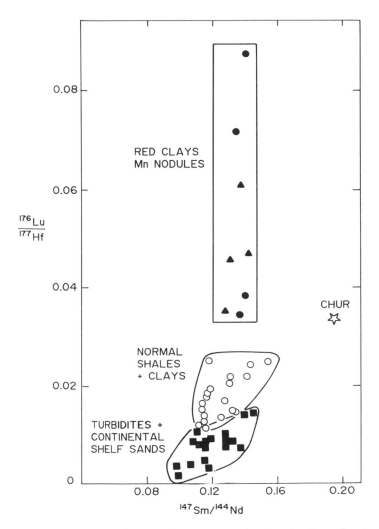

Fig. 9.8. Plot of Lu/Hf *versus* Sm/Nd ratio in different sediment types, showing that large fractionations in Lu/Hf are not accompanied by significant changes in Sm/Nd. After Patchett *et al.* (1984).

sedimentary system. Patchett *et al.* (1984) plotted Lu/Hf ratios against Sm/Nd for different marine sediment types (Fig. 9.8). While $^{147}Sm/^{144}Nd$ ratios are more-or-less constant at ca. 0.12–0.14 in most analysed samples, $^{176}Lu/^{177}Hf$ is strongly fractionated between sandstones and clays. Patchett *et al.* attributed this fractionation to the very strong affinity of Hf for the zircon phase, which because of its resistance to mechanical and chemical attack becomes enriched in sand-grade sediments. Hf is correspondingly depleted in the fine-grained clay fraction.

In contrast, the rare earth elements, including Lu, are either dispersed in major rock-forming minerals or

concentrated in relatively easily altered accessories such as allanite, apatite, monazite and sphene, so that they become enriched in the fine fraction and absorbed onto clay particles and organic material. The sorting of marine sediments according to grain size (Fig. 9.9) then yields low Lu/Hf sands and turbidites on the continental shelf and continental slope, medium Lu/Hf shales and clays, and very high Lu/Hf red clays and Mn nodules in the deep ocean where terrigenous sediment is lacking.

Patchett *et al.* (1984) pointed out that the fractionation of Lu/Hf in different marine sediment reservoirs could be important in testing the extent of sediment recycling into the mantle. If subducted

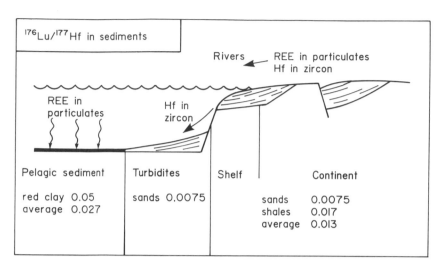

Fig. 9.9. Schematic cross-section of a passive (Atlantic-type) continental margin to show fractionation of $^{176}Lu/^{177}Hf$ ratios between sands, shales and pelagic sediments. After Patchett *et al.* (1984).

sediment has the same composition as bulk crust then it will be difficult to test whether the continents have grown through geological time, or whether their mass has remained constant (section 4.4). However, if subducted sediment has fractionated Lu/Hf ratios with respect to the bulk crust, then over time, mantle reservoirs which receive such an input should develop distinctive Hf isotope compositions.

Patchett *et al.* calculated the Hf and Nd isotopic compositions of mantle reservoirs 2 Byr after receiving subducted sediment of different characteristics (Fig. 9.10). They demonstrated that any individual sediment type subducted (e.g. red clay, average pelagic sediment, or turbidites) would yield distinctive isotopic compositions after 2 Byr of residence in the mantle. If the OIB array is a mixing line between depleted mantle and old subducted sediment, then it follows that pelagic and turbidite sediment would need to be mixed in regular and predictable proportions to generate the strong positive correlation between Hf and Nd in OIB sources. The most enriched of such sources (e.g. Tristan da Cunha) would need to contain ca. 2% of such a sediment mixture in their mantle source.

Patchett *et al.* argued that some present-day island arcs (e.g. Antilles and Banda) are almost exclusively subducting continental terrigenous sediment, while at other island arcs only young basic volcanic material, poor in zircons, is available for recycling into the mantle. If past subduction zones were as non-uniform as those at the present, then thorough mixing of the

subducted materials in the mantle would be required to generate the OIB Hf–Nd isotope correlation. Yet such mixing would eliminate the very heterogeneity in the OIB source which generates the array. Hence, Patchett *et al.* concluded that subduction of continental detritus into the mantle is not the dominant form of mantle recycling.

Salters and Hart (1991) presented additional data for OIB samples in order to establish the locations of the end-member components proposed by Zindler and Hart (1986) to explain mantle isotope systematics. The results (Fig. 9.11) confirm that EMI and EMII are strongly collinear. On the other hand, the isolation of HIMU below the main trend suggests a possible connection with the DMMII component, attributed by Johnson and Beard (1993) to ancient depleted spinel peridotite. Hence the HIMU signature may represent recycled shallow oceanic lithosphere. Other evidence for this model is discussed in section 6.6.

Island-arc basalts (IAB) offer a means of monitoring the composition of material actually being recycled into the mantle, in order to test theoretical models such as Patchett *et al.* (1984). Hf–Nd isotope data were presented by White and Patchett (1984) for IAB sampling depleted and enriched sources (solid symbols in Fig. 9.12). These data fall within the field of OIB samples, showing that old sedimentary material presently being subducted into the mantle has appropriate Hf–Nd systematics to explain the composition of OIB magmas. This may imply that

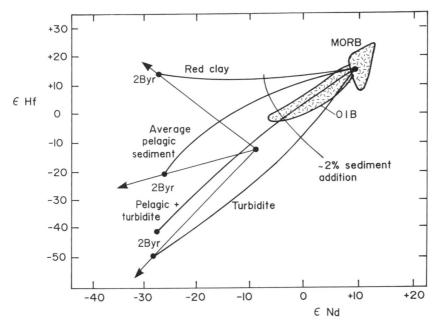

Fig. 9.10. ϵ Hf *versus* ϵ Nd diagram showing MORB and OIB data relative to various mixing models for sediment recycling into the mantle: red clay, average pelagic sediment, turbidites, and a unique mixture of turbidite and pelagic components. Only the latter is consistent with OIB data. After Patchett *et al.* (1984).

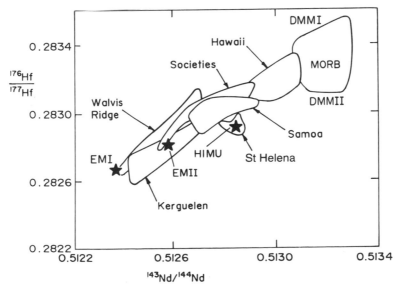

Fig. 9.11. Hf–Nd isotope diagram for oceanic volcanics showing fields for geochemically important ocean islands, along with the estimated compositions of end-members. Modified after Salters and Hart (1991).

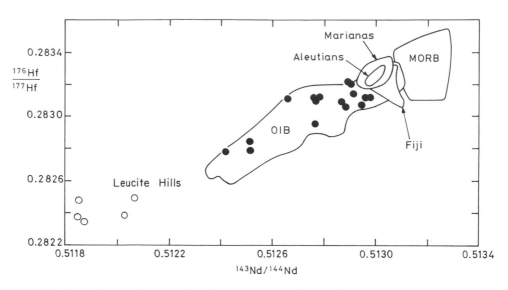

Fig. 9.12. Hf–Nd isotope plot showing a variety of magma types relative to the OIB and MORB arrays from Fig. 9.11: Named fields = arcs with depleted signatures; (●) = enriched island arc basalts; (○) = Leucite Hills volcanics. After Salters and Hart (1991).

non-extreme mixtures of the sediment types in Fig. 9.10 can explain the composition of OIB sources with moderate ease. However, extreme sediment signatures such as pure pelagic material are not consistent with the Hf data. This therefore seems to exclude this material as an exclusive source of the EMI component in OIB (section 6.6).

An alternative mechanism for generating enriched OIB sources is foundering of sub-continental lithosphere (Zindler and Hart, 1986). The latter type of reservoir may be represented by volcanics from the Leucite Hills, which also lie along the same trajectory in Fig. 9.12 as the OIB array. The EMI component is the most likely to be generated by foundering of sub-continental lithosphere. In contrast, EMII can best be explained by sediment subduction. Because these components are more-or-less collinear on the Hf–Nd isotope diagram, the processes which generate these enriched source compositions must have a similar effect on Sm/Nd and Lu/Hf fractionation.

9.2 The La–Ce and La–Ba systems

^{138}La exhibits branched decay; by β emission to ^{138}Ce and by electron capture to ^{138}Ba. The La–Ce decay scheme is a potentially useful isotopic tracer, and the La–Ba scheme may form a useful geochronometer, but their application has been greatly hindered by the very low abundance of the parent

isotope (0.089% of natural lanthanum) and its very long half-life (over 100 Byr, totalled between both routes). Nevertheless, both methods have recently been applied to geological problems.

In addition to the general problems mentioned above, counting determinations of the La decay constants are hampered by the low energy of emitted particles. Hence, early measurements, particularly of the β decay branch, were scattered. To overcome this problem, the counting experiments actually measure the γ decay of isomers (excited states) of the product nuclide, rather than the isobaric decay process itself.

A further complication for counting experiments is the hygroscopic nature of La_2O_3, the material usually used in these studies. Transformations to the hydroxide or carbonate result in weight gains of 17% and 2.5% respectively, but in ten out of twelve recent counting determinations on La, no volatile data were reported (Tanaka and Masuda, 1982). However, despite a large variation in absolute values, all counting determinations since 1970 yield a ratio for β / electron capture decay constants near 0.51 . In addition, two recent counting experiments on anhydrous La oxide yielded average values for the β decay constant of 2.29 and 2.22 × 10^{-12} respectively (Sato and Hirose, 1981; Norman and Nelson, 1983). The La–Ce system was the first of the two methods to be applied geologically, but since the La–Ba case is simpler it will be discussed first.

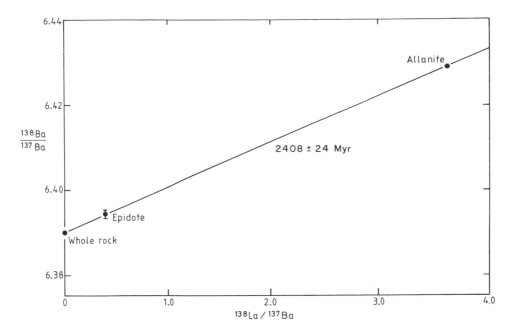

Fig. 9.13. La–Ba isochron diagram for a sample of Amitsoq gneiss, western Greenland. The isochron yields an age of 2408 ± 24 Myr, which dates a metamorphic event rather than the age of the rock. The initial ratio of 6.3897 is invariant in whole-rock systems. After Nakai *et al.* (1986).

9.2.1 La–Ba geochronology

^{138}Ba, the daughter product of the electron capture decay of ^{138}La, is also the most abundant isotope of barium, making up 88% of the natural element. In view, therefore, of the very low abundance of the parent nuclide, significant variations in the abundance of ^{138}Ba are only found in REE-rich and Ba-poor minerals. The first geological measurements were made by Nakai *et al.* (1986) on epidote, allanite and sphene from Precambrian rocks (Fig. 9.13). Nakai *et al.* ratioed ^{138}Ba against ^{137}Ba to yield the following decay equation:

$$\frac{^{138}\text{Ba}}{^{137}\text{Ba}} = \left(\frac{^{138}\text{Ba}}{^{137}\text{Ba}}\right)_I + \frac{^{138}\text{La}}{^{137}\text{Ba}} \cdot \frac{\lambda_{\text{E.C.}}}{\lambda_{\text{total}}} (e^{\lambda_{\text{total}}t} - 1) \qquad [9.3]$$

Ba isotope analyses were normalised to a ^{136}Ba/^{137}Ba ratio of 0.6996.

Nakai *et al.* found that all analysed whole-rock samples had ^{138}Ba/^{137}Ba initial ratios within error of 6.3897, attributed to the very low ^{138}La/^{137}Ba ratios of all such materials. Since the Ba isotope ratio of whole-rock systems is effectively invariant over time, a La–Ba mineral age can be based simply on the analysis of one or more La-enriched minerals. Using the electron capture decay constant of 4.44×10^{-12} yr^{-1} determined by Sato and Hirose

(1981), Nakai *et al.* obtained relatively good agreement between the La–Ba and Sm–Nd ages of a pegmatite from Mustikkamaki (Finland) and Amitsoq gneiss from Greenland.

9.2.2 La–Ce geochronology

The relative harmony between counting and geological determinations of the La electron capture decay branch was not matched by the La β decay route to cerium. This branch is beset by much larger analytical problems, but has more geochemical applications. Tanaka and Masuda (1982) determined the first La–Ce isochron, ratioing ^{138}Ce against ^{142}Ce and normalising to a ^{136}Ce/^{142}Ce ratio of 0.0172 . The decay equation is:

$$\frac{^{138}\text{Ce}}{^{142}\text{Ce}} = \left(\frac{^{138}\text{Ce}}{^{142}\text{Ce}}\right)_I + \frac{^{138}\text{La}}{^{142}\text{Ce}} \cdot \frac{\lambda_\beta}{\lambda_{\text{total}}} (e^{\lambda_{\text{total}}t} - 1) \qquad [9.4]$$

Because the half-lives of both decay branches are so long, they have very little effect on each other. For example, simplifying the equation to:

$$\frac{^{138}\text{Ce}}{^{142}\text{Ce}} = \left(\frac{^{138}\text{Ce}}{^{142}\text{Ce}}\right)_I + \frac{^{138}\text{La}}{^{142}\text{Ce}} (e^{\lambda_\beta t} - 1) \qquad [9.5]$$

only causes a 0.5 % over-estimate in age.

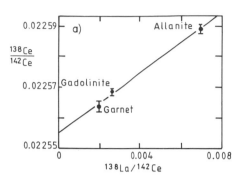
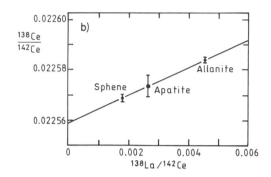

Fig. 9.14. La–Ce isochrons for (a) the Lovbole, and (b) the Mustikkamaki pegmatites of Finland. When combined with Sm–Nd data they imply La β decay constants of $2.70 \pm 0.25 \times 10^{-12}$ and $2.93 \pm 0.41 \times 10^{-12}$ yr^{-1} respectively. After Masuda *et al.* (1988).

Two further technical problems are encountered in Ce isotope analysis. One is the extreme size of the ^{140}Ce peak relative to the small ^{136}Ce and ^{138}Ce peaks (e.g. ^{140}Ce/^{136}Ce = 464.65). Collision of the ^{140}Ce ion beam with gas molecules in the vacuum system causes down-mass peak tailing whose effect on the small peaks must be corrected.

A second major problem is the isobaric interference of ^{138}Ba onto ^{138}Ce. ^{138}Ba is six times more abundant than any other natural Ba isotope which could be used to monitor Ba interference. Therefore any interference correction for ^{138}Ba (even if near zero) will amplify detector noise six-fold. The solution to this problem is to analyse Ce as the oxide species CeO$^+$. Because barium is divalent, the BaO$^+$ species is very unfavourable, so, providing overall Ba levels are kept low by good chemistry, the Ba interference can be taken to be zero without correction. Analysing Ce as the oxide introduces other isobaric interference problems, but these are easily overcome by good chemistry (section 2.1.2).

Tanaka and Masuda (1982) attempted to date separated minerals from the Bushveld pluton, but because of the geological similarity between La and Ce, a limited range of La/Ce ratios was available and the isochron had a large analytical error. Further isochron determinations were made by Masuda *et al.* (1988). However, using the β decay constant of 2.29×10^{-12} yr^{-1} obtained by Sato and Hirose (1981), the La–Ce isochrons gave old ages outside error of the Sm–Nd method. If ages are taken from the Sm–Nd data then the La–Ce isochron slopes can be used to make a geological decay constant determination. Using the Bushveld result and mineral isochrons from two Finnish pegmatites (Fig. 9.14), Masuda *et*

al. calculated an average β decay constant of 2.77×10^{-12} yr^{-1}, about 20% higher than the counting determinations.

In an attempt to provide further geological constraints on the La β decay constant, Dickin (1987a) determined a La–Ce isochron on a suite of Lewisian whole-rock gneisses from northwest Scotland. ^{138}Ce was ratioed against ^{136}Ce, since this gives rise to more manageable isotope ratios, but an equivalent normalising factor was used for fractionation correction. It was argued that whole-rock REE systems were more advantageous for calibrating the decay constant than mineral systems, due to the greater resistance of the former to metamorphic resetting. The original sample suite contained two basic granulites and four intermediate-to-acid granulites. Using the decay constant of Sato and Hirose (1981), the La–Ce isochron age of 2.99 Byr was in good agreement with the Lewisian Sm–Nd age of 2.91 Byr (section 4.1.3).

However, further work on the Sm–Nd systematics of Lewisian gneisses has shown that the samples analysed by Dickin (1987a) comprise two suites with different geological history. Basic gneisses preserve 2.9 Byr crustal formation ages, while intermediate-to-acid whole-rock gneisses have been reset by granulite-facies metamorphism at 2.60 Byr (section 4.1.3). Using the decay constant of Sato and Hirose, the four intermediate-to-acid granulites yield a La–Ce age of 2.65 ± 0.3 Byr (2σ) which is in agreement with the Sm–Nd result on this suite. No meaningful La–Ce age can be calculated on the basic granulites alone.

The most recent geological determination of the La β decay constant (Makishima *et al.*, 1993) has also supported the value of Sato and Hirose. Two La–Ce

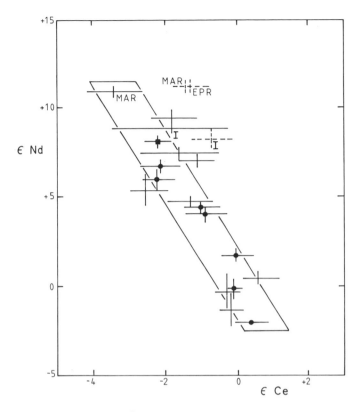

Fig. 9.15. Plot of ϵ Nd against ϵ Ce (part per 10^4 deviation from Bulk Earth isotope composition) for basic lavas. Error bars represent 2 SDM within-run precision. Solid circles are OIB samples used to calculate the best-fit mantle array, whose width is defined by average within-run Ce isotope error. Solid square is 60 Myr Skye plateau lava, relative to calculated value of 60 Myr Bulk Earth. Dotted error bars denote samples whose within-run error bars do not overlap the OIB mantle array. After Dickin (1988).

mineral isochrons were determined on Archean granites from western Australia, which gave more-or-less concordant U–Pb zircon and Rb–Sr mineral isochron ages. Sm–Nd mineral suites also yielded ages within error of the other methods, but with higher error due to large MSWD values. Using a decay constant of 2.29×10^{-12} yr^{-1}, the granites gave La–Ce mineral ages of 2.76 ± 0.41 and 2.69 ± 0.38 Byr, which agree well with U–Pb ages of 2.665 and 2.692 Byr respectively.

9.2.3 Ce isotope geochemistry

Since La and Ce are light rare earth elements (LREE), Ce isotope data form a tracer for time-integrated LREE enrichment or depletion of geological reservoirs. Similarly, Nd isotope data are a tracer for time-integrated fractionation between the middle

REE. Therefore, a combination of Ce and Nd isotope data provides a unique control of the time-integrated light-to-middle REE evolution of complex geological reservoirs in the mantle or crust. Together they may form a powerful petrogenetic tool.

Ce isotope analyses of eight ocean island basalts (OIB) from the Atlantic, Pacific and Indian oceans were presented by Dickin (1987b). When plotted against published Nd isotope analyses the data defined a linear array which fell within error of the meteoritic Bulk Earth point of Shimizu *et al.* (1984). The linearity of this 'mantle array' can be attributed to the coherent behaviour of the REE during processes of mantle evolution.

Subsequently, Tanaka *et al.* (1987) reported Ce isotope data on eleven more ocean island, ocean ridge and island arc basalts (Fig. 9.15). The majority of these were very consistent with the data of Dickin

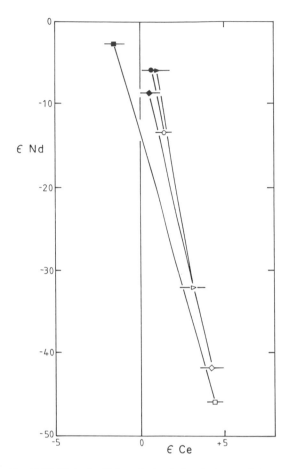

Fig. 9.16. Plot of ϵ Nd against ϵ Ce for Archean plutons, showing greater scatter of initial ratios (solid symbols) than present-day compositions (open symbols). After Tanaka *et al.* (1987).

group of rocks which underwent Ce and Nd isotope evolution starting at the Bulk Earth isotope composition. If these rocks all had linear rare earth profiles, then their Ce–Nd isotope compositions must lie on a linear array whose slope is solely a function of the relative decay constants of the parent nuclides, irrespective of age. Ideally, if such a rock suite were analysed for Ce and Nd isotope composition, the relative decay constants could be calculated without the need to know any concentration or age information.

Tanaka *et al.* attempted to apply this model in practice, and determine the La β decay constant by the analysis of four unrelated continental rocks from around the world with approximately linear REE profiles. An almost perfect linear array was found, but unfortunately this must be attributed to coincidence, since the calculated initial ratios of these samples are actually *more* dispersed than their present-day compositions (Fig. 9.16). Therefore the linearity of this particular data set is coincidental, and its slope cannot be used to determine the La β decay constant. This concept is probably destined to remain a theoretical construct, since the principal difficulty in determining the decay constant is not determination of the La/Ce ratio, but the Ce isotope ratio itself.

A combination of Ce and Nd isotope data was used by Dickin *et al.* (1987) as a tool to study mixing relations during crustal contamination of continental magmas. Twelve Tertiary igneous rocks from Skye in northwest Scotland were analysed for Ce isotope composition. These were compared with theoretical mixing models based on analysed crustal end-members. The Ce *versus* Nd isotope diagram (Fig. 9.17) provides one way to evaluate the merits of different mixing models for analysed lavas. This immediately allows the exclusion of the trondhjemitic mixing line as a relevant petrogenetic model. However, the data allow a more elegant test of the end-members involved in magma mixing.

By taking the Ce isotope compositions of competing crustal end-members, linear mixing lines can be projected back through the analysed product lavas on diagrams of Ce isotope ratio against 1/Ce and La/Ce to model the elemental composition of the mantle-derived precursor prior to contamination. By performing the same calculation for the Sm/Nd data, a complete model light REE profile can be determined for each precursor. These model profiles (dashed) are compared in Fig. 9.18 with a variety of REE profiles for lavas whose isotopic compositions are effectively uncontaminated (solid lines). In this way it was possible to show that the granitic sheet end-member is

(1987b), but three samples lay outside of error of a best-fit mantle array calculated by Dickin (1988). Tanaka *et al.* (1987; 1988) attributed these outliers to incoherent behaviour of the La–Ce and Sm–Nd systems during the evolution of the depleted upper mantle. However, the fact that Nd, Sr and Pb isotope systems in the Ce-anomalous samples were perfectly normal led Dickin (1988) to suggest that larger analytical errors might be responsible. In contrast, Makisima and Masuda (1994) found high ϵ Ce in MORB and rejected the low value of Tanaka *et al.* (1987).

Because of the gradual variation of chemical properties along the lanthanide series, chondrite-normalised REE patterns for large rock reservoirs tend to define approximately linear profiles. Tanaka *et al.* (1987) considered the behaviour of an idealised

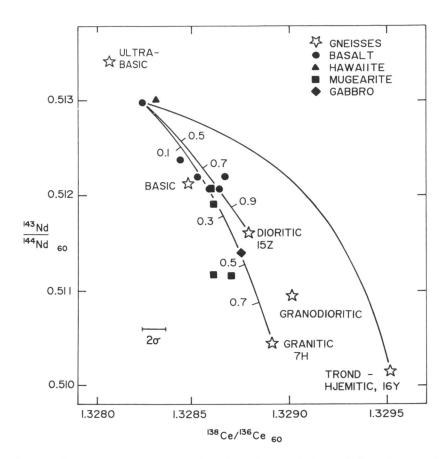

Fig. 9.17. Plot of initial Nd *versus* Ce isotope composition of Skye lavas, relative to Archean basement gneisses at 60 Myr. The distribution of contaminated lava compositions excludes mixing of mantle-derived and trondhjemitic crustal melts. After Dickin *et al.* (1987).

better able to explain the isotopic composition of the contaminated lavas than the intermediate tonalitic gneiss end-member.

The oceans represent another environment where combined Ce and Nd isotope analysis can be used to study mixing processes. Ce–Nd isotope data for Atlantic and Pacific ferromanganese nodules were presented by Tanaka *et al.* (1986) and Amakawa *et al.* (1991). These isotope ratios are argued to be indicative of the composition of the ocean water from which the nodules grew. In the light of Nd and Sr isotope data (section 4.5), the Ce–Nd data were expected to reflect mixing between continental and hydrothermal fluxes into the ocean system. However, the data did not lie on a single mixing line between reasonable MORB and continental end-members (Fig. 9.19).

Tanaka *et al.* (1986) and Elderfield (1992) attributed this behaviour to the shorter residence

time of Ce in seawater (ca. 100 yr) relative to Nd (ca. 500 yr, Elderfield and Graves, 1982). This behaviour causes the effective mixing line between the two sources to have a shallower slope than predicted. For example, the Pacific Ocean is dominated by the hydrothermal flux, while the Atlantic is dominated by the continental flux. Given the ca. 1000 yr mixing time of the ocean system, Nd undergoes better homogenisation between the two oceans than Ce, as shown in Fig. 9.19. Amakawa *et al.* (1992) disputed this interpretation, based on an ϵ Ce value for MORB (-1.3), which is high relative to the Ce–Nd mantle array (Fig. 9.15).

9.3 The K–Ca system

The K–Ca couple was actually the first isotopic system to be suggested as a geochemical tracer for

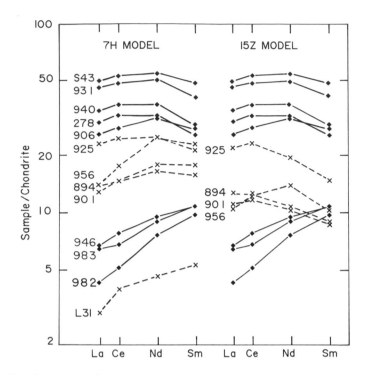

Fig. 9.18. LREE profiles for isotopically uncontaminated lavas (solid lines) compared to model LREE profiles for mantle-derived precursors of contaminated lavas (dashed). The 7H model yields more consistent profiles for the two lava types. After Dickin *et al.* (1987).

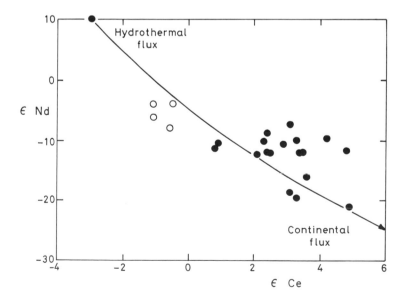

Fig. 9.19. Plot of ϵ Nd against ϵ Ce showing data from Pacific (O) and Atlantic (●) manganese nodules, relative to a simple mixing line between MORB and continental fluxes. After Elderfield (1992).

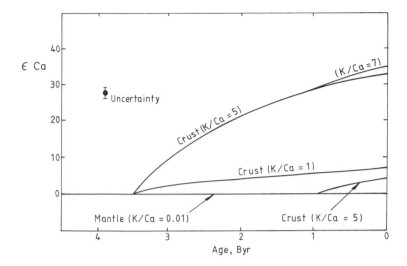

Fig. 9.20. Plot of Ca isotope evolution against time in terms of ϵ units (part per 10^4 deviation from the constant mantle composition). Growth lines are shown for intermediate crust (K/Ca = 1) and granitic crust (K/Ca = 5). After Marshall and DePaolo (1982).

granite petrogenesis (Holmes, 1932). However, this was on the assumption that the major isotope of potassium, ^{41}K, was the radioactive nuclide. (Fortunately this is not really the case or the Earth would have melted from the heat.) When it was realised that ^{40}K was actually the radioactive nuclide, the idea of pursuing the K–Ca system was abandoned, since it was anticipated that radiogenic ^{40}Ca would be swamped by the dominant non-radiogenic ^{40}Ca component. The method finally became viable with the development of modern high-precision mass spectrometers, but has not been widely applied.

Russell et al. (1978) used Ca isotope analysis to investigate mass-dependent fractionation processes, but the first geochronological application of the method was made by Marshall and DePaolo (1982). Because of the large relative differences between Ca nuclide masses, isotope ratios must be corrected for natural and instrumental mass fractionation using a more complex procedure than the simple linear law (section 2.2.2). In practice an exponential mass fractionation correction was used in the two studies mentioned above. Marshall and DePaolo quoted their Ca isotope data as $^{40}Ca/^{42}Ca$ ratios, corrected to a value of 0.31221 for the non-radiogenic $^{42}Ca/^{44}Ca$ ratio.

A variety of meteorites, lunar samples, and mantle-derived materials was analysed by Russell et al. (1978) and Marshall and DePaolo (1982). When age-corrected to yield initial Ca isotope ratios at various

times between 1.3 and 4.6 Byr ago, all of the measurements fell within analytical uncertainty of a $^{40}Ca/^{42}Ca$ ratio of 151.016. This tells us that, because of its very low K/Ca ratio, the Earth's mantle demonstrates negligible growth of radiogenic Ca with time.

Rather than quoting raw isotope ratios, Ca isotope compositions can be reported in terms of ϵ units (part per 10^4 deviation from the mantle composition). However this is more a matter of convenience than necessity, in view of the zero Ca isotope evolution of the mantle reservoir with time. In contrast, granitic crustal reservoirs have high K/Ca ratios of around 5–10 which can display appreciable ^{40}Ca growth with time. For reservoirs more than 1 Byr old these may give rise to isotope ratios outside the error of the mantle value (Fig. 9.20). Because of the relatively short half-life of ^{40}K compared to the age of the Earth, isotopic growth lines are curved in this diagram.

Bearing in mind the branched decay of ^{40}K, we can substitute into the general decay equation [1.10] to derive the following isochron equation for the K–Ca system:

$$\frac{^{40}Ca}{^{42}Ca} = \left(\frac{^{40}Ca}{^{42}Ca}\right)_I + \frac{^{40}K}{^{42}Ca} \cdot \frac{\lambda_\beta}{\lambda_{total}} (e^{\lambda_{total} t} - 1) \qquad [9.6]$$

The branching ratio of β to total decays is 0.8952, and the total decay constant is 5.543×10^{-10} yr^{-1} (section 10.1).

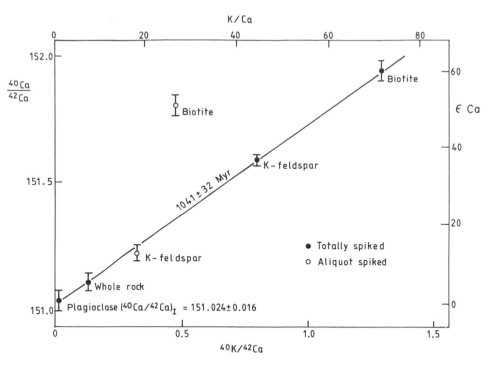

Fig. 9.21. K–Ca isochron plot for separated minerals from the Pikes Peak batholith. Note erroneous results of spiking aliquots (○) rather than the whole dissolution (●). After Marshall and DePaolo (1982).

Marshall and DePaolo tested the K–Ca system as a dating tool by analysing a small suite of separated minerals from the Pikes Peak batholith of Colorado. Plagioclase, whole-rock, K-feldspar and biotite define an isochron array (Fig. 9.21), whose slope yields an age of 1041 ± 32 Myr (2σ). This is within error of other age determinations on this largely un-metamorphosed pluton. The initial ratio of the Pikes Peak batholith (151.024) is within error of the mantle value.

One severe analytical problem that was encountered during this work (other than the mass fractionation behaviour mentioned above) was that samples divided into aliquots before mixing with spike gave erroneous K/Ca ratios. Marshall and DePaolo speculated that this might have been due to partial potassium precipitation from the rock solutions. It is avoided by spiking the whole sample before dissolution.

Marshall and DePaolo (1989) went on to apply the K–Ca method as a petrogenetic tracer in a study of Cenozoic plutons from the western US. Granites emplaced into Paleozoic crust on the continental margin had a similar range of Ca isotope ratios to island-arc volcanics, from values within error of the mantle composition to just outside error (see also Nelson and McCulloch, 1989). In contrast, granites emplaced into Lower Proterozoic basement showed larger Ca isotope enrichments which were correlated with ε Nd (Fig. 9.22). Hence, the Ca isotope ratios of the plutons must be inherited from the crustal source at depth.

Marshall and DePaolo compared the ε Ca–ε Nd compositions of the plutons with crustal evolution models shown by curves in Fig. 9.22. Given a crustal ^{147}Sm/^{144}Nd ratio of 0.1 (determined using the known Nd isotope signature of Colorado basement), Ca–Nd isotope evolution curves were drawn for different crustal K/Ca ratios. If we assume the granites to be total crustal melts then data in Fig. 9.22 imply a K/Ca ratio of about 1 in the source. However, this is higher than most estimates for bulk crust, so a more likely explanation is that ε Ca ratios were fractionated during the melting processes by preferential extraction of the most fusible components of the crust. The usefulness of these constraints for petrogenetic modelling suggest that Ca isotopes have promise as a tracer for crust–mantle mixing, and may find more exploitation in the future.

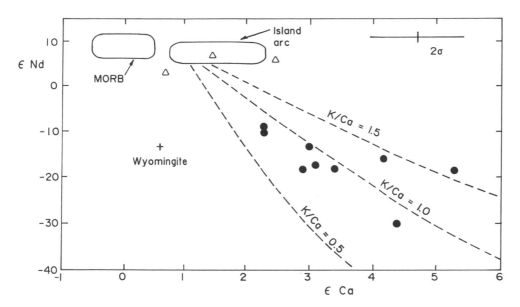

Fig. 9.22. Plot of ϵ Nd against ϵ Ca showing compositions of Cenozoic granitoids emplaced into young basement ($\triangle$) and old basement ($\bullet$), relative to island-arc volcanics and MORB. Curves show model K/Ca ratios for basement with $^{147}Sm/^{144}Nd = 0.1$. After Marshall and DePaolo (1989).

References

Amakawa, H., Ingri, J., Masuda, A. and Shimizu, H. (1991). Isotopic compositions of Ce, Nd and Sr in ferromanganese nodules from the Pacific and Atlantic Oceans, the Baltic and Barents Seas and the Gulf of Bothnia. *Earth Planet. Sci. Lett.* **105**, 554–65.

Amakawa, H., Shimizu, H. and Masuda, A. (1992). Reply to comment (by Elderfield, H.) on 'Isotopic compositions of Ce, Nd and Sr in ferromanganese nodules from the Pacific and Atlantic Oceans, the Baltic and Barents Seas and the Gulf of Bothnia'. *Earth Planet. Sci. Lett.* **111**, 563–5.

Beard, B. L. and Johnson, C. M. (1993). Hf isotope composition of late Cenozoic basaltic rocks from northwest Colorado, USA: new constraints on mantle enrichment processes. *Earth Planet. Sci. Lett.* **119**, 495–509.

DePaolo, D. J. (1981). Neodymium isotopes in the Colorado Front Range and crust–mantle evolution in the Proterozoic. *Nature* **291**, 193–6.

Dickin, A. P. (1987a). La–Ce dating of Lewisian granulites to constrain the ^{138}La β-decay half-life. *Nature* **325**, 337–8.

Dickin, A. P. (1987b). Cerium isotope geochemistry of ocean island basalts. *Nature* **326**, 283–4.

Dickin, A. P. (1988). Mantle and crustal Ce/Nd isotope systematics. *Nature* **333**, 403–4.

Dickin, A. P., Jones, N. W., Thirlwall, M. F. and Thompson, R. N. (1987). A Ce/Nd isotope study of crustal contamination processes affecting Palaeocene magmas in Skye, northwest Scotland. *Contrib. Mineral. Petrol.* **96**, 455–64.

Dixon, D., McNair, A. and Curran, S. C. (1954). The natural radioactivity of lutetium. *Phil. Mag.* **45**, 683–4.

Elderfield, H. (1992). The Ce–Nd–Sr isotope systematics of seawater: comment on 'Isotopic compositions of Ce, Nd and Sr in ferromanganese nodules from the Pacific and Atlantic Oceans, the Baltic and Barents Seas and the Gulf of Bothnia' (by Amakawa, H., Ingri, J., Masuda, A. and Shimizu, H.) *Earth Planet. Sci. Lett.* **111**, 557–61.

Elderfield, H. and Graves, M. J. (1982). The rare earth elements in seawater. *Nature* **296**, 214–9.

Faure, G. (1977). *Principles of Isotope Geology.* Wiley, 464 p.

Fujimaki, H., Tatsumoto, M. and Aoki, K. (1984). Partition coefficients of Hf, Zr and REE between phenocryst phases and groundmass. *Proc. 14th Lunar Planet. Sci. Conf., J. Geophys. Res.* **89** (supp.), B662–72.

Gruau, G., Chauvel, C., Arndt, N. T. and Cornichet, J. (1990). Aluminum depletion in komatiites and garnet fractionation in the early Archean mantle: Hafnium isotopic constraints. *Geochim. Cosmochim. Acta* **54**, 3095–101.

Herr, W., Merz, E., Eberhardt, P. and Signer, P. (1958). Zur Bestimmung der β Halbwertzeit des ^{176}Lu durch den nachweis von radiogenem ^{176}Hf. *Z. Natur.* **13a**, 268–73.

Holmes, A. (1932). The origin of igneous rocks. *Geol. Mag.* **69**, 543–58.

Johnson, C. M. and Beard, B. L. (1993). Evidence from hafnium isotopes for ancient sub-oceanic mantle beneath the Rio Grande rift. *Nature* **362**, 441–4.

Kinny, P. D., Compston, W. and Williams, I. S. (1991). A reconnaissance ion-probe study of hafnium isotopes in zircons. *Geochim. Cosmochim. Acta* **55**, 849–59.

Makishima, A. and Masuda, A. (1994). Ce isotope ratios of N-type MORB. *Chem. Geol.* **188**, 1–8.

Makishima, A., Nakamura, E., Akimoto, S., Campbell, I. H. and Hill, R. I. (1993). New constraints on the ^{138}La β-decay constant based on a geochronological study of granites from the Yilgarn Block, Western Australia. *Chem. Geol.* (*Isot. Geosci. Section*) **104**, 293–300.

Marshall, B. D. and DePaolo, D. J. (1982). Precise age determination and petrogenetic studies using the K–Ca method. *Geochim. Cosmochim. Acta* **46**, 2537–45.

Marshall, B. D. and DePaolo, D. J. (1989). Calcium isotopes in igneous rocks and the origin of granite. *Geochim. Cosmochim. Acta* **53**, 917–22.

Masuda, A., Shimizu, H., Nakai, S., Makishima, A. and Lahti, S. (1988). ^{138}La β-decay constant estimated from geochronological studies. *Earth Planet. Sci. Lett.* **89**, 316–22.

Nakai, S., Shimizu, H. and Masuda, A. (1986). A new geochronometer using lanthanum-138. *Nature* **320**, 433–5.

Nelson, D. R. and McCulloch, M. T. (1989). Petrogenetic applications of the ^{40}K–^{40}Ca radiogenic decay scheme – a reconnaissance study. *Chem. Geol.* (*Isot. Geosci. Section*) **79**, 275–93.

Norman E. B. and Nelson M. A. (1983). Half-life and decay scheme of ^{138}La. *Phys. Rev. C* **27**, 1321–4.

Patchett, P. J. (1983). Hafnium isotope results from Mid-ocean ridges and Kerguelen. *Lithos* **16**, 47–51.

Patchett P. J., Kouvo O., Hedge C. E. and Tatsumoto M. (1981). Evolution of continental crust and mantle heterogeneity: Evidence from Hf isotopes. *Contrib. Mineral. Petrol.* **78**, 279–97.

Patchett P. J. and Tatsumoto M. (1980a). A routine high-precision method for Lu–Hf isotope geochemistry and chronology. *Contrib. Mineral. Petrol.* **75**, 263–7.

Patchett P. J. and Tatsumoto, M. (1980b). Lu–Hf total-rock isochron for the eucrite meteorites. *Nature* **288**, 571–4.

Patchett P. J. and Tatsumoto M. (1980c). Hafnium isotope variations in oceanic basalts. *Geophys. Res. Lett.* **7**, 1077–80.

Patchett P. J. and Tatsumoto M. (1981). Lu/Hf in chondrites and definition of a chondritic hafnium growth curve. *Lunar Planet. Sci.* **XII**, 822–4, Lunar Planet. Inst.

Patchett P. J., White W. M., Feldmann H., Kielinczuk S. and Hofmann A. W. (1984). Hafnium/rare earth element fractionation in the sedimentary system and crustal recycling into the Earth's mantle. *Earth Planet. Sci. Lett.* **69**, 365–78.

Pettingill, H. S. and Patchett, P. J. (1981). Lu–Hf total rock age for the Amitsoq gneisses, West Greenland. *Earth Planet. Sci. Lett.* **55**, 150–6.

Russell W. A., Papanastassiou D. A. and Tombrello T. A. (1978). Ca isotope fractionation on the Earth and other solar system materials. *Geochim. Cosmochim. Acta* **42**, 1075–90.

Salters, V. J. and Hart, S. R. (1989). The hafnium paradox and the role of garnet in the source of mid-ocean-ridge basalts. *Nature* **342**, 420–2.

Salters, V. J. and Hart, S. R. (1991). The mantle sources of ocean ridges, islands and arcs: the Hf-isotope connection. *Earth Planet. Sci. Lett.* **104**, 364–80.

Sato J. and Hirose T. (1981). Half-life of ^{138}La. *Radiochem. Radioanal. Lett.* **46**, 145–52.

Shimizu, H., Tanaka, T. and Masuda, A. (1984). Meteoritic ^{138}Ce/^{142}Ce ratio and its evolution. *Nature* **307**, 251–2.

Stevenson, R. K. and Patchett, P. J. (1990). Implications for the evolution of continental crust from Hf isotope systematics of Archean detrital zircons. *Geochim. Cosmochim. Acta* **54**, 1683–97.

Tanaka, T. and Masuda, A. (1982). The La–Ce geochronometer: a new dating method. *Nature* **300**, 515–18.

Tanaka, T., Shimizu, H., Kawata, Y. & Masuda, A. (1987). Combined La–Ce and Sm–Nd isotope systematics in petrogenetic studies. *Nature* **327**, 113–17.

Tanaka, T., Shimizu, H., Kawata, Y. and Masuda, A. (1988). Reply to: Dickin, A. P. (1988). Mantle and crustal Ce/Nd isotope systematics. *Nature* **333**, 403–4.

Tanaka, T., Usui, A. and Masuda, A. (1986). Oceanic Ce and continental Nd: multiple sources of REE in oceanic ferromanganese nodules. *Terra Cognita* **6**, 114 (abstract).

Tatsumoto, M., Unruh, D. M. and Patchett, P. J. (1981). U–Pb and Lu–Hf systematics of Antarctic meteorites. *Nat. Inst. Polar Res. Tokyo.*

White, W. M. and Patchett, P. J. (1984). Hf–Nd–Sr isotopes and incompatible element abundances in island arcs: implications for magma origins and crust–mantle evolution. *Earth Planet. Sci. Lett.* **67**, 167–85.

Zindler, A. and Hart, S. R. (1986). Chemical geodynamics. *Ann. Rev. Earth Planet. Sci.* **14**, 493–571.

10 K–Ar and Ar–Ar dating

10.1 The K–Ar dating method

Potassium is one of the eight most abundant chemical elements in the Earth's crust, and a major constituent of many rock-forming minerals. However, the radioactive isotope, ^{40}K, makes up only 0.012% of total potassium, so it effectively falls in the low ppm concentration range. 89.5% of ^{40}K decays lead to ^{40}Ca, but in most rocks this is swamped by common (non-radiogenic) ^{40}Ca, which makes up 97% of total calcium. Because variations in radiogenic ^{40}Ca abundance are very limited in most rock systems, this method has a restricted application as a dating tool (section 9.3). Only 11% of ^{40}K decays lead to ^{40}Ar, but since this is a rare gas, the radiogenic component is dominant. It makes up 99.6% of atmospheric argon, equal to 0.93% of dry air by volume.

Decay to ^{40}Ar is by three different routes (section 1.3.1), two of which involve capture of an orbital electron by the nucleus. Positron emission constitutes only 0.01% of decays to ^{40}Ar. The electron capture (EC) decay constant, which can therefore be taken to represent all of the routes from ^{40}K to ^{40}Ar, now has a recommended value of 0.581×10^{-10} yr^{-1}, equivalent to a half-life of 11.93 Byr (*IUGS Subcommission on Geochronology;* Steiger and Jager, 1977). This was based on a weighted mean of the six best counting determinations, evaluated by Beckinsale and Gale (1969).

Decay to ^{40}Ca is by emission of a β particle, and the β decay constant is now taken to be 4.962×10^{-10} yr^{-1}, equivalent to a half-life of 1.397 Byr. The sum of the decay constants for the two branches yields the total ^{40}K decay constant of 5.543×10^{-10} yr^{-1}, equivalent to a half-life of 1.25 Byr.

The fraction of ^{40}K atoms that decay into ^{40}Ar is given by the expression $(\lambda_{EC} /[\lambda_{EC} + \lambda_{\beta}])$. Hence, substituting into the general decay equation [1.10], the growth of ^{40}Ar in a K-bearing rock or mineral can be written as:

$$^{40}\text{Ar}_{\text{total}} = {}^{40}\text{Ar}_{\text{I}} + \frac{\lambda_{\text{EC}}}{\lambda_{\text{total}}} \cdot {}^{40}\text{K}(e^{\lambda_{\text{total}}t} - 1) \qquad [10.1]$$

However, if the system was completely outgassed of Ar at the time of formation, the initial Ar term disappears, and the equation is simplified to:

$$^{40}\text{Ar}^* = \frac{\lambda_{\text{EC}}}{\lambda_{\text{total}}} \cdot {}^{40}\text{K}(e^{\lambda_{\text{total}}t} - 1) \qquad [10.2]$$

where ^{40}Ar* signifies radiogenic argon only. As will be seen below, this is the situation which is normally assumed in K–Ar dating.

10.1.1 Analytical techniques

The isotopic composition of naturally occurring potassium has been found to be effectively constant in all types of rock throughout the Earth, with a few minor exceptions (e.g. Garner *et al.*, 1976). Therefore the ^{40}K content of a mineral or rock is usually found by straightforward chemical analysis for total potassium, followed by multiplication by 1.2×10^{-4} to derive the concentration of the radioactive isotope. Various methods can be used to determine potassium, including x-ray fluorescence, neutron activation analysis, and isotope dilution. A commonly used method is flame photometry, a form of optical emission spectrometry which is especially suitable for the alkali metals. This technique probably yields K determinations approaching 1% accuracy if carefully used. It may be less accurate than other methods (e.g. isotope dilution) but is relatively quick and uses inexpensive equipment. The method is described in detail by Vincent (1960).

Argon trapped in a geological sample is released and purified in an argon extraction line, 'spiked' with an enriched isotope, and then fed into a mass spectrometer for isotopic analysis (Fig. 10.1). Sam-

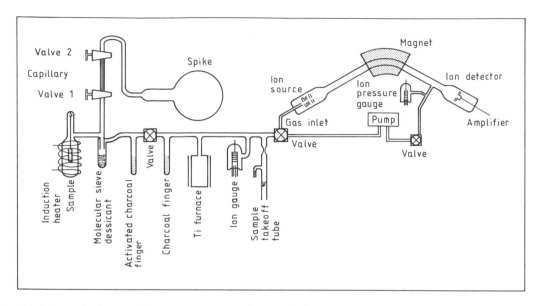

Fig. 10.1. Schematic diagram of an argon extraction line coupled to a static gas mass spectrometer. After Dalrymple and Lanphere (1969).

ples must have the minimum possible surface area for absorption of atmospheric argon, therefore mineral separates or whole-rock chips are not powdered. After loading the sample(s) in the extraction line, the whole line, and especially the sample itself, must be baked under vacuum to extract all possible atmospheric argon from the system. Next, after isolating the pump, the sample is manoeuvred into a disposable molybdenum crucible, which is positioned in a radio-frequency induction furnace. The crucible is heated to ca. 1400 °C, whereupon the sample melts and releases all of the trapped gases. These consist mostly of H_2O and CO_2, with a very small amount of argon and other rare gases. All gases except the rare gases can be removed by reaction with titanium vapour in a Ti sublimation pump or by using a zeolite 'getter'. Activated charcoal fingers may be used for temporary absorption of gases during their manipulation.

Highly enriched ^{38}Ar spike is usually stored in a large glass reservoir bulb. This is connected to a length of capillary tube of fixed volume between two valves with low dead-space (Fig. 10.1). The capillary is opened to the reservoir while valve 1 is closed. Valve 2 is then closed, and the known volume of spike between the two valves is added to the sample by opening valve 1. Because the reservoir pressure falls with each gas withdrawal, successive spike aliquots contain smaller and smaller fractions of ^{38}Ar. However, aliquots are periodically calibrated by

mixing with a known volume of atmospheric argon and performing an isotope dilution analysis. The amount of ^{38}Ar spike added to each sample is determined by noting its order in the sequence and interpolating between the calibration runs (Lanphere and Dalrymple, 1966). The amounts of argon released from a typical sample of a few hundred milligrams are very small, generally less than 10^{-6} cc (cm^3) at STP (standard temperature and pressure = 25 °C at 1 atm). For this reason the isotopic analysis is performed statically; in other words, the entire sample is fed into the mass spectrometer at once, after isolation from the pumps. The ratio $^{40}Ar/^{36}Ar$ in the air may be measured between unknown samples as a check on the calibration of the machine, and normally has a value of 295.5 ± 0.5.

Two different types of mass spectrometer are in common use. Modern rare-gas machines tend to be very similar to solid source machines, with a high accelerating potential of several kV, and peak-switching by changing the magnetic field. The problem with this type of machine is that the high velocities of the ions makes them implant into metal components in the vacuum system whenever these are struck by the ion beam. Such ions diffuse back out of the metal surfaces during the next sample analysis, and this memory effect must be carefully corrected. The effect may be reduced by polishing metal components which the beam is likely to strike. Many

Peak size

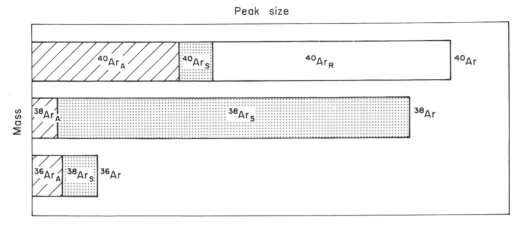

Fig. 10.2. Schematic argon isotope mass spectrum showing fractions of each peak due to radiogenic Ar (white), spike (stipple) and atmospheric contamination (hatch). Size fractions are not shown to scale. After Dalrymple and Lanphere (1969).

older instruments use a low accelerating potential of a few hundred volts and a small permanent-field magnet. The accelerating potential is then switched to focus different nuclides into the collector. This type of machine suffers from very little memory effect but is capable of much poorer precision in the measurement of isotope ratios. Since the source is gaseous, there is no problem of mass-dependent fractionation in either type of machine.

A typical argon isotope mass spectrum is shown in Fig. 10.2. The presence of any ^{36}Ar signal shows that common or non-radiogenic argon is present. In fact this is almost inevitable because of the great difficulty of removing all atmospheric argon from the system. If the sample was completely outgassed at the time of its formation, so that it contains no inherited non-radiogenic Ar, then the measured ^{36}Ar peak can be multiplied by 295.5 to correct the ^{40}Ar peak for atmospheric argon:

$$^{40}\text{Ar*} = {}^{40}\text{Ar}_{\text{total}} - 295.5\ {}^{36}\text{Ar} \qquad [10.3]$$

A value of 0.063% of atmospheric argon is similarly subtracted from the ^{38}Ar peak. The ^{40}Ar and ^{36}Ar peaks must also be corrected for small fractions of these isotopes in the spike. The amount of radiogenic ^{40}Ar* in the sample is then found by comparison with the size of the net ^{38}Ar peak, formed by a known quantity of spike. (In other words, this is an isotope dilution determination.) Given the abundances of ^{40}Ar and ^{40}K in the sample, the age is calculated by

re-arranging equation 10.2:

$$t = \frac{1}{\lambda_{\text{total}}}\ \ln\left[\frac{{}^{40}\text{Ar*}}{{}^{40}\text{K}} \cdot \frac{\lambda_{\text{total}}}{\lambda_{\text{EC}}} + 1\right] \qquad [10.4]$$

It is necessary to assume that the sample contained no initial argon (usually called 'excess' argon), because this might have a ^{40}Ar/^{36}Ar ratio different from atmospheric argon, leading to a mixture with indeterminate ^{40}Ar/^{36}Ar ratio which could not be corrected for atmospheric contamination. In addition, K–Ar ages depend on closed-system behaviour of the sample for K and Ar throughout its history.

The ^{36}Ar/^{40}Ar ratio must be analysed to very high precision because the atmospheric Ar correction magnifies any errors in this measurement by nearly 300. The importance of this effect is shown in Fig 10.3, where the effect of errors in the measurement of ^{36}Ar are shown in terms of the resulting error in calculated age (Cox and Dalrymple, 1967). Once atmospheric contamination exceeds 70% of total argon, errors in ^{36}Ar have serious consequences on the age measurement. This correction is not a problem for old and/or K-rich samples, but is the principal limitation to dating young material.

Since 1967, great improvements of measurement precision have been made, allowing the dating of very young rocks. However, Mussett and Dalrymple (1968) showed that volcanic rocks contain 'locked-in' atmospheric (non-radiogenic) argon, some of which cannot be removed even by baking in a vacuum. Hence, even with a low-blank analytical

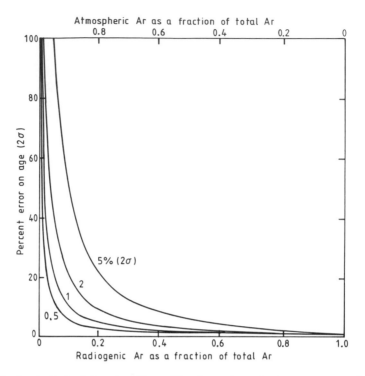

Fig. 10.3. Error magnification in K–Ar dating (*y* axis) resulting from atmospheric argon contamination. Curves are calculated for 0.5, 1, 2 and 5 % errors in the measurement of ^{36}Ar. After Cox and Dalrymple (1967).

system, a small residual atmospheric fraction is almost unavoidable in terrestrial lavas.

10.1.2 Inherited argon and the K–Ar isochron diagram

Since ^{36}Ar is used as a monitor of atmospheric contamination, there is no facility in K–Ar dating to correct for initial argon incorporated into minerals or rocks at the time of crystallisation. Hence, it must be assumed to be absent. However, Damon and Kulp (1958) showed that this assumption could break down, by finding evidence for initial or 'excess' argon in beryl, cordierite and tourmaline. Since these minerals all have a ring structure, it was initially assumed that the stacking of rings created channels in which excess argon inherited from fluids could reside. Damon and Kulp suggested that this problem might also occur in hornblende, where partial vacancy of the alkali-cation site might provide a location for excess argon.

However, excess argon was subsequently also found in pyroxenes by Hart and Dodd (1962). Since the pyroxene structure does not have any suitable voids for argon accommodation, Hart and Dodd argued that it must be located in crystal dislocations and defect structures. This implies that excess argon is a product of the environment of crystallisation rather than the host mineral. Hart and Dodd noted that their analysed pyroxenes were from originally deep-seated rocks, unlike the volcanic or shallow intrusive rocks normally used in K–Ar dating. Hence they warned that excess argon might be a common feature in samples from deep-seated (plutonic) environments.

The occurrence of excess argon was extended to submarine lavas by Dalrymple and Moore (1968). They dated glassy pillow rims and whole-rock pillow cores from flows at 500–5000 m depth on the northeast ridge of Kilauea volcano, Hawaii. Various geological lines of evidence suggested a historical age for the samples, but K–Ar ages of up to 43 Myr were found. Furthermore, a series of samples from rim to interior of one pillow (from 2590 m depth) showed an inverse correlation of apparent K–Ar age with distance from the rim (Fig. 10.4). The results were attributed to entrapment of initial or excess argon

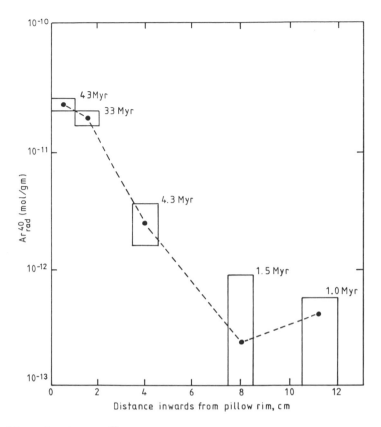

Fig. 10.4. Contents of (excess) radiogenic ^{40}Ar in submarine pillows from Hawaii, plotted against inward distance from the pillow rim. Apparent K–Ar ages for each sample are noted in Myr. After Dalrymple and Moore (1968).

which was inherited from the mantle source by the magmas. Dalrymple and Moore concluded that because these magmas were quenched under substantial hydrostatic pressure, inherited argon was not completely outgassed at the time of eruption, as usually occurs in terrestrial lavas.

However, even some subaerially erupted lavas were subsequently found to contain inherited argon. McDougall *et al.* (1969) encountered measurable radiogenic ^{40}Ar contents in historical age subaerial basalts from New Zealand. Lavas shown by ^{14}C dating of wood inclusions to be less than 1 kyr old nevertheless gave K–Ar ages up to 465 kyr. This led McDougall *et al.* to consider whether such cases of inherited argon could be detected and/or corrected.

They proposed that the raw ^{40}Ar signal (uncorrected for atmospheric contamination) be divided by ^{36}Ar and plotted against the K/Ar ratio to form an isochron diagram analogous to Rb–Sr (Fig. 10.5). This is achieved by expanding the initial Ar term in equation [10.1] to include both atmospheric and excess components (X), and by dividing throughout by ^{36}Ar:

$$\left(\frac{^{40}\text{Ar}}{^{36}\text{Ar}}\right)_{\text{total}} = \left(\frac{^{40}\text{Ar}}{^{36}\text{Ar}}\right)_{\text{atm}} + \left(\frac{^{40}\text{Ar}}{^{36}\text{Ar}}\right)_{\text{X}}$$

$$+ \frac{^{40}\text{K}}{^{36}\text{Ar}} \cdot \frac{\lambda_{\text{EC}}}{\lambda_{\text{EC}} + \lambda_{\beta}} (e^{\lambda_{\text{total}}t} - 1) \quad [10.5]$$

This equation has the form:

$$y = 295.5 + c + xm \quad [10.6]$$

When a suite of samples is analysed from a single completely-outgassed system such as a lava flow, the c term is equal to zero. Therefore, the analysed points, when plotted on an isochron diagram, should define a straight line with an intercept of 295.5, whose slope yields the age of eruption. In fact, this array is merely a mixing line between the samples and atmospheric argon. When the atmospheric correction is performed

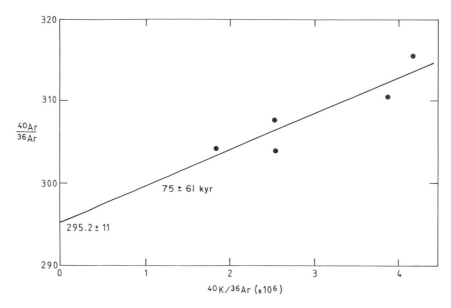

Fig. 10.5. K–Ar isochron plot for a lava of historical age from Mount Wellington, New Zealand, showing a best-fit slope age of 75 kyr. After McDougall *et al.* (1969).

on a single analysis, we effectively make 295.5 the origin and determine the slope.

In the case studied by McDougall *et al.* (1969), the lavas are of approximately zero age. Hence, the analyses which make up their 'isochrons' (e.g. Fig. 10.5) represent trapped argon in the magma variably mixed with atmospheric argon. McDougall *et al.* speculated that the trapped argon might originate from partially digested crustal xenocrysts.

Roddick and Farrar (1971) considered the case of a geologically old sample suite displaying both inherited argon and atmospheric contamination (Fig. 10.6). With inherited and radiogenic argon only, the array ABC is defined, but if variable atmospheric contamination occurs, a scatter (DEF) may result. In principle, a good linear array on the K–Ar isochron diagram should indicate that both the age and initial Ar isotope ratio are meaningful. However, it may be possible for the slope of the line to swing round in a systematic way due to complex mixing processes, so that it yields a good array of meaningless slope. Nevertheless, the isochron diagram is a useful test of K–Ar data where inherited Ar is suspected. This may be a particular problem in plutonic rocks.

Lanphere and Dalrymple (1976) drew a distinction between excess and inherited argon. They defined the former as argon 'incorporated into rocks and minerals by processes other than *in-situ* decay of ^{40}K' and the

latter as argon which 'originates within mineral grains by decay of ^{40}K prior to the rock-forming event'. The second definition seems in fact to be a special case of the former and it would appear that most authors use the two terms interchangeably. These authors were the first to use the ^{40}Ar/^{39}Ar method (see below) to identify excess argon in rocks (Lanphere and Dalrymple, 1971). Their 1976 paper followed up this work by studying the effects of excess argon on the ^{40}Ar/^{39}Ar ages of rock-forming minerals.

10.1.3 Argon loss

The K–Ar method is unique amongst the major radiometric dating methods in having a gaseous daughter product. This means that the K–Ar system reacts differently than couples such as Rb–Sr in response to thermal and hydrothermal events. Because argon is a non-reactive gas, it displays limited partition into fluids, so that the K–Ar system may be more resistant than Rb–Sr to hydrothermal metamorphism. On the other hand, no mineral phase preferentially takes up argon when it is lost from the mineral where it was originally produced. This means that in K–Ar dating, whole-rock analysis confers no additional resistance to metamorphic re-setting (as it does for the Rb–Sr method). On the contrary, in K–

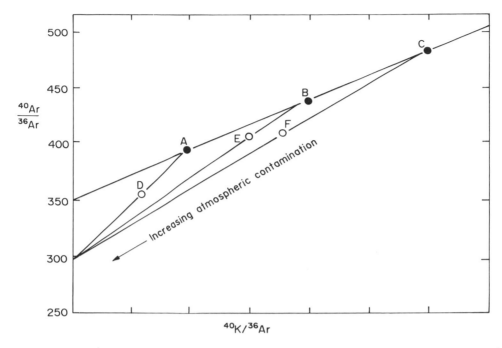

Fig. 10.6. Schematic K–Ar isochron diagram to show the effect of mixing inherited and radiogenic argon (A, B, C), coupled with variable atmospheric contamination (D, E, F). After Roddick and Farrar (1971).

Ar analysis a whole-rock sample is only as resistant to re-setting as its least retentive phase. Consequently whole-rock K–Ar analysis is a last resort, when all mineral phases in the rock are too fine-grained for mineral separation.

A good comparison of argon loss from different minerals during a thermal event is provided by contact metamorphism associated with the Eldora stock in the Colorado front ranges (Hart, 1964). The 54 Myr-old quartz monzonite stock is intruded into ca. 1350 Myr-old amphibolites and schists. Hart analysed biotite, hornblende and K-feldspar at increasing distances from the intrusive contact (Fig. 10.7), and found that despite the limited extent of petrographic alteration, K–Ar mineral ages were reset at large distances from the stock. Hornblende displayed good Ar retention properties, with loss of argon confined primarily to within ten feet (ca. 3m) of the contact. However, coarse biotites were largely reset at distances up to 1000 ft (300 m) from the contact, while K-feldspars had lost a substantial fraction of argon even 20 000 feet (6 km) from the contact. The latter can hardly be said to be within the thermal aureole of the stock, and reflects the now widely accepted view that K-feldspars may lose argon

by diffusion even at ambient temperatures. Biotite Rb–Sr systems were only slightly more resistant to disturbance than K–Ar.

10.1.4 Calibrating the geomagnetic reversal time-scale

One of the most important applications of the K–Ar method has been to calibrate the magnetic reversal time-scale defined by sea floor magnetic anomaly 'stripes'. The amount of ocean floor material recovered which is fresh (unaltered) enough for dating is limited, so most attention has been focussed on dating terrestrial sections (such as basic lavas) which yield a good magnetostratigraphy. The K–Ar method is really the only geochronometer capable of dating young basic rocks. Since its establishment, the reversal time-scale has been subject to almost continuous revision, and some landmarks are reviewed here.

Pioneering work was performed by Cox et al. (1963) on 0–3 Myr-old lavas from California, and by McDougall and Tarling (1964) on 0–3 Myr-old lavas from the Hawaiian islands. Cox et al. used K–Ar dates on sanidine, obsidian, biotite, and whole-rocks,

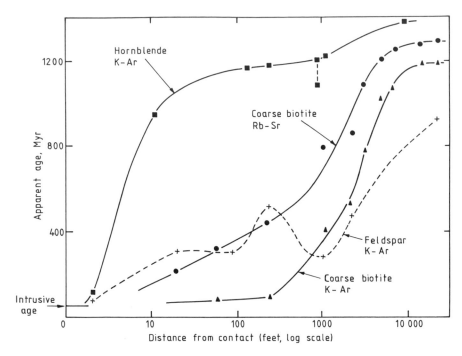

Fig. 10.7. Plot of apparent mineral ages against outward distance from the contact of the Eldora stock, Colorado. After Hart (1964).

while McDougall and Tarling worked on basalt whole-rocks. Good agreement between the two data sets confirmed that the reversal time-scale is due to worldwide changes in the polarity of the Earth's magnetic field, and not due to post-crystallisation alteration phenomena, as had been suggested by some workers.

A comprehensive compilation of data for 354 terrestrial lavas (mostly from ocean islands) was used by Mankinen and Dalrymple (1979) to precisely constrain the polarity time-scale for the last 5 Myr, using the new K–Ar decay constants (Steiger and Jager, 1977). Not all of the available data is in perfect agreement, therefore Mankinen and Dalrymple used a statistical technique to calculate the most probable ages of the three most recent polarity epoch boundaries, such that the standard deviation of apparent dating inconsistencies is minimised (Fig. 10.8).

Unfortunately, the geological record of terrestrial lavas is too fragmented to extend the technique of detailed epoch-boundary dating back beyond 5 Myr. Therefore, Heirtzler *et al.* (1968) used sea floor magnetic anomaly patterns to extend the terrestrial time-scale to ca. 80 Myr by extrapolation. They took as a fixed point an age of 3.35 Myr for the epoch

boundary between Gilbert (reversed) and Gauss (normal), based on Doell *et al.*'s (1966) Sierra Nevada data. Fortunately, the later value for this fixed point (3.4 Myr above) did not move significantly. Heirtzler *et al.* extrapolated from the present day, through the Gilbert–Gauss point, to calibrate the older part of the time-scale, by assuming a constant spreading rate for the South Atlantic ridge over the last 80 Myr. This may seem a crude assumption, but Heirtzler *et al.* justified it on the basis of the good correlation between S Atlantic and N Pacific spreading rates (Fig. 10.9).

During the next two decades, improvements to the reversal time-scale were achieved by adding additional fixed points. LaBrecque *et al.* (1977) made the first major revision by using two fixed points to avoid the extreme extrapolation of Heirtzler *et al.* (1968). The younger point was again the Gilbert (R)–Gauss (N) boundary, while the older (marine anomaly 29), was tied by magneto-stratigraphy of Paleocene limestones near Gubbio, Italy, to the Cretaceous–Tertiary (K–T) boundary. Tying the reversal time-scale to the geological column was advantageous in harmonising the time-scales, but problematical in that it revealed the poor constraints on the K–T boundary itself. For example, sources cited by LaBrecque *et al.*

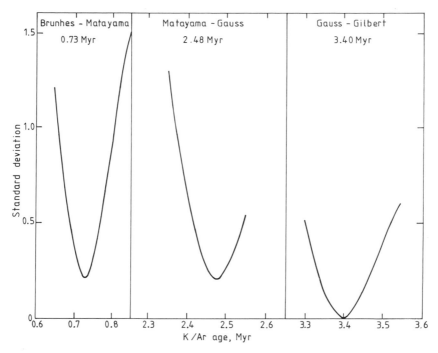

Fig. 10.8 Standard deviation of apparent dating inconsistencies as a function of 'trial' values for polarity boundaries. The best estimate of each boundary age is where error is a minimum. After Mankinen and Dalrymple (1979).

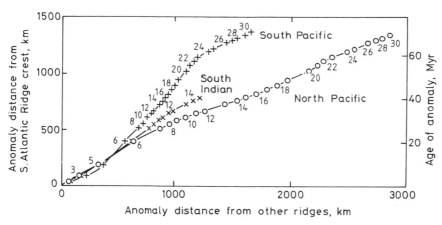

Fig. 10.9. Plot of distance from the ridge to a given anomaly in the South Atlantic, against the corresponding distance for three other ridges. After Heirtzler *et al.* (1968).

(1977) for a 65 Myr K–T boundary age were a poster published by Elsevier (van Eysinga, 1975) and a paper by van Hinte (1976). The latter is merely a citation of Berggren (1972), whose only constraints were a minimum of 57.1 ± 3 from Belgium and a maximum of 68.1 ± 4 from SE England. Fortunately the 65 Myr age for the K–T boundary has proved so robust

in the long run that it has not even been affected by changes in the decay constants used! (This is a coincidence of course.)

Ness *et al.* (1980) adjusted K–Ar dates to the then new decay constants recommended by the IUGC (section 10.1), and proposed a time-scale based on two additional fixed points (total now four). This time-scale was revised by Lowrie and Alvarez (1981), who used magneto-stratigraphy in the 25 – 130 Myr-old Gubbio limestones of Italy to interpolate between 11 fixed points. The fixed points are bio-stratigraphic stage boundaries, for which the dates of Ness et al. were used. However, the large number of fixed points generated a kinked line, due to errors in the absolute age calibration of some points. Harland et al. (1982) ironed out some of these inflections to produce a useful 'working time-scale' (Fig. 10.10). However, Harland's approach has probably now reached its technical limit, since the absolute ages of the fixed points, based mainly on K–Ar glauconite ages, are not reliable enough for more accurate calibration.

Fresh progress in calibrating the reversal time-scale was dependent on new high-precision radiometric dating. This was achieved by analysing volcanic ash layers within terrestrial magneto-stratigraphic succes-

sions. Berggren et al. (1985) provided preliminary data by K–Ar analysis of biotites, and in the process reduced the apparent age of the Eocene–Oligocene boundary from ca. 38 Myr to 36.5 Myr. However, this work was in turn superseded by the single-crystal laser-fusion Ar–Ar method (section 10.3). By analysing single biotite and feldspar crystals from fresh volcanic tephra deposits, Swisher and Prothero (1990) reduced analytical errors to ca. 0.25 Myr (2σ), and determined an Eocene–Oligocene boundary age of 34 Myr. These improvements in precision make it realistic to extend the time-scale based on terrestrial volcanics back from 3 Myr to 35 Myr, because the errors on individual dates no longer exceed the lengths of individual polarity intervals. A preliminary example of this approach is demonstrated by McIntosh et al. (1992). Doubtless, much more detailed sections will be constructed in the future.

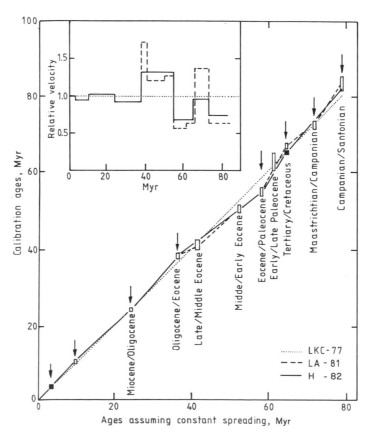

Fig. 10.10. Diagram to show the fixed points which constrain the reversal time-scales of LaBrecque et al. (LKC–77, ■); Lowrie and Alvarez (LA81, ▢); and Harland et al. (H82, ↓). Inset shows calculated consequences for spreading rate on the South Atlantic Ridge. After Harland et al. (1982).

10.2 The ^{40}Ar–^{39}Ar dating technique

10.2.1 Irradiation

^{39}K may be converted to ^{39}Ar by irradiation with fast neutrons in an n,p (neutron capture, proton emission) reaction, permitting the K determination for a K–Ar age to be made as part of the argon isotope analysis:

$$^{39}_{19}K + n \rightarrow\, ^{39}_{18}Ar + p$$

The method was first applied by Wanke and Konig (1959) using a counting technique to detect both ^{39}Ar ($t_{1/2} = 269$ yr) and ^{41}Ar ($t_{1/2} = 2$ hours) produced by neutron activation of ^{40}Ar. However, this method does not permit an atmospheric correction to be made, since ^{36}Ar cannot be adequately measured.

The comparatively long half-life of ^{39}Ar means that it can be regarded as a stable isotope for mass spectrometric analysis, which was first applied to ^{40}Ar–^{39}Ar dating by Merrihue and Turner (1966). It is interesting to note, however, that the concept of combined irradiation and mass spectrometric analysis was applied to the I–Xe system in meteorite studies five years earlier (section 15.2.2).

The production of ^{39}Ar from ^{39}K during the irradiation is expressed as:

$$^{39}Ar = ^{39}K\, \Delta t \int_{\min e}^{\max e} \phi_e\, \sigma_e\, de \qquad [10.7]$$

where Δt is the irradiation time, ϕ_e is the flux density of neutrons with energy e, and σ_e is the capture cross-section of ^{39}K for neutrons of energy e. The production must be integrated over the total range of neutron energies, which is a very difficult calculation in practice. Therefore, the normal procedure is to use a sample of known age as a flux monitor.

Taking the K–Ar decay equation [10.2], which is reproduced here:

$$^{40}Ar* = \frac{\lambda_{EC}}{\lambda_{total}} \cdot ^{40}K(e^{\lambda_{total} t} - 1)$$

and dividing through on both sides by equation [10.7], yields:

$$\frac{^{40}Ar*}{^{39}Ar} = \boxed{\frac{\lambda_{EC}}{\lambda_{total}} \cdot \frac{^{40}K}{^{39}K \Delta t \int \phi_e \sigma_e\, de}}(e^{\lambda_{total} t} - 1) \qquad [10.8]$$

However, the boxed term is the same for sample and standard. Therefore it is customary to refer to it as a single quantity whose reciprocal, J, can be evaluated as a constant (Mitchell, 1968). Hence, for the standard:

$$J = \frac{e^{\lambda t} - 1}{^{40}Ar*/^{39}Ar} \qquad [10.9]$$

where t is known. Rearranging equation [10.9] for samples of unknown age yields:

$$t = \frac{1}{\lambda} \cdot \ln\left[J\left(\frac{^{40}Ar*}{^{39}Ar}\right) + 1 \right] \qquad [10.10]$$

In order to obtain an accurate value of J for each unknown sample, several standards need to be run, representing known spatial positions relative to the unknown samples within the reactor core (Mitchell, 1968). Hence J values for each of the samples can be interpolated.

10.2.2 Corrections

During the irradiation of ^{39}K, interfering Ar isotopes are generated by neutron reactions from calcium and other potassium isotopes (Fig. 10.11). Brereton (1970) and Dalrymple and Lanphere (1971) made detailed studies of the magnitude of these effects and their correction. However, it appears in practice that many workers have simply ignored the interferences.

Mitchell (1968) suggested that acceptable results could be obtained without interference correction on minerals over 1 Myr old, provided that K/Ca was greater than 1. In such circumstances, a simple atmospheric correction may be considered adequate:

$$\frac{^{40}Ar*}{^{39}Ar} = \left(\frac{^{40}Ar}{^{39}Ar}\right)_{meas} - 295.5\left(\frac{^{36}Ar}{^{39}Ar}\right)_{meas} \qquad [10.11]$$

Turner (1971a) showed that Ar interferences could be kept to a minimum by variation of certain irradiation parameters. The principal interferences which must be considered (Fig. 10.11) are:

$$^{40}K \quad n, p \quad \rightarrow\, ^{40}Ar$$
$$^{40}Ca \quad n, n\alpha \rightarrow\, ^{36}Ar$$
$$^{42}Ca \quad n, \alpha \quad \rightarrow\, ^{39}Ar$$

Other interferences occur but may be omitted as insignificant.

The approaches suggested by Turner were:

1) Optimisation of neutron dose according to age (Fig. 10.12a) to maximise ^{39}Ar production, without generating significant artificial ^{40}Ar from ^{40}K. (K content is not considered as a factor because the intended target and interfering one are both K isotopes.)

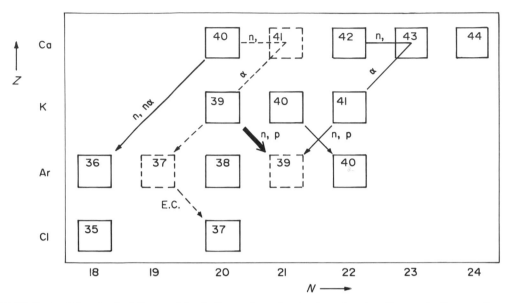

Fig. 10.11. Part of the chart of the nuclides in the region of potassium showing the production reaction (heavy arrow) and major interfering reactions (solid) during ^{40}Ar–^{39}Ar activation. The dashed reaction to ^{37}Ar is the interference monitor (see text). Data from Mitchell (1968).

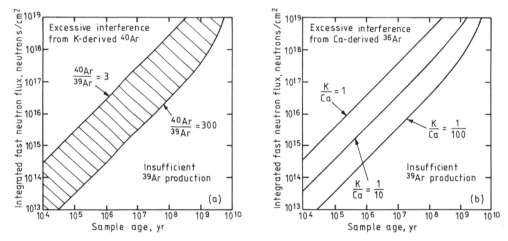

Fig. 10.12. Optimisation of neutron dose for (a) K content, and (b) K/Ca ratio. Hatched area in (a) and bold lines in (b) indicate regions of acceptable compromise between sufficient ^{39}Ar production and minimal ^{40}Ar or ^{36}Ar interference in typical rocks. These areas are almost coincident in the two plots. After Turner (1971a).

2) Optimisation of sample size according to age and K content in order to obtain the total ^{40}Ar and ^{39}Ar yields necessary to achieve the desired counting statistics during mass spectrometric analysis.

3) The K/Ca ratio in the sample also dictates an optimum neutron dose to generate enough ^{39}Ar without significant interfering ^{36}Ar (Fig.

10.12b). However, the optimum values partly overlap with those prescribed by criterion 1.

Theoretically, very young rocks can be activated with less than 1% interferences by following these rules. However, this may require an immense sample size. In practice, a better alternative may be to use more irradiation but apply corrections. The complete correction formula (in terms of Ar isotope ratios) is:

$$\frac{^{40}Ar^*}{^{39}Ar} = \left[\frac{^{40}Ar}{^{39}Ar} - 295.5 \cdot \frac{^{36}Ar}{^{39}Ar} + 295.5 \cdot \frac{^{37}Ar}{^{39}Ar} \cdot \left(\frac{^{36}Ar}{^{37}Ar}\right)_{Ca} - \left(\frac{^{40}Ar}{^{39}Ar}\right)_K\right] \Big/ \left[1 - \frac{^{37}Ar}{^{39}Ar} \cdot \left(\frac{^{39}Ar}{^{37}Ar}\right)_{Ca}\right] \qquad [10.12]$$

where $^{37}Ar/^{39}Ar$ is the interference monitor ratio measured for the unknown, which must be corrected for ^{37}Ar decay from the time of irradiation till analysis ($t_{1/2} = 35$ days); and where $(^{36}Ar/^{37}Ar)_{Ca}$, $(^{39}Ar/^{37}Ar)_{Ca}$ and $(^{40}Ar/^{39}Ar)_K$ are production ratios of Ar isotopes from the subscripted elements. These production ratios are determined by irradiating pure salts of Ca and K respectively in the reactor of interest, and are characteristic of the neutron flux of that reactor. Values for these production ratios measured by different authors for different reactors have typical ranges of 2.1–2.7, 6.3–30 and 0.006–0.031 respectively (Dalrymple and Lanphere, 1971).

10.2.3 Step heating

Because the potassium signature of a sample is converted *in-situ* to an argon signature by the 40–39 technique, it is possible to liberate argon in stages from different domains of the sample and still recover full age information from each step. Merrihue and Turner (1966) demonstrated the effectiveness of this 'step heating' technique in their original Ar–Ar dating study of meteorites, adapting the method from its previous application to I–Xe analysis of meteorites (section 15.2.2).

The great advantage of the step heating technique over the conventional 'total fusion' technique is that progressive outgassing allows the possibility that anomalous sub-systems within a sample may be identified, and, ideally, excluded from an analysis of the 'properly behaved' parts of the sample. This can apply to both separated minerals and whole-rock samples. Most commonly the technique is used to understand samples which have suffered argon loss, but it may also be a help in interpreting samples with inherited argon.

In the case of partially disturbed systems, the domains of a sample which are most susceptible to diffusional argon loss (such as the rim of a crystal) should be outgassed at relatively low temperature, whereas domains with tightly-bound argon (which are most resistant to disturbance) should release argon at higher temperature. In order to understand the history of disturbed samples, results of the step heating analysis are normally presented in one of two ways: as a K–Ar isochron diagram, analogous to a suite of samples analysed by conventional K–Ar; or as an age spectrum plot.

Step heating results from the meteorite Bjurbole (Merrihue and Turner, 1966) are plotted on an isochron diagram in Fig. 10.13. The straight line array indicates a simple one-stage closed-system history for the meteorite. The initial $^{40}Ar/^{36}Ar$ ratio may be only partially meaningful, since it is a mixture of initial Ar and atmospheric contamination.

The isochron plot is useful where inherited argon is suspected, but the age spectrum plot is more helpful to evaluate argon loss. To construct the spectrum plot, the size of each gas release at successively higher temperature is measured by the magnitude of the ^{39}Ar ion beam produced. Each gas release can then be plotted as a bar, whose length represents its volume as a fraction of the total ^{39}Ar released from the sample, and whose value on the y axis is the corrected $^{40}Ar/^{39}Ar$ ratio from equation [10.12]. The latter is proportional to age, which is sometimes plotted on a log scale, and sometimes linear. Determination of a reliable crystallisation age from the spectrum plot depends on the identification of an age 'plateau'. A rigorous criterion for a plateau age is the identification of a series of adjacent steps which together comprise more than 50% of the total argon release, each of which yields an age within 2 standard deviations of the mean (Dalrymple and Lanphere, 1974; Lee *et al.* 1991). However, plateaus have been 'identified' in many instances on the basis of weaker evidence.

The age spectrum plot displays the ideal nature of the K–Ar system in tektite glasses. These are objects which were completely melted during flight through the atmosphere and then rapidly quenched on landing. Young tektites which have not been affected by weathering yield perfect plateaus (Fig. 10.14). However, the most useful application of the 40–39 method is on samples with a complex geological history that *does* involve secondary argon loss.

10.2.4 Argon loss events

In order to assess the usefulness of 40–39 dating on disturbed systems, Turner *et al.* (1966) applied the method to chondritic meteorites. Many of these yield conventional K–Ar ages below 4.5 Byr, with U–He ages clustering around 500 Myr (Anders, 1964). Step heating results from whole-rock samples of the Bruderheim and Colby meteorites generated complex

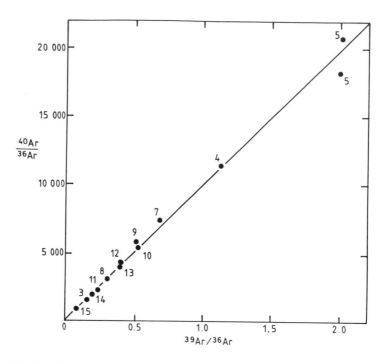

Fig. 10.13. Step heating data for the Bjurbole meteorite presented on the Ar–Ar isochron diagram. Numbers by data points signify temperatures ($\times$100 °C) for each release step. After Merrihue and Turner (1966).

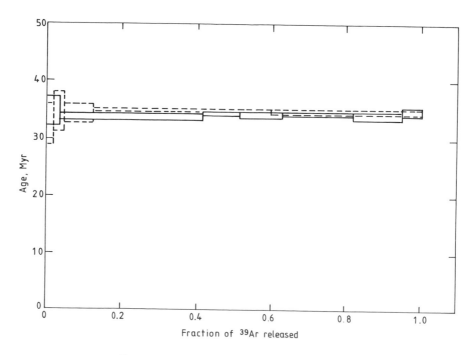

Fig. 10.14. Ideal $^{40}Ar/^{39}Ar$ age spectra for two Texas tektites. After York (1984).

plateaus (Fig. 10.15), which were attributed to argon loss at ca. 500 Myr (Turner *et al.*, 1966; Turner, 1968).

Turner *et al.* suggested that when these meteorites were heated (possibly in a collision between planetessimals), Ar loss occurred from the surface of mineral grains, and its transport within grains was by volume diffusion. This argon loss model is illustrated schematically in Fig. 10.16. Turner *et al.*

argued that step heating analysis of the sample in the vacuum system would mimic the natural thermal event, so that domains near the surface of minerals, which had suffered geological disturbance, would outgas first in the experiment. In contrast, domains near the cores of grains would be resistant to geological disturbance, and would also outgas at the highest temperatures in the laboratory. In order to test this model, Turner *et al.* then calculated

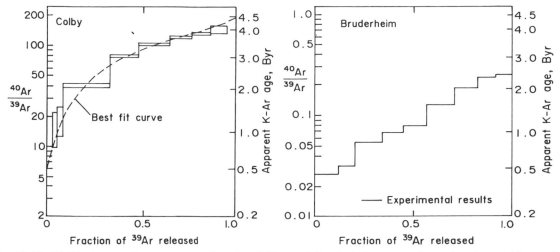

Fig. 10.15. 40–39 argon release patterns for the Colby and Bruderheim meteorites, showing evidence for disturbance after formation. Colby, with ca. 50% Ar loss, still yields an estimate of the age of formation; Bruderheim, with ca. 80% loss, yields only the age of disturbance. After Turner (1968).

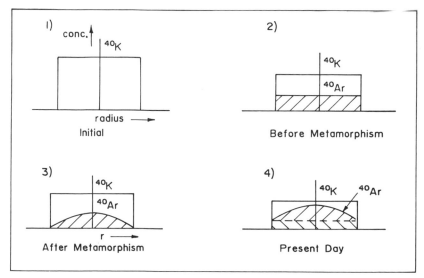

Fig. 10.16. Schematic illustration of the geological history of a mineral grain in a partially disturbed meteorite. 1) At 4500 Myr; 2) 500 Myr ago, before thermal event; 3) immediately after thermal event; 4) present day. After Turner (1968).

theoretical age spectrum plots, assuming that the meteorites were formed at 4500 Myr and metamorphosed in a single event at 500 Myr. These were then compared with observational data.

Turner *et al.* modelled argon loss from a system with different size-distributions of spherical grains. Fig. 10.17a predicts the effect of argon loss from grains of uniform size. A more realistic argon loss model for a multi-grain sample assumes a log–normal size distribution of mineral grains. This generates a family of solutions according to the amount of grain-size variation allowed. Fig. 10.17b shows the results for standard deviation (σ[log radius]) = 0.33, equivalent to 66% of grains falling within a factor of two from the mean radius. Numbers on the curves indicate the fraction of Ar lost. When loss is less than 0.2 (20%) the high-temperature plateau still records a good crystallisation age but the metamorphic age is badly constrained. When the fraction of Ar lost is greater than 0.8 the curves become concave upward and yield a more precise metamorphic age, but the formation age is completely lost. Bruderheim displays the latter pattern (loss = ca. 0.9), while Colby is intermediate (loss = ca. 0.6). The best fit for these meteorites is actually found for σ[log radius] = 0.2.

The good fit of analysed and modelled plateaus in Fig. 10.17 suggests that the diffusional loss mechanism is an accurate model for thermal disturbance of meteorites. This might be expected, since both the

geological event and laboratory measurements were based on heating of anhydrous phases in a vacuum. Turner (1972) demonstrated a similar good fit to the diffusional model for experimental data from a lunar anorthosite. In contrast, terrestrial 40–39 dating generally involves hydrated minerals such as biotite and hornblende. In this case, the diffusional argon loss mechanism may not provide such an accurate model.

A terrestrial example of a discrete argon loss event is provided by 40–39 analysis of the Eldora stock (Berger, 1975), for which conventional K–Ar data were described above. Fig. 10.18 shows age spectrum plots for hornblendes, biotites and feldspars in the vicinity of the stock.

Of the three minerals studied, hornblende (Fig. 10.18a and b) displays the type of pattern most similar to Turner's thermal diffusion degassing model, but this resemblance may be misleading. The most distant sample (not shown) yields an excellent plateau age of ca. 1400 Myr. Samples at 1130, 950, 248 and 34 ft (ca. 350, 290, 75 and 10 m) display serious Ar loss from the outside of grains, but approach the 'true' age in the highest-temperature fractions. However, Berger recognised that this pattern might reflect alteration to biotite, rather than diffusional Ar loss from hornblende. This interpretation is supported by dating experiments on synthetic hornblende–biotite mixtures (Rex *et al.*, 1993). Another problematical observation is that the sample 11 ft (3.5 m) from the contact

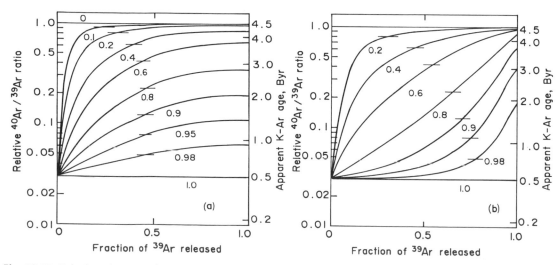

Fig. 10.17. Calculated argon release patterns for a sample composed of spherical mineral grains, with initial age of 4.5 Byr and disturbance at 0.5 Byr. a) Assuming uniform size distribution; b) log–normal size distribution with σ = 0.33. Numbers indicate fraction of argon lost. Horizontal bars show the average (total release) age for each curve. After Turner (1968).

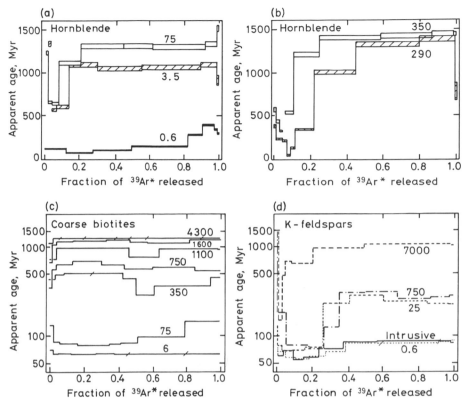

Fig. 10.18. Ar–Ar age spectrum plots for mineral phases at different distances from the Eldora stock. Figures indicate distance in metres. a), b) hornblende, c) biotite, d) K-feldspar. Successive release steps with identical ages are separated by slashes. After Berger (1975).

displays an intermediate 'false' plateau of high quality. Finally, the sample 2 ft (0.6 m) from the contact displays a saddle-shaped pattern, in which the lowest-age fraction approaches the age of metamorphism.

Coarse biotite (Fig. 10.18c) behaves somewhat differently. Its maximum age at infinite distance from the stock (1250 Myr) is lower than the hornblende age. At intermediate distances the spectra are irregular, but show a general decrease in 'plateau' age as the stock is approached. Hence, it appears that biotites can be partially but uniformly outgassed, possibly because of enhanced diffusion parallel to the cleavage. Finally, K-feldspar suffers irregular and disastrous Ar losses, as is known from conventional K–Ar analysis (Fig. 10.18d).

Berger concluded that hornblendes were able to generate plateaus of high quality which were nevertheless meaningless. This may make hornblende a dangerous material on which to base geological interpretations of age, in the absence of independent

confirmatory evidence. Berger argued that partially reset biotites were always identifiable by their irregular patterns, making biotite a reliable basis for age interpretation. The exact meaning of the plateaus in the biotite and hornblende samples distant from the stock is equivocal, since the country-rocks are paragneisses with a long history of thermal events. Subsequent studies have indeed generated many examples of meaningless plateaus in hornblende, and more rarely, in biotite.

10.2.5 ^{39}Ar recoil

The Ar–Ar dating technique was found to be particularly useful for dating small whole-rock samples of lunar material, especially fine-grained mare basalts. The dashed profile in Fig. 10.19 shows a typical release pattern (Turner and Cadogan, 1974), attributed to 8% radiogenic Ar loss from K-rich sites with low Ar retentivity. However, other samples showed either a sharp decrease in apparent age in the

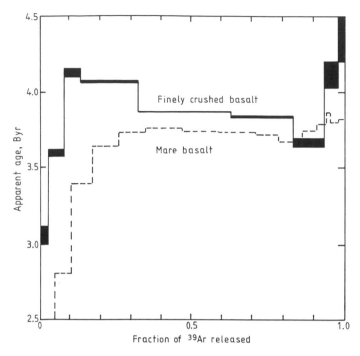

Fig. 10.19. The effect of fine crushing on a 40–39 age spectrum, due to ^{39}Ar recoil. Dashed profile = analysed rock chip of a lunar mare basalt. Solid profile = similar sample activated after fine powdering. After Turner and Cadogan (1974).

high-temperature fractions, or, particularly in fine-grained rocks, a progressive decrease in apparent age over most of the gas release. The latter examples led workers to suspect Ar redistribution within the sample, possibly during the irradiation process.

It was proposed by Mitchell (in Turner and Cadogan, 1974) that recoil of ^{39}Ar during the n,p reaction from ^{39}K could cause small-scale redistribution of this nuclide. Turner and Cadogan calculated that this effect could deplete argon from the surface of a K-bearing mineral to a mean depth of 0.08 μm (Fig. 10.20). In order to test the practical effects of this process on fine-grained material, they powdered a sample of medium-grained ferrobasalt to a grain size of 1–10 μm before irradiation. This was thought to bring ca. 10% of K-bearing lattice sites to within 0.1 μm of a grain boundary, whereupon ^{39}Ar could recoil out of the lattice. It was anticipated that the ^{39}Ar released would enter low-K minerals such as plagioclase, pyroxene and ilmenite, leading to an old apparent age during low-temperature release (K-bearing minerals) and a young apparent age during high-temperature release.

Results of this experiment (Fig. 10.19) show that while abnormally old ages were produced at low temperature, the data approached the 'true' plateau age at intermediate temperatures. Turner and Cadogan argued that ^{39}Ar released by recoil must have been lost from the sample altogether, rather than absorbed by low-K phases. This is probably due to the fact that adjacent grains are in less intimate contact in a powdered sample than a fine-grained rock sample. The unusually high ages in the highest temperature fraction (Fig. 10.19) were tentatively attributed to an incorrect Ca correction, due to ^{37}Ar recoil during the n, α reaction from ^{40}Ca. This transformation should result in four times more recoil than proton emission from ^{39}Ar.

The problem of ^{39}Ar recoil was found to be particularly severe in attempts to apply ^{40}Ar–^{39}Ar dating to the authigenic sedimentary mineral glauconite (e.g. Foland et al., 1984). This is probably due to the very small grain size of the glauconite crystallites which make up the grains of a pellet. However, Smith et al. (1993) showed that this problem might be overcome by encapsulating glauconite grains in small glass ampoules prior to irradiation. The recoil products can then be collected for analysis, in order to correct the Ar release from the rest of the grain.

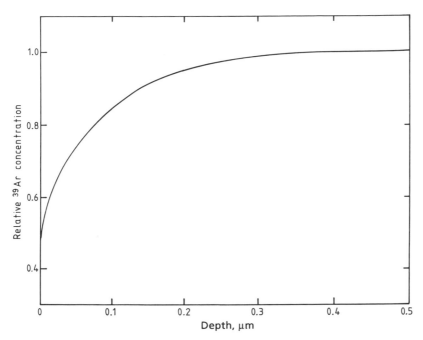

Fig. 10.20. Plot showing calculated drop in ^{39}Ar concentration at the surface of a K-bearing mineral due to recoil, in response to bombardment with an isotropic neutron flux. After Turner and Cadogan (1974).

10.2.6 Dating paleomagnetism

Paleomagnetic measurements are a vital tool in the reconstruction of ancient plate tectonic motions, by comparison of 'apparent polar wander paths' (APWPs) for different continental fragments. One essential step in the construction of an APWP 'track' for a given continent is the dating of the time when magnetic remanence was inherited by the rock. However, the magnetic remanence is relatively easily overprinted because it has a comparatively low blocking temperature.

The dating of magnetic remanence took a major step forward when York (1978) showed from theoretical principles that the processes of thermal de-magnetisation and argon loss from a mineral grain were related. This is because they are both almost exclusively the products of thermal kinetics, in contrast to ^{87}Sr loss (for example) which may be dependent on the presence or absence of aqueous fluids. Hence, the 40–39 method is an ideal tool to date paleomagnetic remanence.

A good example of such work is provided by the oldest reliable Ar–Ar ages for terrestrial rocks (Lopez Martinez *et al.*, 1984), on the Barberton komatiites. Although the time of eruption was constrained by Sm–Nd dating, knowledge of their subsequent thermal history was required to interpret the paleomagnetic intensity study of Hale (1987).

Analyses were performed on whole-rock powders, which were irradiated alongside the '3 GR' hornblende standard to determine neutron fluxes. Fig. 10.21 shows age spectrum data from the best sample analysed. The spectrum consists of three separate sections. At low temperatures (600–800 °C) and high temperatures (>1100 °C) argon loss was observed, resulting in low ages. However, at intermediate temperatures (925–1035 °C) a very stable plateau was observed, from which a best age of 3486 ± 6 Myr (2σ) was obtained. The integrated (total fusion) age of 3336 Myr was significantly younger, due to the effects of the low- and high-temperature steps.

The top half of Fig. 10.21 reports Ca/K ratios, calculated from the measured ^{37}Ar/^{39}Ar ratio (section 10.2.2), which characterise the argon reservoir sampled by each gas release. This helps to identify the mineral phases in the sample which gave rise to different parts of the age spectrum. By microprobe analysis, the authors were able to deduce that the mineral giving rise to the age plateau was metamorphic tremolite, while the low-temperature, low-Ca/K phase was stilpnomelane. The high-Ca/K phase

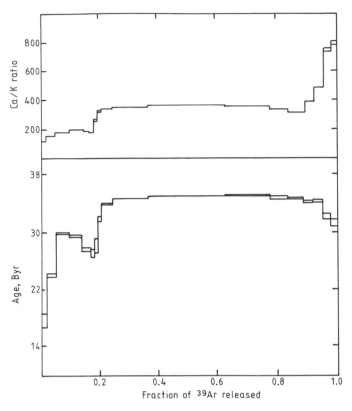

Fig. 10.21. Age and Ca/K spectra from Barberton komatiite sample B40A. Error boxes where visible are 1σ. After Lopez Martinez *et al.* (1984).

may represent pyroxene relics of the original igneous mineralogy.

A K–Ar isochron diagram was plotted (Fig. 10.22) in order to examine the composition of the non-radiogenic end-member, and test for inherited argon. In this case the isochron diagram was plotted in the alternative form $^{36}Ar/^{40}Ar$ *versus* $^{39}Ar/^{40}Ar$ (Turner, 1971b). This representation helps to curtail the strong correlation between the two ordinates which occurs with the conventional K–Ar isochron diagram, making error estimates easier. An initial $^{40}Ar/^{36}Ar$ ratio of 281 ± 18 (2σ) is calculated from the inverse of the y axis intercept, after expansion of analytical errors to absorb a small amount of geological scatter. This is within error of the atmospheric value of 295.5, so insignificant initial argon was probably present. These data are from a sample which was stored in vacuum between irradiation and analysis. This was found to be necessary to prevent a strong absorption of atmospheric argon by the sample. The reciprocal of the x intercept yields the radiogenic $^{40}Ar/^{39}Ar$ ratio, equivalent to an age of 3489 ± 68 Myr (2σ). This is

almost identical to the plateau age.

Age spectrum results from the better of two basaltic komatiite samples are shown in Fig. 10.23. In contrast to the komatiites, these samples display significant excess argon in the low- and high-temperature gas releases, with an integrated age of 3778 Myr. Nevertheless, the best plateau age of 3447 Myr is in close agreement with the best komatiite results. The saddle-shaped form is well known in samples containing excess argon.

The Ca/K plot for the basaltic komatiite suggests that the plateau is due to hornblende, while the disturbed parts of the spectrum are again related to stilpnomelane. Lopez Martinez *et al.* speculated that K–Ar systematics in this mineral might have been disturbed during oxidation from ferro- to ferric-stilpnomelane. Since the plateau ages are in all cases identified with metamorphic minerals, they must be dating a thermal event which occurred less than 100 Myr after eruption. Hale (1987) has tentatively identified this event as the intrusion of the nearby Threespruit granitoid pluton.

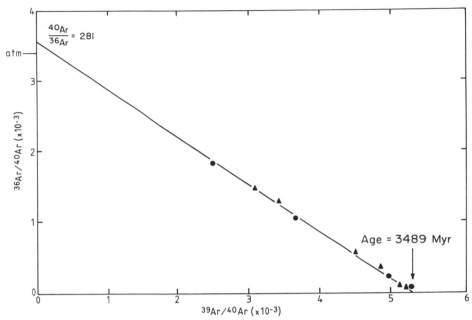

Fig. 10.22. Argon–argon isotope plot for plateau segments of two 40–39 runs (●, ▲) on komatiite B40A. After Lopez Martinez *et al.* (1984).

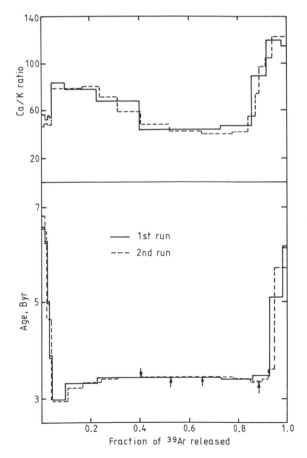

Fig. 10.23. Age and Ca/K spectra for two runs on a basaltic komatiite showing low- and high-temperature emission from high Ca/K domains. Arrows separate successive gas releases with identical ages. After Lopez Martinez *et al.* (1984).

10.2.7 Thermochronometry

The thermal history of meteorites is best interpreted in terms of short-lived thermal events, such as their initial cooling and any subsequent collisions. However, Turner (1969) recognised that in different circumstances, slow cooling from a single event could yield an age spectrum rather similar to that produced by episodic thermal events. The concept is illustrated by the age spectrum of a biotite from the La Encrucijada pluton in Venezuela (Fig. 10.24). Unfortunately, this approach can rarely be used on terrestrial rocks because the low-temperature part of the profile, which is critical in the determination of a precise cooling rate, becomes 'corrupted' by minor diffusional loss of argon at ambient temperatures over geological time.

A more useful approach to quantify the cooling history of crustal rocks is the blocking temperature concept (section 3.3.2), whose theoretical basis was examined by Dodson (1973). Argon loss from a mineral can be described by the thermal diffusion

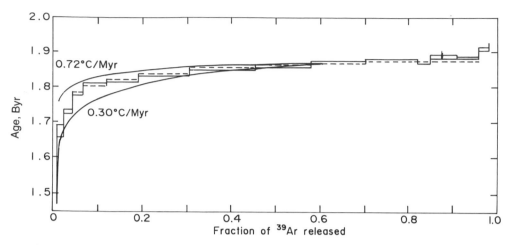

Fig. 10.24. ^{40}Ar/^{39}Ar age spectrum of La Encrucijada biotite, Venezuela, compared to predicted cooling curves based on modelling of Ar diffusion in biotite. After York (1984).

coefficient:

$$D = D_0\, e^{-E/RT} \qquad [10.13]$$

where D_0 is the thermal diffusivity of the mineral, E is the activation energy of argon diffusion, R is the gas constant and T is absolute temperature. The exponent causes D to be a very strong function of temperature. Therefore a small temperature drop can cause a transition from a state where Ar loss by diffusion is rapid to a state where Ar loss by diffusion is very slow. This relatively sharp transition constitutes the process of blocking. The blocking temperature T_B is defined by Dodson (1973) as follows:

$$T_B = \frac{E}{R\,\ln(A\tau D_0/a^2)} \qquad [10.14]$$

where A is a geometrical parameter which takes account of the crystal form of the argon-bearing mineral (55, 27 or 9 for a sphere, cylinder or sheet respectively), a is the length of the average diffusion pathway from the interior to the surface of the grain, and τ is the cooling time constant. The latter is in turn defined as:

$$\tau = RT_B^2/E(-C)_B \qquad [10.15]$$

where $(-C)_B$ is the cooling rate at the blocking temperature T_B. Hence, substituting equation [10.15] into [10.14] yields:

$$T_B R = E/\ln\left[A \cdot \frac{RT_B^2}{E(-C)_B} \cdot \frac{D_0}{a^2}\right] \qquad [10.16]$$

A method to calculate blocking temperatures from Ar–Ar spectrum plots was proposed by Buchan *et al.*

(1977) and developed by Berger and York (1981a). A plateau age must be available on a mineral from a slowly cooled terrane. For each heating step in the plateau, the volume of radiogenic ^{40}Ar released in a given time is used to calculate D/a^2. For planar minerals such as biotite the diffusion equation has the following general form (e.g. Harrison and McDougall, 1981):

$$\frac{D}{a^2} = \frac{(qf)^2}{t} \qquad [10.17]$$

where f is the fractional loss of argon, t is the heating time, and q is a geometric factor.

The results are plotted on a log scale against the reciprocal temperature of each step, forming an Arrhenius plot (e.g. Fig. 10.25). If diffusional Ar loss obeys the Arrhenius law as expected then the steps in the plateau should define a straight line whose slope is the activation energy E and whose y intercept is the frequency factor D_0/a^2. These values enable equation [10.16] to be solved, provided the cooling rate $(-C)_B$ at the blocking temperature (T_B) can be estimated. Fortunately, the temperature solution has a weak dependency on cooling rate, such that an order of magnitude change in this value only causes a 10% change in the calculated blocking temperature. Because T_B appears on both sides of equation [10.16], it must be solved iteratively, but it converges quickly. The power of this technique is that the mineral blocking temperature is calculated directly on the dated material, rather than having to depend on generalised blocking temperatures for different mineral types from the literature, which may not be

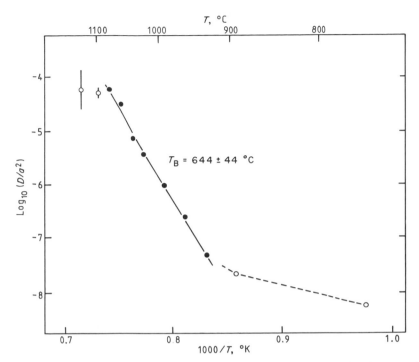

Fig. 10.25. Arrhenius plot for hornblende from a Grenville diorite, Haliburton Highlands, Ontario. The blocking temperature was determined from the array of 7 solid data points. Error is 1σ. After Berger and York (1981a).

applicable to the specific cooling conditions under study.

Berger and York (1981a) applied this 'thermochronometry' method to a study of post-orogenic cooling of the Grenville Province of southern Ontario, Canada. Plutonic ages in the Grenville belt vary from 1.0 to 2.7 Byr, but most K–Ar dates fall below 1 Byr, and are attributed to uplift and cooling after collisional orogeny (Harper, 1967). Berger and York studied dioritic and gabbroic plutons from the Haliburton Highlands, both to determine a detailed cooling curve for the area and to interpret paleomagnetic data on these rocks.

Typical 40–39 profiles from Haliburton diorites which yielded reasonable plateaus are shown in Fig. 10.26. These samples are plotted on Arrhenius plots in Figs. 10.25 and 10.27. The hornblende displays a relatively simple array in Fig. 10.25, although low-temperature and high-temperature points must be excluded from the regression. K-feldspar displays coherent low-temperature behaviour, but high-temperature data are irregular, possibly due to disruption of the lattice above 900 °C. The most unusual behaviour was demonstrated by biotite, which in many cases gave rise to two heating pulses which

defined sub-parallel arrays (such as in Fig. 10.24). Nevertheless, the blocking temperatures calculated from the two segments were usually within error. Berger and York speculated that the break in regular behaviour was due to structural breakdown.

Results from all of the analysed minerals are shown on a diagram of blocking temperature against age (Fig. 10.28). Points without error bars failed a reliability criterion which required both the age plateau and Arrhenius correlation line to have four or five statistically well-fitting data points. The data show a clear picture of fairly rapid cooling (ca. 5 °C/ Myr) from the hornblende blocking temperature of ca. 700 °C at 980 Myr to the biotite blocking temperature of ca. 380 °C at 900 Myr. Thereafter, the data are mainly from plagioclase, which displays considerable scatter. Berger and York's original interpretation (solid line in Fig. 10.28) called for very slow cooling (under 1 °C/Myr) for a further 300 Myr. However, in a study of gabbro from the Hastings Basin of the Grenville (ca. 80 km east of the Haliburton Highlands), Berger and York (1981b) recognised that the apparent slow cooling curve after 900 Myr might really represent a more recent thermal event.

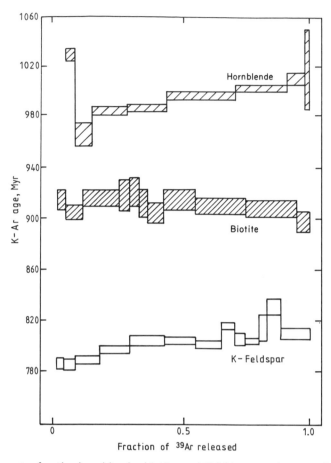

Fig. 10.26. 40–39 age spectra for the hornblende, biotite and K-feldspar analyses used to determine blocking temperatures in Figs. 10.25 and 10.27. Modified after Berger and York (1981a).

The latter interpretation of the plagioclase data was supported by Hanes *et al.* (1988) on the basis of 40–39 dating of the Elzevir and Skootamata plutons of the Hastings basin. Hanes *et al.* analysed three plagioclases which displayed a range of rather mediocre plateau ages between 400 and 600 Myr. Variations in ^{37}Ar/^{39}Ar with the fraction of ^{39}Ar released were used as an index of the Ca/K ratio of different domains within the minerals. A pronounced hump in the middle of these profiles indicated that the analysed plagioclases were multi-phase systems. They display two different types of alteration which may be of different age. Hanes *et al.* suggested that scattered coarse epidote and muscovite alteration might have formed at high temperatures soon after plutonism, while fine-grained sericitic alteration probably represents an event younger than 400 Myr.

The evidence for structural breakdown in biotite points to a possible weakness in the thermochrono-

metry method of Berger and York. Because biotite is a hydrous mineral, diffusional loss of Ar during vacuum heating may not accurately mimic Ar loss in nature under (probable) hydrothermal conditions, as suggested by Giletti (1974). This problem was confirmed by Gaber *et al.* (1988), who showed large divergences in argon diffusivity between hydrothermal and vacuum heating experiments on biotite and hornblende.

The susceptibility of these hydrous minerals to structural breakdown during vacuum heating has largely discredited the application of step heating analysis to determine their blocking temperatures. Instead, there has been a return to the use of 'literature' values for blocking temperature. However, these must be used with caution, since an experimental study by Harrison *et al.* (1985) showed that biotite blocking temperatures are compositionally dependent. For the case of 56% annite, this study

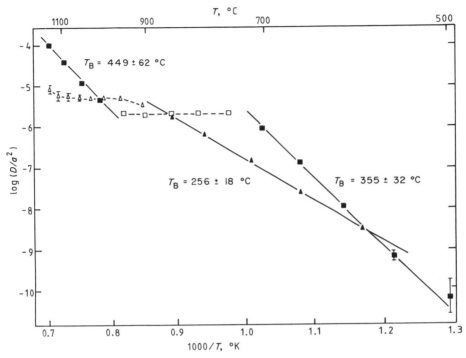

Fig. 10.27. Arrhenius plot for Grenville K-feldspar (▲, △) and biotite (■, ⬜) dated in Fig. 10.26. Blocking temperature calculations were based on solid data points only. Errors are 1σ. After Berger and York (1981a).

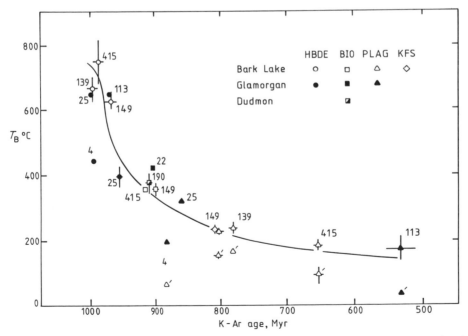

Fig. 10.28. Plot of calculated mineral blocking temperatures from 40–39 data against plateau ages by this method to show a model for crustal cooling after the Grenville orogeny. Solid and open symbols indicate minerals from different plutons. After Berger and York (1981a).

indicated blocking temperatures of 350 °C and 310 °C respectively for cooling rates of 100 °C/Myr and 10 °C/Myr. More iron-rich compositions exhibit greater diffusivities, leading to lower blocking temperatures.

10.2.8 K-feldspar thermochronometry

In contrast to the failure of the thermochronometry method in hydrous minerals, increasing attention has been focussed on the anhydrous mineral K-feldspar. Because of the unpredictable effects of perthitic exsolution, K-feldspars have very variable blocking temperatures, which must be determined for individual dated samples by the step heating method. Heizler *et al.* (1988) demonstrated this technique in determining the cooling curve of the Chain of Ponds pluton, NW Maine. They obtained quite distinct blocking temperatures and ages on three feldspar separates, establishing a post-Appalachian cooling curve from 330 °C to 180 °C (solid symbols in Fig. 10.29). However, other workers (e.g. Foland, 1974) have suggested that diffusion in K-feldspars is controlled by microstructural domains of variable size, rather than on a whole-grain scale. This would involve a different treatment of the step heating data.

Adopting the micro-domain model, Lovera *et al.*

(1989) reinterpreted the data of Heizler *et al.* by breaking each Ar–Ar analysis into a series of sub-plateaus with distinct ages and blocking temperatures. They attributed these sub-plateaus to diffusional domains within each feldspar grain, varying in size by two orders of magnitude. This model was tested by comparing measured step heating data with model spectra based on different domain size-distributions, as applied by Turner *et al.* (1966) to meteorite studies (section 10.2.4). The variable domain-size model was shown to fit the experimental data much better than the uniform model, thus confirming its usefulness. The result of this approach was that each analysis yielded a separate but overlapping cooling curve *segment* (open symbols in Fig. 10.29), rather than a single point on the cooling curve. Further experiments on single feldspar crystals by Lovera et al. (1991) showed that domains of varying size are an intrinsic property of alkali feldspars, which therefore cannot be separated by hand-picking of material for analysis.

The very variable domain size of K-feldspars explains their reputation in K–Ar studies of having such poor argon retentivity as to be useless as a dating tool. The smallest of the domains do indeed suffer argon loss at near ambient temperatures, but the larger domains, sampled in a step heating analysis, can have blocking temperatures as high as biotite. Hence, it is concluded that K-feldspar analysis has a lot of potential in the future for low-temperature thermochronometry (Hanes, 1991). An example of this type of application is in studying the history of sedimentary basins (e.g. Harrison and Burke, 1989).

10.3 Laser probe dating

The application of the laser probe to K–Ar dating is now becoming an important technique, but surprisingly, the method was slow in development. Megrue (1967) pioneered the use of laser ablation for rare gas analysis, but did not apply the method to geochronology until six years later (Megrue, 1973). This study made use of the laser probe in order to date small clasts in a polymict lunar breccia. After activation, spots 100 μm in diameter were irradiated with single pulses from a ruby laser. Each pulse ablated a pit about 30 μm deep, equivalent to about 1 μg of rock, representing a miniature total-fusion analysis of the exposed surface. The aggregate gas fraction from several nearby spots was gettered and cryogenically trapped, before admission to the mass spectrometer for analysis. (Typical equipment is shown in Fig. 10.30.) Analysis of ten different clasts revealed two

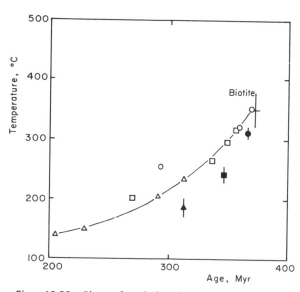

Fig. 10.29. Plot of calculated K-feldspar blocking temperature against plateau age to model cooling of the Chain of Ponds pluton, NW Maine. Solid symbols show the single-plateau analysis of Heizler *et al.* (assuming uniform domain size). Open symbols show the sub-plateau analysis of Lovera *et al.* (assuming variable domain size). After Harrison (1990).

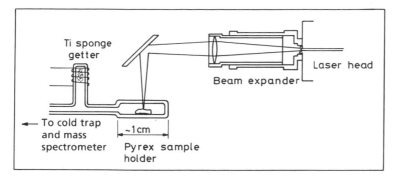

Fig. 10.30. Schematic illustration of laser ablation Ar–Ar dating equipment. After York *et al.* (1981).

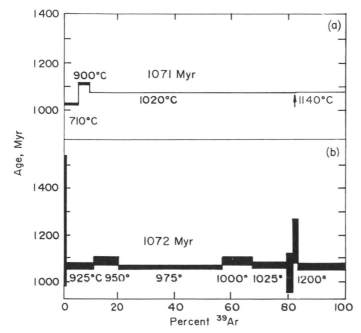

Fig. 10.31. Step heating results for the Hb3GR standard. a) Conventional; b) laser single grain. Quoted ages are average (integrated release) ages. After Layer *et al.* (1987).

arrays of data on a K–Ar isochron diagram with ages of approximately 3.7 and 2.9 Byr.

York *et al.* (1981) developed the laser microprobe technique by showing that a defocussed continuous wave laser could be used to perform step heating analysis in a manner analogous to conventional 40–39 dating. The technique was demonstrated on a whole-rock sample of slate from the Kidd Creek mine, near Timmins, Ontario. The laser beam was focussed to generate a spot 0.6 mm in diameter, which caused progressive argon release from the surface after a few minutes, using a 1 watt power setting. The laser step heating analysis produced results consistent with

conventional step heating of the same sample, representing the timing of a thermal event which opened the K–Ar system in the slate.

The low sensitivity of the MS–10 mass spectrometer used by York *et al.* (1981) limited application of the method, but in subsequent development a purpose-built continuous laser system was coupled to a high-sensitivity mass spectrometer. Layer *et al.* (1987) tested this system by analysing the hornblende standard Hb3GR. This is known from previous step heating analysis (Turner, 1971a) to yield a perfect plateau age (Fig. 10.31a). After activation, single grains up to 0.5 mm across were heated within the laser beam for 30 seconds at increasing power levels. After each heating episode, argon was gettered and then analysed. Excellent plateaus were generated (e.g. Fig. 10.31b), and the integrated release ages fell within error of the conventional step heating result.

In order to test the laser step heating method on a slowly-cooled geological system, Layer *et al.* analysed biotites from the Trout Lake batholith, NW Ontario. Laser step heating of a small (0.25 mm) biotite grain yielded a result (not shown) which was identical to a

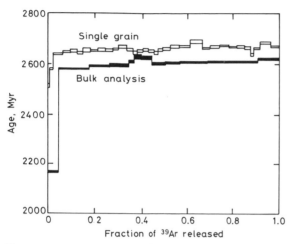

Fig. 10.32. Laser single grain and conventional step heating results for biotites from the Trout Lake batholith, Ontario. After Layer *et al.* (1987).

conventional step heating analysis on 13 mg of biotite. However, laser step heating of a large (1 mm) biotite from the same hand specimen yielded a significantly older age (Fig. 10.32). The total release age for this grain (2654 ± 5 Myr) was closer to the U–Pb intrusive age for the batholith of 2699 ± 2 Myr. This shows that the larger biotites probably closed earlier during metamorphic cooling.

Wright *et al.* (1991) developed this study on the Trout Lake batholith by using the laser step heating method to date a range of single biotite grains of different sizes. Only grains having a regular shape similar to a thin cylinder were analysed. After measurement of the grain radii and activation in the reactor, each specimen was subjected to laser step heating analysis, during which the whole grain was bathed in the laser beam at increasing intensities. For samples displaying normal plateaus, their integrated ages were plotted against grain size (Fig. 10.33). The results display a positive correlation between grain size (cylindrical radius) and integrated age. This was attributed to diffusional argon loss during the original cooling history of the batholith. Wright *et al.* speculated that the scatter of data points to the right of the main array might represent large grains which were either damaged during sample crushing or consisted of natural aggregates of smaller sub-domains.

The positive correlation in Fig. 10.33 is explained

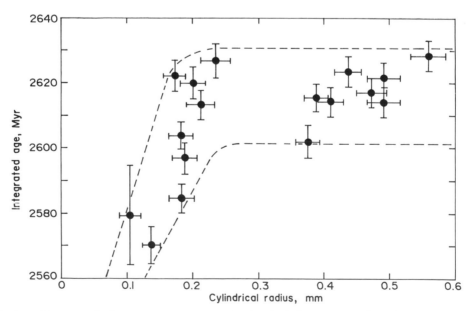

Fig. 10.33. Plot of integrated release age against grain size for single biotite grains from the Trout Lake batholith, Ontario. After Wright *et al.* (1991).

by the larger surface area / volume ratio of smaller grains, resulting in a lower effective blocking temperature than larger grains. Geological determinations of Ar diffusion in biotite (e.g. Onstott *et al.*, 1989) can be used to calculate the size-dependence of blocking temperature (0.1 mm = 275 °C; 0.23 mm = 295 °C). Hence, the array in Fig. 10.33 translates into a cooling curve for the batholith of temperature against time. This yields a calculated cooling rate between 295 and 275 °C of about 0.33 °C/Myr. The high-temperature cooling curve of the pluton can be calculated between the older biotite ages and the U–Pb zircon age of 2700 Myr (with a blocking temperature estimated at around 750 °C). This segment of the cooling curve is much steeper, at around 5 °C/Myr.

Lee *et al.* (1990) tested the laser step heating method on biotite and hornblende grains which had suffered a thermal disturbance long after initial cooling. The sample consisted of baked Archean gneiss adjacent to an Early Proterozoic dyke, and both minerals were analysed by three methods: conventional step heating; single grain laser step heating; and laser spot dating. Biotite ages for the three methods clustered closely around 2050 Myr, interpreted as the time of dyke intrusion. On the other hand, the three techniques produced very different results for hornblende. Conventional step heating of a multi-grain population and laser spot dating generated very variable ages (Fig. 10.34), while laser step heating generated a good plateau, with an apparent age of 2430 Myr. However, this does not correspond to a known geological event.

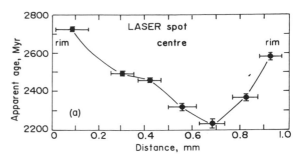

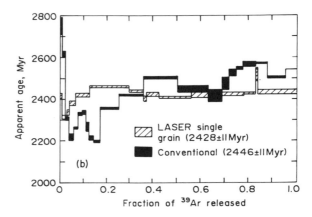

Fig. 10.34. Comparison of spot and step heating ages for a disturbed hornblende sample. a) Profile of laser spot ages across a single grain; b) laser and conventional step heating profiles. After Lee *et al.* (1990; 1991).

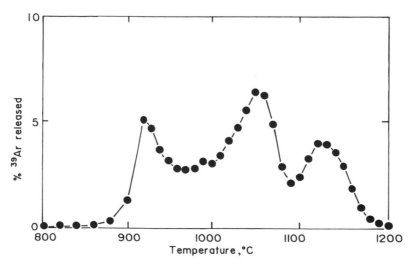

Fig. 10.35. Argon release pattern observed in response to heating of the hornblende standard Mmhb–1. After Lee *et al.* (1991).

Lee *et al.* (1991) speculated that the plateau could result from mixing of argon from different domains in the mineral before release. Heating experiments on the hornblende standard Mmhb–1 showed that argon was released in three principal pulses (Fig. 10.35). The first of these, at 930 °C, was correlated with the onset of structural breakdown at the margins of grains. However, the main phase of breakdown occurred at 1050 °C, forming a strong fabric parallel to cleavage and accompanied by the breakdown of titanite lamellae in the crystal. Finally, at 1130 °C the grains melted. The laser step heating plateau in Fig. 10.34 was formed by argon release between 960 and 1250 °C, suggesting that it may result from argon homogenisation in the grain during structural breakdown. Therefore, although laser step heating is a very powerful technique, it is desirable to check data from disturbed systems by a second technique such as laser spot dating.

References

Anders, E. (1964). Meteorite ages. *Rev. Mod. Phys.* **34**, 287–325.

Beckinsale, R. D. and Gale, N. H. (1969). A reappraisal of the decay constants and branching ratio of ^{40}K. *Earth Planet. Sci. Lett.* **6**, 289–94.

Berger, G. W. (1975). ^{40}Ar/^{39}Ar step heating of thermally overprinted biotite, hornblende and potassium feldspar from Eldora, Colorado. *Earth Planet. Sci. Lett.* **26**, 387–408.

Berger, G. W. and York. D. (1981a). Geothermometry from ^{40}Ar/^{39}Ar dating experiments. *Geochim. Cosmochim. Acta* **45**, 795–811.

Berger, G. W. and York. D. (1981b). ^{40}Ar/^{39}Ar dating of the Thanet gabbro, Ontario: looking through the metamorphic veil and implications for paleomagnetism. *Can. J. Earth Sci.* **18**, 266–73.

Berggren, W. A. (1972). A Cenozoic time-scale – some implications for regional geology and paleobiogeography. *Lethaia* **5**, 195–215.

Berggren, W. A., Kent, D. V. and Flynn, J. J. (1985). Jurassic to Paleogene: Part 2. Paleogene geochronology and chronostratigraphy. In: Snelling, N. J. (Ed.), *The Chronology of the Geological Record. Geol. Soc. Mem.* **10**, Blackwell, pp. 141–95.

Brereton, N. R. (1970). Corrections for interfering isotopes in the ^{40}Ar/^{39}Ar dating method. *Earth Planet. Sci. Lett.* **8**, 427–33.

Buchan, K. L., Berger, G. W., McWilliams, M. O., York, D. and Dunlop, D. J. (1977). Thermal overprinting of natural remanent magnetization and K/Ar ages in metamorphic rocks. *J. Geomag. Geoelectr.* **29**, 401–10.

Cox, A. and Dalrymple, G. B. (1967). Statistical analysis of geomagnetic reversal data and the precision of potassium–argon dating. *J. Geophys. Res.* **72**, 2603–14.

Cox, A., Doell, R. R. and Dalrymple, G. B. (1963). Geomagnetic polarity epochs and Pleistocene geochronology. *Nature* **198**, 1049–51.

Dalrymple, G. B. and Lanphere, M. A. (1969). *Potassium–Argon Dating.* Freeman, 258 p.

Dalrymple, G. B. and Lanphere, M. A. (1971). ^{40}Ar/^{39}Ar technique of K–Ar dating: a comparison with the conventional technique. *Earth Planet. Sci. Lett.* **12**, 300–8.

Dalrymple, G. B. and Lanphere, M. A. (1974). ^{40}Ar/^{39}Ar age spectra of some undisturbed terrestrial samples. *Geochim. Cosmochim. Acta* **38**, 715–38.

Dalrymple, G. B. and Moore, J. G. (1968). Argon 40: excess in submarine pillow basalts from Kilauea Volcano, Hawaii. *Science* **161**, 1132–5.

Damon, P. E. and Kulp, L. (1958). Excess helium and argon in beryl and other minerals. *Amer. Miner.* **43**, 433–59.

Dodson, M. H. (1973). Closure temperature in cooling geochronological and petrological systems. *Contrib. Mineral. Petrol.* **40**, 259–74.

Doell, R. R., Dalrymple, G. B. and Cox, A. (1966). Geomagnetic polarity epochs: Sierra Nevada data, 3. *J. Geophys. Res.* **71**, 531–41.

Foland, K. A. (1974). Ar40 diffusion in homogeneous orthoclase and an interpretation of Ar diffusion in K-feldspar. *Geochim. Cosmochim. Acta* **38**, 151–66.

Foland, K. A., Linder, J. S., Laskowski, T. E. and Grant, K. (1984). ^{40}Ar–^{39}Ar dating of glauconies: measured ^{39}Ar recoil loss from well-crystallized specimens. *Chem. Geol. (Isot. Geosci. Section)* **46**, 241–64.

Gaber, L. J., Foland, K. A. and Corbato, C. E. (1988). On the significance of argon release from biotite and amphibole during ^{40}Ar/^{39}Ar vacuum heating. *Geochim. Cosmochim. Acta* **52**, 2457–65.

Garner, E. L., Machlan, L. A. and Barnes, I. L. (1976). The isotopic composition of lithium, potassium, and rubidium in some Apollo 11, 12, 14, 15, and 16 samples. *Proc. 6th Lunar Sci. Conf.* Pergamon, pp. 1845–55.

Giletti, B. J. (1974). Diffusion related to geochronology. In: Hofmann, A. W., Giletti, B. J., Yoder, H. S. and Yund, R. A. (Eds.), *Geochemical Transport and Kinetics.* Carnegie Inst. Wash., pp. 61–76.

Hale, C. J. (1987). The intensity of the geomagnetic field at 3.5 Ga: paleointensity results from the Komati Formation, Barberton Mountain Land, South Africa. *Earth Planet. Sci. Lett.* **86**, 354–64.

Hanes, J. A. (1991). K–Ar and ^{40}Ar/^{39}Ar geochronology: methods and applications. In: Heaman, L. and Ludden, J. N. (Eds.), *Short Course Handbook on Applications of Radiogenic Isotope Systems to Problems in Geology.* Min. Assoc. Canada, pp. 27–57.

Hanes, J. A., Clark, S. J. and Archibald, D. A. (1988). An ^{40}Ar/^{39}Ar geochronological study of the Elzevir batholith and its bearing on the tectonothermal history of the southwestern Grenville Province, Canada. *Can. J. Earth Sci.* **25**, 1834–45.

Harland, W. B., Cox, A. V., Llewellyn, P. G., Pickton, C. A. G., Smith, A. G. and Walters, R. (1982). *A Geologic Time Scale 1982.* Cambridge Univ. Press., 131 pp.

Harper, C. T. (1967). On the interpretation of potassium–

argon ages from Precambrian shields and Phanerozoic orogens. *Earth Planet. Sci. Lett.* **3**, 128–32.

Harrison, T. M. (1990). Some observations on the interpretation of feldspar ^{40}Ar/^{39}Ar results. *Chem. Geol. (Isot. Geosci. Section)* **80**, 219–29.

Harrison, T. M. and Burke, K. (1989). ^{40}Ar/^{39}Ar thermochronology of sedimentary basins using detrital feldspars: examples from the San Joaquin Valley, California, Rio Grande Rift, New Mexico, and North Sea. In: Naeser, N. D. and McCulloh, T. H. (Eds.), *Thermal History of Sedimentary Basins*, Springer-Verlag, pp. 141–55.

Harrison, T. M., Duncan, I. and McDougall, I. (1985). Diffusion of ^{40}Ar in biotite: Temperature, pressure and compositional effects. *Geochim. Cosmochim. Acta* **49**, 2461–8.

Harrison, T. M. and McDougall, I. (1981). Excess ^{40}Ar in metamorphic rocks from Broken Hill, New South Wales: implications for ^{40}Ar/^{39}Ar age spectra and the thermal history of the region. *Earth Planet. Sci. Lett.* **55**, 123–49.

Hart, S. R. (1964). The petrology and isotopic–mineral age relations of a contact zone in the Front Range, Colorado. *J. Geol.* **72**, 493–525.

Hart, S. R. and Dodd, R. T. (1962). Excess radiogenic argon in pyroxenes. *J. Geophys. Res.* **67**, 2998–9.

Heirtzler, J. R., Dickson, G. O., Herron, E. M., Pitman, W. C. and LePichon, X. (1968) Marine magnetic anomalies, geomagnetic field reversals, and motions of the ocean floor and continents. *J. Geophys. Res.* **73**, 2119–36.

Heizler, M. T., Lux, D. R. and Decker, E. R. (1988). The age and cooling history of the Chain of Ponds and Big Island Pond plutons and the Spider Lake granite, west-central Maine and Quebec. *Amer. J. Sci.* **288**, 925–52.

LaBrecque, J. L., Kent, D. V. and Cande, S. C. (1977). Revised magnetic polarity time scale for Late Cretaceous and Cenozoic time. *Geology* **5**, 330–5.

Lanphere, M. A. and Dalrymple, G. B. (1966). Simplified bulb tracer system for argon analysis. *Nature* **209**, 902–3.

Lanphere, M. A. and Dalrymple, G. B. (1971). A test of the ^{40}Ar/^{39}Ar age spectrum technique on some terrestrial materials. *Earth Planet. Sci. Lett.* **12**, 359–72.

Lanphere, M. A. and Dalrymple, G. B. (1976). Identification of excess ^{40}Ar by the ^{40}Ar/^{39}Ar age spectrum technique. *Earth Planet. Sci. Lett.* **32**, 141–8.

Layer, P. W., Hall, C. M. and York, D. (1987). The derivation of ^{40}Ar/^{39}Ar age spectra of single grains of hornblende and biotite by laser step-heating. *Geophys. Res. Lett.* **14**, 757–60.

Lee, J. K. W., Onstott, T. C., Cashman, K. V., Cumbest, R. J. and Johnson, D. (1991). Incremental heating of hornblende in vacuo: implications for ^{40}Ar/^{39}Ar geochronology and the interpretation of thermal histories. *Geology* **19**, 872–6.

Lee, J. K. W., Onstott, T. C. and Hanes, J. A. (1990). An ^{40}Ar/^{39}Ar investigation of the contact effects of a dyke intrusion, Kapuskasing Structural Zone, Ontario. *Contrib. Mineral. Petrol.* **105**, 87–105.

Lopez Martinez, M., York, D., Hall, C. M. and Hanes, J. A. (1984). Oldest reliable ^{40}Ar/^{39}Ar ages for terrestrial rocks: Barberton Mountain komatiites. *Nature* **307**, 352–4.

Lovera, O. M., Richter, F. M. and Harrison, T. M. (1989). The ^{40}Ar/^{39}Ar thermochronometry for slowly-cooled samples having a distribution of domain sizes. *J. Geophys. Res.* **94**, 17917–35.

Lovera, O. M., Richter, F. M. and Harrison, T. M. (1991). Diffusion domains determined by ^{39}Ar released during step heating. *J. Geophys. Res.* **96**, 2057–69.

Lowrie, W. and Alvarez, W. (1981). One hundred million years of geomagnetic polarity history. *Geology* **9**, 392–7.

McDougall, I., Polach, H. A. and Stipp, J. J. (1969). Excess radiogenic argon in young subaerial basalts from the Auckland volcanic field, New Zealand. *Geochim. Cosmochim. Acta* **33**, 1485–520.

McDougall, I. and Tarling, D. H. (1964). Dating geomagnetic polarity zones. *Nature* **202**, 171–2.

McIntosh, W. C., Geissmen, J. W., Chapin, C. E., Kunk, M. J. and Henry, C. D. (1992). Calibration of the latest Eocene–Oligocene geomagnetic polarity time scale using ^{40}Ar/^{39}Ar dated ignimbrites. *Geology* **20**, 459–63.

Mankinen, E. A. and Dalrymple, G. B. (1979). Revised geomagnetic polarity time scale for the interval 0 to 5 m.y. B.P. *J. Geophys. Res.* **84**, 615–26.

Megrue, G. H. (1967). Isotopic analysis of rare gases with a laser microprobe. *Science* **157**, 1555–6.

Megrue, G. H. (1973). Spatial distribution of ^{40}Ar/^{39}Ar ages in lunar breccia 14301. *J. Geophys. Res.* **78**, 3216–21.

Merrihue, C. and Turner, G. (1966). Potassium–argon dating by activation with fast neutrons. *J. Geophys. Res.* **71**, 2852–7.

Mitchell, J. G. (1968). The argon-40/argon-39 method for potassium–argon age determination. *Geochim. Cosmochim. Acta* **32**, 781–90.

Mussett, A. E. and Dalrymple, G. B. (1968). An investigation of the source of air Ar contamination in K–Ar dating. *Earth Planet. Sci. Lett.* **4**, 422–6.

Ness, G., Levi, S. and Couch, R. (1980). Marine magnetic anomaly time-scales for the Cenozoic and Late Cretaceous: a precis, critique and synthesis. *Rev. Geophys. Space Phys.* **18**, 753–70.

Onstott, T. C., Hall, C. M. and York, D. (1989). ^{40}Ar/^{39}Ar thermo-chronometry of the Imataca complex, Venezuela. *Precambrian Res.* **42**, 255–91.

Rex, D. C., Guise, P. G. and Wartho, J.-A. (1993). Disturbed ^{40}Ar/^{39}Ar spectra from hornblendes: thermal loss or contamination? *Chem. Geol. (Isot. Geosci. Section)* **103**, 271–81.

Roddick, J. C. and Farrar, E. (1971). High initial argon ratios in hornblendes. *Earth Planet. Sci. Lett.* **12**, 208–14.

Smith, P. E., Evensen, N. M. and York, D. (1993). First successful ^{40}Ar/^{39}Ar dating of glauconies: argon recoil in single grains of cryptocrystalline material. *Geology* **21**, 41–4.

Steiger, R. H. and Jager, E. (1977). IUGS Subcommission on Geochronology: convention on the use of decay constants in geo- and cosmochronology. *Earth Planet. Sci. Lett.* **36**, 359–62.

Swisher, C. C. and Prothero, D. R. (1990). Single-crystal ^{40}Ar/^{39}Ar dating of the Eocene–Oligocene transition in North America. *Science* **249**, 760–2.

Turner, G. (1968). The distribution of potassium and argon in chondrites. In: Ahrens, L. H. (Ed.) *Origin and Distribution of the Elements.* Pergamon, pp. 387–97.

Turner, G. (1969). Thermal histories of meteorites by the ^{39}Ar–^{40}Ar method. In: Millman, P. M. (Ed.) *Meteorite Research.* Reidel, pp. 407–17.

Turner, G. (1971a). Argon 40–argon 39 dating: the optimisation of irradiation parameters. *Earth Planet. Sci. Lett.* **10**, 227–34.

Turner, G. (1971b). ^{40}Ar/^{39}Ar ages from the lunar maria. *Earth Planet. Sci. Lett.* **11**, 169–91.

Turner, G. (1972). ^{40}Ar–^{39}Ar age and cosmic ray irradiation history of the Apollo 15 anorthosite, 15415. *Earth Planet. Sci. Lett.* **14**, 169–75.

Turner, G. and Cadogan, P. H. (1974). Possible effects of ^{39}Ar recoil in ^{40}Ar/^{39}Ar dating. *Proc. 5th Lunar Sci. Conf.,* pp. 1601–15.

Turner, G., Miller, J. A. and Grasty, R. L. (1966). Thermal history of the Bruderheim meteorite. *Earth Planet. Sci. Lett.* **1**, 155–7.

van Eysinga, F. W. B. (1975). *Geological Time Table.* (3rd. Edn). Elsevier.

van Hinte, J. E. (1976). A Cretaceous time scale. *Amer. Assoc. Petroleum Geol. Bull.* **60**, 498–516.

Vincent, E. A. (1960). Analysis by gravimetric and volumetric methods, flame photometry, colorimetry and related techniques. In: Smales, A. A. and Wager, L. R. (Eds.), *Methods in Geochemistry.* pp. 33–80.

York, D. (1978). A formula describing both magnetic and isotopic blocking temperatures. *Earth Planet. Sci. Lett.* **39**, 89–93.

York, D. (1984). Cooling histories from ^{40}Ar/^{39}Ar age spectra: implications for Precambrian plate tectonics. *Ann. Rev. Earth Planet. Sci.* **12**, 383–409.

York, D., Hall, C. M., Yanase, Y., Hanes, J. A. and Kenyon, W. J. (1981). ^{40}Ar/^{39}Ar dating of terrestrial minerals with a continuous laser. *Geophys. Res. Lett.* **8**, 1136–8.

Wanke, H. and Konig, H. (1959). Eine neue methode zur Kalium–Argo–Alterbestimmung und ihre Anwendung auf Steinmeteorite. *Z. Natur.* **14a**, 860–6.

Wright, N., Layer, P. W. and York, D. (1991). New insights into thermal history from single grain ^{40}Ar/^{39}Ar analysis of biotite. *Earth Planet. Sci. Lett.* **104**, 70–9.

11 Rare gas geochemistry

The group of elements known as the rare, inert or noble gases possess unique properties which make them important in isotope geology. The low abundance of these rare gases allows them to sensitively record several types of nuclear process, even including rare nuclear fission reactions. (In contrast, the relatively larger abundance of other fission products such as the 'rare' earths swamps fissiogenic production). Another property of these gases is their inertness, which allows unique insights into the Earth's interior because of their lack of interaction with other materials. Finally, as isotopic tracers, they can give information about the degassing history of the mantle, the formation age of the atmosphere, and about mixing relationships between different mantle reservoirs.

11.1 Helium

Helium has two isotopes, ^4He and ^3He. The former was recognised by Rutherford (1906) to be the α decay product of actinide elements, and comprised the first radiometric dating method. However, the great diffusivity of helium made the method very susceptible to thermal disturbance and it has therefore been abandoned in all but the most specialised applications (e.g. Wernicke and Lippolt, 1993).

Non-radiogenic ^3He was first discovered in nature by Alvarez and Cornog (1939). Alvarez and Cornog estimated (using a cyclotron) that atmospheric helium had a $^3He/^4He$ ratio ten times greater than natural oil-well gases from the Earth's crust. Aldrich and Nier (1948) confirmed this observation by mass spectrometric measurements, and determined atmospheric and well-gas $^3He/^4He$ ratios of ca. 1.2×10^{-6} and 1×10^{-7} respectively. They concluded that there must be independent sources of the two isotopes, one of which could be primordial.

11.1.1 Mass spectrometry

Mass spectrometric analysis of helium is broadly similar to argon isotope analysis in K–Ar dating (section 10.1.1). However, in rare gas tracer studies there are no 'extra' isotopes available to allow accurate correction for atmospheric contamination. Therefore it is critical to minimise the *extent* of this contamination during helium extraction and analysis. Uncertainties in the atmospheric 'blank' may contribute the principal error in helium isotope analysis, especially for rock samples. Well-gas samples, being larger, are less susceptible to atmospheric contamination during analysis, but may have come from an open system in the natural environment. In the case of rock analysis, absorbed atmospheric helium is usually driven off by overnight heating at 200–300 °C. The sample gas may then be extracted by melting the rock or by crushing under vacuum. A combination of both techniques (e.g. Kurz and Jenkins, 1981) provides an extra check against the possibility of atmospheric contamination, both in the laboratory and the environment.

Two steps are necessary to reduce blank levels in the mass spectrometer for all rare gas analyses. One is to polish all internal surfaces of a metal instrument to minimise gas absorption into the vacuum system walls. Another is to reduce the internal surface area of the instrument as much as possible, for example by boring the flight tube out of a solid piece of steel rather then using welded pipe. A low internal volume also yields better sensitivity for very small samples.

All rare gas analyses are performed in the static gas mode (i.e. with vacuum pumps isolated). As a result, hydrogen tends to build up in the instrument so that its molecular ions HD^+ and H_3^+ cause isobaric interferences onto $^3He^+$. Therefore, the vacuum system in some older machines contains a small titanium 'getter', designed to absorb H_2 released inside the instrument (Clarke *et al.*, 1969). Nevertheless the peak composed of HD and H_3 may still be much larger than ^3He, and it is essential to separate them by mass. This can be done by making use of the 0.006 atomic mass unit difference between the species

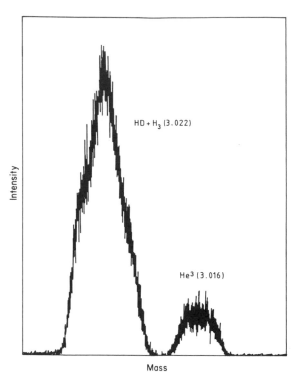

HD + H₃ (3.022)

He³ (3.016)

Intensity

Mass

Fig. 11.1. Scan of peaks in the region of mass 3 during a helium isotope analysis to show machine resolution. Masses are quoted relative to $^{12}C = 12.000$. After Lupton and Craig (1975).

which results from their different nuclear binding energies (Fig. 11.1). In order to achieve this separation at mass 3, a resolution of one mass unit in 600 is necessary, which can be achieved with an instrument of ca. 25 cm radius (Clarke *et al.*, 1969; Kurz and Jenkins, 1981).

In order to measure the very large intensity difference between ^{3}He and ^{4}He signals, it is most convenient to measure the former on a multiplier detector and the latter by Faraday detector. These can only be used in the static collection mode if a branched flight tube is available, because of the extreme divergence of the mass 3 and 4 ion beams (Lupton and Craig, 1975). Alternatively, peak switching is performed by changing the accelerating potential or magnetic field (e.g. Clarke *et al.*, 1969; Poreda and Farley, 1992).

11.1.2 Helium production

In order to determine whether primordial helium is an important constituent in the Earth, it is necessary to determine the ^{3}He/^{4}He ratio of primordial solar system helium, and the production ratio in nuclear and cosmogenic processes. A good indication of the composition of primordial helium is provided by the ^{3}He/^{4}He ratio of $(2 - 4) \times 10^{-4}$ measured in gas-rich carbonaceous chondrites (Pepin and Signer, 1965). These meteorites have such high primordial gas contents that their composition is not significantly perturbed by cosmogenic helium (a product of cosmic-ray spallation effects). In contrast, most ^{3}He in iron meteorites is cosmogenic.

Early calculations of the nuclear ^{3}He/^{4}He production ratio in igneous rocks were made by Morrison and Pine (1955). Radiogenic production of ^{4}He is obvious, since the α particle is synonymous with a ^{4}He nucleus. However, 'nucleogenic' ^{3}He can also be generated by neutron bombardment of light atoms. Radioactive decay of uranium generates a neutron flux in rocks by two mechanisms. Spontaneous fission is a minor source, but by far the dominant source of neutrons is the collision of α particles with nuclei of light elements. Some of these neutrons reach epithermal energies, where they can induce the (n, α) reaction on lithium. The tritium thus produced decays to ^{3}He:

$$^6Li + n \rightarrow {}^3H + \alpha$$
$$^3H \rightarrow {}^3He + \beta \qquad (t_{1/2} = 12 \text{ yr}).$$

Kunz and Schintlmeister (1965) calculated that ^{3}He generation by this reaction is at least three orders of magnitude more efficient than all other neutron-induced reactions. Given the uranium (+ thorium) and lithium content of a rock, the ^{3}He/^{4}He yield can be calculated (Gerling *et al.*, 1971). The results are consistent with the range of $(1 - 3) \times 10^{-8}$ measured empirically in old granites. The calculations were also confirmed by experimental irradiation of ultrabasic rocks in a reactor (Tolstikhin *et al.*, 1974). Some other possible sources of ^{3}He (via tritium) are:

$$^{238}U \rightarrow \text{fission products} + 2 \times 10^{-4} \cdot {}^3H,$$
$$^7Li + \alpha \rightarrow {}^8Be + {}^3H,$$
$$^7Li + \gamma \rightarrow {}^4He + {}^3H.$$

However, Mamyrin and Tolstikhin (1984) calculated total ^{3}He/^{4}He production ratios of 8×10^{-12}, $< 7 \times 10^{-9}$ and ca. 10^{-13} respectively from these reactions, making them insignificant compared with the main (n,α) reaction. It was concluded from these observations and calculations that no nuclear process has been discovered which is capable of generating ^{3}He/^{4}He ratios significantly greater than 10^{-8} in normal rocks. However, uranium ores generate lower ratios, while Li-rich *minerals* generate abnormally high ratios.

Another mineral in which high $^3He/^4He$ ratios have been observed is diamond. Values up to 3×10^{-4} were interpreted by Ozima and Zashu (1983) as indicative of primordial mantle reservoirs, but have been attributed by later workers to either nucleogenic or cosmogenic ^3He production. For example, Lal *et al.* (1987) attributed high $^3He/^4He$ ratios in alluvial diamonds from Zaire to cosmogenic production while exposed at the surface.

On the other hand, Kurz *et al.* (1987) and Zadnik *et al.* (1987) measured $^3He/^4He$ as high as 1.4×10^{-3} in diamonds mined directly from kimberlite pipes at depths of ca. 26 m and 200 m respectively. Since cosmic rays cannot penetrate to such depths, these helium signatures were attributed to nucleogenic production. Evidence for this interpretation came from the observation of isotopic variability within individual diamonds, and the determination of $^3He/^4He$ ratios higher than solar in the latter study. In both cases, ^3He production was attributed to the (n, α) reaction on lithium. For this process to occur, the diamond and its inclusions must be irradiated by neutrons from outside the crystal, so that radiogenic ^4He production in the diamond itself is suppressed.

In-situ cosmogenic helium production in terrestrial rocks was proposed by Jeffrey and Hagan (1969), but was not identified unambiguously until work by Kurz (1986a) and Craig and Poreda (1986). In a detailed helium isotope study of subaerial lavas from Haleakala volcano, Kurz discovered very high $^3He/^4He$ ratios in step heating studies of some near-surface 0.5 – 0.8 Myr-old alkali basalts. Low-temperature gas releases from samples within 0.5 m of the weathered surface yielded $^3He/^4He$ ratios over 10^{-3} (Fig. 11.2). These values are even higher than primordial meteoritic or solar-wind helium. In contrast, step heating of samples from a similar stratigraphic horizon that were buried under ca. 160 m of younger flows yielded MORB-like helium (see below). Helium released by crushing of phenocrysts also gave a MORB signature in both the buried and surface samples. Therefore, Kurz argued that crushing released magmatic helium from vesicles, but that step heating of old surface samples released dispersed cosmogenic helium from the rock matrix. Young surface samples such as the 1790 flow on Haleakala do not show these effects, ruling out anthropogenic bomb tritium as the source of the ^3He.

Kurz (1986b) went on to examine cosmogenic ^3He production as a function of depth below the surface of a lava flow. Spallation reactions caused by cosmo-

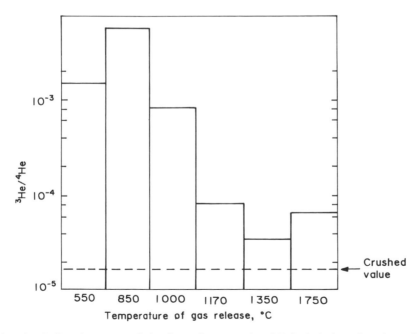

Fig. 11.2. Step heating helium isotope analysis of a surface sample of Haleakala lava showing a large cosmogenic component in low-temperature release steps. After Kurz (1986a).

genic neutrons are the dominant source of ^{3}He at the surface, but neutrons are attenuated exponentially downwards. Nevertheless, ^{3}He abundances showed less attenuation with depth than expected. This was attributed to production by cosmic-ray muons, which have a greater penetration depth than neutrons. Muon capture by nuclei causes neutron emission, which in turn produces ^{3}He via the (n, α) reaction on lithium. The depth dependence of different production routes for ^{3}He is summarised in Fig. 11.3 (Lal, 1987).

Cosmogenic isotopes represent a useful tool for determining exposure ages of rock surfaces (section 14.6). However, the great diffusivity of helium may be a problem in using ^{3}He in such studies. For example, Cerling (1989) showed that helium was often not quantitatively retained in quartz, the most widely used material in surface exposure dating. On the other

hand, ^{21}Ne displays cosmogenic production with an attenuation depth similar to ^{3}He (Sarda et al., 1993). The lower diffusivity of neon may therefore make it more widely useful in surface exposure dating (section 11.4).

11.1.3 Terrestrial primordial helium

Accurate determinations of the atmospheric ^{3}He/^{4}He ratio were made by Mamyrin et al. (1970) and Clarke et al. (1976), yielding ratios of 1.40×10^{-6} and 1.38×10^{-6}. Because atmospheric helium is universally used as a mass spectrometric standard, it is convenient to express ^{3}He/^{4}He ratios in unknown samples relative to the atmospheric ratio in the form $R_{\text{unknown}}/R_{\text{air}}$ (R/R_a). However, because cosmogenic ^{3}He production in the atmosphere is difficult to quantify accurately, it is not possible to prove the existence of a primordial helium source in the Earth *simply* by the fact that the atmosphere is two orders of magnitude richer in ^{3}He than radioactive production in rocks.

Stronger evidence of a primordial helium signature in the Earth was provided by Clarke et al. (1969), when they discovered that deep water from the Pacific ocean was enriched in ^{3}He by up to 20% relative to the atmosphere. However, Sheldon and Kern (1972) and Lupton and Craig (1975) hypothesised that this could conceivably be due to temporary weakening of the Earth's magnetic field, during which the atmospheric ^{3}He/^{4}He ratio was elevated by greater cosmic ray penetration.

Convincing evidence of primordial helium in the Earth was first provided by Mamyrin et al. (1969), who found ^{3}He/^{4}He ratios of ten times the atmospheric ratio in thermal fluids from the Kuril Islands. Subsequently, ^{3}He/^{4}He ratios as high as 20 times atmospheric were found in hot springs from Iceland (Mamyrin et al., 1972). Even higher R/R_a values have been found in oceanic volcanic rocks: up to 26 in subglacial basaltic glasses from the neovolcanic zones of Iceland (Fig. 11.4) and up to 32 in basaltic glass from Loihi Seamount off Hawaii (Kurz et al., 1982). In contrast, Craig and Lupton (1976) found a relatively narrow range of R/R_a ratios around 9 in MORB glasses from different ocean basins. Subsequent data (e.g. Fig. 11.4) have confirmed the narrow helium isotope range in MORB, relative to the large variations in plumes.

The intermediate helium isotope composition of MORB, between atmospheric and plume sources, were explained by partial outgassing of primordial helium from the upper mantle, followed by radiogenic

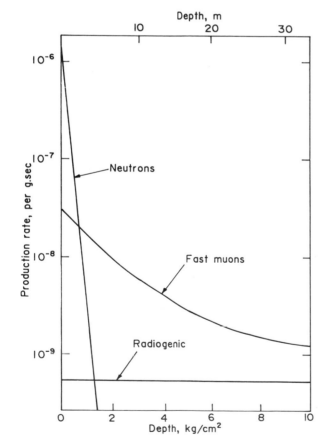

Fig. 11.3. Calculated production rates for ^{3}He by different processes as a function of depth, expressed as kg / cm^2, which is approximately equal to 1/3 depth in m. After Lal (1987).

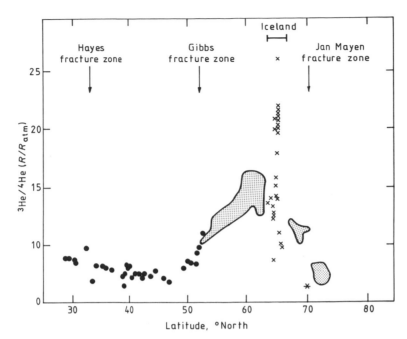

Fig. 11.4. Plot of helium isotope ratios along the Mid Atlantic Ridge as a function of latitude showing the primordial ^{3}He signature of the Iceland plume relative to MORB. After Kurz *et al.* (1985).

helium production. This caused the upper mantle to develop a lower ^{3}He/^{4}He composition than the un-degassed lower mantle, where radiogenic production is swamped by primordial helium. This partial degassing or 'two reservoir' model for the mantle was originally proposed to explain argon isotope systematics (Hart *et al.*, 1979), and was applied to helium by Kaneoka and Takaoka (1980).

Unfortunately Kaneoka and Takaoka based their case on rare gas compositions in phenocrysts from Haleakala volcano, Hawaii, which were subsequently shown to be contaminated with atmospheric argon and cosmogenic helium (Fisher, 1983; Kurz, 1986a). However, the OIB data cited above are reliable indicators of primordial helium because in all cases they come from rocks which have been shielded from *in-situ* cosmogenic production. Hence the two-reservoir model for mantle helium has been widely accepted.

On the other hand, Fisher (1985) argued that noble gases from oceanic island basalts do not require an undepleted (non-degassed) mantle source. His case was based primarily on the different abundances of non-radiogenic rare gases in MORB and OIB. Thus, if OIB come from the un-degassed source, we would expect them to contain more helium than MORB glasses from the degassed upper mantle; however,

OIB glasses actually have ten times *less* ^{3}He than MORB (Fisher, 1985). This evidence appears to pose a problem for the two-reservoir model but is not definitive, due to the poorly constrained behaviour of rare gases during the melting process. For example, the dynamics of mantle convection and melt segregation under ridges must be different from plumes, and ridge magmas probably collect helium from a greater volume of mantle during the melting process (section 13.3.2). Hence, most workers have taken the isotopic evidence in favour of the two-reservoir model as definitive, and over-riding any problems involving rare gas abundances.

The existence of a primordial helium reservoir in the Earth was challenged more recently by Anderson (1993), who attributed this signature to the subduction of cosmic (interplanetary) dust particles. These particles were found to accumulate in ocean-floor sediments by Merrihue (1964). Cosmic dust has ^{3}He/^{4}He ratios similar to gas-rich meteorites (ca. 3×10^{-4}), and a small fraction of these particles falls to Earth without burning up in the atmosphere (Nier and Schlutter, 1990). Hence, ocean-floor sediments develop a 'primordial' helium isotope signature (Fig. 11.5a).

The rare gases in cosmic dust particles may be encapsulated in magnetite grains, which are rela-

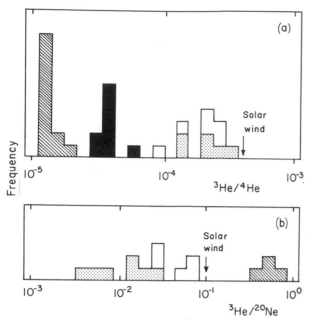

Fig. 11.5. Histograms of a) $^3He/^4He$ and b) $^3He/^{20}Ne$ in cosmic dust particles (stipple) and ocean-floor sediments (white) compared with the rare gas composition of MORB (hatched) and OIB (black). The solar wind composition is shown for reference. Data from Allegre *et al.* (1993).

tively resistant to thermal degassing (Matsuda *et al.*, 1990). Therefore, the cosmic helium in ocean-floor sediments might survive the subduction process and be transported into the deep mantle. In contrast, *atmospheric* rare gases trapped in ocean-floor sediments are very susceptible to thermal degassing. Staudacher and Allegre (1988) argued that subduction-related volcanism is at least 98% efficient in scavenging these atmospheric gases from subducted sediments before they can reach the deep mantle.

Because cosmic dust may survive the 'subduction barrier' against atmospheric rare gases, it has the potential to deliver helium with a primordial signature into the deep mantle. This possibility has been recognised by several workers (e.g. Allegre *et al.*, 1993), but Anderson (1993) took this model a step further by attributing *most* of the primordial helium signal in mantle hot-spots to subducted cosmic dust. Because of the far-reaching consequences of this model, it is important to examine it quantitatively.

Anderson calculated that only 0.4% of the total incoming ^3He flux to the Earth is incorporated into ocean-floor sediment. This provides a potential ^3He flux into the mantle of 8×10^3 cm^3/yr (at standard temperature and pressure), which is 300 times lower

than the estimated rate of ^3He escape from ocean ridges (2.5×10^6 cm^3/yr). It is now agreed by most workers (although not by Anderson) that the composition of the MORB source is buffered by material supply from mantle plumes (the OIB source). Therefore, if the depleted mantle (MORB source) has reached a steady state for helium, then the loss of ^3He must be replaced by deep mantle plumes. But if these plumes are fed by ^3He subduction at the present rate, they can only account for 0.3% of this ^3He. This problem could be overcome by invoking much higher ^3He subduction rates in the past, followed by storage in the lower mantle for a few Byr. Such high rates of extra-terrestrial input might be possible during the intense meteorite bombardment of early Earth history (as recorded on the Moon). However, this effect might be partially offset by hotter conditions in subducted slabs during early Earth history (section 4.4.3).

Anderson cited evidence from neon isotopes in support of this cosmic primordial helium model. However, Allegre *et al.* (1993) used these data to place upper limits on the amount of cosmic ^3He which can enter plume sources. They noted that the $^3He/^{20}Ne$ ratio in cosmic dust is one to two orders of magnitude lower than $^3He/^{20}Ne$ in the upper mantle (Fig. 11.5b). Furthermore, helium has a much greater diffusivity than neon, which would promote its preferential degassing from cosmic dust grains during subduction (Hiyagon, 1994). Therefore it appears that subduction of cosmic dust cannot contribute more than a small fraction of the mantle ^3He budget without causing excessive enrichment of ^{20}Ne in submarine glasses. Hence, the evidence continues to favour a primordial helium reservoir in the Earth.

11.1.4 Helium and heat

The decay of ^{232}Th, ^{235}U and ^{238}U generates six, seven and eight ^4He atoms respectively, but these isotopes also contribute the bulk of radioactive heat to the Earth (the other source being ^{40}K). Hence, helium and heat generation should be related. Assuming a K/U ratio of 10^4 from MORB analyses (Jochum *et al.*,1983), and a Th/U ratio of 4, O'Nions and Oxburgh (1983) calculated that 10^{12} atoms of ^4He would be generated in the mantle per joule of heat production. They then calculated the concentration of U necessary to generate the observed helium and heat fluxes (Fig. 11.6).

O'Nions and Oxburgh determined that in the oceans, the radiogenic helium flux (which comprises 88% of the total) can be sustained by an average U

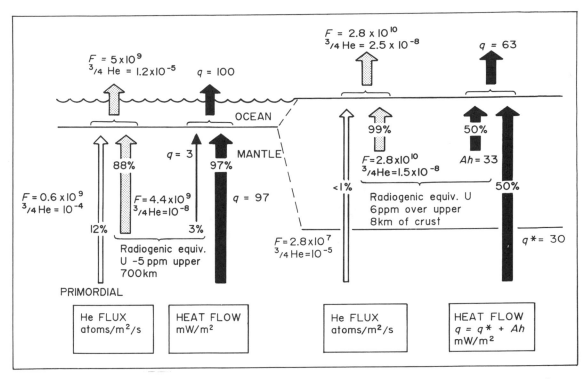

Fig. 11.6. A comparison of calculated fluxes of helium (F) and heat (q) in average oceanic and continental areas. After O'Nions and Oxburgh (1983).

concentration of 1.5 ppb in the whole mantle, or 5 ppb in the upper 700 km of the mantle. However, this quantity of uranium (and equivalent Th and K) can only generate 3% of the observed oceanic heat flux. Hence they concluded that 97% of the oceanic heat flux must come from the lower mantle or core, compared with only 12% primordial helium from this source.

The excess heat may come from core crystallisation, but O'Nions and Oxburgh judged it inadequate to sustain the heat flow over Earth history. Therefore, they proposed that a boundary layer inhibits upward transport of helium from the 'primordial' reservoir much more effectively than the transport of heat. They envisaged this boundary layer at 700 km depth, separating the upper and lower mantle. Other workers have preferred the core–mantle boundary, but recent work suggests that the core may have a limited helium budget (Matsuda *et al.*, 1993).

99% of the continental helium flux is radiogenic, and can be sustained by a U equivalent concentration of 6 ppm in the upper 8 km of the crust. This can also explain 50% of the continental heat flux. Hence, the

other 50% of continental heat flow must be sub-continental, whereas less than 1% (primordial + radiogenic) of the continental helium flux comes from the mantle. In this case it is clear that the continental crust is a boundary layer. Mantle-derived heat can be carried across it conductively, but mantle-derived helium only leaks through the crust in certain discrete areas. These are normally areas of active magmatism.

Well-gas studies demonstrate the local nature of mantle helium transport through the crust. Oxburgh *et al.* (1986) have shown that sedimentary basins which result from crustal loading, such as the Alpine Molasse basin, yield helium with very low R/R_a values around 0.05, whereas sedimentary basins formed by extensional tectonics, such as the Rhine Graben and the Pannonian basin of Hungary, may yield helium with much higher R/R_a values around unity (Fig. 11.7a). The huge 'Panhandle' gas field in the southern US is particularly interesting. It is one of the world's largest gas fields, and has helium contents of up to 2%. In the south, the reservoir is draped over uplifted Proterozoic–Paleozoic basement, and in this region R/R_a values as low as 0.06 have been measured (Fig. 11.7b). In contrast, the northern part of the reservoir

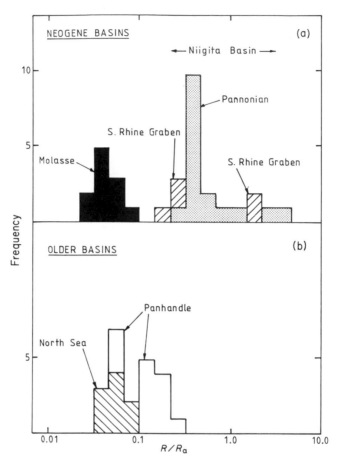

Fig. 11.7. Histograms showing variation in R/R_a values (on a log scale) in different types of sedimentary basin. After Oxburgh *et al.* (1986).

is in an area of recent igneous activity. Here, R/R_a values up to 0.2 have been measured, corresponding to 2% MORB-type helium (Oxburgh *et al.*, 1986).

11.1.5 Helium and volatiles

The comparison of helium isotope data with other rare gas tracers will be discussed below. However, helium isotope compositions can also be used to place important constraints on the interpretation of other volatile species, the most important of which are carbon dioxide and methane.

Carbon fluxes in the Earth are difficult to constrain because of the reactivity of this element. However, Marty and Jambon (1987) argued that if carbon abundance could be tied to ^3He in major mantle products such as MORB, then helium fluxes might be usable as a measure of the carbon flux in a variety of

environments. They collated $C/^3He$ data for MORB from different ocean basins and found that they defined a relatively narrow range with an average $C/^3He$ ratio of ca. 2×10^9. In a parallel study, Jambon and Zimmermann (1987) showed that the $C/^3He$ ratios measured by heating MORB glass and by crushing of vesicles were similar (Fig. 11.8), suggesting that the measured ratios are indicative of the basaltic magma itself, and are not severely fractionated relative to one another during eruption. This is attributed to the similar solubilities of helium and carbon dioxide in basaltic magma. Taken together, these pieces of evidence suggest that the measured ratio is typical of the $C/^3He$ flux from the upper mantle on a world-wide scale.

O'Nions and Oxburgh (1988) took these deductions further by arguing that the oceanic upper mantle flux of $C/^3He$ could be applied also to mantle-derived

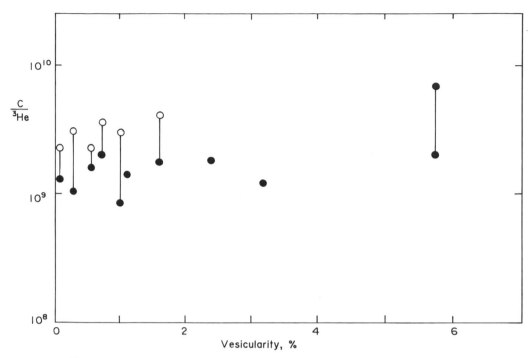

Fig. 11.8. Plot of C/^{3}He ratio against glass vesicularity for MORB samples. (O) = heated glass; (●) = crushed vesicles. After Marty and Jambon (1987).

volatile fluxes through the continental lithosphere. They examined as an example the Pannonian basin of Hungary, which has an R/R_a value as high as 6, suggesting that mantle-derived helium comprises up to 90% of the total helium flux in parts of the basin. This is attributed to its extensional tectonic setting. Measurements of helium abundance for a major aquifer in the Pannonian basin were used by Martel *et al.* (1989) to estimate a mantle ^{3}He flux of 8×10^4 atoms/m^2/s for the basin as a whole, which is greater than the globally averaged oceanic flux (Fig. 11.6). Hence, O'Nions and Oxburgh concluded that if extensional zones such as the Pannonian basin make up a significant fraction of the lithosphere then they make a major contribution also to the global carbon inventory of the crust.

O'Nions and Oxburgh sought to expand these conclusions still further by arguing that C/K ratios are equal in major Earth reservoirs, and hence that magmatic C/K fluxes are similar. By tying potassium in its turn to ^{3}He fluxes, they attempted to quantify the importance for continent building of basaltic rift magmatism such as seen in the Pannonian basin. The calculations can be seen in Fig. 11.9. Diagonal contours define the total fraction of the continental mass which could be attributed to continental rift

volcanism for different ^{3}He fluxes. If the ^{3}He flux for the Pannonian is typical, then such magmatic provinces could have contributed 5 – 10% of the total continental mass, over Earth history, if they cover a corresponding area of the continental surface at any given time. Hence, they may be a significant agency in the continent-building process.

11.1.6 Helium and non-volatiles

Although helium isotope ratios provide the best evidence for a primordial gas reservoir in the Earth, this single isotope ratio cannot provide enough degrees of freedom to constrain the complex mixing processes expected to occur in the mantle. Hence, various attempts have been made to compare helium isotope signatures with other isotope ratios in oceanic volcanics, in order to provide extra constraints on mantle processes.

One such approach is the comparison of helium and strontium isotope data (Kurz *et al.*, 1982; Lupton, 1983). MORB defines a restricted range of compositions on a plot of helium isotope ratio against ^{87}Sr/^{86}Sr, but ocean islands are widely scattered (Fig. 11.10). While Loihi defines the most primordial helium composition, some ocean islands such as

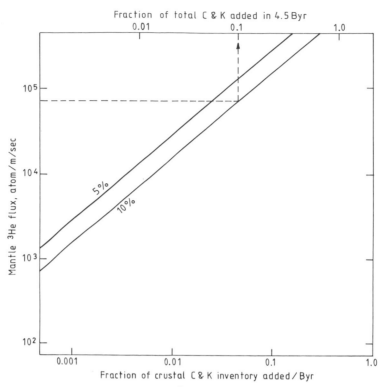

Fig. 11.9. Diagram linking continent-building action to ^{3}He flux for continental lithosphere undergoing extensional tectonics. Diagonal lines show the effect of extending the model to different fractions of the continental surface, at any given time. Dashed line indicates the helium flux of the Pannonian basin. Modified after O'Nions and Oxburgh (1988).

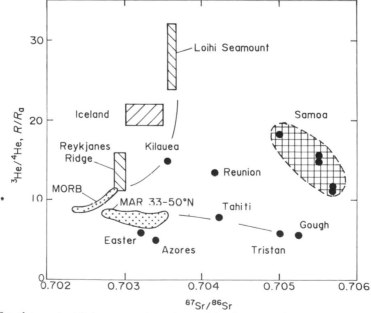

Fig. 11.10. Plot of ^{3}He/^{4}He against Sr isotope ratio to show mixing between the MORB reservoir and primordial and recycled plume sources. After Lupton (1983).

Tristan, Gough and the Azores have ^{3}He/^{4}He ratios lower (more radiogenic) than MORB. This requires a component of radiogenic helium from a long-lived U or Th-rich source, which can most easily be satisfied by the recycling of oceanic crust and sediments into the mantle, as inferred from lithophile isotope data (section 6.6).

Combined helium and strontium isotope data suggest that while plumes represent sources which are enriched with respect to MORB, they are of two types (Kurz *et al.*, 1982). One kind, such as Hawaii and Iceland, samples rare gases leaking from a primordial reservoir which may be the deep mantle. The other type, such as Tristan, Gough and the Azores, samples only recycled material, although these plumes are no less well defined kinematically. Fig. 11.10 also suggests the existence of a third type of plume source, exemplified by Samoa. This might be due to the mixing of primordial and recycled helium. Such a scenario would be feasible if the recycled material sinks down through the mantle until it accumulates on a boundary layer on top of the primordial helium source, often regarded by geochemists as the lower mantle. New data from Samoa continues to support this mixing model (Farley *et al.*, 1992).

The MORB field in Fig. 11.10 breaks into two lobes with geographical boundaries. The main field trends slightly towards the primordial source, while the Mid Atlantic Ridge between 33 and 50 °N defines a subsidiary field with more radiogenic helium, consistent with contamination by the nearby Azores plume. This evidence suggests that plumes break into trains of blobs which locally contaminate the MORB source. According to the nature of the plume material contaminating any given ridge segment, the MORB array can trend towards either the primordial or recycled plume type.

11.2 Argon

Initial, 'excess' or 'inherited' argon is normally regarded as a problem to be avoided in K–Ar and Ar–Ar dating (section 10.1.2). However, the isotopic composition of this argon can be used as a powerful geochemical tracer, especially when used alongside other rare gas data.

Atmospheric contamination is a much more serious problem in the isotopic analysis of 'heavy' rare gases than it is for helium. This is because atmospheric helium has a low abundance due to its escape from the atmosphere, whereas the heavy rare gases have accumulated in the atmosphere over Earth history.

Stringent baking of both equipment and samples is needed to drive off absorbed atmospheric gas before analysis. Because of this problem, the first clear evidence for inherited argon was provided by the analysis of beryl, whose ring-type structure accommodates unusually large quantities of initial argon, swamping the effects of atmospheric contamination. 'Excess' argon was first found in beryl by Aldrich and Nier (1948) and studied in more detail by Damon and Kulp (1958). The latter workers discovered Archean beryls containing more than 99% of excess argon and with ^{40}Ar/^{36}Ar ratios as high as 10^5.

The ^{40}Ar contents of beryl were observed to decrease over geological time (Fig. 11.11), leading Damon and Kulp to propose an extensive early degassing of the Earth in the Archean, decreasing exponentially towards the present. The beryl data shown in Fig. 11.11 were also used by Fanale (1971) to support a more extreme model of catastrophic early degassing of the Earth. He argued that they were not consistent with models of constant degassing intensity through Earth history, such as proposed by Turekian (1964).

Schwartzman (1973) supported the early degassing model using Ar isotope data from the 2.7 Byr-old Stillwater complex. Because this is a basic magma it can yield more direct information about the Archean mantle than beryl-bearing pegmatites with a potentially large crustal input. A Stillwater pyroxene had an excess ^{40}Ar/^{36}Ar ratio of at least 17 900, corresponding to a calculated maximum ^{36}Ar/silicon ratio of 3×10^{-11} for the 2.7 Byr-old mantle source. The ratio of atmospheric ^{36}Ar to mantle silicon at the present day is 1×10^{-10}, so the Earth was apparently outgassed to at least 70% of its present extent by the end of the Archean.

Ocean-floor basalt glasses are an important source of information about the rare gas budget of the present-day mantle because the high water pressure at the site of eruption retains initial magmatic argon in the sample (section 10.1.2). Furthermore, rapid quenching reduces contamination by atmospheric argon dissolved in seawater. In contrast, the crystalline cores of basalt pillows are largely outgassed of magmatic rare gases and contaminated with atmospheric gases during crystallisation (Fisher, 1971).

Hart *et al.* (1979) used the maximum ^{40}Ar content of 3×10^{-6} ml/g in ocean-floor basalt glasses as an estimate of the present-day ^{40}Ar concentration in the upper mantle source (UM). By subtracting this value from the average ^{40}Ar concentration in the Bulk Silicate Earth (estimated from its K abundance), and comparing this to the total ^{40}Ar budget of the

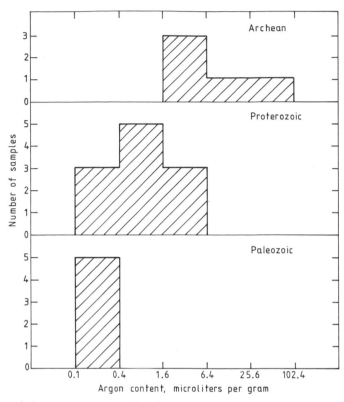

Fig. 11.11. Histograms of the argon content of beryl samples grouped by age to show decrease in ^{40}Ar contents with time. After Damon and Kulp (1958).

atmosphere and crust, Hart *et al.* were able to calculate the mass of mantle which must be outgassed. This calculation is shown in equation [11.1], where square brackets denote concentrations:

$$\text{mass of mantle outgassed} = \frac{^{40}\text{Ar}_{\text{atm}} + {}^{40}\text{Ar}_{\text{crust}}}{[^{40}\text{Ar}_{\text{BSE}}] - [^{40}\text{Ar}_{\text{UM}}]}$$

[11.1]

Using this equation, a potassium abundance of 660 ppm in the Bulk Silicate Earth implies that only 25% of the total mass of the mantle need be outgassed.

Hart *et al.* proposed that the upper mantle was also thoroughly degassed of ^{36}Ar, so that subsequent radiogenic ^{40}Ar production generated high ^{40}Ar/^{36}Ar ratios of up to 16 000. In contrast, the lower mantle was not significantly degassed in argon, such that radiogenic ^{40}Ar production was swamped by the primordial component, allowing only a modest rise in ^{40}Ar/^{36}Ar ratio above the atmospheric value. Hart *et al.* noted the similarity of these model predictions to the ^{40}Ar/^{36}Ar ratios observed in ocean-floor glasses from ridges and mantle plumes respectively (Fig.

11.12). This suggested to them that these data were not seriously perturbed by atmospheric contamination.

The concept of a relatively less degassed or 'undepleted' mantle reservoir with respect to heavy rare gases was strongly contested by Fisher (1983; 1985). He argued that the low ^{40}Ar/^{36}Ar ratios measured in plume environments were a result of atmospheric contamination. This argument was initially aimed at data on xenolithic inclusions in lavas (e.g. Kaneoka and Takaoka, 1980). These appeared to contain primordial argon and helium, but were subsequently shown to be contaminated by atmospheric argon and cosmogenic helium (section 11.1.3). Submarine glasses may be more resistant to such effects, but the interpretation of these data has nevertheless provoked intense controversy.

In order to distinguish between atmospheric contamination and un-degassed mantle sources it is necessary to take stringent experimental precautions. Samples and equipment are thoroughly baked before analysis, and frequent blanks are determined to verify the effectiveness of these procedures. Contamination

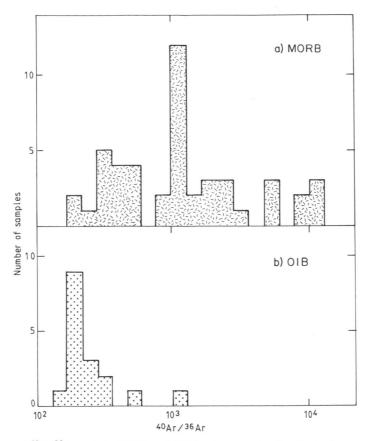

Fig. 11.12. Histograms of $^{40}Ar/^{36}Ar$ ratio in MORB and OIB to show contrasting isotopic composition. After Hart *et al.* (1979).

of glasses by argon from the atmosphere or seawater can be assessed by looking for isotopic heterogeneities within the sample. One approach is to compare the products of thermal degassing with the gases released from vesicles by crushing (e.g. Hart *et al.*, 1983). Another approach is to perform the rare gas analysis by step heating (e.g. Staudacher and Allegre, 1982). This procedure is demonstrated for a MORB sample in Fig. 11.13. Absorbed atmospheric rare gases are released in the low-temperature heating steps, allowing an estimate to be made of the severity of contamination effects. Staudacher *et al.* (1986) used both step heating and crushing to make the most rigorous search for sample contamination.

Hart *et al.* (1983; 1985) monitored the effectiveness of these precautions using rare gas abundance patterns (Fig. 11.14). The radiogenic $^{4}He/^{40}Ar$ ratio of a closed-system mantle (ca. 2) is estimated from the abundance ratio of the parents (U/K). This should correspond to the present-day signature of a primor-

dial rare gas reservoir in the earth. In contrast, seawater is depleted in He/Ar ratio, due to helium escape from the atmosphere. Most MORB glasses are enriched in He/Ar ratio, which is attributed to volatile fractionation during the degassing of MORB magmas. However, a few MORB glasses, along with 'plume' glasses from the flank of Kilauea (Kyser and Rison, 1982) showed strong seawater contamination effects (Fig. 11.14). Hart *et al.* (1983) showed that some of these effects were due to sample alteration, since repeat analysis yielded a markedly different He/Ar ratio. However, they suggested that other samples might have been contaminated as magmas, during emplacement through the thickened oceanic crust at the base of the volcano.

In order to further evaluate the argon isotope signature of MORB-type and un-degassed mantle sources, Allegre *et al.* (1983) and Staudacher *et al.* (1986) compared argon and helium isotope ratios in glasses from ocean ridges and from the Hawaiian

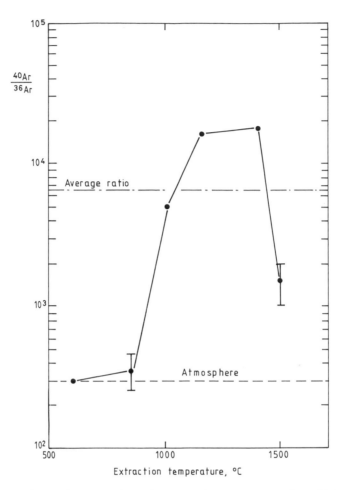

Fig. 11.13. Plot of measured argon isotope ratio against extraction temperature for MORB glass showing probable atmospheric contamination of the low-temperature (< 1000 °C) fractions. After Staudacher and Allegre (1982).

plume. Loihi Seamount gave the highest ^{3}He/^{4}He ratios for uncontaminated mantle-derived materials (presented in inverted form as the ^{4}He/^{3}He ratio in Fig. 11.15). Since this helium signature of Loihi glasses is regarded as primordial, Allegre *et al.* interpreted the low argon isotope ratio in these samples as likewise indicative of an un-degassed mantle source.

In Fig. 11.15 the MORB field has a forked shape, attributed by Staudacher *et al.* (1986) to mixing with primordial and atmospheric helium reservoirs. Both mixing branches show very strong curvature, attributed to the very low argon content of the degassed MORB reservoir, relative to both atmosphere and plume reservoirs. In contrast, a dunite xenolith from Loihi lay far off the MORB–plume mixing line. The argon signature of this sample is consistent with a source in oceanic lithosphere. The high ^{3}He content of this sample cannot be cosmogenic (as for some other xenoliths), since it is submarine; therefore, it was attributed to diffusion of helium from the host magma into the xenolith before eruption.

Allegre *et al.* (1983) observed that MORB glasses also fell on reasonable mixing lines between degassed and un-degassed (plume) sources on a plot of ^{40}Ar/^{36}Ar against strontium isotope ratio (Fig. 11.16). They argued that these mixing lines had a different trajectory than would be expected from surficial contamination processes, and therefore that the isotope data were indicative of processes in the mantle source. However, because rare gases may be decoupled from lithophile element systems, the trajectory of mixing lines may not be a reliable indicator of the end-member compositions.

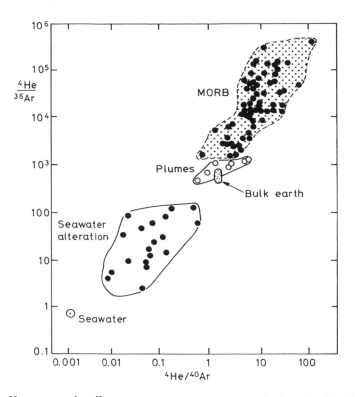

Fig. 11.14. Plot of ^{4}He/^{36}Ar against ^{4}He/^{40}Ar showing the compositions of submarine basalt glasses in relation to seawater and the calculated Bulk Earth composition. (○) = Reykjanes Ridge glasses analysed by Hart *et al.*, 1983. After Hart *et al.* (1985).

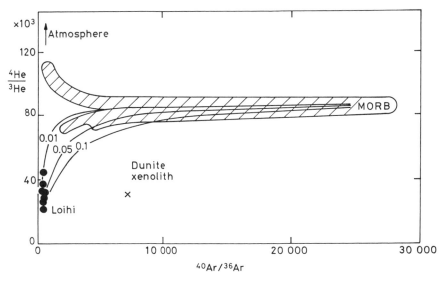

Fig. 11.15. Plot of helium *versus* argon isotope ratio for submarine glasses from MORB and plume environments. Mixing lines are shown for different ^{3}He/^{36}Ar ratios in Loihi relative to MORB sources. After Staudacher *et al.* (1986).

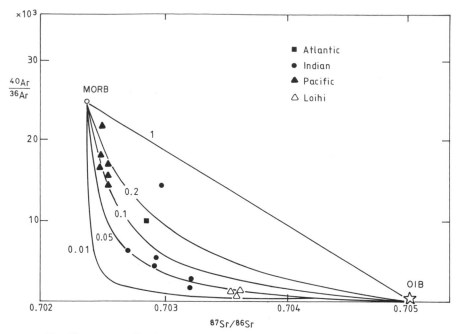

Fig. 11.16. Plot of $^{40}Ar/^{36}Ar$ against $^{87}Sr/^{86}Sr$ for submarine basalt glasses, suggesting coherent mixing between MORB and OIB source mantle. Labels indicate the $^{36}Ar/^{86}Sr$ ratio in the MORB end-member relative to OIB for different mixing lines. After Allegre *et al.* (1986).

Patterson *et al.* (1990) argued that helium isotope systematics in Loihi glasses might also be decoupled from argon, due to the extreme difference in He/Ar ratio between the end-members (as demonstrated, for example, by the hyperbolic form of proposed mixing lines in Fig. 11.15). They pointed out that seawater has between two and four orders of magnitude more ^{36}Ar than Loihi glasses, but two orders of magnitude less ^{3}He. Hence, argon isotope ratios in Loihi magmas might have been contaminated by seawater without affecting their helium signature. In MORB glasses, variable contamination of this kind generates correlations between $^{40}Ar/^{36}Ar$ ratio and $1/^{36}Ar$ abundances, indicative of simple mixing between atmospheric and mantle argon (Fisher, 1986). However, Loihi data do not display such a correlation. Hence, the atmospheric contamination model is hard to evaluate critically in plume environments.

Farley and Craig (1994) made a new examination of this problem, based on helium and argon measurements in olivine phenocrysts from a tholeiitic plume basalt. In this sample from the Juan Fernandez hotspot, Farley and Craig were able to demonstrate a positive correlation between ^{4}He and ^{40}Ar abundances released from fluid inclusions by crushing of phenocrysts. This correlation line is inconsistent with

significant atmospheric contamination of these isotopes, since the atmosphere has negligible ^{4}He. Therefore, the correlation must be attributed to variable gas inventories sampled from a mantle source with constant $^{4}He/^{40}Ar$ ratio.

In contrast, Farley and Craig observed no correlation between ^{4}He and ^{36}Ar abundances. Therefore, they argued that ^{36}Ar abundances must have been perturbed by atmospheric contamination. However, since even gas rich samples showed this behaviour, analytical blank was an unlikely cause. Instead, Farley and Craig attributed the effect to contamination of the magma by seawater argon within the oceanic crust. Hence, they argued that the mantle source sampled by the Juan Fernandez hotspot had a minimum $^{40}Ar/^{36}Ar$ ratio equal to the maximum observed $^{40}Ar/^{36}Ar$ ratio of 7700. However, this is not a lower limit on the argon isotope ratio of the lower mantle (undegassed reservoir), since the Juan Fernandez plume may have been contaminated with radiogenic argon during its ascent through the MORB source.

If this seawater contamination model is correct, it also casts doubt on the reliability of basaltic glasses as samples of primordial argon from the lower mantle. However, it does not disprove the 'two-reservoir' model for mantle rare gases; it simply implies that the $^{40}Ar/^{36}Ar$ of the deep mantle cannot be determined directly. Neon isotope evidence may help to narrow the range of possible values by acting as a monitor for atmospheric contamination (section 11.4).

11.3 Xenon

Xenon is a heavy rare gas with nine stable isotopes (Fig. 11.17). Reynolds (1960) first demonstrated variations in the abundance of ^{129}Xe in meteorites, produced from the extinct nuclide ^{129}I (section 15.2.2). Four other isotopes are fission products of both ^{238}U (Wetherill, 1953), and the extinct nuclide ^{244}Pu (Kuroda, 1960). It is convenient to ratio xenon isotope abundances against ^{130}Xe, which is non-radiogenic and is also shielded from spallation production (Staudacher and Allegre, 1982).

Because Xe is a tracer for two extinct nuclides, meteorite 'xenology' is a powerful tool for studying the condensation of the solar system (Reynolds, 1963). However, 'terrestrial xenology' is also a powerful tool for understanding terrestrial differentiation (Staudacher and Allegre, 1982). The first evidence for excess ^{129}Xe in the Earth (relative to atmospheric xenon) was found in CO_2 well gases from Harding County, New Mexico (Butler et al., 1963). This evidence has such far-reaching implications that no less than four subsequent studies have been devoted to Harding County well gases (Boulos and Manuel, 1971; Phinney et al., 1978; Smith and Reynolds, 1981; Staudacher, 1987).

Studies of xenon isotope data from granitic rocks (Butler et al., 1963) showed that unlike the heavy xenon isotopes, ^{129}Xe is not generated in significant amounts (relative to non-radiogenic xenon) by fission or neutron activation reactions (Fig. 11.18). Therefore Butler et al. concluded that the ^{129}Xe excess in mantle-derived gases must have been due to decay, soon after the formation of the Earth, of extinct ^{129}I. The presence of this extinct nuclide ($t_{1/2} = 16$ Myr) in the Earth demonstrates that its accretion must have occurred within a few Myr of meteorites, which also commonly display ^{129}Xe anomalies (section 15.2.2).

Harding County well gases also contain anomalies in 131, 132, 134 and 136 xenon, relative to the atmosphere. The early work by Butler et al. (1963) and Hennecke and Manuel (1975) suggested that some of this was plutonogenic. However, more precise analysis by Phinney et al. (1978) showed that the excess abundances of ^{131}Xe, ^{132}Xe and ^{134}Xe relative to ^{136}Xe in well gas are a better match to the calculated isotope production from spontaneous fission of ^{238}U rather than ^{244}Pu (Fig. 11.19).

Further advances in terrestrial xenology required the well-gas data to be put into the wider perspective of major terrestrial reservoirs. Technical developments, allowing the xenon analysis of submarine glasses, made this possible. Staudacher and Allegre (1982) found that MORB glasses displayed a coherent relationship between excesses of ^{129}Xe and other heavy xenon isotopes. This was confirmed by subsequent work (Staudacher, 1987), which showed that the most ^{129}Xe-enriched well-gas analyses lay on the same correlation line as MORB (Fig. 11.20a).

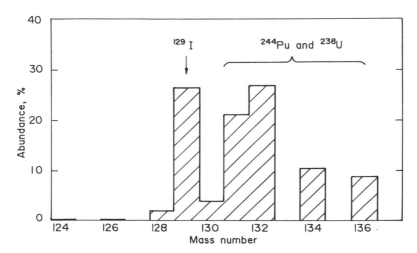

Fig. 11.17. Histogram showing the abundance of xenon isotopes in terrestrial rocks, along with the production route of radiogenic nuclides.

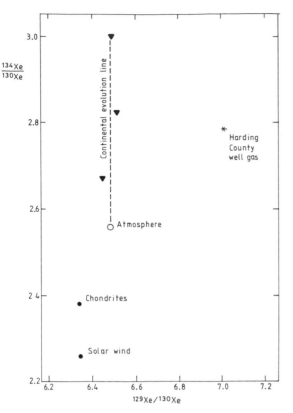

Fig. 11.18. Plot of fissiogenic ^{134}Xe/^{130}Xe against iodogenic ^{129}Xe/^{130}Xe for granites and well gases analysed prior to 1978. Modified after Staudacher and Allegre (1982).

Well-gas and MORB compositions were also coherent on a plot of ^{40}Ar/^{36}Ar against ^{129}Xe/^{130}Xe (Fig. 11.20b). This provides strong evidence that the Harding County well gases sample an upper mantle reservoir similar to the depleted MORB source.

Staudacher and Allegre (1982) argued that the ^{129}Xe excesses in MORB could be explained in the same way as the radiogenic Ar enrichments in MORB. If the degassing process which depleted the upper mantle in ^{36}Ar occurred very early in Earth history, then primordial rare gases including ^{130}Xe would also have been lost, while ^{129}I still remained in the mantle. The high I/Xe ratios generated by this process allowed the small amount of ^{129}I remaining in the upper mantle to generate measurable excesses in ^{129}Xe after its decay. Hence, upper mantle degassing occurred much earlier than lithophile depletion (by crust formation), although these two processes seem to affect roughly similar fractions of the mantle.

Analysis of xenon isotope compositions in glasses from Loihi Seamount and the Reykjanes Ridge gave

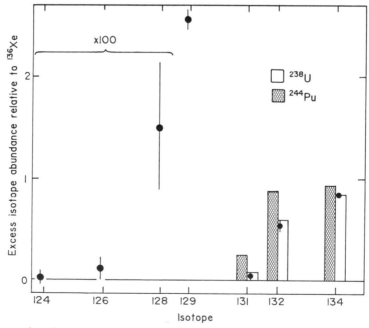

Fig. 11.19. Plot of excess abundances of xenon isotopes in well gas relative to the atmosphere, ratioed against excess ^{136}Xe. The data are compared with modelled production of fissiogenic xenon from ^{238}U and ^{244}Pu. After Phinney et al. (1978).

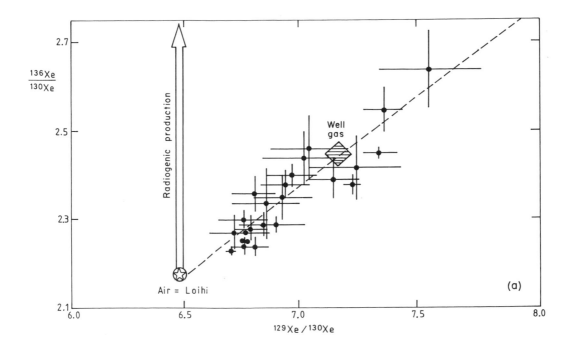

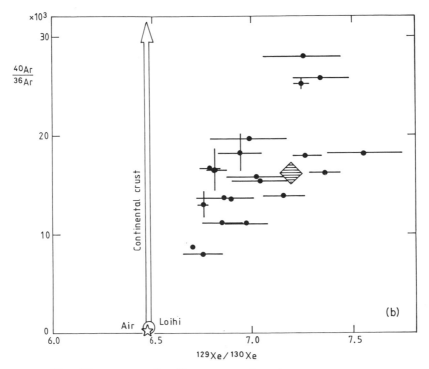

Fig. 11.20. Plots of (a) $^{136}Xe/^{130}Xe$, and (b) $^{40}Ar/^{36}Ar$ against $^{129}Xe/^{130}Xe$ for MORB glasses, compared to other terrestrial components. After Staudacher (1987).

results within error of atmospheric rare gases, but the origin of this signature has been widely disputed. Allegre *et al.* (1983) and Hart *et al.* (1983) attributed it to a less-degassed mantle source, in which radiogenic ^{129}Xe and ^{40}Ar are swamped by large primordial rare gas contents. Alternatively, Ozima *et al.* (1985) proposed that atmospheric xenon is recycled into the deep mantle. However, Staudacher and Allegre (1988) argued that subduction-related volcanism scavenges atmospheric gases from subducted sediments before they can reach the deep mantle. Evidence from the volatile element boron seems to support this contention (section 14.3.4). Finally, Patterson *et al.* (1990) and Craig (1994) argued that the xenon isotope ratios of Loihi glasses (along with argon) are due to atmospheric contamination, introduced directly into plume magmas from seawater. In this case the xenon isotope signature of the lower mantle (like argon) is somewhat indeterminate. Nevertheless, Farley and Poreda (1993), after demonstrating the use of neon isotopes as a monitor of atmospheric contamination, proposed a 'corrected' xenon isotope composition for the plume end-member which was similar to the Loihi/atmosphere values.

Additional evidence for the xenon isotope evolution of mantle reservoirs has been obtained from the analysis of 'coated' diamonds (Ozima and Zashu, 1991). The coats of these diamonds contain relatively large rare gas contents, and are thus suited to isotopic analysis. Ozima and Zashu found xenon isotope ratios identical to the MORB correlation line (Fig. 11.21), suggesting that the same evolution processes gave rise to the mantle sources of MORB and diamonds. In the light of argon and neon isotope evidence (sections 11.2 and 11.4), the MORB array is best attributed to atmospheric contamination of magmas with radiogenic xenon. Similarly, Ozima and Zashu attributed the diamond array to mixing between a radiogenic xenon source and atmospheric xenon contamination (represented by the cores).

The heavy xenon isotope signatures in diamonds can be used to test for plutonogenic or uranogenic production, as demonstrated above for well gases. Ozima and Zashu showed that production from ^{238}U yields the best fit to the xenon data on diamond coats, which in turn places a tight constraint on the time of formation of these coats. Uranogenic xenon (unlike iodogenic and plutonogenic) grows in the mantle over Earth history. Therefore, if the rare gases in diamonds have been isolated from the silicate upper mantle for the past few Byr, they should have developed less ^{136}Xe than young MORB samples (dashed line in Fig. 11.21). The fact that MORB, diamond coats and well gases all lie on the same array (if not a coincidence) suggests that all of these samples were formed relatively recently in geological time. This presents a problem, since other studies have indicated ancient formation ages for diamonds (section 4.2). However, it is possible that old diamond cores were overgrown by younger coats shortly before or during kimberlite magmatism (ca. 100 Myr ago).

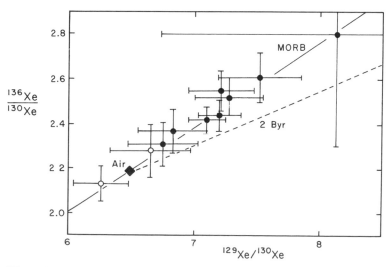

Fig. 11.21. Plot of ^{136}Xe/^{130}Xe against ^{129}Xe/^{130}Xe for coated diamonds, compared with the MORB correlation line of Staudacher (1987). (○) = cores; (●) = coats. After Ozima and Zashu (1991).

11.4 Neon

It is appropriate to discuss neon last amongst the rare gases, since problems discussed above concerning the Earth's primordial rare gas composition reach their most acute form in the interpretation of neon data. Of its three isotopes, ^{21}Ne and ^{22}Ne are nucleogenic and their variations are well understood. In contrast, ^{20}Ne is not known to be nucleogenic, and the causes of its variation in the Earth have been hotly debated. However, a solution to this problem may resolve the ambiguities which remain in the interpretation of argon and xenon isotope signatures from mantle plumes.

The principal nuclear reactions which generate neon isotopes are n,α reactions on ^{24}Mg and ^{25}Mg, which produce ^{21}Ne and ^{22}Ne respectively. Subsidiary reactions are α,n reactions on ^{18}O and ^{19}F which produce ^{21}Ne and ^{22}Na, the latter undergoing β decay to ^{22}Ne. The α,n reaction on ^{17}O to yield ^{20}Ne is unimportant, due to the low abundance of the parent. All α particles are derived from the U-series decay chains, while the neutrons are mostly produced by secondary reactions from α particles. These reactions were first studied by Wetherill (1954), and

have been refined in subsequent work (e.g. *see* Kennedy *et al.*, 1990). The net result of these reactions is to yield a trend towards lower $^{20}Ne/^{22}Ne$ and higher $^{21}Ne/^{22}Ne$ ratios which is most clearly seen in uranium-rich rocks such as granites. Fig. 11.22 shows isotopic data for gas wells from Alberta, Canada, plotted on the commonly used neon three-isotope diagram. The data form a linear array which was attributed to mixing between atmospheric and nucleogenic neon. This is consistent with helium isotope data for these gases, which show a strong radiogenic signature with no mantle-derived component.

Isotopic analysis of exposed terrestrial rocks has also demonstrated the cosmogenic production of ^{21}Ne (Marty and Craig, 1987). This isotope is generated by spallation reactions on Mg, Na, Si and Al, generating a sub-horizontal array on the three-isotope plot. By analysing all three isotopes, the cosmogenic component can be resolved from trapped (magmatic) neon and nucleogenic neon. Graf *et al.* (1991) applied the cosmogenic neon method to quartz separates from Antarctic rocks which had previously been dated by ^{26}Al and ^{10}Be (section 14.6). They demonstrated coherent behaviour of ^{21}Ne with these other cosmo-

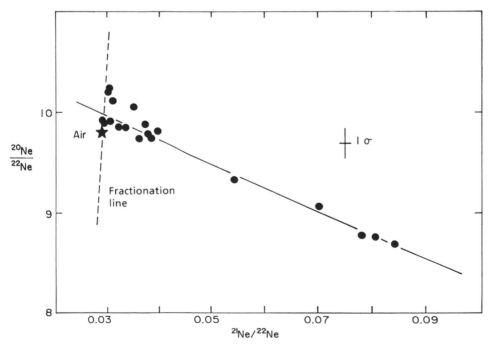

Fig. 11.22. Neon three-isotope correlation diagram showing well gases from the Alberta basin on a mixing line between atmospheric and nucleogenic neon. After Kennedy *et al.* (1990).

genic isotopes, suggesting that neon will be a useful tool in determining cosmic exposure ages of surficial rocks.

The first evidence for non-atmospheric neon in the mantle was presented by Craig and Lupton (1976) on samples of MORB and volcanic gases. These are enriched in ^{20}Ne, as well as nucleogenic ^{21}Ne, relative to ^{22}Ne contents. Subsequently, Harding County well gas was also found to have a composition well removed from atmosphere (Phinney *et al.*, 1978). These ^{20}Ne-enriched components were attributed to exotic primordial rare gas components in the Earth, possibly representing solar neon.

In contrast, Kyser and Rison (1982) speculated that ^{20}Ne enrichment in the analysed samples might be due to mass fractionation of neon from an original mantle composition similar to the atmosphere. They compared a compilation of mantle-derived neon analyses with neon data from geothermal gases in Japan (Nagao *et al.*, 1979). These gases display a mass-fractionation trend (Fig. 11.23) which is explained by preferential diffusion of light neon through the soil to the sampling sites. However, this process also caused marked mass fractionation across the whole argon, krypton and xenon mass spectra, which is not seen in mantle-derived samples relative to atmosphere. Therefore, it is concluded that the neon analyses of

mantle-derived samples cannot be attributed to diffusionally induced fractionation.

Elevated ^{20}Ne abundances were also found in diamonds (Fig. 11.24) by Honda *et al.* (1987) and Ozima and Zashu (1988; 1991). These analyses disprove the ^{20}Ne 'enrichment' model of Kyser and Rison, since diamonds represent *in-situ* solid samples of the mantle. Therefore, Ozima and Zashu reversed the mass fractionation argument of Kyser and Rison, suggesting that diamonds sample a solar neon reservoir in the Earth, whereas the present-day atmosphere has been *depleted* in ^{20}Ne by mass fractionation. They argued that bombardment of the early Earth by radiation caused the massive blow-off of a primitive solar-type atmosphere, leaving a residue enriched in heavy neon. However, this argument is subject to the same objection as the Kyser and Rison model. Neon fractionations of the proposed magnitude between mantle and atmosphere should be accompanied by fractionation of the non-radiogenic isotopes of argon and krypton. In fact, gas-rich MORB glasses have the same ^{38}Ar/^{36}Ar and krypton isotope ratios as the atmosphere (Sarda, *et al.*, 1985; Staudacher *et al.*, 1989). Therefore, gross neon isotope fractionation of the atmosphere relative to mantle is unlikely.

Additional insights into terrestrial neon systematics

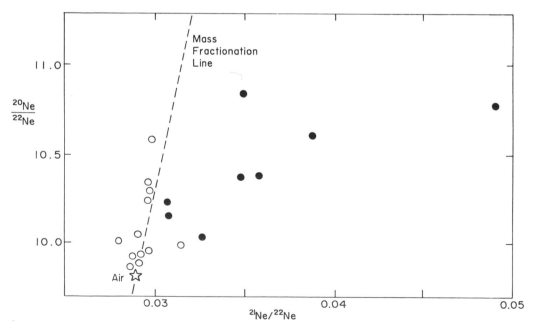

Fig. 11.23. Comparison of mantle samples (●), and mass-fractionated geothermal gases (○), on a neon three-isotope diagram. After Kyser and Rison (1982).

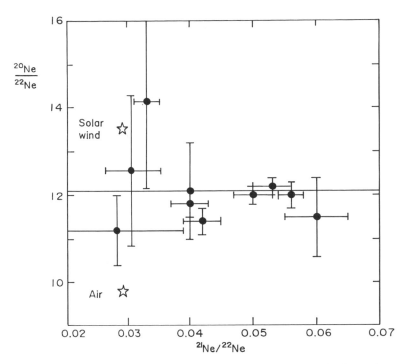

Fig. 11.24. Neon isotope data for diamonds compared with the composition of other solar system components. (The solar-wind composition was determined by analysis of Lunar soil.) For data sources, see text.

have been sought from the analysis of submarine basaltic glasses. Sarda *et al.* (1988), demonstrated the existence of a MORB correlation line passing through the atmosphere point, which has been confirmed by subsequent data (Marty, 1989; Hiyagon *et al.*, 1992). This array (Fig. 11.25) can be explained by three-component mixing of solar-type, atmosphere-type and nucleogenic neon. Sarda *et al.* (1988) also determined neon isotope ratios in several Loihi glasses which fell within error of atmosphere. They believed that these signatures represented a genuine primordial mantle signature, consistent with this reservoir being the principal source of atmospheric neon.

In this model, the primordial Earth would have had a planetary neon signature, which might be achieved by averaging the isotopic compositions of gas-rich meteorites, of which the most commonly recognised are termed neon A, B and C (Black and Pepin, 1969; Black, 1972). The source of the solar-type neon could then be regarded as the vestige of a distinct primordial neon reservoir in the upper mantle (Sarda *et al.*, 1988).

However, subsequent analyses of submarine basalt glasses from Loihi and nearby Kilauea revealed a wider range of neon isotope ratios, stretching from the atmospheric composition towards ^{20}Ne-enriched compositions (Honda *et al.*, 1991; Hiyagon *et al.*, 1992). The enriched end of this array approaches the solar wind composition, but the array has a slope intermediate between the pure mass-fractionation line and the MORB correlation line (Fig. 11.26). Ultra-mafic xenoliths from Reunion and two different localities in the Samoan islands also define neon isotope arrays with slopes intermediate between Hawaii and MORB (Staudacher *et al.*, 1990; Poreda and Farley, 1992).

In order to explain their data, Honda *et al.* (1991; 1993) and Hiyagon *et al.* (1992) attributed all neon in the Earth's interior to mixing between solar and nucleogenic isotopes. The sloping arrays were then attributed to variable atmospheric contamination of this solar mantle neon, and the Loihi neon samples originally analysed by Sarda *et al.* (1988) were attributed almost entirely to atmospheric contamination.

The slopes of data arrays on the three-isotope plot can be expressed by the notation $\delta^{20}Ne/\delta^{21}Ne$ in order to compare neon with other rare gas data. This is done by forcing a regression line through the

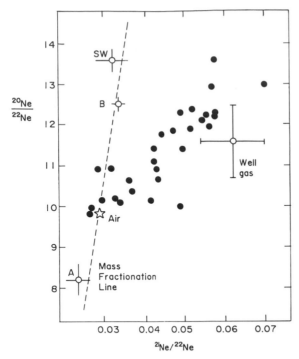

Fig. 11.25. Compilation of MORB neon data (●) on a three-isotope plot. Also, SW = solar wind and A, B = planetary neon compositions from meteorites. For data sources, see text.

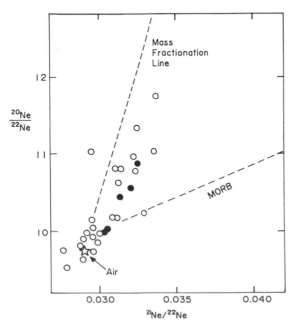

Fig. 11.26. Compilation of Hawaiian neon data on a three-isotope plot. (○) = Loihi; (●) = Kilauea. For data sources, see text.

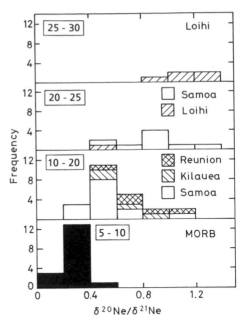

Fig. 11.27. Histograms of $\delta^{20}Ne/\delta^{21}Ne$, as a function of helium R/R_a ratio for submarine glasses and ultramafic xenoliths. After Poreda and Farley (1992).

atmosphere composition, whereupon the gradient is equal to the ratio $\delta^{20}Ne/\delta^{21}Ne$. Poreda and Farley (1992) presented these results in the form of histograms (Fig. 11.27), categorised according to $^{3}He/^{4}He$ ratio (expressed as R/R_a). The results suggest a correlation between neon and helium isotope data, which can be explained by the addition of a nucleogenic–radiogenic neon–helium component to a variably degassed primordial component. Therefore, the near-vertical array of Hawaiian neon isotope analyses measured by Honda et al. (1991) shows that this plume does indeed sample a relatively undegassed source. Hence, the neon isotope data continue to support the 'two reservoir' model for mantle rare gases.

Farley and Poreda (1993) developed the 'atmospheric contamination' model for neon, suggesting that the $^{20}Ne/^{22}Ne$ ratio could be used to monitor and correct atmospheric contamination in other rare gases such as argon. However, because the end-members will have different rare gas abundance ratios (e.g. Ne/Ar), mixing will generate hyperbolic rather than linear arrays, leading to somewhat greater uncertainty in the calculation of uncontaminated end-members.

In order to constrain the curvature of mixing lines,

Farley and Poreda examined Ne–Ar isotope systematics in MORB glasses. The best estimate for $^{40}Ar/^{36}Ar$ in the MORB reservoir remains the maximum value near 30 000 found in a 'popping rock' from the Mid Atlantic Ridge (Staudacher *et al.*, 1989). The very high volatile contents of these magmas probably rendered them relatively immune to atmospheric

contamination. On the other hand, most MORB analyses lie between mixing lines with $^{22}Ne/^{36}Ar$ ratios (*r*) of 0.06–0.6 in atmospheric relative to mantle components.

Many OIB analyses fall within similar bounds (Fig. 11.28) due to contamination of plumes by upper mantle material prior to mixing with atmospheric rare gases. Based on the slightly different Ne/Ar ratios in OIB and MORB, Farley and Poreda estimated that simple two-component mixing between plume and atmospheric rare gases should fall between mixing lines with relative $^{22}Ne/^{36}Ar$ ratios of 0.07 and 0.7. If we fit these curves to the OIB samples with lowest argon isotope ratio, then we can calculate a 'revised' $^{40}Ar/^{36}Ar$ ratio for primordial terrestrial argon. This has a value between 1000 and 3500 (of which Farley and Poreda preferred the latter), compared with the value of 400 proposed by Allegre *et al.* (1983). The higher value would then imply substantial degassing (and consequent radiogenic growth) of this 'primordial' argon reservoir.

The atmospheric contamination model successfully explains the neon isotope arrays in submarine glasses, but it brings us back to the problem of explaining the origin of the atmospheric neon signature. Many workers continue to support neon fractionation during the burn-off of an early terrestrial atmosphere of solar composition (Ozima and Zashu, 1988; 1991). However, an alternative proposed by Marty (1989) is that the atmosphere was formed by late accretion of

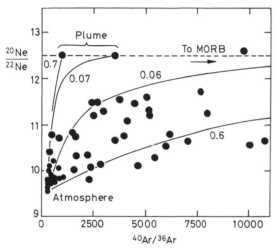

Fig. 11.28 Plot of neon *versus* argon isotope compositions in submarine glasses from ocean islands, showing possible mixing lines due to atmospheric contamination of the rare gas inventories of MORB and plume magmas. Modified after Farley and Poreda (1993).

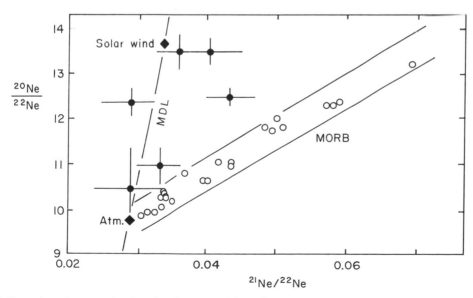

Fig. 11.29. Neon three-isotope plot showing the composition of cosmic dust particles (●) relative to the MORB array and the mass discrimination line (MDL) through solar and atmospheric neon. After Allegre *et al.* (1993).

gas-rich meteorites with planetary neon (e.g. neon A in Fig. 11.25). These could have been accreted to the surface of the Earth subsequent to the formation of the mantle with a solar neon budget. This model is consistent with the fact that neon is the only rare gas to show gross heterogeneity of non-radiogenic isotope ratios in different solar-system bodies.

Allegre *et al.* (1993) attempted to reinstate the planetary model for mantle neon by invoking a new mechanism to explain high $^{20}Ne/^{22}Ne$ ratios of solar type in the earth's mantle. They attributed these signatures to the subduction of cosmic dust particles accumulated in deep-sea sediments. These dust particles become implanted with neon from the solar wind during their exposure in space. Analysis of this material in the atmosphere and in deep sea sediment reveals $^{20}Ne/^{22}Ne$ ratios which span the range between atmospheric and solar compositions (e.g. Nier and Schlutter, 1990). Matsuda *et al.* (1990) suggested that these particles could survive the 'noble gas subduction barrier' and deliver cosmic neon to the deep mantle. This could explain the high $^{20}Ne/^{22}Ne$ ratios of submarine glasses without having to invoke solar-type primordial neon in the earth (Fig. 11.29). However, experimental studies by Hiyagon (1994) suggested that neon would be completely extracted from cosmic dust within 3 years at 500 °C, which is insufficient to sustain this model. Hence, the solar neon model for the Earth is now generally accepted.

References

Aldrich, L. T. and Nier, A. O. (1948). The occurrence of He^3 in natural sources of helium. *Phys. Rev.* **74**, 1590–4.

Allegre, C. J., Sarda, P. and Staudacher, T. (1993). Speculations about the cosmic origin of He and Ne in the interior of the Earth. *Earth Planet. Sci. Lett.* **117**, 229–33.

Allegre, C. J., Staudacher, T., Sarda, P. and Kurz, M. (1983). Constraints on evolution of Earth's mantle from rare gas systematics. *Nature* **303**, 762–6.

Alvarez, L. W. and Cornog, R. (1939). Helium and hydrogen of mass 3. *Phys. Rev.* **56**, 613.

Anderson, D. L. (1993). Helium-3 from the mantle: primordial signal or cosmic dust? *Science* **261**, 170–6.

Black, D. C. (1972). On the origins of trapped helium, neon and argon isotopic variations in meteorites, II. Carbonaceous chondrites. *Geochim. Cosmochim. Acta* **36**, 377–94.

Black, D. C. and Pepin, R. O. (1969). Trapped neon in meteorites. *Earth Planet. Sci. Lett.* **6**, 395–405.

Boulos, M. S. and Manuel, O. K. (1971). The xenon record of extinct radioactivities in the Earth. *Science* **174**, 1334–6.

Butler, W. A., Jeffery, P. M., Reynolds, J. H. and Wasserburg, G. J. (1963). Isotopic variations in terrestrial xenon. *J. Geophys. Res.* **68**, 3283–91.

Cerling, T. E. (1989). Dating geomorphologic surfaces using cosmogenic ^{3}He. *Quat. Res.* **33**, 148–56.

Clarke, W. B., Beg, M. A. and Craig, H. (1969). Excess ^{3}He in the sea: evidence for terrestrial primordial helium. *Earth Planet. Sci. Lett.* **6**, 213–20.

Clarke, W. B., Jenkins, W. J. and Top, Z. (1976). Determination of tritium by mass-spectrometric measurement of ^{3}He. *Int. J. Appl. Rad. Isot.* **27**, 515–22.

Craig, H. (1994). Noble gases in the mantle and atmosphere. In: Lanphere, M. A., Dalrymple, G. B. and Turrin, B. D. (Eds.), *Abs. 8th Int. Conf. on Geochron, Cosmochron. & Isot. Geol., U. S. Geol. Surv. Circ* **107**, p. 70.

Craig, H. and Lupton, J. E. (1976). Primordial neon, helium, and hydrogen in oceanic basalts. *Earth Planet. Sci. Lett.* **31**, 369–85.

Craig, H. and Poreda, R. J. (1986). Cosmogenic ^{3}He in terrestrial rocks: the summit lavas of Maui. *Proc. Natl. Acad. Sci. USA* **83**, 1970–4.

Damon, P. E. and Kulp, L. (1958). Excess helium and argon in beryl and other minerals. *Amer. Miner.* **43**, 433–59.

Fanale, F. P. (1971). A case for catastrophic early degassing of the Earth. *Chem. Geol.* **8**, 79–105.

Farley, K. A. and Craig, H. (1994). Atmospheric argon contamination of ocean island basalt olivine phenocrysts. *Geochim. Cosmochim. Acta* **58**, 2909–17.

Farley, K. A., Natland, J. H. and Craig, H. (1992). Binary mixing of enriched and undegassed (primitive?) mantle components (He, Sr, Nd, Pb) in Samoan lavas. *Earth Planet. Sci. Lett.* **111**, 183–99.

Farley, K. A. and Poreda, R. J. (1993). Mantle neon and atmospheric contamination. *Earth Planet. Sci. Lett.* **114**, 325–39.

Fisher, D. E. (1971). Incorporation of Ar in East Pacific basalts. *Earth Planet. Sci. Lett.* **12**, 321–4.

Fisher, D. E. (1983). Rare gases from the undepleted mantle? *Nature* **305**, 298–300.

Fisher, D. E. (1985). Noble gases from oceanic island basalts do not require an undepleted mantle source. *Nature* **316**, 716–18.

Fisher, D. E. (1986). Rare gas abundances in MORB. *Geochim. Cosmochim. Acta* **50**, 2531–41.

Gerling, E. K., Mamyrin, B. A., Tolstikhin, I. N. and Yakovleva, S. S. (1971). Isotope composition of helium in some rocks. *Geokhimiya* **1971**(10), 1209–17.

Graf, T., Kohl, C. P., Marti, K. and Nishiizumi, K. (1991). Cosmic-ray-produced neon in Antarctic rocks. *Geophys. Res. Lett.* **18**, 203–6.

Hart, R., Dymond, J. and Hogan, L. (1979). Preferential formation of the atmosphere–sialic crust system from the upper mantle. *Nature* **278**, 156–9.

Hart, R., Dymond, J., Hogan, L. and Schilling, J. G. (1983). Mantle plume noble gas component in glassy basalts from Reykjanes Ridge. *Nature* **305**, 403–7.

Hart, R., Hogan, L. and Dymond, J. (1985). The closed-system approximation for evolution of argon and helium in the mantle, crust and atmosphere. *Chem. Geol. (Isot. Geosci. Section)* **52**, 45–73.

Hennecke, E. W. and Manuel, O. K. (1975). Noble gases in

CO_2 well gas, Harding County, New Mexico. *Earth Planet. Sci. Lett.* **27**, 346–55.

Hiyagon, H. (1994). Retention of solar helium and neon in IDPs in deep sea sediment. *Science* **263**, 1257–9.

Hiyagon, H., Ozima, M., Marty, B., Zashu, S. and Sakai, H. (1992). Noble gases in submarine glasses from mid-ocean ridges and Loihi seamount: constraints on the early history of the Earth. *Geochim. Cosmochim. Acta* **56**, 1301–16.

Honda, M., McDougall, I. and Patterson, D. B. (1993). Solar noble gases in the Earth: The systematics of helium–neon isotopes in mantle derived samples. *Lithos* **30**, 257–65.

Honda, M., McDougall, I., Patterson, D. B., Doulgeris, A. and Clague, D. A. (1991). Possible solar noble-gas component in Hawaiian basalts. *Nature* **349**, 149–51.

Honda, M., Reynolds, J. H., Roedder, E. and Epstein, S. (1987). Noble gases in diamonds: occurrences of solar-like helium and neon. *J. Geophys. Res.* **92**, 12507–21.

Jambon, A. and Zimmermann, J. L. (1987). Major volatiles from an Atlantic MORB glass: a size fraction analysis. *Chem. Geol.* **62**, 177–89.

Jeffrey, P. M. and Hagan, P. J. (1969). Negative muons and the isotopic composition of the rare gases in the Earth's atmosphere. *Nature* **223**, 1253.

Jochum, K. P., Hofmann, A. W., Ito, E., Seufert, H. M. and White, W. M. (1983). K, U and Th in mid-ocean ridge basalt glasses and heat production, K/U and K/Rb in the mantle. *Nature* **306**, 431–6.

Kaneoka, I. and Takaoka, N. (1980). Rare gas isotopes in Hawaiian ultramafic nodules and volcanic rocks: constraints on genetic relationships. *Science* **208**, 1366–8.

Kennedy, B. M., Hiyagon, H. and Reynolds, J. H. (1990). Crustal neon: a striking uniformity. *Earth Planet. Sci. Lett.* **98**, 277–86.

Kunz, W. and Schintlmeister, I. (1965). *Tabellen der Atomekerne, Teil II, Kernreaktionen*. AkademieVerlag, 1022 p.

Kuroda, P. K. (1960). Nuclear fission in the early history of the Earth. *Nature* **187**, 36–8.

Kurz, M. D. (1986a). Cosmogenic helium in a terrestrial rock. *Nature* **320**, 435–9.

Kurz, M. D. (1986b). *In-situ* production of terrestrial cosmogenic helium and some applications to geochronology. *Geochim. Cosmochim. Acta* **50**, 2855–62.

Kurz, M. D., Gurney, J. J., Jenkins, W. J. and Lott, D. E. (1987). Helium isotopic variability within single diamonds from Orapa kimberlite pipe. *Earth Planet. Sci. Lett.* **86**, 57–68.

Kurz, M. D. and Jenkins, W. J. (1981). The distribution of helium in oceanic basalt glasses. *Earth Planet. Sci. Lett.* **53**, 41–54.

Kurz, M. D., Jenkins, W. J. and Hart, S. R. (1982). Helium isotopic systematics of oceanic islands and mantle heterogeneity. *Nature* **297**, 43–6.

Kurz, M. D., Meyer, P. S. and Sigurdsson, H. (1985). Helium isotopic systematics within the neovolcanic zones of Iceland. *Earth Planet. Sci. Lett.* **74**, 291–305.

Kyser, T. K. and Rison, W. (1982). Systematics of rare gas isotopes in basaltic lavas and ultramafic xenoliths. *J. Geophys. Res.* **87**, 5611–30.

Lal, D. (1987). Production of ^{3}He in terrestrial rocks. *Chem. Geol. (Isot. Geosci. Section)* **66**, 89–98.

Lal, D., Nishiizumi, K., Klein, J., Middleton, R. and Craig, H. (1987). Cosmogenic ^{10}Be in Zaire alluvial diamonds: implications to ^{3}He excess in diamonds. *Nature* **328**, 139–41.

Lupton, J. E. (1983). Terrestrial inert gases: isotope tracer studies and clues to primordial components in the mantle. *Ann. Rev. Earth Planet. Sci.* **11**, 371–414.

Lupton, J. E. and Craig. H. (1975). Excess ^{3}He in oceanic basalts: evidence for terrestrial primordial helium. *Earth Planet. Sci. Lett.* **26**, 133–9.

Mamyrin, B. A., Anufriyev, G. S., Kamenskiy, I. L. and Tolstikhin, I. N. (1970). Determination of the composition of atmospheric helium. *Geochem. Int.* **7**, 498–505.

Mamyrin, B. A. and Tolstikhin, I. N. (1984). *Helium Isotopes in Nature*. Elsevier, 273 p.

Mamyrin, B. A., Tolstikhin, I.N., Anufriyev, G. S. and Kamenskiy, I. L. (1969). Anomalous isotopic composition of helium in volcanic gases. *Dokl. Akad. Nauka SSSR* **184**, 1197–9.

Mamyrin, B. A., Tolstikhin, I.N., Anufriyev, G. S. and Kamenskiy, I. L. (1972). Isotopic composition of helium in Icelandic hot springs. *Geokhimiya* **1972**(11), 1396.

Martel, D. J., Deak, J., Dovenyi, P., Horvath, F., O'Nions, R. K., Oxburgh, E. R., Stegena, L. and Stute, M. (1989). Leakage of helium from the Pannonian basin. *Nature* **342**, 908–12.

Marty, B. (1989). Neon and xenon isotopes in MORB: implications for the Earth–atmosphere evolution. *Earth Planet. Sci. Lett.* **94**, 45–56.

Marty, B. and Craig, H. (1987). Cosmic-ray-produced neon and helium in the summit lavas of Maui. *Nature* **325**, 335–7.

Marty, B. and Jambon, A. (1987). C/^{3}He in volatile fluxes from the solid Earth: implications for carbon geodynamics. *Earth Planet. Sci. Lett.* **83**, 16–26.

Matsuda, J., Murota, M. and Nagao, K. (1990). He and Ne isotopic studies on the extraterrestrial material in deep-sea sediments. *J. Geophys. Res.* **95**, 7111–17.

Matsuda, J., Sudo, M., Ozima, M., Ito, K., Ohtaka, O. and Ito, E. (1993). Noble gas partitioning between metal and silicate under high pressures. *Science* **259**, 788–90.

Merrihue, C. (1964). Rare gas evidence for cosmogenic dust in modern Pacific red clay. *Ann. N. Y. Acad. Sci.* **119**, 351–67.

Morrison, P. and Pine, J. (1955). Radiogenic origin of the helium isotopes in rocks. *Ann. N. Y. Acad. Sci.* **62**, 69–92.

Nagao, K., Takaoka, N. and Matsubayashi, O. (1979). Isotopic anomalies of rare gases in the Nigorikawa geothermal area, Hokkaido, Japan. *Earth Planet. Sci. Lett.* **44**, 82–90.

Nier, A. O. and Schlutter, D. J. (1990). Helium and neon in stratospheric particles. *Meteoritics* **25**, 263–7.

O'Nions, R. K. and Oxburgh, E. R. (1983). Heat and helium in the Earth. *Nature* **306**, 429–36.

O'Nions, R. K. and Oxburgh, E. R. (1988). Helium volatile fluxes and the development of continental crust. *Earth Planet. Sci. Lett.* **90**, 331–47.

Oxburgh, E. R., O'Nions, R. K. and Hill, R. I. (1986). Helium isotopes in sedimentary basins. *Nature* **324**, 632–5.

Ozima, M., Podosek, F. A. and Igarashi, G. (1985). Terrestrial xenon isotope constraints on the early history of the Earth. *Nature* **315**, 471–4.

Ozima, M. and Zashu, S. (1983). Primitive helium in diamonds. *Science* **219**, 1067–8.

Ozima, M. and Zashu, S. (1988). Solar-type Ne in Zaire cubic diamonds. *Geochim. Cosmochim. Acta* **52**, 19–25.

Ozima, M. and Zashu, S. (1991). Noble gas state of the ancient mantle as deduced from noble gases in coated diamonds. *Earth Planet. Sci. Lett.* **105**, 13–27.

Patterson, D. B., Honda, M. and McDougall, I. (1990). Atmospheric contamination: a possible source for heavy noble gases in basalts from Loihi Seamount, Hawaii. *Geophys. Res. Lett.* **17**, 705–8.

Pepin, R. O. and Signer, P. (1965). Primordial rare gases in meteorites. *Science* **149**, 253–65.

Phinney, D., Tennyson, J. and Frick, U. (1978). Xenon in CO_2 well gas revisited. *J. Geophys. Res.* **83**, 2313–19.

Poreda, R. J. and Farley, K. A. (1992). Rare gases in Samoan xenoliths. *Earth Planet. Sci. Lett.* **113**, 129–44.

Reynolds, J. H. (1960). Determination of the age of the elements. *Phys. Rev. Lett.* **4**, 810.

Reynolds, J. H. (1963). Xenology. *J. Geophys. Res.* **68**, 2939–56.

Rutherford, E. (1906). The production of helium from radium and the transformation of matter. In: Rutherford, E. *Radioactive Transformations*. Yale Univ. Press, pp. 187–93.

Sarda, P., Staudacher, T. and Allegre, C. J. (1985). $^{40}Ar/^{36}Ar$ in MORB glasses: constraints on atmosphere and mantle evolution. *Earth Planet. Sci. Lett.* **72**, 357–75.

Sarda, P., Staudacher, T. and Allegre, C. J. (1988). Neon isotopes in submarine basalts. *Earth Planet. Sci. Lett.* **91**, 73–88.

Sarda, P., Staudacher, T., Allegre, C. J. and Lecomte, A. (1993). Cosmogenic neon and helium at Reunion: measurement of erosion rate. *Earth Planet. Sci. Lett.* **119**, 405–17.

Schwartzman, D. W. (1973). Argon degassing models of the Earth. *Nature Phys. Sci.* **245**, 20–1.

Sheldon, W. R. and Kern, J. W. (1972). Atmospheric helium and geomagnetic field reversals. *J. Geophys. Res.* **77**, 6194–201.

Smith, S. P. and Reynolds, J. H. (1981). Excess ^{129}Xe in a terrestrial sample as measured in a pristine system. *Earth Planet. Sci. Lett.* **54**, 236–8.

Staudacher, T. (1987). Upper mantle origin for Harding County well gases. *Nature* **325**, 605–7.

Staudacher, T. and Allegre, C. J. (1982). Terrestrial xenology. *Earth Planet. Sci. Lett.* **60**, 389–406.

Staudacher, T. and Allegre, C. J. (1988). Recycling of oceanic crust and sediments: the noble gas subduction barrier. *Earth Planet. Sci. Lett.* **89**, 173–83.

Staudacher, T., Kurz, M. D. and Allegre, C. J. (1986). New noble-gas data on glass samples from Loihi Seamount and Hualalai and on dunite samples from Loihi and Reunion Island. *Chem. Geol.* **56**, 193–205.

Staudacher, T., Sarda, P. and Allegre, C. J. (1990). Noble gas systematics of Reunion Island, Indian Ocean. *Chem. Geol.* **89**, 1–17.

Staudacher, T., Sarda, P., Richardson, S. H., Allegre, C. J., Sagna, I. and Dmitriev, L. V. (1989). Noble gases in basalt glasses from a Mid-Atlantic Ridge topographic high at 14 °N: geodynamic consequences. *Earth Planet. Sci. Lett.* **96**, 119–33.

Tolstikhin, I. N., Mamyrin, B. A., Khabarin, L. V. and Erlikh, E. N. (1974). Isotopic composition of helium in ultrabasic xenoliths from volcanic rocks of Kamchatka. *Earth Planet. Sci. Lett.* **22**, 75–84.

Turekian, K. K. (1964). Outgassing of argon and helium from the Earth. In: Brancazio, P. and Cameron, A. G. W. (Eds.), *The Origin and Evolution of Atmospheres and Oceans*. Wiley, pp. 74–83.

Wernicke, R. S. and Lippolt, H. J. (1993). Botryoidal hematite from the Schwarzwald (Germany): heterogeneous uranium distributions and their bearing on the helium dating method. *Earth Planet. Sci. Lett.* **114**, 287–300.

Wetherill, G. W. (1953). Spontaneous fission yields from uranium and thorium. *Phys. Rev.* **82**, 907–12.

Wetherill, G. W. (1954). Variations in the isotopic abundances of neon and argon extracted from radioactive materials. *Phys. Rev.* **96**, 679–83.

Zadnik, M. G., Smith, C. B., Ott, U. and Begemann, F. (1987). Crushing of a terrestrial diamond: $^3He/^4He$ higher than solar meteorites. *Meteoritics* **22**, 541–2.

12 U-series dating

12.1 Secular equilibrium and disequilibrium

The intermediate nuclides in the uranium and thorium decay series have very short half-lives in comparison to their parents, and are usually ignored in the Pb isotope dating methods. However, their short half-lives make these nuclides useful for dating Pleistocene geological events which are too old to be well-resolved by the radiocarbon method and too young to be well-resolved by decay schemes with long half-lives. The manner in which U-series nuclides can fill this 'dating gap' is shown in Fig. 12.1. Generally they are most useful to date events of similar age to their half-life.

A distinctive property of the U-series nuclides which sets them apart from the other dating schemes is that the radiogenic daughters are also radioactive. Hence, in a uranium-bearing system which has been undisturbed for a few million years, a state of 'secular' equilibrium becomes established between the abundances of successive parent and daughter nuclides in the U and Th decay chains, such that the decay rate (or 'activity') of each daughter nuclide in the chain is equal to that of the parent:

$$\text{Activity} = \lambda_0 n_0 = \lambda_1 n_1 = \lambda_2 n_2 = \lambda_N n_N \qquad [12.1]$$

where λ_0 is the decay constant and n_0 is the number of atoms of the original parent, λ_1 and n_1 are the decay constant and abundance of the 1st daughter, and so on. It follows that the abundance of each nuclide will be directly proportional to its half-life (i.e. inversely proportional to its decay constant). The relevant parts of the decay chains are shown in Fig. 12.2.

During geological processes such as erosion, sedimentation, melting or crystallisation, different nuclides in the decay series can become fractionated

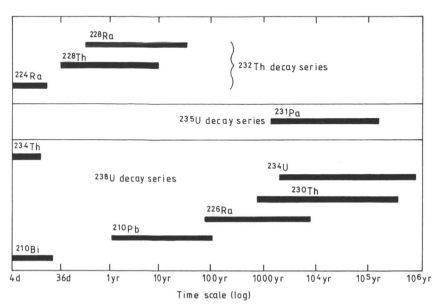

Fig. 12.1. Diagram showing the dating ranges of different nuclides within the three U-series decay chains to show their utility. After Potts (1987).

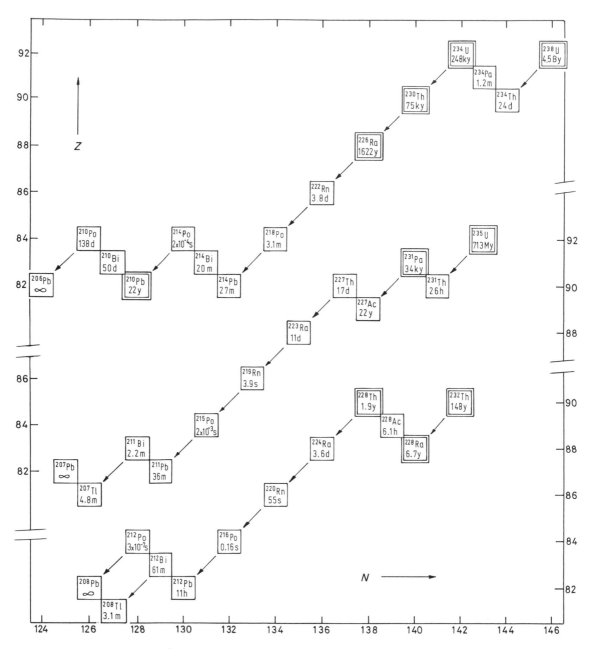

Fig. 12.2. Part of the chart of the nuclides, in term of Z against N, to show species in the Th- and U-series decay chains and their half-lives. Useful species are indicated by double boxes. Before the chemistry of the decay series nuclides was investigated, many of the species were given provisional names which are now obsolete. The only one of these much used today is ionium (^{230}Th). The latter nomenclature will be used here.

relative to one another, due to variations in their chemistry or the structural site they occupy. This results in a state of secular disequilibrium. Such a situation can be utilised in two different ways as a

chronological tool, called respectively the daughter-excess and daughter-deficiency dating methods.

In the daughter-excess method, a deposit is formed with an excess of the daughter beyond the level which

Table 12.1. *U-series dating methods*

Method	Measurement	$t_{1/2}$, kyr	Range, kyr	Application
Daughter excess				
$^{234}U-^{238}U$	^{234}U decay	245.0	< 1500	Coral
^{230}Th	^{230}Th "	75.4	< 500	Deep-sea sedimentation rates
^{231}Pa	^{231}Pa "	32.5	< 200	"
^{210}Pb	^{210}Pb "	0.022	< 0.1	Recent sedimentation
Daughter deficiency				
$^{230}Th-^{234}U$	^{230}Th accum.	75.4	< 500	Marine & fresh-water carbonate; volcanics
$^{231}Pa-^{235}U$	^{231}Pa "	32.5	< 200	"
$^{226}Ra-^{238}U$	^{226}Ra "	1.6	< 10	Closed-system test for ^{230}Th

can be sustained by the abundance of its parent nuclide. Over time, the 'unsupported' daughter decays back until secular equilibrium is restored. If the original fractionation can be determined then the age of the deposit can be calculated by the progress of decay of the excess.

In the daughter-deficiency method, chemical fractionation during the formation of a deposit causes it to take up a radioactive parent but effectively none of its daughter. The age of the deposit can then be determined by measuring the growth of the daughter, up to the point when its abundance is within error of secular equilibrium. Using high-precision mass spectrometric data (section 12.2.2) the useful range may be up to seven half-lives, but other factors may impose lower limits. Table 12.1 summarises some of the more important U-series dating methods.

12.2 Analytical methods

As noted above, the atomic abundance of a U-series nuclide in secular equilibrium is proportional to its half-life. Therefore, the very variable half-lives of the U-series radionuclides causes them to have extreme abundance ratios. Until recently, this has discouraged mass spectrometric determination of U-series nuclides for dating. In contrast, species in secular equilibrium have equal activities (by definition), so radioactive counting is an obvious method for their determination. The traditional technique for measurement of U-series nuclides has been α spectrometry. Counting techniques utilising β and γ particles are not favoured because of the low energies of β transitions

and the complexity of γ-ray spectra (Yokoyama and Nguyen, 1980).

12.2.1 Alpha spectrometry

Because of the very short penetration range of α particles in matter, samples to be counted must be made into thin films. They are normally placed under vacuum in a gridded ion chamber, which is a type of gas ionisation chamber with a short dead-time. If the applied potential between cathode and anode is within a certain range, the electrical pulses generated by α particle emission will be proportional in size to the kinetic energy of those particles. The output can then be fed to a multi-channel analyser in order to register count rates as a function of energy level. To obtain 1σ counting errors of 1%, total counts of 10^4 are required on each peak ($\sigma = \sqrt{n}$). To achieve this, counting times of at least a week are required for most natural samples. Recoil effects gradually contaminate the counter over time with U-series nuclides, raising its background. Hence the counter has a finite effective life, which is shortened if higher-than-normal count rates are measured.

Given the low abundances of the nuclides to be measured in natural materials (part per trillion to part per million range), and the need for a thin source, chemical purification is essential. This normally involves dissolution of the sample in HNO_3 (carbonates) or HF (silicates) followed by anion exchange separation (section 2.1.3). Anion exchange is also used to separate U from Th, since some important α emissions have overlapping energies. For example, the second most abundant decay energy

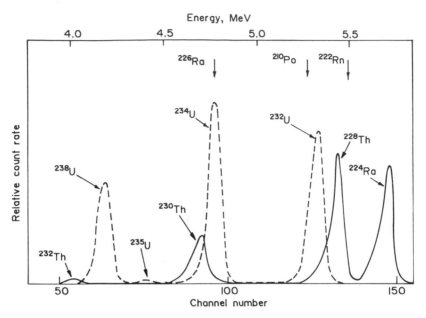

Fig. 12.3. Energy levels of α emissions from spiked U and Th extracts showing mutual interferences. After Kaufman and Broecker (1965).

of ^{234}U (4.72 MeV) almost exactly coincides with the primary ^{230}Th peak (4.68 MeV), given a channel width of ca. 0.07 MeV (Fig. 12.3).

Since chemical extractions are not expected to give a 100% yield, the sample is 'spiked' before chemistry with a known quantity of artificially enriched isotopes, allowing an isotope dilution determination of the sample abundances (section 2.2.6). A widely used U–Th spike is ^{232}U ($t_{1/2} = 72$ yr), which has been allowed to naturally generate its daughter, ^{228}Th ($t_{1/2} = 1.9$ yr). The short half-life of the latter nuclide means that it will be in secular equilibrium with its parent in ca. 20 years (Ivanovich, 1982a). Alternatively, ^{229}Th and ^{236}U may be used, which are separately manufactured. These have much longer half-lives (6 and 70 kyr respectively).

12.2.2 Mass spectrometry

Uranium-series dating by mass spectrometry represents one of the missed opportunities of 1970s isotope geology, since the analytical equipment available at that time was equal to this task, but was not applied until the late 1980s. This omission can be explained by a communications gap between workers in the two fields, and by exaggerated estimates of the problems which might be posed by large nuclide-abundance ratios. The gap was closed in two stages, by Chen *et*

al. (1986) who made the first precise mass spectrometric analysis on ^{234}U, and Edwards *et al.* (1987) who made the first ^{230}Th measurements. These workers showed that mass spectrometric U-series dating offered great improvements in precision over the best α counting determinations.

Edwards *et al.* avoided the difficulty of measuring large ^{238}U/^{234}U ratios by measuring ^{235}U/^{234}U instead. Since ^{238}U/^{235}U has a constant ratio of 137.88 in normal rocks, the conversion is simple. Furthermore, by analysing pure corals with a low detrital ^{232}Th content (see below) the ^{232}Th/^{230}Th abundance ratio was as low as 1.1 (compared to typical ratios of over 250 000 in silicate rocks). These techniques allowed Edwards *et al.* to determine the age of a typical Pleistocene coral to a precision of 123 ± 1.5 kyr (2σ), compared to an α counting determination of 129 ± 9 kyr.

Edwards *et al.* loaded both U and Th (separately) on graphite-coated single rhenium filaments, and analysed as the metal species. Under these conditions the cleanliness of the Th chemical separation is critical. This is demonstrated by the greater ionisation efficiency for very small Th samples, (with negligible ^{232}Th) compared to samples with a larger total Th content (Fig. 12.4). Li *et al.* (1989) demonstrated an alternative approach to Th ionisation which is less demanding of chemical purity. This

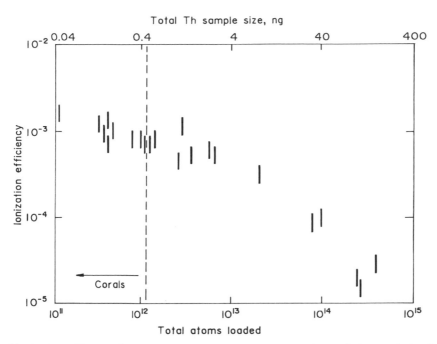

Fig. 12.4. Plot of ionisation efficiency for Th isotope analysis against the total size of Th sample loaded. The typical sample size of clean corals is shown. After Edwards *et al.* (1987).

uses double-filament beads with a very hot centre filament to promote the formation of Th metal ions. Edwards *et al.* used a double 236–233 uranium spike to correct for within-run U fractionation (section 2.2.7), whereas Li *et al.* normalised U isotope data to the natural 235/238 ratio. Both groups used a single ^{229}Th spike for thorium isotope analysis; short half-life spike nuclides cannot be used due to their extremely low abundances.

Mass spectrometric measurements on uraninite samples in secular equilibrium can be used to determine the half-lives of U-series nuclides. The half-life of ^{234}U can be determined very accurately relative to the well-constrained ^{238}U half-life by measurement of ^{234}U/^{238}U. Using this technique, Ludwig *et al.* (1992) determined a half-life of 245.3 ± 0.14 kyr, within error of the previously used value of 244.6 ± 0.7 kyr from α spectrometry (de Bievre *et al.*, 1971). The ^{230}Th half-life can also be determined by analysis of uraninite using a mixed ^{229}Th–^{236}U spike. However, this determination has larger error, since it incorporates the errors of spike calibration based on gravimetric U and Th standards. (Many laboratories calibrate their spike against uraninite, which would lead to a circular argument in this case.) So far, the half-life of 75.4 ± 0.6 kyr

from α counting has not been superseded (Meadows *et al.*, 1980).

In addition to a ten-fold improvement in precision over α counting, mass spectrometry also allows a ten-fold reduction in sample size. The latter improvement allows higher-resolution sampling, as demonstrated by Li *et al.* (1989) in the analysis of thin depositional bands of a speleothem (cave deposit). These technical improvements are causing mass spectrometry to oust α counting as the principal analytical tool in U-series dating. However, they also throw the emphasis of U-series dating work back onto sample collection and preparation, since open-system behaviour of samples becomes more obvious with improvements in analytical precision. These problems will be discussed below.

12.3 **Daughter excess**

12.3.1 ^{234}U

^{238}U decays via two very short-lived intermediates to ^{234}U (Fig. 12.2). Since ^{234}U and ^{238}U have the same chemical properties, it might be thought that they would not be fractionated in geological processes. However, Cherdyntsev and co-workers (1965, 1969) showed that such fractionation does occur. In fact, natural waters show a considerable range in $^{234}U/^{238}U$ activity from unity (secular equilibrium) to values of ten or more (e.g. Osmond and Cowart, 1982). Cherdyntsev *et al.* (1961) attributed these fractionations to radiation damage of crystal lattices, caused both by α emission and by recoil of parent nuclides. In addition, radioactive decay may leave ^{234}U in a more soluble +6 charge state than its parent (Rosholt *et al.*, 1963). These processes together enable preferential leaching of the two very short-lived intermediates and the longer-lived ^{234}U nuclide into groundwater (termed the 'hot atom' effect). The short-lived nuclides have a high probability of decaying into ^{234}U before they can be adsorbed onto a substrate, and ^{234}U is itself stabilised in surface waters as the soluble UO_2^{++} ion, due to the generally oxidising conditions prevalent in the hydrosphere.

The variety of weathering conditions prevailing in the terrestrial environment leads to very variable $^{234}U/^{238}U$ activity ratios in fresh-water systems. However, the long residence time of uranium in seawater (>300 kyr, Ku *et al.*, 1977) maintains seawater $^{234}U/^{238}U$ within narrow limits, corresponding to an activity ratio of 1.14 ± 0.03, (2σ), (Goldberg and Bruland, 1974). A major uranium

sink in the oceans is calcium carbonate, with which uranium is co-precipitated. This is deposited in shallow water by marine organisms and in deep water as an authigenic mineral (i.e. by direct chemical precipitation). At the time of deposition, this material takes on the 'daughter excess' character of seawater, but once isolated the excess decays away until secular equilibrium with the parent is regained (Fig. 12.5). Given an estimate of the original $^{234}U/^{238}U$ fractionation, and given subsequent closed system behaviour, the system can be used as a dating tool until it returns to within analytical error of secular equilibrium.

Unfortunately, many problems are encountered in the practical application of this method. As noted above, the variable uranium isotope fractionations observed in fresh-water systems preclude its application there. In addition, pelagic sediments are ruled out by open-system behaviour of uranium after deposition (Ku, 1965), while mollusc shells also tend to take up uranium after deposition (Kaufman *et al.*, 1971). However, the method has been applied with reasonable success to the dating of corals (e.g. Thurber *et al.*, 1965).

The decay of excess ^{234}U can be expressed by the fundamental decay equation [1.5]. Although this equation was derived in section (1.4) for atomic abundances, it is also true for activities (by dividing

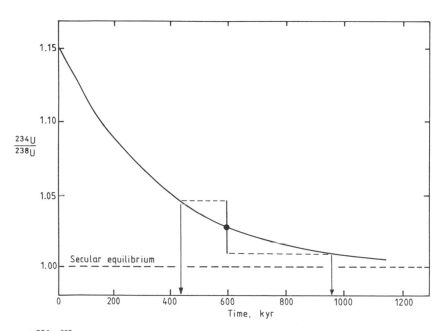

Fig. 12.5. Plot of $^{234}U/^{238}U$ activity against time showing the return to secular equilibrium after isolation from seawater. Arrows show the effect of typical α counting errors on age uncertainty.

both sides by the decay constant, e.g. $\lambda^{234}U$):

$$\frac{n}{\lambda} = \frac{n_0}{\lambda} \, e^{-\lambda t} \qquad [12.2]$$

$$A = A_0 e^{-\lambda t} \qquad [12.3]$$

In order to date a carbonate sample by the decay of excess ^{234}U (Fig. 12.5), we can substitute into equation [12.3] to yield:

$$^{234}U^{x}_{present} = {}^{234}U^{x}_{initial} \, e^{-\lambda_{234}t} \qquad [12.4]$$

where 'x' signifies excess activities above secular equilibrium, and 'initial' signifies the activity at the time of precipitation.

Henceforth in this chapter, all nuclide quantities will be presented in terms of activities unless otherwise stated. However, absolute activities are not as readily measurable as activity ratios, so it is convenient to divide through by ^{238}U activities. But because of the very long half-life of ^{238}U, the activity of $^{238}U_{present}$ is the same as $^{238}U_{initial}$. So:

$$({}^{234}U^{x}/{}^{238}U)_{present} = ({}^{234}U^{x}/{}^{238}U)_{initial} \, e^{-\lambda_{234}t} \qquad [12.5]$$

Since these quantities are in the form of activities, the excess $^{234}U/^{238}U$ activity is equal to the total activity ratio minus one (that part corresponding to secular equilibrium). So:

$$\left(\frac{{}^{234}U}{{}^{238}U}\right)^{total}_{present} - 1 = \left[\left(\frac{{}^{234}U}{{}^{238}U}\right)^{total}_{initial} - 1\right] e^{-\lambda_{234}t} \qquad [12.6]$$

Hence, if we assume that the initial activity ratio of the sample is given by present-day seawater, we can calculate the age of a coral simply by measuring the present-day activity ratio. Chen et al. (1986) showed that modern seawater in the Pacific and Atlantic oceans had a homogeneous level of ^{234}U activity, with values of 1.143 and 1.144 respectively. Given the > 300 kyr residence time of uranium in seawater (Ku et al., 1977), this gives us a strong expectation that the activity ratio should have been close to this value within the 1.2 Myr theoretical range of the method.

Unfortunately, the large analytical uncertainties of α counting (normally > 2%) limited the usefulness of the ^{234}U method in the past. For example, measurement errors of 2% on isotope ratios at 600 kyr led to uncertainties of ca. +350 / −150 kyr (Fig. 12.5). Therefore, the ^{234}U method was superseded by the more precise ^{230}Th deficiency method within the 500 kyr range of the latter technique (section 12.4.1), while above this range the ^{234}U method was itself unusable.

With the advent of mass spectrometric analysis the ^{230}Th deficiency method remains more powerful

below 500 kyr, but the ^{234}U method allows the possibility of dating back to 1 Myr with tolerable precision. For example, Moore et al. (1990) obtained a single ^{234}U age of 750 ± 13 kyr for a submerged coral terrace off Maui (Hawaiian Islands), which they claimed to be consistent with the age of this terrace calculated from subsidence rates of the extinct Haleakala volcano.

Ludwig et al. (1991) made a more detailed study of submerged coral terraces off NW Hawaii. Comparison of ^{234}U ages with terrace depth led to a subsidence curve which is approximately linear for the last 500 kyr, at a rate of 2.6 mm/yr (Fig. 12.6). Small undulations on the subsidence curve shown in Fig. 12.6 represent the calculated effect of eustatic sea-level fluctuations. These cause development of coral terraces by periodically neutralising subsidence (to create a sea-level 'stand') and then exacerbating subsidence, to drown the reef.

The good fit of data points to a linear subsidence model in Fig. 12.6 provides evidence of the reliability of the $^{234}U/^{238}U$ dating method, including the closed-system assumption for uranium systems in coral. This is attributed to the good preservation of submarine coral systems. In contrast, coral which has suffered fresh-water percolation is very susceptible to open-system behaviour. For example, Bard et al. (1991) found that many coral specimens over 50 kyr old which had been dated by ^{230}Th at Lamont Doherty had calculated initial ^{234}U activities above the seawater value of 1.14 (Fig. 12.7). These corals come from raised terraces on the island of Barbados, which is undergoing net tectonic uplift with time.

Since the evidence from Hawaii is consistent with a constant ^{234}U activity in seawater (at least for the past 500 kyr), the high apparent initial ratios found by Bard et al. must be attributed to open-system behaviour of uranium. Unfortunately this subaerial uranium redistribution will also affect the ^{230}Th ages calculated from these samples. For example, the sample with a 528 kyr apparent age in Fig. 12.7 comes from a lower terrace on Barbados, and hence must be younger than the samples with apparent ages of 230 and 418 kyr.

12.3.2 ^{230}Th

Uranium and thorium display very different chemistries in the hydrosphere. As noted above, uranium tends to exist as soluble UO_2^{++} ions in the oxidising environment normally found on the Earth's surface; hence it has a relatively long residence time in natural water bodies. In contrast, thorium is readily adsorbed

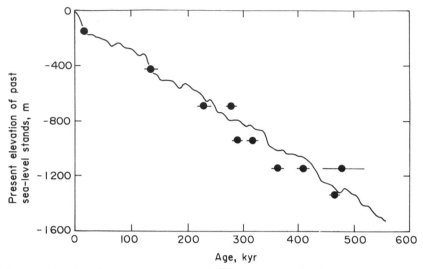

Fig. 12.6. Plot of terrace depth against mass spectrometric ^{234}U age for corals off NW Hawaii showing the good fit to a cooling subsidence curve (modulated by eustatic variations). After Ludwig *et al.* (1991).

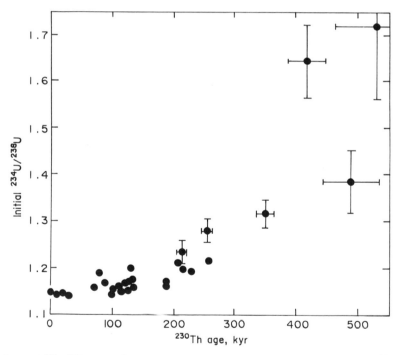

Fig. 12.7. Plot of initial ^{234}U/^{238}U activities in Barbados coral terraces, against calculated ^{230}Th ages, to show apparent initial ratios above the seawater value. After Bard *et al.* (1991).

onto the surface of detrital grains and has a very short residence time in natural waters (e.g. ca. 350 yr in the oceans; Goldberg and Koide, 1962).

The differing behaviour of U and Th causes fractionation between them during the formation of different sediment types, leading to systems out of secular equilibrium. As noted above, ^{238}U decays via two very short-lived intermediates to ^{234}U in seawater. This in turn decays to ^{230}Th, but the latter is almost immediately adsorbed onto the sediment surface. This process continually enriches the sediment surface in ^{230}Th. Because it is preferentially enriched on the sediment surface, relative to its (^{234}U) parent, ^{230}Th is 'unsupported' and out of secular equilibrium. However, after isolation from the sediment–water interface, this unsupported ^{230}Th begins to decay back to secular equilibrium with its parent. Hence, this method should allow the dating of sedimentary deposition.

In early work on sediment dating by Piggot and Urry (1939, 1942), ^{230}Th activity was monitored using its short-lived daughter, ^{226}Ra, which reaches secular equilibrium with ^{230}Th after only 10 kyr or so. This approach was adopted because of the much greater ease of radioactive counting of ^{226}Ra decay, compared to ^{230}Th decay. However, Kroll (1954) demonstrated that this method was problematical, due to the greater mobility of Ra than Th in the sediment system. This problem was overcome by advances in α counting, which allowed the direct measurement of ^{230}Th decay (Isaac and Picciotto, 1953).

Thorium adsorption onto detrital grains is so much more effective than uranium adsorption that for young sediments the uranium-supported component (i.e. the component in secular equilibrium) can be effectively ignored. In other words:

$$^{230}\text{Th}_{\text{excess}} \approx \,^{230}\text{Th}_{\text{total}} \qquad [12.7]$$

Therefore we can use the method as a dating tool by means of the simple decay equation:

$$^{230}\text{Th}_{\text{present}} = \,^{230}\text{Th}_{\text{initial}} \, e^{-\lambda_{230}t} \qquad [12.8]$$

Since the ^{230}Th excess method is used to study sedimentation, it is convenient to formulate t in terms of sediment depth, D, (in a core) and sedimentation rate, R:

$$t = D/R \qquad [12.9]$$

If we substitute this into equation [12.8] and take the natural log of both sides, we obtain:

$$\ln(^{230}\text{Th}_\text{P}) = \ln(^{230}\text{Th}_\text{I}) - D(\lambda_{230}/R) \quad [12.10]$$

This corresponds to the equation for a straight line:

$$y = c - xm \qquad [12.11]$$

Hence, if the natural log of the present-day ^{230}Th activity is plotted against depth in the core, the sedimentation rate can be obtained from the

reciprocal of the slope (solid line in Fig. 12.8):

$$R = -\lambda_{230} \frac{D}{\ln^{230}\text{Th}_\text{P} - \ln^{230}\text{Th}_\text{I}} \qquad [12.12]$$

Although the effects of U-supported ^{230}Th may be negligible near the sediment surface, this component becomes increasingly important as the system approaches secular equilibrium with increasing burial depth (dashed line in Fig. 12.8). Two possible sources of U-supported ^{230}Th may be present. Detrital grains are expected to contain ^{230}Th which is in secular equilibrium with ^{234}U and ^{238}U, even at the sediment surface. In contrast, authigenic minerals such as calcite are expected to contain no ^{230}Th at the sediment surface, but develop increasing ^{230}Th activities with depth, until this reaches secular equilibrium with their uranium inventory.

Two steps are normally taken to correct ^{230}Th sediment activities for the component in secular equilibrium. Firstly, the carbonate fraction is removed by mineral separation; then the excess ^{230}Th activity in the bulk clay fraction is calculated by subtracting ^{234}U activity (in secular equilibrium with

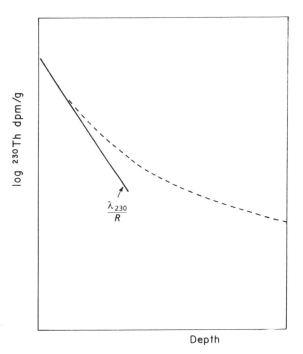

Fig. 12.8. Schematic plot of log ^{230}Th activity (decays per minute per gram) against depth to show behaviour expected in a core formed by a constant sedimentation rate. Solid line = young sediments; dashed line = older sediments.

the ^{230}Th daughter) from total ^{230}Th activity (e.g. Ku, 1976).

$$^{230}\text{Th}_{\text{excess}} = {}^{230}\text{Th}_{\text{total}} - {}^{234}\text{U} \qquad [12.13]$$

The corrected (excess) activities determined in this way are substituted into equations [12.10] and [12.12] above to determine sedimentation rates. Since the concentration of ^{230}Th in the oceans is expected to be constant through time, and the adsorption process is expected to be of constant efficiency, the initial concentration of ^{230}Th in the detrital sediment fraction should be constant. If the bulk sedimentation rate (R) remains constant with time then excess ^{230}Th activity will decrease as a log function with depth. Fig. 12.9a shows data from a Caribbean core which fit this model (Ku, 1976), yielding a linear fit (of log activity against depth). The regression slope yields a sedimentation rate R of 25 ± 1 mm/kyr for the last 300 kyr. Unfortunately, not all cores yield such linear results, so other U-series techniques have been applied in order to improve the dating

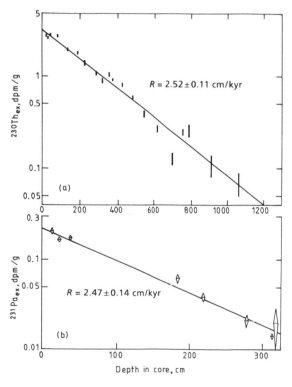

(a)

$R = 2.52 \pm 0.11$ cm/kyr

(b)

$R = 2.47 \pm 0.14$ cm/kyr

Depth in core, cm

Fig. 12.9. Plots of (a) excess ^{230}Th activity, and (b) excess ^{231}Pa activity, against depth in a sediment core, yielding two independent estimates of average sedimentation rate. After Ku (1976).

confidence. One of these is the ^{231}Pa method (Fig. 12.9b), discussed next.

12.3.3 ^{231}Pa

^{235}U decays *via* a short-lived intermediate, ^{231}Th, to ^{231}Pa. The chemistry of protoactinium is similar to that of thorium in that it is preferentially adsorbed onto the surfaces of clay minerals. Hence, the ^{231}Pa method can be used analogously to the ^{230}Th method to date sedimentation rates. However, the much lower abundance of ^{235}U, relative to ^{238}U, and the shorter half-life of ^{231}Pa (34.3 kyr) lead to larger analytical errors. Therefore, this method is most useful as a concordancy test on ^{230}Th data to check for closed-system conditions (e.g. Fig. 12.9b).

12.3.4 ^{230}Th–^{232}Th

Picciotto and Wilgain (1954) claimed that a drawback in using the method of absolute ^{230}Th activity was the need to assume a constant precipitation rate of this species (per sediment mass), which was difficult to check. To avoid this problem, these authors suggested using ^{232}Th as a reference isotope to normalise for variable absolute levels of adsorbed Th. They justified this approach on the basis that ^{230}Th and ^{232}Th ($t_{1/2} = 14$ Byr) are chemically identical, so that they should be removed from seawater at the same rate. Because ^{232}Th has such a long half-life, it suffers no significant decay within the dating range of ^{230}Th. Therefore, if we assume that initial ^{230}Th/^{232}Th activities at the sediment surface remain constant at any given locality through time, we can divide both sides of equation [12.8] by ^{232}Th (where x signifies excess activities):

$$\left(\frac{^{230}\text{Th}^{\text{x}}}{^{232}\text{Th}}\right)_{\text{present}} = \left(\frac{^{230}\text{Th}^{\text{x}}}{^{232}\text{Th}}\right)_{\text{initial}} e^{-\lambda_{230}t} \qquad [12.14]$$

Applying this to the activity *versus* depth plot we obtain:

$$\ln\left(\frac{^{230}\text{Th}^{\text{x}}}{^{232}\text{Th}}\right)_{\text{P}} = \ln\left(\frac{^{230}\text{Th}^{\text{x}}}{^{232}\text{Th}}\right)_{\text{I}} - D\frac{\lambda_{230}}{R} \qquad [12.15]$$

Picciotto and Wilgain pointed out that, for this method to work, effectively all of the Th in the sediment must have been chemically precipitated, and not be detrital. However, 30% or more of the total ^{232}Th budget in a pelagic sediment is normally within the detrital phases (Goldberg and Koide, 1962). Consequently, Ku *et al.* (1972) have argued that the effect of dividing by ^{232}Th is very similar to the effect

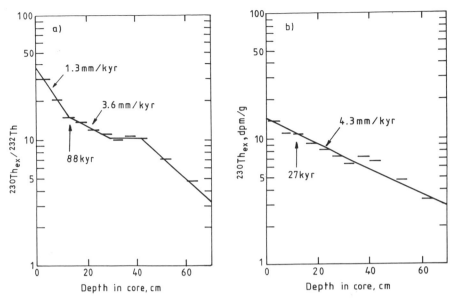

Fig. 12.10. Thorium isotope results from the ZEP 15 core (Mid Atlantic Ridge) showing interpretations of sedimentation history using a) the $^{230}Th/^{232}Th$ method and b) the simple ^{230}Th method. After Ku (1976).

of dividing by the non-carbonate (i.e. detrital) fraction in the analysed sample. If the detrital fraction in the sediment is constant then this does not cause a problem, but if it varies with depth, then this will perturb the initial $^{230}Th/^{232}Th$ ratios and hence lead to erroneous ages and sedimentation rates. This problem is illustrated in Fig. 12.10 with data for a core from the Mid Atlantic Ridge. The $^{232}Th/^{230}Th$ plot (Fig. 12.10a) yields an age for the 12 cm deep horizon (arrowed) which is more discordant from the ^{14}C age of 17 kyr than the simple ^{230}Th plot (Fig. 12.10b).

In order to reduce the perturbing effect of the detrital component on $^{230}Th/^{232}Th$ ages, Goldberg and Koide (1962) used a technique by which authigenic minerals and adsorbed Th were leached from the detrital component with hot hydrochloric acid. This led them also to adopt a different correction for U-supported ^{230}Th. On the assumption that no detrital ^{230}Th component was leached, they excluded the component in secular equilibrium. Instead they corrected for U-supported ^{230}Th in the authigenic (carbonate) component, which is expected to grow with time. This is equivalent to the ^{230}Th daughter-deficiency method, and will be dealt with in detail below (section 12.4.1). If the immediate parent (^{234}U) is assumed to be in equilibrium with ^{238}U (an approximation) then the growth of U-supported Th is given by equation [12.25] (see p. 319). This is

subtracted from total ^{230}Th activity to determine excess ^{230}Th:

$$^{230}Th_{excess} = {}^{230}Th_{total} - {}^{238}U(1 - e^{-\lambda_{230}t}) \qquad [12.16]$$

Ku(1976) claims that this method also has drawbacks, since in fact some thorium leaks from detrital phases during the acid leach process.

12.3.5 $^{231}Pa-^{230}Th$

The similarity in the chemistry of Pa and Th prompted Sackett (1960) and Rosholt et al. (1961) to suggest their use in conjunction as a dating tool. Three factors suggested that the adsorbed initial $^{230}Th/^{231}Pa$ activity ratio should be a constant (≈ 11) defined by the production ratio of the two species: the isotope ratio of their parents is relatively constant in seawater (as demonstrated by the concordance of ^{231}Pa and ^{230}Th dates); they are both adsorbed rapidly compared to their half-lives; and direct river-borne contribution of ^{231}Pa and ^{230}Th is negligible (Scott, 1968). In this case, equation [12.8] can be divided by the corresponding equation for protoactinium, yielding:

$$\left(\frac{^{230}Th}{^{231}Pa}\right)^{excess}_{P} = \left(\frac{^{230}Th}{^{231}Pa}\right)^{excess}_{I} e^{-(\lambda_{230}-\lambda_{231})t} \qquad [12.17]$$

Assuming that the U-supported component is in

secular equilibrium in the detrital phase, excess ^{231}Pa and ^{230}Th are defined as follows:

$$^{231}Pa_{excess} = {}^{231}Pa_{total} - {}^{235}U \qquad [12.18]$$

$$^{230}Th_{excess} = {}^{230}Th_{total} - {}^{234}U \qquad [12.19]$$

Equation [12.17] can then be solved for t by assuming the initial ratio to be 11. The early work of Sackett (1960) and Rosholt et al. (1961) appeared to bear out the assumption, but subsequent work yielded variable excess ^{230}Th/^{231}Pa activities at the sediment surface. Sediments often have surface ratios much higher than 11 (e.g. Sackett, 1964), while manganese nodules may have ratios much lower than 11 (e.g. Sackett, 1966). Hence, it is concluded that variable fractionation between ^{231}Pa and ^{230}Th occurs during sedimentation, rendering the method useless.

12.3.6 ^{230}Th sediment stratigraphy

In view of the difficulties described above, it may be concluded that all of the ^{230}Th methods should be regarded as semi-quantitative as far as absolute dating is concerned. However, ^{230}Th data may be a powerful tool for stratigraphic correlation of quaternary sediments. An example of this application is provided by the study of Scholten et al. (1990) on a 5m core from the Norwegian Sea near Jan Mayen (Fig. 12.11). Excess ^{230}Th activity data fitted an average decay curve equivalent to a sedimentation rate of 1.9 cm/kyr, in reasonable agreement with the rate of 1.6 cm/kyr calculated from oxygen isotope stratigraphy. However, the data display large short-term variations superimposed on the mean decay curve.

Traditionally, variations of this type have been attributed to changes in sedimentation rate. However, this is clearly impossible for some segments of core 23059, which define a positive slope of excess activity against depth. In order to examine these short-term activity variations, Scholten et al. corrected the data for radioactive decay since burial (using the mean decay curve), and then ratioed these initial (excess) ^{230}Th concentrations against ^{232}Th to correct for variable carbonate contents. The resulting values display variations with depth which are correlated with δ^{18}O (Fig. 12.12). Scholten et al. attributed these variations to the influence of climatic factors on ^{230}Th deposition rate. Climatic changes affect the productivity of plankton, and hence the amount of sinking organic matter.

The biogenic particle flux was argued by Mangini and Diester-Haas (1983) to control the downward flux of radionuclides off NW Africa, and hence ^{230}Th activity variations in sediment cores. Therefore Scholten et al. argued that the low initial excess ^{230}Th/^{232}Th levels in isotope stages 2 and 6 (Fig. 12.12) were due to a widespread reduction of biogenic paleo-productivity during these cold periods. This allows the opportunity of correlating ^{230}Th variations between different sites in an ocean system. Similar results may be obtained using ^{231}Pa/^{230}Th activity ratios (Kumar et al., 1993), and using the cosmogenic isotope ^{10}Be (section 14.3.3).

12.3.7 ^{210}Pb

Within the ^{238}U decay chain, the daughter product of ^{226}Ra is the gas ^{222}Rn. This escapes into the

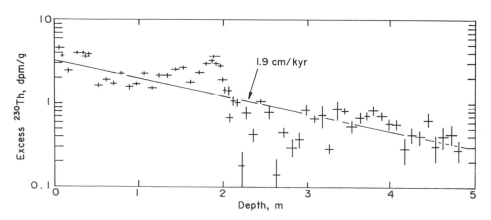

Fig. 12.11. Plot of excess ^{230}Th activity (on a log scale) against depth in core 23059 from the Norwegian Sea. Regression line indicates average sedimentation rate. After Scholten et al. (1990).

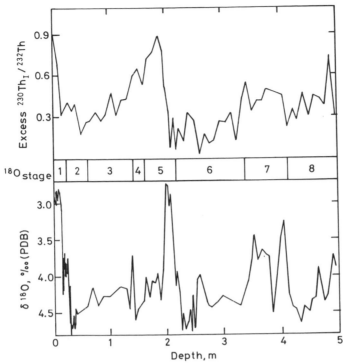

Fig. 12.12. Comparison of the depth-dependence of excess initial ^{230}Th/^{232}Th and δ^{18}O in core 23059. Numbered intervals are stages based on ^{18}O stratigraphy. Stages 1 and 5 represent the present and the 120–130 kyr interglacials. After Scholten *et al.* (1990).

atmosphere from the whole land surface. However, ^{222}Rn has a half-life of only three days, and is followed by four intermediates with half-lives of minutes to seconds, ultimately yielding longer-lived ^{210}Pb. This is estimated to remain in the upper atmosphere for a few days, before the majority returns to earth in precipitation. Thereafter, unsupported ^{210}Pb decays away with a half-life of 22.3 yr. The use of ^{210}Pb was first suggested as a tool to date snow accumulation by Goldberg (1963). However, it can also be used to date very recent fresh-water and marine sedimentation (e.g. Krishnaswamy *et al.*, 1971; Koide *et al.*, 1972) because ^{210}Pb has an aqueous residence time of only a year or two before adsorption onto sediment.

If the ^{210}Pb concentration in newly precipitated snow or sediment remains more-or-less constant with time at a given locality (as expected), then the system will behave exactly the same as the ^{230}Th excess method. We can then use ^{210}Pb activity at the present-day surface to determine initial ^{210}Pb, and solve for the age of a buried ice or sediment sample:

$$^{210}\text{Pb} = {}^{210}\text{Pb}_{\text{initial}}\ e^{-\lambda_{210}t} \qquad [12.20]$$

As with ^{230}Th, if we plot the log of ^{210}Pb activity against depth, the slope yields the sedimentation rate. The first application of the method was to snow chronology (Crozaz *et al.*, 1964). The calculated sedimentation rate of snow at the South Pole in water equivalents (6 ± 1 cm/yr) compared well with a rate determined from yearly 'ice varves'.

The short half-life of ^{210}Pb also makes it ideally suited to the dating of historical-age sediments. For example, the method has become an important tool in studying the history of heavy metal pollution of coastal waters and lakes. Bruland *et al.* (1974) used the method in a study of metal pollution of the Santa Monica basin off Los Angeles. A log plot of total ^{210}Pb activity against depth (Fig. 12.13) yields a linear fit at shallow depths, but the profiles flatten out at ca. 8 cm depth due to the effect of ^{210}Pb supported by ^{226}Ra. However, this can be corrected by subtracting ^{226}Ra activity in order to calculate excess ^{210}Pb activities:

$$^{210}\text{Pb}_{\text{excess}} = {}^{210}\text{Pb}_{\text{total}} - {}^{226}\text{Ra} \qquad [12.21]$$

When the data are plotted in this form the usable range of the method is extended to ca. 150 yr. For the

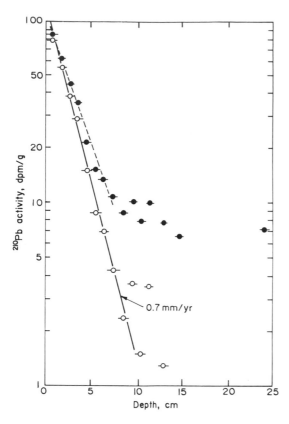

Fig. 12.13. Plot of ^{210}Pb activity against depth in recent sediments of the Santa Monica basin. Solid symbols: total ^{210}Pb activity, including ^{226}Ra-supported fraction. Open symbols: excess ^{210}Pb only. After Bruland *et al.* (1974).

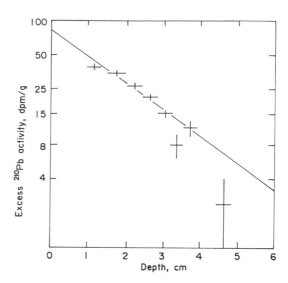

Fig. 12.14. Plot of excess ^{210}Pb activity against depth in a sub-alpine pond from Yosemite National Park, California. After Shirahata *et al.* (1980).

Santa Monica basin the corrected (excess) ^{210}Pb data yield a sedimentation rate of 0.7 mm/yr (Fig. 12.13).

A particularly appropriate application of the ^{210}Pb method is to studies of anthropogenic contamination of sediments. Shirahata *et al.* (1980) applied the method to a remote sub-alpine pond in Yosemite National Park, in order to assess the regional atmospheric fallout of Pb from car exhausts. A sedimentation rate of 0.6 mm/yr was calculated from ^{210}Pb data (Fig. 12.14). Bioturbation of the sediment was ruled out because all bomb-produced radionuclides remained within ca. 2 cm of the sediment surface. Total Pb concentrations in the sediment were found to increase four-fold over the past 100 years, and this change was accompanied by a change in ^{206}Pb/^{207}Pb ratio from a natural local value of 1.15 to an exotic value of 1.2 . The latter was typical of the sources of Pb ore used in the United States for the

manufacture of leaded gasoline (prior to its withdrawal).

Despite these achievements with the ^{210}Pb method, caution must be exercised in the interpretation of data, since more recent studies (e.g. Santschi *et al.*, 1983; Benoit and Hemond, 1991) have shown that ^{210}Pb can be re-mobilised from the surfaces of sediment grains into sediment pore-waters, and thence into the overlying water column. Benoit and Hemond showed from theoretical modelling that ^{210}Pb redistribution could occur by pore-water diffusion, without the need for particle re-working.

12.4 Daughter-deficiency methods

12.4.1 ^{230}Th: theory

The tendency described above for thorium adsorption onto clay minerals leads to low Th levels in ground waters, in contrast to their moderate U levels. Thus, when biogenic or authigenic calcite is formed it tends to contain appreciable U concentrations (a few ppm) but negligible Th. This leads to a situation where ^{230}Th is strongly deficient relative to its parent, ^{234}U. The subsequent regeneration of ^{230}Th can then be used as a dating tool.

The first application of this technique was made as early as 1926 by Khlapin, who used short-lived ^{226}Ra as a measure of ^{230}Th activity. Khlapin assumed that the ^{234}U parent taken up by calcite was itself in secular equilibrium with ^{238}U, and that Th uptake was negligible. Under these conditions, we can treat ^{230}Th production from ^{234}U as if it were derived directly from ^{238}U. To calculate net ^{230}Th accumulation, we must then subtract the fraction which has decayed to ^{226}Ra. Substituting into the relevant Bateman equation [1.13], the abundance (not activity) of ^{230}Th after time t is given as follows:

$$n^{230}\text{Th} = \frac{\lambda_{238}}{\lambda_{230} - \lambda_{238}} \cdot n^{238}\text{U}_\text{I}(e^{-\lambda_{238}t} - e^{-\lambda_{230}t})$$

[12.22]

where I signifies the initial ratio. But these abundances may be easily converted into activities by dividing by the relevant decay constants:

$$\frac{^{230}\text{Th}}{\lambda_{230}} = \frac{\lambda_{238}}{\lambda_{230} - \lambda_{238}} \cdot \frac{^{238}\text{U}_\text{I}}{\lambda_{238}}(e^{-\lambda_{238}t} - e^{-\lambda_{230}t})$$ [12.23]

Now cancelling λ_{238} and multiplying both sides by λ_{230}:

$$^{230}\text{Th} = \frac{\lambda_{230}}{\lambda_{230} - \lambda_{238}} \cdot {}^{238}\text{U}_\text{I}(e^{-\lambda_{238}t} - e^{-\lambda_{230}t})$$ [12.24]

However, because of the very long half-life of ^{238}U relative to the other species, its activity is effectively constant over time. Therefore ^{238}U$_\text{I}$ can be approximated by ^{238}U, $e^{-\lambda_{238}t}$ is approximately 1 and $\lambda_{230} - \lambda_{238}$ is approximately λ_{230}, which, then cancels to yield:

$$^{230}\text{Th} = {}^{238}\text{U}(1 - e^{-\lambda_{230}t})$$ [12.25]

Finally, dividing through by ^{238}U yields the decay equation which can be used for dating:

$$\frac{^{230}\text{Th}}{^{238}\text{U}} = 1 - e^{-\lambda_{230}t}$$ [12.26]

However, it was noted above that ^{234}U and ^{238}U in natural waters are very rarely in secular equilibrium. This introduces a complication into the decay equation, since there is an extra contribution to ^{230}Th by excess ^{234}U until the latter has decayed away. ^{230}Th production by excess ^{234}U (x) is given by an equation analogous to [12.24]:

$$^{230}\text{Th}^\text{x} = \frac{\lambda_{230}}{\lambda_{230} - \lambda_{234}} \cdot {}^{234}\text{U}_\text{I}^\text{x}(e^{-\lambda_{234}t} - e^{-\lambda_{230}t})$$ [12.27]

But excess ^{234}U can only be conveniently measured as a ratio against ^{238}U. Therefore we divide both sides of equation [12.27] by ^{238}U activity. This is effectively constant over time due to its long half-life, so that

present and initial ^{238}U activities are interchangeable:

$$\left(\frac{^{230}\text{Th}}{^{238}\text{U}}\right)^\text{x} = \frac{\lambda_{230}}{\lambda_{230} - \lambda_{234}} \cdot \left(\frac{^{234}\text{U}}{^{238}\text{U}}\right)_\text{I}^\text{x}(e^{-\lambda_{234}t} - e^{-\lambda_{230}t})$$

[12.28]

But the excess activity ratio is equal to the total activity ratio minus one (corresponding to secular equilibrium). So:

$$\left(\frac{^{230}\text{Th}}{^{238}\text{U}}\right)^\text{x} = \frac{\lambda_{230}}{\lambda_{230} - \lambda_{234}} \cdot \left[\left(\frac{^{234}\text{U}}{^{238}\text{U}}\right)_\text{I} - 1\right] \cdot (e^{-\lambda_{234}t} - e^{-\lambda_{230}t})$$

[12.29]

We can substitute equation [12.6] into this equation in order to convert initial 234/238 activities to the present day measured activities (P):

$$\left(\frac{^{230}\text{Th}}{^{238}\text{U}}\right)^\text{x} = \frac{\lambda_{230}}{\lambda_{230} - \lambda_{234}} \cdot \left[\left(\frac{^{234}\text{U}}{^{238}\text{U}}\right)_\text{P} - 1\right] \cdot \frac{(e^{-\lambda_{234}t} - e^{-\lambda_{230}t})}{e^{-\lambda_{234}t}}$$

[12.30]

But the final term simplifies to yield:

$$\left(\frac{^{230}\text{Th}}{^{238}\text{U}}\right)^\text{x} = \frac{\lambda_{230}}{\lambda_{230} - \lambda_{234}} \cdot \left[\left(\frac{^{234}\text{U}}{^{238}\text{U}}\right)_\text{P} - 1\right] \cdot (1 - e^{-(\lambda_{230} - \lambda_{234})t})$$

[12.31]

Finally, adding the ^{230}Th production from equilibrium and excess ^{234}U, (equations [12.26] and [12.31]), we obtain:

$$\frac{^{230}\text{Th}}{^{238}\text{U}} = 1 - e^{-\lambda_{230}t}$$

$$+ \frac{\lambda_{230}}{\lambda_{230} - \lambda_{234}} \cdot \left[\frac{^{234}\text{U}}{^{238}\text{U}} - 1\right] \cdot (1 - e^{-(\lambda_{230} - \lambda_{234})t})$$ [12.32]

This equation could be used directly to solve ages, but it has become normal procedure to rearrange it by multiplying through by ^{238}U/^{234}U. This yields:

$$\frac{^{230}\text{Th}}{^{234}\text{U}} = \frac{1 - e^{-\lambda_{230}t}}{^{234}\text{U}/{}^{238}\text{U}}$$

$$+ \frac{\lambda_{230}}{\lambda_{230} - \lambda_{234}} \cdot \left[1 - \frac{1}{^{234}\text{U}/{}^{238}\text{U}}\right] \cdot (1 - e^{-(\lambda_{230} - \lambda_{234})t})$$

[12.33]

This equation was plotted as an isochron diagram (Fig. 12.15), by Kaufman and Broecker (1965). Effectively, the calibration line for ^{234}U/^{238}U = 1 (secular equilibrium) yields the age in terms of ^{230}Th build-up, while the near-vertical isochron lines apply

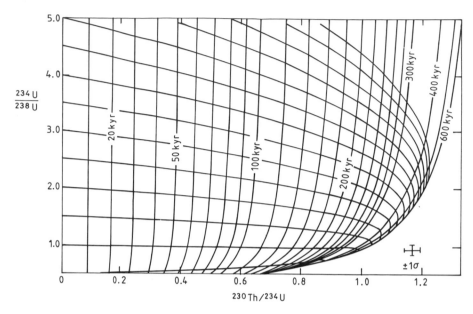

Fig. 12.15. Th–U isochron diagram for systems containing no ^{232}Th. Labelled, steeply-dipping lines are isochrons; lateral lines are growth lines. Error bar shows typical uncertainty for α spectrometry. After Kaufman and Broecker (1965).

the correction for non-equilibrium U isotope compositions. As can be seen on the diagram, this correction is unnecessary for samples less than ca. 30 kyr old. The maximum dating range of the ^{230}Th method is ca. 300 kyr by α counting, but this may be extended to over 400 kyr by mass spectrometry.

12.4.2 ^{230}Th: applications

The ^{230}Th–^{234}U method is applicable to the dating of any closed-system carbonate which is free from contamination by initial detrital thorium. It can provide far better precision for coral dating than the ^{234}U–^{238}U method alone. This is illustrated in Fig. 12.16 by a compilation of high-precision α spectrometry data for un-recrystallised corals (Veeh and Burnett, 1982). It can be seen that typical measurement errors in ^{234}U/^{238}U ratio lead to age uncertainties of over 100%, whereas errors in ^{230}Th/^{234}U lead to age errors of only 10%.

The precision of ^{230}Th coral dating has been further enhanced by the mass spectrometric method, as demonstrated by the analysis of live reef-forming corals from the Vanuatu arc, east of Australia (Edwards *et al.*, 1988). In these specimens, ^{230}Th ages were compared with historical ages based on yearly growth bands. The latter are about 1 cm wide, and can be accurately counted in specimens at least

200 years old. ^{230}Th ages were determined with errors as low as $\pm$ 3 yr (2σ), and were in excellent agreement with the historical age of the corals (Fig. 12.17). ^{230}Th dating of corals between 9000 and 40 000 yr old has been used very effectively to calibrate the radiocarbon time-scale (section 14.1.5).

^{230}Th dating of corals has also been very important in constraining Pleistocene sea-level variations with time. This is possible because sea-level highs during interglacial periods become marked by coral terraces which are stranded as sea-level falls again. Fig. 12.18 shows a compilation (from Moore, 1982) of α counting ^{230}Th dates on corals. In general there is a good correlation between sea-level highs and solar radiation received by the Earth (insolation). This supports Milankovitch (1941), who suggested that changes in Pleistocene climate were largely due to changes of insolation, caused by variations in the Earth's orbit ('Milankovitch forcing').

Unfortunately, the error bars on α counting ages are so large that they are almost of equal magnitude to the Pleistocene climatic cycle. This is illustrated in Fig. 12.19 by comparison of α spectrometry dates with the calculated insolation curve at the time of high sea-level stand VII (also known as 5e). A 129 kyr coral terrace date by Ku (in Edwards *et al.*, 1987) is centred on the insolation high, corresponding to an interglacial. However, determinations by Bloom *et al.*

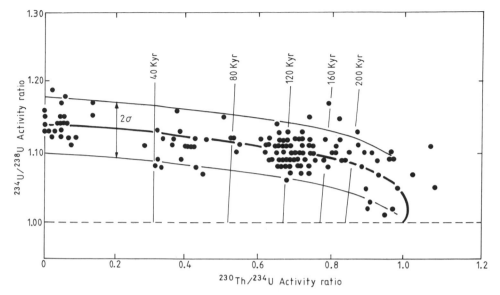

Fig. 12.16. Plot of $^{234}U/^{238}U$ *versus* $^{230}Th/^{234}U$ activity ratio for un-recrystallised coral data with 2σ errors better than 5% . Heavy near-horizontal curve shows evolution of activity ratios with time, starting from present-day seawater composition. Vertical lines are isochrons. After Veeh and Burnett (1982).

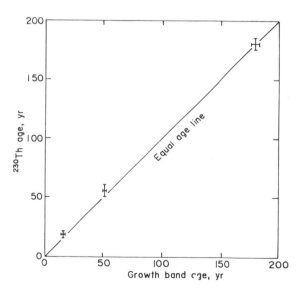

Fig. 12.17. Comparison of mass spectrometric ^{230}Th ages of living corals with historical ages based on annual growth bands. After Edwards *et al.* (1988).

(1974) bracket two insolation highs and are actually centred on the intervening low. Mass spectrometric dates on the same samples were in all three cases within error of the α counting determination, but are now centred unequivocally in the insolation high at 127 kyr (Edwards *et al.*, 1987). This implies that the 140 kyr VIIa terrace ages are probably spurious.

Another important approach for constraining Pleistocene sea-level variations is ^{230}Th dating of speleothem (stalactites, stalagmites etc.) from submarine caves. These formations grow during periods of low sea-level stand, when they are exposed subaerially to percolating calcareous solutions. When sea-level rises and they become drowned, growth stops, and an erosional hiatus is formed. The densely crystalline form of speleothem deposits is conducive to good closed-system behaviour, so that this material is ideal for U-series dating. Therefore, drowned speleothem and coral terraces form a complementary couple for Pleistocene sea-level studies.

The first mass spectrometric dating study on submarine cave deposits was made by Li *et al.* (1989) on a sample from 15 m depth in a Bahamas 'Blue Hole'. A detailed sequence of U-series age determinations on the 12 cm-thick flowstone showed carbonate deposition over 280 kyr (Fig. 12.20). Within this period, there were four internal hiatuses corresponding to sea-level stands above −12 m (relative to present-day sea-level). These data are consistent with the 'orbitally tuned' record of oxygen isotope variations in ocean water, termed SPECMAP (Imbrie *et al.*, 1984).

A notable feature of the new data set is the evidence

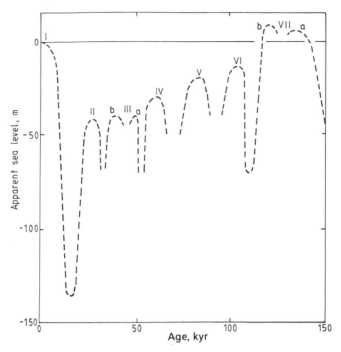

Fig. 12.18. Composite curve of paleo sea-level relative to present-day sea-level for the Late Pleistocene period. High sea-level stands are denoted by roman numerals. After Moore (1982).

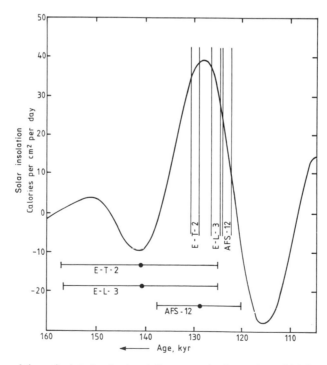

Fig. 12.19. Comparison of the calculated solar insolation curve in the region of high sea-level stand VII with age data for coral terraces by α counting (wide error bars) and mass spectrometry (thin bars). Note that ages increase to the left. After Edwards *et al.* (1987).

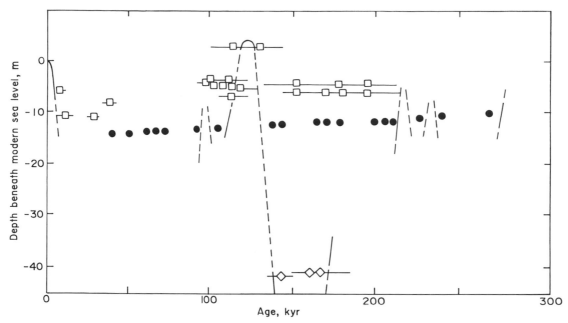

Fig. 12.20. Pleistocene sea-level curve for the Bahamas (dashed lines), based on U-series ages on drowned speleothems. (●) = mass spectrometric data. (◇) = speleothem and (▢) = coral terrace ages by α spectrometry. After Li *et al.* (1989).

for low sea-level at 140 kyr, consistent with the α counting determinations of Gascoyne *et al.* (1979). Winograd (1990) challenged this evidence on the grounds of apparent conflict with older α counting ages for corals (e.g. Bloom *et al.*, 1974). However, Lundberg *et al.* (1990) showed that if careful attention is given to the quoted analytical error limits of good-quality α counting and mass spectrometric data, then there is no conflict between the speleothem and coral ages. In this connection, it is important to remember that the 1σ error limits traditionally quoted for α spectrometry correspond to only 68% confidence that the 'true' age is within the quoted limits. The availability of high-quality mass spectrometric data would seem to be the perfect time to switch to 2σ errors (95% confidence), such as are used in other fields of isotopic dating.

The ^{230}Th deficiency method can also be used to date terrestrial authigenic carbonates such as tufas and speleothems, where the initial U isotope composition is unknown. Indeed, mass spectrometric analysis may even allow U-series calibration of annual growth banding, identified by microscopic variations in luminescence (Baker et al., 1993). Speleothems may be used to study karst processes and other geomorphological and climatic events. For example, Gascoyne *et al.* (1983) presented speleothem

data as a histogram of frequency against time (based on ^{230}Th dates). Because glacial freezing inhibits percolation of mineralised waters (and hence carbonate deposition), glacial advances in local areas can be mapped with time by a diminution in speleothem frequency (Fig. 12.21).

Speleothems can also be used to calibrate climatic variations directly, by measuring δ^{18}O signatures and U-series ages on the same cave deposits. For example, Winograd *et al.* (1992) made a combined U-series and stable isotope study on the calcite lining of a water-filled cavern in Nevada (USA) called Devils Hole. The results of this study suggested similar glacial cycles to the SPECMAP oceanic record, but were not directly correlated with the *timing* of cycles in the SPECMAP record. This led Winograd *et al.* to question the SPECMAP model, in which the oceanic stable isotope record is tuned to fit the Earth's orbital variations (Milankovitch forcing). However, other workers (e.g. Imbrie *et al.*, 1993) have attributed the mismatch to the more complex relationship between δ^{18}O and paleoclimate in continental groundwater, compared to seawater.

One of the most interesting U-series applications is the dating of human bones and cultural deposits. In the past, the large sample size requirements of α counting precluded direct analysis of such material.

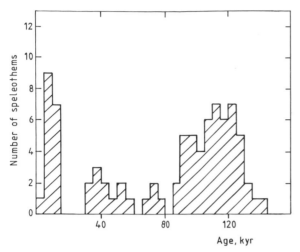

Fig.12.21. Plot of speleothem frequency from NW England against ages determined by U-series dating. The gap between 20 and 30 kyr corresponds to the peak of the last glaciation. After Schwarcz and Blackwell (1983).

Instead, ages were based on speleothem deposits which pre-dated and post-dated a 'cultural' layer (e.g. Schwarcz, 1989). However, with the advent of mass-spectrometric U-series analysis, it may sometimes be possible to date tooth fragments directly. This approach is based on the tendency of dentine and enamel to take up uranium from the surrounding environment shortly after deposition. McDermott *et al.* (1993) showed that U-series analysis gave ages concordant with electron spin resonance (ESR) ages on bovoid dental fragments, assuming that uranium uptake occurred soon after deposition. Hence

McDermott *et al.* were able to confirm ESR dates for the appearance of early modern humans at least 100 kyr ago.

12.4.3 ^{230}Th: dirty calcite

Because fossil bones may be encased by subsequent tufa deposits, U-series analysis of such material has been very useful for dating Pleistocene human and animal remains (e.g. Schwarcz and Blackwell, 1991). However, the most interesting tufas are often impure, for the very reason that if they contain bones they will probably contain other detrital material. This introduces initial ^{230}Th, which if not corrected for, may cause serious errors in calculated ages.

In cases where detrital contamination is minor, the same laboratory technique may be used as for clean material: the sample is leached with dilute nitric acid in an attempt to dissolve the carbonate fraction without disturbing the detrital component. This may diminish the contamination to a level where it is swamped by other errors. However, the detrital component is not usually inert in nitric acid, but often contains a certain fraction of loosely bound uranium and thorium which is removed by the leaching process. The extent of this leakage may be monitored by measuring the activity of ^{232}Th. If this reaches a level of more than a few % of ^{230}Th activity then it may be necessary to correct the carbonate data for leaching of radionuclides from the contaminating detrital phase (Ku and Liang, 1984).

U-series data for dirty calcites are best visualised on an isochron diagram. The most common form involves ratioing both ^{230}Th and ^{234}U against ^{232}Th (Fig. 12.22). If all U and Th isotopes are leached from

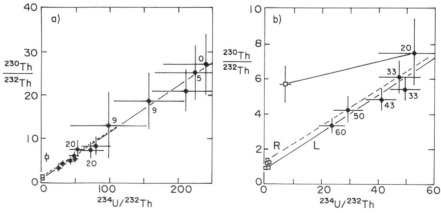

Fig. 12.22. U–Th isochron diagrams showing results from leaching of artificial mixtures of calcite and mud with 5–7 M nitric acid. (●, L) = leachates; (□, R) = residues from leaching; numbers indicate % of mud in the sample. Diagram (b) is a blow-up of the lower left corner of (a). After Przybylowicz *et al.* (1991).

the residue with equal efficiency then a cord joining the leachate and residue points can be interpreted as an isochron line. The slope will then yield the ^{230}Th/^{234}U ratio of the carbonate component, which can be used to calculate the sample age in the same way as for clean material (Fig. 12.15).

One problem with the data presentation in Fig. 12.22 is that the two variables become very highly correlated as the ^{232}Th fraction diminishes in size. The large error bars should therefore be represented by elongated error ellipses (e.g. Fig. 12.24) rather then rectangular error boxes. Similarly, a regression program utilising correlated errors should be used to calculate isochron slopes. An alternative data presentation, utilised by Kaufman (1971) but not shown here, involves plotting ^{230}Th/^{234}U against ^{232}Th/^{234}U. On this diagram the age of the sample is represented by the intercept on the y axis, and increasing detrital contamination is indicated by displacement away from the y axis. Similar plots have been used more widely in U-series studies of silicate systems (section 13.3.3.).

Przybylowicz et al. (1991) performed leaching experiments on artificial mixtures of pure calcite speleothem and mud in order to test the reliability of the leaching method in dating dirty calcites (e.g. Fig. 12.22). The results show that residues are displaced slightly (occasionally substantially) above the array of leachate compositions. This is probably due to slight preferential leaching of uranium relative to thorium from the detrital phase during the leaching process, and may yield apparent ages somewhat below the true value. Schwarcz and Latham (1989) argued that this problem could be diminished by regressing leachate analyses alone. In this case it is no longer necessary to assume a lack of differential isotopic fractionation during the leaching process. Isotopic fractionation is permitted, provided that the amount of such fractionation is the same in all samples. This has the effect of shifting the isochron line sideways in Fig. 12.22b but not changing its slope.

A similar type of correlation diagram can also be used to correct the ^{234}U/^{238}U ratio for detrital contamination. Fig. 12.23 shows these data for the experiment described above. However, this is less important since the ^{230}Th age is only weakly dependent on the ^{234}U/^{238}U ratio in samples less than 300 kyr in age.

An application of the 'leach–leach' technique to dating natural mixtures is shown in Fig. 12.24 for contaminated travertines which enclose the 'Mousterian cultural layer' at Tata, Hungary (Schwarcz and Skoflek, 1982). Regression of four leachates leads to

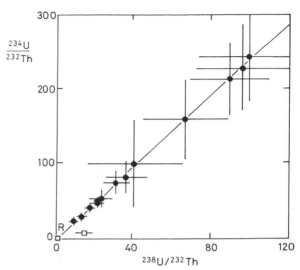

Fig. 12.23. Correlation plot to correct for detrital perturbation of uranium isotope ratios in impure carbonates. (●) = leachates; (□) = residues. After Przybylowicz et al. (1991).

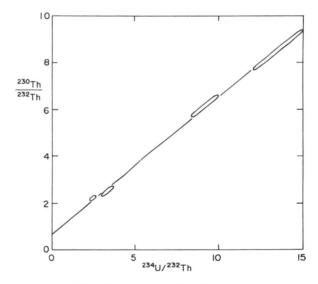

Fig. 12.24. ^{230}Th/^{232}Th versus ^{234}U/^{232}Th isochron diagram for leachates of contaminated travertine from Tata, Hungary. Ellipses portray correlated error limits. Modified after Schwarcz and Latham (1989).

an age of 101 ± 4 kyr for carbonate enclosing the cultural layer, which is bracketed between the ages of 78 ± 5 and 118 ± 37 kyr in overlying and underlying clean travertine layers.

In order to achieve a high-precision result from the leach–leach technique it is desirable to leach three or more samples with variable detrital contents from the horizon to be dated. However, under these circumstances it is possible that different samples might show variable degrees of isotopic fractionation during leaching. This problem can be avoided by total digestion of a suite of variably contaminated samples from the same deposit. If they all contain the same detrital component, i.e. have the same initial $^{230}Th/^{232}Th$ ratio, and have remained as closed-systems, then they will define a perfect isochron line.

Bischoff and Fitzpatrick (1991) tested the relative performance of the total dissolution, leach–leach and leachate–residue methods on a series of artificial mixtures of natural detritus and carbonate. (They also tested the effect of leaching with different acid strengths.) Typical results showed that the total dissolution method was superior to the other two techniques for artificial mixtures (Fig. 12.25). The total dissolution method is also more versatile, in that it can be applied to the dating of any type of Pleistocene material with homogeneous initial ratio and closed-system behaviour (Luo and Ku, 1991). However, in the dating of dirty calcite it may be difficult to obtain a large enough range of detrital–carbonate variations to define a good regression line. Furthermore, in the pursuit of such a range of mixtures, samples with variable initial ratio may be analysed. Therefore, the total dissolution, leach–leach and leach–residue methods may all be viable alternatives for dating dirty calcite in different circumstances.

12.4.4 ^{231}Pa

The build-up of ^{231}Pa in carbonates can be used as a dating tool in a way analogous to ^{230}Th. The immediate parent of ^{231}Pa is assumed to be always in equilibrium with its parent (^{235}U) due to its short half-life of 26 hours. Hence, the age relationship is analogous to the simple form of equation [12.26] for ^{230}Th build-up:

$$\frac{^{231}Pa}{^{235}U} = 1 - e^{-\lambda_{231}t} \qquad [12.34]$$

In practice, ^{231}Pa activity is usually measured *via* its short-lived daughter products ^{227}Ac or ^{227}Th. Additionally, ^{235}U activity may be determined by measuring ^{238}U, since the 238/235 activity ratio is constant in most materials at 21.7. However, this illustrates a major drawback of the ^{231}Pa accumulation method. Because ^{235}U is so much less abundant than ^{238}U, measured count rates for ^{231}Pa or its daughters will be twenty times lower than for ^{230}Th. Since counting statistics are the major source of uncertainty in U-series dating, the ^{230}Th method is much preferred to ^{231}Pa in practice. The ^{230}Th and ^{231}Pa dating equations [12.33] and [12.34] can be combined into one (Ivanovich, 1982b). However, there does not seem to be much merit in this, since no simplifications are achieved. The principal application of the ^{231}Pa technique is as a concordancy test for ^{230}Th dates, and this can most simply be achieved using the two

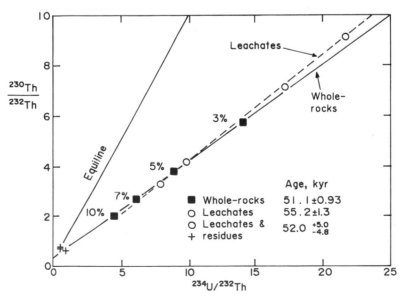

Fig. 12.25. U–Th isochron diagram for artificial mixtures analysed by total sample dissolution (■); leaching (○); and residue (+). After Bischoff and Fitzpatrick (1991).

methods independently. Lalou *et al.* (1993) adopted this approach to date ocean-ridge hydrothermal deposits.

References

Baker, A., Smart, P. L., Edwards, R. L. and Richards, D. A. (1993). Annual growth banding in a cave stalagmite. *Nature* **364**, 518–20.

Bard, E., Fairbanks, R. G., Hamelin, B., Zindler, A. and Hoang, C. T. (1991). Uranium–234 anomalies in corals older than 150,000 years. *Geochim. Cosmochim. Acta* **55**, 2385–90.

Benoit, G. and Hemond, H. F. (1991). Evidence for diffusive redistribution of ^{210}Pb in lake sediments. *Geochim. Cosmochim. Acta* **55**, 1963–75.

Bischoff, J. L. and Fitzpatrick, J. A. (1991). U-series dating of impure carbonates: an isochron technique using total-sample dissolution. *Geochim. Cosmochim. Acta* **55**, 543–54.

Bloom, A. L., Broecker, W. S., Chappell, J. M. A., Matthews, R. K. and Mesolella, K. J. (1974). Quaternary sea level fluctuations on a tectonic coast: new ^{230}Th/^{234}U dates from the Huon Peninsula, New Guinea. *Quat. Res.* **4**, 185–205

Bruland, K. W., Bertine, K., Koide, M. and Goldberg, E. D. (1974). History of metal pollution in Southern California coastal zone. *Envir. Sci. Tech.* **8**, 425–32.

Chen, J. H., Edwards, R. L. and Wasserburg, G. J. (1986). ^{238}U, ^{234}U and ^{232}Th in seawater. *Earth Planet. Sci. Lett.* **80**, 241–51.

Cherdyntsev, V. V. (1969). *Uranium 234.* Atomizdat, Moskva. [Translation by Schmorak, J. Israel Prog. Sci. Trans. 1971, 234 p.]

Cherdyntsev, V. V., Kazachevskii, I. V. and Kuz'mina, E. A. (1965). Dating of Pleistocene carbonate formations by the thorium and uranium isotopes. *Geochem. Int.* **2**, 794–801.

Cherdyntsev, V. V., Orlov, D. P., Isabaev, E. A. and Ivanov, V. I. (1961). Isotopic composition of uranium in minerals. *Geochemistry* **10**, 927–36.

Crozaz, G., Picciotto, E. and DeBreuck, W. (1964). Antarctic snow chronology with Pb210. *J. Geophys. Res.* **69**, 2597–604.

de Bievre, P., Lauer, K. F., Le Duigou, Y., Moret, H., Muschenborn, G., Spaepen, J., Spernol, A., Vaninbroukx, R. and Verdingh, V. (1971). In: Hurrell, M. L. (Ed.), *Proc. Int. Conf. Chem. Nucl. Data,* Inst. Civil Eng. Lond., pp. 221–5.

Edwards, R. L., Chen, J. H. and Wasserburg, G. J. (1987). ^{238}U–^{234}U–^{230}Th–^{232}Th systematics and the precise measurement of time over the past 500,000 years. *Earth Planet. Sci. Lett.* **81**, 175–92.

Edwards, R. L., Taylor, F. W. and Wasserburg, G. J. (1988). Dating earthquakes with high-precision thorium230 ages of very young corals. *Earth Planet. Sci. Lett.* **90**, 371–81.

Faure, G. (1977). *Principles of Isotope Geology.* Wiley, 464 p.

Gascoyne, M., Benjamin, G. J., Schwarcz, H. P. and Ford, D. C. (1979). Sea-level lowering during the Illinoian glacia-tion: evidence from a Bahama 'Blue Hole'. *Science* **205**, 806–8.

Gascoyne, M., Schwarcz, H. P. and Ford, D. C. (1983). Uranium-series ages of speleothem from Northwest England: correlation with Quaternary climate. *Phil. Trans. Roy. Soc. Lond. B* **301**, 143–64.

Goldberg, E. D. (1963). Geochronology with Pb–210. In: *Radioactive Dating.* I.A.E.A., Vienna, pp. 121–31.

Goldberg, E. D. and Bruland, K. (1974) Radioactive geochronologies. In: Goldberg, E. D. (Ed.) *The Sea.* vol. 5, Wiley Interscience, pp. 451–89.

Goldberg, E. D. and Koide, M. (1962). Geochronological studies of deep sea sediments by the ionium/thorium method. *Geochim. Cosmochim. Acta* **26**, 417–50.

Imbrie, J., Hays, J. D., Martinson, D. G., McIntyre, A., Mix, A. C., Morley, J., Pisias, N., Prell, W. and Shackleton, N. J. (1984). The orbital theory of Pleistocene climate: support from a revised chronology of the marine δ^{18}O record. In: Berger, A. L. *et al.* (Eds.), *Milankovitch and Climate, Part 1.* Reidel, pp. 269–305.

Imbrie, J., Mix, A. C. and Martinson, D. G. (1993). Milankovitch theory viewed from Devils Hole. *Nature* **363**, 531–3.

Isaac, N. and Picciotto, E. (1953). Ionium determination in deep-sea sediments. *Nature* **171**, 742–3.

Ivanovich, M. (1982a). Spectroscopic methods. In: Ivanovich, M. and Harmon, R. S. (Eds), *Uranium Series Disequilibrium Applications to Environmental Problems.* Oxford Univ. Press, pp. 107–44.

Ivanovich, M. (1982b). Uranium series disequilibria applications in geochronology. In: Ivanovich, M. and Harmon, R. S. (Eds), *Uranium Series Disequilibrium Applications to Environmental Problems.* Oxford Univ. Press, pp. 56–78.

Kaufman, A. (1971). U-series dating of Dead Sea carbonates. *Geochim. Cosmochim. Acta* **35**, 1269–81.

Kaufman, A. and Broecker, W. S. (1965). Comparison of Th230 and C14 ages for carbonate materials from lakes Lahontan and Bonneville. *J. Geophys. Res.* **70**, 4039–54.

Kaufman, A., Broecker, W. S., Ku, T. L. and Thurber, D. L. (1971). The status of U-series methods of mollusc dating. *Geochim. Cosmochim. Acta* **35**, 1155–83.

Khlapin, V. G. (1926). *Dokl. Akad. Nauk. SSSR* **178**.

Koide, M., Soutar, A. and Goldberg, E. D. (1972). Marine geochronology with Pb210. *Earth Planet. Sci. Lett.* **14**, 442–6.

Krishnaswamy, S., Lal, D., Martin, J. M. and Meybek, M. (1971). Geochronology of lake sediments. *Earth Planet. Sci. Lett.* **11**, 407–14.

Kroll, V. St. (1954). On the age determination in deep-sea sediments by radium measurements. *Deep-sea Res.* **1**, 211–15.

Ku, T. L. (1965). An evaluation of the U^{234}/U^{238} method as a tool for dating pelagic sediments. *J. Geophys. Res.* **70**, 3457–74.

Ku, T. L. (1976). The uranium series methods of age determination. *Ann. Rev. Earth Planet. Sci.* **4**, 347–79.

Ku, T. L., Bischoff, J. L. and Boersma, A. (1972). Age studies of Mid-Atlantic Ridge sediments near 42 °N and 20 °N. *Deep-sea Res.* **19**, 233–47.

Ku, T. L., Knauss, K. G. and Mathieu, G. G. (1977). Uranium in open ocean: concentration and isotopic composition. *Deep-Sea Res.* **24**, 1005–17.

Ku, T. L. and Liang, Z. C. (1984). The dating of impure carbonates with decay-series isotopes. *Nucl. Instr. Meth. in Phys. Res. A* **223**, 563–71.

Kumar, N., Gwiazda, R., Anderson, R. F. and Froelich, P. N. (1993). $^{231}Pa/^{230}Th$ ratios in sediments as a proxy for past changes in Southern Ocean productivity. *Nature* **362**, 45–8.

Lalou, C., Reyss, J. L. and Brichet, E. (1993). Actinide-series disequilibrium as a tool to establish the chronology of deep-sea hydrothermal activity. *Geochim. Cosmochim. Acta* **57**, 122–131.

Li, W. X., Lundberg, J., Dickin, A. P., Ford, D. C., Schwarcz, H. P., McNutt, R. H. and Williams, D. (1989). High-precision mass-spectrometric uranium-series dating of cave deposits and implications for palaeoclimate studies. *Nature* **339**, 534–6.

Ludwig, K. R., Simmons, K. R., Szabo, B. J., Winograd, I. J., Landwehr, J. M., Riggs, A. C. and Hoffman, R. J. (1992). Mass-spectrometric ^{230}Th–^{234}U–^{238}U dating of the Devils Hole calcite vein. *Science* **258**, 284–7.

Ludwig, K. R., Szabo, B. J., Moore, J. G. and Simmons, K. R. (1991). Crustal subsidence rate off Hawaii determined from $^{234}U/^{238}U$ ages of drowned coral reefs. *Geology* **19**, 171–4.

Lundberg, J., Ford, D. C., Schwarcz, H. P., Dickin, A. P. and Li, W. X. (1990). Dating sea level in caves: reply. *Nature* **343**, 217–18.

Luo, S. and Ku, T. L. (1991). U-series isochron dating: a generalised method employing total-sample dissolution. *Geochim. Cosmochim. Acta* **55**, 555–64.

Mangini, A. and Diester-Haass, L. (1983). Excess Th-230 in sediments off NW Africa traces upwelling in the past. In: Suess, A. E. and Thiede, J. (Eds), *Coastal Upwelling.* Plenum. Part A, pp. 455–70.

McDermott, F., Grun, R., Stringer, C. B. and Hawkesworth, C. J. (1993). Mass-spectrometric U-series dates for Israeli Neanderthal/early modern hominid sites. *Nature* **363**, 252–5.

Meadows, J. W., Armani, R. J., Callis, E. L. and Essling, A. M. (1980). Half-life of ^{230}Th. *Phys. Rev. C* **22**, 750–4.

Milankovitch, M. M. (1941). Canon of insolation and the ice age problem. Koniglich Serbische Akademie, Beograd. [Translation, Israel Prog. Sci. Trans., Washington D. C.]

Moore, J. G., Clague, D. A., Ludwig, K. R. and Mark, R. K. (1990). Subsidence and volcanism of the Haleakala Ridge, Hawaii. *J. Volc. Geotherm. Res.* **42**, 273–84.

Moore, W. S. (1982). Late Pleistocene sea-level history. In: Ivanovich, M. and Harmon, R. S. (Eds.), *Uranium Series Disequilibrium Applications to Environmental Problems,* Oxford Univ. Press, pp. 481–96.

Osmond, J. K. and Cowart, J. B. (1982). Ground water. In: Ivanovich, M. and Harmon, R. S. (Eds.), *Uranium Series Disequilibrium Applications to Environmental Problems,* Oxford Univ. Press, pp. 202–45.

Picciotto, E. G. and Wilgain, S. (1954). Thorium determination in deep-sea sediments. *Nature* **173**, 632–3.

Piggot, C. S. and Urry, W. D. (1939). The radium content of an ocean bottom core. *J. Wash. Acad. Sci.* **29**, 405–15.

Piggot, C. S. and Urry, W. D. (1942). Time relations in ocean sediments. *Bull. Geol. Soc. Am.* **53**, 1187–210.

Potts, P. J. (1987). *Handbook of Silicate Rock Analysis.* Blackie, 602 p.

Przybylowicz, W., Schwarcz, H. P. and Latham, A. G. (1991). Dirty calcites. 2. Uranium-series dating of artificial calcite–detritus mixtures. *Chem. Geol. (Isot. Geosci. Section)* **86**, 161–78.

Rosholt, J. N., Emiliani, C., Geiss, J., Koczy, F. F. and Wangersky, P. J. (1961). Absolute dating of deep-sea cores by the Pa231/Th230 method. *J. Geol.* **69**, 162–85.

Rosholt, J. N., Shields, W. R. and Garner, E. L. (1963). Isotopic fractionation of uranium in sandstone. *Science* **139**, 224–6.

Sackett, W. M. (1960). The protoactinium–231 content of ocean water and sediments. *Science* **132**, 1761–2.

Sackett, W. M. (1964). Measured deposition rates of marine sediments and implications for accumulation rates of extraterrestrial dust. *Ann. N. Y. Acad. Sci.* **119**, 339–46.

Sackett, W. M. (1966). Manganese nodules: thorium–230: protoactinium–231 ratios. *Science* **154**, 646–7.

Santschi, P. H., Li, Y. H., Adler, D. M., Amdurer, M., Bell, J. and Nyffeler, U. P. (1983). The relative mobility of natural (Th, Pb and Po) and fallout (Pu, Am, Cs) radionuclides in the coastal marine environment: results from model ecosystems (MERL) and Narragansett Bay. *Geochim. Cosmochim. Acta* **47**, 201–10.

Scholten, J. C., Botz, R., Mangini, A., Paetsch, H., Stoffers, P. and Vogelsang, E. (1990). High resolution $^{230}Th_{ex}$ stratigraphy of sediments from high-latitude areas (Norwegian Sea, Fram Strait). *Earth Planet. Sci. Lett.* **101**, 54–62.

Schwarcz, H. P. (1989). Uranium series dating of Quaternary deposits. *Quat. Int.* **1**, 7–17.

Schwarcz, H. P. and Blackwell, B. (1985). Dating methods of Pleistocene deposits and their problems: II. Uranium series disequilibrium dating. *Geosci. Can. Reprt* **2**, 9–17.

Schwarcz, H. P. and Blackwell, B. (1991). Archaeological applications. In: Ivanovich, M. and Harmon, R. S. (Eds.), *Uranium Series Disequilibrium Applications to Environmental Problems.* 2nd Edn, Oxford Univ. Press, pp. 513–52.

Schwarcz, H. P. and Latham, A. G. (1989). Dirty calcites. 1. Uranium-series dating of contaminated calcite using leachates alone. *Chem. Geol. (Isot. Geosci. Section)* **80**, 35–43.

Schwarcz, H. P. and Skoflek, I. (1982). New dates for the Tata, Hungary archaeological site. *Nature* **295**, 590–1.

Scott, M. R. (1968). Thorium and uranium concentrations and isotope ratios in river sediments. *Earth Planet. Sci. Lett.* **4**, 245–52.

Shirahata, H., Elias, R. W., Patterson, C. C. and Koide, M. (1980). Chronological variations in concentrations and isotopic compositions of anthropogenic atmospheric lead in sediments of a remote subalpine pond. *Geochim. Cosmochim. Acta* **44**, 149–62.

Thurber, D. L., Broecker, W. S., Blanchard, R. L. and

Potratz, H. A. (1965). Uranium-series ages of Pacific atoll coral. *Science* **149**, 55–8.

Veeh, H. H. and Burnett, W. C. (1982). Carbonate and phosphate sediments. In: Ivanovich, M. and Harmon, R. S. (Eds.), *Uranium Series Disequilibrium Applications to Environmental Problems*. Oxford Univ. Press, pp. 459–80.

Winograd, I. J. (1990). Dating sea level in caves: comment. *Nature* **343**, 217–8.

Winograd, I. J., Coplen, T. B., Landwehr, J. M., Riggs, A. C., Ludwig, K. R., Szabo, B. J., Kolesar, P. T. and Revesz, K. M. (1992). Continuous 500,000-year climate record from vein calcite in Devils Hole, Nevada. *Science* **258**, 284–7.

Yokoyama, Y. and Nguyen H. V. (1980). Direct and non-destructive dating of marine sediments, manganese nodules, and corals by high resolution γ-ray spectrometry. In: Goldberg, E. D., Horibe, Y. and Saruhashi, K. (Eds.), *Isotope Marine Chemistry*. Uchida Rokkaku, Ch. 14.

13 U-series geochemistry of igneous systems

U-series dating of sedimentary rocks was discussed in the previous chapter. These isotopes can also be used as dating tools for igneous rocks; however, their application as isotopic tracers is probably more important. The short half-lives of the decay series nuclides makes them ideally suited to studies of magma segregation from the mantle and magma evolution in the crust, since these processes operate over similar time periods. With a half-life of 75.4 kyr, ^{230}Th is by far the most important of these geological tracers, and will be the main focus of this chapter. However, attention will also be given to other shorter-lived isotopes used in conjunction with thorium. Note that all isotopic abundances of U-series nuclides referred to in this chapter are expressed as activities, unless specifically stated to be atomic.

Until recently, all U-series measurements on igneous rocks were made by α spectrometry, as for sedimentary rocks (section 12.2). Following the application of mass-spectrometry to U-series dating of carbonates, it was quickly applied to igneous systems (Goldstein et al., 1989). However, analysis of ^{230}Th is made more difficult in silicate rocks by their very large atomic ^{232}Th/^{230}Th ratios. For example, a basalt with a typical Th/U concentration ratio of 4, and in secular equilibrium, will have a ^{232}Th/^{230}Th atomic ratio of 240 000. For a single-sector mass-spectrometer with an abundance sensitivity of 1 ppm at 2 a.m.u. (section 2.2.3), the ^{232}Th peak will then generate a peak tail at mass 230 which is one-quarter of the size of the ^{230}Th peak. However, McDermott et al. (1993) showed that accurate ^{230}Th data can still be obtained under these conditions if the exponentially curved baseline shape is carefully interpolated under the ^{230}Th peak. Alternatively, peak-tailing can be reduced using an energy filter (section 2.2.3).

13.1 Geochronology of volcanic rocks

In some ways, U-series systems in igneous rocks are simpler than carbonates, because ^{234}U and ^{238}U are always effectively in secular equilibrium. On the other hand, they are more complex than pure carbonates in that they invariably contain initial Th at the time of cooling. Hence, a U–Th isochron diagram must always be used to date igneous rocks by the ^{230}Th method.

13.1.1 The U–Th isochron diagram

After time t, the net ^{230}Th activity in a silicate sample is the sum of ^{230}Th growth from U decay and the residue of partially decayed initial ^{230}Th. In other words we sum equations [12.8] and [12.25]:

$$^{230}\text{Th}_P = {}^{230}\text{Th}_I\, e^{-\lambda_{230}t} + {}^{238}\text{U}(1 - e^{-\lambda_{230}t}) \qquad [13.1]$$

It is convenient to divide through by ^{232}Th, whose activity is effectively constant between t initial and the present:

$$\left(\frac{^{230}\text{Th}}{^{232}\text{Th}}\right)_P = \left(\frac{^{230}\text{Th}}{^{232}\text{Th}}\right)_I e^{-\lambda_{230}t} + \frac{^{238}\text{U}}{^{232}\text{Th}}(1 - e^{-\lambda_{230}t})$$

$$[13.2]$$

This is the equation for a straight line, and is plotted on a diagram of ^{230}Th/^{232}Th against ^{238}U/^{232}Th (Fig. 13.1) which is analogous to the Rb–Sr isochron diagram. As in the Rb–Sr system, a suite of cogenetic samples of the same age define a linear array whose slope yields the age. However, because the ^{230}Th daughter product is itself subject to decay, this leads to more complex isotope systematics.

The evolution of igneous systems on the U–Th isochron diagram depends on their composition with respect to a state of secular equilibrium. Samples in secular equilibrium must, by definition, have equal ^{230}Th and ^{238}U activities. Hence, they must have equal ^{230}Th/^{232}Th and ^{238}U/^{232}Th activity ratios in Fig. 13.1. Such samples lie on a slope of unity in this diagram, called the 'equiline' by Allegre and Condomines (1976).

Now considering a suite of rock or mineral samples; at the time of their crystallisation they have

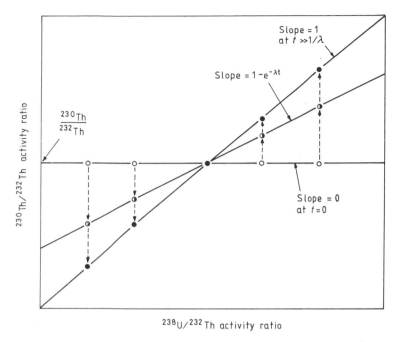

Fig. 13.1. Isotopic evolution of igneous rocks on the ^{230}Th/^{232}Th *versus* ^{238}U/^{232}Th isochron diagram. Symbols: (○): time $t = 0$; (half-filled): samples after elapsed time t; (●): samples after effectively infinite time ($t \gg 1/\lambda$).

variable U/Th ratios but a constant (initial) ^{230}Th/^{232}Th activity ratio, forming a horizontal line in Fig. 13.1. The point of intersection of this array with the equiline must by definition remain invariant; since it starts its closed-system evolution in secular equilibrium it must remain so. All other samples in the rock suite evolve with time. Those to the right of the invariant point are daughter (^{230}Th) deficient relative to ^{238}U, and ^{230}Th therefore builds up with time. They move vertically upwards until they also reach the equiline (secular equilibrium). Those to the left of the equiline have daughter (^{230}Th) excess relative to the parent. They move vertically downwards with time until they reach the equiline. The more initial thorium that is present, the higher the intersection between the initial composition and the equiline. Conversely, when no initial Th is present (as in pure carbonates), evolution begins along the x axis. Hence, in a cogenetic suite which has not reached equilibrium the initial Th isotope ratio is given by the intersection of the isochron array with the equiline. It is not the composition at the y axis.

The first use of this system to date igneous minerals was made by Cerrai *et al.* (1965). Kigoshi (1967) tested the method by dating three igneous rocks of different age; a Cretaceous granite (effectively of infinite age), a 35.7 kyr-old pumice (dated by ^{14}C on a

wood inclusion) and a lava of historical (effectively zero) age. His results (Fig. 13.2) demonstrated the method to be effective. The old granite samples lie on the equiline, the pumice samples yield a U–Th isochron age of 38 kyr, and the historical lava yields a zero slope.

To achieve a high-precision age from the U–Th method, a reasonable spread of ^{238}U/^{232}Th ratios is needed within each sample suite. Kigoshi carried over the leaching techniques of carbonate dating in order to maximise this spread of U/Th ratios. However, this is a potentially dangerous technique, since in the leaching process disequilibrium may be introduced between different parents and daughters in the decay series. For example, ^{230}Th may be preferentially leached relative to ^{238}U from radiation-damaged lattice sites, yielding spuriously old ages. This is called the 'hot atom effect', and is the same process (section 12.3.1) which gives rise to variable ^{234}U/^{238}U ratios in natural waters. Kigoshi attempted to counteract this problem by measuring ^{238}U and ^{232}Th activities *via* their short-lived daughter products ^{234}Th and ^{228}Th respectively, which should also occupy radiation-damaged sites.

Taddeucci *et al.* (1967) avoided the complexities of the 'hot atom' effect by using conventional physical separation and total dissolution of minerals to date

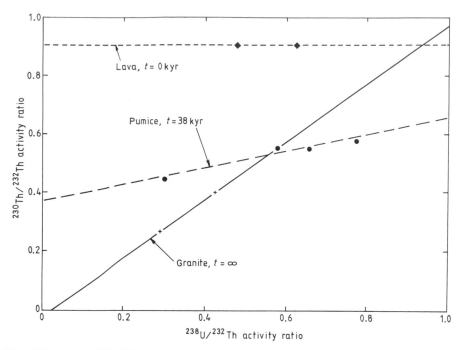

Fig. 13.2. $^{230}Th/^{232}Th$ *versus* $^{238}U/^{232}Th$ isochron diagram for three suites of leachates from whole-rock samples. After Kigoshi (1967).

five rhyolitic tuffs from the Mono Craters of California. However, U–Th dates on phenocryst–glass pairs did not display good agreement with other methods. For example, the hornblende–glass pair gave an apparent U–Th age of only 1 kyr, whereas K–Ar dating yielded an age of ca. 7 kyr, and ^{14}C dating of undisturbed lake sediments near the volcano gave a minimum eruptive age of 2.2 kyr.

Allegre (1968) subsequently showed that separated phenocryst phases from one of Taddeucci *et al.*'s rocks defined an isochron age of 25 kyr (Fig. 13.3). The implication of this discordancy is that the different analytical methods are dating different events. U–Th ages on phenocryst minerals probably date their crystallisation in a magma chamber, whereas the K–Ar method dates the time of eruption (if outgassing of volatiles during eruption was effective). The hornblende–glass age is meaningless, since these two systems did not close at the same time. The discordance between dating methods is therefore caused by the relatively long residence period of magma in the chamber, after phenocryst formation.

In contrast to the Mono Craters case, Allegre and Condomines (1976) and Condomines and Allegre (1980) were able to achieve good linearity of phenocryst and whole-rock points in dating studies

of the Irazu volcano (Costa Rica) and Stromboli volcano (Italy). This implies that in these systems phenocryst growth only briefly preceded eruption. On the other hand, Capaldi and Pece (1981) claimed to find gross Th isotope disequilibrium between different mineral phases in modern lavas from Etna, Vesuvius and Stromboli. This led Capaldi *et al.* (1982) to completely write off the U–Th method as a dating tool. However, in repeat analyses of the samples from the same Etna and Vesuvius lavas, Hemond and Condomines (1985) were unable to find mineralogical disequilibrium of Th isotope ratios. This suggests that Capaldi *et al.* (1982) over-reacted when they dismissed the method. It is true that there are quite a number of instances where Th isotope disequilibrium has been found on a mineralogical scale (Capaldi *et al.*, 1985). However, if phenocryst phases are screened by petrographic examination to exclude entrained xenocrysts then many young lavas can yield accurate U–Th crystallisation ages (e.g. Condomines *et al.*, 1982).

13.2 Magma chamber evolution

Just as initial Sr isotope compositions are useful as a geochemical tracer when using the Rb–Sr method,

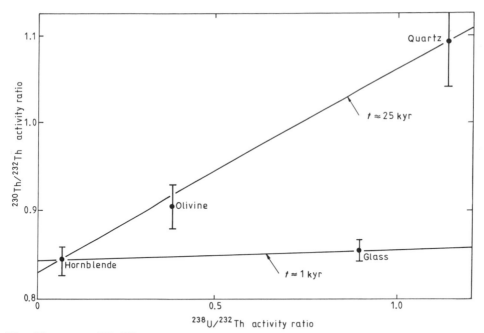

Fig. 13.3. ^{230}Th/^{232}Th *versus* ^{238}U/^{232}Th isochron diagram for hornblende–olivine–quartz phenocryst assemblages and glass matrix from a Mono Craters rhyolite (California), showing isotopic discordancy between phenocrysts and glass. After Allegre (1968).

initial Th isotope compositions are also a useful product of the U–Th method. One area where they have proved particularly valuable is in studying magma chamber evolution.

In the same way that Th isotope evolution in a volcanic rock can be used to date crystallisation, Th isotope evolution between successive eruptions can be used to date the residence of magma in a chamber. We can use the same equation as for the rock system, except that what we input as the 'final composition' on the left-hand side is actually the initial Th activity ratio of a magma at the time of eruption (E), while the Th activity ratio on the right-hand side is the composition of the magma in the chamber at the time of influx (I), simplistically, from the mantle. The quantity 't' then represents the residence time of the magma batch in the chamber (all nuclide ratios in activities):

$$\left(\frac{^{230}\text{Th}}{^{232}\text{Th}}\right)_{\text{E}} = \left(\frac{^{230}\text{Th}}{^{232}\text{Th}}\right)_{\text{I}} e^{-\lambda_{230}t} + \frac{^{238}\text{U}}{^{232}\text{Th}}\left(1 - e^{-\lambda_{230}t}\right)$$

[13.3]

Allegre and Condomines (1976) preferred to refer all times to the present, introducing T as the age of influx into the chamber and t as the time of eruption. It is best to first rearrange equation [13.3] to gather the exponent terms in one place:

$$\left(\frac{^{230}\text{Th}}{^{232}\text{Th}}\right)_{\text{E}} = \left[\left(\frac{^{230}\text{Th}}{^{232}\text{Th}}\right)_{\text{I}} - \frac{^{238}\text{U}}{^{232}\text{Th}}\right] e^{-\lambda_{230}t} + \frac{^{238}\text{U}}{^{232}\text{Th}}$$

[13.4]

Then:

$$\left(\frac{^{230}\text{Th}}{^{232}\text{Th}}\right)_{t} = \left[\left(\frac{^{230}\text{Th}}{^{232}\text{Th}}\right)_{T} - \frac{^{238}\text{U}}{^{232}\text{Th}}\right] e^{\lambda_{230}(t-T)} + \frac{^{238}\text{U}}{^{232}\text{Th}}$$

[13.5]

Just as we make a closed-system assumption in the case of rock dating, so we must assume that the magma in the chamber remains a closed-system to U and Th during its evolution and eruption. This assumption should not be upset by Rayleigh crystal fractionation, since the distribution coefficients of both U and Th are so low that both elements normally remain entirely in the liquid.

13.2.1 The Th isotope evolution diagram

The ^{230}Th evolution of magmas can be shown on U–Th isochron diagrams, but it is also convenient to display isotopic evolution on a plot of Th activity ratio against time (Fig. 13.4). This plot is analogous

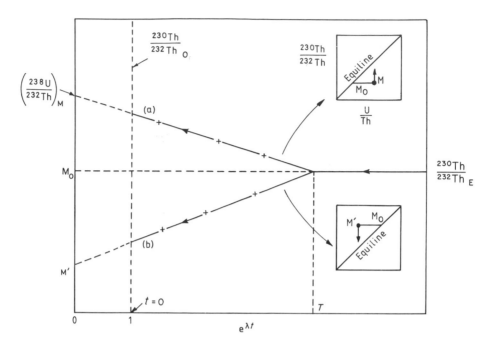

Fig. 13.4. Schematic diagram of Th isotope evolution against time for two closed-system magma chambers: displaying a) daughter deficiency; and b) daughter excess. Plotted data are initial Th activities at the time of eruption (E). Insets show U/Th fractionation and subsequent evolution on U–Th isochron diagrams. After Condomines *et al.* (1982).

to the evolution diagrams of Sr or Nd isotope composition against time (e.g. section 4.2), but because the ^{230}Th half-life is short relative to the time periods under study, the x axis must be calibrated in log time.

A magma body in secular equilibrium must maintain its ^{230}Th/^{232}Th activity ratio. Hence, such evolution is described by a horizontal line in Fig. 13.4. If the system undergoes U/Th fractionation (horizontal displacement from M_0 to M or M′ on the inset diagrams) then the ^{230}Th/^{232}Th activity ratio of the magma will evolve over time to regain secular equilibrium. A magma enriched in ^{238}U/^{230}Th (to the right of the equiline on the inset, with activity M) defines a line of slope (a) on the main diagram. Similarly, a magma depleted in ^{238}U/^{230}Th (activity ratio M′ on the inset) defines an evolution line (b) on the main diagram. If we extrapolate the growth lines to $e^{\lambda t} = 0$, then we can see from equation [13.4] that the y ordinate in the main diagram describes the ^{238}U/^{232}Th activity ratio of the evolving magma.

The U–Th isotope system is a very powerful tool for studying magma chamber evolution because the 75.4 kyr half-life of ^{230}Th is the same order of

magnitude as magma chamber events. However, a significant data-base is needed to unravel the history of most volcanoes, which involve repeated magma injection and eruption events. A simple scenario of this type is illustrated in Fig. 13.5. In this case a primary mantle source in secular equilibrium supplies a series of magma batches over a period of time which have a constant, but disequilibrium ^{238}U/^{232}Th activity ratio generated by the partial melting process (see below). After a period of magma evolution in a high-level chamber (sloping line), the chamber is emptied by eruption and re-filled, causing a kick back to the starting composition. However, in the real world, mixing of magmas of different ages is likely to occur, and this will generate a more complex pattern of magma evolution.

Early studies (Allegre and Condomines, 1976; Condomines and Allegre, 1980) lacked sufficient data to resolve the magmatic history of long-lived volcanoes, and their results were ambiguous. However, a later study by Condomines *et al.* (1982) provided enough data to interpret the history of the Etna volcano in Sicily. Thirteen mineral–whole-rock U–Th isochrons were determined, along with analyses of recent lavas. The results were plotted on a Th

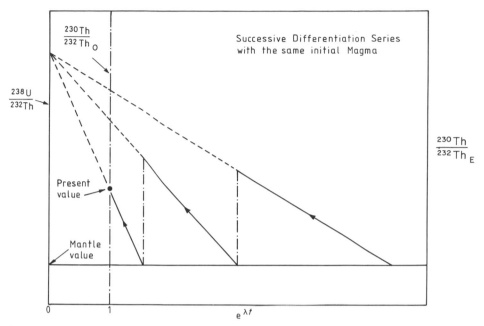

Fig. 13.5. Schematic illustration of the thorium isotope evolution of a periodically tapped and re-filled magma chamber with constant U/Th ratio. After Condomines *et al.* (1982).

isotope evolution diagram (Fig. 13.6a), and an isochron diagram (Fig. 13.6b).

On the Th isotope evolution diagram (Fig. 13.6a) the Etna data provide evidence for four episodes of eruption and magma replenishment, the last three of which (numbered) tie in with dates of major caldera collapse events. Between these events, small magma tappings monitor Th isotope evolution in the high-level chamber. However, ^{230}Th/^{232}Th ratios fall too rapidly during these periods to be explained by closed-system evolution, given the observed range of U/Th ratios (hatched band on the left-hand axis). Therefore, Condomines *et al.* invoked a magma mixing model to explain these steep trends, suggesting that the sub-horizontal evolution line represented a deep, long-lived alkali basalt reservoir which continually supplied magma to higher levels, where a low- ^{230}Th/^{232}Th tholeiitic component was added intermittently.

Initial ratios of lavas at the time of eruption are plotted on a U–Th isochron diagram in Fig. 13.6b. The data suggest mixing between an old magma nearly in secular equilibrium (M_0) and a young one substantially out of equilibrium (M_1). Condomines *et al.* argued that the low ^{230}Th/^{232}Th component (M_1) could not be a crustal contaminant, since this should be close to secular equilibrium, whereas Fig. 13.6 indicates it to be far from equilibrium. However, this does not exclude the possibility of sediment contamination of the mantle source of these magmas. The straight line (1) represents an instantaneous mixing model whereas the curved lines (2, 3) model progressive mixing over a time interval. Present-day ratios (of old lavas) are not plotted on Fig. 13.6b because they do not yield any intelligible information about magma evolution.

13.3 Melting models

13.3.1 Ocean island volcanics

Volcanoes located on continental crust appear to show complex ^{230}Th evolution patterns, probably because mantle-derived magmas become trapped during their rise through low-density sialic basement. In contrast, oceanic volcanoes might be expected to show simpler behaviour, since oceanic crust is easily punctured by rising magma. Therefore, oceanic volcanics provide a window to study Th isotope processes in the mantle.

The first detailed U-series measurements on oceanic volcanics were made by Oversby and Gast (1968) on recent ocean island lavas. This study revealed disequilibrium between isotopes of the ultra-incompatible elements, radium and thorium. Oversby and Gast suggested that these fractionations

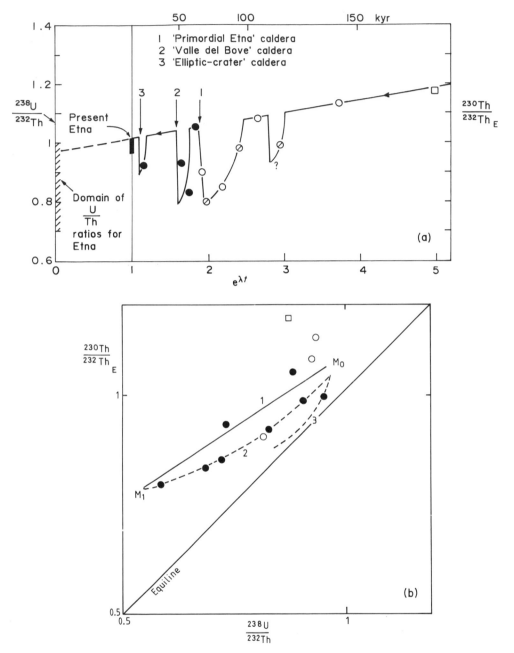

Fig. 13.6. U–Th data for the Etna volcano. a) History of the Etna volcano on a thorium isotope evolution diagram to show four magma influx–mixing–eruption events. Symbols indicate erupted products at different stages of volcano evolution. b) U–Th activity data for Etna lavas at the time of eruption showing possible mixing processes (symbols as in a). M_0 = old magma; M_1 = young magma. After Condomines *et al.* (1982).

were probably inherited from the melting event in the mantle source of the magmas. ^{230}Th activities were observed to be higher than ^{238}U, suggesting that thorium was a more incompatible element than uranium. This conclusion was supported over ten years later (Condomines *et al.*, 1981) in studies of MORB magma genesis (see below).

In view of the short half-lives of ^{226}Ra, Oversby

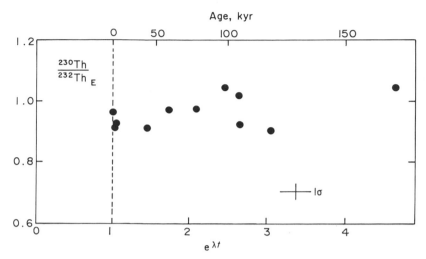

Fig. 13.7. Plot of erupted (initial) Th activity ratios against time for Piton de la Fournaise volcano (Reunion Island) showing constant magma composition over time, within error. After Condomines *et al.* (1988).

and Gast attributed disequilibrium of this species to rapid ascent of ocean island magmas from the source area (< 10 000 yr). This conclusion has also been supported by radium analysis of MORB (section 13.6). In addition, rapid ascent of ocean island magmas has been supported by studies of their Th isotope chemistry over time. For example the volcanoes of Mauna Kea (Hawaii), Marion Island (SW Indian Ocean) and Piton de la Fournaise (Fig. 13.7) all show constant initial $^{230}Th/^{232}Th$ ratios (within error) over the last 250 kyr (Newman *et al.*, 1984; Condomines *et al.*, 1988). This suggests that magma transport from the melting zone to the surface probably occurred within a few kyr, without storage in a deep crustal reservoir. Therefore, the calculated initial Th isotope ratio for each eruption is probably very close to the source value.

13.3.2 Ocean ridge processes

Mid ocean ridges present the minimum crustal thickness which must be traversed by ascending mantle-derived magmas. Therefore, in this environment we should have the best opportunity to see back through the processes of magma evolution during ascent, to study source processes and chemistry. However, some workers (e.g. O'Hara and Mathews, 1981) have suggested that MORB magmas spend many eruptive cycles in periodically re-filled, periodically tapped magma chambers under the ridge, which then grossly perturb the incompatible element and isotopic signatures of the product magmas. In this

case, they argued, it would be almost impossible to 'invert' the data (see section 7.2) to reconstruct source chemistry.

The short half-life of ^{230}Th has provided a powerful tool to test these models of MORB magma evolution. In the first detailed study of MORB samples, Condomines *et al.* (1981) found that fresh, young crystalline basalts and glasses from the 'FAMOUS' area on the Mid Atlantic Ridge (37 °N) had a narrow range of Th isotope and U/Th ratios. Because these samples were all less than 5 kyr old, their present Th isotope compositions can be taken as initial ratios at the time of eruption.

When plotted on a U–Th isochron diagram, the FAMOUS data fall well to the left of the equiline, showing them to be far from isotopic equilibrium. U–Th disequilibrium must have been inherited during the melting process (Fig. 13.8), because U and Th are both ultra-incompatible elements, and cannot be fractionated from each other during Raleigh crystallisation in a magma chamber. If the residence time of magma in such chambers was more than a few tens of kyr, U–Th activities would again reach equilibrium (Fig. 13.8). Because this has not happened, we can deduce that the transport of magma from the melting zone to the surface was relatively rapid, which is not consistent with prolonged evolution in an open-system magma chamber. This conclusion has subsequently been supported by evidence of isotopic disequilibrium of even shorter-lived U-series nuclides in MORB glasses (section 13.6.1).

In addition to the FAMOUS area, Condomines *et*

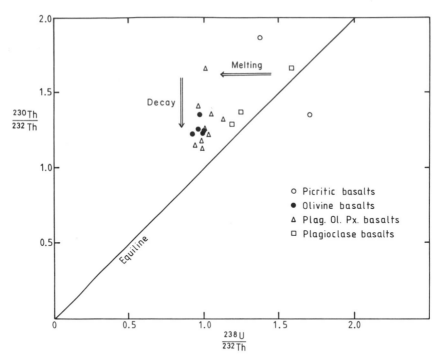

Fig. 13.8. U–Th isochron diagram showing analyses of young lavas from the FAMOUS area of the Mid Atlantic Ridge, to the left of the equiline. Arrows show the effects of partial melting and radioactive decay. After Condomines *et al.* (1981).

al. (1981) showed that other young ridge basalts and OIB also fell on the left side of the equiline (Fig. 13.8). Hence in all of these cases, melts were enriched in Th/U relative to the source. This implies greater incompatibility of Th over U during melting. Allegre and Condomines (1982) expressed this Th/U fractionation during melting by the quantity 'k':

$$k = \frac{(^{238}U/^{232}Th)_{magma}}{(^{238}U/^{232}Th)_{source}} \qquad [13.6]$$

However, it is more useful to express this ratio by its reciprocal, termed 'r' as used by McKenzie (1985a). In addition to facilitating the algebra, this formulation avoids confusion of 'k' with 'κ' (the atomic $^{232}Th/^{238}U$ ratio):

$$r = \frac{(^{238}U/^{232}Th)_{source}}{(^{238}U/^{232}Th)_{magma}} \qquad [13.7]$$

However, the source is assumed to be on the equiline, so its $^{238}U/^{232}Th$ activity is equal to its $^{230}Th/^{232}Th$ activity. Furthermore, if the analysed sample was extracted from the source in less than a few thousand years, then the source Th isotope activity is equal to

that measured in the magma:

$$\left(\frac{^{238}U}{^{232}Th}\right)_{source} = \left(\frac{^{230}Th}{^{232}Th}\right)_{source} = \left(\frac{^{230}Th}{^{232}Th}\right)_{magma} \qquad [13.8]$$

Substituting into equation [13.8] we obtain:

$$r = \frac{(^{230}Th/^{232}Th)_{magma}}{(^{238}U/^{232}Th)_{magma}} = \left(\frac{^{230}Th}{^{238}U}\right)_{magma} \qquad [13.9]$$

This is represented in Fig. 13.9 by the gradient of lines which project to the origin. Hence we can determine U/Th fractionation during melting from a U-series analysis of the magmatic product.

Allegre and Condomines noted that the U/Th fractionation factor must correlate with the degree (fraction) of partial melting. Since MORB and OIB samples display similar fractionation factors (Fig. 13.9), it was argued that they represent similar degrees of melting. Hence, the differences in chemistry and long-lived isotopic systems between MORB and OIB must reflect differences in source chemistry rather than melting regime.

Corroboration of the Mid Atlantic Th isotope results came from a similar study on the East Pacific

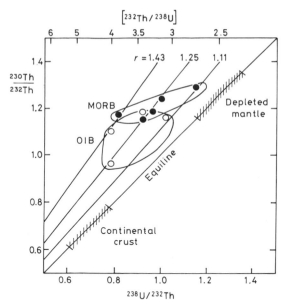

Fig. 13.9. U–Th isochron diagram showing fields for MORB and OIB relative to the Th/U fractionation factor during melting (r). Modified after Allegre and Condomines (1982).

Rise by Newman *et al.* (1983). These data lie to the left of the equiline, and fall within error of unaltered basalts from the FAMOUS area, but with more scatter, particularly in $^{238}U/^{232}Th$ activity. Since U and Th are both ultra-incompatible elements, Newman *et al.* recognised that it was very difficult to generate the required U/Th fractionations at the degrees of melting normally expected for MORB (ca. 10%). They suggested that under these conditions, a U–Th-rich accessory phase might be required to explain the data.

Thompson *et al.* (1984) reversed this problem, arguing that the incompatible element signatures of MORB rocks, including U–Th disequilibrium data, could only be generated by very low degrees of mantle melting. They noted that such an explanation was consistent with observations for other isotope systems on MORB glasses (e.g. Cohen *et al.*, 1980), which showed the observed Rb/Sr, Sm/Nd and U/Pb ratios in MORB magmas to be fractionated relative to the ratios required in the source to generate observed isotope compositions (see section 6.3.2).

This line of argument was developed by McKenzie (1985a), who performed calculations to determine the maximum percentage of partial melting which was consistent with the observed U/Th fractionations. For this purpose he assumed that Th was perfectly

incompatible (i.e. its bulk distribution coefficient between solid and liquid, D, is zero), and that there was no magma residence time in a ridge chamber before eruption. In order to generate a $^{230}Th/^{238}U$ enrichment (r) of 1.25 using a batch melting model and a D value of 0.005 for uranium, the maximum degree of melting permitted was 2%. However, this result is not consistent with major element considerations, which require MORB to be a large degree (ca. 15%) melt of the mantle. Hence, McKenzie ruled out the simple batch melting model for MORB genesis, and adopted instead the dynamic melting model of Langmuir *et al.* (1977).

In this model, melts are extracted simultaneously from a vertical thickness of perhaps 60 km of mantle (horizontal lines in Fig. 13.10). The melts ascend quickly to the surface in a conduit, mixing as they go (solid vertical lines in Fig. 13.10). Meanwhile the source itself moves slowly upwards through the melting zone (dashed vertical lines). At any given point the source contains only 2% melt (porosity), but as it moves upwards, and melts are tapped off, it becomes more and more depleted in incompatible elements. If melts mix equally from the whole melting zone, then the effect of dynamic partial melting on incompatible-element abundances is similar to batch melting. This is because (in the extreme case) the source is completely exhausted of these elements by

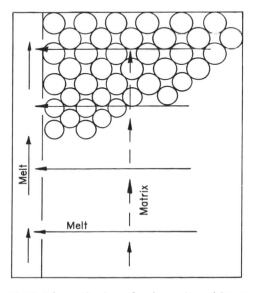

Fig. 13.10. Schematic view of a dynamic melting model for the generation of MORB. See text for discussion. After McKenzie (1985a).

the time it reaches the top of the melting zone. In other words, incompatible-element extraction is 100% efficient. However, for short-lived *nuclides*, the two melting models can yield different results.

If the rate of mantle upwelling is rapid relative to the half-life of the nuclide in question (e.g. ^{230}Th), then this nuclide behaves like a stable element. In this case, dynamic melting will yield an aggregate melt similar to batch melting, and the 15% melt fraction necessary to explain major-element data cannot satisfy the Th isotope data. However, if the rate of mantle upwelling is very slow relative to the ^{230}Th half-life, then ^{230}Th which is removed from the source at the base of the melting column is replenished in the source as it ascends by decay from residual uranium (which is less incompatible than thorium). Consequently, as upwelling progresses, the ^{238}U/^{232}Th activity of the source increases, but it remains on the equiline. After extraction of all U and Th from the source (15% aggregate melt) the ^{238}U/^{232}Th activity of the melt will be the same as the source (Fig. 13.11), but the ^{230}Th/^{238}U activity ratio (r) is still the same as in the first increment of melting at the base of the melting zone.

In between these two extremes (simple batch and dynamic melting), it is possible to determine the rate of mantle upwelling which will yield the observed r value in MORB, given the bulk distribution coefficients (D) for U and Th. Williams and Gill (1989)

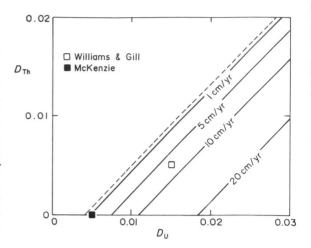

Fig. 13.12. Diagram showing the relationship between calculated mantle upwelling rate (cm/yr) and bulk distribution coefficients (D) for U and Th, assuming a dynamic melting model with 2% porosity and yielding a ^{230}Th/^{238}U activity (r) of 1.2 in MORB. After Williams and Gill (1989).

presented these relationships in diagrammatic form, based on the equations developed by McKenzie (1985a). This information is shown in Fig. 13.12. The values of D chosen by McKenzie (1985a) lead to a calculated mantle upwelling rate of only 1 cm/yr. In contrast, Williams and Gill (1989) argued for a much higher D value for uranium (Fig. 13.12). This allows a more rapid rate of upwelling (ca. 7 cm/yr).

These two different upwelling rates make very different predictions about mantle processes under the ridge. A value of 1 cm/yr is less than the rate of plate motion, which led McKenzie (1985b) to argue that the melting zone under the ridge is funnel-shaped. Trace elements are then extracted from a wide swath of mantle near the base of the funnel, while major elements are dominated by the melt extracted from the apex of the funnel (Fig. 13.13). Because the trace-element extraction zone is larger, this lower domain dominates the U–Th systematics of the melt, which acts like a small-degree batch melt of the original mantle source.

In contrast, with the higher upwelling rate of Williams and Gill (1989), melt extraction occurs from a vertical slice under the ridge, yielding a result closer to the simple dynamic model. However, the consequence is that the ^{230}Th/^{232}Th activity of the erupted products is substantially higher than the source (Fig. 13.11), while the ^{238}U/^{232}Th activity is similar to the source. O'Nions and McKenzie (1993) pointed out

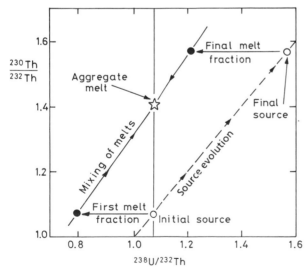

Fig. 13.11. Consequences of very slow mantle upwelling under the ridge for the Th isotope systematics of MORB magmas. Note that the rate of *magma* upwelling is very rapid relative to the ^{230}Th half-life.

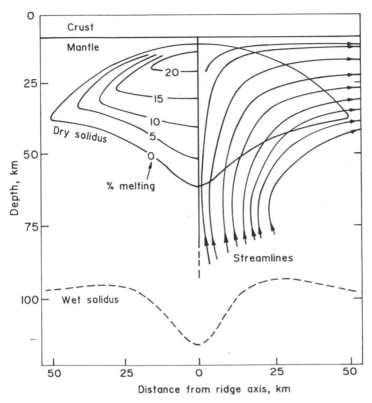

Fig. 13.13. Schematic diagram of mantle convection streamlines under a mid ocean ridge to show a large volume of mantle sampled for trace elements, relative to a small volume under the ridge apex sampled for major elements. After Galer and O'Nions (1986).

that in this case, the Th/U ratio of the source (sections 6.4.2 and 13.4) should be determined not from the Th isotope ratio of MORB (equation [13.9] above), but directly from the elemental U/Th ratio of MORB. Hence, this model predicts that short-lived isotopes *are* fractionated by the melting process under ridges, but stable incompatible elements are *not* fractionated under these conditions. However, by adopting this model, we are no longer able to explain the discrepancies between the trace-element and isotopic signatures of long-lived isotopic systems (section 6.3.2). Therefore, the present author prefers the funnel-shaped melting model shown in Fig. 13.13.

Additional evidence for trace-element fractionation at the base of the melting column comes from experimental studies on mineral melt partition coefficients. As noted above, ^{230}Th/^{238}U enrichment (increased r) is only possible if thorium is more incompatible than uranium. However, clinopyroxene, the principal U- and Th-bearing mineral in the spinel

peridotite upper mantle, has solid/liquid partition coefficients (D) which are *greater* for Th than U (Beattie, 1993a). Therefore, melting in the spinel peridotite stability field (above 70 km depth) cannot generate the observed Th/U fractionations.

On the other hand, Beattie (1993b) showed that garnet has solid/liquid partition coefficients which *are* suitable to generate the observed ^{230}Th/^{238}U enrichments. This was confirmed by LaTourrette *et al.* (1993), who determined values of D_{Th}/D_U equal to ca. 0.1 for garnet. Hence, the Th/U fractionations observed in MORB must originate from melting at greater than 70 km depth, and the resulting liquids must be transported to the surface quickly, before substantial ^{230}Th decay. Both of these requirements are met by the melting model in Fig. 13.13. However, this model encounters greater problems when shorter-lived isotopes such as ^{226}Ra and ^{231}Pa are considered. These make more extreme demands on the melting conditions, which will be discussed in section 13.6.1.

13.3.3 The Th–U isochron diagram

During the course of ^{226}Ra analysis of MORB samples (section 13.6), Reinitz and Turekian (1989) introduced a different presentation of the U–Th isochron diagram, which has also proved useful in the discussion of Ra–Th data (e.g. Rubin and Macdougall, 1990). We will take the opportunity here to discuss this format, and compare it with the conventional isochron plot for U–Th data, before applying it in section 13.6 to Ra–Th data.

Reinitz and Turekian plotted ^{230}Th/^{238}U activity (corresponding to the gradient in the normal U–Th isochron diagram) against ^{232}Th/^{238}U activity. In the case where there is no initial thorium present, the age is given by the intercept on the y axis (equation [12.26]):

$$\left(\frac{^{230}\text{Th}}{^{238}\text{U}}\right)_P = 1 - e^{-\lambda_{230}t} \qquad [13.10]$$

This plot is demonstrated using the MORB data of Reinitz and Turekian (1989) in Fig. 13.14 (solid symbols). A regression through these data points yields an intercept of 0.751, corresponding to an apparent age of 151 kyr. The equiline has an intercept of unity (infinite age), and forms a horizontal array independent of Th/U ratio. The intersection between the regression line and the equiline defines an invariant point (I). The zero-age Th/U fractionation array runs from the origin through point I and, as time passes, samples on this array evolve along vertical lines towards the equiline (Fig. 13.14). The slope of the zero-age array yields the estimated initial ^{230}Th/^{232}Th activity of the suite, indicative of the mantle source composition. For the data of Reinitz and Turekian this ratio is 2.0.

When the MORB data of Newman et al. (1983) are plotted on the Th–U diagram (Fig. 13.14; open symbols) they fall close to a zero-age line, with a much lower ^{230}Th/^{232}Th activity (ca. 1.22). This is

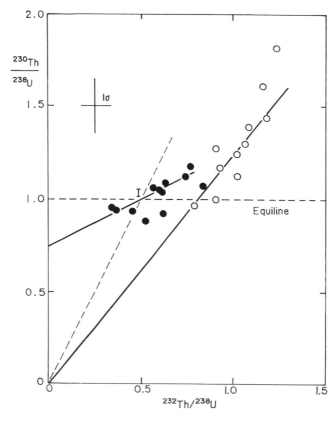

Fig. 13.14. Th–U isochron plot showing data for the East Pacific Rise. Solid symbols: data of Rubin and Macdougall (1988) and Reinitz and Turekian (1989); open symbols: data of Newman et al. (1983). Modified after Reinitz and Turekian (1989).

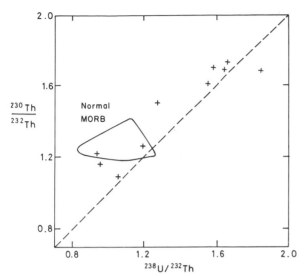

Fig. 13.15. Comparison of the field of ^{232}Th-enriched MORB data (+, Rubin and Macdougall, 1988) with 'normal' Th isotope data for MORB on a U–Th isochron diagram. After Williams and Gill (1989).

strange, because the two suites came from similar areas of the East Pacific Rise (EPR). Reinitz and Turekian suggested that the data of Newman et al. had been upset, either by some geological process, or by analytical error. However, the data of Newman et al. are consistent with more recent α counting data for the EPR (Ben Othman and Allegre, 1990) and mass spectrometric data for the Juan de Fuca and Gorda ridges (Goldstein et al., 1989; 1991). In contrast, the data of Reinitz and Turekian are consistent only with the small data set of Rubin and Macdougall (1988). The latter are compared with the 'normal' ridge data in a conventional U–Th isochron diagram (Fig. 13.15). Williams and Gill (1989) suggested that the data of Rubin and Macdougall might represent ^{232}Th-depleted melts derived from the LIL-depleted top of the mantle melting column under the ridge. This might be caused by multiple melting pulses affecting the same mantle source (Rubin and Macdougall, 1992).

13.3.4 U–Th model age dating

The consistency of mass spectrometric Th isotope data from the crest of the Juan de Fuca and Gorda ridges prompted Goldstein et al. (1991) to apply the method as a dating tool for young off-axis MORB glasses. In order to be sure that these older MORB glasses remained as closed-systems to U and Th, and

were not affected by sea-floor alteration, the samples were subjected to chemical screening, in addition to the normal processes of hand picking and surface leaching of the glass chips (Goldstein et al., 1989; 1991). Sensitive chemical screening was provided by boron analysis, since fresh glasses have boron levels of ca. 1 ppm, while altered glasses have boron contents more than an order of magnitude higher (Spivak and Edmond, 1987). In addition, samples were analysed for ^{234}U/^{238}U, since alteration may cause this ratio to increase or decrease (Macdougall et al., 1979). However, this test is probably less sensitive, since seawater has a ^{234}U/^{238}U activity ratio only slightly above unity.

It is not practical to date MORB samples using the U–Th isochron technique, since glasses have a narrow range of U–Th ratios, and phenocryst phases cannot be used. Therefore, a model age dating technique must be applied. In this approach the initial activity ratio of the off-axis samples is estimated by analysing on-axis samples from the same ridge segment. Goldstein et al. demonstrated this approach for three ridge segments, of which two Juan de Fuca ridge segments are shown in Fig. 13.16.

Goldstein et al. (1991) compared model U–Th ages from samples on both sides of the ridge axis with the spreading rate determined from magnetic data. Generally, the fit was found to be quite good, although some asymmetry is apparent in the U–Th ages. Because both ridges show the same sense of asymmetry, it is possible that this is a real feature of the spreading geometry. However, there is, as yet, insufficient data to be confident of this interpretation. On the other hand, complexity in the age structure of the ridge is suggested by low apparent U–Th ages just below the summit of the Endeavour segment. Goldstein et al. attributed these ages to young lavas erupted near the ridge summit which flowed down over its flanks.

13.3.5 Genesis of oceanic granites

Work on Icelandic lavas by Sigmarsson et al. (1991) has shown that Th isotope data can also be used to investigate the genesis of felsic rocks in oceanic environments. Two alternative models for these rocks involve either direct fractionation of mafic magmas or partial melting of pre-existing mafic crust. However, because of the young age of this crust, conventional radiogenic tracers such as Sr and Nd cannot resolve these models. In contrast, the short half-life of ^{230}Th may allow the solution of this problem.

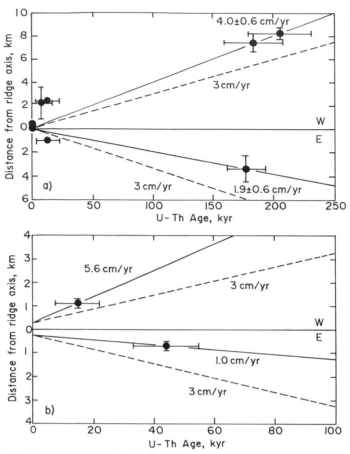

Fig. 13.16. Plot of U–Th model ages compared with distance from the axis of the Juan de Fuca ridge, East Pacific: (a) the Endeavour segment; and (b) the Southern segment. Magnetically determined spreading rates are shown by dashed lines. After Goldstein *et al.* (1991).

Rift-zone tholeiites from Iceland are displaced to the left of the equiline by partial melting, but after 0.5 Myr, when their excess ^{230}Th activity has decayed back to secular equilibrium, they are displaced downwards to a different range of Th activity ratios on the equiline (Fig. 13.17). Consequently, historical-age felsic melts produced by anatexis of Icelandic crust will have a lower ^{230}Th/^{232}Th activity than direct magmatic differentiates of juvenile basic magma. Sigmarsson *et al.* found that dacitic volcanics from the Hekla volcano in southern Iceland had ^{230}Th/^{232}Th activities too low to be derived from contemporaneous mafic magma, but consistent with the melting of older crust. Hekla rhyolites were then modelled by magmatic fractionation of the dacite melts (Fig. 13.17).

Th isotope studies of volcanic systems from Iceland have also been able to show the effect of contamina-

tion of basic magmas by crust a few Myr old (Hemond *et al.*, 1988; Sigmarsson *et al.*, 1991). In major volcanic centres, where the oxygen isotope composition of the crust has been overprinted by meteoric hydrothermal convection, Th activity ratios also correlate with δ^{18}O. An example of this behaviour is shown in Fig. 13.18 for contaminated quartz tholeiites from the active rift zones of Iceland.

13.4 Mantle evolution

The above discussions have focussed on the use of disequilibrium conditions to make deductions about short-period processes of magma generation and evolution. However, the ^{230}Th/^{232}Th activity ratios of primary mantle-derived magmas can also be a tracer for the atomic U/Th ratio of the source reservoir at the time of volcanism. However, this is

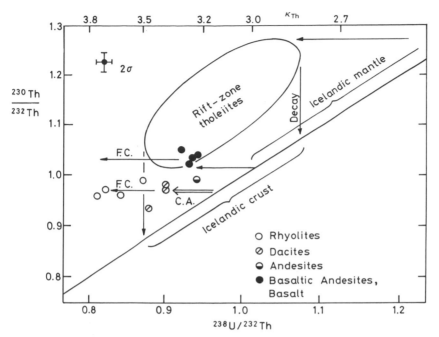

Fig. 13.17. U–Th isochron diagram showing the consequences of fractional crystallisation (F.C.) *versus* crustal anatexis (C.A.) for the production of felsic melts in Hekla volcano, Iceland. After Sigmarsson *et al.* (1991).

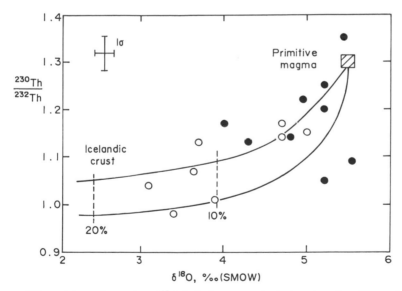

Fig. 13.18. Diagram of Th activity ratio against $\delta^{18}O$ (relative to SMOW) showing the effect of contamination of mafic magmas by young crust. (●) = olivine tholeiite; (○) = quartz tholeiite. After Hemond *et al.* (1988).

normally quoted in its reciprocal form as the atomic $^{232}Th/^{238}U$ ratio, κ. The $^{232}Th/^{230}Th$ activity ratio, equal to the $^{232}Th/^{238}U$ activity ratio, is converted to κ by multiplying by the ratio of the half-lives (3.134). The κ value is, in turn, almost identical to the overall

Th/U weight ratio. This geochemical tracer can be used in the study of long-term mantle evolution processes. As such, it is most useful in conjunction with other radiogenic tracers.

Condomines *et al.* (1981) demonstrated that

MORB and OIB samples formed a mantle array on a Th activity ratio *versus* Sr isotope ratio plot which is analogous to the Nd *versus* Sr isotope mantle array (section 6.3.2). This data set has subsequently been augmented by additional analyses, but the correlation remains strong (Fig. 13.19). The strength of the mantle correlation line provides powerful evidence for the rapid ascent of most oceanic magmas through the crust, since a long pre-eruptive history would tend to cause downward migration of points off the mantle array. This, indeed, is exactly what appears to have happened to some of the Icelandic data in Fig. 13.19. On the other hand, some altered ocean-floor basalts from the FAMOUS area of the Mid Atlantic Ridge fall above the Th–Sr isotope array due to contamination with ^{230}Th-enriched seawater. Data for the East Pacific Rise from Rubin and Macdougall (1988) and Reinitz and Turekian (1989) also fall well above the array, but this has been attributed to multiple melting pulses (Rubin and Macdougall, 1992).

Condomines *et al.* (1981) used the Bulk Earth Sr isotope composition calculated from the Nd–Sr isotope mantle array (0.70475) to calculate a Bulk Earth κ value of 3.55, in reasonable agreement with the κ value of ca. 3.7 in chondrites. However, this was probably a coincidence, since Allegre *et al.* (1986) subsequently argued from Pb isotope evidence that the Bulk Earth κ value is ca. 4.2 (section 6.4.2). In addition the so-called Nd–Sr mantle array has now become a 'mantle disarray' (section 6.3.2), composed of mixing lines between several different mantle

components. In this situation, the stro_ the Th–Sr mantle array is surprising. F_ be attributed to the relatively close similarities between the Rb–Sr and U_ The new Bulk Earth κ value leads to _ ^{87}Sr/^{86}Sr ratio of 0.7055 (Fig. 13.19).

U–Th isotope data yield an 'instantaneous' κ value for the source; in other words the value pertaining in the source at the time of volcanism. However, we can also calculate a 'time-integrated' κ value for a given reservoir, such as the upper mantle, from Pb isotopes (section 6.4.2). The 'time-integrated' ratio is the average ratio over Earth history for a sample from that reservoir.

Instantaneous and time-integrated κ values were first compared in ocean island volcanics by Oversby and Gast (1968), and revealed a paradox which has been confirmed in more recent work on MORB and OIB. For example, the instantaneous κ value for North Atlantic MORB is ca. 2.5, while the time-integrated value is ca. 3.75 . Galer and O'Nions (1985) argued that the simplest solution to this paradox is to postulate that Pb recently extracted from the MORB source only had a brief residence time there, and spent most of Earth history in a reservoir with a Th/U value similar to Bulk Earth. Using a Bulk Earth κ value of 3.9 (based on chondrites) they calculated that an average Pb residence time of only 600 Myr in a Th-depleted MORB reservoir ($\kappa =2.5$) would give the time-integrated κ value of 3.75 needed to explain MORB

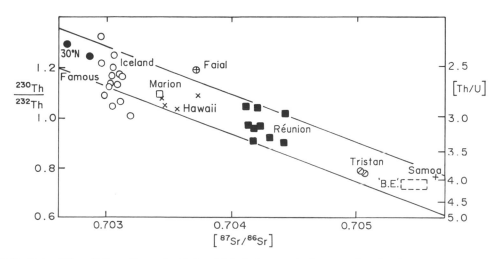

Fig. 13.19. Plot of Th activity ratio against [atomic] Sr isotope ratio for oceanic volcanics. OIB and MORB samples show a strong mantle array, except for Icelandic basalts with a long pre-eruptive history. After Condomines *et al.* (1988).

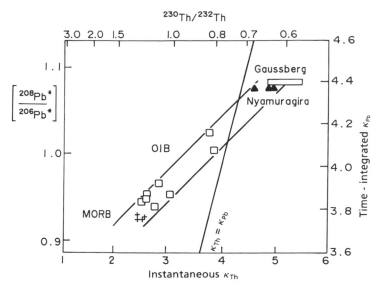

Fig. 13.20. Plot of time-integrated *versus* instantaneous κ for MORB (+), OIB (☐), and two continental alkali volcanoes lying to the right of the geochron. After Williams and Gill (1992).

Pb isotope compositions (section 6.4.2). Revision of instantaneous κ in MORB to 2.58 (O'Nions and McKenzie, 1993) does not significantly affect this residence time.

The short residence time of Pb in the MORB reservoir implied very effective buffering of the upper mantle by a Bulk Earth reservoir. This made extreme demands on mantle evolution models, and Galer and O'Nions were forced to propose exchange between the upper and lower mantle on a massive scale. However, the higher Bulk Earth κ value of 4.2 proposed by Allegre *et al.* (1986) allows the residence time of Pb in the MORB source to be increased to 1600 − 1800 Myr and reduces the mass of the high-κ reservoir required to buffer the upper mantle to only 10% over Earth history. This buffering probably occurs by mixing of enriched mantle plumes into the convecting asthenosphere, since OIB lie on a 'mantle array' between MORB and the U–Th geochron (representing the Bulk Earth). This is shown on a plot of instantaneous *versus* time-integrated κ (Fig. 13.20).

Williams and Gill (1992) proposed that the enriched reservoir which buffers the U–Th systematics of OIB and MORB is sub-continental lithosphere. They argued that the signature of this component is represented by alkali basalts from Nyamuragira volcano in Eastern Zaire and Gaussberg volcano, Antarctica (see also Williams *et al.*, 1992). Both of these volcanoes display low $^{230}Th/^{232}Th$ activity

ratios, yielding instantaneous κ values of ca. 5, well above the Bulk Earth value. The fact that these data lie on an extension of the mantle array supports the contention that they represent one of the enriched components which buffer the asthenosphere. However, there is also good evidence for the role of subducted sediment as an enriched component which buffers the upper mantle. Evidence for this process is discussed below.

13.5 Subduction zone processes

The first comprehensive study of Th isotope systematics in subduction-related magmas was undertaken by Hemond (1986). This work showed a distribution clustering near the equiline on the U–Th isochron diagram (Fig. 13.21a), but with strong departures from the Th–Sr isotope mantle array (Fig. 13.21b). The data set was augmented by Gill and Williams (1990) and McDermott and Hawkesworth (1991), with similar findings.

The incoherent behaviour of Th–Sr systematics (Fig. 13.21b) suggests more complex processes at subduction zones, compared to ridges and plumes. However, some systematic variations emerge on the U–Th isochron diagram (Fig. 13.21a), where continental arcs tend to lie on the U-depleted side of the equiline, while oceanic arcs lie to the U-enriched side. High elemental U/Th ratios (low Th/U) have long been known in the island arc tholeiite series (Jakes

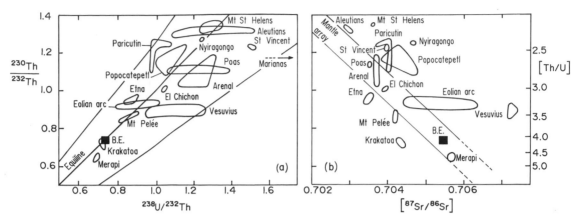

Fig. 13.21. Isotopic data for subduction-related volcanics, on (a) the U–Th isochron diagram, and (b) the Th–Sr isotope diagram. After Condomines *et al.* (1988).

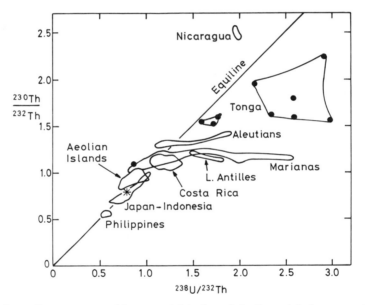

Fig. 13.22. U–Th isochron diagram summarising recent data for subduction-related magmas. Star = average crust. After Hawkesworth *et al.* (1991).

and Gill, 1970), so the distribution of these data to the right of the equiline is not surprising. This effect is seen most clearly in LIL-depleted arcs such as Tonga and the Marianas (Fig. 13.22), but is small or absent in associated back-arc volcanics. For these reasons, the effect is best explained by U metasomatism from the subducted slab into the melting zone in the overlying mantle wedge.

The slab-derived uranium flux can be seen clearly if the overlying wedge is LIL-depleted (e.g. Tonga, Marianas), but in arcs with a more enriched wedge,

the uranium flux may be swamped by U and Th derived by normal partial melting processes. Hence, continental arcs, underlain by enriched mantle lithosphere, tend to show U/Th behaviour similar to within-plate basalts. This effect can be seen by plotting the $^{238}U/^{230}Th$ activity ratio ($1/r$ in section 13.3) against the total Th content of the rock, thus showing that arcs with strong U/Th enrichment are characterised by low total Th contents (McDermott and Hawkesworth, 1991). The slight upward displacement of all arcs in Fig. 13.23 relative to MORB was

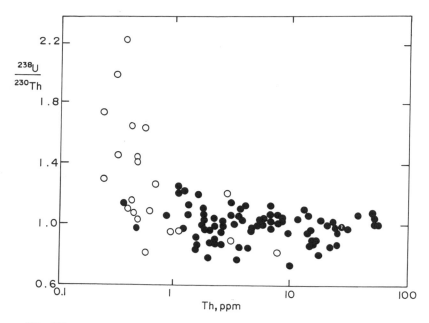

Fig. 13.23. Plot of $^{238}U/^{230}Th$ activity (= U/Th fractionation from the equiline) against total Th content of arc volcanics. Open symbols are from the Tonga and Mariana arcs. After McDermott and Hawkesworth (1991).

attributed to the effects of residual amphibole in the melting zone. This retains some thorium, causing the magma to show slight relative enrichment in U/Th ratios.

Magmas to the right of the equiline in Fig. 13.22 must reflect U/Th fractionation shortly before eruption, in order to preserve isotopic disequilibrium. Independent evidence for the role of slab-derived fluids in this process comes from correlated $^{10}Be/^9Be$ and $^{238}U/^{230}Th$ ratios in the Southern Volcanic Zone of the Andes (Sigmarsson *et al.*, 1990). Because ^{10}Be is a cosmogenic isotope, it can only be introduced into the melting zone of arc magmas by subduction of ocean-floor crust and sediment (section 14.3.4). Therefore, a positive correlation between $^{10}Be/^9Be$ and $^{238}U/^{230}Th$ ratios (Fig. 13.24) suggests a similar location for uranium enrichment of arc magmas. Several of the Southern Volcanic Zone samples have $^{238}U/^{230}Th$ ratios close to the equiline (in common with other continental arcs), but these all have elevated cosmogenic beryllium. If we project these samples back to zero cosmogenic beryllium then we can estimate a $^{238}U/^{230}Th$ ratio of 0.8–0.9 in the wedge-derived component, which is typical of asthenosphere-derived magmas from other tectonic environments.

A second distinct feature of subduction-related magmas on the U–Th isochron diagram is the

extension of some arc suites to ^{232}Th-enriched compositions towards the origin (e.g. Philippines, Indonesia, Fig. 13.22). Because these suites lie close to the equiline, it appears that U/Th fractionation was relatively ancient. This is confirmed on a plot of κ_{Pb} against κ_{Th} in Fig. 13.25. Correlated increases in these values must reflect ancient U–Th fractionation events, since κ_{Pb} is controlled by long-lived U and Th isotopes. The best explanation of these data is contamination of arc magmas by partial melts of subducted sediments, which have appropriate compositions on both the U–Th isochron diagram and κ–κ diagrams (McDermott and Hawkesworth, 1991). Some sediment inevitably escapes the melting process, and is then recycled into the deep mantle.

Processes of fluid metasomatism and sediment contamination in arcs can be summarised on a Th activity *versus* Sr isotope plot (Fig. 13.26). On this diagram, the Indonesian arc defines a hyperbolic array which is consistent with two-component mixing between normal depleted mantle and subducted sediment. In contrast, Tongan data trend upwards towards altered MORB and marine carbonate, from whence the U-enriched fluid flux is probably derived. Final note must be made regarding the Nicaraguan data. These have very high ^{230}Th contents, reminiscent of a metasomatised source, but actually fall on the U-depleted side of the equiline (Fig. 13.22). This

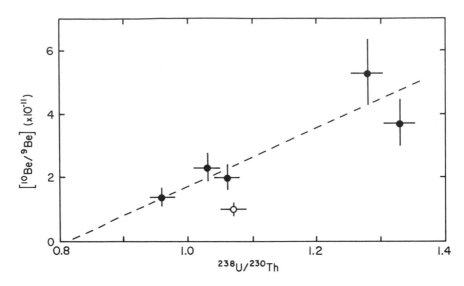

Fig. 13.24. Plot of $^{10}Be/^9Be$ ratio against $^{238}U/^{230}Th$ activity ratio in the Southern Volcanic Zone of the Andes, showing correlated enrichments. The open circle indicates a sample that has been perturbed by contamination in the continental crust. Revised, after Sigmarsson *et al.* (1990).

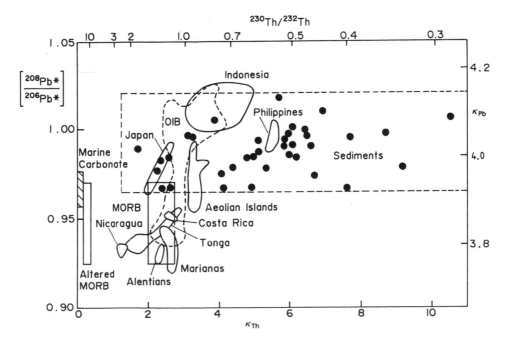

Fig. 13.25. Plot of κ_{Th} against κ_{Pb} to show correlated high values in some arcs and in ocean-floor sediments (•). Fields for OIB, MORB, altered MORB, and marine sediments are shown for reference. After McDermott and Hawkesworth (1991).

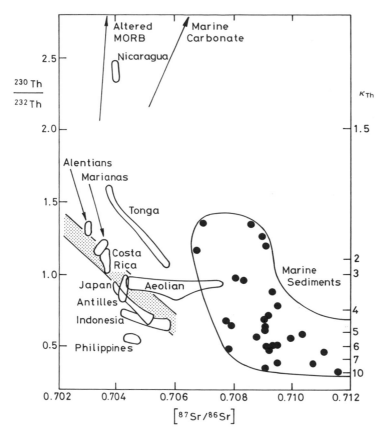

Fig. 13.26. Th–Sr isotope diagram showing possible mixing models to explain the departure of arc magmas from the mantle array of MORB and OIB compositions (shaded). After McDermott and Hawkesworth (1991).

unusual signature is probably best attributed to a source which suffered U metasomatism some period prior to magma generation.

13.6 Short-lived U-series isotopes

In addition to ^{230}Th, there are five shorter-lived nuclides in the U and Th decay schemes which may be useful in the study of igneous systems. These are shown in Fig. 12.2, but for convenience are summarised below. The equilibration times (t_{Eq}) shown represent the maximum useful range of each species, based on the assumption that its activity will be within error of secular equilibrium after 5 half-lives. Each arrow represents a decay transition, but only nuclides relevant to volcanic systems are shown. Disequilibrium between short-lived nuclei in volcanic rocks was discovered very early (e.g. Joly, 1909), but has only recently been subjected to detailed study. ^{210}Pb was observed to be out of secular equilibrium

^{238}U $>>>>$ ^{230}Th $>$ ^{226}Ra $>>>>>>$ ^{210}Pb $>>>$ ^{206}Pb

$t_{1/2}$, yr		75 200	1600	22
t_{Eq}, yr		400 000	8000	100

^{235}U $>>$ ^{231}Pa $>>>>>>>>>$ ^{207}Pb

$t_{1/2}$, yr	34 300
t_{Eq}, yr	170 000

^{232}Th $>$ ^{228}Ra $>>$ ^{228}Th $>>>>>>$ ^{208}Pb

$t_{1/2}$, yr	5.77	1.91
t_{Eq}, yr	30	10

with ^{230}Th in ocean island lavas by Oversby and Gast (1968). This demonstrated the occurrence of Th/Pb fractionation within 100 yr of eruption. However, the chemistry of Pb is so different from the other nuclides that it is difficult to use in petrogenetic interpretations. It will not be discussed further here. [228]Ra and [228]Th can be used to measure very short-period changes in magma chemistry, but have only rarely been found out of isotopic equilibrium.

Of the other short-lived nuclides, [226]Ra has been the most widely used. It has traditionally been measured by counting (sometimes *via* its shorter-lived decay products), but despite its short half-life, recent advances have allowed its measurement by mass spectrometry (e.g. Cohen and O'Nions, 1991). This permits [226]Ra abundances in the femtogram range (10^{-9} ppm) to be determined to better than 1% precision. [231]Pa abundances in igneous rocks are too low to measure by alpha spectrometry because of the low abundance of the parent isotope, [235]U. However, with the advent of mass spectrometry, [231]Pa measurements are also possible in the femtogram range, and this nuclide shows promise as a useful geochemical and chronological tool (Goldstein *et al.*, 1993).

13.6.1 [226]Ra

In view of its short equilibration time of 8000 yr, [226]Ra is useful in studies of geologically rapid magmatic processes. However, a disadvantage is the lack of a longer-lived radium isotope to normalise against in order to exclude chemical fractionation. Williams *et al.* (1986) proposed that this problem might be overcome by using barium as a proxy for a stable radium isotope. For this to be useful, the two elements must have similar distribution coefficients, so that they behave the same in melting and fractionation processes. Rubin and Macdougall (1990) applied this concept to MORB glasses from the East Pacific Rise. The data are shown on an 'isochron' diagram of Ra/Th activity ratio against the [Ba/Th] weight ratio in Fig. 13.27. Because this diagram plots daughter/parent on the *y* axis, it is analogous to the [230]Th–[238]U isochron diagram (Fig. 13.14).

If a magma suite is extracted from a source in secular equilibrium and with homogeneous trace-element chemistry, then all zero-age lavas should lie on a straight line through the origin, corresponding to a constant ratio of [226]Ra activity/Ba concentration. Distribution along the line is due to [226]Ra and Ba fractionation relative to Th, presumed to be during partial melting. The intersection of this zero-age fractionation line with the equiline defines the Ba/Th ratio of the source. After separation from the source, magmas evolve vertically towards the equiline. Since the half-life of [226]Ra is much less than the

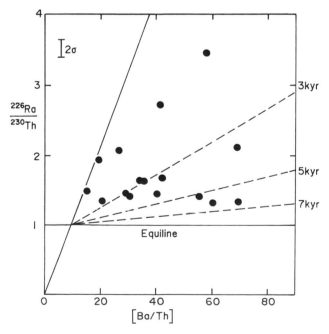

Fig. 13.27. Ra/Th–[Ba/Th] 'isochron' diagram for MORB glasses from the East Pacific Rise. After Rubin and Macdougall (1990).

parent (^{230}Th), the latter can be considered as effectively stable over the time periods under consideration (< 10 kyr). Therefore, for a theoretical sample with zero initial thorium, the decay equation is analogous to equation [13.10]:

$$\frac{^{226}\text{Ra}}{^{230}\text{Th}} = 1 - e^{-\lambda_{226}t} \qquad [13.11]$$

Therefore, ages are given by the intersection of isochrons on the y axis. However, since all samples have Ba/Th greater than the source, isotopic evolution is by decay of excess ^{226}Ra. Therefore, it is convenient to show the ages as isochron lines for a given source composition (Fig. 13.27).

A problem with this evidence is the apparently abnormal U–Th analyses obtained on the same samples (section 13.3.3). However, the radium data have subsequently been confirmed by mass spectrometric ^{226}Ra analysis of MORB glasses from the Juan de Fuca, Gorda, and East Pacific ridges (Volpe and Goldstein, 1993). Glasses from the ridge axis yield excess ^{226}Ra activities of up to 2.5 times greater than ^{230}Th activity. In contrast, off-axis glasses from these ridges yield ^{226}Ra/^{230}Th activity ratios of 1.00, as expected from their greater age. This gives us confidence that radium has not been mobilised in these glasses by sea-floor alteration, and that the disequilibrium data represent processes in the mantle.

If barium is an accurate analogue for stable radium, then the Th–Ra–Ba method can be used in conventional isochron dating of magma fractionation events. Reagan et al. (1992) applied this method to the dating of anorthosite phenocryst growth from phonolitic magmas of the Mount Erebus volcano, Antarctica. The data are used here to demonstrate the use of the conventional format isochron diagram for radium (analogous to the U–Th isochron diagram). This is based on the equation:

$$\frac{(^{226}\text{Ra})_{\text{P}}}{[\text{Ba}]} = \frac{(^{226}\text{Ra})_{\text{I}}}{[\text{Ba}]} e^{-\lambda_{226}t} + \frac{^{230}\text{Th}}{[\text{Ba}]}(1 - e^{-\lambda_{226}t}) \qquad [13.12]$$

where square brackets denote concentrations. This is analogous to equation [13.2] for the U–Th system.

Because anorthoclase readily takes up divalent but not trivalent ions it has Th/Ba ratios of effectively zero. The isochron age is then determined by the glass points, yielding crystallisation ages of ca. 2.5 kyr for samples from two recent eruptions (Fig. 13.28). However, it may be dangerous to rely on two-point phenocryst–glass ages without other supporting evidence (section 13.1.1). A more complete example of a Th–Ra–[Ba] isochron was provided by Schaefer

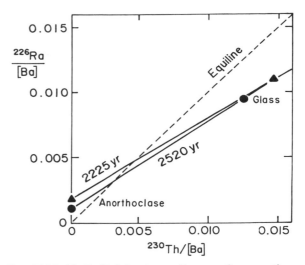

Fig. 13.28. Th–Ra/[Ba] isochron diagram for anorthoclase–glass pairs from the 1984 and 1988 phonolite eruptions of Mt. Erebus, Antarctica. After Reagan et al. (1992).

et al. (1993) on the 1985 pumice eruption of Nevado del Ruiz volcano, Columbia. In this case the glass point was colinear with three different mineral phases (and the whole-rock), yielding a best-fit age of 6.1 ± 0.5 kyr. This result was interpreted as the average age of an extended period of crystal fractionation, rather than a discrete magmatic differentiation event.

Volpe and Goldstein (1993) showed that within a given ocean-ridge segment, ^{230}Th/[Ba] ratios were constant in lavas of different ages (below 10 kyr). This implies that elemental fractionation effects were constant over this time. Hence, it is reasonable to also assume constant ^{226}Ra/^{230}Th ratios at the time of eruption. In this case, ^{226}Ra/^{230}Th ratios can be used to determine model ages for very young ocean-ridge basalts, in a similar way to the model age approach developed for U–Th data (section 13.3.4). Using this approach, the variation of ^{226}Ra/^{230}Th ratio in a suite of lavas from the axial valley of the Juan de Fuca ridge allowed the calculation of relative age differences of up to 1200 yr between different eruptions.

Evidence for ^{226}Ra/^{230}Th disequilibrium in MORB places even tighter constraints on melting models beneath ridges than were provided by Th/U data. ^{226}Ra/^{230}Th ratios in MORB glasses are as high as 2.5, despite the short 8000 yr equilibration time of ^{226}Ra. The dynamic melting model of McKenzie (1985a) can explain the Ra–Th data at very low porosities, assuming a Ra distribution coefficient of zero, because it assumes instantaneous melt extraction from all levels in the melting column. In this model, most of the radium must come from the base of the melting column, because ultra-incompatible elements are very efficiently extracted at low degrees of melting. This must be situated at > 70 km depth to allow melting in the garnet stability zone (section 13.3.2). However, it not known whether melts can ascend from 80 km depth with sufficient velocity to avoid decay of the excess ^{226}Ra inventory of the magma.

To avoid this problem, Spiegelman and Elliott (1993) proposed that the porosity of the melting system increases from a value of zero at the base of the melting column, to a maximum of ca. 0.5 % at the top of the melting column. This has the effect of holding back the complete release of ultra-incompatible elements, so that some of the Ra is derived from shallower levels in the melting column. The ^{226}Ra extracted from these shallower levels can then reach the surface without undergoing significant decay. This model gives mathematically acceptable solutions to the Th–U and Ra–Th data, but it is questionable as to whether melt extraction can begin at the porosities of less than 0.1% required in this model. An alternative approach (Rubin and Macdougall, 1988; Qin, 1992) is to invoke disequilibrium melting.

All of the melting models discussed above, and in section 13.3.2, assume complete equilibrium between crystals and melt throughout the melting process. This means that the inventory of trace elements within each mineral grain must be homogenised by diffusion before any melt is removed from contact with the surface of the grain. Hence ultra-incompatible elements can be stripped out from the entire grain by very small degrees of melting. However, Qin argued that if the melting rate is of the same order of magnitude as the volume diffusion rate of cations in mineral grains (e.g. garnet and cpx), then incompatible elements will be stripped out from grains in layers, like the shells of an onion.

Qin (1992) argued that disequilibrium melting is almost unavoidable in the generation of ^{226}Ra excesses, since the secular equilibration time of this nuclide (8000 yr) is comparable with the 10^4 yr diffusional equilibrium time of cations in a mineral grain at the temperature of basaltic melting (section 6.2.2). Qin took this argument further, by suggesting that differing rates of volume diffusion for different cations could cause *diffusional fractionation* of U-series nuclides and other incompatible-element couples. However, this overlooks the fact that the source spends a minimum of 100 kyr in the melting zone, for a 50 km-deep melting zone upwelling at 5 cm/yr.

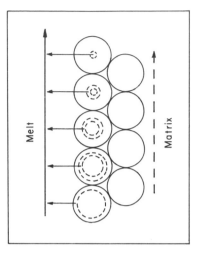

Fig. 13.29. Schematic illustration of a disequilibrium melting model to explain retention of ultra-incompatible elements in the solid phase of the melting column under a ridge. Times for melts and matrix to rise through the melting zone may be ca. 10 kyr and 100 kyr respectively.

These figures suggest that disequilibrium melting cannot be maintained over the whole depth of the melting column under ridges, but might occur at the base of the column in such as way as to hold back complete release of ultra-incompatible elements into the melt. This model is illustrated in Fig. 13.29, and would have the effect of retaining Ra in the solid phase until shallower depths, without having to invoke melt extraction from the source at ultra-low porosities.

13.6.2 ^{231}Pa

The first precise ^{231}Pa analyses of igneous rocks were made by Goldstein *et al.* (1993) on MORB glasses from the East Pacific Rise and the Juan de Fuca and Gorda ridges. On-axis samples yielded ^{231}Pa/^{235}U activity ratios of up to 2.9, while off-axis samples returned to secular equilibrium over ca. 150 kyr. Protoactinium data cannot be plotted on an isochron diagram because no suitable stable analogue to ^{231}Pa has yet been identified. However, Goldstein *et al.* argued that within single ridge segments, U–Pa model ages could be determined in a manner analogous to U–Th and Th–Ra model ages. By assuming a zero age in the sample from each axial valley with highest ^{231}Pa/^{235}U activity, Goldstein *et al.* calculated model ages for other axial and off-axis samples which were within error of their U–Th model ages.

The ^{231}Pa enrichments in on-axis samples were similar to ^{226}Ra enrichments in the same samples (Volpe and Goldstein, 1993), and much greater than ^{230}Th enrichment relative to ^{238}U. Therefore, the order of incompatibility of the U-series elements is Pa ≈ Ra ≫ Th > U. Similar arguments apply to the generation of excess ^{231}Pa and ^{226}Ra activities, and it is presently unclear whether they are due to equilibrium dynamic melting at low porosity (< 1 %), or whether a disequilibrium melting model must be invoked.

13.6.3 ^{228}Ra

The first detailed application of very short-lived species to magma evolution was made by Capaldi et al. (1976) on the volcanoes of Etna and Stromboli. However, one of the most interesting applications of these species is to the study of carbonatite magma genesis, as exemplified in Oldoinyo Lengai volcano, in the East African Rift of Tanzania. The 1960 and 1988 carbonatite eruptions from this volcano were studied, respectively, by Williams et al. (1986) and Pyle et al. (1991). Both of these eruptions showed strong disequilibrium between ^{228}Ra ($t_{1/2} = 5.77$ yr) and its parent ^{232}Th. The 1988 eruption also demonstrated ^{228}Th disequilibrium ($t_{1/2} = 1.91$ yr); however, it was not possible to test for this phenomenon in the 1960 eruption, since the samples reached secular equilibrium in the twenty years between sampling and analysis! The so-called '1963' eruption of Williams et al. (1986) is also excluded from this discussion because of uncertainty about its eruption age (Williams et al., 1988).

It is now generally agreed that carbonatites are formed by the evolution of peralkaline magmas in conditions of strong CO_2 enrichment, probably involving the segregation of immiscible droplets of carbonate magma from a silicate magma host (e.g. Pyle et al., 1991). The discovery of ^{228}Ra disequilibrium in the Oldoinyo carbonatites suggests that the segregation process probably occurred shortly before eruption. Over this time-scale, ^{226}Ra ($t_{1/2} = 1620$ yr) can be treated effectively as a stable isotope. Therefore the ^{228}Ra decay equation can be divided by ^{226}Ra to yield an isochron relation analogous to equation[13.2] (Capaldi et al., 1976):

$$\left(\frac{^{228}\text{Ra}}{^{226}\text{Ra}}\right)_P = \left(\frac{^{228}\text{Ra}}{^{226}\text{Ra}}\right)_I e^{-\lambda_{228}t} + \frac{^{232}\text{Th}}{^{226}\text{Ra}}(1 - e^{-\lambda_{228}t})$$

[13.13]

However, Williams et al. (1986) preferred to use the alternative isochron diagram, where ages are defined by the intercept on the left-hand axis, using the following equation, analogous to equation [13.10]:

$$\left(\frac{^{228}\text{Ra}}{^{232}\text{Th}}\right)_P = 1 - e^{-\lambda_{228}t}$$

[13.14]

This activity ratio is plotted in Fig. 13.30 against ^{226}Ra/^{232}Th activity. The zero-age line passes through the origin and represents a Ra/Th fractionation line. The slope of this line is the ^{228}Ra/^{226}Ra activity ratio, equal in turn to the ^{232}Th/^{238}U activity ratio, since ^{226}Ra and ^{228}Ra are in separate decay chains. For Oldoinyo Lengai, this ratio was found to be unity. Radioactive decay of a system located on the fractionation line causes it to move vertically downwards towards the equiline and permits an age to be assigned.

The initial ratio of the 1960 carbonatite is plotted on Fig. 13.30. If this magma is attributed to a single instantaneous event which caused Ra/Th fractionation, then this event occurred seven years before eruption. However, more complex models are possible. For example, if segregation occurred in two events, ^{228}Ra formed in the first event might decay before the second event (Fig. 13.30). In this case, the time from the second event to eruption is less than seven years. Alternatively, carbonatite segregation might have occurred over a period of time. In Fig. 13.31, this process is modelled by drawing Ra/Th growth curves for different rates of Ra enrichment in the carbonatite magma (relative to the conjugate silicate liquid). These growth curves are then calibrated by determining the time necessary for enrichment of the effectively stable ^{226}Ra isotope. A simple model of constant enrichment rate yields a calculated duration of this process for the 1960 magma of ca. 18 years, if this occurred immediately prior to eruption (Fig. 13.31).

13.6.4 ^{228}Th

The addition of ^{228}Th data can potentially allow selection between short-term differentiation models such as those outlined above, because of its dependence on very recent events. These data were collected by Pyle et al. (1991) for the 1988 carbonatite of Oldoinyo Lengai, but due to analytical difficulties, Ra/Th activity ratios could not be measured. Therefore, the advantage of the combined systems was lost. However, data for Mt Etna measured by Capaldi et al. (1976) and re-interpreted by Cortini (1985) can be used to demonstrate the method.

The effects of a hypothetical Ra/Th fractionation event are shown in Fig. 13.32. A reservoir previously

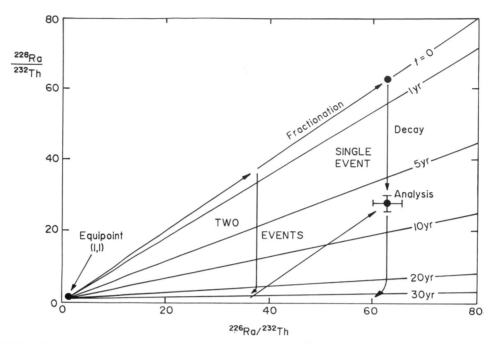

Fig. 13.30 Ra–Th isochron diagram showing single-event and two-event models for the age of the 1960 Oldoinyo Lengai carbonatite since radium enrichment. After Williams *et al.* (1986).

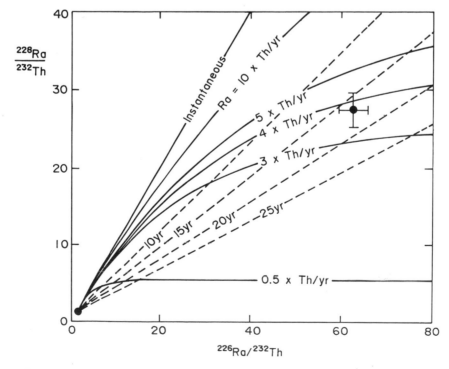

Fig. 13.31. Ra–Th isochron diagram showing continuous radium enrichment models for the 1960 carbonatite of Oldoinyo Lengai. For explanation, see text. After Williams *et al.* (1986).

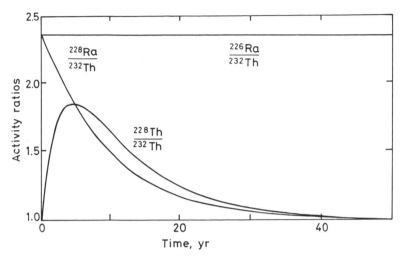

Fig. 13.32. Schematic illustration of the effects of Ra/Th fractionation on isotope systematics of a system previously in secular equilibrium. After Capaldi *et al.* (1976).

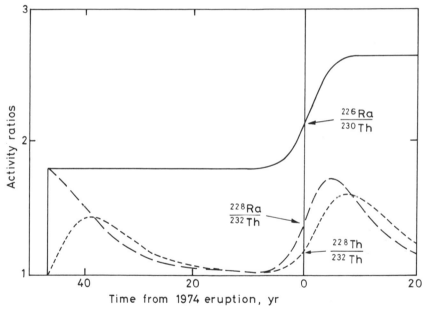

Fig. 13.33. Model of two-stage radium enrichment proposed to explain Ra and Th isotope systematics of the 1974 eruptions of Etna volcano. For discussion see text. Modified after Cortini (1985).

in secular equilibrium (e.g. a long-lived magma body) becomes enriched in radium relative to thorium, due to some form of differentiation event. After this enrichment event, excess ^{226}Ra and ^{228}Ra activities decay away at different rates. ^{228}Th is initially in equilibrium with its ultimate parent (^{232}Th), but subsequently builds to a peak and then decays as it approaches equilibrium with its immediate parent (^{228}Ra).

The single-stage model illustrated above can explain any two sets of isotope ratios in the 1974 eruption of Mt Etna, but cannot satisfy all three ratios. Cortini (1985) therefore developed more complex models in an attempt to explain all of the data. Fig. 13.33 shows a model involving two radium-enrichment events. The first occurs ca. 50 yr before eruption. After the short-lived species have returned to secular equilibrium, a second Ra-enrichment event occurs over a duration of about ten years. Eruption to produce the 1974 lavas occurs during this second (continuous) enrichment event. Doubtless, these short-lived systems will find increasing application to future eruptions of per-alkaline volcanic systems.

References

Allegre, C. J. (1968). ^{230}Th dating of volcanic rocks. *Earth Planet. Sci. Lett.* **5**, 209–10.

Allegre, C. J. and Condomines, M. (1976). Fine chronology of volcanic processes using ^{238}U–^{230}Th systematics. *Earth Planet. Sci. Lett.* **28**, 395–406.

Allegre, C. J. and Condomines, M. (1982). Basalt genesis and mantle structure studied through Th–isotopic geochemistry. *Nature* **299**, 21–4.

Allegre, C. J., Dupre, B. and Lewin, E. (1986). Thorium/uranium ratio of the Earth. *Chem. Geol.* **56**, 219–27.

Beattie, P. (1993a). The generation of uranium series disequilibria by partial melting of spinel peridotite: constraints from partitioning studies. *Earth Planet. Sci. Lett.* **117**, 379–91.

Beattie, P. (1993b). Uranium–thorium disequilibria and partitioning on melting of garnet peridotite. *Nature* **363**, 63–5.

Ben Othman, D. and Allegre, C. J. (1990). U–Th isotopic systematics at 13°N East Pacific Ridge segment. *Earth Planet. Sci. Lett.* **98**, 129–37.

Capaldi, G., Cortini, M., Gasparini, P. and Pece, R. (1976). Short-lived radioactive disequilibria in freshly erupted volcanic rocks and their implications for the pre-eruption history of a magma. *J. Geophys. Res.* **81**, 350–8.

Capaldi, G., Cortini, M. and Pece, R. (1982). Th isotopes at Vesuvius: evidence for open system behaviour of magma-forming processes. *J. Volc. Geotherm. Res.* **14**, 247–60.

Capaldi, G., Cortini, M. and Pece, R. (1985). On the reliability of the ^{230}Th^{238}U dating method applied to young volcanic rocks–(reply). *J. Volc. Geotherm. Res.* **26**, 369–76.

Capaldi, G. and Pece, R. (1981). On the reliability of the ^{230}Th–^{238}U dating method applied to young volcanic rocks. *J. Volc. Geotherm. Res.* **11**, 367–72.

Cerrai, E., Dugnani Lonati, R., Gazzarini, F. and Tongiorgi, E. (1965). Il methodo iono–uranio per la determinazione dell'eta dei minerali vulcanici recenti. *Rend. Soc. Mineral. Ital.* **21**, 47–62

Cohen, R. S., Evensen, N. M., Hamilton, P. J. and O'Nions, R. K. (1980). U–Pb, Sm–Nd and Rb–Sr systematics of mid-ocean ridge basalt glasses. *Nature* **283**, 149–53.

Cohen, A. S. and O'Nions, R. K. (1991). Precise determination of femtogram quantities of radium by thermal ionization mass spectrometry. *Anal. Chem.* **63**, 2705–8.

Condomines, M. and Allegre, C. J. (1980). Age and magmatic evolution of Stromboli volcano from ^{230}Th–^{238}U disequilibrium data. *Nature* **288**, 354–7.

Condomines, M., Hemond, Ch. and Allegre, C. J. (1988). U–Th–Ra radioactive disequilibria and magmatic processes. *Earth Planet. Sci. Lett.* **90**, 243–62.

Condomines, M., Morand, P. and Allegre, C. J. (1981). ^{230}Th–^{238}U radioactive disequilibria in tholeiites from the FAMOUS zones (Mid-Atlantic Ridge, 36° 50′ N): Th and Sr isotopic geochemistry. *Earth Planet. Sci. Lett.* **55**, 247–56.

Condomines, M., Tanguy, J. C., Kieffer, G. and Allegre, C. J. (1982). Magmatic evolution of a volcano studied by ^{230}Th–^{238}U disequilibrium and trace elements systematics: the Etna case. *Geochim. Cosmochim. Acta* **46**, 1397–416.

Cortini, M. (1985). An attempt to model the timing of magma formation by means of radioactive disequilibria. *Chem. Geol. (Isot. Geosci. Section)* **58**, 33–43.

Galer, S. J. G. and O'Nions, R. K. (1985). Residence time of thorium, uranium and lead in the mantle with implications for mantle convection. *Nature* **316**, 778–82.

Galer, S. J. G. and O'Nions, R. K. (1986). Magma genesis and the mapping of chemical and isotopic variations in the mantle. *Chem. Geol.* **56**, 45–61.

Gill, J. B. and Williams, R. W. (1990). Th isotope and U-series studies of subduction-related volcanic rocks. *Geochim. Cosmochim. Acta* **54**, 1427–42.

Goldstein, S. J., Murrell, M. T. and Janecky, D. R. (1989). Th and U isotopic systematics of basalts from the Juan de Fuca and Gorda Ridges by mass spectrometry. *Earth Planet. Sci. Lett.* **96**, 134–46.

Goldstein, S. J., Murrell, M. T., Janecky, D. R., Delaney, J. R. and Clague, D. A. (1991). Geochronology and petrogenesis of MORB from the Juan de Fuca and Gorda ridges by ^{238}U–^{230}Th disequilibrium. *Earth Planet. Sci. Lett.* **107**, 25–41 & **109**, 255–72 (erratum).

Goldstein, S. J., Murrell, M. T. and Williams, R. W. (1993). ^{231}Pa and ^{230}Th chronology of mid-ocean ridge basalts. *Earth Planet. Sci. Lett.* **115**, 151–9.

Hawkesworth, C. J., Hergt, J. M., McDermott, F. and Ellam, R. M. (1991). Destructive margin magmatism and the contributions from the mantle wedge and subducted crust. *Australian J. Earth Sci.* **38**, 577–94.

Hemond, Ch. (1986). *Geochimie Isotopique du Thorium et du Strontium dans la Serie Tholeiitique d'Islande et dans des Series Calco-alcalines Diverses.* These 3eme Cycle, Université Paris VII, 151 pp.

Hemond, Ch. and Condomines, M. (1985). On the reliability of the ^{230}Th–^{238}U dating method applied to young volcanic rocks – discussion. *J. Volc. Geotherm. Res.* **26**, 365–9.

Hemond, Ch., Condomines, M., Fourcade, S., Allegre, C. J., Oskarsson, N. and Javoy, M. (1988). Thorium, strontium and oxygen isotopic geochemistry in recent tholeiites from Iceland: crustal influence on mantle-derived magmas. *Earth Planet. Sci. Lett.* **87**, 273–85.

Jakes, P. and Gill, J. B. (1970). Rare earth elements and the island arc tholeiitic series. *Earth Planet. Sci. Lett.* **9**, 17–28.

Joly., J. (1909). On the radioactivity of certain lavas. *Phil. Mag.* **18**, 577.

Kigoshi, K. (1967). Ionium dating of igneous rocks. *Science* **156**, 932–4.

Langmuir, C. H., Bender, J. F., Bence, A. E. and Hanson, G. N. (1977). Petrogenesis of basalts from the FAMOUS area: Mid-Atlantic Ridge. *Earth Planet. Sci. Lett.* **36**, 133–56.

LaTourrette, T. Z., Kennedy, A. K. and Wasserburg, G. J. (1993). Thorium–uranium fractionation by garnet: evidence for a deep source and rapid rise of oceanic basalts. *Science* **261**, 739–42.

Macdougall, J. D., Finkel, R. C., Carlson, J. and Krishnaswami, S. (1979). Isotopic evidence for uranium exchange during low-temperature alteration of oceanic basalt. *Earth Planet. Sci. Lett.* **42**, 27–34.

McDermott, F., Elliott, T. R., van Calsteren, P. and Hawkesworth, C. J. (1993). Measurement of $^{230}Th/^{232}Th$ ratios in young volcanic rocks by single-sector thermal ionisation mass spectrometry. *Chem. Geol. (Isot. Geosci. Section)* **103**, 283–92.

McDermott, F. and Hawkesworth, C. (1991). Th, Pb, and Sr isotope variations in young island arc volcanics and oceanic sediments. *Earth Planet. Sci. Lett.* **104**, 115.

McKenzie, D. (1985a). $^{230}Th–^{238}U$ disequilibrium and the melting processes beneath ridge axes. *Earth Planet. Sci. Lett.* **72**, 149–57.

McKenzie, D. (1985b). The extraction of magma from the crust and mantle. *Earth Planet. Sci. Lett.* **74**, 81–91.

Newman, S., Finkel, R. C. and Macdougall, J. D. (1983). $^{230}Th–^{238}U$ disequilibrium systematics in oceanic tholeiites from 21°N on the East Pacific Rise. *Earth Planet. Sci. Lett.* **65**, 17–33.

Newman, S., Finkel, R. C. and Macdougall, J. D. (1984). Comparison of $^{230}Th–^{238}U$ disequilibrium systematics in lavas from three hot spot regions: Hawaii, Prince Edward and Samoa. *Geochim. Cosmochim. Acta* **48**, 315–324.

O'Hara, M. J. and Mathews, R. E. (1981). Geochemical evolution in an advancing, periodically replenished, periodically tapped, continuously fractionated magma chamber. *J. Geol. Soc. Lond.* **138**, 237–77.

O'Nions, R. K. and McKenzie, D. (1993). Estimates of mantle thorium/uranium ratios from Th, U and Pb isotope abundances in basaltic melts. *Phil. Trans. Roy. Soc. Lond. A* **342**, 65-77.

Oversby, V. M. and Gast, P. W. (1968). Lead isotope compositions and uranium decay series disequilibrium in recent volcanic rocks. *Earth Planet. Sci. Lett.* **5**, 199–206.

Pyle, D. M., Dawson, J. B. and Ivanovich, M. (1991). Short-lived decay series disequilibria in the natrocarbonatite lavas of Oldoinyo Lengai, Tanzania: constraints on the timing of magma genesis. *Earth Planet. Sci. Lett.* **105**, 378–96.

Qin, Z. (1992). Disequilibrium partial melting model and its implications for trace element fractionations during mantle melting. *Earth Planet. Sci. Lett.* **112**, 75–90.

Reagan, M. K., Volpe, A. M. and Cashman, K. V. (1992). ^{238}U- and ^{232}Th-series chronology of phonolite fractionation at Mount Erebus, Antarctica. *Geochim. Cosmochim. Acta* **56**, 1401–7.

Reinitz, I. and Turekian, K. K. (1989). $^{230}Th–^{238}U$ and $^{226}Ra–^{230}Th$ fractionation in young basaltic glasses from the East Pacific Rise. *Earth Planet. Sci. Lett.* **94**, 199–207.

Rubin, K. H. and Macdougall, J. D. (1988). ^{226}Ra excesses in mid-ocean-ridge basalts and mantle melting. *Nature* **335**,

Rubin, K. H. and Macdougall, J. D. (1990). Dating of neovolcanic MORB using ($^{226}Ra/^{230}Th$) disequilibrium. *Earth Planet. Sci. Lett.* **101**, 313–22.

Rubin, K. H. and Macdougall, J. D. (1992). Th–Sr isotopic relationships in MORB. *Earth Planet. Sci. Lett.* **114**, 149–57.

Schaefer, S. J., Sturchio, N. C., Murrell, M. T. and Williams, S. N. (1993). Internal ^{238}U-series systematics of pumice from the November 13, 1985, eruption of Nevado del Ruiz, Colombia. *Geochim. Cosmochim. Acta* **57**, 1215–19.

Sigmarsson, O., Condomines, M. and Fourcade, S. (1992). Mantle and crustal contribution in the genesis of recent basalts from off-rift zones in Iceland: constraints from Th, Sr and O isotopes. *Earth Planet. Sci. Lett.* **110**, 149–62.

Sigmarsson, O., Condomines, M., Morris, J. D. and Harmon, R. S. (1990). Uranium and ^{10}Be enrichments by fluids in Andean arc magmas. *Nature* **346**, 163–5.

Sigmarsson, O., Hemond, Ch., Condomines, M., Fourcade, S. and Oskarsson, N. (1991). Origin of silicic magma in Iceland revealed by Th isotopes. *Geology* **19**, 621–4.

Spiegelman, M. and Elliott, T. (1993). Consequences of melt transport for uranium series disequilibrium in young lavas. *Earth Planet. Sci. Lett.* **118**, 1–20.

Spivak, A. J. and Edmond, J. M. (1987). Boron isotope exchange between seawater and the oceanic crust. *Geochim. Cosmochim. Acta* **51**, 1033–43.

Taddeucci, A., Broecker, W. S. and Thurber, D. L. (1967). ^{230}Th dating of volcanic rocks. *Earth Planet. Sci. Lett.* **3**, 338–42.

Thompson, R. N., Morrison, M. A., Hendry, G. L. and Parry, S. J. (1984). An assessment of the relative roles of crust and mantle in magma genesis: an elemental approach. *Phil. Trans. Roy. Soc. Lond. A* **310**, 549–99.

Volpe, A. M. and Goldstein, S. J. (1993). $^{226}Ra–^{230}Th$ disequilibrium in axial and off-axis mid-ocean ridge basalts. *Geochim. Cosmochim. Acta* **57**, 1233–41.

Williams, R. W., Collerson, K. D., Gill, J. B. and Deniel, C. (1992). High Th/U ratios in subcontinental lithospheric mantle: mass spectrometric measurement of Th isotopes in Gaussberg lamproites. *Earth Planet. Sci. Lett.* **111**, 257–68.

Williams, R. W. and Gill, J. B. (1989). Effects of partial melting on the uranium decay series. *Geochim. Cosmochim. Acta* **53**, 1607–19.

Williams, R. W. and Gill, J. B. (1992). Th isotope and U-series disequilibria in some alkali basalts. *Geophys. Res. Lett.* **19**, 139–42.

Williams, R. W., Gill, J. B. and Bruland, K. W. (1986). Ra–Th disequilibria systematics: timescale of carbonatite magma formation at Oldoinyo Lengai volcano, Tanzania. *Geochim. Cosmochim. Acta* **50**, 1249–59.

Williams, R. W., Gill, J. B. and Bruland, K. W. (1988). Ra–Th disequilibria: timescale of carbonatite magma formation at Oldoinyo Lengai volcano, Tanzania. *Geochim. Cosmochim. Acta* **52**, 939. Reply to Gittins, J. (1988). Comment on 'Ra–Th disequilibria systematics: timescale of carbonatite magma formation at Oldoinyo Lengai volcano, Tanzania'. *Geochim. Cosmochim. Acta* **52**, 937.

14 Cosmogenic nuclides

The Earth undergoes continuous bombardment by cosmic rays from the galaxy. These are atomic nuclei (mainly protons) travelling through interstellar and interplanetary space at relativistic speeds. The net flux of cosmic-ray energy intercepted by the Earth is low: equivalent in intensity to visible starlight. However, the energy of each particle is very high, averaging several billion electron volts (the kinetic energy of a gas molecule at 10 000 °K is about one electron volt). Cosmic rays can therefore interact strongly with matter.

Cosmic rays generate unstable nuclides in two principal ways: by direct bombardment of target atoms, and by the agency of cosmic-ray-generated fast neutrons. The latter are produced by the collision of cosmic rays with atmospheric molecules and slowed by further collisions to thermal kinetic energies. These 'thermal' neutrons are able to interact with the nuclei of stable atoms, causing transformations to radioactive nuclei. The 'cosmogenic' nuclides thus produced can be used as dating tools and as radioactive tracers.

The measurement of cosmogenic nuclides falls into two developmental stages. Early work, almost entirely on ^{14}C, was by radioactive counting. More recently, accelerator mass spectrometry has revolutionised the field of cosmogenic nuclides, allowing ^{14}C measurement on very small samples and allowing the utilisation of several other cosmogenic nuclides for the first time.

14.1 Carbon-14

The collision of cosmic-ray-produced thermal neutrons with nitrogen nuclei has a reasonable probability of generating radiocarbon by an n,p reaction:

$$^{14}_{7}N + n \rightarrow {}^{14}_{6}C + p$$

Oxidation to carbon dioxide follows rapidly, and the radioactive CO_2 joins the carbon cycle. It may be absorbed photosynthetically by plants, or it may exchange with CO_2 in water and ultimately be deposited as carbonate.

^{14}C decays by beta emission back to ^{14}N with a half-life of ca. 5700 yr. Hence, atmospheric ^{14}C activity is the result of an equilibrium between cosmogenic production and radioactive decay. During their life-time, living tissues will exchange CO_2 with the atmosphere, and hence remain in radioactive equilibrium with it. However, on death this exchange is expected to stop, whereupon the ^{14}C in the tissue decays with time. If the initial level of ^{14}C activity in a carbon sample at death (A_0) can be predicted, and if it has subsequently remained a closed system, then by measuring its present level of activity (A), its age (t) can be determined. This can be expressed as the radioactive decay law (from equation [1.5]):

$$A = A_0 \, e^{-\lambda t} \qquad [14.1]$$

The idea of using radiocarbon as a dating tool was conceived by W.F. Libby, for which he received the Nobel prize for Chemistry in 1960. The early history of the field is described by Kamen (1963) and a twenty-five-year review was given by Ralph and Michael (1974). Taylor (1987) has written an account of its archaeological applications.

The Earth's magnetic field deflects incoming charged particles so that the equatorial cosmic-ray flux is four times less than the polar flux (Fig. 14.1). Therefore, one of the first questions which Libby and his co-workers investigated was whether the present-day activity of ^{14}C was uniform over the Earth's surface. No latitude dependence was found in modern wood (Anderson and Libby, 1951), and the average specific activity found was 15.3 disintegrations per minute per gram (dpm/g). Hence, geographical homogenisation of ^{14}C in the atmosphere before uptake by plants appears to be a justifiable assumption.

More recent data which demonstrate the rate of atmospheric ^{14}C homogenisation were provided accidentally by atmospheric nuclear explosions. Fig. 14.2 shows the level of ^{14}C at different locations around the world after the addition of excess ^{14}C from atmospheric tests (Libby, 1970). World-wide atmospheric homogenisation occurs after only two or three

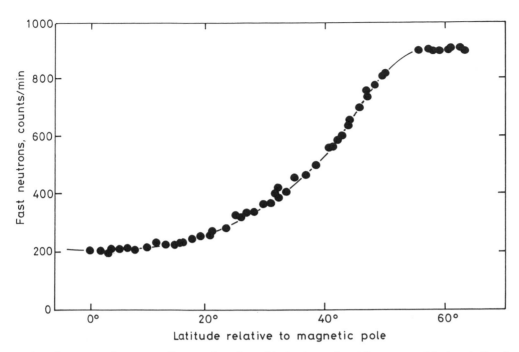

Fig. 14.1. Plot of cosmogenic neutron flux as a function of latitude to show the geographical variation in cosmic ray-intensity. After Simpson (1951).

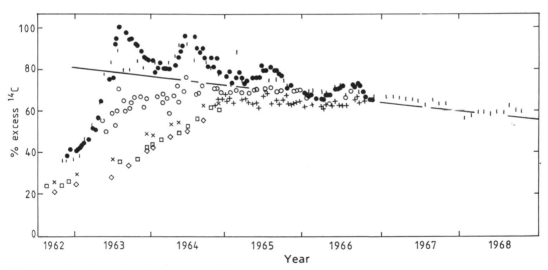

Fig. 14.2. Excess (bomb-produced) atmospheric ^{14}C measured at different localities on the globe, during and after the peak of atmospheric nuclear testing. Localities: (□) 71 ° N; (I) Mojave Desert, 36 ° N; (o) 9 ° N; (×) 18 ° S; (+) 21 ° S; (◊) 38 ° S; (•) 41 ° S. After Libby (1970).

years. The recovery rate of the Mojave Desert samples after 1965 suggests that the time for buffering of the atmosphere by surface ocean water is somewhat longer (17 years), but this is still very short relative to the ^{14}C half-life.

Libby (1952) also assumed that the atmosphere had a constant ^{14}C activity through time, as a result of equilibrium between a constant rate of production and decay. Hence, the ^{14}C activity of recent organic tissue was taken to be equal to the initial activity in the past. A closed-system assumption was justified on the basis that complex organic molecules cannot exchange carbon with the environment after death. However, such exchange can occur in many carbonates, making them less reliable as dating material. Libby determined a ^{14}C half-life of 5568 ± 30 yr from a weighted mean of the four most precise laboratory counting determinations, all of which clustered closely around the mean. The above-mentioned assumptions were supported (Arnold and Libby, 1949) by a good concordance between ^{14}C dates and historical ages for nine test samples (Fig. 14.3).

In the natural reduction of CO_2 to carbon by photosynthesis, and during laboratory preparation for analysis (e.g. combustion of carbon to CO_2), isotopic fractionation between carbon isotopes can occur. This is due to the weaker bonds, and hence greater reactivity, of the lighter isotope (section 2.2.2). In order to assess the fractionation between ^{14}C and ^{12}C in natural and laboratory processes, Craig (1954) proposed that the ^{13}C/^{12}C ratio of samples be

measured by mass spectrometry. Because fractionation is mass-dependent, ^{14}C/^{12}C fractionation will be twice as great as ^{13}C/^{12}C fractionation. The latter is expressed relative to the PeeDee belemnite (PDB) standard (Craig, 1957):

$$\delta^{13}C = \left[\frac{(^{13}C/^{12}C)_{sample}}{(^{13}C/^{12}C)_{PDB}} - 1 \right] \cdot 10^3 \quad [14.2]$$

This fractionation factor can be directly converted into a correction to the ^{14}C age using Fig. 14.4 (Mook and Streurman, 1983). In this diagram, normal δ^{13}C compositions for various types of sample are shown. Because 'modern wood' is established as the reference point for calibrating the efficiency of ^{14}C counting equipment, age corrections must be applied rative to this type of material (Fig. 14.4), which has a normal or 'calibration' value of δ^{13}C = −25 per mil (relative to PDB). In marine carbonates, this effect is offset by the 400 yr ^{14}C age of ocean surface water, which must be subtracted.

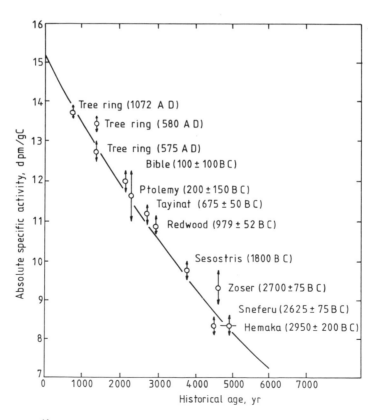

Fig. 14.3. Plot of measured ^{14}C activity (disintegrations per minute per gram of carbon) in archaeological samples of known age against predicted activity based on modern wood and a 5568 yr half-life. After Libby (1952).

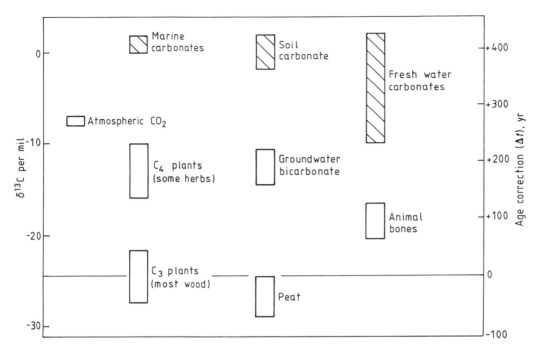

Fig. 14.4. Carbon isotope fractionation effects in different materials, and necessary corrections to calibrated ^{14}C ages for C_3 plants (wood). Carbonates are hatched. After Mook and Streurman (1983).

14.1.1 ^{14}C measurement by counting

The development of the radiocarbon method went hand in hand with the development of low-level counting techniques. The specific activity of ^{14}C is small, yielding a maximum count-rate of 13.6 decays per minute per gram (dpm/g) for modern wood, but only 0.03 dpm/g for a sample 50 kyr old. Furthermore, the maximum β energy is low (156 keV), so that in a solid source of non-zero thickness a significant fraction of particles would be absorbed by other carbon atoms in the sample.

Libby's early determinations of ^{14}C activity were on samples of solid carbon using a 'screen wall' geiger counter. However, this method was soon replaced by the analysis of CO_2 in a gas counter (de Vries and Barendsen, 1953). CO_2 is very readily prepared, and in the gas counter there is no risk of losing counts (due to absorption) before the particles reach the detector.

Unfortunately the natural background level of activity which will be measured by a gas counter (cosmic rays and gamma emission from natural materials) is far larger than the level of activity from the sample itself. Hence, two screening techniques are used (Fig. 14.5). The first is a thick wall of material which itself has a low level of activity (e.g. 'old' lead).

The second component is an array of geiger tubes arranged immediately round the gas proportional counter. The geiger tubes are electronically connected in anti-coincidence to the proportional counter. If a high-energy particle, such as a cosmic ray, enters the shielding, it will trigger the geiger tubes at almost the same time as the proportional counter, and the two signals will cancel out. The dramatic effects of these shielding techniques on the background were demonstrated by Ralph (1971) using a counter filled with 'dead' CO_2 made from anthracite coal.

Count rates (dpm) were:

No shielding:	1500
Shielded by iron and mercury:	400
Shielded and with anti-coincidence counters turned on:	8

Since the work described above (and during its progress), Libby's assumptions and half-life value have been re-examined. However, it was decided to continue to publish radiocarbon ages using Libby's atmospheric composition and half-life (Godwin, 1962). These are called 'conventional ages'. Correction factors are subsequently applied to determine a true 'historical' age. We will now re-examine two of Libby's assumptions.

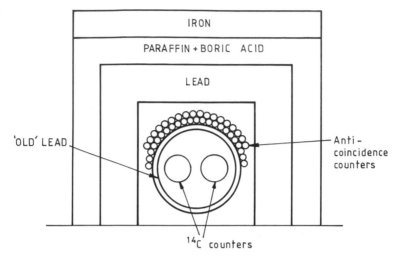

Fig. 14.5. Components in passive and active screening of a CO_2 gas counter. After Mook and Streurman (1983).

14.1.2 Closed-system assumption

Loss of carbon from a system during its geological life-time is not usually a problem in radiocarbon dating. However, contamination with extraneous environmental carbon may be a major problem. To exclude such contamination, rigorous sample preparation procedures have been developed.

When dating wood or charcoal for archaeological purposes it is desirable to determine the time when the tree was cut down. Hence, it is only necessary to exclude post-mortem exchange with the environment. For this, an Acid–Alkali–Acid leaching treatment referred to as the AAA treatment has been found to be effective (Olsson, 1980). The three steps are:

1) Leach with 4% HCl at 80 °C for 24 hours to remove sugars, resins, soil carbonate and infiltrated humic acids.
2) Leach with up to 4% NaOH at up to 80 °C for at least 24 hours to remove infiltrated tannic acids (this step also removes part of the lignin).
3) Repeat step 1 to remove any atmospheric CO_2 absorbed during the alkali step.

The overall process removes about 50% of the original carbon.

When dating tree rings for calibration studies (see below), the objective is quite different. In this case it is essential to sample only material laid down in the year of growth corresponding to the annual ring. This requires that all material deposited during the subsequent life of the tree (e.g. lignin) must be leached away. This is accomplished by inserting a

step 1a into the above procedure in which the wood chips are bleached by progressive addition of an almost equal weight of sodium perchlorate powder in dilute acetic acid at 70 °C. The procedure removes up to 75% of the carbon, leaving a residue of pure cellulose for analysis (Mook and Streurman, 1983).

When dating bones, all of the inorganic carbonate fraction must be removed by leaching with very dilute HCl because this fraction invariably exchanges carbon with groundwater. The organic carbon fraction in the bone is in the form of collagen, which is resistant to post-mortem exchange. Different methods for treatment of bones are described by Olsson *et al.* (1974). Leaching with acid has also been shown to improve the accuracy of radiocarbon ages on corals (see below).

14.1.3 Initial ratio assumption

As radiocarbon measurements became more precise, systematic age discrepancies between historical material and radiocarbon dates began to suggest that the level of ^{14}C activity in the atmosphere had varied with time. The first evidence for such temporal variations in ^{14}C activity was provided by Suess (1955), who found that 20th century wood showed a 2% depletion in activity relative to 19th century wood. This was attributed to dilution of radioactive carbon by 'dead' carbon introduced into the atmosphere by burning fossil fuel (nuclear tests later drove the equilibrium in the other direction by adding ^{14}C to the atmosphere). Subsequently, de Vries (1958) found that late-17th century wood had ca. 2% higher activity than 19th century wood. These two 'anomalies' are sometimes called the 'Suess' and 'de Vries' effects.

The discovery of secular variations in ^{14}C activity has provoked various models which attempt to explain this record. Forbush (1954) observed that the 11-year cycle of sunspot activity was inversely correlated with cosmic-ray intensity. This is because high levels of solar activity (marked by increased sunspot activity) cause an increase in the solar wind of ionised particles, which extends the Sun's magnetic field and deflects galactic cosmic rays away from the Earth. Calculations by Oeschger *et al.* (1970) suggest that the stratospheric cosmic-ray flux may be nearly doubled at solar minima, relative to maxima.

Because historical records are available for sunspot frequency, this provided a means of charting cosmic-ray intensity, and hence ^{14}C production, over the last 1500 years. Stuiver (1961) performed these calculations, and suggested that a sunspot minimum in the late-17th century could explain the 'de Vries effect' ^{14}C activity maximum at that time. This was confirmed by Stuiver (1965) using more detailed ^{14}C data (Fig. 14.6).

Extension of the ^{14}C activity curve back well before the time of Christ revealed large long-term variations, in addition to the short-term effects attributed to changes in solar cosmic-ray production (Suess, 1965). Elsasser *et al.* (1956) had predicted that if the strength of the Earth's magnetic field displayed secular variations, as suggested by Thellier (1941 and following), then this would have affected the paleo cosmic-ray flux incident on the atmosphere, and hence ^{14}C production. However, strong evidence of a causal relationship with the Earth's field strength was not established until Bucha and Neustupny (1967) provided more extensive paleomagnetic intensity measurements. These data revealed sinusoidal variations in the Earth's magnetic field strength which were of a global nature and which matched the sinusoidal deviations between radiocarbon and absolute ages.

By modelling the effect of paleomagnetic intensity variations on ^{14}C activity, Bucha and Neustupny were able to match the deviations between tree-ring and radiocarbon time-scales almost exactly (Fig. 14.7). A comparison with historically dated wood showed a very similar result, except that this curve was translated upwards by ca. 100 yr. This can be attributed to the average time delay between wood growth and utilisation. Because the model of Bucha and Neustupny linked the long time-period deviations between radiocarbon and absolute ages to variations in the global magnetic field, it also implied that the

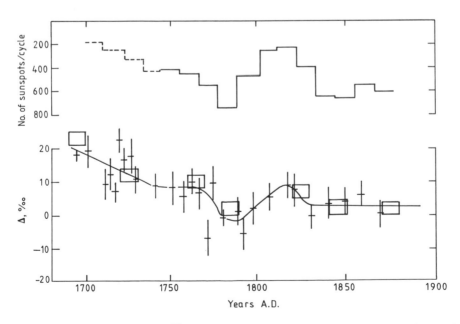

Fig. 14.6. Plots of sunspot activity and relative ^{14}C activity, expressed as Δ, parts per mil, to show coherent anti-correlation in the 17th and 18th centuries. A best-fit curve is drawn using two ^{14}C data sets (error boxes and error crosses). After Stuiver (1965).

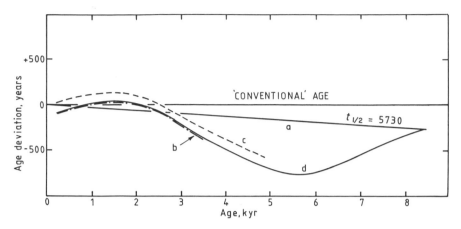

Fig. 14.7. Plot of age deviation between 'conventional' radiocarbon ages (half-life = 5568 yr) and other age determinations. a) Radiocarbon method using 5730 yr half-life: b) dendrochronology time-scale; c) historical time-scale; d) radiocarbon method using 5730 yr half-life and correction for variations in Earth's magnetic field intensity. After Bucha and Neustupny (1967).

deviations should be of a systematic world-wide nature. Hence it gave grounds for the establishment of very precise calibration sequences, which could then be used for world-wide correction of 'conventional' radiocarbon ages to calendar ages.

14.1.4 Dendrochronology

It was quickly realised that the most accurate way to calibrate the 'conventional' ^{14}C time-scale for initial ^{14}C variations was to integrate radiocarbon dates with tree-ring chronologies. Great efforts have been expended in this task over the last 30 years.

The longest dendrochronology calibration range has been achieved using the stunted bristlecone pine. When this work began the species was known as *Pinus aristata*. However, it is now usually termed *Pinus longaeva*. The semi-desert habitat of this tree gives rise to great longevity and good preservation after death. Ferguson (1970) erected a continuous master chronology reaching back over 7000 yr, based on several living trees and 17 specimens of dead wood from the White Mountains of eastern-central California (Fig. 14.8). This suite now extends nearly 8700 years (to 6700 B.C.), and includes the oldest living tree at ca. 4600 years old (Ferguson and Graybill, 1983).

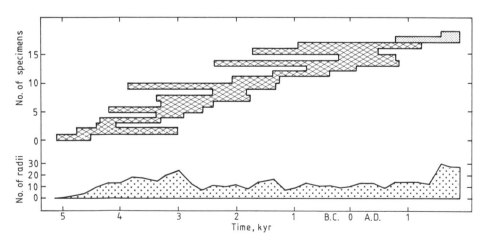

Fig. 14.8. A 'master' tree-ring chronology based on living and dead specimens of Bristlecone pine with overlapping age ranges. Upper chart shows range of each specimen. Lower chart shows total number of radii from which raw data were derived. After Ferguson (1970).

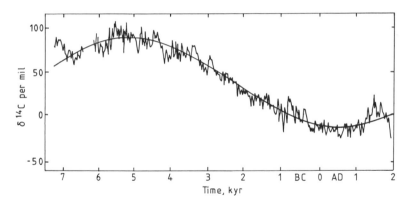

Fig. 14.9. Changes in atmospheric ^{14}C activity in the last 9000 years, presented in the form of isotopic fractionation per mil, based on 'continuous' Bristlecone pine and 'floating' European oak chronologies. After Bruns *et al.* (1983).

Suess (1970) presented a data set of 315 radiocarbon measurements for bristlecone pine from Ferguson's collection, and used this to construct a continuous calibration curve from 5200 years B.C. to the present. One of the prominent features of this curve was the presence of numerous 'wiggles' with wavelengths of 100–300 years, superimposed on the longer-term variations discussed above. Suess attracted much criticism because his calibration curve was drawn by eye through the measured points, rather than using a statistical curve fit. Many other workers (as late as Pearson *et al.*, 1977) maintained that the second order 'wiggles' identified by Suess were an artifact of statistical uncertainties in the data, and had no real meaning. However, this was an unrealistic model, since the known 'de Vries effect' wiggles of the 17th century A.D. were of similar magnitude. The reality of the 'Suess' wiggles in the ancient radiocarbon record (3500 years B.C.) was finally established by De Jong *et al.* (1979). These wiggles are seen superimposed on long-term ^{14}C variations in the 9000 yr calibration curve shown in Fig. 14.9.

The convoluted shape of the calibration curve introduces ambiguities to ^{14}C dating within many periods, since a single radiocarbon age can correspond to more than one historical age. These ambiguities may sometimes be resolved by applying historical constraints (e.g. section 14.2.1). Alternatively, they may be avoided in the dating of wood if a sample spanning more than about 50 growth rings can be dated. This ring sequence then forms a small 'floating' calibration curve which can be 'wiggle matched' with the known calibration curve to yield a much more accurate time-span for the growth of the sample wood. Suess and Strahm (1970) demonstrated this technique when they dated a floating tree ring

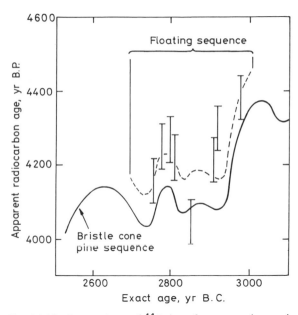

Fig. 14.10. Comparison of ^{14}C data for a wood sample and the calibration curve to show the application of 'wiggle matching'. The dashed line is the proposed fit to the measured data, shown with error bars. After Suess and Strahm (1970).

sequence from Auvernier (Switzerland) against the Bristlecone pine calibration curve (Fig. 14.10). This procedure allowed the age uncertainties on the Auvernier material to be reduced from hundreds of years to decades.

In order to obtain the highest quality calibration curve, it is desirable to analyse samples of single tree rings. The small size of the bristlecone pine trees limits the precision which can be obtained, because of the limited amount of sample for analysis. However, the ability to 'wiggle match' floating tree ring chronologies with the continuous bristlecone pine calibration has permitted more detailed calibration of the curve using larger trees (e.g. De Jong *et al.*, 1979).

In Europe, the most important tree for detailed calibration purposes is the oak (*Quercus petraea*). This is partly because the oak very rarely has missing annual growth rings, due to its manner of growth. In contrast, the widespread Alder may lack up to 45% of its annual rings (Huber, 1970). The oak is also ideal because it is a long-lived, large tree which displays good resistance to decay after death. In North America, the last 1500 yr of the ^{14}C time-scale has been calibrated in great detail (Stuiver and Pearson, 1986) using large trees such as the Douglas fir (*Pseudotsuga menziesii*) and Giant redwood (*Sequoia gigantea*). This curve was adopted as a new international standard at the 1985 Radiocarbon business meeting (Mook, 1986). Part of this curve is used below (section 14.2.1).

Even more exact dates are possible if the floating chronology comes from an area geographically near to the calibration chronology. Having 'wiggle matched' the radiocarbon data to obtain historical ages with uncertainties of a few decades, the widths of the tree rings themselves are then matched between the floating and calibration material, to obtain an exact date. However, this procedure is only possible if the two chronologies come from areas with the same weather pattern, thus giving rise to similar growth variations.

Hillam *et al.* (1990) used this procedure to date a Neolithic wooden walkway from Somerset (England) to the probable year of its construction (3806 B.C.). This age is based on the fact that ten timbers had bark surfaces with ages of 3807/3806 B.C., and must therefore have been cut down in that calendar year. A single sample with a bark age of 3800 B.C. probably represents a later repair to the walkway. This work suggests that as dendrochronologies are completed for more areas of the world, it should increasingly be possible to date large wood samples to the exact age of their felling.

Comparatively large (20 per mil) ^{14}C variations in wood from single sunspot cycles have been claimed by some workers (e.g. Baxter and Farmer, 1973; Fan *et al.*, 1986). However, atmospheric ^{14}C variations on this time-scale are not consistent with the experimental data of Stuiver and Quay (1981). The latter workers modelled small (4 per mil) ^{14}C variations over sunspot cycles which are at the limits of measurement precision.

14.1.5 Uranium series calibration

Many attempts have been made to extend the calibration of the ^{14}C time-scale beyond the limit of dendrochronology. These have mainly used varved lake sediments (e.g. Tauber, 1970) or ice cores (e.g. Hammer *et al.*, 1986). However, these methods usually involve long interpolations between dated points, based on estimated sedimentation or precipitation rates. Consequently they are not very accurate.

Bard *et al.* (1990a) took a major step in extending the radiocarbon calibration by using mass spectrometric U–series analysis (section 12.2.2). This method was used to assign absolute ages to Barbados corals previously analysed for ^{14}C. In view of uncertainties about closed-system behaviour in carbonates, the method was tested by analysis of samples less than 10 kyr old. These gave ages in good agreement with the dendrochronology time-scale, after applying a 400 yr correction for ^{14}C equilibration between atmospheric and seawater reservoirs. Results for older samples were presented on a plot of Δ^{14}C activity (relative to modern wood) against U–Th age (Fig. 14.11). Samples in the range 10–15 kyr gave Δ^{14}C activities well within error of those predicted from geomagnetic field strength data. Samples older than 15 kyr initially gave more scattered data (open circles in Fig. 14.11). However, repeat analysis of the ^{14}C measurements after strong acid leaching gave more consistent results (Bard et al., 1990b, 1993).

Mazaud *et al.* (1991) compared the coral data with a ^{14}C production model based on an improved geomagnetic intensity record. The good agreement between the coral data and the predicted ^{14}C activity curve means that long-term activity variations in the atmosphere and hydrosphere can be explained largely by variable cosmogenic production (in response to secular variations in the magnetic field). Hence, climatic effects, which can affect the ^{14}C equilibrium between atmospheric and marine carbonate reservoirs, must play a subordinate role. However, Stuiver *et al.* (1991) argued that climate could have a second-order effect on atmospheric ^{14}C/^{12}C activity ratios by releasing ^{12}C from oceanic carbonate through changes in ocean circulation. There is no evidence for this process within the range of the dendrochronological time-scale, but it may have occurred during the major climatic re-adjustments of the last deglaciation.

Edwards *et al.* (1993a) tested this proposal by means of a more detailed coral study for the period 8–14 kyr B.P. They found a markedly rapid decrease in ^{14}C activity between 12 and 11 kyr B.P., which they

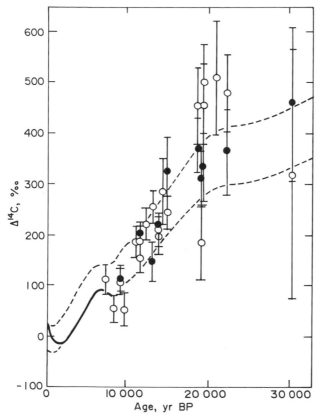

Fig. 14.11. Plot of Δ ^{14}C relative to modern wood of samples dated by dendrochronology (heavy curve) and U–Th analysis ($\bullet$, $\circ$). Solid and open symbols signify leached and un-leached radiocarbon analyses respectively. Dashed lines show the envelope of ^{14}C activity predicted from a theoretical cosmogenic model. After Bard *et al.* (1990a,b).

attributed to dilution of atmospheric ^{14}C by 'dead' carbon. This was attributed to changes in the circulation of the North Atlantic ocean, which may also have caused the glacial re-advance of the Younger Dryas event. Hence, it was argued that climatic changes *can* perturb the overall control of the geomagnetic field on atmospheric ^{14}C activity for short periods of time. Further detailed work on corals is required to test this model.

14.2 Accelerator mass spectrometry

Mass spectrometry is potentially a powerful alternative to radioactive counting in the determination of cosmogenic nuclides because it utilises every atom of this nuclide in the sample. In contrast, counting determinations utilise only the small number of atoms which decay during the measurement experiment. If decay rates are very high (corresponding to half-lives of less than a thousand years) then the latter method

may be most efficient. However, for longer-lived nuclides, mass spectrometry has the ability to out-perform counting.

Cosmogenic nuclides are characterised by very low abundances relative both to other isotopes of the same element and to isobaric interferences from other elements. The former problem is exemplified by the fact that even modern carbon, with the highest ^{14}C/^{12}C ratio (1.2×10^{-12}), would cause unacceptably large peak tails in the mass spectrum of a 'conventional' machine such as normally used by geologists. Such machines typically have 'abundance sensitivities' (peak tail at one mass unit distance) of 10^{-6} at the uranium mass, which may decrease to ca. 10^{-9} at the mass of carbon.

Abundance sensitivity might be improved sufficiently to measure ^{14}C in a 'conventional' mass spectrometer by increasing its size, introducing electrostatic filters, and by increasing the accelerating potential and magnet current. The latter approaches respectively filter and overwhelm the spread of

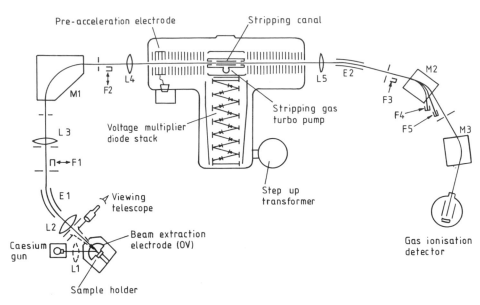

Fig. 14.12. Schematic illustration of the Toronto accelerator mass spectrometer showing typical features of such instruments. M1–M3: magnetic analysers; E1–E2: electrostatic analysers; L1–L5: electrostatic lenses. F1–F5: Faraday cups; After Kieser *et al.* (1986).

energies of the ions emitted by the source. Accelerator mass spectrometers usually have all of these features (Fig. 14.12), but they are not central to accelerator mass spectrometry (AMS). The principal attributes of the tandem accelerators used in AMS are the charge-exchange process, which removes molecular interferences, and the very high ion energies achieved, which allow energy-loss detectors to resolve atomic isobars.

It has been suggested by Lal (1988) that the principal impetus for the development of AMS was the fact that accelerators became available for this purpose as their applications in physics diminished. As with many techniques, AMS began as a method looking for an application, but quickly took off as a useful tool in its own right. In the future, laser-induced resonance ionisation may replace the role of accelerators in excluding isobaric interferences (e.g. Labrie and Reid, 1981), but at the present AMS is the dominant method.

The essence of the tandem accelerator is the initial acceleration of negative ions by a positive potential in the megavolt range, followed by charge exchange of the ion beam, after which positive ions are accelerated back to zero potential. During the charge-stripping process, isobars of different elements often behave differently, allowing their subsequent separation, while molecular ion isobaric interferences are

destroyed. Charge stripping may be performed by passing the ion beam through an electron-stripping gas (e.g. argon), through a thin graphite film, or (in very high energy accelerators) a thin metal foil. Experience with carbon has shown that charge stripping to a 3+ state is often most effective, since triple-charged CH_2 molecular ions have not been observed (Litherland, 1987). This avoids the need for a high-resolution magnetic analyser to resolve molecular ions by their mass defect.

14.2.1 Radiocarbon dating by AMS

Determination of ^{14}C, ^{26}Al and ^{129}I can be performed on low-energy tandem accelerators (Litherland, 1980; 1987). In these cases the atomic isobars, ^{14}N, ^{26}Mg and ^{129}Xe do not form stable negative ions, so that complete separation occurs in the source (e.g. Purser *et al.*, 1977). However, separation of the atomic $^{14}C^-$ ion from the molecular ion $^{12}CH_2^-$ depends on the charge-stripping stage of the tandem accelerator.

Most ^{14}C analyses by AMS are presently performed on a solid graphite sample. A typical preparation method is the catalytic reduction of CO_2. However, Bronk and Hedges (1987) have experimented with a CO_2 gas ion source. In order to achieve a ^{14}C ion beam of 15 ions per second from modern carbon, an 'intense' ^{12}C ion beam of 2 μA must be generated. This is normally achieved using a caesium sputter source, which ejects negative carbon ions by bombarding the sample with Cs^+ from an ion gun. The efficiency of AMS radiocarbon measurement is illustrated by the fact that it yields the same count rate from 1 mg of carbon as the β count rate from a whole gram of carbon. Nevertheless, a 55 kyr-old carbon sample yielding a 2 μA ^{12}C beam has a ^{14}C count rate of only one ion per minute (corresponding to a $^{14}C/^{12}C$ ratio of ca. 1.2×10^{-15}).

The sputter source generates an ion beam with variable ion energies. After acceleration to a few tens of kilovolts, these must be 'cleaned up' using an electrostatic analyser before the beam is ready for the accelerator. In addition, it is necessary to split the major and minor ion beams with a magnetic analyser before the accelerator, in order to minimise scattering of the ^{14}C beam by collision of the ^{12}C beam with gas molecules.

In ^{14}C dating, the most effective charge-stripping medium is provided by a higher gas pressure in the central ultra-high-voltage 'stripping canal' of the tandem accelerator (Fig. 14.12). Differential pumping of the acceleration tubes at either end of the tandem generator can maintain a pressure 5000 times lower here than in the stripping canal (Litherland, 1987). The charge-stripping process generates a range of charge states in the positive ion beam, such that only ca. 50% of ions have the selected charge. Therefore, the accelerator system must be calibrated

against standards of known $^{14}C/^{12}C$ ratio before unknown samples are run. Production of $^{14}C^{3+}$ using a 3 MV accelerator is ideal for radiocarbon measurement, but $^{14}C^{2+}$ ions from a 1.4 MV accelerator can also be used (Lee et al., 1984).

The very high energy of the positive ion beam (normally > 1 MeV) allows the use of ionisation counters which measure the energy of collected ions as well as their abundance. This provides a final means of resolving any residual molecular ions in the ^{14}C beam (possibly generated by recombination after the accelerator). Fig. 14.13 shows that the molecular ion beams of ^{13}C and ^{12}C are barely significant in modern carbon, but dominant in 47 kyr-old carbon.

A good example of the application of AMS to radiocarbon analysis is provided by the dating of the Shroud of Turin (Damon et al., 1989). This was believed to be possibly the burial cloth of Christ, although its sudden appearance in the 1350s raised the probability that it was instead a medieval 'icon'. The advent of AMS analysis provided the opportunity to perform an absolute date on the fabric using a total of only 7 cm^2 (150 mg) of cloth. This was divided between university laboratories at Arizona, Oxford and Zurich. A major worry was that the fabric might have been contaminated during its long history as a relic. Therefore, a variety of different cleaning procedures was performed by each group on different sub-samples of cloth. In addition, each group dated three control fabric samples of different known ages

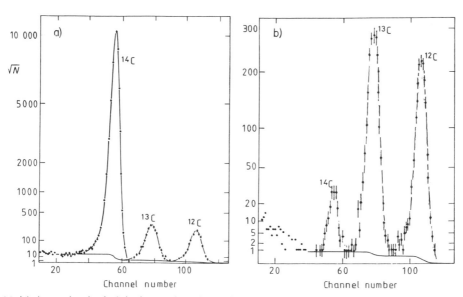

Fig. 14.13. Multi-channel pulse-height (energy) analysis of radiocarbon dating samples from an ionisation detector. a) Modern carbon; b) 47.4 kyr-old carbon. Typical error bars are shown. After Litherland (1987).

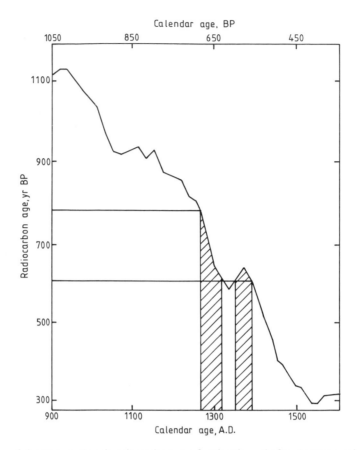

Fig. 14.14. Translation of the 'conventional' radiocarbon age for the Shroud of Turin into a calendar age. Age limits are at the 95% confidence level (2σ), including the estimated error of the calibration curve. Of the two possible calendar age ranges (shaded), the more recent is excluded by historical data. After Damon *et al.* (1989).

to form a general impression of the between-lab. reproducibility.

The results from the three laboratories, collated independently at the British Museum, were in good agreement, although the shroud itself yielded poorer reproducibility than the controls, possibly due to contamination. The 'conventional' radiocarbon age of the shroud (B.P.) was translated into a calendar age using the calibration curve of Stuiver and Pearson (1986). The 95% confidence limits on the conventional age include an inflection in the calibration curve which technically creates two possible calendar age ranges, A.D. 1262–1312 and 1353–1384 (Fig. 14.14). However, since the shroud went on view in the 1350s, the latter range can fortunately be excluded on historical grounds. It is concluded that the linen of the shroud was derived from cotton which grew in A.D. 1290 ± 25 years, and it is therefore a medieval artifact.

14.3 Beryllium-10

Cosmic rays interact directly with nitrogen and oxygen atoms in the atmosphere, causing spallation (fragmentation) into the light atoms Li, Be and B. Amongst these, one of the nuclides produced is the unstable isotope ^{10}Be. Cosmogenic ^{10}Be can also be generated in the surface layer of exposed rocks by *in-situ* production. However, this subject will be dealt with under section 14.6.

^{10}Be decays by pure β emission to ^{10}B. It was first observed in naturally occurring material by radioactive counting (Arnold, 1956). However, even at that time Arnold recognised that mass spectrometry might supplant radioactive counting as a method for the determination of ^{10}Be. This is because the relatively long half-life of 1.51 Myr (Hofmann *et al.*, 1987) makes counting a very inefficient process for ^{10}Be analysis of natural samples. For example, McCorkell *et al.* (1967) used 1200 tonnes of ice-water to make ^{10}Be (and ^{26}Al) measurements by β counting on Greenland ice. In contrast, Raisbeck *et al.* (1978) made the first AMS measurement on similar material using only 10 kg of ice-water.

Analytically, ^{10}Be determination by AMS is similar to ^{14}C, but involves some additional complications. Because Be does not form stable negative ions, the BeO species must be used, upon which the isobaric interference of ^{10}BeO$^-$ is a serious problem. This is overcome by passing the ion beam through an absorber gas (in front of the detector) whose pressure is adjusted to completely stop ^{10}B transmission. The high ion velocity of the ^{10}Be beam generated by AMS allows this species to pass through to the detector, which consists of a gas ionisation counter in front of a surface barrier detector which finally absorbs the ion beam. The first detector measures energy loss (ΔE) of the ions as they collide with gas molecules in the chamber (this is related to atomic number). The second detector measures residual energy. Using this bivariate discriminant, ^{10}Be ions can be resolved from all other signals (Fig. 14.15) to yield a very low

background. ^{10}Be contents of samples are measured relative to added ^{9}Be spike, and normalised for machine mass discrimination by frequent standard analysis.

14.3.1 ^{10}Be in the atmosphere

^{10}Be enters the hydrological cycle by attachment to aerosols, from which it is scavenged by precipitation. Consequently, it has a very short (ca. 1 week to 2 yr) residence time in the atmosphere and, unlike ^{14}C, is not homogenised within the atmosphere prior to its 'fallout'.

It was originally intended (Raisbeck *et al.*, 1979) that ^{10}Be analysis of rainwater would allow accurate constraints to be placed on the global average ^{10}Be production rate. However, two factors complicate a determination of the ^{10}Be flux in rainwater. One is the tendency for comparatively Be-rich soil particulates to be caught up in the atmosphere and cause secondary contamination of rain (Stensland *et al.*, 1983). Once this effect is removed, individual

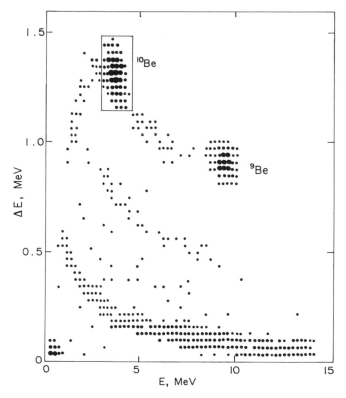

Fig. 14.15. Plot of ΔE against E for a typical geological sample, to show resolution of ^{10}Be from other species. Dot size indicates the number of counts in each bin (smallest = 1 count). After Brown *et al.* (1982).

depositional events still turn out to have very variable ^{10}Be contents (Brown *et al.*, 1989).

One way of gauging the effect of soil re-suspension on ^{10}Be abundances is to compare them with ^{7}Be data. The latter species has relatively similar atmospheric production rates to ^{10}Be, but much lower levels in soils due to its very short (53 day) half-life. It is measured by γ counting. An alternative way of assessing the effects of soil re-suspension is to compare continental and oceanic ^{10}Be deposition rates. Brown *et al.* (1989) used both of these approaches in an analysis of ^{10}Be precipitation in Hawaii and the continental US to test its behaviour at temperate latitudes. Average ^{10}Be contents of Hawaiian rain, in which atmospheric soil suspension is negligible, are very similar to ^{7}Be in rain from Illinois in the US (Fig. 14.16), but in both cases a few events yield very large relative contents.

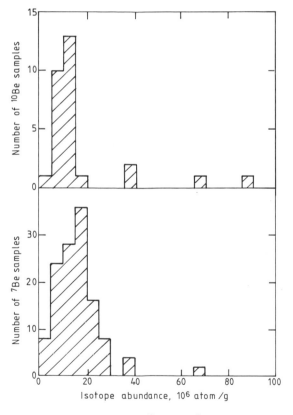

Fig. 14.16. Histograms of ^{10}Be and ^{7}Be concentration respectively in rainfall from Mauna Loa, Hawaii and Bondville, Illinois. The former are weekly rainfall aggregates, the latter represent individual showers. After Brown *et al.* (1989).

The variability of ^{10}Be contents in individual rain showers makes it difficult to determine accurate annual fluxes for mid-latitudes. A summary of these data as a function of latitude (Fig. 14.17) shows the variability of these estimates (Brown *et al.*, 1992). However, at tropical latitudes the estimates of annual ^{10}Be flux are more reproducible. The latter are in good agreement with a global ^{10}Be flux estimate of 10^6 atom/cm^2/yr, based on cosmic-ray intensity as a function of latitude (Lal and Peters, 1967). Therefore, at the present time, the curve in Fig. 14.17 represents the best available estimate of the atmospheric ^{10}Be flux.

Atmospheric ^{10}Be accumulates in snow and ice, but its half-life is too long to date such deposits. However, it can be used as a tracer of climatic changes and to understand the processes modulating cosmogenic ^{14}C production in the atmosphere. The first detailed study of this type was made by Raisbeck *et al.* (1981a) on a 906 m-long ice core from the Dome C station, eastern Antarctica. This core had been dated on the basis of oxygen isotope correlation with ^{14}C-dated marine sediments. A detailed analysis of the top 40 m of core revealed a ^{10}Be maximum around 1700 A.D. which correlated with the ^{14}C de Vries effect maximum (section 14.1.3) and the 'Maunder' sunspot minimum at this time (Eddy, 1976). Consequently these data supported the model of solar modulation of cosmic-ray intensity.

Subsequent studies of Greenland ice cores from the Camp Century and Milcent stations (Beer *et al.*, 1984) confirmed the ^{10}Be maximum associated with the Maunder sunspot minimum (Fig. 14.18). In addition, Beer *et al.* (1985) performed a Fourier transform analysis of very detailed isotopic data from the Milcent core. This revealed second order ^{10}Be variations with a 9–11 year cycle time equal to the 'sunspot cycle', both before and during the Maunder minimum. Beer *et al.* concluded from these findings that solar activity continues to vary, even when sunspots are not actually visible.

In contrast to the good agreement between ^{14}C and ^{10}Be data for the last 500 yr, a reconnaissance study of the whole Dome C core (Raisbeck *et al.*, 1981a) revealed a strong correlation of ^{10}Be with δ^{18}O, which was attributed to the climatic effect of the last ice age. No significant correlation was observed with geomagnetic field strength, which reached a minimum value 6000 years ago and has been argued to control long-term ^{14}C variations. Yiou *et al.* (1985) suggested a partial solution to this problem by attributing the ^{10}Be maximum during the last glaciation to lower precipitation at that time. This would sweep the same

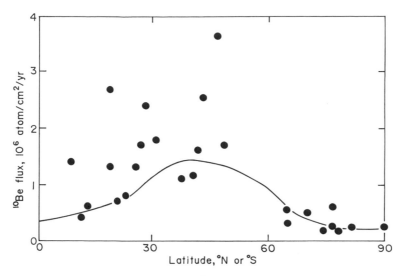

Fig. 14.17. Summary of empirical estimates of annual ^{10}Be flux in rainfall, as a function of latitude. These are compared with a theoretical model based on cosmic-ray intensity (shown by the curve). After Brown *et al.* (1992).

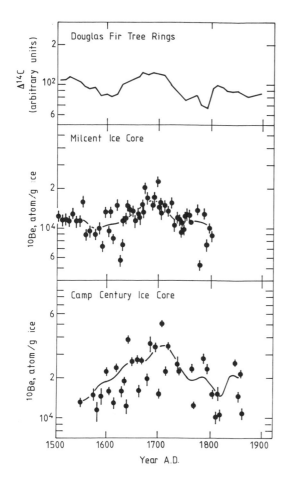

amount of ^{10}Be out of the atmosphere, but concentrate it in a lower volume of ice, causing the ^{10}Be record to be compressed.

The long-term Antarctic Dome C data were again matched by results from Camp Century (Fig. 14.19). Beer *et al.* (1988) suggested two alternatives to explain the lack of correlation between the ^{10}Be and geomagnetic intensity data. One idea was that ^{10}Be ice-core data are recorded at high latitudes where the field strength has a weak influence on atmospheric cosmic-ray intensity. An alternative suggestion was that the long-term ^{10}Be and ^{14}C activity variations are not caused by geomagnetic field changes but by a complex interplay of climatic effects and solar cosmic-ray modulations. However, recent U–Th calibration of the radiocarbon time-scale (section 14.1.5) suggests that ^{14}C variations up to 30 kyr old can largely be explained by variations in the Earth's magnetic field. Climatic effects play a subordinate role.

14.3.2 ^{10}Be in soil profiles

Beryllium is partitioned very strongly from rain water onto the surface of soil particles such as clay minerals.

Fig. 14.18. Plot of ^{10}Be variations in the Milcent and Camp Century ice cores during the last 500 yr, compared with age-corrected ^{14}C levels in tree rings. Curves through ^{10}Be data represent ca. 30 yr running averages. After Elmore *et al.* (1987).

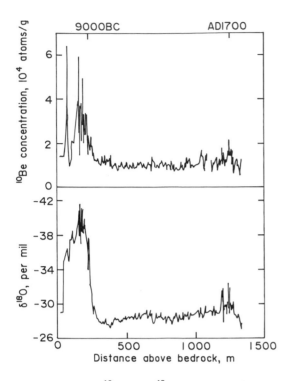

Fig. 14.19. Plots of ^{10}Be and $\delta\,^{18}$O variations in the Camp Century ice core (Greenland) over a depth of 1400 m, corresponding to ca. 10 kyr. After Beer *et al.* (1988).

If we assume that ^{10}Be adsorption is perfect, and that a given soil section is developed by weathering of rock or rock debris without the addition or removal of sediment during the weathering process, then the soil section should contain a complete inventory of all deposited ^{10}Be that has not yet decayed. This process offers the opportunity of dating a soil profile by measuring the total accumulation of ^{10}Be in the section, but it is apparent that there are a large number of assumptions.

The ^{10}Be inventory of a soil profile in Virginia represents a case where beryllium uptake appears to be nearly 100% efficient (Pavich *et al.*, 1985). ^{10}Be activities display a smooth decay curve against depth (Fig. 14.20a), with a total inventory of 9×10^{11} atom/cm^2. We can compare this value with a theoretical inventory, N, assuming 100% uptake over a given period of time. This is given by the equation:

$$N = \frac{q}{\lambda}(1 - e^{-\lambda t}) \qquad [14.3]$$

where q is the input flux from rainfall and t is the accumulation time. If the profile is infinitely old, relative to the 1.5 Myr half-life of ^{10}Be, then it will reach saturation, where the input flux from rainfall is balanced by the rate of decay (λ). Equation [14.3] then simplifies to $N = q/\lambda$. Given an annual deposition flux of 1.3×10^6 atom/cm^2 at this latitude, the

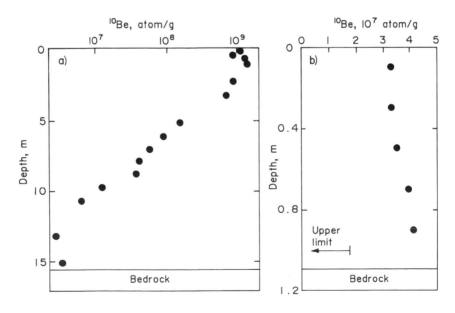

Fig. 14.20. Plots of ^{10}Be activity as a function of depth in two soil profiles. a) Virginia Piedmont, after Pavich *et al.* (1985); b) Orinoco Basin, after Brown *et al.* (1992).

saturation inventory will be 3×10^{12} atom/cm^2, which is three times the observed inventory. The discrepancy can be explained by loss of ^{10}Be-enriched soil from the top of the profile by erosion, and its replacement at the bottom of the profile by weathering of bedrock with no ^{10}Be. Solving equation [14.3] for t (using the observed inventory) yields the residence time of ^{10}Be in the profile, equal to 800 kyr.

Very different ^{10}Be behaviour is demonstrated by a soil profile from the Orinoco Basin (Fig. 14.20b), which has a total inventory of only 5×10^9 atom/cm^2. Assuming an annual ^{10}Be flux for this latitude of 0.4×10^6 atom/cm^2 yields a ^{10}Be residence period in the soil profile of only 12 kyr (because t is short, relative to the ^{10}Be half-life, it can be approximated by N/q). This low value is best explained by 'breakthrough' of ^{10}Be from the base of the profile by leaching (Brown *et al.*, 1992).

Differences in ^{10}Be retention between the two cases described above can be understood in the light of laboratory experiments on beryllium partition between soil and water (You *et al.*, 1989). These studies showed that beryllium retention on soil particles is strongly pH-dependent, with distribution coefficients of ca. 10^5 in neutral conditions (pH 7), but less than 100 at pH 2. Hence the more acidic conditions in tropical soils are less favourable for beryllium retention.

In alkaline soils, beryllium mobility within the soil profile may be very limited, and in these conditions ^{10}Be may be used as a stratigraphic tool. An example is provided by a ^{10}Be study of Chinese loess, in which carbonate-rich conditions yield a pH value of 8

(Chengde *et al.*, 1992). The profile was dated magnetically back to 800 kyr, and represents the products of wind-borne deposition through varying climatic conditions. Chengde *et al.* tuned their profile to the Quaternary climatic record provided by sea-floor ^{18}O variations. They concluded that during arid periods, rapid loess deposition was accompanied by high fluxes of ^{10}Be, adsorbed onto wind-blown particles. These sections were interspersed with wetter periods with lower depositional fluxes.

14.3.3 ^{10}Be in the oceans

Marine sediments were some of the first materials to be successfully analysed for ^{10}Be, since they have concentrations measurable by β counting. The objective was to use ^{10}Be as a dating tool for oceanic sediments. However, early studies, which simply compared ^{10}Be abundances at various depths with theoretical cosmogenic production rates, were unreliable. A more rigorous study was made by Tanaka and Inoue (1979) on paleomagnetically dated sediment cores from the Pacific ocean. These workers showed that absolute ^{10}Be concentrations were variable from site to site, but that values at a given depth, relative to the sediment surface, were consistent with a theoretical decay path (Fig. 14.21). The good agreement between the concentration data and the theoretical reference line suggests that cosmogenic ^{10}Be production has been constant to within about 30% over the last 2.5 Myr.

Early studies of the behaviour of stable ^{9}Be in the oceans suggested that it was one of the class of

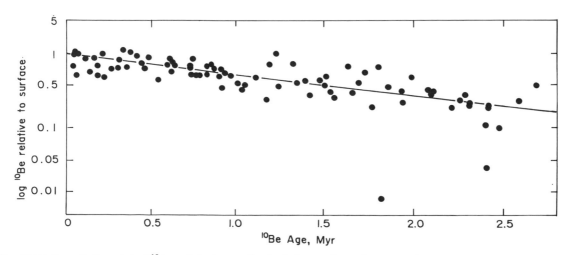

Fig. 14.21. Compilation plot of ^{10}Be activity (normalised relative to the sediment surface) against burial age (depth) for five cores from the North Pacific. After Tanaka and Inoue (1979).

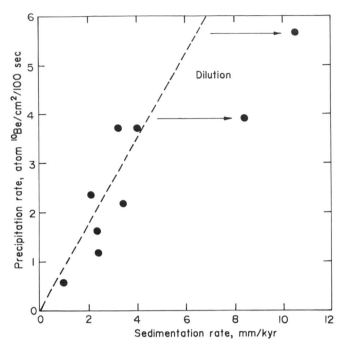

Fig. 14.22. Plot of ^{10}Be precipitation rate against particulate sedimentation rate for the different Pacific Ocean sites. Dashed line represents a constant ^{10}Be concentration in different cores. Points to the right of this correlation line are attributed to dilution by excess (coarse) clastic material. After Tanaka and Inoue (1979).

elements which very quickly precipitates from seawater (Merrill *et al.*, 1960). Tanaka and Inoue (1979) tested a model in which fine particulates are the principal carrier of ^{10}Be by plotting the ^{10}Be precipitation rate against sedimentation rate (Fig. 14.22). The good positive correlation in most of the data suggested that the particulate model was valid, so the net ^{10}Be deposition flux at any given locality is dependent on sedimentation rate and not the cosmogenic flux. ^{10}Be deposition rates are seen to vary by a factor of three above and below the theoretically predicted ^{10}Be flux based on nuclear cross-sections (Reyss *et al.*, 1981). Tanaka and Inoue argued that the transport of fine particulates by ocean currents could generate the observed large variations in ^{10}Be depositional flux at different localities. This has been largely confirmed by subsequent AMS studies of sediment cores and of ocean water itself.

A detailed understanding of the aqueous ^{10}Be system requires a consideration of the oceanic residence time. Merrill *et al.* (1960) estimated the residence time of beryllium using equation [14.4] (Goldberg and Arrhenius, 1958):

$$\text{residence time} = \frac{\text{total oceanic inventory}}{\text{total rate of introduction}} \quad [14.4]$$

This equation holds for a steady-state (equilibrium) system, which is approximated if the flux is constant for three residence times. For ^{10}Be the equation can conveniently be calculated per unit area:

$$\text{residence time} = \frac{\text{total water column budget/unit area}}{\text{supply flux/unit area}}$$
$$[14.5]$$

Merrill *et al.* determined a residence time for ^{9}Be attached to particulate matter of 150 yr, but they estimated a longer residence time of 570 yr for the dissolved beryllium budget.

The first estimate of the soluble ^{10}Be budget of the oceans was made by Yokoyama *et al.* (1978) based on the ^{10}Be/^{9}Be ratios of manganese nodules. By assuming that these incorporated dissolved beryllium directly from seawater, and using published ^{9}Be abundances in the oceans, they calculated the oceanic ^{10}Be budget as 2×10^9 atoms/g. Almost identical concentrations were determined by Raisbeck *et al.* (1980) in the first direct ^{10}Be determinations on deep ocean waters, but their estimated residence time (630 yr) differed markedly from that of Yokoyama *et al.* (1600 yr) due to the use of different cosmogenic flux estimates. Raisbeck *et al.* used their own estimate

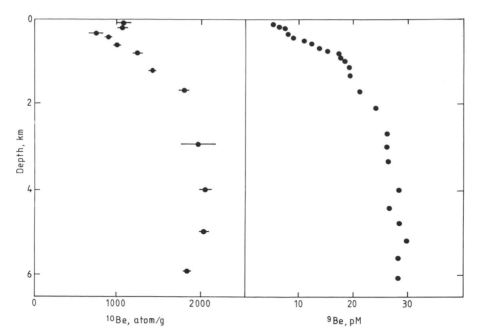

Fig. 14.23. Concentrations of [10]Be (atom/g) and [9]Be (picomol, pM) plotted as a function of water depth in the open ocean of the east Pacific. After Kusakabe *et al.* (1987).

of the cosmogenic [10]Be flux, based on one year's rain from a single locality in France, uncorrected for re-suspension of soil. This can now be seen to be an over-estimate. Using the theoretical production rate of Reyss *et al.* (1981), both studies lead to a soluble [10]Be residence time in the oceans of ca. 1200 yr.

Arnold (1958) divided the behaviour of elements such as beryllium in the oceanic system into three categories: soluble/sorbed ions; pelagic particulate-controlled ions; and inshore particulate-controlled ions. All three categories can be seen to control [10]Be. Despite its tendency to be adsorbed onto particulates, dissolved [10]Be has a longer residence time in the oceans than similar adsorbable species such as [230]Th (section 12.3.2). This difference can be explained by a nutrient-like behaviour in beryllium (Measures and Edmond, 1982). Berillium concentration profiles in Pacific ocean water (Fig. 14.23) show strong depletion near the surface where they are adsorbed onto organic matter, but relative enrichment at depth due to the breakdown of dead organic matter as it falls through the water column (Kusakabe *et al.*, 1987). The new Pacific data yield a deep water [10]Be concentration of 2×10^9 atom/g which is in agreement with the earlier results quoted above. However, Atlantic ocean water displays a different signature which may reflect the large river water input into this ocean basin.

The behaviour of [10]Be in the near-shore environ-ment is very different from the open ocean, as proposed in Arnold's model. Sediment cores from the continental rises off western Africa and western North America show [10]Be accumulation rates at least an order of magnitude larger than the theoretical cosmogenic flux (Mangini *et al.*, 1984; Brown *et al.*, 1985). This is attributed to the lateral transport of large quantities of [10]Be, by ocean currents, into areas of high deposition (due to continental run-off or excess biological production). In these localities the transported [10]Be is effectively scavenged and carried to the bottom.

Similar excess [10]Be sedimentation rates have been seen in freshwater lakes (Raisbeck *et al.*, 1981b). In this case, soil erosion in the drainage basin which supplied the lake caused the introduction of ber-yllium-rich sediment. Lundberg *et al.* (1983) proposed that the excess [10]Be was introduced dominantly on organic matter rather than silicate particles, consistent with the nutrient-like behaviour of [10]Be proposed for the oceanic system.

14.3.4 [10]Be in magmatic systems

Perhaps the most important application of [10]Be as a geological tracer is in studies of the relationship between sediment subduction and island arc volcanism. In a reconnaissance study, Brown *et al.* (1982) demonstrated ^{10}Be concentrations in island arc volcanics ($2.7 \times 10^6 - 6.9 \times 10^6$ atom/g) which were generally much higher than the levels seen in a control group of continental and oceanic flood basalts. Brown *et al.* argued against high-level contamination of the analysed volcanics on the grounds that the short half-life of ^{10}Be renders it extinct in all but surficial deposits, while ^{10}Be levels in rain-water are too low to cause the observed enrichments. In contrast, it has long been known (e.g. Arnold, 1956) that pelagic sediments have very large ^{10}Be contents of over 10^9 atoms/g. Brown *et al.* therefore attributed their data to the involvement of subducted ocean floor sediment in the source area of island arc magmatism.

Subsequent studies (e.g. Tera *et al.*, 1986) confirmed the general observation of high ^{10}Be in island arc volcanics and low ^{10}Be in the non-arc control group (Fig. 14.24). Detailed studies were also undertaken to assess the effects of weathering on the ^{10}Be contents of lavas. Analyses of material collected during or immediately after eruption were shown to contain the same range of ^{10}Be contents as historical lavas. Contamination of non-arc samples by radiogenic ^{10}Be was only observed in severely altered samples. *In-situ* cosmogenic production of ^{10}Be in lavas at depths of over 1 cm was excluded by the low ^{10}Be abundances in 16 Myr-old Columbia River basalts.

A surprising result of the detailed study of Tera *et al.* was that several arcs have ^{10}Be levels as low as the maximum of 1×10^6 atoms/g in the control group. These included all samples from the Mariana, Halmahera (Moluccan) and Sunda arcs. To explain these observations, Tera *et al.* suggested four requirements for a positive ^{10}Be signal in arc

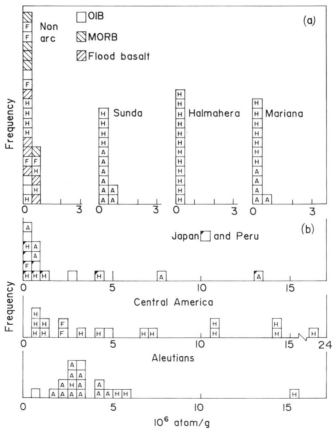

Fig. 14.24. Histograms of ^{10}Be abundance in volcanic rocks. a) Non-arc control group and low-^{10}Be arcs; b) high-^{10}Be arcs. Symbols: A = active volcano; H = historic flow; F = fresh sample collected during or immediately after eruption. After Tera *et al.* (1986).

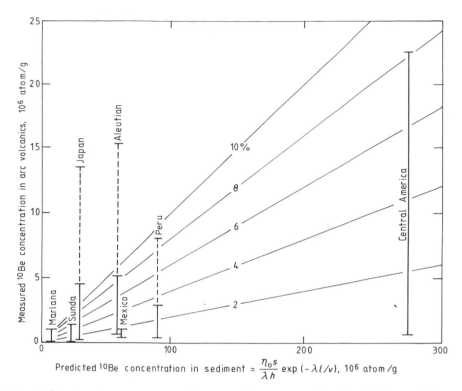

Fig. 14.25. Plot of ^{10}Be data for seven arcs against a model parameter for the efficiency of ^{10}Be supply to arc magma sources. η_o = ^{10}Be abundance of sediment; s = Plio-Pleistocene sedimentation rate; h = sediment thickness; l = distance from trench to magma source; v = plate velocity. ^{10}Be signals are modelled for different bulk percentage sediment contributions to magmas. After Tera *et al.* (1986).

volcanics:

1) adequate ^{10}Be inventory in trench sediments;
2) subduction rather than accretion of uppermost Be-enriched sediments;
3) incorporation of sediment in the magma source area;
4) transport time from sedimentation to magma source area < 10 Myr.

Failure of any one of these criteria could preclude a positive ^{10}Be signal. However, they did not observe simple correlations between ^{10}Be and geophysical parameters such as age of the subducting plate.

In an attempt to harmonise their data from different arcs, Tera *et al.* plotted ^{10}Be against a complex quantity involving sedimentation rate, sediment thickness, plate velocity, and distance from trench to magma source (Fig. 14.25). Since the volcanic front is always located about 100 km above the seismic plane, the latter quantity is inversely proportional to the dip of the Benioff zone. The contrast between the ^{10}Be-rich Central American data

and other arcs is explained by the high sedimentation rate and steep subduction angle of the former. However, both the Japanese and Aleutian arcs have a single ^{10}Be-rich data point (shown by the dashed range in Fig. 14.25) which does not fit the model.

A further step in rationalising the ^{10}Be systematics of arc volcanics was achieved by considering the data relative to non-cosmogenic (^{9}Be) abundances (Monaghan *et al.*, 1988; Morris and Tera, 1989). Within the different minerals of a single rock sample, ^{10}Be is normally strongly correlated with total Be content (Fig. 14.26), implying that radiogenic and non-radiogenic Be were mixed before magmatic differentiation occurred. This further strengthens the arguments against surficial contamination of the lavas by ^{10}Be, and also argues against crustal ^{10}Be assimilation by magmas. The enriched Be contents of the groundmass, relative to phenocrysts, show that Be behaves as an incompatible element during magmatic differentiation.

In contrast to the mineral systematics, most whole-rock samples analysed by Morris and Tera did not

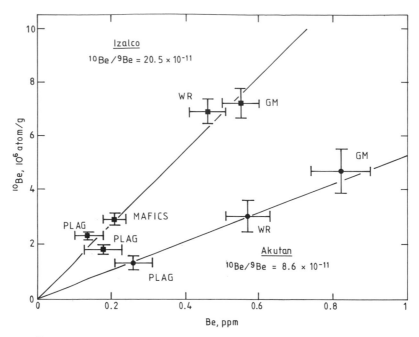

Fig. 14.26. Plot of ^{10}Be against total Be content for separated mineral phases, plus whole-rock (WR) and groundmass (GM) from two young lavas. Samples are from the Izalco volcano (Central America) and Akutan (Aleutians). After Morris and Tera (1989).

show a strong correlation between ^{10}Be and total Be. This is consistent with the fact that ^{10}Be/^{9}Be ratios *are* correlated with absolute ^{10}Be abundance (Fig. 14.27). These findings suggest that most of the rocks analysed, which were basalts, did not have their ^{10}Be contents perturbed by magmatic differentiation, so that the interpretations of Tera *et al.* (1986) still hold. However, some andesites lie significantly to the right of the main trend (Fig. 14.27), including the Japanese and Aleutian samples with abnormally high ^{10}Be contents in Fig. 14.25 (Tera *et al.*, 1986).

Further constraints on the timing of subduction-related processes were obtained by combining ^{10}Be/^{9}Be and ^{238}U/^{230}Th data. Sigmarsson *et al.* (1990) observed that these ratios were correlated in the Southern Volcanic Zone of the Andes (section 13.5). Based on this correlation, and the much shorter half-life of ^{230}Th than ^{10}Be, Sigmarsson *et al.* suggested that the time-scale for dehydration, melting and eruption of these arc magmas was probably less than 20 kyr.

A further step in understanding subduction-zone processes was achieved by comparing ^{10}Be/^{9}Be and boron/beryllium ratios in arc lavas (Morris *et al.*, 1990). Several arcs displayed a strong positive correlation between these two variables (Fig. 14.28),

despite the fact that the beryllium isotopes and boron have different distributions in the subducted slab. ^{10}Be is concentrated in the uppermost sediment layers and diminishes rapidly downwards, ^{9}Be is distributed throughout the sediment column, whereas B is principally concentrated in the hydrothermally altered basaltic crust. Hence, there is no *a priori* reason why these three species should display coherent behaviour in arc volcanics. The fact that they do behave coherently in widely separated volcanoes along the length of an arc suggested to Morris *et al.* a very thorough homogenisation mechanism for Be and B during the process of subduction-related magma genesis. While such a process could occur in the solid state, it is easiest to conceive of the mixing of fluids driven off from different parts of the subducted slab. The convergence of all of the correlation lines at the origin points to complete stripping out of all boron from the subduction zone, with no long-term residence of this element in the mantle.

The observed correlation between ^{10}Be/^{9}Be ratio and elemental B/Be ratio suggests that the latter may represent a useful proxy for the former. This is important because it widens the applicability of beryllium data. Firstly, the elemental B/Be ratio can be used as a tracer of the slab component in arcs with

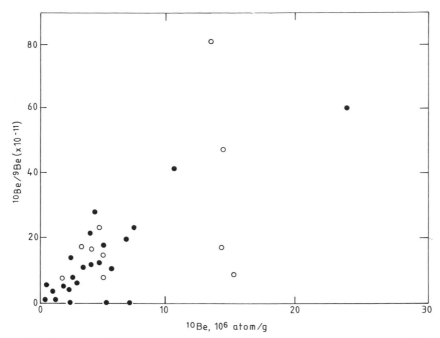

Fig. 14.27. Plot of $^{10}Be/^9Be$ against ^{10}Be abundance for basalts (●) and evolved rocks (○) from different arc and non-arc environments. After Morris and Tera (1989).

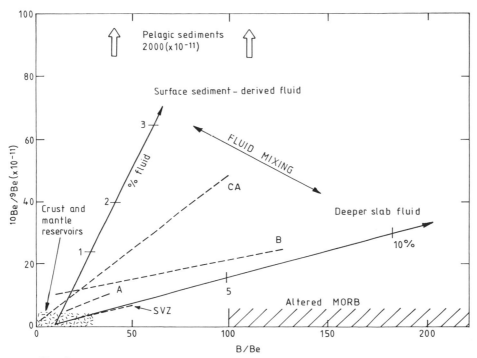

Fig. 14.28. Plot of $^{10}Be/^9Be$ against elemental B/Be ratio, showing correlation lines for arc volcanics (dashed) relative to possible mixing end-members. Numbered ticks denote calculated percentages of slab-derived fluids necessary to generate observed arrays by contamination of the mantle source. Arcs: CA = Central American; A = Aleutians; B = Bismark arc, New Guinea; SVZ = Southern Volcanic Zone of the Andes. After Morris et al. (1990).

low subduction rates, where ^{10}Be is extinct by the time of eruption. Secondly, elemental ratios can be measured with less sophisticated analytical equipment such as ICP-MS. These advantages were demonstrated by Edwards *et al.* (1993b) in a study of basaltic lavas from the Indonesian arc. Edwards *et al.* were able to combine B/Be ratios with other radiogenic isotope systems in order to uniquely specify the Pb, Sr and Nd isotope signatures of the slab-derived component, which could be modelled by a 80%–20% mixture of basaltic crust and Indian Ocean sediment. This signature was also distinguishable from enriched and depleted reservoirs in the mantle wedge. The use of elemental B/Be data made these deductions possible despite the fact that ^{10}Be abundances were at baseline, showing this nuclide to be extinct in the analysed lavas.

14.4 Chlorine-36

^{36}Cl is analogous to ^{10}Be in its atmospheric production, in this case by spallation of ^{40}Ar rather than ^{14}N, and like ^{10}Be it is quickly swept from the atmosphere by precipitation. However, unlike ^{10}Be, ^{36}Cl is not

removed from groundwater by adsorption onto particulates, but remains in the aqueous medium as it travels through geological strata. This fact, coupled with its relatively short half-life of 0.301 Myr, makes ^{36}Cl potentially very useful in the dating or tracing of Quaternary groundwater systems. Cosmogenic ^{36}Cl can also be generated in the surfaces of exposed rocks by *in-situ* production. However, this subject will be covered under section 14.6.

The principal obstacle in AMS analysis of ^{36}Cl is isobaric interference by ^{36}S. This forms abundant negative ions and is not removed by the charge-stripping process. It can be resolved by its lower energy loss in the gas counter, but this is most effective at energy levels above 48 MeV, requiring an accelerator of at least 6 MV potential. This rules out ^{36}Cl analysis with lower-energy (2 MV) tandetrons (Wolfli, 1987). A 'time-of-flight' analyser may also be used before the gas counter (Fig. 14.29) in order to eliminate peak tailing from the relatively very large ^{35}Cl and ^{37}Cl ion beams, which are not adequately resolved by the preceding magnetic and electrostatic analysers in the system. Time-of-flight analysis can only be performed on pulsed ion beams, which are

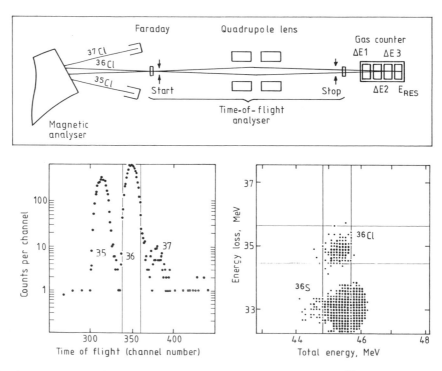

Fig. 14.29. Analyser segment and output data of an AMS instrument designed for ^{36}Cl determination, showing the use of time-of-flight analysis to resolve ^{36}Cl from ^{35}Cl and ^{37}Cl and energy loss detection to resolve from ^{36}S. After Wolfli (1987).

controlled by pulsing the sputter source. The analysis relies on the fact that lighter masses are accelerated to slightly higher velocities than heavier ones, so that after traversing a distance of a metre or so, they arrive at the detector a few nanoseconds earlier. Hence ^{36}Cl is resolved from both ^{36}S, ^{35}Cl and ^{37}Cl (Fig. 14.29).

The first use of ^{36}Cl as a hydrological tracer was not based on the cosmogenic isotope at all, but on anthropogenic bomb-produced ^{36}Cl. This resulted from seven large tests conducted on the sea surface from 1952 to 1958, which caused neutron activation of marine chlorine. Profiles of anthropogenic ^{36}Cl against time were determined in a Greenland ice core (Elmore et al., 1982), in Canadian groundwater (Bentley et al., 1982) and in a soil profile from New Mexico (Phillips et al., 1988). All of these measurements showed a very sharp spike in ^{36}Cl, with a duration of 15–20 years (Fig. 14.30). It is anticipated that in the near future anthropogenic ^{36}Cl will be a powerful hydrological tracer, replacing bomb-produced tritium as the latter becomes extinct.

In contrast to anthropogenic ^{36}Cl, cosmogenic ^{36}Cl is applied to the dating of ancient groundwater, hundreds of kyr in age. For simple sedimentary aquifers this has been quite successful. In a study of the Great Artesian Basin of eastern Australia, Bentley et al. (1986) analysed ^{36}Cl/total Cl ratios in 26 groundwater samples up to 800 km from the recharge area. The relatively simple hydrodynamics of the basin allows quite accurate theoretical calculation of the ages of these waters, which are compared in Fig. 14.31 with $^{36}Cl/Cl$ ages since isolation from the atmosphere. The good correlation observed between the two dating methods is encouraging for further application of the ^{36}Cl method as a groundwater geochronometer.

Slightly more complex systematics were found by Phillips et al. (1986) in a study of the Milk River aquifer in southern Alberta, Canada. ^{36}Cl and total Cl concentrations in the aquifer were very variable, but $^{36}Cl/Cl$ ratios decreased smoothly down the hydrological gradient, yielding maximum residence ages of 2 Myr at the distal end of the aquifer. A much lower hydrodynamic age of 0.5 Myr calculated for this aquifer may be a result of overlooking the effects of Quaternary glacial epochs, when water movement would have been restricted. Alternatively, the ^{36}Cl age may be too old due to mixing with old 'dead' chlorine percolating out of shaley strata which bound the aquifer above and below. ^{129}I data (section 14.5) suggest that the latter process has indeed occurred, but the true age is probably between the two extremes.

The ^{36}Cl method has been more problematical in

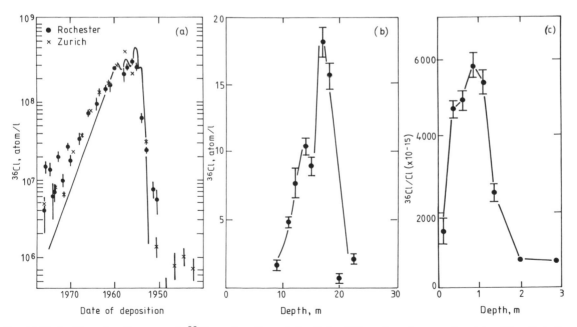

Fig. 14.30. Profiles of anthropogenic ^{36}Cl as a function of depth in different environments. a) Ice (Dye 3 station, central south Greenland); b) groundwater (Borden landfill, Ontario); c) desert soil (New Mexico). After Gove (1987) and Fabryka-Martin et al., (1987).

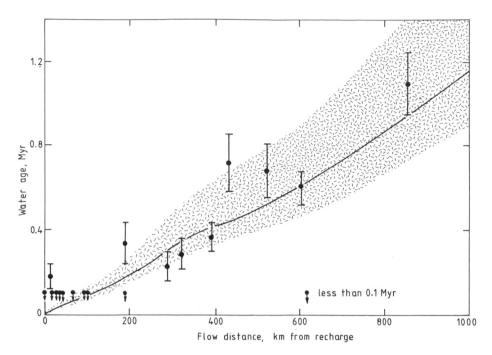

Fig. 14.31. Plot of ^{36}Cl/Cl age against hydrologically calculated age (solid curve) since recharge for groundwater of the Great Artesian Basin, Queensland, Australia. Error bars = ^{36}Cl ages; Stippled area = uncertainty on hydrological age. After Bentley *et al.* (1986).

studying groundwater ages in igneous rocks, due to the complex hydrological relationships between fracture systems, and to interference by local radiogenic ^{36}Cl production. These problems have been evaluated in a case study of the Stripa granite, Sweden, which is intruded into meta-sedimentary country rocks termed leptite. The Stripa granite has unusually high uranium contents of ca. 40 ppm, causing a substantial neutron flux which generates ^{36}Cl by the n,γ reaction on ^{35}Cl. The U content of the leptite, at 5 ppm, also generates significant, if much lower levels of radiogenic ^{36}Cl.

Analysis of Stripa groundwaters (Fig. 14.32) yields values which fall between the ^{36}Cl/Cl ratios predicted for *in-situ* radiogenic isotope production in the two dominant rock types (Andrews *et al.*, 1989). These values are so high that they exceed and swamp normal cosmogenic ^{36}Cl/Cl ratios. Anthropogenic levels could be even higher, but the lack of any tritium signal in the Stripa groundwaters precludes significant involvement. Hence, it is concluded that ^{36}Cl in Stripa groundwater results from *in-situ* radiogenic production in rocks. ^{36}Cl is therefore only a viable dating method for waters in uranium-poor rocks such as sedimentary aquifers.

14.5 Iodine-129

There are over 100 cosmogenic isotopes with masses over 40 and half-lives over one year, which are therefore potentially useful geochemical tracers or dating tools (Henning, 1987). However, most of these elements are metals, and they are not suited to AMS analysis due to the difficulty of forming negative ions. One of the few heavy isotopes to have found significant application is ^{129}I, which is formed in modest abundance in the atmosphere by spallation of Xe, and which, as a non-metal, forms good negative ion beams.

^{129}I analysis by AMS is relatively straightforward, since the only isobaric interference (^{129}Xe) does not form stable negative ions (Elmore *et al.*, 1980). The principal interference is ^{127}I, which at isotope ratios above 10^{12} forms a peak tail that must be removed by time-of-flight analysis in addition to magnetic and electrostatic analysers. The ^{129}I/^{127}I detection limit under these conditions is about 10^{-14}.

As in the case of ^{36}Cl, the ^{129}I tracer has been used to study the entry of anthropogenic material into natural systems. In a study of a marine sediment core from the continental slope off Cape Hatteras (North

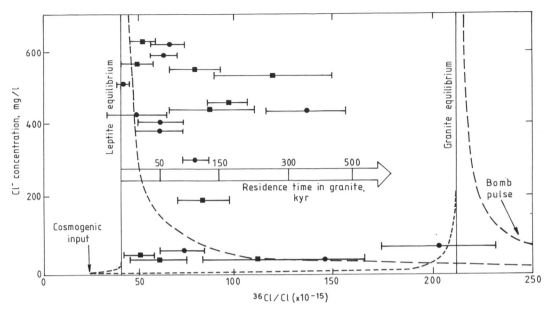

Fig. 14.32. Plot of ^{36}Cl/Cl ratios in Stripa groundwaters against total Cl contents, relative to models for radiogenic, cosmogenic, and anthropogenic production (and mixing between them, shown by dashed lines). After Andrews *et al.* (1989).

Carolina), Fehn *et al.* (1986) found ^{129}I/^{127}I levels at the sediment surface to be two orders of magnitude higher than the relatively constant levels at depth (Fig. 14.33). A zone of intermediate ^{129}I/^{127}I ratios just below the surface was attributed to bioturbation. It would appear that the chemical behaviour of ^{129}I in the ocean system is intermediate between ^{10}Be and ^{36}Cl, in that it is strongly adsorbed onto organic matter, but when this breaks down much of the iodine is lost back to the pore-water. The bulk of such iodine is subsequently re-mixed with seawater, but some is retained in the sediment pile.

^{129}I has a much longer half-life (15.7 Myr) than the other scientifically useful cosmogenic nuclides, so that it is applicable to much older systems. However, its geological applications are complicated by the significant radiogenic iodine generated by *in-situ* uranium fission. This is examined in case studies of the Great Artesian Basin of Australia and the Stripa granite of Sweden (Fabryka-Martin *et al.*, 1985; 1989).

Groundwaters in the Great Artesian Basin range up to ca. 1 Myr in age on the basis of hydrological and ^{36}Cl evidence (above), so that negligible decay of ^{129}I is expected. Therefore, in the absence of contamination by radiogenic iodine or extraneous water sources, ^{129}I/^{127}I ratios should be constant across the basin. Analytical data bear out this

prediction to a reasonable extent (Fig. 14.34), consistent with the very low uranium content of the aquifer rocks and the hydrostatic overpressure of the artesian basin relative to potential contaminant water bodies. Since the near-surface water is itself old relative to human activity, there is no anthropogenic signature. However, excess ^{129}I/^{127}I ratios (above normal cosmogenic levels) were seen in water of ca. 150 and 500 kyr age (Fig. 14.34). Fabryka-Martin *et al.* attributed the latter result to contamination by radiogenic ^{129}I from granitic basement, which forms the floor of the aquifer at its distal end. The cause of the other high value is unknown.

Very different conditions were found in studies of the Stripa granite groundwater (Fabryka-Martin *et al.*, 1989). In this case radiogenic ^{129}I is present at levels two orders of magnitude higher than cosmogenic iodine. The ^{129}I systematics at Stripa can be seen most clearly when plotted against ^{36}Cl/Cl (Fig. 14.35). Except for one shallow water sample with a prominent anthropogenic ^{36}Cl signature, the data form an array trending from estimated meteoric recharge towards a pure radiogenic component. This array could result from mixing between two end-members, but it could also result from variable, but correlated, production of radiogenic ^{129}I and ^{36}Cl in the granite, since both are controlled by the uranium content of the rock. Hence, the main role for ^{129}I is to

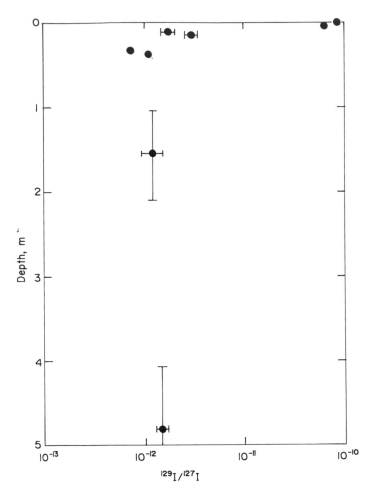

Fig. 14.33. Plot of ^{129}I/^{127}I ratios as a function of depth for a sediment core from 1 km water depth off Cape Hatteras, North Carolina, showing anthropogenic iodine within the top 10 cm of section. After Fehn *et al.* (1986).

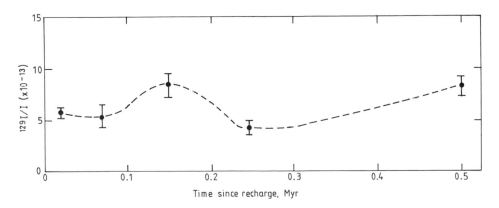

Fig. 14.34. Plot of ^{129}I/^{127}I against estimated age of waters (since recharge) in the Great Artesian Basin of Australia. After Fabryka-Martin *et al.* (1985).

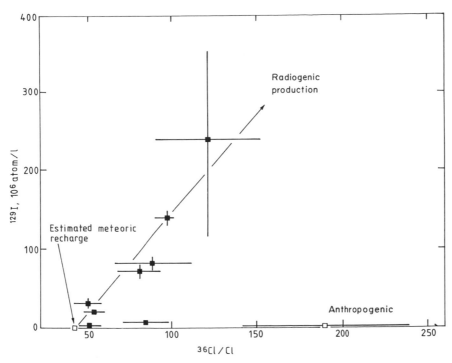

Fig. 14.35. Plot of absolute ^{129}I abundance against the ^{36}Cl/total Cl ratio for groundwaters from depths up to 1200 m in the Stripa mine, Sweden. Cosmogenic, radiogenic and anthropogenic signatures are shown. After Fabryka-Martin *et al.* (1989).

gauge *in-situ* radiogenic perturbation of ^{36}Cl ages in groundwater systems.

14.6 Aluminium-26

14.6.1 Meteorite exposure ages

Atmospheric production of ^{26}Al is much lower than ^{14}C or ^{10}Be because the progenitor, ^{40}Ar, constitutes only about 1% of atmospheric gases. *In-situ* production at the Earth's surface is also low, due to atmospheric attenuation of cosmic rays. Consequently, the first studies of cosmogenic ^{26}Al were in studies of the Moon and meteorites. After the break-up of their parent bodies, meteorite fragments are exposed to intense cosmic-ray bombardment during their travel through space, causing substantial ^{26}Al production. If these fragments have cosmic exposure ages of at least a few Myr, then their surfaces will reach saturation in ^{26}Al production ($t_{1/2} = 0.7$ Myr).

After falling to Earth, atmospheric shielding protects meteorite fragments from significant further ^{26}Al production, and decay of this inventory can then be used to determine a terrestrial residence age.

Because the abundances of ^{26}Al in meteorites are comparatively high, they do not demand accelerator mass spectrometry. Instead, they are measured using non-destructive γ counting, by putting the whole meteorite fragment in a large shielded detector. Attenuation of γ particles by the sample itself is corrected by empirical modelling (e.g. Evans *et al.*, 1979).

In the first large-scale survey of Antarctic meteorites, Evans *et al.* (1979) compared the ^{26}Al activities in these samples with those in 'falls' (observed at impact), with a zero terrestrial residence. The falls had a moderately well-defined range of ^{26}Al activities for a given compositional class of meteorites (Fig. 14.36). The outlier from this class can be attributed to a failure to reach ^{26}Al saturation, due to the short cosmic-ray exposure history of the fragment. In contrast, Antarctic meteorites ranged to substantially lower activities, indicating significant terrestrial residence ages in several cases. Unfortunately, these values are only semi-quantitative, due the relatively long half-life of ^{26}Al, and due to uncertainties in production. The latter problem arises because of the poor penetrative capacity of the low-energy cosmic

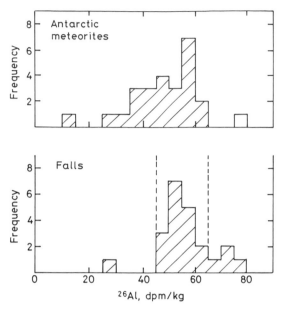

Fig. 14.36. Histogram of ^{26}Al activities in Antarctic meteorites, compared with American 'falls'. After Evans et al. (1979).

rays which generate ^{26}Al, making the cosmic production rate susceptible to depth within the fragment.

Although ^{26}Al represents a good reconnaissance tool for terrestrial age determination (e.g. Evans and Reeves, 1987), ^{36}Cl provides a more precise method (Nishiizumi et al., 1979). This arises from its shorter (0.3 Myr) half-life, and from more accurately known saturation values, due to its generation by penetrative high-energy cosmic rays. However, the analysis is more technically demanding, and must be performed by accelerator mass spectrometry. Results of a large ^{36}Cl study of Antarctic meteorite ages are presented in Fig. 14.37 (Nishiizumi et al., 1989a).

The high quality of terrestrial ^{36}Cl ages for Antarctic meteorites has led to their use to determine the half-life of another cosmogenic nuclide, ^{41}Ca. With a half-life of only 0.1 Myr, this shows great promise as a precise dating tool, but an efficient AMS analysis method has only recently been developed (Fink et al. 1990). Another problem limiting the application of ^{41}Ca has been uncertainty in the half-life. To solve this problem, Klein et al. (1991)

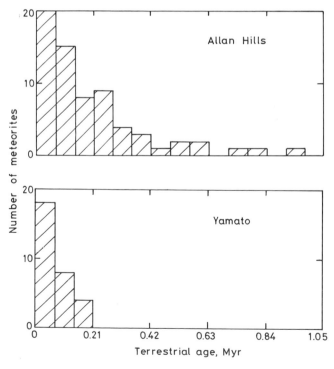

Fig. 14.37. Histograms of calculated terrestrial residence age for Antarctic meteorites from the Allan Hills and Yamato Mountains areas, based on the decay of cosmogenic ^{36}Cl. After Nishiizumi et al. (1989a).

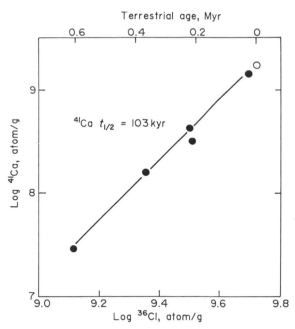

Fig. 14.38. Plot of ^{41}Ca *versus* ^{36}Cl activity in Antarctic meteorites (●) and a 'fall' (○), yielding a 'geological' half-life determination for ^{41}Ca. Calculated ^{36}Cl ages are also shown. After Klein *et al.* (1991).

performed ^{41}Ca analyses on aliquots of Antarctic iron meteorites which had already been dated by ^{36}Cl. The results (Fig. 14.38) reveal a strong linear correlation between abundance of the two species, whose slope corresponds to the ratio of the half-lives. Taking a ^{36}Cl half-life of 301 ± 4 kyr yields a precise ^{41}Ca half-life of 103 ± 7 kyr. This nuclide promises to be a major dating tool in the future.

14.6.2 Terrestrial exposure ages

Because of atmospheric attenuation of cosmic rays, most terrestrial materials have $^{26}Al/^{27}Al$ ratios less than 10^{-14}. However, in some aluminium-poor minerals such a quartz, the content of (non-cosmogenic) ^{27}Al may be as low as few ppm, so that after a few thousand years exposure to cosmic rays, $^{26}Al/^{27}Al$ ratios of 10^{-11} to 10^{-13} may be generated, within the measurement capability of AMS. These data can then be used to calculate exposure ages of terrestrial rock surfaces.

The principal obstacle to AMS analysis of ^{26}Al is the formation of sufficient negative Al ions during sputtering, which is only about 25% efficient (Middleton and Klein, 1987). The metal species

must be used rather than the oxide because the latter suffers a severe interference from MgO. However, Mg does not form negative metal ions at all, so there are no isobaric interferences on the metal-ion signal.

The principal application of ^{26}Al as a geochronometer is in the measurement of rock exposure ages. It would be possible to use this nuclide alone for this purpose, but in view of the many possible permutations of exposure and erosion history, the use of two nuclides with different half-lives provides a more powerful constraint on these models. The normal choice is to combine ^{26}Al measurements ($t_{1/2} = 0.705$ Myr) with ^{10}Be ($t_{1/2} = 1.51$ Myr).

The atmospheric $^{26}Al/^{10}Be$ production ratio has been determined as about 4×10^{-3} by sampling from high-flying aircraft, but the *in-situ* production ratio in quartz has been measured as 6 (Nishiizumi *et al.*, 1989b). Because the atmospheric ^{10}Be production rate is comparatively high, great care must be taken to ensure that rock samples to be used for exposure dating are not contaminated by the atmospheric or so-called 'garden variety' of ^{10}Be (Nishiizumi *et al.*, 1986). Because of its resistance to chemical weathering, quartz is comparatively resistant to contamination by garden variety ^{10}Be. This, along with its low ^{27}Al content makes it an excellent material for exposure dating. In quartz, ^{10}Be is derived from spallation of ^{16}O, while ^{26}Al is produced by spallation of ^{28}Si and mu-meson capture by the same species. Most of this production occurs in the top half-metre of the rock surface, but limited ^{26}Al production can occur at depths of up to 10 m (Middleton and Klein, 1987).

There are two limiting models for the interpretation of surface exposure data (e.g. Nishiizumi *et al.*, 1991a). These are illustrated on a plot of $^{26}Al/^{10}Be$ ratio against absolute count rate (Fig. 14.39). The upper curve shows the effect of increasing exposure age, for the case where erosion rate is zero. The lower curve shows the effect of different steady-state erosion rates, for the case where exposure age is infinite (relative to half-life). At the point of saturation (after a few half-lives), the $^{26}Al/^{10}Be$ ratio is 2.88. The relatively small separation between the curves for the two end-member models is due to the limited (factor of two) difference between the half-lives of ^{26}Al and ^{10}Be.

Application of the Al–Be exposure method is demonstrated in Fig. 14.39 using data from nunataks of the Allan Hills area of Antarctica (Nishiizumi *et al.*, 1991a). The results display a range of Al/Be ratios close to the steady-state erosion curve. However, the zero-erosion-rate (variable exposure age) model

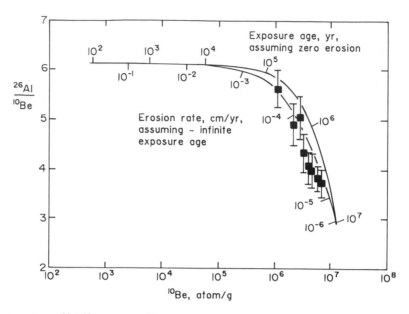

Fig. 14.39. Plot of analysed $^{26}Al/^{10}Be$ against ^{10}Be abundance (corrected to production at sea-level) for Allan Hills quartz samples, used as a measure of minimum exposure age and/or maximum erosion rate. After Nishiizumi *et al.* (1991a).

cannot be ruled out. The lowest $^{26}Al/^{10}Be$ yields the strongest constraint, representing a minimum exposure age of 1.4 Myr or a maximum erosion rate of 0.24 mm/kyr. Samples to the left of the erosion line in Fig. 14.39 may be explained by burial under ice for some period in the past. During times of burial, points move downwards to the left, due to the greater rate of decay of ^{26}Al relative to ^{10}Be.

Another interesting example of the ^{26}Al–^{10}Be method is provided by surface-exposure dating of rocks at Meteor Crater, Arizona (Nishiizumi *et al.*, 1991b). Samples were taken from the upper few cm of large blocks in the ejecta blanket of the impact. The lithology of these blocks shows that they were from strata buried to more than 10 m before the impact. On the other hand, their large size suggests that they were unroofed of any overlying ash blanket soon after the impact event. ^{26}Al and ^{10}Be abundances were calibrated against the empirical production rates of Nishiizumi *et al.* (1989b), and yield a consensus of ages around 50 kyr, along with a few younger ages (e.g. Monument rock) which may indicate more recent exposure above the ash blanket. These ages are also supported by ^{36}Cl exposure ages on the same samples (Fig. 14.40; Phillips *et al.*, 1991).

Technically, the 50 kyr age shown in Fig. 14.40. represents a minimum exposure age, assuming zero erosion. However, sampled surfaces were found to be

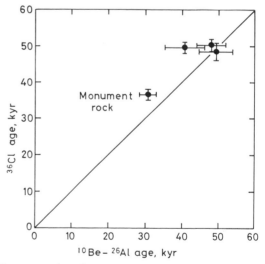

Fig. 14.40. Plot of minimum exposure ages for selected Meteor Crater samples showing concordancy between results from different cosmogenic isotopes. After Nishiizumi *et al.* (1991b).

coated with 'rock varnish', which takes thousands of years to develop. Therefore, erosion was probably negligible in the arid climate of Arizona, and the measured ages are probably close to the time of impact. Recent work with the stable rare gases suggests that cosmogenic production of neon may also represent a useful tool for surface exposure dating when combined with Al and Be (section 11.4).

References

Anderson, E. C. and Libby, W. F. (1951). World-wide distribution of natural radiocarbon. *Phys. Rev.* **81**, 64–9.

Andrews, J. N., Davis, S. N., Fabryka-Martin, J., Fontes, J. C., Lehmann, B. E., Loosli, H. H., Michelot, J-L., Moser, H., Smith, B. and Wolf, M. (1989). The *in-situ* production of radioisotopes in rock matrices with particular reference to the Stripa granite. *Geochim. Cosmochim. Acta* **53**, 1803–15.

Arnold, J. R. (1956). Beryllium-10 produced by cosmic rays. *Science* **124**, 584–5.

Arnold, J. R. (1958). Trace elements and transport rates in the ocean. *2nd UN Conf. on Peaceful Uses of Atomic Energy* **18**, 344–6.

Arnold, J. R. and Libby, W. F. (1949). Age determinations by radiocarbon content: Checks with samples of known age. *Science* **110**, 678–80.

Bard, E., Arnold, M., Fairbanks, R. G. and Hamelin, B. (1993). ^{230}Th–^{234}U and ^{14}C ages obtained by mass spectrometry on corals. *Radiocarbon* **35**, 191–9.

Bard, E., Hamelin, B., Fairbanks, R. G. and Zindler, A. (1990a). Calibration of the ^{14}C timescale over the past 30,000 years using mass spectrometric U–Th ages from Barbados corals. *Nature* **345**, 405–10.

Bard, E., Hamelin, B., Fairbanks, R. G., Zindler, A., Mathieu, G. and Arnold, M. (1990b). U/Th and ^{14}C ages of corals from Barbados and their use for calibrating the ^{14}C timescale beyond 9000 years B.P. *Nucl. Instr. Meth. in Phys. Res. B* **52**, 461–8.

Baxter, M. S. and Farmer, J. G. (1973). Radiocarbon: short-term variations. *Earth Planet. Sci. Lett.* **20**, 295–9.

Beer, J., Andree, M., Oeschger, H., Siegenthaler, U., Bonani, G., Hofmann, H., Morenzoni, E., Nessi, M., Suter, M., Wolfli, W., Finkel, R. and Langway, C. (1984). The Camp Century ^{10}Be record: implications for long-term variations of the geomagnetic dipole moment. *Nucl. Instr. Meth. in Phys. Res. B* **5**, 380–4.

Beer, J., Oeschger, H., Finkel, R. C., Castagnoli, G., Bonino, G., Attolini, M. R. and Galli, M. (1985). Accelerator measurements of ^{10}Be: the 11 year solar cycle. *Nucl. Instr. Meth. in Phys. Res. B* **10**, 415–18.

Beer, J., Siegenthaler, U., Bonani, G., Finkel, R. C., Oeschger, H., Suter, M. and Wolfli, W. (1988). Information on past solar activity and geomagnetism from ^{10}Be in the Camp Century ice core. *Nature* **331**, 675–9.

Bentley, H. W., Phillips, F. M., Davis, S. N., Gifford, S., Elmore, D., Tubbs, L. E. and Gove, H. E. (1982). Thermonuclear ^{36}Cl pulse in natural water. *Nature* **300**, 737–40.

Bentley, H. W., Phillips, F. M., Davis, S. N., Habermehl, M. A., Airey, P. L., Calf, G. E., Elmore, D., Gove, H. E. and Torgerson, T. (1986). Chlorine 36 dating of very old groundwaters, 1, The Great Artesian Basin, Australia. *Water Resour. Res.* **22**, 1991–2002.

Bronk, C. R. and Hedges, R. E. M. (1987). A gas source for radiocarbon dating. *Nucl. Instr. Meth. in Phys. Res. B* **29**, 45–9.

Brown, E. T., Edmond, J. M., Raisbeck, G. M., Bourles, D. L., Yiou, F. and Measures, C. I. (1992). Beryllium isotope geochemistry in tropical basins. *Geochim. Cosmochim. Acta* **56**, 1607–24.

Brown, L., Klein, J. and Middleton, R. (1985). Anomalous isotopic concentrations in the sea off Southern California. *Geochim. Cosmochim. Acta* **49**, 153–7.

Brown, L., Klein, J., Middleton, R., Sacks, I. S. and Tera, F. (1982). ^{10}Be in island-arc volcanoes and implications for subduction. *Nature* **299**, 718–20.

Brown, L., Stensland, G. J., Klein, J. and Middleton, R. (1989). Atmospheric deposition of ^{7}Be and ^{10}Be. *Geochim. Cosmochim. Acta* **53**, 135–42.

Bruns, M., Rhein, M., Linick, T. W. and Suess, H. E. (1983). The atmospheric ^{14}C level in the 7th millennium BC. *P.A.C.T. (Physical And Chemical Techniques in Archaeology)* **8**, 511–16.

Bucha, V. and Neustupny, E. (1967). Changes in the Earth's magnetic field and radiocarbon dating. *Nature* **215**, 261–3.

Chengde, S., Beer, J., Tungsheng, L., Oeschger, H., Bonani, G., Suter, M. and Wolfli, W. (1992). ^{10}Be in Chinese loess. *Earth Planet. Sci. Lett.* **109**, 169–77.

Craig, H. (1954). Carbon-13 in plants and the relationships between carbon-13 and carbon-14 variations in nature. *J. Geol.* **62**, 115–49.

Craig, H. (1957). Isotopic standards for carbon and oxygen and correction factors for mass-spectrometric analysis of carbon dioxide. *Geochim. Cosmochim. Acta* **12**, 133–49.

Damon, P. E. and 20 others (1989). Radiocarbon dating of the Shroud of Turin. *Nature* **337**, 611–15.

de Jong, A. F. M., Mook, W. G. and Becker, B. (1979). Confirmation of the Suess wiggles: 3200–3700 BC. *Nature* **280**, 48–9.

de Vries, H. (1958). Variation in concentration of radiocarbon with time and location on Earth. *Proc. Konikl. Ned. Akad. Wetenschap B* **61**, 94–102.

de Vries, H. and Barendsen, G. W. (1953). Radiocarbon dating by a proportional counter filled with carbon dioxide. *Physica* **19**, 987–1003.

Eddy, J. A. (1976). The Maunder minimum. *Science* **192**, 1189–202.

Edwards, R. L., Beck, J. W., Burr, G. S., Donahue, D. J., Chappell, A. L., Bloom, A. L., Druffel, E. R. M. and Taylor, F. W. (1993a). A large drop in atmospheric $^{14}C/^{12}C$ and reduced melting in the Younger Dryas, documented with ^{230}Th ages of corals. *Science* **260**, 962–8.

Edwards, C. M. H., Morris, J. D. and Thirlwall, M. F. (1993b). Separating mantle from slab signatures in arc lavas using B/Be and radiogenic isotope systematics. *Nature* **362**, 530–3.

Elmore, D., Conard, N. J., Kubik, P. W., Gove, H. E., Wahlen, M., Beer, J. and Suter, M. (1987). ^{36}Cl and ^{10}Be profiles in Greenland ice: dating and production rate variations. *Nucl. Instr. Meth. in Phys. Res. B* **29**, 207–10.

Elmore, D. and 11 others. (1980). Determination of ^{129}I using tandem accelerator mass spectrometry. *Nature* **286**, 138–40.

Elmore, D., Tubbs, L. E., Newman, D., Ma, X. Z., Finkel, R., Nishiizumi, K., Beer, J., Oeschger, H. and Andree, M. (1982). ^{36}Cl bomb pulse measured in a shallow ice core from Dye 3, Greenland. *Nature* **300**, 735–7.

Elsasser, W., Ney, E. P. and Winckler, J. R. (1956). Cosmic-ray intensity and geomagnetism. *Nature* **178**, 1226–7.

Evans, J. C., Rancitelli, L. A. and Reeves, J. H. (1979). ^{26}Al content of Antarctic meteorites: implications for terrestrial ages and bombardment history. *Proc. 10th Lunar Planet. Sci. Conf.* 1061–72.

Evans, J. C. and Reeves, J. H. (1987). ^{26}Al survey of Antarctic meteorites. *Earth Planet. Sci. Lett.* **82**, 223–30.

Fabryka-Martin, J., Bentley, H., Elmore, D. and Airey, P. L. (1985). Natural iodine-129 as an environmental tracer. *Geochim. Cosmochim. Acta* **49**, 337–47.

Fabryka-Martin, J., Davis, S. N. and Elmore, D. (1987). Applications of ^{129}I and ^{36}Cl in hydrology. *Nucl. Instr. Meth. in Phys. Res. B* **29**, 361–71.

Fabryka-Martin, J., Davis, S. N., Elmore, D. and Kubik, P. W. (1989). In-situ production and migration of ^{129}I in the Stripa granite, Sweden. *Geochim. Cosmochim. Acta* **53**, 1817–23.

Fan, C. Y., Chen, T. M., Yun, S. X. and Dai, K. M. (1986). Radiocarbon activity variation in dated tree rings grown in Mackenzie delta. *Radiocarbon* **28**, 300–5.

Fehn, U., Holdren, G. R., Elmore, D., Brunelle, T., Teng, R. and Kubik, P. W. (1986). Determination of natural and anthropogenic ^{129}I in marine sediments. *Geophys. Res. Lett.* **13**, 137–9.

Ferguson, C. W. (1970). Dendrochronology of Bristlecone pine, *Pinus aristata*. Establishment of a 7484-year chronology in the White Mountains of eastern-central California, USA. In: I. U. Olsson (Ed.), *Radiocarbon Variations and Absolute Chronology, Proc. 12th Nobel Symp.* Wiley, pp. 571–93.

Ferguson, C. W. and Graybill, D. A. (1983). Dendrochronology of Bristlecone pine: a progress report. *Radiocarbon* **25**, 287–8.

Fink, D., Middleton, R., Klein, J. and Sharma, P. (1990). ^{41}Ca measurement by accelerator mass spectrometry and applications. *Nucl. Instr. Meth. in Phys. Res. B* **47**, 79–96.

Forbush, S. E. (1954). Worldwide cosmic-ray variations, 1937–1952. *J. Geophys. Res.* **59**, 525–42.

Godwin, H. (1962). Half-life of radiocarbon. *Nature* **195**, 984.

Goldberg, E. D. and Arrhenius, G. O. S. (1958). Chemistry of Pacific pelagic sediments. *Geochim. Cosmochim. Acta* **13**, 153–212.

Gove, H. E. (1987). Tandem-accelerator mass-spectrometry measurements of ^{36}Cl, ^{129}I and osmium isotopes in diverse natural samples. *Phil. Trans. Roy. Soc. Lond. A* **323**, 103–19.

Hammer, C. U., Clausen, H. B. and Tauber, H. (1986). Ice-core dating of the Pleistocene/Holocene boundary applied to a calibration of the ^{14}C time scale. *Radiocarbon* **28**, 284–91.

Henning, W. (1987). Accelerator mass spectrometry of heavy elements: ^{36}Cl to ^{205}Pb. *Phil. Trans. Roy. Soc. Lond. A* **323**, 87–99.

Hillam, J., Groves, C. M., Brown, D. M., Baillie, M. G. L., Coles, J. M. and Coles, B. J. (1990). Dendrochronology of the English Neolithic. *Antiquity* **64**, 210–20.

Hofmann, H. J., Beer, J., Bonani, G., Von Gunten, H. R., Raman, S., Suter, M., Walker, R. L., Wolfli, W. and Zimmermann, D. (1987). ^{10}Be half-life and AMS-standards. *Nucl. Instr. Meth. in Phys. Res. B* **29**, 32–6.

Huber, B. (1970). Dendrochronology of central Europe. In: I. U. Olsson (Ed.), *Radiocarbon Variations and Absolute Chronology, Proc. 12th Nobel Symp.* Wiley, pp. 233–5.

Kamen, M. D. (1963). Early history of carbon-14. *Science* **140**, 584–90.

Kieser, W. E., Beukens, R. P., Kilius, L. R., Lee, H. W. and Litherland, A. E. (1986). Isotrace radiocarbon analysis–equipment and procedures. *Nucl. Instr. Meth. in Phys. Res. B* **15**, 718–21.

Klein, J., Fink, D., Middleton, R., Nishiizumi, K. and Arnold, J. (1991). Determination of the half-life of ^{41}Ca from measurements of Antarctic meteorites. *Earth Planet. Sci. Lett.* **103**, 79-83.

Kusakabe, M., Ku, T. L., Southon, J. R., Vogel, J. S., Nelson, D. E., Measures, C. I. and Nozaki, Y. (1987). The distribution of ^{10}Be and ^{9}Be in ocean water. *Nucl. Instr. Meth. in Phys. Res. B* **29**, 306–10.

Labrie, D. and Reid, J. (1981). Radiocarbon dating by infrared laser spectroscopy. *Appl. Phys.* **24**, 381–6.

Lal, D. (1988). In-situ-produced cosmogenic isotopes in terrestrial rocks. *Ann. Rev. Earth Planet. Sci.* **16**, 355–88.

Lal, D. and Peters, B. (1967). Cosmic-ray produced radioactivity on the Earth. In: *Handbook of Physics.* **46/2**. Springer, pp. 551–612.

Lee, H. W., Galindo-Uribarri, A., Chng, K. H., Kilius, L. R. and Litherland, A. E. (1984). The ^{12}CH$_2^{+2}$ molecule and radiocarbon dating by accelerator mass spectrometry. *Nucl. Instrum. Meth. in Phys. Res. B* **5**, 208–10.

Libby, W. F. (1952). *Radiocarbon Dating.* University of Chicago Press, 124 p.

Libby, W. F. (1970). Ruminations on radiocarbon dating. In: I. U. Olsson (Ed.), *Radiocarbon Variations and Absolute Chronology, Proc. 12th Nobel Symp.* Wiley, pp. 629–40.

Litherland, A. E. (1980). Ultra-sensitive mass spectrometry with accelerators. *Ann. Rev. Nucl. Part. Sci.* **30**, 437–73.

Litherland, A. E. (1987). Fundamentals of accelerator mass spectrometry. *Phil. Trans. Roy. Soc. Lond. A* **323**, 5–21.

Lundberg, L., Ticich, T., Herzog, G. F., Hughes, T., Ashley, G., Moniot, R. K., Tuniz, C., Kruse, T. and Savin, W. (1983). ^{10}Be and Be in the Maurice River – Union Lake system of Southern New Jersey. *J. Geophys. Res.* **88**, 4498–504.

Mangini, A., Segl, M., Bonani, G., Hofmann, H. J., Morenzoni, E., Nessi, M., Suter, M., Wolfli, W. and Turekian, K. K. (1984). Mass-spectrometric ^{10}Be dating of deep-sea sediments applying the Zurich tandem accelerator. *Nucl. Instr. Meth. in Phys. Res. B* **5**, 353–8.

Mazaud, A., Laj, C., Bard, E., Arnold, M. and Tric, E. (1991). Geomagnetic field control of ^{14}C production over the last 80 kyr: implications for the radiocarbon time-scale. *Geophys Res. Lett.* **18**, 1885–8.

McCorkell, R., Fireman, E. L. and Langway, C. C. (1967). Aluminium-26 and Beryllium-10 in Greenland Ice. *Science* **158**, 1690–2.

Measures, C. I. and Edmond, J. M. (1982). Beryllium in the water column of the central North Pacific. *Nature* **297**, 513.

Merrill, J. R., Lyden, E. F. X., Honda, M. and Arnold, J. R. (1960). The sedimentary geochemistry of the beryllium isotopes. *Geochim. Cosmochim. Acta* **18**, 108–29.

Middleton, R. and Klein, J. (1987). ^{26}Al: measurement and applications. *Phil. Trans. Roy. Soc. Lond. A* **323**, 121–43.

Monaghan, M. C., Klein, J. and Measures, C. I. (1988). The origin of ^{10}Be in island-arc volcanic rocks. *Earth Planet. Sci. Lett.* **89**, 288–98.

Mook, W. G. (1986). Business meeting: Recommendations/resolutions adopted by the Twelfth International Radiocarbon conference. *Radiocarbon* **28**, 799.

Mook, W. G. and Streurman, H. J. (1983). Physical and chemical aspects of radiocarbon dating. *P.A.C.T.* (Physical And Chemical Techniques in Archaeology) **8**, 31–55.

Morris, J. D., Leeman, W. P. and Tera, F. (1990). The subducted component in island arc lavas: constraints from Be isotopes and B–Be systematics. *Nature* **344**, 31–6.

Morris, J. D. and Tera, F. (1989). ^{10}Be and ^{9}Be in mineral separates and whole-rocks from island arcs: implications for sediment subduction. *Geochim. Cosmochim. Acta* **53**, 3197–206.

Nishiizumi, K., Arnold, J. R., Elmore, D., Ferraro, R. D., Gove, H. E., Finkel, R. C., Beukens, R. P., Chang, K. H. and Kilius, L. R. (1979). Measurements of ^{36}Cl in Antarctic meteorites and Antarctic ice using a van de Graaff accelerator. *Earth Planet. Sci. Lett.* **45**, 285–92.

Nishiizumi, K., Elmore, D. and Kubik, P. W. (1989a). Update on terrestrial ages of Antarctic meteorites. *Earth Planet. Sci. Lett.* **93**, 299–313.

Nishiizumi, K., Kohl, C. P., Arnold, J. R., Klein, J., Fink, D. and Middleton, R. (1991a). Cosmic-ray produced ^{10}Be and ^{26}Al in Antarctic rocks: exposure and erosion history. *Earth Planet. Sci. Lett.* **104**, 440–54.

Nishiizumi, K., Kohl, C. P., Shoemaker, J. R., Arnold, J. R., Klein, J., Fink, D. and Middleton, R. (1991b). In-situ ^{10}Be and ^{26}Al exposure ages at Meteor Crater, Arizona. *Geochim. Cosmochim. Acta* **55**, 2699–703.

Nishiizumi, K., Lal, D., Klein, J., Middleton, R. and Arnold, J. R. (1986). Production of ^{10}Be and ^{26}Al by cosmic rays in terrestrial quartz *in-situ* and implications for erosion rates. *Nature* **319**, 134–6.

Nishiizumi, K., Winterer, E. L., Kohl, C. P., Klein, J., Middleton, R., Lal, D. and Arnold, J. R. (1989b). Cosmic ray production rates of ^{10}Be and ^{26}Al in quartz from glacially polished rocks. *J. Geophys. Res.* **94**, 17,907–15.

Oeschger, H., Houtermans, J., Loosli, H. and Wahlen, M. (1970). The constancy of cosmic radiation from isotope studies in meteorites and on the Earth. In: Olsson, I. U. (Ed.) *Radiocarbon Variations and Absolute Chronology, Proc. 12th Nobel Symp.* Wiley, pp. 471–98.

Olsson, I. U. (1980). ^{14}C in extractives from wood. *Radiocarbon* **22**, 515–24.

Olsson, I. U., El-Daoushy, M. F. A. F., Abdel-Mageed, A. I. and Klasson, M. (1974). A comparison of different methods for pretreatment of bones. *Geol. Foren. Stockh. Forhandl.* **96**, 171–81.

Pavich, M. J., Brown, L., Valette-Silver, J. N., Klein, J. and Middleton, R. (1985). ^{10}Be analysis of a Quaternary weathering profile in the Virginia Piedmont. *Geology* **13**, 39–41.

Pearson, G. W., Pilcher, J. R., Baillie, M. G. L. and Hillam, J. (1977). Absolute radiocarbon dating using a low altitude European tree-ring calibration. *Nature* **270**, 25–8.

Phillips, F. M., Bentley, H. W., Davis, S. N., Elmore, D. and Swanick, G. B. (1986). Chlorine 36 dating of very old groundwater 2. Milk River aquifer, Alberta, Canada. *Water Resour. Res.* **22**, 2003–16.

Phillips, F. M., Mattick, J. L. and Duval, T. A. (1988). Chlorine 36 and tritium from nuclear weapons fallout as tracers for long-term liquid and vapour movement in desert soils. *Water Resour. Res.* **24**, 1877–91.

Phillips, F. M., Zreda, M. G., Smith, S. S., Elmore, D., Kubic, P. W., Dorn, R. I. and Roddy, D. J. (1991). Age and geomorphic history of Meteor Crater, Arizona, from cosmogenic ^{36}Cl and ^{14}C in rock varnish. *Geochim. Cosmochim. Acta* **55**, 2695–8.

Purser, K. H., Liebert, R. B., Litherland, A. E., Beukens, R. P., Gove, H. E., Bennett, C. L., Clover, M. R. and Sondheim, W. E. (1977). An attempt to detect stable N-ions from a sputter ion source and some implications of the results for the design of tandems for ultra-sensitive carbon analysis. *Rev. Phys. Appl.* **12**, 1487–92.

Raisbeck, G. M., Yiou, F., Fruneau, M., Lieuvin, M. and Loiseaux, J. M. (1978). Measurements of ^{10}Be in 1,000- and 5,000-year-old Antarctic ice. *Nature* **275**, 731–3.

Raisbeck, G. M., Yiou, F., Fruneau, M., Loiseaux, J. M., Lieuvin, M. and Ravel, J. C. (1979). Deposition rate and seasonal variations in precipitation of cosmogenic ^{10}Be. *Nature* **282**, 279–80.

Raisbeck, G. M., Yiou, F., Fruneau, M., Loiseaux, J. M., Lieuvin, M., Ravel, J. C. and Lorius, C. (1981a). Cosmogenic ^{10}Be concentrations in Antarctic ice during the past 30,000 years. *Nature* **292**, 825–6.

Raisbeck, G. M., Yiou, F., Fruneau, M., Loiseaux, J. M., Lieuvin, M., Ravel, J. C., Reyss, J. M. and Guichard, F. (1980). ^{10}Be concentration and residence time in the deep ocean. *Earth Planet. Sci. Lett.* **51**, 275–8.

Raisbeck, G. M., Yiou, F., Lieuvin, M., Ravel, J. C., Fruneau, M. and Loiseaux, J. M. (1981b). ^{10}Be in the environment: some recent results and their application. *Proc. Symp. Accel. Mass Spectrom.* Argonne Natl. Lab., pp. 228–43.

Ralph, E. K. (1971). Carbon-14 dating. In: Michael, H. N. and Ralph, E. K. (Eds.) *Dating Techniques for the Archaeologist.* M.I.T. Press, pp. 1–48.

Ralph, E. K. and Michael, H. N. (1974). Twenty-five years of radiocarbon dating. *Amer. Scient.* **62**, 553–60.

Reyss, J. L., Yokoyama, Y. and Guichard, F. (1981). Production cross sections of ^{26}Al, ^{22}Na, ^{7}Be from argon and of ^{10}Be, ^{7}Be from nitrogen: implications for the production rates of ^{26}Al and ^{10}Be in the atmosphere. *Earth Planet. Sci. Lett.* **53**, 203–10.

Sigmarsson, O., Condomines, M., Morris, J. D. and Harmon, R. S. (1990). Uranium and ^{10}Be enrichments by fluids in Andean arc magmas. *Nature* **346**, 163–5.

Simpson, J. A. (1951). Neutrons produced in the atmosphere by cosmic radiations. *Phys. Rev.* **83**, 1175–88.

Stensland, G. J., Brown, L., Klein, J. and Middleton, R. (1983). Beryllium-10 in rain. *EOS* **64**, 283 (abstract).

Stuiver, M. (1961). Variations in radiocarbon concentration and sunspot activity. *J. Geophys. Res.* **66**, 273–6.

Stuiver, M. (1965). Carbon-14 content of 18th- and 19th-century wood: variations correlated with sunspot activity. *Science* **149**, 533–4.

Stuiver, M., Braziunas, T. F., Becker, B. and Kromer, B. (1991). Climatic, solar, oceanic and geomagnetic influences on late-glacial and Holocene atmospheric ^{14}C/^{12}C change. *Quat. Res.* **35**, 1–24.

Stuiver, M. and Pearson, G. W. (1986). High-precision radiocarbon time-scale calibration from the present to 500 BC. In: Stuiver, M. and Kra. R. S. (Eds), *Calibration Issue, 12th International Radiocarbon Conference. Radiocarbon* **28**, 805–38.

Stuiver, M. and Quay, P. D. (1981). Atmospheric ^{14}C changes resulting from fossil fuel CO_2 release and cosmic-ray flux variability. *Earth Planet. Sci. Lett.* **53**, 349–62.

Suess, H. E. (1955). Radiocarbon concentrations in modern wood. *Science* **122**, 415–7.

Suess, H. E. (1965). Secular variations of the cosmic-ray-produced carbon 14 in the atmosphere and their interpretations. *J. Geophys. Res.* **70**, 5937–52.

Suess, H. E. (1970). Bristlecone-pine calibration of the radiocarbon time- scale 5200 B. C. to the present. In: I. U. Olsson (Ed.), *Radiocarbon Variations and Absolute Chronology, Proc. 12th Nobel Symp.* Wiley, pp. 303–11.

Suess, H. E. and Strahm, C. (1970). The Neolithic of Auvernier, Switzerland. *Antiquity* **44**, 91–9.

Tanaka, S. and Inoue, T. (1979). ^{10}Be dating of North Pacific sediment cores up to 2.5 million years B.P. *Earth Planet. Sci. Lett.* **45**, 181–7.

Tauber, H. (1970). The Scandinavian varve chronology and C14 dating. In: Olsson, I. U. (Ed.) *Radiocarbon Variations and Absolute Chronology, Proc. 12th Nobel Symp.* Wiley, pp. 173–96.

Taylor, R. E. (1987). *Radiocarbon Dating, an Archaeological Perspective.* Academic Press, 212 p.

Tera, F., Brown, L., Morris, J., Sacks, I. S., Klein, J. and Middleton, R. (1986). Sediment incorporation in island-arc magmas: inferences from ^{10}Be. *Geochim. Cosmochim. Acta* **50**, 535–50.

Thellier, E. O. (1941). Sur la verification d'une methode permettant de determiner l'intensite du champ magnetique terrestre dans le passe. *Compte Rendu Acad. Sci. Paris* **212**, 281.

Wolfli, W. (1987). Advances in accelerator mass spectrometry. *Nucl. Instr. Meth. in Phys. Res. B* **29**, 1–13.

Yiou, F., Raisbeck, G. M., Bourles, D., Lorius, C. and Barkov, N. I. (1985). ^{10}Be in ice at Vostok Antarctica during the last climatic cycle. *Nature* **316**, 616–17.

Yokoyama, Y., Guichard, F., Reyss, J. L., Van, N. H. (1978). Oceanic residence times of dissolved beryllium and aluminium deduced from cosmogenic tracers ^{10}Be and ^{26}Al. *Science* **201**, 1016–17.

You, C. F., Lee, T. and Li, Y. H. (1989). The partition of Be between soil and water. *Chem. Geol.* **77**, 105–18.

15 Extinct radionuclides

15.1 Definition

An 'extinct radionuclide' is understood to be one that was formed by a process of stellar nucleosynthesis prior to the coalescence of the solar system, and which has subsequently decayed away to zero. Most extinct nuclides have very short half-lives, but a few have long half-lives in the millions of years range. These may have persisted in solar-system materials at high enough concentrations to generate observable variations in the isotopic composition of daughter products. These parent–daughter pairs are of interest to cosmochemists because they may be able to provide information about the origin of the solar system and its early history.

The production rate of an arbitrary solar-system nuclide as a function of time is shown schematically in Fig. 15.1. After the 'big bang', nucleosynthetic production in stars proceeds at a rate p which may be steady or very variable (depending on the process), during the lifetime of the galaxy (T). However, prior to condensation in the new solar nebula, it is anticipated that much or all solar-system matter was out of nucleosynthetic 'circulation' for a time period in some form of interstellar cloud. The time between last nucleosynthesis ('star death') and major condensation ('glob formation') is termed Δ (Fig. 15.1). Determination of Δ for different extinct nuclides may reveal information about the process which led to solar-system coalescence. Therefore, it is one of the major goals of isotope cosmochemistry.

In order to derive useful information from the daughter products of extinct radionuclides, it is necessary to study material which has not been significantly re-worked during the life of the solar-system. Hence, chondritic meteorites, which appear to

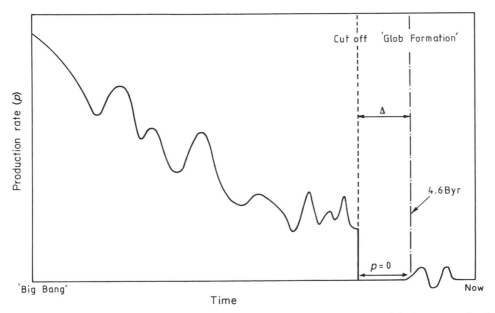

Fig. 15.1. Schematic illustration of the variation in production rate (p) of a given nuclide between the 'big bang' and termination of nucleosynthesis, followed by a period 'Δ' prior to solar-system coalescence. After Wasserburg (1985).

Table 15.1. *Extinct radionuclides*

Parent	Daughter	Decay mode	Mean life (Myr)	Half-life (Myr)	λ (yr^{-1})
^{244}Pu	various	fission	120	82	8.5×10^{-9}
^{129}I	^{129}Xe	β	23	16	4.3×10^{-8}
^{247}Cm	^{235}U	$3\alpha, 2\beta$	22.5	15.6	4.4×10^{-8}
^{107}Pd	^{107}Ag	β	9.4	6.5	1.1×10^{-7}
^{53}Mn	^{53}Cr	β	5.3	3.7	1.9×10^{-7}
^{60}Fe	^{60}Ni	2β	2.1	1.5	4.7×10^{-7}
^{26}Al	^{26}Mg	β	1.0	0.7	9.8×10^{-7}
^{41}Ca	^{41}K	β	0.15	0.1	6.7×10^{-6}

most nearly represent the original accretion components of the solar system, are the main objects of study. However, for some nuclide pairs, iron and stony meteorites, which were subject to early planetary differentiation processes, may also be useful. Cosmic-ray bombardment is one process to which meteorites are particularly susceptible. This can cause nuclear transformations, which must be excluded as a mechanism for generating daughter-product anomalies before the latter are attributed to extinct parents.

Most of the scientifically important extinct radio-nuclides with half-lives over 10^5 years are shown in Table 15.1 in order of decreasing stability, and will be discussed below. The extinct nuclide ^{146}Sm ($t_{1/2} = 103$ Myr) was examined in section 4.1, but will not be discussed here because it places only weak constraints on solar-system condensation. In Table 15.1, mean lives ($1/\lambda$) are quoted in addition to half-lives because they are helpful in understanding the production history of extinct nuclides.

15.2 Species present in the early solar system

15.2.1 Extant actinides

The age of the universe is estimated at ca. 15 Byr from nucleosynthetic models which predict the ages of the oldest stars. Hence, the age of the galaxy at the time of solar-system condensation was ca. 10 Byr. This is also the lifetime of a typical star like the Sun. However, large stars have much shorter lifetimes, which may even be less than 1 Myr in duration. Since the solar-system coalesced from the debris of 'dead' stars, it is theoretically possible that any given atom in the solar-system could have been processed through only one previous star, or through numerous previous stars.

This indeterminacy leads to uncertainty in the production rate of solar-system nuclides over the life of the galaxy. This uncertainty is particularly severe for r-process nuclides, which are only generated in the supernova explosions which terminate the life-histories of large stars (section 1.2). The uncertainty in r-process production is, in turn, a major source of uncertainty in the determination of Δ; in fact the two are inextricably linked. Therefore, it is essential, in estimating Δ for a given nuclide, to constrain its production history. The best constraints on long-term r-process production rates are provided by long-lived 'extant' nuclides. Therefore the extant actinides, ^{235}U, ^{238}U and ^{232}Th will be examined before discussion of individual extinct nuclides.

The gulf of unstable nuclides between the end of the s-process nucleosynthetic ladder and the actinide elements (section 1.2.2) means that these nuclides can only be generated by the r-process. The seed nuclei for this process are clearly the nuclei at the top of the s-process ladder, but these nuclides, especially Pb, have small neutron capture cross-sections, creating a nucleosynthetic barrier. As a result, the production ratios of the actinides are constrained to be close to unity. This factor is critical in using them to model r-process production rates over the life time of the galaxy.

We begin with the abundances of these nuclides in chondritic meteorites. Normalising to ^{235}U $= 1$, ^{238}U and ^{232}Th have present-day abundances of ca. 138 and 520 respectively in carbonaceous chondrites. Correcting for decay to initial abundances at 4.55 Byr and re-normalising to ^{235}U $= 1$, we obtain ratios of 3.45 and 8.18 for ^{238}U and ^{232}Th respectively, but these are still higher than estimated production ratios (relative to ^{235}U) of 0.66 and 1.27 respectively (Broecker, 1986). We will now use these data to test two extreme production models (Fig. 15.2).

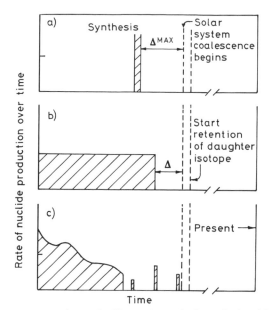

Fig. 15.2. Schematic illustration of the relationship between production models and Δ calculation. a) Single supernova event yielding maximum value of Δ; b) constant 'continuous' production followed by a short Δ period; c) complex variation in production rate ('granular model'). After Wasserburg and Papanastassiou (1982).

If all uranium formation was attributed to a single supernova event (Fig. 15.2a), then we could calculate the apparent timing of this event by the subsequent decay of short-lived ^{235}U (half-life ca. 700 Myr) relative to longer-lived ^{238}U (half-life ca. 4500 Myr). The calculation based on uranium isotopes alone may be somewhat more reliable than that involving thorium because no chemical fractionation can have occurred during solar-system coalescence. The calculated time of the event is 2.1 Byr before solar-system coalescence. Evidence for the presence of the short-lived actinide ^{244}Pu in the early solar system rules out a model with such a large Δ value.

The other extreme model involves 'continuous' supernova production (Fig. 15.2b). Taken to its limit this is an impossibility, since each supernova event terminates the evolution of a star, and the scattered debris must be incorporated in a new star before nucleosynthesis can continue. However, for nuclides with half-lives of hundreds of Myr, a supernova frequency as low as 100 Myr in the production history of an element will be a close approximation to continuous production. Under this model, the abundance of an unstable nuclide builds up until it reaches a level where the rate of synthesis is equalled by the rate of decay. This point of saturation is reached sooner in short-lived than long-lived nuclides (Fig. 15.3).

The growth curves in Fig. 15.3 can be presented in the form of isotope ratios (Fig. 15.4) of ^{238}U and ^{232}Th against ^{235}U. The time at which the curves intersect the primordial solar-system composition

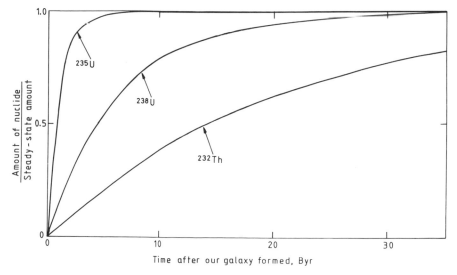

Fig. 15.3. Contrasting rates of approach of U and Th isotopes to the steady-state abundance in a model of 'continuous' supernova actinide production. After Broecker (1986).

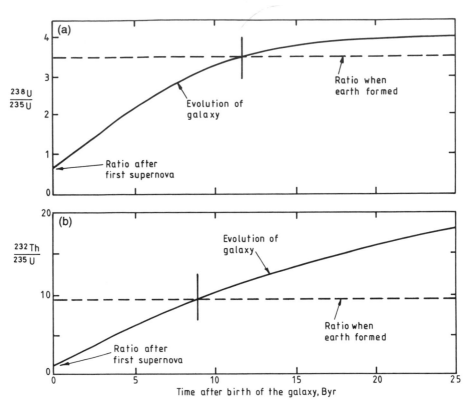

Fig. 15.4. Growth curves for $^{238}U/^{235}U$ and $^{232}Th/^{235}U$ in a 'continuous' supernova production model, relative to the primordial solar-system value. After Broecker (1986).

(calculated above) yields a crude estimate for the age of the galaxy at solar-system coalescence. Adding 4.55 Byr, we obtain estimates for the age of the galaxy of 16.5 and 13.5 Byr from Figs. 15.4a and 15.4b respectively. Given the many assumptions made, these agree surprisingly well with the ca. 15 Byr age of the universe calculated from the oldest stars, which therefore validates the 'continuous supernova' model (to some extent). Under this model, the extant actinides provide a very poor constraint on the value of Δ. However, that is the role of the *extinct* nuclides.

In between the two extreme models described above, there is an infinite number of intermediate models where discrete production events of variable size occur at variable intervals (e.g. Fig. 15.2c). These are termed 'granular' models. Ideally, we would like to use the actinide data to put an upper limit on the 'granularity' of these models. Many workers have used rather arbitrary models involving a combination of 'continuous' production and late 'discrete' events. However, Trivedi (1977) suggested that it was more reasonable to assume supernova events at regular

intervals. He proposed a simple model in which 50 supernova spikes of equal size were equally spaced with an interval of 140 Myr (assuming a galactic age at solar-system formation of 7 Byr). Relative to a theoretical 'stable' actinide, this means that the most recent supernova products would undergo 50-fold dilution by relatively 'cold' material generated in previous events. This model represents a useful yardstick for comparison with the dilution factors proposed for extinct nuclides.

15.2.2 I–Xe

The nuclide ^{129}I has a half-life of 16 Myr, and decays by β emission to ^{129}Xe (Fig. 15.5). Wasserburg and Hayden (1955) searched for ^{129}Xe anomalies, but were unsuccessful. However, ^{129}Xe excesses were eventually demonstrated by Reynolds (1960), making this the first extinct nuclide to be 'found'.

If the excess ^{129}Xe signatures discovered by Reynolds were due to the decay of now-extinct ^{129}I in the meteorite samples, it would be expected that

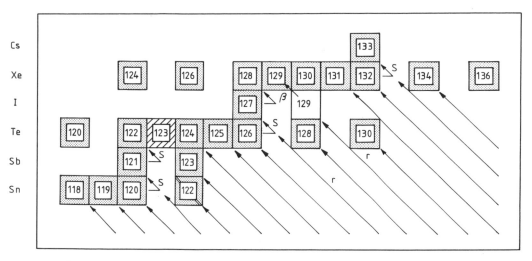

Fig. 15.5. Part of the chart of the nuclides in the region of iodine showing nucleosynthetic production and radioactive decay routes.

stable ^{127}I would still be present. Therefore, to test this model, it was necessary to look for correlations between the abundance of ^{127}I and excess ^{129}Xe in each sample. To do this by chemical analysis would have been laborious and inaccurate. However, Jeffrey and Reynolds (1961) conceived of an elegant means of making the ^{129}Xe$_{excess}$/^{127}I measurement in a single mass spectrometric analysis. By irradiating whole-rock samples of the Richardton chondrite with slow neutrons in a reactor, it was possible to generate the stable isotope ^{128}Xe from ^{127}I by the following n, γ and β decay reactions:

$$^{127}\text{I} + \text{n} \rightarrow {}^{128}\text{I} + \gamma$$
$$^{128}\text{I} \rightarrow {}^{128}\text{Xe} + \beta + \bar{\nu}$$

Hence, the I/Xe ratio could be determined by isotopic analysis of xenon alone.

Jeffrey and Reynolds made a further technical advance in their method of sample analysis. Rather than one-step outgassing of xenon from each meteorite sample by melting it (to produce a single data point), they outgassed the sample in a series of increasing temperature steps, admitting each new gas-release separately to the mass spectrometer for analysis. It is interesting to note that a very similar neutron irradiation and 'step-heating' mass spectrometric method was applied to K–Ar geochronology five years later, revolutionising it to the ^{40}Ar–^{39}Ar method (section 10.2).

It is convenient to display the xenon isotope data on a plot somewhat analogous to an Ar–Ar isochron diagram (section 10.2.3). Jeffrey and Reynolds

demonstrated a correlation between ^{129}Xe and ^{128}Xe abundance, ratioing both of these against the non-radiogenic ^{132}Xe isotope (e.g. Fig. 15.6). If the efficiency of the activation process is calibrated, the excess ^{128}Xe abundance translates into the abundance of the non-radioactive iodine isotope, ^{127}I. Similarly, because every ^{129}I atom has by now been converted to ^{129}Xe by radioactive decay, the excess ^{129}Xe abundance translates into the ^{129}I abundance at the time when meteorite samples were isolated from a common reservoir. Hence, the slope of any array of data points observed in this diagram (^{129}Xe$_{excess}$/^{128}Xe$_{excess}$) has no age significance. Instead, it indicates the initial ^{129}I/^{127}I ratio when the meteorite cooled to the point where its minerals became closed to diffusional loss of xenon into space. Meanwhile, the intercept of the 'iso-concentration of ^{129}I' line on the y axis represents the non-radiogenic ^{129}Xe/^{132}Xe ratio.

The slope of the array in Fig. 15.6 corresponds to an initial ^{129}I/^{127}I ratio of 1×10^{-4}. Subsequent work on a wide variety of meteorites has confirmed this value with only small variations. This suggests that ^{129}I was widely distributed through the solar nebula (e.g. Podosek, 1970; Wasserburg *et al.*, 1977; Niemeyer, 1979). Excess ^{129}Xe is also found in terrestrial rocks and magmas, but in the Earth it is not correlated with ^{127}I abundance. This occurrence must represent the outgassing of noble gases from a deep Earth reservoir which also once contained ^{129}I (section 11.3).

The ^{129}I/^{127}I ratio calculated from meteorite studies has been used as a 'model age' chronometer to

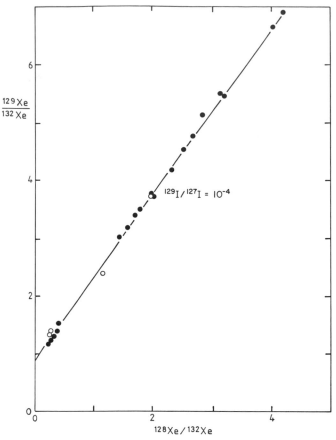

Fig. 15.6. Xe–Xe plot for stepwise degassed samples of the Richardton meteorite showing the line of 'iso-concentration' of ^{129}I. Solid and open symbols indicate gas fractions released above and below 1100 °C respectively. After Hohenberg *et al.* (1967).

measure the time-interval (Δ) between last nucleosynthesis and coalescence of the solar nebula. However, as with model ages in general, there are several major assumptions which must be made in any attempt to calculate a possible Δ value. Estimates must be made of:

1) the ratio of ^{129}I/^{127}I originally produced by nucleosynthesis;
2) the rate of nucleosynthesis over time, prior to the Δ period; and
3) in a granular model, any dilution of the last addition of radioactively 'hot' nucleosynthetic material by 'cold' material from earlier events.

These problems are best examined by comparing some extreme solutions which were summarised by Wasserburg (1985).

It has been traditionally assumed that iodine is generated by the r-process (Fig. 15.5). Production ratios (p_{127}/p_{129}) can only be determined by theoretical calculation; hence there are large uncertainties. However, they are generally assumed to be near unity. For example, values which have been used in the literature are unity (Wasserburg *et al.*, 1960; Wasserburg, 1985), 1.3 (Cameron, 1962; quoted by Hohenberg, 1969) and 2.9 +1/−2 (Seeger *et al.*, 1965; quoted by Schramm and Wasserburg, 1970).

The model originally conceived by Reynolds (1960) involved synthesis of iodine with a ^{129}I/^{127}I ratio of unity in a single event. This would decay to a ratio of 10^{-4} over ca. 12 half-lives, yielding a maximum Δ value of ca. 200 Myr (Fig. 15.7a). However, if all iodine was generated in a single supernova (i.e. zero dilution of 'hot' supernova iodine by 'cold' ^{127}I), then all other r-process elements would have to have formed at this time, which is incompatible with actinide evidence.

The other extreme model assumed more-or-less

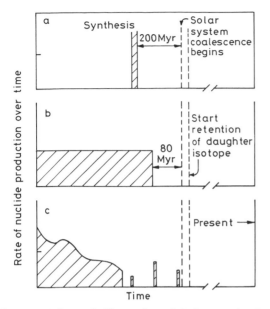

Fig. 15.7. Schematic illustration of iodine production models and consequent Δ calculation. a) Single supernova event yielding maximum value of Δ (ca. 200 Myr); b) constant 'continuous' production followed by a period Δ of ca. 80 Myr; c) complex variation in production rate ('granular model') yielding indeterminate Δ value. After Wasserburg and Papanastassiou (1982).

constant supernova activity throughout the galaxy. If their products were kept mixed then r-process production of solar-system material might be regarded as relatively constant (Dicke, 1969). Under this model, the total number of atoms of stable ^{127}I after time T (at the termination of nucleosynthesis, Fig. 15.1) is defined as:

$$n_T^{127} = p^{127} T \qquad [15.1]$$

where p is the average production rate. Similarly, the total number of ^{129}I atoms after time T can be approximated as:

$$n_T^{129} = p^{129}/\lambda = p^{129} \cdot \text{mean life} \qquad [15.2]$$

Dividing equation [15.2] by [15.1] we obtain:

$$\left(\frac{n^{129}}{n^{127}}\right)_T = \frac{p_{129}}{p_{127}} \cdot \frac{\text{mean life}}{T} \qquad [15.3]$$

Therefore, assuming a production ratio of unity in a model where iodine is formed by frequent and well-mixed supernovae over a period of 10 Byr, the ^{129}I/^{127}I ratio when nucleosynthesis is interrupted is 2.3×10^{-3}. This would take nearly five half-lives to

decay to a value of 1×10^{-4}, yielding a Δ value of ca. 80 Myr (Fig. 15.7b). However, this model has a major conceptual problem. It is very sensitive to the contributions of iodine from 'late' supernovae near the end of the nucleosynthetic period. If these form a significant fraction of the total iodine budget, then they are apt to de-stabilise the smooth growth model, giving rise to a 'granular' model (Wasserburg and Papanastassiou, 1982), illustrated in Fig 15.7c.

In reality, consideration of the rate of supernova occurrence in the whole galaxy (about one every 100 years) relative to the size of the galaxy, suggests that in our corner of the galaxy, a granular model is almost inevitable, relative to the relatively short half-life of ^{129}I. In this case, the most critical quantity is the dilution factor for the last addition of hot iodine (^{129}I/^{127}I ≈ 1) by cold or nearly-cold iodine from earlier events (^{129}I/^{127}I ≈ 0). A dilution factor of 100 has been proposed by Cameron and Truran (1977). Coupled with a production ratio of unity, this would yield a Δ value of ca. 110 Myr. However, as the dilution factor approaches 10^4, Δ can approach zero. Not until the review of Wasserburg (1985) was it explicitly pointed out that such 'extreme' solutions are possible.

Podosek (1970) used the slightly variable initial ^{129}I/^{127}I ratios of 0.7×10^{-4} to 1.3×10^{-4} measured in different meteorites to calculate the relative cooling times of different meteorites, by assuming that 'hot' and 'cold' iodine sources were homogenised in the solar nebula. It is not necessary to know the iodine isotope production ratio or the dilution factor of 'hot' by 'cold' iodine in the solar nebula for this calculation. However, the assumption of initial ^{129}I/^{127}I homogeneity in the solar nebula is not yet proven. Crabb et al. (1982) argued instead that the variations in initial ratio are due to imperfect mixing of iodine from different sources (variable dilution factors). More recently, Podosek's research group has at times felt the evidence to be inconclusive (Bernatowicz et al., 1988), and at other times argued that the 'ages' have genuine time-significance (Swindle et al., 1991).

Still more uncertainties may be present in modelling the time-scale of iodine nucleosynthesis and solar-system condensation. Because iodine is not greatly separated from the s-process nucleosynthetic pathway, there remains a possibility that ^{129}I might be produced in less extreme environments than supernovae. Furthermore, it must not be forgotten that ^{127}I is certainly generated by the s-process, so that uncertainties about the relative s- and r-process contribution to total iodine production are also present. Taking all of the uncertainties together, it is clear that ^{129}I cannot place tight constraints on the relative timing of nucleosynthesis and solar-system condensation (contrary to early claims). For this we must turn to other systems.

15.2.3 Pu–Xe

^{244}Pu has a half-life of 82 Myr. The clearest evidence of extinct ^{244}Pu in meteorite materials is provided by fission products, most notably a large excess abundance of 132, 134 and 136 xenon, which has been matched to the signature of laboratory Pu fission products (Fig. 15.8). ^{244}Pu is always compared to the abundance of other actinide elements in drawing conclusions about solar-system origins.

^{244}Pu is most conveniently ratioed against ^{238}U, but this involves elemental as well as isotopic abundances, and the former are susceptible to chemical fractionation after condensation of the nebula. Hence, representative analysis has been very difficult. The first determination of the initial ^{244}Pu/^{238}U ratio, based on meteoritic phosphate (Wasserburg *et al.*, 1969), yielded a value of ca. 0.035. However, this value may not be representative of the bulk (whole-rock) meteorite due to partition effects. Whole-rock analysis of different meteorites has yielded a large range of values, but the best consensus is for a chondritic value below 0.007 (Fig. 15.9).

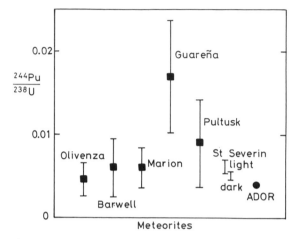

Fig. 15.9. Summary of Pu/U ratios for whole-rock samples of various meteorites. After Hagee *et al.* (1990).

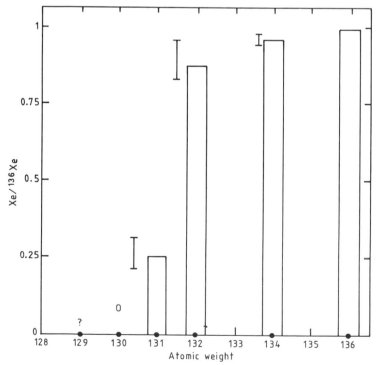

Fig. 15.8. Histogram of measured meteoritic Xe isotope abundance ratios relative to ^{136}Xe, compared to ratios observed in laboratory fission products (tie bars). After Wasserburg and Papanastassiou (1982).

Plutonium data alone are not able to apply tight constraints to Δ, for the same reasons as were given for iodine. However, if we assume that ^{129}I and ^{244}Pu were added at the same time, then we can use the data together to constrain Δ. For example, a continuous model yields an equilibrium ^{244}Pu/^{238}U ratio of 0.018, which can decay to 0.007 after a Δ period of ca. 100 Myr. This is in reasonable agreement with the continuous model for ^{129}I, which yields a Δ value of 80 Myr. Similarly, if we assume a granular model in which both elements undergo 50-fold dilution of the last r-process addition with cold material, both plutonium and iodine yield Δ values of ca. 130 Myr. Both of these results argue against very late addition of very dilute material (which was able to explain the iodine data alone). On the other hand, if ^{129}I is not an r-process nuclide, then its addition does not have to be accompanied by ^{244}Pu, and these constraints disappear.

15.2.4 Al–Mg

The nuclide ^{26}Al has a half-life of 0.72 Myr and decays to ^{26}Mg (Fig. 15.10). The discovery of extinct ^{26}Al was a much more difficult task than ^{129}I, and was only made possible by the fall of the Allende meteorite in February 1969. This is an agglomerate of fine-grained debris and chondrules which also contains 'inclusions' with a refractory mineralogy which appear to be very early condensation products from the solar nebula.

Analyses of Mg isotope ratios in minerals separated from Allende inclusions demonstrate mass-fractionation-dependent variations in ^{25}Mg/^{24}Mg ratio. These are normally of the order of a few parts per mil (Fig. 15.11), but some inclusions (e.g. EK1-04-1 and C1) show larger effects. These were termed FUN samples (showing Fractionation and Unknown Nuclear anomalies) by Wasserburg *et al.* (1977). Because ^{27}Al was always a nuclide of comparatively low abundance, it was at first very difficult to demonstrate radiogenic ^{26}Mg abundances outside error of mass fractionation processes.

The first evidence for radiogenic ^{26}Mg was demonstrated by Lee *et al.*, (1976). Subsequently,

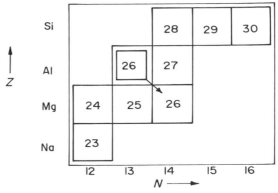

Fig. 15.10. Part of the chart of the nuclides in the region of Mg and Al showing decay of extinct ^{26}Al to ^{26}Mg.

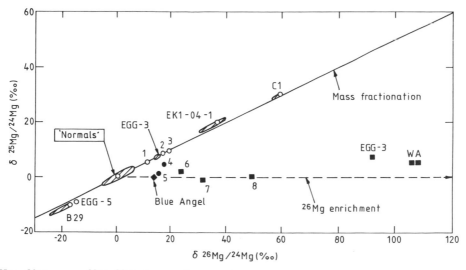

Fig. 15.11. ^{25}Mg/^{24}Mg *versus* ^{26}Mg/^{24}Mg isotope diagram showing deviations of Allende inclusions from the normal solar-system value, in parts per mil (δ). These may be due to mass fractionation in the solar nebula (open symbols) or decay of extinct ^{26}Al (solid symbols). After Wasserburg and Papanastassiou (1982).

larger ^{26}Mg anomalies were found by very careful hand-picking of inclusion- and alteration-free plagioclase grains from the WA Allende inclusion. By thus maximising the Al/Mg ratio of the analysed material, an excess ^{26}Mg abundance of 97 parts per mil was found in one anorthite grain (Lee *et al.*, 1977). Ion microprobe studies of selected areas of individual crystals allowed material of even higher Al/Mg ratio to be analysed, yielding ^{26}Mg enrichments of up to 500 per mil (Bradley *et al.*, 1978; Hutcheon *et al.*, 1978).

It is convenient to display Al–Mg data on an isotope ratio *versus* element ratio plot somewhat analogous to an isochron diagram (e.g. Fig. 15.12). Hence, ^{26}Mg/^{24}Mg is plotted against an Al/Mg ratio. In a conventional isochron diagram, the ratio plotted on the abscissa would be ^{26}Al/^{24}Mg. However, since every ^{26}Al atom has been converted to ^{26}Mg by radioactive decay at the present day, the ^{27}Al/^{24}Mg ratio is plotted instead. Therefore, the slope of any array of data points observed in this diagram has no age significance, but indicates the initial ^{26}Al/^{27}Al ratio in the sample suite at the time when sub-systems were isolated from a common reservoir.

The data shown in Fig. 15.12 for separated minerals from the Allende EGG-3 inclusion display a good ^{26}Mg/^{27}Al correlation, yielding a well-defined initial ^{26}Al/^{27}Al ratio of 4.9×10^{-5} (Armstrong *et al.*,

1984). A similar array for the WA inclusion gave a ^{26}Al/^{27}Al ratio of $5.1 \pm 0.6 \times 10^{-5}$, (Lee *et al.*, 1977). However, in the inclusion USNM 3529-26 (Armstrong *et al.*, 1984), minerals from different parts of the inclusion yielded different initial ratios, with a higher value in the core (3.8×10^{-5}) than the rim (2.3×10^{-5}). This spatial variation is best explained by Mg loss, particularly from the margins of the inclusion, during a later metamorphic event. The event may have resulted from the heat output of ^{26}Al decay itself.

Subsequent studies of Mg isotope ratio in Allende inclusions have made use of the ion microprobe for direct analysis of material with very low common Mg contents. Data on Allende melilites and spinels by Steele and Hutcheon (1979) yield an initial ^{26}Al/^{27}Al ratio of 2 (± 1) $\times 10^{-5}$, although data on the refractory mineral hibonite (from the same inclusion) imply a ratio as high as 8×10^{-5}. Nevertheless, these results are broadly in line with the data on separated minerals.

The great significance of ^{26}Al for cosmochemistry is its short half-life of 0.72 Myr. Because this is only 4% of the ^{129}I half-life, its presence in the early solar system constrains a nucleosynthetic event to have occurred much more imminently before the coalescence of the solar system. Classical nucleosynthetic models (e.g. Arnett, 1969; Truran and Cameron,

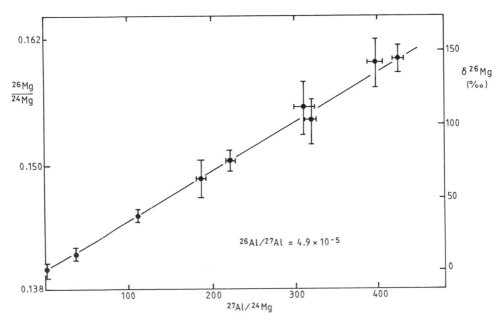

Fig. 15.12. Plot of δMg against Al/Mg for the Allende EGG-3 inclusion showing best-fit line of constant initial ^{26}Al/^{27}Al ratio with a value of 4.9×10^{-5}. After Armstrong *et al.* (1984).

1978) attribute ^{26}Al to explosive carbon burning in the envelope of a supernova, and predict a ^{26}Al/^{27}Al production ratio of ca. 10^{-3}. It would take only 3 Myr for this ratio to decay to the value of 5×10^{-5} ratio found in several meteorite samples. Lee *et al.* (1976) suggested that this was due to late addition to the solar nebula of freshly synthesised material from a nearby nova or supernova explosion. Cameron and Truran (1977) pointed out that such an explosion in the vicinity of a condensing solar nebula was a very unlikely coincidence unless the supernova itself triggered the collapse of an interstellar cloud to form the solar system. This model is illustrated in Fig. 15.13.

For a time, the 'supernova trigger' model for solar-system coalescence was widely accepted. However, spectral data from the High Energy Astronomical Observatory satellite (HEAO 3) revealed a γ line (Fig. 15.14) due to ^{26}Al decay from a diffuse galactic source (Mahoney *et al.*, 1984). An average galactic ^{26}Al/^{27}Al

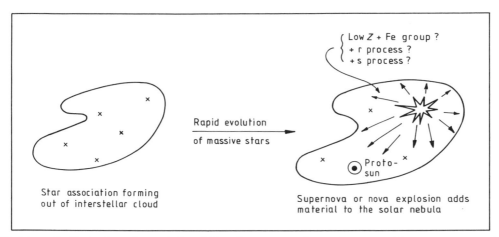

Fig. 15.13. Schematic illustration of a model in which solar-system collapse is promoted by a supernova which also seeds it with short-lived nuclides. After Wasserburg (1985).

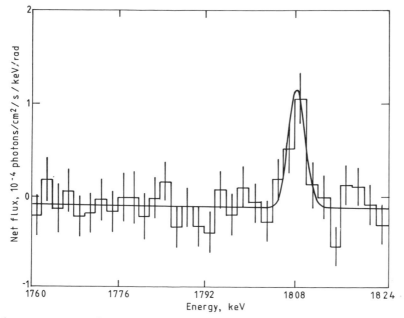

Fig. 15.14. Galactic γ-ray spectrum showing a peak at 1808 keV which is attributed to decay of excited ^{26}Mg, itself produced by the decay of ^{26}Al in interstellar space. After Mahoney *et al.* (1984).

ratio of ca. 10^{-5} was determined, remarkably close to that deduced for the early solar-system from Allende inclusions. The high ^{26}Al abundance in the galaxy means that supernovae, which are rare, cannot be the principal sources. Indeed, recent experimental studies (Champagne *et al.*, 1983) suggest that Red Giants can generate Al with a 26/27 production ratio of unity.

Although astronomical observations rule out the supernova model, isotopic data still imply late injection of ^{26}Al into the pre-solar cloud. Wasserburg (1985) suggested that this could be caused by rapidly evolving stars within the interstellar cloud, followed by rapid separation and condensation on a ca. 1 Myr time-scale (Wasserburg, 1985). On the other hand, Clayton (1975, 1979) has argued that most isotopic anomalies observed in meteorites, including extinct radionuclide signatures, are inherited from pre-solar dust grains. If this was true for ^{26}Al, then both Al isotope and Al/Mg ratios must have been inherited intact from these pre-solar materials. This implies that many meteorite mineral phases (e.g. Ca–Al inclusions in Allende) are also pre-solar. Wasserburg (1985) contested this argument on mineralogical grounds, believing that most or all of the analysed meteorite phases crystallised within the solar system.

Evidence against a pre-solar age for Ca–Al inclusions was provided by the discovery of ^{26}Mg excesses in a clast from the chondrite Semarkona (Hutcheon and Hutchison, 1989). These excesses were correlated with Al/Mg ratio, implying a ^{26}Al/^{27}Al ratio of 7×10^{-6} at the time of crystallisation. Hutcheon and Hutchison argued that the mineralogy and REE chemistry of the clast were the result of igneous processes, implying a planetary, rather than nebular origin. If these inferences are correct, it implies that some planetary differentiation was occurring only 2 Myr after the formation of the refractory Allende inclusions. The heat liberated by ^{26}Al decay might itself have contributed to the melting of small planetary bodies.

All of the samples described above (inclusions and clasts) are now generally thought to have formed during solar-system condensation. However, evidence has recently been found for the preservation of pre-solar grains in the Murchison carbonaceous chondrite. These grains are composed of silicon carbide, and have exotic rare gas signatures which match the abundance patterns expected in carbon-burning Red Giants (Lewis *et al.*, 1990). This means that they have escaped significant heating during solar-system coalescence. Zinner *et al.* (1991) found large excesses of ^{26}Mg in some of these grains, equivalent to initial ^{26}Al/^{27}Al ratios from 10^{-5} up to nearly unity. These

ratios probably date from the time of expulsion of the grains, in the solar wind of a Red Giant, into interstellar space. It is possible that this solar wind (rather than a supernova, for example) triggered the collapse of a giant molecular cloud to form the solar system (Nuth, 1991).

15.2.5 Pd–Ag

^{107}Pd decays by β emission to ^{107}Ag with a half-life of 6.5 Myr. The only objects with a high enough Pd/Ag ratio to yield measurable variations in ^{107}Ag abundance are iron meteorites. The first successful discovery of radiogenic ^{107}Ag was made by Kelly and Wasserburg (1978) on the Santa Clara meteorite. They deduced that ca. 10 Myr might have elapsed between a nucleosynthetic event and the coalescence and differentiation of iron-cored small planets. Subsequent work has revealed radiogenic ^{107}Ag in several meteorites (Chen and Wasserburg, 1984; 1990), of which the best data are from the Gibeon meteorite. These results yield an initial ^{107}Pd/^{108}Pd ratio of 2.3×10^{-5} (Fig. 15.15). Since ^{107}Pd *versus* Ag correlations are observed in bodies which have clearly been melted since accretion of the solar system, the isotopic signatures must reflect the presence of 'live' short-lived nuclides in the early solar system. They cannot have been inherited from pre-solar grains (Clayton, 1975).

Wasserburg and Papanastassiou (1982) drew attention to the relative similarity between the initial iodine, aluminium and palladium ratios of ca. 1×10^{-4}, 0.5×10^{-4} and 0.2×10^{-4} respectively. They pointed out that if the additions of 'hot' material occurred comparatively early (e.g. $\Delta =$ ca. 200 Myr), as suggested by Schramm and Wasserburg (1970), then their different half-lives would have attenuated the short-lived nuclides to very different degrees by the time of solar-system condensation. In contrast, the comparatively similar abundance ratios actually observed may imply a late addition with similar degrees of dilution by cold material. However, the more recent discovery of galactic ^{26}Al prompted Wasserburg (1985) to draw back, and admit that 'it is possible that the value of ca. 10^4 for the abundance of ^{26}Al, ^{107}Pd and ^{129}I is a coincidence'.

15.2.6 Mn–Cr

^{53}Mn decays to ^{53}Cr with a half-life of 3.7 Myr. For several reasons, this extinct nuclide reinforces the evidence from ^{26}Al and ^{107}Pd for the early history of the solar system. Birck and Allegre (1985; 1988)

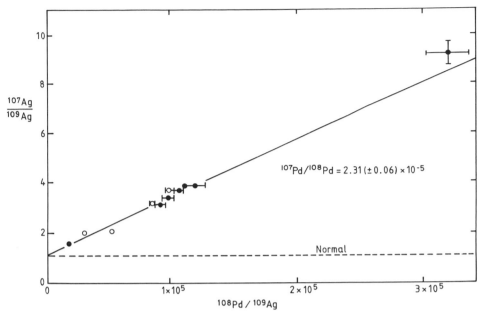

Fig. 15.15. Ag isotope *versus* Pd–Ag diagram showing evidence of extinct ^{107}Pd in iron meteorites. The best-fit line to seven metal samples from the Gibeon meteorite (•) yields an initial ^{107}Pd/^{108}Pd ratio of 2.3×10^{-5}. Open symbols denote other group IVA iron meteorites. After Wasserburg (1985).

found correlated variations in ^{53}Cr/^{52}Cr and Mn/Cr ratio in several meteorites, including Allende, Murchison, Indarch and the pallasite meteorite Eagle Station. Results for the latter are shown in Fig. 15.16, and indicate an initial ^{53}Mn/^{55}Mn ratio of 2.3×10^{-6}. Since the mixed silicate–iron mineralogy of this meteorite indicates a high-temperature origin, the Mn–Cr isotope systematics must result from *in-situ* decay of ^{53}Mn in differentiated planetary bodies. Hence ^{53}Mn provides additional evidence for nucleosynthetic processes immediately before coalescence of the solar system.

^{53}Mn is part of the iron group of elements (section 1.2) which are thought to be synthesised in large stars shortly before supernova explosion. Hence it was argued (e.g. Rotaru *et al.*, 1992) that a supernova did indeed shortly pre-date solar-system condensation, as initially proposed on the basis of ^{26}Al. From the 2.7 Myr half-life of ^{53}Mn, it would follow that such an event occurred ca. 20 Myr before planetary differentiation. However, there are other possible routes for the synthesis of ^{53}Mn (Birck and Allegre, 1985). Furthermore, curium isotope evidence (or the lack of it; see below) argues against a late r-process addition to the solar nebula. Therefore, although there is good evidence for a late nucleosynthetic addition to the solar nebula, its source remains in doubt.

In addition to the evidence from radiogenic ^{53}Cr, abundances of non-radiogenic ^{54}Cr may also provide constraints on the origin of solar-system materials. Rotaru *et al.* (1992) found that differential dissolution of chondrite samples by acid leaching resulted in a variety of ^{54}Cr abundances against other Cr isotopes. Furthermore, the Orgueil meteorite did not display any phase with a solar (= terrestrial) ^{54}Cr abundance. Rotaru *et al.* attributed these observations to the mixing of multiple nucleosynthetic components in the formation of the solar-system. Apparently the different sources of Cr were concentrated in different mineral types, so that they could be sampled by leaching the resulting aggregation of cosmic dust particles. Other Fe-group elements (e.g. Fe, Ti) apparently do not show these effects because material from different sources was concentrated in the same mineral types. Rotaru *et al.* coupled this evidence with the radiogenic ^{53}Cr data and suggested that as much as a few per cent of the solar nebula might result from material newly synthesised a few Myr before planet formation.

15.2.7 Fe–Ni

^{60}Fe decays to ^{60}Ni *via* ^{60}Co with a half-life of 1.5 Myr. Shukolyukov and Lugmair (1993a) found ^{60}Ni

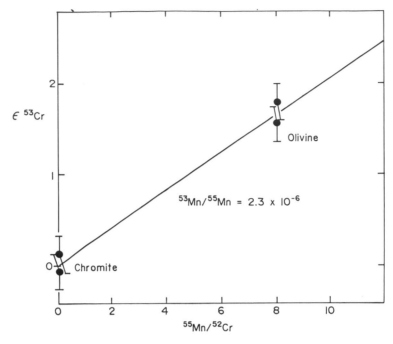

Fig. 15.16. Mn–Cr correlation diagram for mineral phases of the Eagle Station meteorite, yielding an initial $^{53}Mn/^{55}Mn$ ratio of 2.3×10^{-6} at the time of igneous differentiation of its parent body. After Birck and Allegre (1988).

excesses as high as 50 epsilon units in the achondrite Chervony Kut, facilitated by the extremely high Fe/Ni ratios in this meteorite (up to 350 000). These are attributed to nickel partition into the cores of differentiated planetessimals. Whole-rock samples of Chervony Kut displayed an 'isochron' correlation between $^{60}Ni/^{58}Ni$ and Fe/Ni ratio, indicating that ^{60}Fe was still alive at the time of differentiation, with a $^{60}Fe/^{56}Fe$ ratio of 3.9×10^{-9}. Whole-rock analysis of the achondrite Juvinas also generated an 'isochron' relation, but with a lower $^{60}Fe/^{56}Fe$ ratio of ca. 4×10^{-10} (Shukolyukov and Lugmair, 1993b). The difference between these extinct nuclide abundance ratios implies a difference in closure age of ca. 4.7 Myr between the two meteorites, if the parent body had a homogeneous $^{60}Fe/^{56}Fe$ ratio. This evidence points to widespread distribution of ^{60}Fe in the solar nebula, and suggests that this nuclide was a major source of heat. Coupled with ^{26}Al, ^{60}Fe is capable of causing planetary melting within a few million years of accretion. Since the production routes for ^{60}Fe are unclear, precise Δ values cannot be calculated. However, the evidence is consistent with Mn–Cr data in suggesting addition of newly synthesised material to the nebula only a few million years before condensation.

15.3 Absent species

15.3.1 Cm–U

^{247}Cm decays to ^{235}U with a half-life of 16 Myr. The significance of this species is that like ^{244}Pu it is formed by the r-process only, but unlike ^{244}Pu it has a comparatively short half-life. Hence the abundance of ^{247}Cm can indicate conclusively whether late additions of material to the nebula (called for above) were r-process products or not. It has been noted that, unlike ^{129}I, ^{26}Al has a heterogeneous distribution, reaching peaks in certain Allende inclusions. Hence, if ^{247}Cm was present in the late ^{26}Al addition, these same samples should be enriched in the daughter product of curium, ^{235}U. The search for ^{247}Cm therefore consists of looking for very small variations in uranium isotope composition. Unfortunately, ^{235}U is normally used as an enriched isotope in U abundance determination by isotope dilution, and hence most laboratories are susceptible to artificial perturbations in this ratio. Perhaps for this reason, numerous claims have been made for excess ^{235}U, but all are suspect.

In an attempt to conclusively resolve this problem, Chen and Wasserburg (1981) made a very careful investigation of U isotope composition using a double 233/236 uranium spike to correct for mass fractionation during analysis (section 2.2.7), on samples for which large ^{26}Al signatures had been demonstrated. This study placed a maximum limit of $\pm 0.4\%$ on uranium isotope variation in these very favourable samples, from which Chen and Wasserburg calculated a maximum initial ^{247}Cm/^{238}U ratio of 1.5×10^{-3}.

This value can be compared with the analogous ^{244}Pu/^{238}U ratio, which sets an upper limit on the dilution factor for late r-process material. Since inferred initial ^{247}Cm abundances are substantially lower than ^{244}Pu, it is deduced that the late ^{26}Al addition to the solar nebula was not r-process material. Wasserburg (1985) argued that the low ^{247}Cm abundance places a strong lower limit on Δ (100 Myr, Wasserburg and Papanastassiou, 1982) for the last r-process addition to the solar nebula. However, it should be pointed out that this argument only excludes late *heterogenous* additions of ^{247}Cm to the solar nebula (accompanying ^{26}Al). This is because it rests on the observed homogeneity of ^{235}U/^{238}U isotope ratios rather than their absolute value.

15.3.2 Ca–K

^{41}Ca decays to ^{41}K with a half-life of only 0.1 Myr. Hence, if ^{41}K excesses were found in solar-system material they would imply a very late addition of hot material to the solar nebula. Such anomalies might be expected in Allende material displaying ^{26}Al signatures. Begemann and Stegmann (1976) sought ^{41}Ca signatures in Allende samples, and believed they had found them. However, subsequent work by Hutcheon *et al.* (1984) attributed this to $(^{40}$Ca^{42}Ca$)^{++}$ dimers, creating a peak which could not be resolved in mass from ^{41}K. These workers set an upper limit on initial ^{41}Ca/^{40}Ca of 8×10^{-9}, equivalent to a minimum Δ value of 1.8 Myr if the ^{41}Ca/^{40}Ca ratio immediately after addition of hot material was 10^{-4}.

When coupled with positive signals from the Mn–Cr and Fe–Ni systems, the Ca–K evidence suggests that late addition of stellar core material to the solar nebula occurred between 2 and 20 Myr before condensation. However, bombardment by stellar wind (carrying ^{26}Al) may have occurred as a distinct later event.

References

Armstrong, J. T., Hutcheon, I. D., and Wasserburg, G. J. (1984). Disturbed Mg isotopic systematics in Allende CAI. In: *Lunar Planet. Sci.* **XV**, Lunar Planet. Inst., 15 (abstract).

Arnett, W. D. (1969). Explosive nucleosynthesis in stars. *Astrophys. J.* **157**, 1369–80.

Begemann, F. and Stegmann, W. (1976). Implications from the absence of ^{41}K anomaly in a Allende inclusion. *Nature* **259**, 549–50.

Bernatowicz, T. J., Podosek, F. A., Swindle, T. D. and Honda, M. (1988). I–Xe systematics in LL chondrites. *Geochim. Cosmochim. Acta* **52**, 1113–21.

Birck, J. L. and Allegre, C. J. (1985). Evidence for the presence of ^{53}Mn in the early solar-system. *Geophys. Res. Lett.* **12**, 745–8.

Birck, J. L. and Allegre, C. J. (1988). Manganese–chromium isotope systematics and the development of the early solar-system. *Nature* **351**, 579–84.

Bradley, J. G., Huneke, J. C. and Wasserburg, G. J. (1978). Ion microprobe evidence for the presence of excess ^{26}Mg in an Allende anorthite crystal. *J. Geophys. Res.* **83**, 244–54.

Broecker, W. (1986). *How to Build a Habitable Planet*. Eldigio Press, Columbia Univ., 291p.

Cameron, A. G. W. (1962). The formation of the sun and the planets. *Icarus* **1**, 13–69.

Cameron, A. G. W. and Truran, J. W. (1977). The supernova trigger for formation of the solar-system. *Icarus* **30**, 447–61.

Champagne, A. E., Howard, A. J. and Parker, P. D. (1983). Nucleosynthesis of ^{26}Al at low stellar temperatures. *Astrophys. J.* **269**, 686–8.

Chen, J. H. and Wasserburg, G. J. (1981). The isotopic composition of uranium and lead in Allende inclusions and meteorite phosphates. *Earth Planet. Sci. Lett.* **52**, 1–15.

Chen, J. H. and Wasserburg, G. J. (1984). The origin of excess ^{107}Ag in Gibeon (IVA) and other iron meteorites. In: *Lunar Planet. Sci.* **XV**, Lunar Planet. Inst., 144 (abstract).

Chen, J. H. and Wasserburg, G. J. (1990). The isotopic composition of Ag in meteorites and the presence of ^{107}Pd in proto-planets. *Geochim. Cosmochim. Acta* **54**, 1729–43.

Clayton, D. D. (1975). Extinct radioactivities: trapped residuals of presolar grains. *Astrophys. J.* **199**, 765–9.

Clayton, D. D. (1979). Supernovae and the origin of the solar-system. *Space Sci. Rev.* **24**, 147–226.

Crabb, J., Lewis, R. S. and Anders, E. (1982). Extinct ^{129}I in C3 chondrites. *Geochim. Cosmochim. Acta* **46**, 2511–26.

Dicke, R. H. (1969). The age of the galaxy from the decay of uranium. *Astrophys. J.* **155**, 123–34.

Hagee, B., Bernatowicz, T. J., Podosek, F. A., Johnson, M. L., Burnett, D. S. and Tatsumoto, M. (1990). Actinide abundances in ordinary chondrites. *Geochim. Cosmochim. Acta* **54**, 2847–58.

Hohenberg, C. M. (1969). Radioisotopes and the history of nucleosynthesis in the galaxy. *Science* **166**, 212–15.

Hohenberg, C. M., Podosek, F. A. and Reynolds, J. H. (1967). Xenon–iodine dating: sharp isochronism in chondrites. *Science* **156**, 233–6.

Hutcheon, I. D., Armstrong, J. T., and Wasserburg, G. J. (1984). Excess in ^{41}K in Allende CAZ: confirmation of a hint. In: *Lunar Planet. Sci.* **XV**, Lunar Planet. Inst., 387–8 (abstract).

Hutcheon, I. D. and Hutchison, R. (1989). Evidence from the Semarkona ordinary chondrite for ^{26}Al heating of small planets. *Nature* **337**, 238–41.

Hutcheon, I. D., Steele, I. M., Smith, J. V. and Clayton, R. N. (1978). Ion microprobe, electron microprobe and cathodoluminescence data for Allende inclusions with emphasis on plagioclase chemistry. In: *Proc. 9th Lunar Planet. Sci. Conf.* Pergamon, pp. 1345–68.

Jeffrey, P. M. and Reynolds, J. H. (1961). Origin of excess ^{129}Xe in stone meteorites. *J. Geophys. Res.* **66**, 3582–3.

Kelly, W. R., and Wasserburg, G. J. (1978). Evidence for the existence of ^{107}Pd in the early solar-system. *Geophys. Res. Lett.* **5**, 1079–82.

Lee, T., Papanastassiou, D. A. and Wasserburg, G. J. (1976). Demonstration of ^{26}Mg excess in Allende and evidence for ^{26}Al. *Geophys. Res. Lett.* **3**, 109–13.

Lee, T., Papanastassiou, D. A. and Wasserburg, G. J. (1977). Aluminum-26 in the early solar-system: Fossil or fuel? *Astrophys. J. (Lett.)* **211**, L107–10.

Lewis, R. S., Amari, S. and Anders, E. (1990). Meteoritic silicon carbide: pristine material from carbon stars. *Nature* **348**, 293–8.

Mahoney, W. A., Ling, J. C., Wheaton, W. A., Jacobsen, A. S. (1984). HEAO-3 discovery of ^{26}Al in the interstellar medium. *Astrophys. J.* **286**, 578–85.

Niemeyer, S. (1979). I–Xe dating of silicate and troilite from IAB iron meteorites. *Geochim. Cosmochim. Acta* **43**, 843–60.

Nuth, J. (1991). Small grains of truth. *Nature* **349**, 18–19.

Podosek, F. A. (1970). Dating of meteorites by the high-temperature release of iodine-correlated Xe129. *Geochim. Cosmochim. Acta* **34**, 341–65.

Reynolds, J. H. (1960). Determination of the age of the elements. *Phys. Rev. Lett.* **4**, 8–9.

Rotaru, M., Birck, J. L. and Allegre, C. J. (1992). Clues to early solar-system history from chromium isotopes in carbonaceous chondrites. *Nature* **358**, 465–70.

Schramm, D. N. and Wasserburg, G. J. (1970). Nucleo-chronologies and the mean age of the elements. *Astrophys. J.* **162**, 57–69.

Seeger, P. A., Fowler, W. A. and Clayton, D. D. (1965).

Nucleosynthesis of heavy elements by neutron capture. *Astrophys. J. Supp.* **11**, 121–66

Shukolyukov, A. and Lugmair, G. W. (1993a). Live iron-60 in the early solar system. *Science* **259**, 1138–42.

Shukolyukov, A. and Lugmair, G. W. (1993b). ^{60}Fe in eucrites. *Earth Planet. Sci. Lett.* **119**, 159–66.

Steele, I. M. and Hutcheon, I. D. (1979). Anatomy of Allende inclusions: mineralogy and Mg isotopes in two Ca–Al-rich inclusions. In: *Lunar Planet. Sci.* **X**, Lunar Planet. Inst., 1166–8 (abstract).

Swindle, T. D., Caffee, M. W., Hohenberg, C. M., Lindstrom, M. M. and Taylor, G. J. (1991). Iodine–xenon studies of petrographically and chemically characterized Chainpur chondrules. *Geochim. Cosmochim. Acta* **55**, 861–80.

Trivedi, B. M. P. (1977). A new approach to nucleocosmo-chronology. *Astrophys. J.* **215**, 877–84.

Truran, J. W. and Cameron, A. G. W. (1978). ^{26}Al production in explosive carbon burning. *Astrophys. J.* **219**, 226–9.

Wasserburg, G. J. (1985). Short-lived nuclei in the early solar-system. In: Black, D. C. and Matthews, M. S. (Eds.), *Protostars and Planets*. Univ. Arizona Press, pp. 703–37.

Wasserburg, G. J., Fowler, W. A. and Hoyle, F. (1960). Duration of nucleosynthesis. *Phys. Rev. Lett.* **4**, 112–14.

Wasserburg, G. J. and Hayden, R. J. (1955). Time interval between nucleogenesis and the formation of meteorites. *Nature* **176**, 130–1.

Wasserburg, G. J., Lee, T. and Papanastassiou, D. A. (1977). Correlated O and Mg isotopic anomalies in Allende inclusions: II magnesium. *Geophys. Res. Lett.* **4**, 299–302.

Wasserburg, G. J. and Papanastassiou, D. A. (1982). Some short-lived nuclides in the early solar-system – a connection with the placental ISM. In: Barnes, C. A., Clayton, D. D. and Schramm, D. N. (Eds.), *Essays in Nuclear Astrophysics*. Cambridge Univ. Press, pp. 77–140.

Wasserburg, G. J., Schramm, D. N. and Huneke, J. C. (1969). Nuclear chronologies for the galaxy. *Astrophys. J. (Lett.)* **157**, L91–6.

Zinner, E., Amari, S., Anders, E. and Lewis, R. (1991). Large amounts of extinct ^{26}Al in interstellar grains from the Murchison meteorite. *Nature* **349**, 51–4.

16 Fission track dating

16.1 Track formation

The spontaneous fission of ^{238}U releases about 200 MeV of energy, much of which is transferred to the two product nuclides as kinetic energy. They travel about 7 µm in opposite directions, leaving a single trail of damage through the medium which is about 15 µm long. Fission fragment tracks were originally observed in cloud chambers and photographic emulsions. Subsequently, Silk and Barnes (1959) produced artificial tracks in muscovite by irradiating uranium-coated flakes in a reactor. The resulting fragment tracks were observed at high magnification under the electron microscope.

'Fission tracks' (Fleischer et al., 1964) are only found in insulating materials. Fleischer et al. (1965a) proposed that the passage of the charged fission fragment causes ionisation of atoms along its path by stripping away electrons (Fig. 16.1a). The positively charged ions then repel each other, creating a cylindrical zone of disordered structure (Fig. 16.1b). This, in turn, causes relaxation stress in the surrounding matrix. It is the resulting 100 Å (10 nm)-wide zone of strain (Fig. 16.1c) which is actually seen under the electron microscope. Conductors do not display fission tracks because the free movement of electrons in their lattice structure neutralises the charged damage zone.

The ability to generate tracks depends on the mass of the ionising particle and the density of the medium. In muscovite, the lowest mass particle which can generate tracks by irradiation is about 30 atomic mass units (a.m.u.). Fission fragments, with masses of ca. 90 and 135 a.m.u. respectively, are well above this threshold, so that they always generate tracks. On the other hand, α particles, the major product of uranium decay, are so far below the critical mass that they cannot create tracks. Neither can they cause track erasure (Fleischer et al., 1965b).

Price and Walker (1962a) showed that when irradiated material was abraded to expose fission tracks at the surface, the damage zone could be preferentially dissolved by mineral acids, leading initially to a very fine channel only 25 Å wide. However, this could be enlarged by further chemical etching to yield a wide pit which was observable

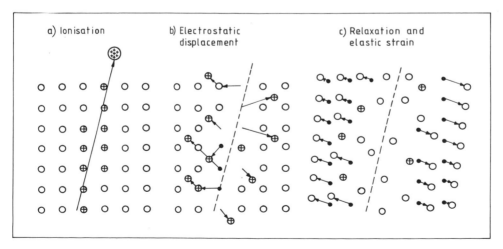

Fig. 16.1. Schematic illustration of the process of formation of a fission track in a crystalline insulating solid. After Fleischer et al. (1975).

under the optical microscope. Price and Walker (1962b) first discovered 'fossil' fission tracks in minerals, created by the spontaneous fission of dispersed uranium atoms. They went on to suggest (Price and Walker, 1963) that their density could be used as a dating tool for geological materials up to a billion years old. This was verified by Fleischer *et al.* (1965a) who obtained dates on artificial and natural glasses and minerals which were in agreement with other methods (Fig. 16.2).

Price and Walker (1963) demonstrated that spontaneous fission of ^{238}U was the only significant source of tracks in most natural materials. Induced fission of ^{235}U by natural thermal neutrons can be ignored, as can cosmic ray-induced fission of uranium. Spallation recoils induced by cosmic rays could, in principle, generate tracks in geological material exposed at the surface for very long time-periods. This is the principal source of tracks in meteorites (e.g. Lal *et al.*, 1969), but atmospheric shielding reduces their abundance to negligible levels in terrestrial rocks (Fleischer *et al.*, 1975). Therefore the total production of spontaneous fission tracks (F_s) per unit volume of rock can be derived from the general decay equation [1.9]:

$$F_s = \frac{\lambda_{\text{fission}}}{\lambda_\alpha}\,^{238}\text{U}\,(e^{\lambda_\alpha t} - 1) \qquad [16.1]$$

The ^{238}U fission decay constant is ca. 7×10^{-17} yr^{-1} ($t_{1/2} = 9.9 \times 10^{15}$ yr; Naeser *et al.*, 1989). There is some disagreement as to its exact value, but it will be seen below that this uncertainty need not enter into geological age determinations. Fissiogenic decay is over a million times lower than the α decay constant of ^{238}U, so it can be ignored in determining the isotopic abundance of uranium through time.

After polishing and etching a surface of the material to be dated, a fraction q of the total tracks will be visible at the surface. Therefore, the measured spontaneous fission track density, ρ_s, will be $q\,F_s$:

$$\rho_s = q\,\frac{\lambda_{\text{fission}}}{\lambda_\alpha}\,^{238}\text{U}\,(e^{\lambda_\alpha t} - 1) \qquad [16.2]$$

Price and Walker recognised that the most effective way of measuring the uranium concentration was to irradiate the sample with neutrons in a reactor, and thereby produce artificial tracks by the induced fission of ^{235}U. Following equation [16.2], the induced track

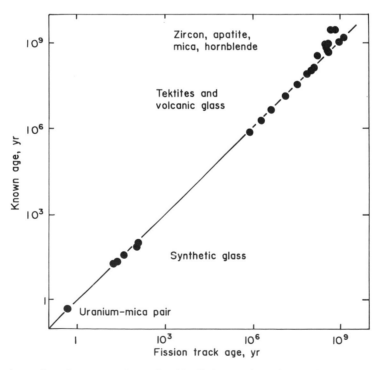

Fig. 16.2. A comparison of specimen ages determined by fission track analysis with those from historical or other radiometric sources. After Fleischer *et al.* (1965a).

density will be:

$$\rho_i = q\,^{235}U\phi\sigma \qquad [16.3]$$

where ϕ is the thermal neutron flux per unit volume and σ is the cross-section of ^{235}U for induced fission by thermal neutrons. If the sample material, including uranium concentration and etching procedure, is identical for these two experiments then the ratio of track densities can be used to solve for t, and then q goes out of the equation and the uranium concentrations are replaced by the $^{238}U/^{235}U$ isotope ratio only:

$$\frac{\rho_s}{\rho_i} = \frac{\lambda_{fission}}{\lambda_\alpha}\cdot\frac{137.88}{\phi\sigma}(e^{\lambda_\alpha t}-1) \qquad [16.4]$$

This can be rearranged to yield an equation in terms of t:

$$t = \frac{1}{\lambda_\alpha}\ln\left(1+\frac{\rho_s}{\rho_i}\cdot\frac{\lambda_\alpha}{\lambda_{fission}}\cdot\frac{\phi\sigma}{137.88}\right) \qquad [16.5]$$

It is possible to determine ϕ and σ directly by using flux monitors such as iron wire or copper foil. However, these types of flux monitors may not react to reactor conditions in exactly the same way as geological material. Therefore, an alternative procedure is to do a fission track analysis of a standard material with known uranium concentration. Fleischer *et al.* (1965a) used fragments of glass microscope-slides to calibrate the Brookhaven graphite reactor in this way. However, this does not avoid the uncertainty of the ^{238}U fission decay constant.

To eliminate both the flux term and the decay constant term, many workers started to use minerals dated by K–Ar as internal standards for the irradiation. Fleischer and Hart (1972) formalised

this system into the 'zeta calibration'. A sample of known age is used to calculate ζ by rearranging equation [16.4] and dividing both sides by the track density ρ_d in a given glass dosimeter:

$$\zeta = \frac{\phi\sigma}{137.88\lambda_{fission}\rho_d} = \frac{e^{\lambda_\alpha t}-1}{\lambda_\alpha(\rho_s/\rho_i)\rho_d} \qquad [16.6]$$

To date an unknown sample, the age equation [16.5] is now modified by substitution of ζ:

$$t = \frac{1}{\lambda_\alpha}\ln\left(1+\frac{\zeta\lambda_\alpha\rho_s\rho_d}{\rho_i}\right) \qquad [16.7]$$

The failure to resolve the decay constant problem can perhaps be attributed to this method, which transfers the uncertainty into the age determination of the geological reference material. Use of such material was recommended for all fission track dating studies by a working group of the IUGS Subcommission on Geochronology (Hurford, 1990). One of the most well-known of these standards is the 28 Myr-old Fish Canyon Tuff, Colorado (Naeser *et al.*, 1981; Hurford and Green, 1983).

16.2 Track etching

Several different types of geological material are suitable for the determination of fission track ages. Fleischer and Price (1964a) tested them with different acid or alkali leaching solutions to determine the most effective for track observation. The precise progress of the etching process depends on the composition of the matrix and the nature, concentration and temperature of the acid. This can give rise to a surprising variation in the appearance of etched tracks in different materials (e.g. Fig. 16.3), and may affect the

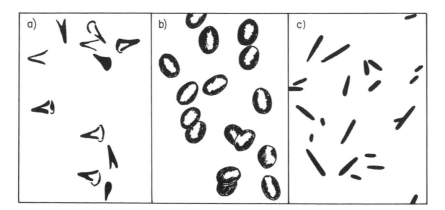

Fig. 16.3. Drawings of etched fission tracks induced by the same source (^{252}Cf) in different material: a) K-feldspar; b) soda-lime glass; c) lexan polycarbonate. Width of each field is 40 μm. From photographs by Fleischer *et al.* (1968).

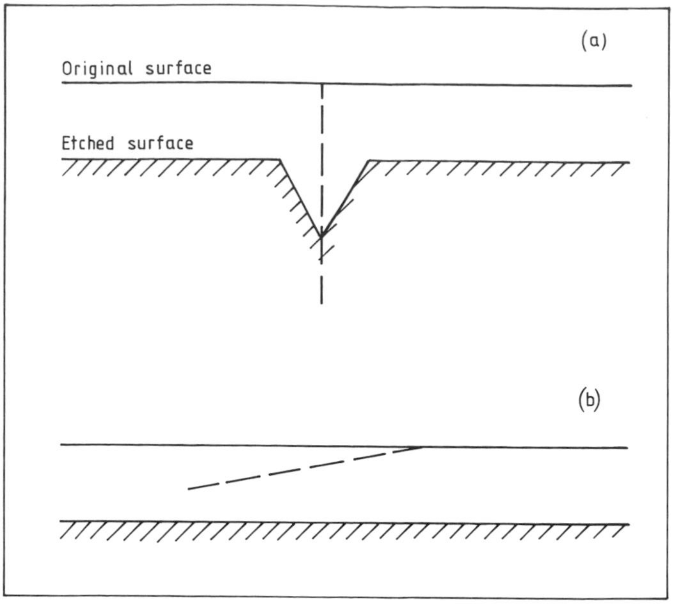

Fig. 16.4. Schematic illustration of the progress of track etching. a) Perpendicular to surface; b) tangential. After Fleischer and Price (1964b).

accuracy of track counting. These problems were discussed by Fleischer and Price (1964b) in an assessment of the fission track dating of glass.

The geometry of an etched track depends on the rate of etching down the axis of the track (from its intersection with the surface), relative to the general rate of attack of the polished surface (Fig. 16.4a). One problem in accurate track counting is to distinguish etched tracks from other features. For example, track pits in glass are at first pointed, but with increased etching time they round out. The optimal etching time is then a compromise between the need to make large enough pits to count quickly, and the tendency for large round-bottomed pits to be confusable with etched porosity. However, this is not such a problem in mineral phases.

Another source of uncertainty for both glass- and mineral-dating is caused by tracks which barely register in the etched surface. For example, tracks which are almost tangential to the surface may be completely erased by etching (Fig. 16.4b). Other tracks may not have intersected the original polished surface, but are exposed by the general attack of the surface during etching. These discrepancies will average out statistically if large numbers of tracks are counted with identical spatial geometry (see below), but may cause large errors when spatial geometry varies. A more detailed discussion of track

formation and track etching is given by Fleischer *et al.* (1975).

Fleischer and Price (1964a) estimated the dating range of fission track analysis with different types of material. Using the criterion that dates of reasonable precision can only be determined when the track density is at least 100 per cm^2, the lower end of the dating range can be estimated for different types of material according to uranium content (Fig. 16.5).

16.3 Counting techniques

Close examination must now be given to the assumptions involved in fission track dating. The first of these, noted above, is that the induced track count is performed on identical material to the spontaneous track count. Several different experimental methods are available which attempt to reach this ideal. Different approaches may be best for different sample material.

16.3.1 Population method

This expression was coined by Naeser (1979a), but was effectively the method adopted by the earliest workers (e.g. Price and Walker, 1963). The term refers to the fact that the spontaneous and induced tracks are counted in different splits or sub-populations of

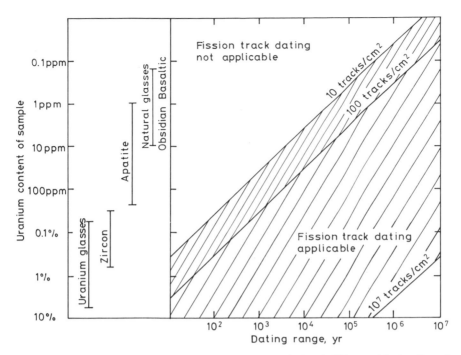

Fig. 16.5. Diagram to show the dating range for fission track analysis of different kinds of geological material according to uranium content. After Wagner (1978).

material, which are nevertheless assumed to sample the same population. This depends on the material having a homogeneous distribution of uranium between the two splits. The method has proved particularly successful for dating glass and apatite, but unsuccessful for sphene and zircon, where uranium distribution is very variable both within and between grains.

The sample is separated into two splits (Fig. 16.6). One is irradiated with thermal neutrons along with the standard (flux-monitor). Both spontaneous and induced tracks are to be registered under spherical (4π) spatial geometry. Therefore, after irradiation of the induced-track split, both splits are mounted in epoxy, ground, polished and etched under identical conditions. This reveals an internal surface of the material and also removes any extraneous superficial tracks generated by uranium-bearing dust particles. Track densities are counted in both splits. The induced-track density is calculated by subtracting the spontaneous-track density (un-irradiated sample) from the total-track density (irradiated sample).

The population method should be statistically tested by counting track densities in numerous grains or glass shards in each split. Alternatively, if a large piece of glass or mineral is available it can be cut or

cleaved so that the two faces to be counted are nearly identical sections through the sample. The latter method was adopted by Price and Walker (1963) in their analyses of muscovite. Price and Walker took the extra precaution of irradiating the split for spontaneous-track counting in a cadmium box (which screens out thermal neutrons) so that both splits should be treated as nearly identically as possible prior to etching. However, this precaution has now been dispensed with.

In the analysis of apatite, pre-irradiation heating of the induced-track split has been found advantageous to erase all spontaneous tracks by thermal annealing (see below). This allows the induced track density to be determined directly in the irradiated split. However, this procedure may be problematical in dating glass because it may affect the etching properties of the irradiated split, leading to systematic track counting errors.

16.3.2 External detector method

In this technique (Fleischer *et al.*, 1965a) the uranium content of the material to be dated is determined by inducing counts in an external detector rather than in the sample material itself. The sample is ground,

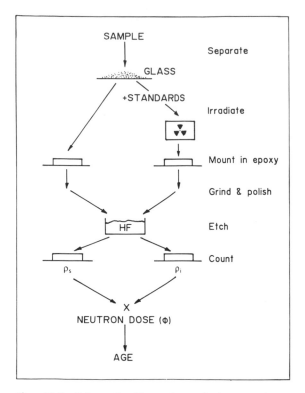

Fig. 16.6. Schematic illustration of the population method of fission track analysis. After Naeser and Naeser (1984).

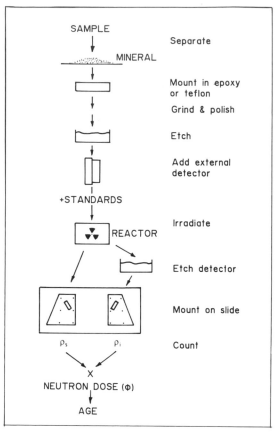

Fig. 16.7. Schematic illustration of the external detector method of fission track analysis, as described by Naeser (1979a). In this version the counting of spontaneous tracks is performed after irradiation, unlike the sequence described in the text. After Naeser and Naeser (1984).

polished, etched and counted, after which a sheet of detector material is placed in intimate contact with the etched surface. This must be done with absolute cleanliness to exclude uranium-bearing dust grains (see above). The external detector is commonly a low-uranium mica or a plastic such as lexan. After irradiation, the external detector is removed from the sample, etched and counted (Fig. 16.7).

The advantage of the external detector method is that both spontaneous and induced tracks are generated by the same sample material. Hence, it is suited to the analysis of material with a very heterogeneous distribution of uranium. The main disadvantage of the method is that the spontaneous and induced tracks are recorded under different spatial geometry conditions (Fig. 16.8). Spontaneous tracks are generated in the interior of the rock, and can therefore be formed by uranium atoms both above and below the etched plane (spherical or 4π geometry). In contrast, tracks induced in the external detector come out from the surface of the analysed material and are therefore generated with approximately one-half the frequency (hemi-spherical or 2π

geometry). Reimer *et al.* (1970) questioned whether the efficiency of induced-track formation is exactly 50%, or whether small biasses are introduced. However, subsequent experiments (discussed by Hurford and Green, 1982) showed that in most cases the ideal efficiency of 50% is achieved.

16.3.3 Re-etching method

This technique, described by Price and Walker (1963), is similar to the external detector method in that a sample is irradiated *after* polishing, etching and counting of spontaneous tracks. However, the sample itself is now re-etched and re-counted to determine the induced-track density by subtraction. As for the external method, induced tracks are formed

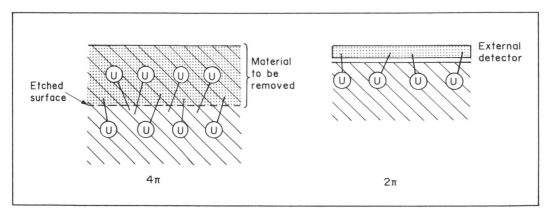

Fig. 16.8. Schematic illustration of the difference between 4π and 2π geometry in track formation.

with only 50% efficiency (2π geometry). The disadvantage of this method is that spontaneous-track pits will be unduly enlarged after the second etch, and may obscure some induced tracks. It is consequently less popular than the external method.

16.3.4 Re-polishing method

This technique (Naeser *et al.*, 1989) is an improvement on the re-etching method, and yields results similar to the 'mirror image' population method (Price and Walker, 1963). The sample is polished, etched and counted for spontaneous-track density. After irradiation it is re-polished to a depth of at least 20 μm to reveal a new internal face with 4π track geometry. This is then etched and counted to determine the induced-track density by subtraction. The method has the advantage that both spontaneous and induced tracks are recorded under identical geometry, and spontaneous tracks are not over-enlarged by double etching. Also, surface contamination during irradiation is not a problem. The spontaneous and induced tracks are not generated by exactly the same sample material, but the two etched surfaces are so close together uranium inhomogeneity in the grain as a whole is unlikely to significantly bias the data. A disadvantage compared with the normal population method is that the two etching steps are performed separately, and may therefore vary slightly in efficiency.

16.4 Detrital populations

An advantage of the external detector method of fission track counting is the ability to determine a separate age from each grain of the population (this

also applies to the less-widely used re-polishing method). This capability is useful if a heterogeneous age population is suspected in the sample, as in the case of sedimentary rocks with mixed provenance (e.g. Hurford and Carter, 1991). One type of such material is represented by volcanic ash (tephra) deposits which have been re-worked in the sedimentary environment. These may be important stratigraphic markers when associated with vertebrate fossil remains (e.g. Kowallis *et al.*, 1986).

Fission track results for individual detrital grains may be presented in histogram form. However, a more quantitative age estimate is possible if errors are assigned to each individual grain determination, so that the data can be presented as a probability density function (Hurford *et al.*, 1984). This function is simply the summation of the Poisson age distributions for each of the individual grain determinations. Fig. 16.9 shows such a plot for zircons from the re-worked El Ocote tephra deposit in Mexico, which displays a bimodal age distribution (Kowallis *et al.*, 1986). The younger peak places a maximum age on the time of sedimentary re-working, and is in agreement with the estimated biostratigraphic age of associated fossil material. These results show that application of the population method to fission track dating of this tephra would yield a meaningless average of the two age populations.

Walter (1989) suggested that additional assessments could be made of the quality of detrital fission track ages if the raw data (spontaneous- *versus* induced-track densities) were plotted for each grain. This yields an isochron diagram (Fig. 16.10) where the slope of each correlation line is proportional to age. These lines should pass through the origin, corresponding to a grain with zero uranium content. The linearity of each correlation line can be used to assess the influence of systematic analytical errors or geological disturbance on the reliability of the best-fit ages.

16.5 Track annealing

From the very beginning of fission track studies (Silk and Barnes, 1959) it has been known that fission

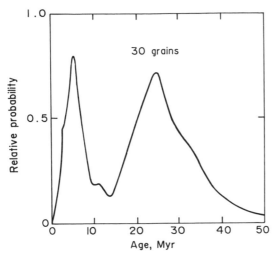

Fig. 16.9. Plot of probability density as a function of age for fission track data on detrital zircons from the re-worked El Ocote tephra from Mexico. After Kowallis *et al.* (1986).

tracks can fade under certain conditions. This was first seen as a result of electron bombardment during microscopy. However, elevated temperatures are the most important cause of track fading or 'annealing'. During this process the displaced ions within the damage track lose their charge and return to their normal lattice positions, after which the track is no longer susceptible to preferential acid attack.

In experiments on track annealing in mica, Fleischer *et al.* (1964) claimed that track annealing progressed by the accumulated 'healing up' of short segments at random points along the length of tracks. However, subsequent work on other materials (e.g. on glass by Storzer and Wagner, 1969) suggested that the healing process occurs principally at the ends of each track, causing a regular and progressive short-ening. As the length of tracks is diminished by healing they have a smaller probability of intersecting the free surface during the etching treatment. Hence, fewer tracks become etched and the apparent track density decreases (Fig. 16.11). This correlation between track length and track density is termed the 'random line segment model' (Fleischer *et al.*, 1975).

Early studies showed that different materials have different degrees of resistance to fission track annealing (Fleischer and Price, 1964a). In addition, however, a temperature–time relationship is found for the annealing process. The higher the temperature, the shorter the time required for complete annealing of tracks in any given material. To examine this

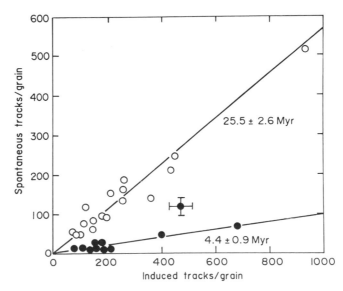

Fig. 16.10. Spontaneous- *versus* induced-track isochron diagram showing data for individual zircon grains from the El Ocote tephra. After Walter (1989).

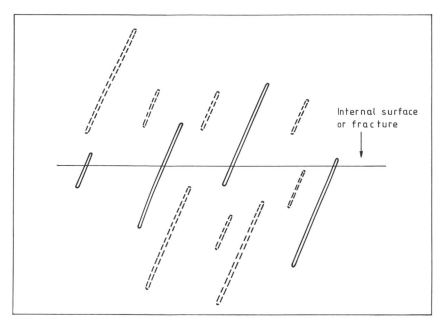

Fig. 16.11. Schematic illustration of the effect of track shortening on the observed density of etched tracks. Short and long tracks are of equal abundance, but the latter have a higher probability of becoming etched. After Laslett *et al.* (1982).

behaviour, Fleischer and Price (1964b) performed laboratory annealing experiments on the mineral indochinite and found that annealing obeyed a Boltzmann's law relation:

$$t = A\, e^{E/kT} \qquad [16.8]$$

where t is the time for track fading, A is a constant, E is the activation energy, k is Boltzmann's constant and T is absolute temperature. Much of the work since this time has been devoted to determining accurate Boltzmann relation annealing curves for different materials, both by laboratory and well-constrained geological studies.

Detailed laboratory experiments were performed on apatite and sphene by Naeser and Faul (1969) and on tektite glass by Storzer and Wagner (1969). These studies showed that annealing is a progressive process. Different degrees of track annealing in different materials each define their own Boltzmann's relation lines when shown on Arrhenius plots of time against reciprocal temperature (Fig. 16.12). This implies that as annealing progresses (as measured by the fraction of tracks lost) it also becomes progressively more difficult (Storzer and Wagner, 1969). Hence, when comparing the annealing properties of different minerals it is necessary to compare equal fractions of track loss, such as 50%

(Fig. 16.12). However, the strong fanning of lines in Fig. 16.12 (reflecting a range of activation energies for annealing) may be due to compositional variations within each type of material.

Following this line of investigation, Storzer and Poupeau (1973) compared laboratory annealing rates (in the same material) for freshly induced tracks and spontaneous tracks which had been partially annealed in nature. They found that as temperature was raised the fresh tracks were initially lost at a much higher rate, but that at a certain 'plateau' temperature the rates of annealing became equal.

Storzer and Poupeau argued that if both spontaneous and induced tracks were subjected to a heat treatment before counting then fission track ages could be corrected for partial annealing in the environment. Track counting must be by the population method; therefore the sample must have a uniform distribution of uranium. After irradiation of the induced-track sample, track counting analysis is performed by stepwise annealing of both spontaneous- and induced-track samples in the laboratory. After each heating step a new surface of both samples is polished, etched and counted.

Results from this procedure are shown in Fig. 16.13 for a North American tektite. Above a certain threshold temperature (ca. 100 °C), induced tracks

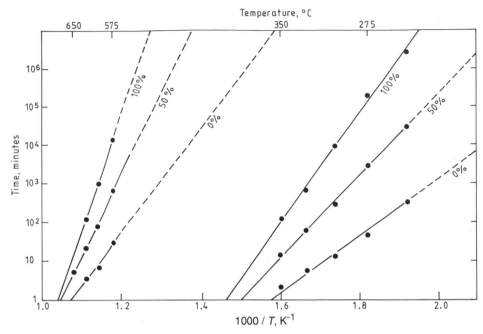

Fig. 16.12. Arrhenius plot to show the coherent progress of annealing in apatite and sphene. After Naeser and Faul (1969).

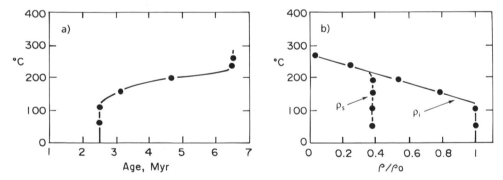

Fig. 16.13. Demonstration of the 'plateau-annealing' technique on a North American tektite. a) Apparent fission track age as a function of temperature step; b) fraction of induced and spontaneous tracks remaining at a given temperature (relative to initial density ρ_0). After Storzer and Wagner (1982).

start to fade, but spontaneous tracks are resistant. Therefore the apparent fission track age increases rapidly with temperature. However, as laboratory heating approaches the temperature at which annealing occurred in the environment, spontaneous tracks also start to fade, and the apparent age therefore reaches a plateau (Fig. 16.13). Storzer and Wagner (1982) argued that this 'plateau-annealing' technique can yield corrected fission track ages in glasses with a precision of $\pm 10\%$ (2σ).

16.6 Uplift and subsidence rates

Wagner and Reimer (1972) demonstrated the usefulness of apatite fission track ages for tectonic studies by applying them to Alpine uplift rates. Subsequently, Wagner et al. (1977) developed this technique by measuring apatite fission track ages over a 3000 m range of vertical relief in the Central Alps. Fission track ages on Alpine apatites do not conform to metamorphic isograds or terrane boundaries, but

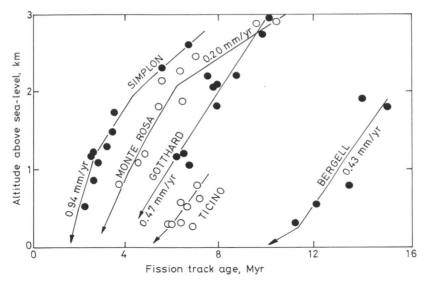

Fig. 16.14. Plot of fission track ages against topographic altitude for Alpine apatites, and deduced apparent uplift rates. After Wagner *et al.* (1977).

display a strong correlation with topographic relief (Fig. 16.14). They clearly represent cooling ages from Alpine metamorphism due to tectonic uplift, and can be used directly to calculate apparent uplift rates over the last few million years.

If a 'freezing-in' or 'blocking' temperature could be calculated for the Alpine apatites then the uplift rates in Fig. 16.14 could be converted into cooling rates. Two problems are faced in this task. The first is that the laboratory experiments show that blocking occurs over a range of temperatures. The second is that this range is itself dependent on the cooling rate. Hence, the argument is to some extent circular. Wagner *et al.* (1977) estimated a temperature for 50% track retention of 100–120 °C, half-way between 0% and 100% annealing temperatures of 60 and 180 °C, at a cooling rate of ca. 20 °C/Myr estimated from Rb–Sr biotite ages.

By combining apatite fission track data with biotite K–Ar and Rb–Sr, muscovite K–Ar, muscovite Rb–Sr and monazite U–Pb data, Wagner *et al.* were able to calculate cooling rates for different regions of the Alps over the last 35 Myr (Fig. 16.15). These results suggest that cooling in the Central Alps (Ticino and Gotthard areas) has been relatively uniform, while that in the East (Bergell) has slowed and the West (Simplon and Monte Rosa) has speeded up in the last few Myr. These conclusions are consistent with Fig. 16.14, and suggest that the Alps have undergone differential geographic uplift through time.

The idea of using a vertical traverse of apatite fission track ages to deduce tectonic histories was applied by Naeser (1979b) to bore-hole studies of sedimentary basins. Naeser proposed that in sedimentary sequences which are at their maximum burial temperature, apparent fission track ages would show a relationship with burial depth similar to Fig. 16.16. At shallow depths, burial heating is insignificant and fission track ages reflect the sediment source (detrital ages). As depth of burial increases, apatites undergo increased thermal annealing, and display decreasing apparent fission track ages, until they finally reach a total annealing zone with zero apparent age. The interval between zero and total annealing is called the Partial Annealing Zone (PAZ).

The upper- and lower-temperature bounds of the PAZ will depend on the age of the sedimentary basin. Naeser (1981) collected fission track age data from sedimentary basins with different burial rates. By making geological estimates of the effective burial (annealing) time in each basin, Naeser was able to make geological determinations of the Boltzmann lines for thermal annealing in apatite. These were confirmed by Gleadow and Duddy (1981) in a study of bore-hole data from the Otway Basin of Victoria, SE Australia.

The effective annealing time at present-day down-hole temperatures was estimated from the burial history of the basin (Fig. 16.17), suggesting that peak temperatures have been maintained for ca. 30 Myr.

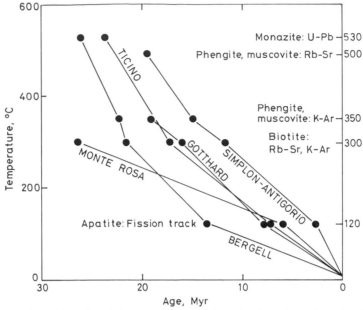

Fig. 16.15. Proposed cooling history for different regions of the Alps, based on time since closure of different radiogenic mineral systems. After Wagner *et al.* (1977).

Using these estimates, annealing properties were determined for the Otway Basin apatites (Fig. 16.18), which were consistent with other bore-hole data (Naeser, 1981). In addition, the Boltzmann line for 50% annealing was consistent with laboratory annealing experiments. However, the temperature *interval* between 0 and 100% annealing was narrower than that predicted by the divergence of Arrhenius relation annealing lines from the laboratory data of Naeser and Faul (1969).

A complicating factor in the analysis of track fading in apatite is the discovery that annealing temperature is compositionally dependent (Green *et al.*, 1985). Fission track analyses were performed on individual apatite grains from a single horizon in an Otway drill hole with a present-day temperature of 92 °C. These conditions result from progressive burial over the last 120 Myr. Chlor-apatite grains were found to give results near the depositional age, whereas fluor-apatites gave ages as low as zero (Fig. 16.19). Hence, when laboratory and geological annealing processes are compared, it is important that the material in the two types of experiment is as near compositionally identical as possible.

Bearing these findings in mind, Green *et al.* (1985) argued that Boltzmann lines for different percentages of track annealing did not have a fan-shaped distribution, but were parallel. This would imply that the activation energy for track fading was

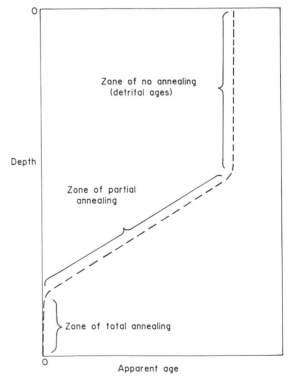

Fig. 16.16. Schematic illustration of the variation of apparent fission track age with depth in bore-hole samples from a sedimentary basin. After Naeser (1979b); Naeser *et al.* (1989).

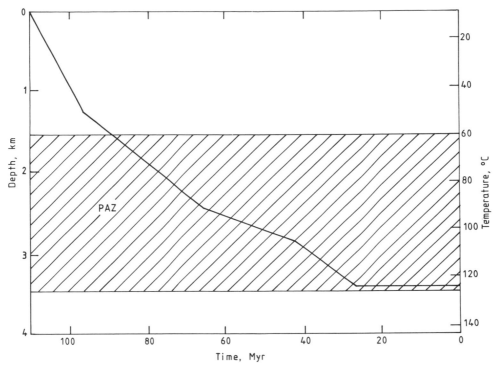

Fig. 16.17. Estimate of the burial history of the Otway Basin, SE Australia, based on a geothermal gradient of 30 °C/km. Shaded zone indicates width of PAZ determined from fission track analysis at different depths. After Gleadow and Duddy (1981).

constant for all tracks in a given sample, although track fading would still occur over a temperature interval, as assumed in the plateau method. However, close examination of the data set of Green *et al.* (not shown here) suggests that Boltzmann lines *are* divergent, although not to the extent suggested by early experiments. This conclusion is supported by more recent comparisons between laboratory data for apatite annealing (Laslett *et al.*, 1987) and drill-hole data from the Otway basin, Australia (Green *et al.*, 1989). Using more sophisticated models for the thermal and burial history of basin development, the laboratory and drill-hole data give coherent results.

In the above discussion, geologically well-known thermal basin histories were used to calibrate the annealing behaviour of apatite tracks. Given this background, fission track data could then be used to study geologically unknown basins. For example, Briggs *et al.* (1981) used this approach in order to compare the thermal histories of two sedimentary basins in the San Joaquin Valley of California. The Tejon oil field is divided into two parts by the seismically active White Wolf fault. One part, the

Basin Block, was a Late Tertiary depocentre which underwent strong subsidence. The other, Tejon Block, was less depressed. Fission track analysis of apatite from bore-holes shows the different geological history of the two blocks (Fig. 16.20).

Naeser *et al.* (1989) used these data, along with Boltzmann annealing lines from other geological locations, to calculate the thermal histories of the two blocks. Given geological evidence that the present down-hole temperatures represent peak values, the temperatures necessary for total annealing can be used to calculate effective heating times of ca. 1 Myr and 10 Myr for the Basin and Tejon block respectively (Fig. 16.21). These results are consistent with geological evidence for much more rapid burial of the Basin block, and do not require a perturbation in geothermal gradient.

16.7 Track length measurements

Because annealing initially causes shortening of tracks rather than their complete erasure, the use of etched track densities to chart the progress of annealing is an indirect approach. Therefore, in samples with a

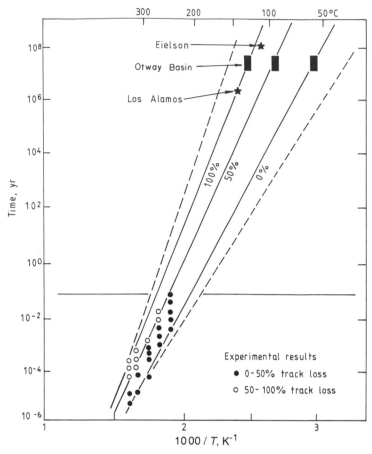

Fig. 16.18. Arrhenius plot for fission track annealing in apatite from Otway Basin bore-holes (blocks), other bore-holes (stars) and laboratory experiments (spots and dashed lines). After Gleadow and Duddy (1981).

complex thermal history, there will be considerable uncertainty in interpreting the apparent fission track age. Fission track data can be used more effectively to study the thermal history of a sample if the apparent age determined from track density is augmented by study of the length of etched tracks. Such studies were first made in dating micas (Maurette *et al.*, 1964; Bigazzi, 1967), but were applied in detail to tektite glasses by Storzer and Wagner (1969).

Etching of fission tracks in glasses tends to yield circular pits because of the smaller difference in structure, and hence etching rate, between tracks and the free surface. Consequently, the progress of annealing in glasses is accompanied by a decrease in the diameter rather than length of etched tracks (Fig. 16.22). Storzer and Wagner suggested that measurements of pit diameter could be used to correct fission track ages in glasses for the effects of annealing.

In some tektites, Storzer and Wagner measured an identical range of pit sizes resulting from the etching of spontaneous and induced fission tracks (Fig. 16.23a). These were interpreted as samples which had not suffered any annealing. In contrast, the range of pit sizes produced by etching of spontaneous tracks was displaced downwards in other samples (e.g. Fig. 16.23b). These samples must have suffered annealing in the geological environment, and gave younger fission track ages. Using the data from Fig. 16.22, Storzer and Wagner corrected these samples for the annealing process, whereupon they gave fission track ages within error of the un-annealed tektites (ca. 0.7 Myr). They speculated that the annealing process was a result of bush-fires in the Australian outback where the tektites were collected.

Under the 'random line segment model' of Fleischer *et al.* (1975) there should be a linear

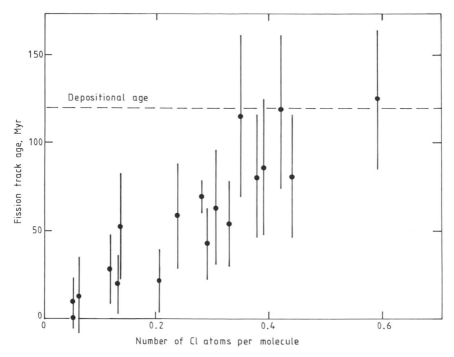

Fig. 16.19. Plot of measured fission track ages in individual apatites from Otway Group sandstones, Australia, to show compositional dependence of track annealing. After Green *et al.* (1985).

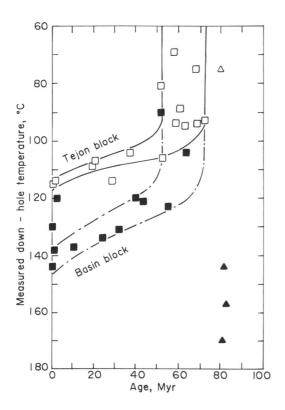

relationship between track density and average track length. However, when we only count track densities, we ignore additional information about the variation in track length on either side of the mean value. This *variation* in track length can yield additional information about the cooling history of a sample, because the longer the residence time of a sample in the partial annealing zone, the larger will be the variation in track length around the mean value.

Track length data can be collected in two different ways. One is to measure the apparent length of tracks which intersect the etched surface. These are termed 'projected track lengths' (e.g. Dakowski, 1978). The apparent length of these tracks is biassed from the true length distribution by three factors. Firstly, truncation by the surface reduces apparent length. Secondly, tracks undergo visual fore-shortening to an extent which depends on their angle to the surface.

Fig. 16.20. A comparison between fission track ages in bore-holes from the Basin and Tejon blocks of the San Joaquin Valley, California. Apatite data (■, □) give thermal history information while zircon data (▲, △) yield provenance ages. After Naeser *et al.* (1989).

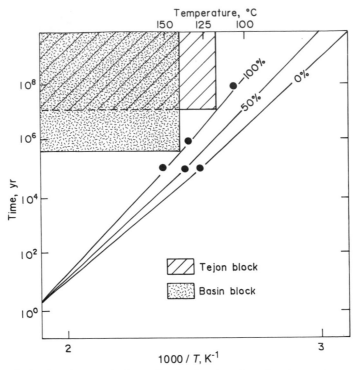

Fig. 16.21. Use of the Arrhenius plot to calculate effective heating times for the total annealing horizon in bore-holes from the Tejon and Basin blocks, San Joaquin Valley, California. After Naeser *et al.* (1989).

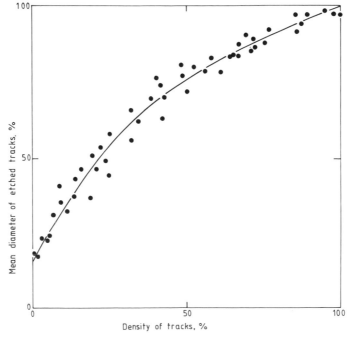

Fig. 16.22. Plot of mean pit diameter against pit density in variably annealed tektite glass samples, relative to the original pit size and density before annealing. After Storzer and Wagner (1969).

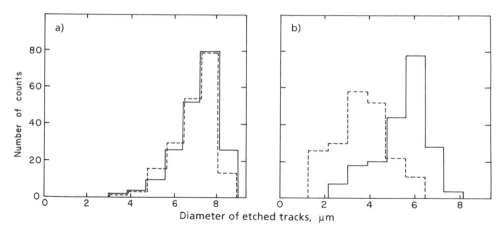

Fig. 16.23. Histograms of etched pit diameter for spontaneous (dashed) and induced (solid) fission tracks. a) Unannealed tektite; b) partially annealed tektite. After Storzer and Wagner (1969).

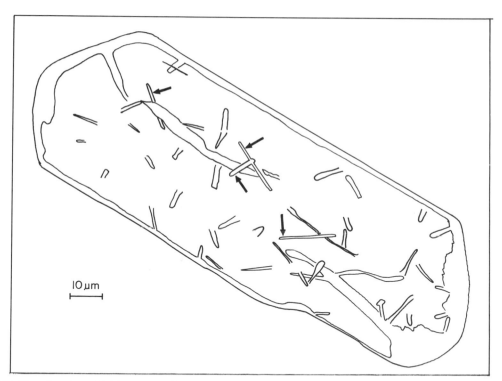

Fig. 16.24. Line drawing of high-contrast features in an etched apatite grain, viewed under dry (non immersion) conditions. Four confined tracks are visible (arrowed). From a photograph by Gleadow *et al.* (1986).

Thirdly, more frequent intersection of long tracks with the surface biases apparent length upwards. The alternative approach is to measure 'confined track length' (Bhandari *et al.*, 1971; Green, 1981; Laslett *et al.* 1982; 1984). These are tracks which do not break the general etched surface, but which become etched by the penetration of acid down a channel inside the mineral which intersects the track (Fig. 16.24). The two most common types are termed 'Track-IN-Track' (TINT) and 'Track-IN-CLEavage' (TINCLE) respectively (Lal *et al.*, 1969; Fleischer *et al.*, 1975).

Measurement of confined tracks which are

horizontal (parallel to the etched surface) leads to the minimum bias from true track length. Such tracks are recognised by the fact that the whole track goes in and out of focus at once when the objective lens is racked up and down. However, even these tracks are affected by three types of bias. One is the more frequent intersection of long tracks with an etching channel. The second is the effect of over-etching (Fig. 16.25a). It is necessary that some tracks become over-etched to ensure that other are not under-etched. Since tracks have effectively zero width before etching, over-etched tracks can be recognised by their non-zero

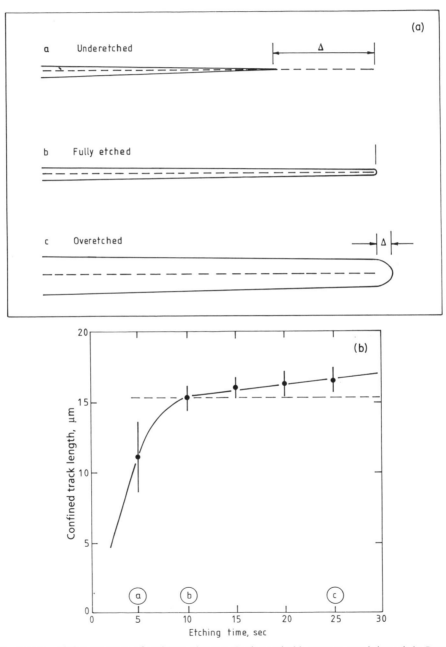

Fig. 16.25. Illustration of the progress of etching. a) For a single track; b) average track length in Durango apatite (Mexico) as a function of etching time in 5M HNO_3 at 21 °C. After Laslett *et al.* (1984).

width, which is almost the same as the excess length (2 × Δ, Fig. 16.25a). The optimum etching time is determined by experiments on incremental etching (Fig. 16.25b). The standard etching time of 20 seconds used by Laslett *et al.* (1984) leads on average to 1 μm of over-etching.

The third cause of bias in confined track lengths is the effect of crystallographic orientation (Green and Durrani, 1977). Compared to un-annealed apatite grains, in which track length is effectively equal in all directions, annealed apatites demonstrate less shortening of tracks parallel to the *c* axis than in other orientations (Fig. 16.26). Therefore, in order to see the whole range of track lengths in an annealed apatite (and thus gain the maximum information), the prismatic section is observed, containing tracks at all angles to the *c* axis (Gleadow *et al.*, 1986).

Laslett *et al.* (1982) and Gleadow *et al.* (1986) compared the results of projected and confined track length in samples with different thermal histories. They argued that while projected track lengths yield only subtle indications of different thermal histories, confined tracks gave clear diagnostic indicators of

thermal history. These can be divided into five types (Fig. 16.27a–e).

Induced tracks (Fig. 16.27a) are the longest and most uniform type (16 ± 1μm, based on several different sample types). Tracks in undisturbed volcanics and rapidly-cooled shallow intrusions are also uniform within a single sample (Fig. 16.27b), but there is some variation between sample means (ca. 14–15.5μm). This can be attributed to limited annealing at near-ambient temperatures over periods of tens of Myr. Tracks in undisturbed basement apatite (Fig. 16.27c) are somewhat shorter (means of 12–14μm), with a skewed distribution attributed to slow cooling from regional metamorphism. Finally, bimodal and mixed distributions (Fig. 16.27 d and e) are attributed to various types of two-stage thermal history, in which pre-existing tracks were partially erased by a thermal event between initial cooling and the present.

Although this approach is a clear advance in the understanding of thermal histories, it lacks quantitative control. For example, the measurement of confined track lengths is made subjective by the

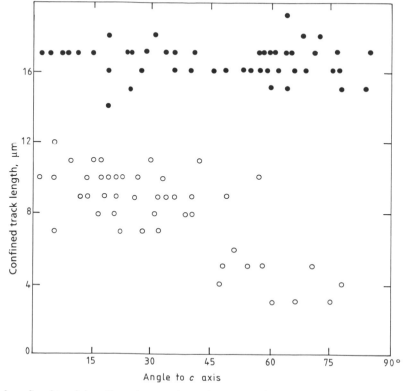

Fig. 16.26. Plot of confined track length against angle to the *c* axis of apatite grains. Solid = un-annealed. Open = annealed at 350 °C for 1 hour. After Laslett *et al.* (1984).

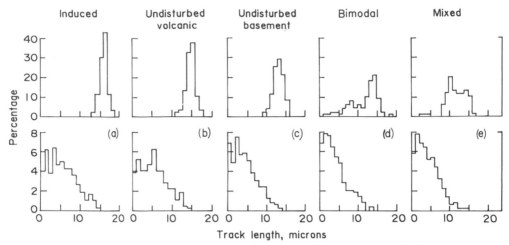

Fig. 16.27. Histograms of track length (as a percentage of total sample) for apatites with different types of thermal history (see text). Top row: horizontal confined tracks; bottom row: projected tracks. After Gleadow *et al.* (1986).

unconstrained depth range over which they can be observed. On the other hand, projected track lengths are biassed, but if they are ratioed against the length of projected *induced* tracks (in a population experiment) then these biasses are cancelled out. In this situation, Wagner (1988) showed that the upper end of the projected track length distribution *can* provide quantitative constraints on thermal histories (despite the poor control provided by the *mean* length of these tracks).

Wagner (1988) compared the percentage of spontaneous tracks in excess of 10μm long with the percentage of induced tracks over 10μm long, to yield the ratio 'c_s/c_i'. A value of 10μm was chosen as a convenient length cut-off; the exact figure is not critical because we are dealing with the *ratio* of spontaneous to induced tracks above this length. Wagner argued that the c_s/c_i ratio defined above could be used to study the lower temperature bound of the PAZ because the longest projected tracks are the most sensitive to incipient annealing. This is illustrated by laboratory annealing experiments (Fig. 16.28) where the fraction of remaining projected tracks over 10μm is compared with the fraction of total tracks remaining, after a given amount of annealing. The fraction of long tracks (c/c_o) is found to decrease much faster than the total track density (ρ/ρ_o).

Wagner *et al.* (1989) presented bore-hole evidence from the Otway Basin in support of a low-temperature cut-off of ca. 60 °C for the PAZ (Fig. 16.29). At bore-hole temperatures above 60 °C, the proportion of projected spontaneous tracks with

lengths above 10μm approached zero. They argued that the residue of long tracks at temperatures between 60 and 90 °C could be attributed to Cl-rich apatites with more resistance to annealing. Hence, a single study (measuring ca. 1000 spontaneous and induced tracks) can be used to estimate the time since a rock passed through both the upper- and lower-temperature boundary of the PAZ. The projected track length measurement is made at the same time as the track density count, using a digitising tablet.

Wagner and Hejl (1991) generalised the above approach by calculating an apparent fission track age for all projected track lengths exceeding a series of cut-off lengths (x) from zero to 15μm (at 1μm-intervals). Relative to the conventional fission track age ($x = 0$), the normalised age (t) at different values of x is given as:

$$t = c_s\rho_s/c_i\rho_i \qquad [16.9]$$

These apparent ages can then be plotted as an 'age-spectrum' diagram. The significant thing about this diagram is that the cut-off for preservation of fission tracks of different lengths is related to temperature, although probably not as a linear function. Hence, the apparent-age diagram is itself a plot of temperature against time, if we can calibrate the cut-off length against temperature. As a preliminary calibration we can adopt the following points:

1) A track length of zero should correspond to the conventional blocking temperature of ca. 120 °C which marks the upper-temperature limit of the PAZ.

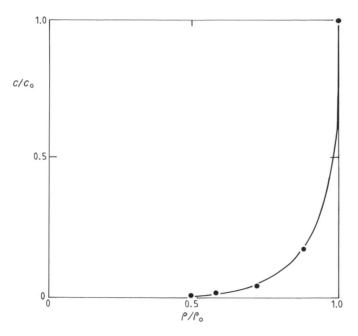

Fig. 16.28. Plot of remaining fraction of tracks over 10 μm (c/c_o) compared to total tracks remaining (ρ/ρ_o) after different degrees of laboratory annealing. After Wagner (1988).

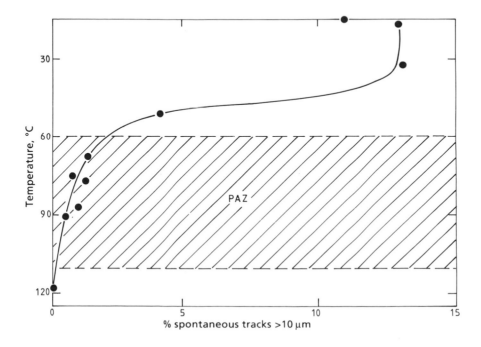

Fig. 16.29. Plot of the percentage of projected spontaneous tracks in apatite with lengths over 10 μm, against down-hole temperature in an Otway Basin drill-hole, to show the low-temperature bound of the partial annealing zone (PAZ). After Wagner et al. (1989).

2) Wagner (1988) assigned a cut-off length of 10μm to the blocking temperature of 60 °C at the lower-temperature limit of the PAZ.

3) A track length of 15μm represents the case of zero annealing, which may be sustained at ambient temperatures of ca. 30 °C.

Hence, a crude temperature calibration has been applied to the cut-off length scale by the present author (Fig. 16.30).

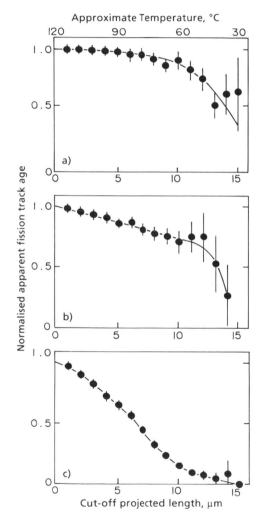

Fig. 16.30. Fission track age-spectrum diagrams for three geological examples: a) Fish Canyon Tuff; b) Alpine (Gotthard Pass) granite; c) paragneiss from 2353 m depth in the Continental Deep Drill hole, Bavaria. For discussion, see text. Modified after Wagner and Hejl (1991).

Wagner and Hejl presented fission track age-spectrum diagrams for three rocks with different thermal histories in order to explore the usefulness of this diagram (Fig. 16.30). The Fish Canyon Tuff from Colorado (Fig. 16.30a) shows the effects of rapid cooling through the PAZ, with minimal loss of short tracks, but does show some loss of long tracks. Therefore, it has either suffered a mild thermal pulse since eruption (for which there is little evidence), or else ambient temperatures are sufficient to cause some annealing. The Gotthard Pass granite from the Alps (Fig. 16.30b) shows the effect of slower cooling through the PAZ, due to a uniform but rapid uplift rate. Finally, a paragneiss from the Continental Deep Drill hole in Bavaria (Fig. 16.30c) shows the behaviour of a sample which is undergoing slow uplift, but is still buried. The sample came from 2353 m depth, with an ambient temperature of 70 °C, near the low-temperature limit (= upper depth limit) of the PAZ.

The above examples suggest that the age-spectrum fission track diagram is likely to be an important tool for studies of cooling processes in the future. They also suggest that the PAZ does not have a sharp cut-off at 60 °C, but a much more gradual low-temperature limit. Additional examples of the use of this technique are given by Grivet *et al.* (1993).

References

Bhandari, N., Bhat, S. G., Rajogopalan, G., Tamhane, A. S. and Venkatavaradan, V. S. (1971). Fission fragment lengths in apatite: recordable track lengths. *Earth Planet. Sci. Lett.* **13**, 191–9.

Bigazzi, G. (1967). Length of fission tracks and age of muscovite samples. *Earth Planet. Sci. Lett.* **3**, 434–8.

Briggs, N. D., Naeser, C. W. and McCulloh, T. H. (1981). Thermal history of sedimentary basins by fission-track dating. *Nucl. Tracks* **5**, 235–7 (abstract).

Dakowski, M. (1978). Length distributions of fission tracks in thick crystals. *Nucl. Track., Det.* **2**, 181–9.

Fleischer, R. L. and Hart, H. R. (1972). Fission track dating: techniques and problems. In: Bishop, W., Miller, J. and Cole, S. (Eds.), *Calibration of Hominoid Evolution*. Scottish Academic Press, pp. 135–170.

Fleischer, R. L. and Price, P. B. (1964a). Techniques for geological dating of minerals by chemical etching of fission fragment tracks. *Geochim. Cosmochim. Acta* **28**, 1705–14.

Fleischer, R. L. and Price, P. B. (1964b). Glass dating by fission fragment tracks. *J. Geophys. Res.* **69**, 331–9.

Fleischer, R. L., Price, P. B., Symes, E. M. and Miller, D. S. (1964). Fission track ages and track-annealing behaviour of some micas. *Science* **143**, 349–51.

Fleischer, R. L., Price, P. B. and Walker, R. M. (1965a). Tracks of charged particles in solids. *Science* **149**, 383–93.

Fleischer, R. L., Price, P. B. and Walker, R. M. (1965b). Effects of temperature, pressure, and ionization on the formation and stability of fission tracks in minerals and glasses. *J. Geophys. Res.* **70**, 1497–502.

Fleischer, R. L., Price, P. B. and Walker, R. M. (1968). Charged particle tracks: tools for geochronology and meteor studies. In: Hamilton, E. and Farquhar, R. M. (Eds), *Radiometric Dating for Geologists.* Wiley Interscience, pp. 417–435.

Fleischer, R. L., Price, P. B. and Walker, R. M. (1975). *Nuclear Tracks in Solids.* University of California Press, 605 p.

Gleadow, A. J. W. and Duddy, I. R. (1981). A natural long-term track annealing experiment for apatite. *Nucl. Tracks* **5**, 169–74.

Gleadow, A. J. W., Duddy, I. R., Green, P. F. and Lovering, J. F. (1986). Confined fission track lengths in apatite: a diagnostic tool for thermal history analysis. *Contrib. Mineral. Petrol.* **94**, 405–15.

Green, P. F. (1981). 'Track-in track' length measurements in annealed apatites. *Nucl. Tracks* **5**, 121–8.

Green, P. F., Duddy, I. R., Gleadow, A. J. W. and Tingate, P. R. (1985). Fission-track annealing in apatite: track length measurements and the form of the Arrhenius plot. *Nucl. Tracks* **10**, 323–8.

Green, P. F., Duddy, I. R., Laslett, G. M., Hegarty, K. A., Gleadow, A. J. W. and Lovering, J. F. (1989). Thermal annealing of fission tracks in apatite 4. Quantitative modelling techniques and extension to geological time-scales. *Chem. Geol. (Isot. Geosci. Section)* **79**, 155–82.

Green, P. F. and Durrani, S. A. (1977). Annealing studies of tracks in crystals. *Nucl. Track Det.* **1**, 33–9.

Grivet, M., Rebetez, M., Ben Ghouma, N., Chambaudet, A., Jonckheere, R. and Mars, M. (1993). Apatite fission-track age correction and thermal history analysis from projected track length distributions. *Chem. Geol. (Isot. Geosci. Section)* **103**, 157–69.

Hurford, A. J. (1990). Standardization of fission track calibration: recommendation by the Fission Track Working Group of the I.U.G.S. Subcommission on Geochronology. *Chem. Geol. (Isot. Geosci. Section)* **80**, 171–8.

Hurford, A. J. and Carter, A. (1991). The role of fission track dating in discrimination of provenance. In: Morton, A. C., Todd, S. P. and Haughton, P. D. W. (Eds.) *Developments in Sedimentary Provenance Studies. Geol. Soc. Spec. Pub.* **57**, 67–78.

Hurford, A. J., Fitch, F. J. and Clarke, A. (1984). Resolution of the age structure of the detrital zircon populations of two Lower Cretaceous sandstones from the Weald of England by fission track dating. *Geol. Mag.* **121**, 269–77.

Hurford, A. J. and Green, P. F. (1982). A users' guide to fission track dating calibration. *Earth Planet. Sci. Lett.* **59**, 343–54.

Hurford, A. J. and Green, P. F. (1983). The ζ age calibration of fission-track dating. *Isot. Geosci.* **1**, 285–317.

Kowallis, B. J., Heaton, J. S. and Bringhurst, K. (1986).

Fission-track dating of volcanically derived sedimentary rocks. *Geology* **14**, 19–22.

Lal, D., Rajan, R. S. and Tamhane, A. S. (1969). Chemical composition of nuclei of $Z > 22$ in cosmic rays using meteoritic minerals as detectors. *Nature* **221**, 33–7.

Laslett, G. M., Gleadow, A. J. W. and Duddy, I. R. (1984). The relationship between fission track length and track density in apatite. *Nucl. Tracks* **9**, 29–37.

Laslett, G. M., Green, P. F., Duddy, I. R. and Gleadow, A. J. W. (1987). Thermal annealing of fission tracks in apatite, 2. A quantitative analysis. *Chem. Geol. (Isot. Geosci. Section)* **65**, 1–13.

Laslett, G. M., Kendall, W. S., Gleadow, A. J. W. and Duddy, I. R. (1982). Bias in measurement of fission-track length distributions. *Nucl. Tracks* **6**, 79–85.

Maurette, M., Pellas, P. and Walker, R. M. (1964). Etude des traces fission fossiles dans le mica. *Bull. Soc. Franc. Miner. Cryst.* **87**, 6–17.

Naeser, C. W. (1979a). Fission-track dating and geological annealing of fission tracks. In: Jager, E. and Hunziker, J. C. (Eds.), *Lectures in Isotope Geology.* Springer-Verlag, pp. 154–69.

Naeser, C. W. (1979b). Thermal history of sedimentary basins: Fission-track dating of subsurface rocks. In: Scholle, P. A., and Schluger, P. R. (Eds.), *Aspects of Diagenesis. Soc. Econ. Paleontol. Mineral. Spec. Pub.* **26**, pp. 109–12.

Naeser, C. W. (1981). The fading of fission tracks in the geologic environment – data from deep drill holes. *Nucl. Tracks.* **5**, 248–50 (abstract).

Naeser, C. W. and Faul, H. (1969). Fission track annealing in apatite and sphene. *J. Geophys. Res.* **74**, 705–10.

Naeser, C. W., Zimmermann, R. A. and Cebula, G. T. (1981). Fission-track dating of apatite and zircon: an interlaboratory comparison. *Nucl. Tracks* **5**, 65–72.

Naeser, N. D. and Naeser, C. W. (1984). Fission-track dating. In: Mahaney, W. C. (Ed.), *Quaternary Dating Methods. Developments in Paleontology and Stratigraphy* **7**. Elsevier, pp. 87–100.

Naeser, N. D., Naeser, C. W. and McCulloh, T. H. (1989). The application of fission-track dating to the depositional and thermal history of rocks in sedimentary basins. In: Naeser, N. D. and McCulloh, T. H. (Eds.), *Thermal History of Sedimentary Basins.* Springer-Verlag, pp. 157–80.

Price, P. B. and Walker, R. M. (1962a). Chemical etching of charged particle tracks in solids. *J. Appl. Phys.* **33**, 3407–12.

Price, P. B. and Walker, R. M. (1962b). Observation of fossil particle tracks in natural micas. *Nature* **196**, 732–4.

Price, P. B. and Walker, R. M. (1963). Fossil tracks of charged particles in mica and the age of minerals. *J. Geophys. Res.* **68**, 4847–62.

Reimer, G. M., Storzer, D. and Wagner, G. A. (1970). Geometry factor in fission track counting. *Earth Planet. Sci. Lett.* **9**, 401–4.

Silk, E. C. H. and Barnes, R. S. (1959). Examination of fission fragment tracks with an electron microscope. *Phil.*

Mag. **4**, 970–2.

Storzer, D. and Poupeau, G. (1973). Ages plateaux de mineraux et verres par la methode des traces de fission. *C. R. Acad. Sci. Paris* **276**, 137–9.

Storzer, D. and Wagner, G. A. (1969). Correction of thermally lowered fission track ages of tektites. *Earth Planet. Sci. Lett.* **5**, 463–8.

Storzer, D. and Wagner, G. A. (1982). The application of fission track dating in stratigraphy: a critical review. In: Odin, G. S. (Ed.), *Numerical Dating in Stratigraphy.* Wiley, pp. 199–221.

Wagner, G. A. (1978). Archaeological applications of fission-track dating. *Nucl. Track. Det.* **2**, 51–63.

Wagner, G. A. (1988). Apatite fission-track geochrono-thermometer to 60 °C: projected length studies. *Chem. Geol. (Isot. Geosci. Section)* **72**, 145–53.

Wagner, G. A., Gleadow, A. J. W. and Fitzgerald, P. G. (1989). The significance of the partial annealing zone in apatite fission-track analysis: projected track length measurements and uplift chronology of the Transantarctic Mountains. *Chem. Geol. (Isot. Geosci. Section)* **79**, 295–305.

Wagner, G. A. and Hejl, E. (1991). Apatite fission-track age-spectrum based on projected track-length analysis. *Chem. Geol. (Isot. Geosci. Section)* **87**, 1–9.

Wagner, G. A. and Reimer, G. M. (1972). Fission-track tectonics: the tectonic interpretation of fission track apatite ages. *Earth Planet. Sci. Lett.* **14**, 263–8.

Wagner, G. A., Reimer, G. M. and Jager, E. (1977). Cooling ages derived by apatite fission-track, mica Rb–Sr and K–Ar dating: the uplift and cooling history of the Central Alps. *Mem. Inst. Geol. Min. Univ. Padova* **30**, 1–27.

Walter, R. C. (1989). Application and limitation of fission-track geochronology to Quaternary tephras. *Quat. Int.* **1**, 35–46.

Appendix A

Chart of the nuclides

(see two following pages)

The chart is broken into six sections. The atomic number, Z is plotted on the ordinate against the neutron number, N, on the abscissa. Mass numbers, A, are ringed. Stable and naturally occurring unstable nuclides are shown, along with a few extinct nuclides of cosmochemical interest. The abundance of stable nuclides is quoted in per cent. (Unstable nuclides with half-lives over 10^{12} yr are included in this category.) Half-lives of naturally occurring unstable nuclides are quoted in seconds, minutes, hours, days and years (s, m, h, d, yr). Data mainly from Lide, D. R. (Ed.) *CRC Handbook of Chemistry and Physics* 1994–1995 (75th Edn), CRC press.

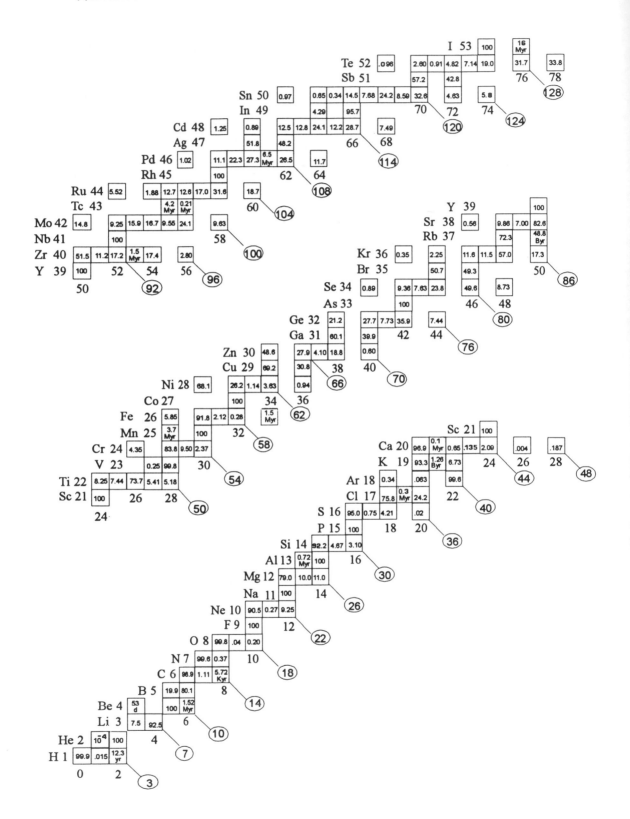

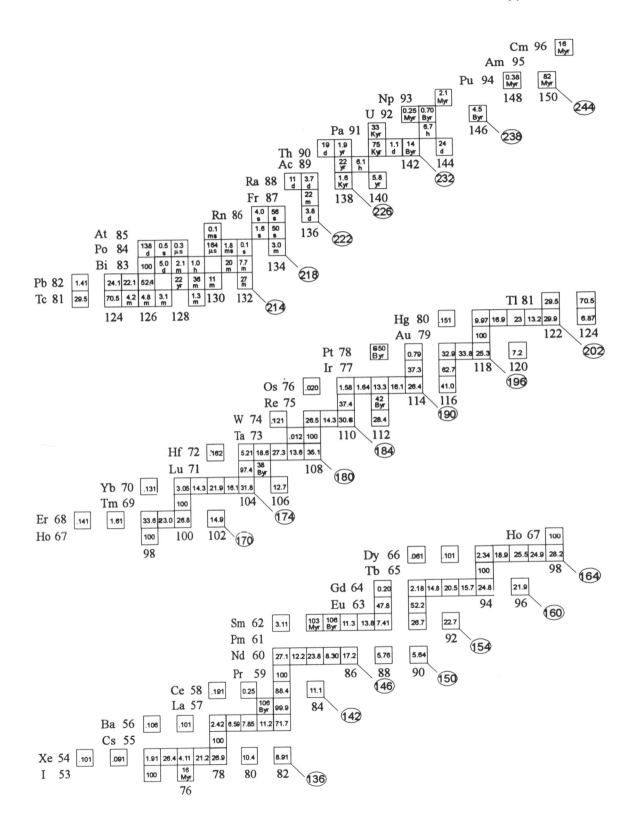

Appendix B

Updated material for the 1997 paperback edition

1.1 The chart of the nuclides

Recent experiments on the Darmstadt heavy ion accelerator, reviewed by Normile (1996), have helped to expand the field of experimentally known unstable nuclides on the neutron-rich side of the path of stability (Fig. 1.1). These new isotopes were manufactured by bombarding target material with ^{238}U, triggering fission reactions which produce a few very neutron-rich nuclei. The products were separated by mass spectrometry and their properties examined using high-energy detectors (e.g. Fig. 14.29). Knowledge about these unstable nuclei will improve our understanding of the nucleosynthetic r-process, which operates within this part of the nuclide chart (Fig. 1.7).

1.2 Nucleosynthesis

Problems have been recognised in the x-process model for generating the light elements Li, Be and B, since cosmic ray spallation cannot explain the observed isotope ratios of these elements in solar system materials. However, Casse *et al.* (1995) proposed that carbon and oxygen nuclei ejected from supernovae can generate these nuclides by collision with hydrogen and helium in the surrounding gas cloud. This process is believed to occur in regions such as the Orion nebula. The combination of supernova production with spallation of galactic cosmic rays can explain observed solar system abundances of Li, Be and B.

3.5 Seawater evolution

Neogene seawater evolution has provided a challenge to geochemists to see what are the finest-scale changes in Sr isotope evolution which can be documented. Dia *et al.* (1992, op. cit.) and Clemens *et al.* (1993, op. cit.) claimed to observe changes in ^{87}Sr/^{86}Sr, correlated with δ ^{18}O, with a periodicity of about 0.1 Myr. However, subsequent work by the same research groups (Henderson *et al.*, 1994; Clemens *et al.*, 1995) failed to reproduce these cycles in three drill cores (including two used in the original work). Instead, the new data fell on a linear evolution path defined by other studies (e.g. Hodell *et al.*, 1990, op. cit.). Hence, the apparent periodicity in the earlier work is attributed to analytical artifacts and does not reflect seawater Sr isotope evolution.

For the data of Dia *et al.* (1992), the analytical artifact was apparently a breakdown in the accuracy of the fractional correction. Thus, Clemens *et al.* (1995) were able to reproduce the temporal periodicity using a linear-law fractionation correction, but this also generated a positive correlation between $^{88}Sr/^{86}Sr$ and fractionation-corrected $^{87}Sr/^{86}Sr$ ratios, indicative of a fractionation bias (section 2.2.2). After correction of this bias, the periodicity disappeared. The data of Clemens *et al.* (1993) were not subject to this bias, since the more accurate exponential law was used. However, Henderson *et al.* (1994) showed that only 3 out of 75 samples analysed by Clemens *et al.* (1993) lay outside 2 σ (95%) confidence limits from the linear evolution path fitted through the data. Since four outliers would be expected at this confidence limit, the apparent periodicity in this data set is probably not statistically significant (Henderson *et al.*, Fig. 1b). Farrell *et al.* (1995) carried out a study with similar sampling density and analytical precision to the above work, but using 455 samples extending over the past 6 Myr. These data constrain the seawater evolution curve to an average confidence limit of ±0.00002 (2 σ). The curve shows undulations with a 1–2 Myr periodicity which are realistic reflections of changing Sr fluxes, given the 2.5 Myr residence time of Sr in the ocean system.

The two principal mechanisms which have been proposed to account for short-term variation in seawater Sr isotope ratio are Himalayan uplift and glacial cycles. Recent work explores these mechanisms.

Blum and Erel (1995) attempted to quantify the amount of radiogenic Sr that could be released by glacial erosion. In order to do this, they used ammonium acetate leaching to analyse the isotopic composition of exchangeable Sr in weathered glacial moraines. Weathered soils from six moraines displayed a negative correlation between the isotopic composition of leachable Sr and the age of the soil (Blum and Erel, Fig. 1). Most notably, a very radiogenic $^{87}Sr/^{86}Sr$ ratio of 0.795 was observed in soil from the youngest (400 yr old) moraine. Blum and Erel used these data to argue that a spike of

radiogenic Sr is released by weathering of moraines immediately after glaciation. Modelling of this spike suggested that it could yield an incremental increase in $^{87}Sr/^{86}Sr$ of 0.00005 for each 100 kyr glacial cycle. If repeated over the Pleistocene period, this could reproduce (within error) the evolution slope of Hodell *et al.* (1990). However, the results of Blum and Erel are heavily weighted by the youngest data point. The older (2 kyr – 300 kyr) soils also display a positive correlation between Sr in exchangeable sites and whole-soil digestion, suggesting that soil composition plays a significant role in controlling the isotopic composition of leachable Sr, as well as soil age. Hence, more data are needed to test the model.

Harris (1995) re-examined mechanisms for the 20 Myr old peak in the rate of change of seawater $^{87}Sr/^{86}Sr$ ratio (Fig. 3.30). He claimed that there is no evidence in the Bengal Fan for increased Himalayan erosion 20 Myr ago. Instead, he suggested that the inferred ^{87}Sr spike in river water at that time was due to the exposure and chemical weathering of metasedimentary rocks with a large budget of leachable ^{87}Sr.

Additional work on the Bengal Fan by Derry and France-Lanord (1996) suggested that relationships between erosion and river water composition are more complex. These workers studied the mineralogical and isotopic stratigraphy of the Bengal Fan over the past 20 Myr using ODP drill core samples. They found two contrasting regimes of clay mineral and Sr isotope distribution (Derry and France-Lanord, Fig. 6). For ages greater than 7 Myr and less than 1 Myr, illite was predominant in the <2 μm clay fraction, with $^{87}Sr/^{86}Sr$ ratios of 0.71–0.72. In contrast, the period 1–7 Myr ago was characterised by high smectite abundances and Sr isotope ratios of 0.72–0.75. Derry and France-Lanord attributed elevated Sr isotope ratios during the 1–7 Myr period to intensified chemical weathering on the Ganges-Brahmaputra flood-plain, in contrast to a regime dominated by physical erosion before and after this period. However, it is significant that these changes do not have an obvious effect on the seawater Sr evolution curve. Therefore, during the period 1–7 Myr ago, the more radiogenic Sr signature in the Ganges-Brahmaputra system must have been largely offset by a 50% drop in the Sr load of the river.

4.1.1 $^{146}Sm–^{142}Nd$

The extinct nuclide ^{146}Sm can place important constraints on models of early Earth differentiation. Therefore, the claim by Harper and Jacobsen (1992, op. cit.) to find terrestrial ^{142}Nd anomalies was tested by Sharma *et al.* (1996). Three sample types were analysed: normal Nd, normal Nd spiked with 30 or 57 ppm ^{142}Nd, and 3.8 Byr old rocks from Isua, West Greenland. The latter included one sample with a claimed 32 ppm excess of ^{142}Nd, and one unknown sample. Samples were run in static mode on a Finnigan MAT 262 machine. Focus positions were shown to have some influence on the measured relative abundances of ^{142}Nd; nevertheless, results for the three sample types were generally consistent with the predicted ^{142}Nd excesses (Sharma *et al.*, Fig. 5). Hence, the existence of ^{142}Nd anomalies in the Earth is confirmed with good confidence.

Jacobsen and Harper (1996) showed that the ^{142}Nd excess can be used with initial ϵ ^{143}Nd to model early differentiation of the silicate Earth. A simple two-stage model assumes Bulk Earth evolution until time T, when Sm/Nd fractionation occurs in some part of the mantle. The fractionated mantle reservoir is subsequently sampled at the time of magmatism. For the 3.81 Byr old Isua rocks, this model yields an age of 4.47 Byr for the fractionation event, 100 Myr after Earth accretion (Jacobsen and Harper, Fig. 7). In contrast, 4 Byr old samples of Acasta gneiss, with calculated initial ^{143}Nd values (ϵ[t]) of +3.5, gave no ^{142}Nd anomaly. This implies either strong Sm/Nd fractionation at a relatively late date (ca. 4.3 Byr ago), or that the high ϵ ^{143}Nd values of these samples are metamorphic artifacts (section A.4.3).

4.3 Model ages and crustal processes

Komatiites have been widely used as probes of mantle Nd isotope signatures in the early Earth, and hence to constrain Early Archean mantle evolution (section 4.4.3). A study by Lahaye *et al.* (1995) has important implications for this data set, because it implies that the initial Nd isotope signatures of many komatiites may have been disturbed by subsequent alteration. Lahaye *et al.* compared calculated initial isotope compositions (ϵ Nd[t]) for whole-rock samples and separated pyroxenes in five komatiite flows from the Abitibi and Barberton belts. Many whole-rocks show small (1–2 ϵ unit) deviations from the pyroxenes, but a few show much larger deviations, up to +5 and –10 units (Lahaye *et al.*, Fig. 7). In view of these results, Nd isotope data on komatiites should ideally be based on a combination of whole-rock and mineral analyses.

Bowring and Housh (1995) augmented the Nd isotope data set for Early Archean depleted mantle with new data from 3.94 to 4.0 Byr old Acasta gneisses. These data do not substantially extend the 'envelope' of most depleted mantle compositions in

the Early Archean, but they are significant because they are based on granitic and tonalitic gneisses, which are normally more resistant to metamorphic disturbance than mafic rocks. Bowring and Housh also reported a very enriched Nd isotope signature (ϵ Nd = –4.8 at 4 Byr) in an amphibolite gneiss from the same suite. However, this signature has probably been disturbed, since it yields an impossibly old T(CHUR) Nd model age of 4.96 Byr.

Moorbath and Whitehouse (1996) observed that the disturbed amphibolite and the granitoid rocks lay an a single Sm–Nd errorchron, with an apparent age of 3.2 ± 0.15 Byr (Moorbath and Whitehouse, Fig. 1). This could be a coincidence; however, other samples of the granitoid suite lie on the same array, suggesting that it may date a Mid-Archean metamorphic event. This implies that the calculated initial ratios of the two granitoid samples with highest apparent values of ϵ Nd (t) might have been slightly disturbed (Moorbath and Whitehouse, Fig. 2). These findings show the importance of having large enough Nd isotope data sets so that their internal consistency can be evaluated as a test for closed system behaviour.

Another area of Nd isotope interpretation which continues to be controversial is the significance of Nd model ages as crustal formation ages. Because mixing between material of different ages can give rise to non-unique interpretations of Nd isotope data, crustal formation ages must be constrained by other evidence, such as U–Pb crystallisation ages. Typically, Nd model ages define the oldest apparent age for a terrane, while a minimum age is defined by U–Pb dating of cross-cutting plutons. The smaller the difference between the average Nd model age and the U–Pb age, the tighter the constraints on the Nd model age as a true crustal extraction age. To illustrate this premise, I will discuss an example from the Grenville Province of Canada, where the interpretation of Nd model ages as crustal formation ages has been disputed.

Dickin and McNutt (1989, op. cit.) reported a step in TDM model ages in the Ontario Grenville belt from values of 2.4–2.7 Byr, interpreted as variably reworked Archean crust, to values around 1.9 Byr, interpreted as an accreted Early Proterozoic arc. In contrast, DeWolf and Mezger (1994) proposed that the 1.9 Byr model ages represent mixtures of 2.7 and 1.5 Byr old components. These alternative interpretations can be tested using U–Pb data (e.g. Corrigan *et al.*, 1994), which yield 1.7–1.74 Byr crystallisation ages for plutonic orthogneisses in the disputed area. Since these orthogneisses cross-cut older foliated gneisses on which the Sm–Nd data were determined, 1.74 Byr is a minimum age for any components involved in mixing. The relative proximity of this minimum age to the 1.9 Byr TDM age of the Proterozoic terrane shows that it is juvenile Early Proterozoic crust, with a minor or negligible Archean component. It is concluded that Nd isotope data remain uniquely powerful for determining crustal formation ages in highly metamorphosed orogenic belts (section A.5.5).

5.1 U–Pb isochrons

The problem of uranium mobility in most rock systems limits U–Pb isochron dating to certain specialised applications. Some areas where the method has been successfully applied are in dating sediment deposition by means of marine carbonate, dating prograde metamorphism by means of garnet porphyroblasts, and dating fresh Quaternary-age volcanic rocks.

Uranium has a seawater residence time four orders of magnitude longer than Pb, leading to a very high μ value of ca. 75 000 (Jahn and Cuvellier, 1994). Because U is also thought to have a higher carbonate/seawater partition coefficient than Pb, the μ value of marine carbonates could be even higher. An upper limit of 230 000 was suggested by Jahn and Cuvellier, corresponding to a U content of 1–3 ppm and Pb content of ca. 1 ppb (part per billion). The highest μ value observed in an ancient biogenic calcite is 50 000 (Smith et al., 1991, op. cit.), but most ancient carbonates have values far lower than this, attributed to open-system behaviour during post-depositional diagenesis and secondary alteration.

Jones et al. (1995) showed that in the Capitan Limestone of New Mexico, some samples had μ values over 3000, which appear to be a primary depositional feature. The sample suite came from a Permian reef complex and consisted of massive abiotic botryoids (ca. 1 cm in size), made of acicular aragonite recrystallised to low-Mg calcite. Microsampling for oxygen and carbon isotope analysis suggested that relatively pristine and diagenetically disturbed domains were intimately interfingered. Samples preserving possible primary μ values had high U (3 ppm), high Sr (3000 ppm) and low Mn contents (<10 ppm). The high μ values allowed the calculation of radiogenic $^{206}Pb*/^{238}U$ ages, as in equation [5.8], after correction for a small common Pb component. Using this method, six texturally well-preserved samples gave an average age of 250 ± 3 Myr which is consistent with the age of deposition (Jones et al., Fig. 2a). These results are somewhat encouraging for U–Pb dating of carbonates because

they show that some sample domains can remain relatively undisturbed during diagenesis. However, the interfingering of pristine and altered domains suggests that reliable sampling of pristine material will be very difficult.

Garnets are important minerals in the geothermometry and barometry of metamorphic rocks. Therefore, the direct dating of this material would allow constraints to be placed on heating and cooling rates during regional metamorphism. Both the Sm–Nd and U–Pb systems have potential for dating metamorphic garnets, but they have different strengths and weaknesses. Garnets grown under amphibolite facies conditions (ca. 550 °C) can give concordant ages of prograde garnet growth from the two methods (section 4.1.4). However, Mezger et al. (1992) argued that the Sm–Nd system was opened at ca. 600 °C (upper amphibolite facies) in all but very large inclusion-free grains. Therefore, they suggested that garnet Sm–Nd ages usually date cooling rather than prograde mineral growth.

Mezger et al. (1991) proposed a higher closure temperature of ca. 800 °C for the U–Pb system in garnet. However, this system suffers from the tendency for uranium to be concentrated in minute inclusions, rather than in the garnet lattice. For example, in the first study of this type, on the Pikwitonei granulite terrane in northern Manitoba, Mezger et al. (1989) attempted to use the U–Pb system to date prograde garnet growth during prolonged Late Archean metamorphism. After correction for a small common Pb component, radiogenic $^{206}Pb*/^{238}U$ and $^{207}Pb*/^{235}U$ ages were calculated using equations [5.8] and [5.9]. Unfortunately, most samples gave discordant ages between the two Pb systems, indicative of open-system behaviour. A later study (DeWolf et al., 1996) showed that this was due to Pb loss from micron-sized monazite inclusions in the garnet grains.

The absence of a non-radiogenic isotope for fractionation normalisation has always degraded Pb isotope analysis relative to other systems, such as Sr and Nd. Getty and DePaolo (1995) proposed that in the dating of Quaternary-age material, $^{207}Pb/^{204}Pb$ variations within an isochron suite are so small that this ratio can be used for fractionation normalisation. This allows a one to two orders of magnitude improvement in the precision with which $^{206}Pb/^{207}Pb$ can be measured, compared with the silica gel method. With this improvement in precision, Quaternary lavas can be dated to ca. 10% uncertainty using $^{238}U–^{206}Pb$ mineral isochrons.

5.3.1 **The geochron**

Based on evidence from extinct nuclides (Chapter 15), Ca–Al inclusions from the Allende chondrite are regarded as the oldest solar system objects, and are therefore of particular interest in dating early Solar System evolution. Inclusions and chondrules both have high U/Pb ratios, but a significant common Pb component rules out direct application of the U–Pb method. Several Pb–Pb dating studies have been performed on mixed suites of chondrules and inclusions. However, Chen and Wasserburg (1981) obtained the first precise age from inclusions alone. A Pb–Pb isochron based solely on inclusions gave an age of 4568 ± 5 Myr, whereas forcing the isochron through the Canyon Diablo point decreased the age to 4559 ± 4 Myr.

Two-point isochrons can be calculated between Canyon Diablo and any individual inclusion, and these are termed model 207/206 ages because they rely on the assumption that initial lead in the inclusion was the same as Canyon Diablo. Tilton (1988) preferred these ages to the isochron age of 4568 Myr, because the 207/206 ages did not appear to be correlated with the ^{206}Pb/^{204}Pb ratio (which measures the amount of common Pb in each sample). However, a re-evaluation of the data set by Allegre *et al.* (1995) did show such a correlation (Allegre *et al.*, Fig. 3). Believing that the discrepancies were due to mobilisation of common Pb from the matrix into the inclusions, Allegre *et al.* obtained more radiogenic data by progressive leaching experiments. These new data points gave 207/206 ages within error of the most radiogenic data of Chen and Wasserburg. They give a best age of 4566 +2/–1 for Allende Ca–Al inclusions, and hence solar system condensation.

The best age for chondrule formation was determined in a different way (Gopel *et al.*, 1994). In this case, phosphate grains in chondrules had very low common Pb contents, and could be dated by the U–Pb concordia method. Most data were concordant, but phosphates from different meteorites gave a large range of apparent ages. Gopel *et al.* speculated that resetting occurred shortly after cooling (presumably during aggregation into larger bodies). This would give rise to Pb loss lines which were parallel to the concordia (Gopel *et al.*, Fig. 2). Hence, the oldest age of 4563 Myr must be considered the best estimate for the time of formation of the oldest chondrules, only 3 Myr after Ca–Al inclusions.

The meaning of the apparent Pb–Pb age of the Earth, relative to meteorites, has been re-examined in several recent papers (Allegre *et al.*, 1995; Galer and Goldstein, 1996; Halliday *et al.*, 1996; Harper and Jacobsen, 1996). It is known that the Bulk Silicate Earth (BSE) has a much higher μ value (ca. 9) than bulk chondrites (ca. 0.7). This difference could be due to Pb volatilisation during terrestrial accretion, or to Pb partition into the core. Comparison of the partition coefficients of Pb and other siderophile–chalcophile elements suggests that core formation was the main cause of high μ values in the BSE (e.g. Galer and Goldstein, 1996). If this is correct, terrestrial Pb isotope evolution can be constrained by the Hf–W extinct nuclide system (section A.15.2.8). Since the Hf–W evidence seems to favour late core formation, a young Pb–Pb age for the Earth should be expected (Halliday *et al.*, 1996). This is consistent with the evidence from Archean galena compositions (Fig. 5.26), which favour a low μ value in very early Earth history. A 100 Myr period of core formation can largely explain the Pb paradox, since it causes the BSE, and hence average MORB, to lie to the right of the geochron (section 6.4.1). However, Pb still displays anomalous behaviour in OIB (Fig. 6.26), due to its abnormal geochemistry in the lithophile system (section 6.6).

5.5 **Pb–Pb dating and crustal evolution**

The Pb and Nd isotope systems can both provide powerful constraints on crustal evolution, but their very different geochemical properties may lead to different uses. The most controversial applications of these systems are to complex crustal growth problems involving possible mixing between sources of different ages. Therefore, some examples will be examined here to explore the different constraints available from Pb and Nd isotope data (see also section A.4.3).

The Pb isotope system comprises two coupled dating systems, whereas Sm–Nd normally offers only one. Therefore Pb–Pb data should theoretically provide more control than Nd on mixing processes between crustal reservoirs of different ages. However, U–Pb isotope systems are susceptible to open system behaviour over geological history, whereas whole-rock Sm–Nd systems are very resistant to such effects. Therefore, when Pb–Pb data are used to study ancient mixing events, the measured isotope ratios cannot normally be corrected to unique initial ratios. Instead, model initial ^{207}Pb/^{204}Pb ratios are usually determined by projecting the Pb–Pb data back along an isochron line corresponding to the age of the mixing event (e.g. Fig. 5.29). Hence, we are left with a single isotopic tracer, analogous to ^{143}Nd/^{144}Nd.

Davis *et al.* (1996) compared the use of initial

$^{207}Pb/^{204}Pb$ and $^{143}Nd/^{144}Nd$ ratios in Late Archean granitoids as tracers of the extent of Early to Mid-Archean basement in the Slave province of NW Canada. They found a relatively good correlation between the two tracers (Davis *et al.* Fig. 5), supporting a model of two-component mixing between juvenile (recently mantle-derived) and old crustal end-members. The mixing line is somewhat curved, indicating a Pb/Nd ratio in the old crustal end-member about three times higher than the juvenile end-member. In principal this makes Pb a more sensitive tracer of hidden crust than Nd. However, evidence from Greenland (Fig. 7.28) suggests that Pb may behave less reproducibly. Hence, the scatter in Fig. 5 (Davis *et al.*) is probably due to non-stoichiometric mixing of Pb rather than Nd.

In crustal provinces where high grade metamorphism occurs much later than magmatism, Pb isotope data may be subject to greater uncertainty. An example is provided by the Grenville Province, where metamorphism at 1.1 Byr occurred up to 1.6 Byr after crustal formation. DeWolf and Mezger (1994) measured Pb isotope ratios on K feldspars from various rock-types in the Grenville Province of Ontario and New York State. Data for Mid-Proterozoic rocks of the Adirondacks and Central Metasedimentary Belt formed a compact array of shallow slope, while data from the Central Gneiss Belt (CGB) defined a steeper but more scattered array (DeWolf and Mezger, Fig. 2). DeWolf and Mezger attempted to use the Pb–Pb data set to test Nd isotope mapping of the extent of Archean crust in the CGB. However, the strong susceptibility of Pb isotope data to open-system behaviour during high grade metamorphism makes Pb a less sensitive tool than Nd for mapping crustal formation ages in the Grenville Province. A combination of Nd isotope and U–Pb data indicates the existence of a juvenile Early Proterozoic terrane in the CGB (section A.4.3). However, meta-igneous rocks in this terrane have ambiguous Pb isotope signatures that cannot be resolved from the Archean or Mid-Proterozoic arrays. This applies both to the 207/204–206/204 (isochron) diagram and to the 208/204–206/204 (thorogenic) diagram (DeWolf and Mezger, Fig. 2).

6.1 Mantle heterogeneity

Over the past couple of decades, the convective structure of the mantle has been much debated, with geophysicists generally advocating a single layer of convection cells in the mantle, while geochemists have generally advocated two-layer convection. This debate may finally have been resolved by evidence that the viscosity of the mantle increases by more than an order of magnitude from top to bottom (e.g. Bunge *et al.*, 1996).

The variable viscosity model gives rise to a convective regime somewhere between the two extremes described above. In principle the lower mantle is part of the main convective system of the mantle, but in practice its high viscosity largely isolates it from the rapidly convecting upper mantle. For example, convective flow in the upper mantle is argued to streak out and homogenise heterogeneities over time, but modelling by Manga (1996) suggests that instead of being streaked out, lower mantle blobs of high viscosity may actually aggregate together over time.

6.5 Mantle reservoirs in isotopic multispace

Entrainment of asthenospheric upper mantle into the melting zone beneath ocean islands has been accepted for some time (section 6.2). However, Hart *et al.* (1992, op. cit.) were the first to propose that major entrainment of a lower mantle component (FOZO) could substantially modify the isotopic composition of plumes. FOZO was initially proposed as a depleted component at a 'focus zone' between the MORB and HIMU sources, but subsequently revised by Hauri *et al.* (1994) to a less depleted signature within the interior of the mantle tetrahedron (Hauri *et al.*, Fig. 17). Modelling by Hauri *et al.* permitted a wide range in the extent of lower mantle entrainment into plumes (between 5% and 90%). However, numerical modelling by Farnetani and Richards (1995) suggested that the centre part of the plume head (which normally undergoes partial melting) would contain less than 10% entrained material.

New views of the viscosity structure of the mantle (section A.6.1) may make simple entrainment models less attractive. If the Earth has a viscous lower mantle, subducted lithosphere may enter the 'body' of the lower mantle and form megaliths there, rather than piling up on the core-mantle boundary, as expected for a low-viscosity lower mantle. The degree of lower mantle homogenisation in these two models will be very different. The low-viscosity model might generate a relatively homogeneous FOZO component, available for entrainment in rising plumes. However, the high-viscosity model is likely to be much messier, giving rise to plumes with complex internal mixing relationships between a variety of sources, some enriched and some depleted relative to Bulk Earth.

Detailed examination of individual hot-spots may offer the best hope to determine the origins of 'non-enriched' lower mantle sources (such as FOZO), and the ways in which these sources are juxtaposed in rising plumes. For example, Thirlwall *et al.* (1994) showed that Icelandic basalts and North Atlantic MORB display sub-parallel but distinct Pb isotope arrays (Thirlwall *et al.*, Fig. 10). Hence, they argued that isotopic heterogeneity in the Iceland plume could not be caused by mixing with MORB. They speculated that the Iceland plume array might represent a 'young HIMU' type source. However, Kerr *et al.* (1995) suggested that the Iceland plume was formed from recycled 'lower oceanic lithosphere'. This would generate a more depleted signature than recycled oceanic crust (the probable origin of HIMU).

Hanan and Graham (1996) argued that MORB Pb–Pb arrays for all three major ocean basins converge on a common point on a plot of $^{207}Pb/^{206}Pb$ versus $^{208}Pb/^{206}Pb$ (Hanan and Graham, Fig. 1). They suggested that this represents a common mantle component 'C' (erroneously amplified by their copy editor to 'carbon'). This component resembles the FOZO composition of Hauri *et al.* (1994) since it has effectively the same Pb isotope ratio, and is some-times, but not always, associated with an enhanced primordial helium signature (relative to typical MORB). However, 'C' differs from FOZO in its effect, since it is regarded as a plume component entrained into the MORB source rather than a lower mantle component entrained into plumes. In this sense, 'C' resembles the recycled oceanic lithosphere component of Kerr *et al.* (1995).

Allegre and Lewin (1995) and Allegre *et al.* (1996) re-examined another source of evidence about mantle mixing processes, based on mantle isochrons. Rb–Sr, Sm–Nd and Lu–Hf mantle isochrons for combined data sets of MORB and OIB yield apparent ages of ca. 1.4, 1.2 and 0.8 Byr respectively (Allegre *et al.*, Fig. 2). These ages are substantially less than two-point isochrons through average oceanic volcanics and the Bulk Earth, which yield ages around 2 Byr, interpreted as the mean crustal extraction age of the crust. In contrast, Allegre *et al.* argued that the slope ages of the MORB–OIB mantle isochrons date the 'stirring time' of the upper mantle, representing the average time for re-homogenisation of recycled oceanic crust. However, these ages are minima, since the parent–daughter pairs are fractionated during extraction of basaltic magmas from their mantle sources. Allegre *et al.* argued that this fractionation cannot explain the deviation of 0.8–1.4 Byr mantle isochrons from the 2 Byr crustal extraction 'isochron'. However, other

workers (e.g. Salters, 1996) suggest that the OIB deviations can be explained by parent–daughter fractionation if plume basalts represent 5% melts of their sources (Salters, Fig. 3). Therefore, it appears that parent–daughter mantle isochrons cannot yield more reliable constraints than previously available on the average mixing time of the mantle.

The Pb–Pb system offers an opportunity to determine mantle isochron ages without the effects of parent–daughter fractionation during partial melting. However, the relatively short half-lives in this system involve isotopic evolution along growth curves, such that the oldest sub-system on a Pb–Pb isochron diagram largely determines the apparent age (Christensen and Hofmann, 1994). Hence, Allegre and Lewin (1995) argued that the (average) stirring time of the upper mantle is effectively half the apparent MORB–OIB Pb–Pb isochron age of 1.9 Byr. It is unclear, however, whether the Pb–Pb array is really dating the mantle stirring time, or a more complex time-integrated relationship between the evolution of the MORB and HIMU components (e.g. as modelled by Christensen and Hofmann, 1994).

6.6 Identification of enriched mantle components

Woodhead *et al.* (1993, op. cit.) measured $\delta^{18}O$ values as high as 7 in basaltic glasses from Pitcairn sea-mounts, and attributed them to sediment contamination of the EMI mantle reservoir. Eiler *et al.* (1995) tested these findings by analysing oxygen isotope compositions of olivine and plagioclase phenocrysts from Pitcairn Island. All olivines had $\delta^{18}O$ values close to 5.2, while plagioclase had δ values near 6.1, consistent with mass fractionation effects at magmatic temperatures. Therefore, the variations seen by Woodhead *et al.* are almost certainly due to local contamination of the glasses, either before or after solidification. Harmon and Hoefs (1995) presented a much larger data set of oxygen isotope determinations for oceanic basalts. They plotted $\delta^{18}O$ against Sr, Nd and Pb isotope ratios in order to explore the meaning of the oxygen isotope variations (Harmon and Hoefs, Fig. 9). Many of these variations can be attributed to local contamination effects (for example in Icelandic lavas). However, a very slight elevation of $\delta^{18}O$ values in the Societies-Samoa data set (relative to MORB) might be due to sediment contamination of the EMII mantle source (Fig. 9c).

Debate continues on whether the enriched mantle components EMI, EMII and HIMU are generated by crustal recycling or foundering of sub-continental lithosphere. Chauvel *et al.* (1995), developing earlier models by various workers, suggested that Pb is removed from subducting oceanic crust by metasomatism and deposited in the overlying mantle wedge. It would then be incorporated into the continental crust via arc magmatism. This non-magmatic movement of Pb can explain the decoupling of Pb from the other isotope systems (including U–Th). However, some fraction of slab-derived Pb is likely to be left behind in the mantle wedge, lowering its μ value. Over time, the wedge should develop unradiogenic Pb signatures, and most likely also migrate into the enriched quadrant of the Sr–Nd isotope diagram (Dickin, 1996). Subduction erosion of such material, combined with the slab, can generate the LoNd array. This is consistent with the observation of Hart *et. al.* (1986, op. cit.) that EMI and HIMU 'appear to be complementary to each other in terms of the enrichments and depletions in various parent/daughter ratios'.

Analysis of dredged basalts from the Afanasy–Nikitin rise, at the southern end of the non-seismic 85° East Ridge, extended the composition of the EMI component to more extreme isotope compositions. Two samples analysed by Mahoney *et al.* (1996) were colinear with an extension of the Pitcairn seamount array in Nd–Pb, Sr–Pb, and Pb–Pb isotope space (Mahoney *et al.*, Fig. 3). In addition, nine analyses quoted from a Russian source (Sushchevskaya *et al.*, unpublished), define an array which is colinear with the data of Mahoney *et al.* (and Pitcairn data) in Nd–Pb and Sr–Pb isotope space. The Russian data are not colinear on a Pb–Pb isochron diagram, but this could be due to analytical fractionation effects. Mahoney *et al.* ruled out *in situ* lithospheric mantle as a source for the A–N data, since, at the time of volcanism, the A–N rise was situated on young oceanic crust, far from any continent. However, the data may not represent a lower mantle source composition, because they are quite distinct from the Crozet hot spot, to which the 85° E ridge and A–N rise are attributed. Hence, Mahoney *et al.* speculated that the A–N component may represent non-plume material entrained for a brief time into the Crozet plume. Nevertheless, this does not detract from the importance of this example as a pointer to the origins of the EMI component recognised elsewhere. In this regard, the data are consistent with an origin from subducted lithosphere.

7.2.3 Lithospheric mantle contamination

Isotopic tracers have been widely used to monitor crustal contamination of continental magmas during their ascent, and to some extent this process can now be quantified. In contrast, the relative importance of lithospheric and asthenospheric mantle sources continues to be a matter of debate. Thermal constraints (McKenzie and Bickle, 1988) suggest that the high melting rates necessary to erupt flood basalt provinces can only be satisfied by melting of mantle plumes. However, lithospheric extension will cause small volume melting of metasomatised lithosphere, generating potassic mafic magmas which may have extreme isotopic compositions. Several workers recognised that these processes might act together, leading to contamination of asthenospheric magmas in the mantle lithosphere, as well as the overlying crust. Good examples of this process come from the Mesozoic flood basalt province of Gondwana (southern Africa, South America and Antarctica).

In the Karroo Province of southern Africa, picrite basalts from Nuanetsi define a Sm–Nd pseudo-isochron with an apparent age of 1 Byr. This was initially interpreted as a mantle isochron dating the age of a magma source in the lithosphere (Ellam and Cox, 1989). However, re-examination of the data (Ellam and Cox, 1991), suggested that the array was a mixing line, formed by contamination of picritic magmas (from a depleted plume source) by potassic mafic magmas from a lithospheric lamproite source (Ellam and Cox, 1991, Fig. 2). Trace element mixing calculations suggested that the lamproite component made up about 20% of the mixed magma, but dominated its Nd and Pb isotope signature. However, osmium contents of the two end-members were about equal (section 8.5). A similar model may explain Nd–Os isotope ratios in the Stillwater Complex, Montana (data of Lambert *et al.*, 1994).

An important aspect of the Gondwana flood basalt province is its geochemical provinciality, which provides strong evidence for lithospheric control of magma chemistry. One example of this provinciality is the identification of high-Ti and low-Ti flood basalt provinces in southern Africa and in the Parana basin of South America (e.g. Hawkesworth *et al.*, 1984, op. cit.). If the same provinciality was found in mafic potassic rocks, this would support the lithospheric contamination model for flood basalts. Such a case was observed for the high-Ti Parana province (e.g. Hawkesworth *et al.*, 1992), but the absence of low-Ti alkali mafic rocks caused major uncertainty about the petrogenesis of low-Ti Parana basalts (Hergt *et al.*, 1991). This problem was solved by the discovery of low-Ti alkali mafic rocks on the flanks of the Parana basin (Gibson *et al.*, 1996).

Comparison of the Nd isotope systematics of high-Ti and low-Ti alkali mafic rocks from the Parana showed that while the ranges of Sm/Nd ratios were identical, Nd isotope compositions were quite distinct (Gibson *et al.*, Fig. 6). This suggests that the lithospheric mantle underlying high-Ti and low-Ti provinces has distinct trace element enrichment ages, possibly reflecting the geographical extent of Late Proterozoic crustal re-working. If the isotopic signatures of the most enriched alkali mafic suites are used as end-members in a mixing model, the isotopic compositions of low-Ti flood basalts can be explained by ca. 20% contamination of a primitive plume end-member in the lithospheric mantle, followed by extensive contamination in the crust (Gibson *et al.*, Fig. 5). However, the high-Ti flood basalts require ca. 50% contamination in the mantle lithosphere, but less crustal contamination. Such models for the interaction of plumes with the mantle lithosphere are likely to be a continuing major focus in geochemical studies of continental basalts.

7.3.3 Flood basalts

The Columbia River Basalt Province has been one of the most difficult to explain by the plume model, since its location above a subduction zone was thought to preclude involvement of the Yellowstone plume, which tracked across the western margin of North America during the Neogene period. Nevertheless, DePaolo (1983, op. cit.) advocated a plume source, and this was supported by Brandon and Goles (1988) on the basis of trace element evidence from Imnaha basalts. Subsequently, Brandon and Goles (1995) re-interpreted published isotopic data in terms of a plume component, identifying component 2 of Carlson and Hart (1988, op. cit.) with the Yellowstone plume rather than subduction-processed upper mantle. This model requires that the plume broke through the subducting slab ca. 14 Myr ago. While this may appear unlikely, it has been supported by recent rare gas evidence (Dodson *et al.*, 1996), which indicates an enhanced primordial helium signature (relative to upper mantle) in olivine phenocrysts from an Imnaha basalt flow. These findings bring the Columbia River Basalt Province into line with the plume model for other terrestrial flood basalt provinces.

8.2 Determination of the Re decay constant

Incomplete sample dissolution and sample-spike homogenisation have been major obstacles in accurate Re–Os dating. In order to overcome these problems,

Shirey and Walker (1995) adopted the Carius tube digestion method, by which samples are dissolved in sealed glass ampoules under high temperature and pressure. Reagents were sulphuric acid and chromium oxide, and the samples were heated to 180 °C. Outer metal safety jackets were used, but the pressure is retained entirely by the sealed glass tube. After reaction, the products are frozen before release.

The Carius tube technique was used in two recent studies of iron meteorites. Smoliar *et al.* (1996) determined isochron slopes on four different classes of iron meteorites. Evidence from extinct nuclides (section 15.2.5) suggests that group IIIA iron meteorites have ages within 5 Myr of the age of angrite meteorites (4558 Myr). Assuming this age for IIIA meteorites, the Re–Os isochron slope gave a Re decay constant of 1.666×10^{-11} yr^{-1}, equivalent to a half-life of 41.6 ± 0.4 Byr. The error estimate was based on the possibility of a 1% error on the spike calibration, but since this would be a systematic error, its magnitude is not well constrained. Using the same half-life, other iron meteorite groups (except IVB) yield ages and initial ratios on a common osmium evolution curve, identified with the evolving solar nebula. However, the trend is mainly defined by the young age of IVA irons, which may reflect open Re–Os systems, since Pd–Ag data for Gibeon (a IVA iron) show this to be an early body (section 15.2.5). Shen *et al.* (1996) made a precise isochron determination on group IIAB irons, yielding a slope which was identical to the IIAB isochron of Smoliar *et al.* (within analytical error). The spread of Re/Os ratios obtained for other groups did not allow precise ages to be determined.

8.4 Mantle osmium

Osmium isotope evidence offers a unique geochemical insight into mantle processes. Hence this method has been the focus of much recent work attempting to understand mantle evolution.

Roy-Barman and Allegre (1994) made a detailed comparison between ocean ridge basalts and abyssal peridotites from various ridges. These two sample types are expected to have the same Os isotope ratios, since they are thought to represent (respectively) the eruptive product and the residue from melting. Samples were leached with oxalic acid (and then HBr) to remove ferro-manganese coatings, which have radiogenic seawater osmium signatures. Leached peridotites from the Mid-Atlantic Ridge (MAR) had a very narrow range of ^{187}Os/^{186}Os ratios from 1.02 to 1.04 (^{187}Os/^{188}Os of 0.122–0.125), in good agreement with previous work. However, leached MORB samples had elevated ratios of 1.07 to 1.11 (^{187}Os/^{188}Os of 0.128–0.133), within the range of North Atlantic OIB (Roy-Barman and Allegre, Fig. 4). Roy-Barman and Allegre interpreted the MORB values as magmatic because of agreement between leached glass and crystalline basalt compositions. They proposed two alternative models to explain the differences between MORBs and peridotites: contamination of MAR basaltic magmas by assimilation of hydrothermally altered oceanic crust; or contamination of the MAR magma source with isotopically enriched osmium from the Azores plume or from enriched streaks in a marble cake mantle (Roy Barman and Allegre, 1996). A definitive choice between these models cannot yet be made. However, contamination of the MORB source by recycled oceanic crust has been invoked previously to explain lithophile isotope ratios (Chapter 6), and has also been invoked to explain MORB Lu–Hf and U–Th signatures (sections A.9.1.3 and A.13.3.2).

Sea-floor contamination processes have been ruled out in OIB samples with more than 0.03–0.05 ppb Os (e.g. Marcantonio et al., 1995). However, the ultimate origin of enriched Os isotope signatures in OIB remains unclear. In a study of drilled lavas from Mauna Loa, Hawaii, Hauri et al. (1996) demonstrated correlations between Os, Sr, Nd and Pb isotope ratios in lavas with over 0.3 ppb osmium, while Bennett et al. (1996) extended the Os–Sr–Pb correlation line to more enriched compositions, reaching a ^{187}Os/^{186}Os ratio of 1.21 (187/188 = 0.145) in samples from the Koolau volcano. This array approaches the composition of a Pitcairn sample (Hauri et al., Fig. 4), thought to approximate the isotopic composition of the EMI mantle component (Reisberg et al., 1993, op. cit.).

Various models were examined to account for the Hawaiian array. Both Hauri et al. (1996) and Bennet et al. (1996) ruled out simple mixing between plume and depleted upper mantle sources, arguing instead that two separate components were available within the plume. Hauri et al. identified these components as an EMI plume signature, entraining lower mantle with unradiogenic osmium (FOZO). They tentatively attributed the plume signature to a mixture of recycled oceanic crust and pelagic sediment. Evidence in support of such mixing was presented by Hauri (1996), who observed correlations between isotopic and major element compositions of Mauna Loa basalts. Hauri attributed these correlations to variable enrichment of the plume source by pods of eclogite derived from recycled crust (Hauri, Fig. 1). However, an alternative origin for EMI which should be given

consideration is recycling of lithospheric mantle, which might contain pyroxenite veins formed above subduction zones.

Osmium data for the Canary Islands (Marcantonio et al., 1995) are representative of an increasing number of OIB analyses with radiogenic osmium but with Pb isotope ratios intermediate between HIMU and EMI (Hauri et al., 1996, Fig. 4). The decoupling of Pb and Os isotope data in these samples has posed a problem for simple mixing lines between DMM and HIMU (e.g. Marcantonio et al., 1995). The Canaries data were tentatively attributed to mixing between DMM and a HIMU–EMI mix (Hart, pers. comm. to Marcantonio et al., 1995). This model is consistent with lithophile isotope evidence, which has suggested an intimate relationship between HIMU and EMI, corresponding to recycled oceanic crust and subcontinental lithosphere (section 6.5).

Alternatively, Walker et al. (1995) and Widom and Shirey (1996) proposed that the radiogenic Os signatures of many plumes could be derived directly from the core. They suggested that a metallic core component could be incorporated into plumes at the core-mantle boundary, such that its high Os content would overprint the isotopic signature of the plume. However, the operation of such a process is speculative, whereas lithospheric recycling is a process which is known to occur. Hence, the present author regards the lithosphere as a more attractive source of the observed isotopic variations.

The behaviour of osmium during subduction has important implications for Os mantle evolution because of the major influence which recycled oceanic crust is believed to have on OIB sources. For lithophile isotope systems, analysis of island arc basalts has played an important part in understanding subduction zone processes and mantle recycling (section 6.7). However, Os isotope analysis of arc volcanics is difficult because of their very low Os contents, typically less than 10 ppt (parts per trillion). To overcome this problem, Brandon et al. (1996) analysed lherzolite and harzburgite xenoliths in arc lavas. These xenoliths have osmium concentrations of 0.1–2.1 ppb, and are believed to sample the mantle wedge which overlies the subduction zone.

Brandon et al. (1996) determined Os isotope ratios in whole-rock xenoliths which ranged to 5% above and below the chondritic composition, from a minimum near the field of abyssal peridotites to a maximum ^{187}Os/^{188}Os ratio of 0.134 (187/186 = 1.12). Os and Nd isotope ratios displayed a broad negative correlation (Brandon et al., Fig. 3), which could be explained by mixing of depleted upper mantle with

5–15% of a subduction-related component derived from 80–95% basalt and 5–20% sediment. This model is supported by the relatively low Al_2O_3 ratios of the xenoliths (especially those from Simcoe, Washington). These define a vertical trend (Brandon *et al.*, Fig. 4), which is distinct from the diagonal enrichment–depletion trend seen in orogenic lherzolites (see below). Brandon *et al.* suggested that their results reflect metasomatic Os transport in a chloride-rich slab-derived fluid or melt. However, more data are needed to verify these findings.

8.4.2 Lithospheric evolution

Xenolith suites are a major source of data about the lithospheric mantle (section 7.1), but since they are not usually cogenetic, they rarely form isochrons. Osmium model ages may be particularly useful in studies of the lithospheric mantle because the Re–Os system can effectively date mantle depletion events, as well as the mantle enrichment events normally dated by lithophile isotope systems. Samples of depleted mantle can be used to calculate conventional osmium model ages, based on the intersection of sample and Bulk Earth evolution lines (e.g. Pearson *et al.*, 1995a, Fig. 11). However, the Re/Os ratio of mantle samples may be disturbed by metasomatic rhenium addition after the initial crust-forming Re depletion event. An alternative approach (Walker *et al.*, 1989a, op. cit.) is to assume that the depletion event generated Re/Os ratios of zero in the samples. Re depletion model ages (T_{RD}) can then be calculated by projecting a horizontal line back to the Os mantle growth curve (Pearson *et al.*, Fig. 11).

Rhenium depletion ages represent minimum ages for lithospheric formation. However, they can be calculated directly from the Os isotope ratios of peridotites, allowing the accumulation of reasonably large date sets (e.g. Carlson and Irving, 1994, Fig. 6). From these data we can deduce that the Wyoming, Kaapvaal and Siberian cratons all possess mantle keels with minimum formation ages of ca. 3 Byr. Where osmium data are available for enriched samples from the same locality, these can further constrain the age of the lithospheric mantle. For example, eclogites from Siberian kimberlites yield Re–Os model ages around 3 Byr, consistent with a 2.9 Byr Re–Os isochron age (Pearson *et al.*, 1995b), and with the (minimum) 3 Byr mantle depletion age.

Reisberg and Lorand (1995) proposed an improved approach to overcome the problem of rhenium metasomatism in dating mantle peridotites. In studies of two orogenic lherzolites (Ronda and East Pyrenees),

they observed good correlations between Os isotope ratios and aluminium contents of whole-rock peridotite samples (Reisberg and Lorand, Fig. 1). Reisberg and Lorand proposed that these plots are isochron analogues, in which aluminium, an immobile and incompatible element, is a proxy for rhenium. Because the Re/Al ratio of the original rock before metasomatism is unknown, the slope cannot be solved directly to yield an age. However, if Re and Al behaved coherently in the original rock, the intercept on the y axis will yield the initial Os isotope ratio in the usual way. This initial ratio can then be used to determine an accurate rhenium depletion model age. The resulting 1.3 Byr model age for Ronda (Reisberg and Lorand, Fig. 2) is in good agreement with Re–Os errorchrons determined on the peridotite and pyroxenite fractions (section 8.4.2). Rhenium mobility for the East Pyrenees samples prevented the determination of a Re–Os isochron, but the 2.3 Byr model age is in reasonable agreement with Nd model ages of Pyrenean lower crustal xenoliths.

Meisel *et al.* (1996) argued that the most fertile (Al-rich) lherzolites from the subcontinental lithosphere could be used to constrain the Os isotope evolution of the 'primitive upper mantle' (PUM). Mantle xenoliths from Kilburn Hole (New Mexico), West Eifel (Germany) and the Baikal Rift (Mongolia) formed a linear array on a plot of Os isotope ratio against Al_2O_3 (Meisel *et al.*, Fig. 1b). The most Al-rich samples were consistent with an estimated Al_2O_3 content of 4.2% in the PUM, and gave an average $^{187}Os/^{188}Os$ ratio of 0.129 (187/186 = 1.08), in good agreement with the most Al-rich of the massive peridotites analysed by Reisberg and Lorand (1995). The new PUM value is in agreement with the average for ordinary chondrites, but is significantly more radiogenic than the mantle value of Allegre and Luck (1980, op. cit.), carbonaceous chondrites, and 'chondritic' initial ratios from the 1.97 Byr old Outokumpu ophiolite of Finland (Walker *et al.*, 1996), and 2.7 Byr old komatiites from the Kambalda area of Western Australia (Foster *et al.*, 1996).

8.7 Seawater osmium evolution

Ravizza (1993) proposed that osmium isotope compositions of ancient seawater could be determined from metalliferous sediments deposited near mid-ocean ridges. Because these sediments have greater rates of deposition than the pelagic clays used previously, the fraction of hydrogenous (seawater derived) osmium dominates over the cosmic dust fraction. Hence, reliable seawater signatures can be determined by bulk analysis of metalliferous sediments without the need for leaching. Twelve bulk samples with ages up to 28 Myr were analysed from DSDP cores near the East Pacific Rise. The results were in good agreement with leached pelagic clay data, and showed a rapid increase in seawater $^{187}Os/^{186}Os$ over the last 15 Myr, but relatively constant ratios between 18 and 28 Myr. Sr isotope ratios from the same samples were in good agreement with published data, but when compared with osmium, they showed significant decoupling between the two systems (Ravizza, Fig. 4). This is not surprising, in view of the very different chemistry of the two elements.

Peucker-Ehrenbrink *et al.* (1995) extended the detailed seawater osmium record to 80 Myr, using a combination of leached and bulk sediment analyses. However, they showed that for slowly deposited pelagic clays, a more gentle leaching procedure was necessary to remove hydrogenous osmium without releasing the cosmic dust fraction. Data from sediments with a relatively high deposition rate were in good agreement with the earlier studies, but data from slowly deposited sediments were less reliable. Leached 80 Myr old sediment showed for the first time a radiogenic osmium isotope signature below the K–T boundary. In addition, gently leached samples on either side of the boundary constrained a sharp drop in Os isotope ratio to the immediate vicinity of the boundary (Peucker-Ehrenbrink *et al.*, Fig. 3). Peucker-Ehrenbrink *et al.* calculated that a meteorite impact at the K–T boundary could have released a pulse of dissolved osmium into the oceans equivalent in size to a 5 Myr duration of the global run-off flux (see below). It should be noted that none of the samples in these seawater osmium studies was age corrected. Such corrections were argued to be insignificant, but will become important as older material is analysed.

In order to understand seawater osmium evolution it is important to quantify the fluxes which control its composition. The principal source of radiogenic osmium is river-borne run-off from old continental crust. However, possible sources of unradiogenic osmium are ultramafic rocks and the dissolution of cosmic dust. Esser and Turekian (1993a) estimated a bulk $^{187}Os/^{186}Os$ ratio of 10.5 for continental crust by the analysis of glacial loess and deltaic river sediments. This value is only slightly higher than present-day seawater (8.6), but it may not reflect the actual composition of 'potentially soluble' osmium delivered to the sea by rivers. Pegram *et al.* (1994) argued that in the oxidising conditions of river water, this potentially soluble component will not actually be

in solution but adsorbed on the ferro-manganese coatings of particulate sediment. When this sediment reaches the sea, reducing conditions may break down the ferro-manganese oxides, releasing osmium to seawater. Pegram *et al.* simulated this release using the peroxide leaching method. In a case study on Oneida Lake, New York State, they found leachable Os isotope ratios to be about 50% higher than bulk sediment ratios.

On a global scale, the $^{187}Os/^{186}Os$ ratio of leachable osmium was found to be very variable, ranging from values of 1.4–7.1 in rivers draining ultramafic rocks to values of 10–20 in more typical drainage basis (Pegram *et al.*, 1994). However, large rivers such as the Mississippi and the Ganges had ratios close to 16, a best estimate for the global average. Given an isotope ratio of 16 for leachable river-borne osmium, and around unity for dissolution of cosmic dust and sea floor hydrothermal alteration, the river-borne flux represents 50% of the global osmium supply to the oceans. Because Fe–Mn oxides in river water and seawater have similar osmium contents, we can deduce that river water and seawater have similar osmium concentrations, even though the absolute values are poorly constrained. Based on an elaboration of this approach, Peucker-Ehrenbrink and Ravizza (1996) estimated a seawater osmium residence time of about 16 kyr. This relatively short residence time (compared to Sr, section 3.5.2), suggests that seawater osmium might undergo short-term isotopic variations. Ferro-manganese crusts may provide a good record of spatial and temporal variations in seawater osmium isotope composition. For example, Burton *et al.* (1996) found a small difference between the composition of recent crusts from the N Atlantic and Pacific–Indian oceans (EOS 77 (46), F321, abstract).

Finally, Esser and Turekian (1993b) reported that osmium isotope analysis of coastal sediments could be used as a tracer of environmental pollution, based on anthropogenic osmium in contaminated sewage sludge.

9.1.3 Hf–Nd systematics of mantle depletion

Salters and Zindler (1995) applied a hot-SIMS method to hafnium isotope analysis of diopside separates from mantle peridotite samples. This involves bombarding the sample on a hot filament with Ar^+ ions to enhance ionisation. Abyssal peridotites and lherzolite xenoliths from New Mexico had Hf–Nd signatures close to the MORB field. However, peridotite xenoliths from Salt Lake Crater, Hawaii, had very radiogenic hafnium, far above the mantle Hf–Nd array (Salters and Zindler, Fig. 2). This isotopic signature is indicative of an incompatible element (Hf)-depleted source, but the trace element patterns of the xenoliths are indicative of incompatible element enrichment. Two alternative depletion-enrichment models were examined in order to explain the data. These involved either ancient (ca. 1 Byr old) or more recent (100 Myr old) depletion of the mantle source, followed by recent metasomatic enrichment. The ancient depletion model seems less likely because such an event should perturb Nd as well as Hf isotope signatures. The subsequent metasomatism would then have to coincidentally bring Nd isotope ratios back to normal upper mantle values. The more recent depletion model involves fairly extreme Lu/Hf fractionation in the sub-oceanic lithosphere, but such fractionation is possible if garnet is involved. Whatever the origin of these xenoliths, it is clear that such materials do not significantly affect OIB magmatism (including Hawaii), since OIB samples display a coherent Hf–Nd mantle array.

Salters (1996) presented an enlarged Hf–Nd data set for MORB samples in order to re-examine the 'hafnium paradox' in ocean ridge magmatism. Trace element fractionation during melting is quantified using the δ value, which compares the actual Lu/Hf and Sm/Nd ratios in MORB samples with the ratios necessary to generate observed isotopic signatures by evolution from a 2 Byr old chondritic source. Almost all MORBs and OIBs have significant positive δ (Lu/Hf) values, which demand melting in the garnet stability zone. In addition, average δ values from four different ridges showed a positive correlation with the water depth over the ridge (Salters, Fig. 6), which in turn is inversely dependent on the rate of magma generation at the ridge. However, one would expect to see the largest δ value (garnet signature) at the shallowest ridge, since melting under these ridges is believed to start at greater depths in the mantle. To avoid this problem, Salters argued that the shape of the melting zone was different under shallow and deep ridges, with a more triangular shape under deep ridges, which would enhance the relative amount of melting in the garnet zone (Salters, Fig. 7). In contrast, Hirschmann and Stolper (1996) argued that the Lu/Hf fractionation in MORB results from melting of garnet pyroxenite (rather than garnet peridotite) in a marble cake mantle. According to this model, the 'garnet signature' from pyroxenite melting is preserved under deep ridges with low melting rates, but under shallow ridges, this signature is diluted by large-scale melting of spinel peridotite. Evidence from other isotopic systems may help to resolve this debate (sections A.8.4 and A.13.3.2).

9.2.3 Ce isotope geochemistry

New Ce isotope analyses on ocean floor manganese nodules from the Atlantic Ocean (Amakawa et al., 1996) gave values which were much less scattered than previous data with large errors. On a plot of ϵ Ce versus Nd (Amakawa et al., Fig. 4), the new data fall close to a mixing line between MORB and continental crust, consistent with the mixing model of Elderfield (1992, op. cit.).

10.2.5 ^{39}Ar recoil

Argon recoil has important implications for minerals whose diffusional history is explained in terms of micro domains. The most important examples are feldspars, which have exsolution lamellae about 0.01–0.3 μm thick, but show little evidence for recoil effects in their plateau ages or Arrhenius plots. The lack of any such evidence led McDougall and Harrison (1988) to speculate that a large fraction of ^{39}Ar recoils occurred at low energy, with very small displacements. Onstott et al. (1995) re-examined this question using theoretical calculations and ion implantation experiments, but these continued to support a mean ^{39}Ar recoil distance of 0.082 μm. The implications were examined for three minerals showing exsolution of K-rich and K-poor lamellae (amphibole, plagioclase and K-feldspar). In all three cases, calculations indicated that ^{39}Ar concentrations would be significantly homogenised and ^{37}Ar almost totally homogenised between adjacent lamellae (e.g. Onstott et al., Fig. 14). Onstott et al. concluded that the lamellae were too small to be the domains controlling volume diffusion of argon in the samples of amphibole and plagioclase studied. The situation for K-feldspar was less clear, but they suggested that some reinterpretation of results might be necessary for the smallest domain sizes of K-feldspar used in thermal history analysis. This question remains controversial (section A.10.2.8).

The small grain size of clay minerals (typically 5–1000 nm thick) makes them very susceptible to ^{39}Ar loss by recoil effects. To combat this process, the technique of encapsulation was developed (Smith et al., 1993, op. cit.), whereby escaping ^{39}Ar is held in an ampoule so that it can be collected and analysed with argon released during laser heating. However, experiments by Dong et al. (1995) suggested that the escaping argon fraction is not lost from illite and smectite by direct recoil, but by thermal degassing of low retentivity sites which have picked up recoiling nuclides. These low retentivity sites are actually the two free surfaces of the clay mineral grain, and the amount of ^{39}Ar loss is inversely proportional to thickness. For example, a grain of illite 100 nm (1000 Å) thick, made of 100 silicate–cation–silicate composite layers, will lose about 1% of its ^{39}Ar (2 out of 200 surfaces); however, a grain 10 nm thick might lose as much as 10% of its ^{39}Ar budget.

Dong *et al.* (1995) argued that ^{39}Ar loss during irradiation may match ^{40}Ar loss during geological heating (diagenesis). They measured the ^{39}Ar signal retained by encapsulation, then subtracted this fraction from the total gas release of the sample (including all encapsulated argon). The resulting 'argon retention age' should be the same as a non-encapsulated age, and was argued to be a better estimate of the time of deposition or early diagenesis than the encapsulated age. However, retention ages measured on Paleozoic clays were older to varying degrees than the non-encapsulated ages, which Dong *et al.* attributed to the effect of different evacuation pressures during irradiation. In view of these uncertainties the method cannot be considered a reliable dating tool.

10.2.8 K-feldspar thermochronometry

Thermochronometry is based on linear Arrhenius relationships observed when ^{39}Ar is released from diffusion domains of uniform size and K content which obey a simple diffusion law. In their development of the K-feldspar method, Lovera *et al.* (1989, op. cit.) showed that non-linear Arrhenius trends produced by conventional step heating could be resolved into separate linear segments by cycling the heating schedule up and down (e.g. Richter *et al.*, 1991, Fig. 2). Subsequent work on a variety of samples showed that these line segments were effectively parallel, indicating relative constancy of diffusional activation energies. The vertical separation between different linear segments on the Arrhenius diagram can then be interpreted in terms of relative domain size. If the effective diffusion radius of the domain which generates the first, low-temperature array is set arbitrarily to be r_0 ('r' is the equivalent of 'a' in section 10.2.7), then the relative diffusion dimension of each subsequent gas release point can be calculated using the following relation:

$$\log (r/r_0) = [\log (D(T)/r^2) - \log (D(T)/r_0^2)] / 2$$

Because we are only interested in the relative diffusion dimension, the exact geometry assumed (slab or sphere, etc.) has little effect on the calculations. Richter *et al.* (1991) proposed that each value of log (r/r_0) should be plotted against cumulative ^{39}Ar release in a manner analogous to an age spectrum plot (Richter *et al.*, Fig. 3b). This plot therefore bridges the gap between the age spectrum diagram and the Arrhenius diagram.

The working of the 'Richter plot' was demonstrated on an analysis from the Quxu pluton, Himalayas. The diagram gave evidence of four plateaus, of which the highest temperature three correspond to plateaus on the age spectrum plot (Richter *et al.*, 1991, Fig. 3a). The lowest temperature plateau had no corresponding age because the first five steps were perturbed by excess ^{40}Ar which probably diffused into the grains after cooling. However, this did not affect the ^{39}Ar released during this part of the analysis, which can still be used to obtain diffusional data. In contrast, the highest temperature step gave an age, but no diffusional information, because it occurred during sample melting. After the gas release is broken into plateaus, the volume fraction of each is calculated, and an iterative programme is used (Lovera, 1992) to determine the actual radius ratios of the domains which will model the observed profile of relative diffusion dimensions. The solution is shown in Richter *et al.*, Fig. 3c. Finally, a cooling curve is determined (iteratively) which will yield the observed age spectrum.

In several experiments on K-feldspars from different orogens, the radius ratio of largest to smallest domains was determined to be between 100 and 500. Lovera *et al.* (1993) performed step heating experiments on the Chain of Ponds pluton, Maine (sample MH-10) in an attempt to find the absolute sizes of these domains. After crushing, K-feldspar grains were separated into four size fractions, averaging 425, 138, 54 and 42 μm diameter. Following irradiation, a full step heating analysis was performed on each size fraction, and the resulting Arrhenius plots compared (Lovera *et al.*, Fig. 6). The results showed a dramatic decrease in argon retentivity between the 138 μm and 54 μm size fractions, attributed to the 'breaking open' of the largest diffusion domain in crushing to 54 μm. Hence, the diameter of this domain was inferred as 50–100 μm. Based on relative domain size ratios, the smallest domain size was then estimated at ca. 0.1 μm.

In an accompanying optical and TEM study, Fitz Gerald and Harrison (1993) attempted to determine the crystallographic identity of the different domain sizes. They tentatively correlated the largest domains with blocks of K-feldspar surrounded by fractured and turbid zones, and the smallest with the 0.1 μm distance between albite exsolution lamellae. However, they were unable to identify any intermediate sized domains. Because the estimated size of the smallest diffusion domains is similar to the ^{39}Ar recoil distance, doubts have been expressed (section A.10.2.5) about the meaning of these data. However, empirically, cooling curves based on this technique are in good agreement with other geochronological evidence. For example, Lee (1995a) demonstrated good agreement of K-feldspar cooling curves with 40–39 muscovite ages and fission track apatite ages in a study of tectonic uplift in the Snake Range, Nevada (e.g. Lee, Fig. 10).

10.3.1 Laser probe dating methods

Laser probe dating was developed using continuous wave infra-red lasers, either defocused for step heating, or focussed for spot analysis. These lasers are effective for heating most samples, but cannot be focussed below a spot size of 50 µm, and are not effective for analysing pale coloured minerals such as feldspar. To overcome these problems, ultraviolet lasers have been introduced (e.g. Kelley et al., 1994). Ultraviolet laser light is obtained by frequency doubling, which is only possible using pulsed lasers. The power available with such a system is much lower, but it is effective for spot ablation because the energy is more efficiently concentrated in a 10 µm diameter spot. The ultraviolet laser ablation microprobe (UVLAMP) offers the opportunity for in situ analysis of thin sections with high spatial precision, both laterally and with depth in the sample. Kelley et al. demonstrated the depth resolution by using a rastered beam to ablate successive 1 µm deep steps into the surface of a K-feldspar grain. The resulting isotopic depth profile shows the diffusion of atmospheric argon into the surface of the grain (Kelley et al., Fig. 4). The method shows great potential for detailed studies of argon diffusion in minerals.

10.3.2 Multi-path diffusion

It has long been known that large biotite grains can lose radiogenic strontium or argon more rapidly than predicted by volume diffusion (e.g. section 3.3.2). This behaviour was also seen in argon diffusion experiments using hydrothermal bombs, and can be explained if biotite has an 'effective diffusion radius' of about 150 µm (e.g. Harrison et al., 1985; op. cit.). This implies that biotite grains consist of domains of ca. 150 µm radius, within which argon moves by volume diffusion, but between which there is a mechanism of enhanced transport along crystallographic defects.

The increased amount of spatial argon data available from the laser probe has allowed models of multi-path diffusion to be tested using natural mineral systems. Phillips and Onstott (1988) made the first such study, based on mantle-derived phlogopites from the Premier kimberlite, South Africa. 'Chrontour' mapping of 75 laser spots revealed 1.2–1.4 Byr apparent ages in the rim of a large grain, rising to a maximum of 2.4 Byr in the core (Phillips and Onstott, Fig. 4). This pattern was attributed to the loss of inherited mantle argon from the rim of the grain during 1.2 Byr-old kimberlite emplacement. However, the shape of the argon loss profile does not fit a volume diffusion model (Phillips and Onstott, Fig. 5). Phillips and Onstott attributed these patterns to enhanced diffusive loss of argon from a wide band (round the rim of the grain) due to structural defects. However, the nature of these defects was unclear.

Lee and Aldama (1992) demonstrated a numerical model which combined volume diffusion and enhanced ('short-circuit') diffusion into a general model termed 'multi-path' diffusion. In this model, the transfer of species from the crystal lattice to high-diffusivity paths, and vice versa, is governed by exchange coefficients (K2 and K1 respectively). The magnitude of these coefficients is critical in determining whether multi-path diffusion will occur in a given situation. Their effects can be examined on an Arrhenius plot (Lee, 1995b, Fig. 6). At high temperatures (>800 °C), volume diffusion is the dominant mechanism for argon loss from a mineral, but a relatively small exchange coefficient (K2) will allow a sudden change from lattice-dominated to short-circuit argon transport as temperature falls, forming a sigmoidally-shaped diffusion curve. However, small values of K2 will inhibit the role of short-circuit diffusion, maintaining a lattice-dominated diffusion mechanism to lower temperatures (<500 °C).

Lee (1995b) re-examined old hydrothermal diffusion experiments on hornblende and biotite at different grain sizes in order to evaluate the case for simple volume diffusion or multi-path diffusion in these minerals. In both cases, enhanced diffusivities for large grain sizes appear to favour multi-path diffusion over simple volume diffusion (Lee, Fig. 4). The similar occurrence of multi-path diffusion in these two minerals may result from similar styles of alteration, both in the laboratory and natural systems. For example, Onstott et al. (1991) suggested that fast diffusion pathways in hydrothermal biotite may be created during the experiment by the breakdown of chlorite layers, whereas Kelley and Turner (1991) attributed fast diffusion pathways in natural hornblende to biotite alteration.

Hodges *et al.* (1994) compared argon diffusion patterns in muscovite and biotite using the laser probe. Single mica grains were analysed from the 1700 Myr-old Crazy Basin monzogranite of central Arizona, a terrane which has been argued to display either very slow cooling or multiple metamorphic events. Laser step heating of both muscovite and biotite gave very similar plateau ages of 1412 and 1410 Myr. However, for muscovite, the gas release steps which form the plateau occurred after melting had begun. Therefore, this plateau undoubtedly results from argon homogenisation during melting, and the similarity to the biotite plateau age is probably a coincidence. These findings are consistent with previous work, where argon homogenisation was shown to be a greater problem in laser step heating than conventional step heating (section 10.3).

Laser spot dating revealed very different age patterns in other muscovite and biotite grains from the same sample. Muscovite spot ages ranged from 1650 Myr in a small core area to 1270 Myr at the rim, with a roughly concentric pattern (Hodges *et al.*, Fig. 3a). This age distribution fits a simple volume diffusion model for cylindrical geometry with an effective diffusion radius similar to the grain size. However, laser spot ages in the biotite grain had a quite different distribution, from 1420 Myr in a large core area to 1150 Myr at the rim (Hodges *et al.*, Fig. 3b). The large area with (reset) 1400 Myr ages, along with a zone of young ages approaching the core, was evidence of the operation of fast diffusion pathways. Hodges *et al.* interpreted the data in terms of small domains with an effective diffusion radius of 150 μm, as postulated in hydrothermal experiments. However, distributed fast diffusion pipes may be a better explanation for the large core area with homogeneous 1400 Myr ages. More precise spot dating with the ultraviolet laser probe may help to distinguish between these alternative models for multi-path diffusion.

Villa (1994) claimed to find evidence for multi-path diffusion in K-feldspar. He performed isothermal heating experiments which appeared to be more consistent with a multi-path diffusion model than the multi-domain size model (section A.10.2.8). However, Lovera *et al.* (1996) questioned the design of these experiments, since the same initial rate of argon loss (and hence diffusivity) was observed at both 580 and 750 °C (solid line, Lovera *et al.*, Fig. 1). In contrast, by modelling the argon release in a previous analysis of the same sample by Villa, Lovera *et al.* demonstrated an Arrhenius relationship of normal slope, but offset to high apparent diffusivities by a rapid heating schedule. This relationship is consistent with volume diffusion from small domains. The lack of consistent variations in activation energy between different domain size fractions (Harrison *et al.*, 1991), is also consistent with a simple diffusion model.

11.1.3 Helium in the mantle

It is now widely understood that heavy rare gases in OIB magmas are very susceptible to shallow-level contamination in the magmatic system (sections 11.2–11.4). However, recent work shows that helium isotope signatures in OIB may also be susceptible to such contamination processes.

Hilton *et al.* (1993a) found a strong correlation between R/R_a value and petrology in submarine volcanic glasses from the Lau back-arc basin, situated behind the Tongan arc. Basaltic samples from the centre of the basin had relatively high helium contents (up to 10 nl/g), and normal MORB-like R/R_a values of 8 (Hilton *et al.*, Fig. 3). However, more differentiated glasses from just behind the magmatic arc had much lower helium contents (<0.2 nl/g), and R/R_a values as low as unity. Based on the correlation between $^3He/^4He$, helium content, and petrology, Hilton *et al.* attributed the lower R/R_a values in differentiated glasses to shallow level contamination, probably due to crustal assimilation by magmas which had been largely degassed of mantle helium.

Hilton *et al.* (1995) observed similar evidence in phenocryst phases from two different lava series on Heard Island, Kerguelen Archipelago. Phenocrysts from the Laurens Peninsula series had a plume-type helium signature with R/R_a of 16–18. However, phenocrysts from the Big Ben series had MORB-like R/R_a values in helium-rich samples, but lower R/R_a values in helium-poor samples (Hilton *et al.*, Fig. 1a). In the latter sample suite, isotopic disequilibrium was observed between olivine and cpx phenocrysts. Therefore, Hilton *et al.* suggested that contamination with radiogenic helium occurred immediately before eruption, probably during phenocryst growth in shallow magma chambers.

Several other hot-spots with R/R_a values lower than MORB and elevated Sr isotope ratios (up to 0.705) have been attributed to crustal recycling into the OIB source reservoir (Fig. 11.10). However, Hilton *et al.* (1995) observed that low R/R_a values in these islands were also associated with low total helium contents (Hilton *et al.*, Fig. 2). Hence, they suggested that the radiogenic helium signatures in these islands also result from contamination within the oceanic lithosphere, rather than sediment recycling into the deep mantle.

A different scenario is seen in Samoan basalts, which have a unique combination of elevated R/R_a and radiogenic Sr (Fig. 11.10). Peridotite xenoliths in these lavas have very radiogenic Sr isotope signatures, attributed to metasomatism by an EMII mantle component (section 6.6.2). However, helium isotope analysis of fluid inclusions from the xenoliths revealed high R/R_a values of around 12 (Farley, 1995a). This was unexpected, since a recycled sediment component should have radiogenic helium with low R/R_a values. However, the fluids also had $C/^3He$ ratios of ca. 3×10^9 which were typical of mantle values, and distinct from ratios of over 10^{11} seen in sediments. The combined evidence suggests that the volatile component of the metasomatic fluid was derived from the deep mantle and only recently mixed with a volatile-poor melt of subducted sediment. Because subducted sediment accumulates radiogenic helium from uranium decay, the high R/R_a value of the metasomatic fluid places limits on the mantle residence time of the sediment since subduction. Based on binary strontium–helium mixing calculations, Farley estimated a residence time of only 10 Myr, suggesting that the sediments were incorporated into the plume from the nearby Tongan trench. Helium is also decoupled from lithophile isotope systems in Hawaii, but Valbracht et al. (1996) speculated that this decoupling also occurred at shallow depths within the oceanic lithosphere, rather than in the plume itself (Valbracht et al., Fig. 8).

Beginning with the work of Galer and O'Nions (1985, op. cit.), it was realised that incompatible elements may have short residence times in the convecting upper mantle (sections 6.4.2 and 13.4). In view of the ease with which helium can escape from any system at elevated temperatures, it represents the ultimate incompatible element, and therefore should have the shortest residence time. Since helium subduction is believed to be negligible, input to the upper mantle is restricted to primordial helium escape from the lower mantle, plus in situ production of radiogenic helium from U-series isotopes.

Kellogg and Wasserburg (1990) assumed a steady state between supply and degassing in order to determine the residence time of helium in the upper mantle. They argued that ridges are the principal sites of helium escape from the upper mantle (whereas hot spots dominate in outgassing the lower mantle). Hence, they used a simple calculation to estimate the residence time of helium in the upper mantle:

$$\tau = \text{mass of upper mantle/rate of outgassing}$$

Based on a depth of 670 km, the upper mantle has a mass of 1×10^{27} g. Also, based on ocean floor pro-

duction at 3.5 km^3/yr and a melting depth of 60 km, the rate of mantle outgassing is estimated at 7×10^{17} g/yr. Hence, Kellogg and Wasserburg calculated a helium residence time of 1.4 Byr in the upper mantle. However, O'Nions and Tolstikhin (1994) estimated a somewhat shorter residence time of 1.1 Byr, based on an upper mantle mass of 1.1×10^{27} g and an outgassing rate of 1×10^{18} g/yr (corresponding to a melting depth of 90 km). These relatively short residence times suggest that the upper mantle has been completely outgassed of primordial helium. However, in view of the extreme volatility of helium, they also represent a minimum for the upper mantle residence times of lithophile elements.

In continental regions, helium isotope signatures have been used to detect the presence of mantle-derived magmas at depth, under sedimentary basins and rift zones. More recently, helium analysis of geothermal fluids was used by Hilton et al. (1993b) and Hoke et al. (1994) to probe the width of the mantle melting zone behind the Andean subduction zone. Both studies revealed high R/R_a values (indicative of a significant fraction of mantle helium) in the magmatic zone centred on the Altiplano. In contrast, R/R_a values below 0.5 were found in the trench zone in front of the magmatic arc, and behind the Eastern Cordillera (Hoke et al., Fig. 5). Hilton et al. found good agreement between helium analysis of geothermal fluids and phenocrysts from nearby volcanos. However, because of the high altitude of these volcanos, special rapid-crushing procedures were necessary to minimise contamination from a large in situ cosmogenic component. Hoke et al. attributed high R/R_a values in the central section of the Andes to thinned lithosphere resulting from subduction erosion. The decrease in R/R_a values behind the Eastern Cordillera probably marks the transition from hot, thin lithosphere above the subduction zone to the thick, cold lithosphere of the Brazilian shield.

11.1.7 Helium in sediments

It is well established that high $^3He/^4He$ ratios in ocean floor sediments reflect the accumulation of interplanetary dust particles (IDPs). However, the question of temporal variability in the IDP flux has only recently been examined (Takayanagi and Ozima, 1987). These authors studied ^3He variability in a 10 m pelagic clay core from the Central Pacific and a 150 m nanno-fossil ooze core from the South Atlantic. The former spanned 0–3 Myr, while the latter, with generally higher sedimentation rates, spanned 0–40 Myr. Sedimentation rates were determined in both cases by paleomagnetism, supplemented in the 3 Myr-old core by ^{10}Be data (section 14.3.3). The observed range of ^{3}He/^{4}He ratios was attributed to mixing of 0.1–1 ppm of IDPs with terrestrial sediment (Takayanagi and Ozima, Fig. 2). However, the ^{3}He content of IDPs is 10 orders of magnitude higher than terrestrial sediment, so the IDPs totally dominate the ^{3}He budget of the samples. In both cores, ^{3}He contents were inversely correlated with sedimentation rate (e.g. Takayanagi and Ozima, Fig. 3). The ^{3}He deposition flux was therefore determined by multiplying the ^{3}He content by the sediment mass accumulation rate (mass is used because ocean floor sediments undergo compaction after deposition). The results (e.g. Fig. 5) suggested flux variations over time, but did not display any overall trend. The average ^{3}He flux over the past 40 Myr was estimated as 1.5 ($\pm$1) $\times$ 10^{-15} ml/cm^2/yr (at STP).

Generally similar results were obtained by Farley (1995b) on a 22 m core of pelagic clay from the central North Pacific, spanning the past 72 Myr. During the Quaternary, the sedimentation rate was high, yielding a ^{3}He flux of about 1.1 $\times$ 10^{-15} ml cm^2/yr, in good agreement with Takayanagi and Ozima (1987). However, in the deeper part of the core, the calculated ^{3}He flux was lower, averaging about 0.5 $\times$ 10^{-15} ml/cm^2/yr. It is not clear whether this represents a real variation in the interplanetary dust flux over time, or a reduction in the retentivity of ^{3}He with depth. Farley also observed no ^{3}He peak at the K–T boundary, indicating that the extra-terrestrial signals from iridium and helium are decoupled. This was attributed to impact-induced vapourisation and outgassing of the K–T bolide.

Marcantonio *et al.* (1995) made a more detailed study of the Quaternary ^{3}He flux, based on a 4 m core of carbonate-rich sediment from the Central Pacific, spanning the last 200 kyr. After correcting for dilution by biogenic carbonate, their ^{3}He/^{4}He data lay on the same mixing line observed by Takayanagi and Ozima (1987) between terrigenous and IDP components. Marcantonio also determined initial excess ^{230}Th activities on the same samples. Normalisation of ^{3}He to ^{230}Th can remove the effects of variable sediment dilution because ^{230}Th is constantly produced in seawater from ^{234}U and is rapidly transported to the ocean floor by adsorption onto sinking particulate matter (section 12.3.2). When plotted against ages from oxygen isotope stratigraphy, ^{3}He and ^{230}Th showed strong covariation, with peak signals during interglacial periods (Marcantonio *et al.*, Fig. 3). These peaks can be explained by redistribution of sediments during interglacial periods. Based on the ratio of ^{230}Th activity to ^{3}He content (Marcantonio *et al.*, Fig. 4), an

average ^{3}He deposition flux of 0.96 $\times$ 10^{-15} ml/cm^2/yr was determined for the past 200 kyr.

Farley and Patterson (1995) made a similar study of Quaternary ^{3}He variation, based on a 9 m core of foram-nanofossil ooze from the flank of the Mid-Atlantic Ridge, spanning the period 250–450 kyr BP. ^{3}He contents were inversely correlated with δ ^{18}O variations, which are interpreted as monitors of glacial–interglacial cycles (Farley and Patterson, Fig. 1). Farley and Patterson noted a proposal by Muller and MacDonald (1995) that Quaternary climate variations might be due to a 100 kyr cycle of variations in the Earth's orbital inclination, causing periodic encounters with a cloud of IDPs which could partially block out solar radiation. Hence, Farley and Patterson speculated that the helium isotope data might be recording a causal relationship between IDP accumulation and climate. However, according to this model interglacial periods should be characterised by the lowest ^{3}He flux, whereas Farley and Patterson found the opposite relationship. Therefore, the evidence appears to favour an alternative explanation that climatically induced variations in sedimentation rate caused apparent variations in the ^{3}He flux which could not be adequately corrected with the available age data for the core.

11.3 Xenon

Porcelli and Wasserburg (1995a) extended the steady-state helium model of Kellogg and Wasserburg (1990) to xenon. They argued that upper mantle xenon is not a residue of early Earth degassing (as previously proposed), but is the result of input from the lower mantle and from subduction of atmospheric xenon, along with *in situ* radiogenic production. The supply of rare gases from lower to upper mantle was attributed to mass transfer in plumes. A fraction of the plume is degassed at the hot spot, but the bulk is mixed into the upper mantle. In the mass transfer model, rare gases are not fractionated from one another, so they are all thought to have the same upper mantle residence time.

Since the upper mantle is argued to be in steady state, it bears no memory of its early history. Hence, ^{129}Xe/^{130}Xe variations in the upper mantle (attributed to extinct iodine) are explained solely by mixing of lower mantle and atmospheric xenon. The atmosphere is at the unradiogenic end of the MORB ^{129}Xe/^{130}Xe range, and is argued to have a source independent from the deep Earth, probably by late accretion of volatile-rich material after degassing of the deep earth. In contrast to this composition, the lower mantle is constrained to be at the radiogenic end of the MORB range.

Variations in ^{136}Xe/^{130}Xe are more complex, because there are components from both extinct plutonium and extant uranium fission (Porcelli and Wasserburg, 1995a, Fig. 4). In response to iodine and plutonium decay, it is thought that lower mantle xenon evolved from an initial solar or planetary composition to the hypothetical composition of present day lower mantle (P). In the steady state model, xenon which escapes from the lower mantle is supplemented by uranium decay in the upper mantle, and then mixed with subducted atmospheric xenon, to reach the upper mantle composition (D). The MORB xenon array is then attributed to mixing between D and the atmosphere, due to contamination of magmas at the sea floor. OIB should define a similar array of gentler slope between the atmosphere and P. However, this prediction cannot yet be tested reliably, due to the difficulty of obtaining definitive non-atmospheric xenon signatures from hot spots. The available non-atmospheric data from Samoan xenoliths do not appear to support the model, since all data appear colinear with the MORB array (Poreda and Farley, 1992, op. cit., Fig. 3). However, this may be due to overprinting of the plume signature by entrainment of upper mantle xenon.

Another isotope which may constrain terrestrial xenon systematics is ^{128}Xe. Well gas data from Caffee *et al.* (1988) define a positive correlation between iodogenic ^{129}Xe/^{130}Xe and non-radiogenic ^{128}Xe/^{130}Xe (Jacobsen and Harper, 1996, Fig. 3). This implies that the (upper) mantle and the atmosphere show relative fractionation of non-radiogenic rare gases, and that this fractionation occurred in early Earth history, when ^{129}I was not extinct (see also Tolstikhin and O'Nions, 1996). This would be consistent with the mass fractionation of atmospheric krypton and xenon relative to solar compositions (Pepin, 1992, Figs. 4 & 5). However, we should note that there is also evidence against rare gas fractionation between the mantle and atmosphere. This comes from the light xenon isotopes, ^{124}Xe and ^{126}Xe, which show atmospheric ratios against ^{130}Xe (Fig. 11.19), and from the identical ratios of krypton and non-radiogenic argon in volatile-rich MORB and the atmosphere (e.g. Hiyagon *et al.*, 1992, op. cit., Fig. 6).

Further uncertainties in Xe systematics arise from the findings of Meshik *et al.* (1995), who suggested that variations in the relative abundances of heavy xenon isotopes could be due to open system conditions during the β decay of uranium fission products. High temperature conditions could allow preferential escape of some short-lived intermediates, affecting the isotopic composition of xenon isotopes, which are the final decay products. In the light of all the above problems, terrestrial xenon systematics remain in considerable doubt.

11.4 Neon

Following acceptance of the solar neon model for the Earth, neon and helium isotope data can now be placed in a unified model. This is because helium does not suffer significant atmospheric contamination, while atmospheric contamination of neon can be corrected by normalising to the solar ^{20}Ne/^{22}Ne ratio. The compositions of depleted mantle (D), lower mantle (P), and meteorites (representing the initial Earth) form a linear growth curve on a diagram of ^{21}Ne/^{22}Ne against ^{4}He/^{3}He (e.g. Porcelli and Wasserburg, 1995b, Fig. 4). This is consistent with the common progenitors U and Th, which produce radiogenic helium and nucleogenic neon. Porcelli and Wasserburg (1995b) also argued that the linear Ne–He growth curve is consistent with the steady state model for mantle rare gases.

12.2.2 U–Th mass spectrometry

Asmerom and Edwards (1995) described a new method for loading Th as the fluoride. When used with a normal double filament technique, this technique improved the ionisation efficiency of large thorium samples (such as igneous rocks). Using the new method, a 200 ng Th sample had an ionisation efficiency of 2×10^{-4}, an order of magnitude better than previous methods (Fig. 12.4).

12.3.1 ^{234}U

The excess ^{234}U method is less precise as a dating tool than ^{230}Th, but for some types of sample it is the only method available. An important example is the dating of ocean floor Fe–Mn crusts (Chabaux *et al.*, 1995). Growth rates of these crusts have been measured by the ^{230}Th daughter excess method, but because the initial thorium isotope ratio of seawater is unconstrained, an absolute date is not possible. However, ^{234}U can be used to obtain absolute ages by projecting excess activities back to the known seawater composition. Such absolute ages are necessary because there is evidence that the surfaces of Fe–Mn crusts have often been lost, due to geological erosion or abrasion during dredging operations.

Chabaux *et al.* (1995) reported a case-study on two Fe–Mn crusts, dredged from 1900 m depth on a

West Pacific seamount. For both crusts the ^{230}Th and ^{234}U methods gave consistent growth rates of ca. 7 mm/Myr (Chabaux et al., Fig. 1). This suggests that closed-system conditions were preserved, and that initial uranium and thorium isotope ratios remained constant (within error) during the 150 kyr period of deposition. By projecting ^{234}U activities back to the seawater value, Chabaux et al. located the original growth surface of the crusts, and calculated that abrasion losses of 0.2–0.4 mm had occurred from the surfaces. Extrapolation to the growth surface also allows calculation of the local ^{230}Th/^{232}Th activity ratio of seawater, which would allow thorium isotope mapping in the oceans.

Excess ^{234}U activity data for planktonic foraminifera would be very useful, since these are the basis of the seawater oxygen isotope record (section 12.4.2). However, forams have low U contents (typically 20 ppb), which tend to be swamped by the U contents of ferro-manganese diagenetic overgrowths. Henderson and O'Nions (1995) showed that dithionite solution (a reducing agent) could be used to clean recent forms in order to recover normal seawater uranium isotope ratios. However, a test on 2 Myr old forams showed excess ^{234}U activities above the seawater value which must have been introduced from pore waters after sedimentation. This suggests that forams do not remain a closed system for uranium, and therefore cannot be used for dating or to constrain the uranium isotope evolution of seawater.

12.3.2 ^{230}Th

The rapid adsorption of ^{230}Th onto particulate matter makes it a very useful oceanographic tracer. Hence, several studies have been directed at understanding its behaviour in seawater, including its ocean residence time. The activity of ^{230}Th in North Atlantic seawater was determined by Cochran et al. (1987) by pumping large volumes of seawater, at different depths, through a filter system designed to scavenge ^{230}Th. Two profiles showed increasing activity with depth, both on particulates and in solution (Cochran et al., Fig. 7). High levels of dissolved ^{230}Th at depth were attributed to attainment of sorption equilibrium between particulates and seawater. Riverine supply of ^{230}Th causes slight enrichment in shallow seawater off Cape Hatteras, but makes a negligible contribution to the total ^{230}Th inventory.

Yu et al. (1996) used these results to make a new estimate of the ^{230}Th residence time in North Atlantic seawater. This value is determined from the ^{230}Th inventory per unit volume of water (n = activity/decay constant), divided by the supply flux per unit volume. Since riverine supply of ^{230}Th is considered insignificant, the supply flux is equal to oceanic ^{234}U decay. Hence, in terms of activities:

$$\tau\ ^{230}\text{Th} = 1/\lambda_{230\text{Th}}\ A_{230\text{Th}}/A_{234\text{U}}$$

Based on profiles of ^{230}Th activity against depth, Yu et al. estimated an average ^{230}Th activity of 0.65 d.p.m./m^3 in the North Atlantic at 25°N. This compares with a ^{234}U activity of 2700 d.p.m./m^3 which is constant throughout the oceans due to the long residence time of uranium. Plugging these values into the above equation gave a τ value of only 26 yr, much shorter than previously estimated.

12.3.5 ^{231}Pa–^{230}Th

When compared to the longer 111 yr residence time of ^{231}Pa, the short seawater residence of ^{230}Th implies significant differences in the oceanic behaviour of these species. Because of its particle-reactive behaviour, very little ^{230}Th can be transported laterally (advected) before it is scavenged and sedimented. In contrast, ^{231}Pa can be advected by ocean currents before it is scavenged in a location with a high sinking particle flux. Yu et al. (1996) showed that these different properties allow ^{231}Pa/^{230}Th activity ratios to be used as monitors of ocean circulation.

The present day Atlantic Ocean is dominated by a 'conveyer belt' which transports North Atlantic Deep Water (NADW) southwards to the Antarctic. Radiocarbon evidence (section A.14.1.6) suggests that NADW has a residence time of 200–300 yr in the Atlantic. Comparison of this value with the ocean residence times of ^{231}Pa and ^{230}Th indicates that about 50% of ^{231}Pa, but only 10% of ^{230}Th produced in the Atlantic will be exported to the Southern Ocean. These predictions are supported by ^{231}Pa/^{230}Th activity measurements on (recent) core tops from ocean floor sediment (Yu et al., Fig. 2a), which show an average activity ratio of only 0.06 in the Atlantic, but 0.17 in the Southern Ocean (relative to a production ratio of 0.09).

Yu et al. made the critical observation that sediments deposited at the time of the last glacial maximum had exactly the same distribution pattern of ^{231}Pa/^{230}Th activity ratios (Fig. 2b). From this observation they concluded that the ocean conveyor belt operated at a very similar rate during the glacial maximum. This casts doubt on a widely-favoured model in which the conveyer belt was thought to have partially or completely ceased during the glacial maximum (section A.14.1.6). Hence, if the new evidence is substantiated by further study, it will have major repercussions on our understanding of paleo-oceanography.

12.3.6 ^{230}Th sediment stratigraphy

The short ocean residence time of ^{230}Th makes this tracer useful in constraining the deposition fluxes of species such as ^{10}Be with longer residence times. Frank *et al.* (1995) illustrated this approach in a study of sediment stratigraphy from the Weddell Sea, Antarctica. Both ^{10}Be abundances and excess initial ^{230}Th activities showed large variations in concentration between glacial periods (stages 2, 4 and 6) and interglacials (stages 1, 5e and 7). Average burial fluxes were calculated for each climatic stage, based on initial radio-nuclide abundances, dry bulk density, and sedimentation rate. These fluxes were positively correlated with sedimentation rate (Frank *et al.*, Fig. 6), and varied by more than an order of magnitude. Since ^{230}Th activities vary to the same extent as ^{10}Be abundance, these variations cannot be attributed principally to boundary scavenging of advected nuclides (although this may play some role). Instead, Frank *et al.* attributed the variations to sediment transport. High fluxes during interglacial periods were attributed to focussing of 'marine snow' (radionuclide-bearing diatoms) by strong bottom currents. Low fluxes during glaciations were attributed to 'bulldozing' of sediments by grounded ice shelves, which replaced young (isotopically hot) sediment by old (isotopically dead) material.

12.4.3 ^{230}Th: dirty calcite

In dirty calcite systems, Th/U isochrons offer some advantages over the more popular U/Th isochron presentation because Th/U isochrons minimise the problem of error correlation between the variables. To show how Th/U isochrons yield the composition of the detrital-free component in a dirty calcite system, Ludwig and Titterington (1994) plotted synthetic data in three dimensions (Ludwig and Titterington, Fig. 3). They also presented a maximum likelihood method for calculation of the best-fit isochron, an approach already applied to some other isotopic systems (section 2.3).

13.3.2 ^{230}Th and ocean ridge processes

Comparison of U-series data with other geochemical tracers may yield fresh insights into magma genesis at ridges and hot spots. Sims *et al.* (1995) used a combination of Sm–Nd and U–Th data to test melting models for MORB and a variety of Hawaiian lavas including tholeiites, alkali basalts and basanites. The quantity α (Sm/Nd) is the relative difference between the Sm/Nd ratio of the rock and the ratio in the source necessary to generate its isotopic composition over geological time (see also section A.9.1.3). Values of α (Sm/Nd) are expected to vary with the degree of melting, as demonstrated by their correlation with silica saturation in Hawaiian lavas. Sims *et al.* also observed a roughly hyperbolic correlation in Hawaiian lavas between α (Sm/Nd) and the ^{230}Th/^{238}U activity ratio (represented by k in their Fig. 3, which is the reciprocal of *r*, section 13.3.2). They attributed this correlation to simple batch melting of a garnet lherzolite source, with melt fractions ranging from 6% (tholeiites) to 0.25% (basanites). However, a dynamic melting model is necessary to explain ^{226}Ra/^{230}Th activity data for Hawaiian tholeiites (Hemond *et al.*, 1994). MORB data display a similar range of Th/U activity ratios to Hawaii, but with much lower α values (Sims *et al.*, Fig. 3). Several possible explanations for the MORB array were examined, but dynamic melting is the most plausible.

A relationship between U-series activity ratios and ridge depth may also throw light on mantle melting processes. In a study of MORB glasses from the Azores plateau, Bourdon *et al.* (1996) demonstrated an inverse correlation between ^{230}Th/^{238}U activity ratio (*r*) and water depth above the axis of the Mid-Atlantic Ridge (Bourdon *et al.*, Fig. 7). In view of the great sensitivity of Th/U disequilibrium to melting porosity, one would not expect to see the greatest disequilibrium in samples from the shallowest part of the ridge, with the highest melting rate. The fact that an inverse correlation is observed led Bourdon *et al.* to suggest that the main control on Th/U disequilibrium across the Azores plateau must be the depth at which melting is initiated. They attributed this to either an enhanced volatile or heat content of the mantle, associated with the Azores plume. An alternative explanation is enhanced contamination of the mantle by eclogite or garnet pyroxenite from the plume (Hirschmann and Stolper, 1996, section A.9.1.3). However, Bourdon *et al.*, argued that this could not account for the Azores observations, since Th/U excesses do not correlate with isotopic or trace element evidence for source enrichment.

13.3.6 ^{230}Th in continental lavas

The discovery that excess ^{230}Th activities can only be generated in the presence of garnet (section 13.3.2) has provided a basis for understanding U series systematics in young continental lavas. Asmerom and Edwards (1995) demonstrated this approach in comparing young (<10 kyr) basalts from the Pincate volcanic field of the Basin and Range province in Mexico, and the 'San Francisco' volcanic field of the Colorado Plateau. Nd isotope and geochemical data are consistent with an asthenospheric origin for basalts of the Basin and Range, but a lithospheric origin for those of the Colorado Plateau.

U series analysis was performed to see whether basalts from these different sources would possess different ^{230}Th signatures, reflecting melting at different depths. The results were consistent with this model (Asmerom and Edwards, Fig. 2). Asthenospheric melts had excess ^{230}Th activities typical of MORB magmas (r = 1.2–1.35), whereas the lithospheric melts had no excess ^{230}Th activity (r = 0.99–1.02). Some possible explanations for equilibrium ^{230}Th/^{238}U activity ratios could be large degrees of mantle melting or slow ascent to the surface. However, both of these possibilities were ruled out by the observation of large excess ^{231}Pa activities in these rocks (^{231}Pa/^{235}U activity = 2.0). Melting in the spinel peridotite field can generate excess ^{231}Pa, but not excess ^{230}Th activity (section A.13.6.2). Therefore, Asmerom and Edwards suggested that alkali basalts of the San Francisco volcanic field were probably generated by shallow melting of subcontinental lithosphere within the spinel peridotite field.

13.4 Mantle evolution

U–Th data for lavas of the Snake River Plain, Idaho, provide additional evidence for the role of plumes in mantle enrichment (Reid, 1995). Two basalts from the Craters of the Moon volcanic field fall near the equiline (r = 0.98–1.01), but when the κ (Th) values of these rocks are compared with κ (Pb) values, the data points represent the greatest deviations yet observed from the mantle array of OIB samples (Reid, Fig. 5). This cannot be explained by crustal contamination or *in situ* decay, but can be explained by a uranium enrichment event which was too recent to affect Pb isotopes, yet sufficiently ancient for ^{234}U to decay to ^{230}Th. Evidence for elevated levels of ^{3}He in the Snake River Plain basalts led Reid to suggest that U enrichment of the lithosphere was caused by CO_2 metasomatism associated with the Yellowstone Plume. A similar model was proposed by Williams and Gill (1992, op. cit.) to explain less extreme U enrichment of the source of Nyiragongo volcano (Reid, Fig. 5).

13.6.1 ^{226}Ra

Chabaux and Allegre (1994) studied correlations between excess ^{226}Ra and ^{230}Th activity in OIB alkali basalts, carbonatites, and arc volcanics in order to compare the melting processes in these environments. In OIB, they found a positive correlation between excess ^{226}Ra and ^{230}Th activities, forming a linear array of negative slope on an activity ratio plot of ^{230}Th/^{226}Ra against ^{230}Th/^{238}U (Chabaux and Allegre, Fig. 1). In addition, a weak correlation was found between ^{230}Th/^{238}U activity ratio and the reciprocal of the estimated buoyancy flux of each plume (Chabaux and Allegre, Fig. 2). Based on these observations, Chabaux and Allegre argued that excess ^{226}Ra activities were generated during mantle melting and not by contamination processes in the crust (a possibility discussed by Hemond *et al.*, 1994). In addition, Chabaux and Allegre inferred from the linear correlation in Fig. 1 that the magmas were transferred from the melting zone to the surface within about two ^{226}Ra half-lives, otherwise this correlation line would have been excessively degraded. However, the data do not allow a definitive choice between the type of melting model (section A.13.6.2).

Carbonatites display an array of opposite slope to OIB in the activity ratio plot of ^{230}Th/^{226}Ra against ^{230}Th/^{238}U (Chabaux and Allegre, Fig. 3). However, these differences are due to the behaviour of the ^{230}Th/^{238}U system (not ^{226}Ra), which implies a reversal in the bulk distribution coefficients of U and Th under the conditions of CO_2 enrichment associated with carbonatite melting. Similarly, arc volcanics display a roughly vertical array in Fig. 3, due to the observation of both excesses and deficiencies of ^{230}Th in arc lavas (section 13.5). In both cases, these units display excess ^{226}Ra activities greater or equal to OIB. This is probably due to the role fluids in both rifting and subduction-related environments, but the details are not yet clear.

13.6.2 ^{231}Pa

The shorter-lived U-series systems (^{231}Pa and ^{226}Ra) can provide additional constraints on rates of melting and magma ascent under ocean ridges. Ideally, such constraints will help to select between the equilibrium porous flow (EPF) and channel flow models. However, the solutions to melting models are non-unique, and the data cannot yet conclusively support or refute a given model.

Lundstrom *et al.* (1994, 1995) argued that MORB data could best be explained by the EPF model of

Spiegelman and Elliot (1993, op. cit.). In this model, the magma is argued to rise through the porous, partially molten medium, remaining in equilibrium with the medium as it rises. The rate of magma transport to the surface is non-instantaneous and controls the degree of isotopic disequilibrium in the shorter-lived species. In contrast, Iwamori (1994) argued that rapid melt extraction and ascent (channel flow) provides a better model for ridges and plumes. Iwamori made a brief examination of two-dimensional models for ridge or plume melting. These were examined in more detail by Richardson and McKenzie (1994). Two dimensional modelling takes into account the decreasing rate of upwelling on moving away from the ridge or plume axis (Fig. 13.13). However, in view of other large uncertainties about melting behaviour, the advantages of two-dimensional over one-dimensional modelling are relatively minor.

Lundstrom *et al.* (1995) compared the results of one dimensional EPF and channel flow models with new and published data from the Juan de Fuca ridge which display large variations in U-series activity ratios. These ratios were plotted on bivariate plots (Lundstrom *et al.*, Fig. 2), in which the denominators on each axis were either the same (Figs. 2a and 2c) or had a constant ratio (Fig. 2b). Under these conditions, mixing processes will generate straight lines. To achieve this configuration, the U/Th ratio was plotted in the form 'r' in Fig. 2b and 1/r in Fig. 2a. Linear arrays of data were found on all three diagrams and attributed to mixing between different magmatic end-members. The predicted evolution of these end-member magmas was also shown as a function of depth (Lundstrom *et al.*, Fig. 1b). The EPF model can best be understood by seeing how it explains the principal features of the data in Figs. 1 and 2 of Lundstrom *et al.* (1995).

In Fig. 2c (Lundstrom, 1995), the product magmas are believed to have the same $^{238}U/^{232}Th$ activity ratio as the source. In other words, no net U/Th fractionation is assumed in this model (unlike Sims *et al.*, 1995, for example). Excess activities of ^{230}Th, ^{231}Pa and ^{226}Ra are attributed solely to radiogenic production in the rising solid after melting has begun. Larger excess ^{230}Th activities are generated in one magmatic end-member by melting a deeper source which is also richer in garnet. The resulting magmas begin with more excess ^{230}Th (due to higher garnet in the source), and are then further enriched by a longer period of ascent in the garnet field (Fig. 1b).

In Fig. 2b, excess ^{231}Pa and ^{230}Th activities are positively correlated, but the correlation line does not go through the origin. In other words, while $^{230}Th/^{238}U$ activities (r) diminish in the shallow magma towards secular equilibrium, excess ^{231}Pa activity remains. This occurs because Pa/U fractionation can be caused by cpx as well as garnet, so magmas generated near the top of the garnet field (with little excess ^{230}Th) continue to show large ^{231}Pa excesses. The effect is amplified by the higher levels of cpx assumed in the deeper garnet-enriched source.

Finally, in Fig. 2a, excess ^{226}Ra and ^{230}Th are anti-correlated. In the EPF model, ^{226}Ra produced in the early stages of melting (at depth) largely decays away during the slow rise of magmas to the surface. However, higher levels of excess ^{226}Ra were maintained in melts from the shallow source (Fig. 1b) by assuming lower degrees of melting. Hence, it can be seen that the Juan de Fuca ridge data are explained in the EPF model by carefully adjusting the initial assumptions of melting depth, source composition and degree of melting to generate the two magmatic end-members. However, in view of the major uncertainties that remain about the composition of the source and the shape of the melting zone (section A.9.1), it seems unlikely that the modelling of Lundstrom *et al.* allows a conclusive choice of EPF above channel flow.

Another area of debate, in relation to EPF versus channel flow, is the means by which melts can re-equilibrate with the solid during their ascent. Assuming that re-equilibration occurs by volume diffusion only, calculations by Spiegelman and Kenyon (1992) and Hart (1993) suggested that melts would be very unlikely to remain in equilibrium with the solid during ascent. However, Lundstrom *et al.* (1995) argued that dissolution and reprecipitation was a viable mechanism by which melts could maintain chemical equilibrium with the solid during ascent by EPF. More work is required on this question.

14.1.6 ^{14}C in the oceans

Early radiocarbon measurements on ocean water showed variable apparent ages that could be explained by the aging of water bodies as they travelled in major ocean currents. For example, a detailed radiocarbon study of the Atlantic Ocean by Broecker *et al.* (1960) gave results consistent with the established oceanographic model for the Atlantic (Fig. 4.37). In this model, tropical waters are transported (advected) to the North Atlantic, where they cool, and because of their high salinity, sink to form North Atlantic Deep Water (NADW). This body of deep salty water flows southward to the Antarctic, and after mixing with

Antarctic bottom water, ultimately reaches the Pacific (e.g. Broecker and Denton, 1989, Fig. 22).

After correction for anthropogenic contamination, tropical waters were found to have a ^{14}C composition of −50 per mil (relative to nineteenth century wood), equivalent to an apparent ^{14}C age of ca. 400 yr. As they flow north, these waters impart the same apparent ^{14}C age to North Atlantic surface water. After mixing with cold Arctic water, they feed NADW with ^{14}C values of about −70 parts per mil. As this deep water moves southward, ^{14}C activities fall progressively as the water body ages, reaching a δ value of −145 in Antarctic bottom water. Upwelling of this old water gives rise to apparent ages up to 1200 yr in Antarctic surface water. Similar upwelling also yields old apparent ages in the North Pacific. These variations in apparent ^{14}C age were summarised for ocean surface water by Bard et al. (1994, Fig. 1).

Much new data on the geochemistry of deep ocean water was provided by the Geochemical Ocean Sections Study (GEOSECS) program, including a new ^{14}C study of 2200 deep water samples by Stuiver et al. (1983). Pre-anthropogenic ^{14}C data (Stuiver et al., Fig. 1b) confirmed that the Antarctic ocean is a mixing zone between young Atlantic water (NADW) and older deep water from Pacific and Indian oceans. If the ^{14}C change from NADW to Pacific water was due entirely to decay (at ca. 11 per mil/century), this would imply a total water age of ca. 1700 yr. However, because of water transport and mixing, the average mixing time in each individual ocean basin is much shorter than this: ca. 200–300 yr in the Atlantic, 500 yr in the Pacific, and only 80 yr in the Antarctic, which is the medium for communication between the other oceans.

Broecker and Denton (1989) suggested that the global system of deep water transport, or ocean 'conveyor belt' played a critical role in controlling climate switches between glacial and interglacial periods. Evidence from elemental tracers suggested that the conveyor belt might have been 'turned off' during glacial periods, which would have tended to amplify the cooling effect in the North Atlantic by preventing the export of cold NADW and the import of warm tropical water. Broecker and Denton also speculated that a similar effect occurred during the Younger Dryas event, a temporary glacial re-advance which interrupted the last deglaciation.

If the ocean conveyer belt was turned off during the Younger Dryas, then the flow of tropical water to the North Atlantic would cease, and the apparent age of North Atlantic surface water might rise. Bard et al. (1994) attempted to test this model by ^{14}C analysis of plant fossils and planktonic forams from terrestrial and oceanic sediment cores in the vicinity of a distinctive stratigraphic marker (the Vedde ash bed). Data from four North Atlantic cores gave an average ^{14}C age 1400 yr older than the terrestrial section, but a 650 yr correction was necessary to account for marine bioturbation. This left a net age difference of about 750 yr (compared to the present day value of 400 yr), which Bard attributed to reduced ventilation of North Atlantic surface water to the atmosphere. Some of this difference could be accounted for by increased sea ice cover during the Younger Dryas, but the preferred model of Bard et al. was cutting off the formation of NADW. However, recent protoactinium evidence presents a problem for this model, since it suggests that the supply of NADW to the Antarctic Ocean was as great during glacial periods as at the present day (section A.12.3.5).

14.3.1 ^{10}Be in the atmosphere

Ever since ^{10}Be records on ice cores have been acquired, their poor correspondence with geomagnetic intensity records has been a problem. The 150 kyr record of ^{10}Be deposition from the Vostok ice core, Antarctica, provides an opportunity for resolving geomagnetic and climatic controls on ^{10}Be deposition, since it covers the whole of the last glacial cycle (Raisbeck et al., 1987). This record contains two ^{10}Be peaks during the last glaciation, whose ages of ca. 35 and 60 kyr, correspond roughly with minima in geomagnetic field strength. However, two problems arise. Firstly, the peaks appear to be much sharper than the magnetic field variations, and secondly, the Earth's magnetic field does not shield the polar regions from the cosmic ray flux, so if the ^{10}Be flux was locally derived, we should not expect to see any correspondence with magnetic field at all.

Mazaud et al. (1994) showed that these problems could be overcome by more sophisticated modelling of the ^{10}Be deposition flux. They optimised agreement between the ice core and geomagnetic records by adjusting two parameters: the rate of ice accumulation through the core, and the fraction of total ^{10}Be deposition contributed by atmospheric transport from lower latitudes. Adjusting the rate of ice accumulation modifies the apparent ^{10}Be flux through time by diluting or concentrating peaks in the signal, at the same time changing the age calibration of the core. Varying the fraction of transported ^{10}Be changes the peak to baseline ratio of ^{10}Be variations. Mazaud et al. were able to provide an excellent fit of the redistributed record to a production model with 25%

local polar flux and 75% transported [10]Be (Mazaud et al., Fig. 2). Their approach is validated by the agreement of their model ice accumulation record with other published evidence. Hence, they demonstrated that variations in [10]Be deposition can be explained by changes in the Earth's magnetic field, consistent with ^{14}C evidence.

14.3.3 [10]Be in the oceans

Marine sediment records offer an alternative approach to studying paleo [10]Be production, with more consistent accumulation rates than ice cores. However, ocean currents may compromise the marine record by redistributing [10]Be between different deposition sites. Lao et al. (1992) overcame this problem by comparing [10]Be abundances with the U-series nuclides [230]Th and [231]Pa. These nuclides have similar ocean chemistry to beryllium, but are produced at a constant rate from uranium in solution. Lao et al. compared [10]Be production at the present day and 20 kyr ago (corresponding to the last glacial maximum when the ^{14}C flux was 140% of its present value). Lao et al. normalised both [10]Be and [231]Pa fluxes against [230]Th. However, based on a comparison of ocean residence times (section A.12.3.5), we can best normalise climatic effects on [10]Be deposition by comparing the [10]Be/[231]Pa ratio at the present day and 20 kyr ago. After excluding one site with abnormal chemistry, 17 sites from the Pacific yield an average [10]Be flux enhancement of 144% during the last glacial maximum, in excellent agreement with [14]C.

Comparisons between the geomagnetic field and [10]Be deposition have also been made by observing their variation over time in a single site. However, the existence of a direct (inverse) relationship between these variables depends on the neutralisation of climatic effects. In the absence of U-series data for normalisation, this may only happen by chance. Thus, Robinson et al. (1995) observed a good inverse relationship between [10]Be and magnetic intensity in an 80 kyr old sediment core from the central North Atlantic. However, Raisbeck et al. (1994) observed no such relationship in a 600–800 kyr old section from the equatorial Pacific ocean.

Comparisons between the Be isotope ratio of modern ocean masses were first made by Ku et al. (1990). These revealed a consistent global pattern of increasing [10]Be/[9]Be ratio along the ocean conveyer belt, with values ranging from 0.6×10^{-7} in NADW, through 1×10^{-7} in the Antarctic, to values as high as 1.6×10^{-7} in deep Pacific water (Ku et al., 1990; summarised in von Blanckenburg et al., 1996, Fig. 2).

This variation implies beryllium ocean residence times similar to those of the ocean water masses. The precise residence time is difficult to constrain because of the different sources of the two isotopes: [10]Be from global rain and [9]Be from river water. However, Ku et al., estimated values of 500 yr in the Atlantic and 1200 yr in the Pacific.

Beryllium isotope ratios in the 10^{-7} range are larger than most cosmogenic isotope ratios, offering the possibility of analysis by conventional (non-accelerator) mass spectrometry. Such a method was developed by Belshaw et al. (1995) and applied by von Blanckenburg et al. (1996) to the beryllium isotope analysis of ocean floor ferro-manganese crusts. Uranium isotope ratios were used to date the surface layer of the crust (used for Be analysis) in order to apply an age correction in cases where the original growth surface had been removed by abrasion (section A.12.3.1). Crusts of various ages (0–300 kyr) gave very consistent initial Be isotope ratios within each ocean basin, consistent with the beryllium composition of overlying deep ocean water. This suggests that the oceans have generally maintained the same circulation pattern over the past 300 kyr. However, more detailed comparisons of glacial and interglacial [10]Be/[9]Be ratios are needed to test for any interruptions to the conveyer belt during glacial maxima.

Ku et al. (1995) made the first comparison of [26]Al and [10]Be abundances in ocean water. They found [26]Al/[27]Al ratios in surface water which were consistent with the expected [26]Al flux from atmospheric production (by spallation of argon). Therefore, contributions of in situ cosmogenic [26]Al from cosmic dust or wind-blown continental dust do not appear to be significant in seawater. However, the [26]Al/[10]Be ratio measured in surface water was an order of magnitude lower than the atmospheric production ratio, which was attributed to the much shorter ocean residence time of [26]Al.

Wang et al. (1996) compared [26]Al, [10]Be and [230]Th records in ocean floor sediments. The authigenic (seawater-derived) fractions of these nuclides were removed from core samples by leaching with NaOH. [10]Be/[9]Be ratios in the leachates were in good agreement with the composition of overlying (North Pacific) deep water, whereas [10]Be/[9]Be ratios in the bulk sediment were 50% lower, due to input from detrital continental beryllium. The [26]Al budget in the sediment column was also in good agreement with the atmospheric production flux and with [26]Al in ocean water (Ku, et al., 1995). Comparison of authigenic [26]Al, [10]Be and excess initial [230]Th revealed [26]Al/[230]Th ratios consistent with the estimated production ratio, but excess abundances of ^{10}Be relative to both the other nuclides (Wang *et al.*, Fig. 5). It was concluded that ^{26}Al and ^{230}Th have similar (very short) residence times in ocean water, whereas the longer residence time of ^{10}Be allows lateral transport (advection) of ^{10}Be into the North Pacific, where it is scavenged by high biogenic production. Similar effects are seen in the ^{231}Pa/^{230}Th system (section 12.3.5), but the effect on ^{10}Be/^{26}Al is larger, due to the longer ocean residence of ^{10}Be. However, ratios of ^{231}Pa/^{230}Th are easier to measure, so both methods are likely to be very useful in paleo-oceanography.

14.6.3 ^{36}Cl exposure dating

The concept of exposure dating using *in situ* produced cosmogenic ^{36}Cl was suggested by Davis and Schaeffer (1955), but could not be effectively applied until the advent of AMS analysis (Phillips *et al.*, 1986). Although developed after the ^{10}Be and ^{26}Al methods, ^{36}Cl offers several advantages. It is applicable to a variety of rock types because it is generated from three parents (K, Ca, and Cl) with different chemistry, while most rocks have low background levels of chlorine. Furthermore, interferences by nucleogenic ^{36}Cl are minimal, and contamination from atmospheric ^{36}Cl is only a problem in severely weathered material.

The three main sources of *in situ* cosmogenic ^{36}Cl are neutron activation of ^{35}Cl and neutron-induced spallation of ^{40}K and ^{40}Ca. A subordinate source is from negative muon capture by ^{40}Ca. The relative importance of these production routes depends on the relative K–Ca–Cl abundances in the target and the degree of shielding by overlying rock. Zreda *et al.* (1991) made an empirical study of spallation production, while Liu *et al.* (1994) studied the depth dependence of the neutron activation reaction. Spallation reactions are caused by fast neutrons, which decrease exponentially with depth. However, activation reactions require (slow) thermal neutrons, which are produced when fast neutrons undergo glancing collisions with substrate atoms. Hence, slow neutrons reach a peak intensity at about 15 cm depth in rocks (Liu *et al.*, Fig. 5). The occurrence of multiple routes makes the calculation of total production rates more complex, but it also offers the possibility of greater age control, as will be shown below.

The simplest scenario for cosmogenic dating is the instantaneous transport of rock from below the cosmic ray penetration depth to the surface, followed by exposure without shielding by other material (such as snow cover) and without significant erosion. In this case, all production routes for ^{36}Cl can be summed.

However, corrections must be made for the latitude-dependence of cosmic ray intensity and the altitude-dependence of atmospheric shielding. The simple scenario may be achieved in relatively young lava flows, which are instantaneously exposed at the surface and have not yet suffered significant weathering. However, in many geological environments, erosion is a significant factor. In this case, a single cosmogenic isotope determination only allows the solution of exposure age at known erosion rate or erosion rate at known exposure age. For example, using a lava flow from Nevada of known age, Shepard *et al.* (1995) were able to estimate the weathering rate based on ^{36}Cl/^{35}Cl ratios (Shepard *et al.*, Fig. 3).

A more complex scenario arises when we want to date the age of a debris-deposit based on the exposure ages of clasts from within the deposit. Zreda *et al.* (1994) considered a model in which sample rocks, initially buried within the deposit, are gradually exposed at the surface by erosion of the fine grained matrix (Zreda *et al.*, Fig. 2). This model applies to the problem of dating glacial moraines based on the exposure ages of boulders on the moraine surface, and to dating meteorite impacts based on blocks exposed on the ejecta blanket. Zreda *et al.* argued that if ^{36}Cl dates are available from several different boulders on a moraine or ejecta blanket, the spread of ages (outside analytical error) could be used to model the exposure history of the deposit.

To demonstrate these effects, Zreda and Phillips (1995) modelled ^{36}Cl/Cl ratios for six boulders buried at depths from zero to 300 g/cm^2 (approximately 1.5 m depth in soil of density equal to 2). Total ^{36}Cl production was attributed 50% to spallation reactions and 50% to neutron activation. The erosion rate was set so that the deepest boulder just reaches the surface at the present day. The modelling results show a wide range of ^{36}Cl/Cl ratios, both below and slightly above the growth curve for zero depth (Zreda and Phillips, Fig. 8.7). The growth curves for deeply buried boulders start with slow production, due to the shielding effect of the overlying matrix. As the boulder approaches the surface, the rate of production accelerates, but the total ^{36}Cl inventory remains well below that of a surface sample. However, boulders which were initially subject to shallow burial can actually show greater total ^{36}Cl production than at the surface, due to the peak of neutron activation production at a depth of 50 g/cm^2.

Zreda and Phillips (1995) modelled the erosion history of ejecta blocks from Meteor Crater in a similar way. However, in this case the spread of apparent exposure ages was based on 1000 model points, randomly buried from zero to a chosen maximum depth (Zreda and Phillips, Fig. 8.10b). The model results were compared with the actual spread of ages in four blocks at Meteor Crater, with a mean age of 49.7 ± 0.9 kyr (Monument Rock was excluded, see Fig. 14.40). The best fit was obtained by assuming burial to a maximum depth of 300 g/cm^2, removed at 30 g/cm^2/kyr, so that all boulders reached the surface within 10 kyr. Such modelling cannot yield a unique solution to the erosion process, and the model results should ideally be compared with a larger set of measured ages. Nevertheless, the modelling does suggest that the four blocks yield exposure ages close to the estimated time of impact.

In principal, the use of two spallogenic nuclides (e.g. ^{36}Cl and ^{10}Be, or ^{10}Be and ^{26}Al) allows the deconvolution of exposure ages and erosion rates. However, in practice, analytical and geological uncertainties make the method quite weak (e.g. Fig. 14.39). A more powerful application of the ^{36}Cl technique would make independent use of the neutron activation route to ^{36}Cl, in comparison with one of the three spallation nuclides (^{36}Cl, ^{10}Be or ^{26}Al). Because spallation and neutron activation yield peak nuclide production at different depths in a geological surface, they should yield a clear resolution of exposure histories involving different rates of erosion (Liu et al., 1994, Fig. 7c). With this objective in mind, Bierman et al. (1995) described a method for isolating the neutron activation component of ^{36}Cl, by releasing chlorine from fluid inclusions within rock samples. Hence, comparison of activation and spallogenic exposure ages should be possible in the near future.

A different application of the ^{36}Cl method was proposed by Stone et al. (1996), using ^{40}Ca spallation as a tool for exposure dating of calcite. The abundance of the target nuclide, along with the relatively low abundance of chlorine in calcite, makes this an analytically favourable method which shows great promise for exposure dating of karst landscapes.

15.2.6 Mn–Cr

New constraints on the cooling of chondrite parent bodies come from ^{53}Mn excesses observed in carbonate fragments from two chondritic meteorites (Endress et al., 1996). These poly-crystalline fragments, mostly of dolomitic composition, are uniformly spread through the meteorites Orgueil and Ivuna. They are interpreted as remnants of carbonate veins which formed during very early aqueous alteration of the meteorite parent body. Five dolomite fragments, analysed by ion microprobe (SIMS), define a relatively linear array on a diagram of Cr isotope composition against Mn/Cr (Endress et al., Fig. 2). The slope of the array corresponds to a ^{53}Mn/^{55}Mn ratio of 2×10^{-6} at the time of carbonate crystallisation. The time required to achieve this ratio, starting from the value of 4.4×10^{-5} in 4566 Myr Allende inclusions, is less than 20 Myr. This indicates that the CI meteorite parent body cooled very rapidly after aggregation to temperatures where liquid water could exist.

15.2.8 Hf–W

^{182}Hf decays to ^{182}W (tungsten) by double β decay with a half-life of 9 Myr. Norman and Schramm (1993) proposed the Hf–W system as an r-process chronometer, but its application was delayed by the technical difficulties of tungsten isotope analysis (similar to osmium, section 8.1). Harper et al. (1991) and Harper and Jacobsen (1996) successfully compared W isotope compositions in different solar-system bodies by N–TIMS analysis, and these results were verified and extended by Lee and Halliday (1995) using magnetic sector ICP–MS. Lee and Halliday presented the combined data set in the form of ϵ ^{182}W values, representing part per 10 000 variations in ^{182}W/^{183}W (Harper and Jacobsen, 1996) or ^{182}W/^{184}W (Lee and Halliday, 1995), relative to a terrestrial tungsten standard, NIST 3163. The results showed ϵ W values within error of zero in a terrestrial lava, a Lunar mare basalt, and bulk powders of the Allende and Murchison chondrites (Lee and Halliday, Fig. 1a). In contrast, sawn blocks from the iron meteorites Arispe, Coya Norte and Toluca all had ϵ values close to –4.

In principal, these variations could be explained by either cosmogenic production, pre-solar grains, or live ^{182}Hf in the early solar system. However, the ^{182}W anomaly in iron meteorites is unique because it represents a deficiency in one isotope, rather than an enrichment. This makes both cosmogenic and pre-solar origins for the anomaly quite unlikely, but it can be readily explained by elemental Hf–W partitioning while ^{182}Hf was still alive in the early solar system. Since W is strongly siderophile, whereas Hf is lithophile, the iron cores of meteorite parent bodies developed very low Hf/W ratios. Furthermore, evidence from the Fe–Ni system (section 15.2.7) shows that these cores formed very early, within about 5 Myr of chondrite formation. Therefore, iron meteorites should preserve a nearly primordial ϵ ^{182}W composition for the solar-system. In contrast, ^{182}W growth would continue in other solar system bodies until ^{182}Hf was extinct.

According to this model, we should expect to see a correlation between ^{182}W/^{184}W and ^{180}Hf/^{184}W in stony meteorites, analogous to the correlation lines defined by other extinct nuclides. A weak correlation of this type was reported by Ireland (1991), based on ion probe W/Hf analysis of zircon from the Simmern chondrite (Ireland, Fig. 1 inset). This correlation line could imply an initial ^{182}Hf/^{180}Hf ratio around 2.5×10^{-4} in the primordial solar system, but unfortunately this is only a maximum value, due to uncorrected molecular ion interferences. The results were further clouded by apparent ^{182}W excesses in two terrestrial zircons and two zircons from the Vaca Muerta meso-siderite, but these results are almost certainly inter-ference effects. Until these results were verified by further analysis, the meaning of the iron meteorite data remained uncertain. However, the data of Lee and Halliday were confirmed by new work (Lee and Halliday, 1996).

Harper and Jacobsen (1996) did not have ^{182}W data on chondrites to use as a benchmark for Solar System W isotope evolution. They based the initial Solar System ^{182}Hf/^{180}Hf value on a predicted supernova production ratio of 2×10^{-5}, which implied a theoretical Bulk Earth ^{182}W value (after Hf extinction) only $0.3 \, \epsilon$ units above iron meteorites. Hence, the measured $4 \, \epsilon$ unit excess in the Bulk Silicate Earth would require Hf/W fractionation while ^{182}Hf was still alive, implying very early terrestrial core formation. In contrast, Lee and Halliday (1995) measured W isotope ratios in two bulk chondrite samples, and found ϵ values of $+4$ relative to iron meteorites. This implies identical ϵ W values in the Bulk Earth and Bulk Silicate Earth, and hence that the Earth's core segregated at least 60 Myr after iron meteorites, when ^{182}Hf was extinct (Lee and Halliday, Fig. 2). However, the chondrite data also imply an unexpectedly high ^{182}Hf/^{180}Hf ratio of 2.6×10^{-4} in the early solar system. Since ^{182}Hf lies off the direct s-process path, and can only be produced by high neutron fluxes, this points to a very late supernova source. The significance will be discussed in section A.15.4.

15.3.2 Ca–K

Recent work by Srinivasan et al. (1994, 1996) has moved ^{41}Ca from a species absent to a species present in the early solar system. Srinivasan et al. looked for ^{41}K, the decay product of ^{41}Ca, in Ca–Al-rich inclusions (CAIs) from the Efremovka CV3 chondrite. These inclusions were regarded as an ideal place to search for ^{41}K anomalies because of their unusually fresh petrography and the known presence of ^{26}Al

anomalies. The pristine nature of the samples is very important, because ^{41}K anomalies are only visible at extreme Ca/K ratios, which would be compromised by re-mobilisation of common potassium.

Srinivasan et al. used the ion microprobe to analyse K isotope ratios in pyroxenes and other high-Ca phases from four different inclusions. Special pre-cautions were taken to resolve the ^{41}K signal from interfering molecular ions. ^{40}CaH$^+$ was resolved by its excess mass at high spectral resolution. However, ^{40}Ca^{42}Ca^{2+} ions cannot be resolved by mass, but were monitored and corrected via the related species ^{40}Ca^{43}Ca^{2+}. The reliability of this correction is demon-strated by the constant ^{41}K/^{39}K ratios determined in terrestrial samples with Ca/K variations spanning 9 orders of magnitude. In contrast, Efremovka inclusions showed a strong correlation between ^{41}K/^{39}K and ^{40}Ca/^{39}K ratio (Srinivasan et al., 1996, Fig. 5). If this correlation results from in situ decay of ^{41}Ca after the inclusions were formed then its slope corresponds to an initial ^{41}Ca/^{40}Ca ratio of 1.4×10^{-8}. Minerals from the same suite of inclusions also gave initial ^{26}Al/^{27}Al ratios typical of other CAIs, around 5×10^{-5}.

In principal, the ^{41}K signal in Efremovka inclusions could be explained by recent cosmogenic production of ^{41}Ca by cosmic ray neutrons. However, Srinivasan et al. cited rare gas isotope evidence for Efremovka which pointed to a low cosmogenic neutron flux during the fragment's 11 Myr exposure history in space. A second alternative is that ^{41}K was inherited as a 'fossil' component from ^{41}Ca decay in pre-solar grains. However, this can be ruled out because the correlation between ^{41}K/^{39}K and Ca/K ratio is ob-served in grains which show clear evidence of crys-tallisation from a liquid. Therefore, Srinivasan et al. concluded that the ^{41}K signal could best be explained by the presence of live ^{41}Ca in the early solar system. It is possible that this isotope could be produced within the solar nebula by bombardment with energetic par-ticles from an early 'active' sun. However, this model cannot produce ^{41}Ca and ^{26}Al in their correct relative abundances. Therefore, the most attractive model attributes both nuclides to a common nucleosynthetic process immediately before solar system condensation.

If we assume simultaneous injection of ^{41}Ca and ^{26}Al, and we know the production ratios relative to stable calcium and aluminium, we can solve for both Δ and the dilution factor of 'hot', freshly injected material by 'cold' pre-existing material. This is the same approach that was attempted for the two longer lived r-process nuclei ^{129}I and ^{244}Pu (section 15.2.3). Unfortunately, similar problems arise here, because the production ratios are poorly constrained. However, the half-life of ^{41}Ca is so short (0.1 Myr), that tight constraints on Δ arise from almost any model. In practice, both supernovae and Red Giants (asymptotic giant branch, or AGB stars) can be made to fit the data, implying a Δ value of 1 Myr or less, and a dilution factor of 100 or more. These models will be discussed further in section A.15.4.

15.4 Conclusions

Extinct nuclide evidence still does not lead to a unique model for the origin of the solar system, but the increasing weight of evidence is narrowing the options. The presence of very short-lived species is best explained by reviving the 'trigger hypothesis' (Cameron et al., 1995), by which a single event caused late nuclide injection into a molecular cloud, and also triggered the collapse of the cloud. The species ^{41}Ca, ^{26}Al and ^{107}Pd could be explained by either a Red Giant or supernova; ^{53}Mn, ^{60}Fe, ^{129}I and ^{182}Hf are best explained by a supernova, while actinides must be produced in supernovae. However, more than one type of supernova may be required to fit all of the production ratios. Wasserburg et al. (1996) argued that the effective frequency by which supernovae could contribute to galactic nuclide production could be much higher than the one per 100 Myr commonly assumed. In addition, they suggested that different types of supernovae could have contributed 'spikes' of different nuclides to the pre-solar nebula. The low level of ^{129}I implies an early supernova source (Δ around 100 Myr). Wasserburg et al. then grouped ^{182}Hf with a late actinide addition from a second supernova (Δ around 10 Myr). Finally, ^{41}Ca, ^{26}Al and ^{60}Fe, require very late addition, possibly from a third supernova source. A model of multiple supernova sources was also proposed by Harper (1996), who envisaged star birth in a large molecular cloud with a complex series of injection and mixing events (Harper, Fig. 1).

It is implicit in the above arguments for late supernova injection ($\Delta = 10$ Myr) that this included ^{247}Cm. Uranium isotope evidence excludes substantial heterogeneous late addition of curium (section 15.3.1), but not homogeneous addition. Therefore, if late r-process additions were substantial (Lee and Halliday, 1995) then they must have been homogeneously mixed into the nebula. Recent evidence (Russell et al., 1996) suggests that ^{26}Al addition to the solar nebula was more widespread than previously thought, and therefore that this and other radionuclides may have been homogeneously distributed.

Russell et al. found Mg isotope anomalies in two CAIs (inclusions) and two Al-rich chondrules from the unmetamorphosed ordinary chondrites Semarkona, Moorabie, Inman and Chainpur. The CAIs gave initial $^{26}Al/^{27}Al$ ratios of 5×10^{-5}, exactly the same as found in inclusions from carbonaceous chondrites. The observation of consistent ratios in inclusions from different chondrite groups suggests that these inclusions might have been fairly widely distributed in the nebula. This suggestion is supported by the detection of extinct ^{26}Al in some chondrules from Inman and Chainpur. These gave lower initial $^{26}Al/^{27}Al$ ratios than the inclusions, around 1×10^{-5}. If the isotopic variation is attributed to ^{26}Al decay in a homogeneous reservoir, the data imply a period of ca. 2 Myr between CAI formation and chondrule formation. In contrast, other chondrules from the same meteorite (Chainpur) show no Al–Mg anomalies, suggesting crystallisation 5 Myr after CAI inclusions. This time span is consistent with U–Pb ages on chondrules and inclusions (section A.5.3.1), thus supporting their origin from a homogeneous reservoir.

16.4 Detrital populations

Galbraith (1988) introduced a new kind of isochron diagram for the presentation of multi-grain fission track data. This differs from other isochron plots used in geology because the two variables plotted are the apparent fission track age of each grain, and the standard error of each grain age (σ). These quantities are plotted in the form of age/σ against 1/σ (Galbraith, Fig. 4). In this plot, the slope of an array indicates the average age of the suite of grains analysed, and this age can be indicated on a calibrated arc (Fig. 4). In practice, Galbraith argued, it is more convenient to normalise the average slope to a horizontal, and plot the y axis on a log scale from +2 to −2 (e.g. Galbraith, Fig. 8). The age of any individual point is then determined by projecting from the zero point on the y axis, through the data point, to the calibrated arc of ages (on a log scale). This 'radial' plot is designed for fission track data sets with a high degree of scatter, either due to mixed detrital ages or variable cooling ages.

16.6 Uplift and subsidence rates

It is well known that heterogeneous fission track cooling ages can be generated from apatite grains in a sedimentary rock, due to variations in the composition of detrital apatites. However, O'Sullivan and Parrish (1995) showed that such variations can also occur within single samples of plutonic rock. Single grain apatite analysis was used to study uplift ages in granitoid rocks from the Coast Ranges of British Columbia. The results were broadly in agreement with earlier fission track analysis of bulk samples by the population technique, but some samples showed a marked spread in apparent fission track ages, such that the ages were correlated with the chlorine content of individual grains (O'Sullivan and Parrish, Fig. 6). O'Sullivan and Parrish suggested that single grain analysis of this sort could augment the information normally available from an uplift age profile (O'Sullivan and Parrish, Fig. 7). Traditional analysis of such a section, by the population method, yields a single age point for each elevation. These are then combined to produce the age profile. Single grain analysis can provide additional constraints on the profile because the chemical dependence of annealing is greatest within the PAZ. Hence, the grain population changes from all young grains below the PAZ, through mixed ages in the PAZ, to all old ages above the PAZ. These types of constraints are similar to those which have been achieved in other uplifted crustal sections by track length analysis.

16.7 Track length measurements

Laslett *et al.* (1994) made a comparison of the usefulness of confined and projected track lengths for thermal history analysis. Confined track lengths bear a simpler relationship to the true track length distribution, but projected tracks (also referred to as semi-tracks) are more numerous. Therefore, Laslett *et al.* performed simulations to test the ability of 2000 projected tracks and 100 confined tracks to recover the true track length variation in a mixed population with lengths of 14.5 and 10 μm. The proportion of shorter tracks (p) was varied in different simulations from 20% to 80%. Results showed that when the shorter tracks made up 60 to 80% of the population, both methods could recover the length of these tracks and the correct value of p with similar error bars (Laslett *et al.*, Fig. 10). However, when the proportion of short tracks fell below 50%, the projected track data were seriously compromised. Therefore, Laslett *et al.* concluded that a relatively small number of confined track measurements can more reliably recover the true track length distribution than a large number of projected track lengths.

Hejl (1995) identified an additional problem in recovering the true length distribution of annealed tracks from projected track lengths. This is due to the existence of unetchable gaps in the middle of tracks, a phenomenon which appears to be more widespread

than previously realised. Hejl proposed that the true lengths of confined tracks could be recovered, despite unetchable gaps, by a double etching procedure. This breaks through the 'unetchable gap' to produce a track with a wide middle (double etched) and a narrow extension (beyond the original gap). The total track length can then be measured (Hejl, Fig. 1). This procedure would be more difficult on projected tracks.

It is now generally agreed that confined tracks are superior to projected tracks for thermal history analysis. However, confined tracks must be counted using fixed selection criteria to exclude the possibility of subjective bias. Laslett *et al.* (1994) recommended the 'bright reflection' criterion, which exploits the property of etched tracks at a low angle to the horizontal (less than about 15%) to show a bright image in reflected light. In contrast, a criterion which requires tracks to be in focus along their entire length is unsuitable because it causes a higher rejection rate for long tracks than for short tracks.

In order to use track length analysis to make quantitative interpretations of thermal histories, Laslett *et al.* (1987, op. cit.) developed a model to predict the degree of track shortening after heating episodes of different intensity. This model was based on isothermal laboratory annealing experiments on large gem-quality apatite samples from Durango, Mexico (Green *et al.*, 1986). Arrhenius (temperature–time) relationships were observed for given fractions of track shortening (Laslett *et al.*, Fig. 5). These are analogous to the Arrhenius relationships for track density (e.g. Fig. 16.18). The isothermal model was extended to more complex thermal histories involving variable temperature conditions by Duddy *et al.* (1988), and applied to geological timescales by Green *et al.* (1989, op. cit.). In their approach, a predicted temperature–time curve is divided into intervals (e.g. 1 Myr each), and after each interval the degree of shortening of existing tracks is calculated. At a constant elevated temperature, track shortening occurs rapidly at first, because the track ends are least energetically stable, but subsequently slows considerably. In addition to the annealing of old tracks, new track formation is simulated at 10 Myr intervals. Tracks formed at each of these intervals define an evolution path of reduced track length against time (Green *et al.*, Fig. 1a). The sum of these evolution paths at the present day forms a histogram of track length distribution (Fig. 1c).

Green *et al.* (1989) modelled the track length distribution expected from several different types of thermal history. This process, of predicting track length distributions from thermal history data, is termed forward modelling. They tested the model results against data from the Otway basin, SE Australia. In the Otway basin case, the thermal history problem is over-constrained because Green *et al.* had access to full track length distribution data for a suite of borehole samples from different depths. Therefore, they were able to simplify their analysis by discarding the information on track length variation, and comparing only predicted and measured mean track lengths. The model results (Green *et al.*, Fig. 5) provide a good fit to the observed data at low temperature, but at high temperatures the data indicate greater resistance to annealing than predicted. This can be attributed to higher average Cl contents in Otway basin apatite, relative to Durango apatite on which the model was based.

A suite of samples at different depths is not always available from one locality, but by using the complete track length distribution (rather than just the mean) it should be possible to test possible thermal histories using a single sample. However, several different model thermal histories might generate track length distributions which fit the observed data set. Therefore, it is desirable to test many different thermal histories in order to see what range of possible histories can generate the observed data. The final objective of this process, to determine which thermal history best explains the observed track length distribution, is termed inversion modelling (e.g. Corrigan, 1991; Gallagher, 1995).

Because initial track development and subsequent track shortening are processes subject to random noise, thermal histories cannot be uniquely determined, but must be based on probabilities. In practice, thermal histories chosen at random (the Monte Carlo method) are used to calculate track length distributions by forward modelling. The results are then tested against the observed (or simulated) track length data set. After a few hundred iterations it is possible to map out a range of possible thermal histories which are consistent with the data set. Within this range of possibilities, the highest probability density defines an optimum, but not necessarily unique, thermal history (e.g. Corrigan, Fig. 8b). Relatively well constrained thermal histories can be projected back to the last temperature maximum, but beyond this time the thermal history is very poorly constrained.

A continuing source of uncertainty in such modelling is the extrapolation of laboratory experiments to geological time-scales. Gallagher compared three alternative annealing models which have been proposed over the past ten years. The first of these uses the laboratory data of Green *et al.* (1986), extrapolated by Laslett (1987, op. cit.). The second model (Carlson, 1990) uses the same data set, coupled with

an *ad hoc* geometrical model, but has been criticised as non-realistic in crystallographic terms (e.g. Green *et al.*, 1993). The third model is based on new laboratory data by Crowley *et al.* (1991). Since the models of Carlson and Crowley *et al.* diverge on opposite sides of the Laslett *et al.* model, the latter still appears to be the most useful for thermal history modelling.

References

Allegre, C. J. and Lewin, E. (1995). Isotopic systems and stirring times of the Earth's mantle. *Earth Planet. Sci. Lett.* **136**, 629–46.

Allegre, C. J., Dupre, B. and Lewin, E. (1996). Three time-scales for the mantle. In: Hart, S. and Basu, A. (Eds), *Earth Processes: Reading the Isotopic Code. Geophys. Monograph* **95**, Amer. Geophys. Union, pp. 99–108.

Allegre, C. J., Manhes, G. and Gopel, C. (1995). The age of the Earth. *Geochim. Cosmochim. Acta* **59**, 1445–56.

Amakawa, H., Nozaki, Y. and Masuda, A. (1996). Precise determination of variations in the $^{138}Ce/^{142}Ce$ ratios of marine ferromanganese nodules. *Chem. Geol.* **131**, 183–95.

Asmerom, Y. and Edwards, R. L. (1995). U-series isotope evidence for the origin of continental basalts. *Earth Planet. Sci. Lett.* **134**, 1–7.

Bard, E., Arnold, M., Mangerud, J., Paterne, M., Labeyrie, L., Duprat, J., Melieres, M. A., Sonstegaard, E. and Duplessy, J. C. (1994). The North Atlantic atmosphere-sea surface ^{14}C gradient during the Younger Dryas climatic event. *Earth Planet. Sci. Lett.* **126**, 275–87.

Belshaw, N. S., O'Nions, R. K. and von Blanckenburg, F. (1995). A SIMS method for $^{10}Be/^9Be$ ratio measurement in environmental materials. *Int. J. Mass Spectrom. Ion Phys.* **142**, 55–67.

Bennett, V. C., Esat, T. M. and Norman, M. D. (1996). Two mantle-plume components in Hawaiian picrites inferred from correlated Os–Pb isotopes. *Nature* **381**, 221–3.

Bierman, P., Gillespie, A., Caffee, M. and Elmore, D. (1995). Estimating erosion rates and exposure ages with ^{36}Cl produced by neutron activation. *Geochim. Cosmochim. Acta* **59**, 3779–98.

Blum, J. D. and Erel, Y. (1995). A silicate weathering mechanism linking increases in marine $^{87}Sr/^{86}Sr$ with global glaciation. *Nature* **373**, 415–8.

Bourdon, B., Langmuir, C. H. and Zindler, A. (1996). Ridge-hotspot interaction along the Mid-Atlantic Ridge between 37° 30' and 40° 30' N: the U–Th disequilibrium evidence. *Earth Planet. Sci. Lett.* **142**, 175–89.

Bowring, S. A. and Housh, T. (1995). The Earth's early evolution. *Science* **269**, 1535–40.

Brandon, A. D. and Goles, G. G. (1988). A Miocene subcontinental plume in the Pacific Northwest: geochemical evidence. *Earth Planet. Sci. Lett.* **88**, 273–83.

Brandon, A. D. and Goles, G. G. (1995). Assessing subcontinental lithospheric mantle sources for basalts: Neogene volcanism in the Pacific Northwest, USA as a test case. *Contrib. Mineral. Petrol.* **121**, 364–79.

Brandon, A. D., Creaser, R. A., Shirey, S. B. and Carlson, R. W. (1996). Osmium recycling in subduction zones. *Science* **272**, 861–4.

Broecker, W. S. and Denton, G. H. (1989). The role of ocean-atmosphere reorganizations in glacial cycles. *Geochim. Cosmochim. Acta* **53**, 2465–501.

Broecker, W. S., Gerard, R., Ewing, M. and Heezen, B. C. (1960). Natural radiocarbon in the Atlantic Ocean. *J. Geophys. Res.* **65**, 2903–31.

Bunge, H. P., Richards, M. A. and Baumgardner, J. R. (1996). Effect of depth-dependent viscosity on the planform of mantle convection. *Nature* **379**, 436–8.

Caffee, M. W., Hudson, G. B., Velsko, C., Alexander, E. C., Huss, G. R. and Chivas, A. R. (1988). Non-atmospheric noble gases from CO_2 well gases. *Lunar Planet. Sci.* **XIX**, 154–5 (abstract).

Cameron, A. G. W., Hoflich, P., Myers, P. C. and Clayton, D. D. (1995). Massive supernovae, Orion gamma rays, and the formation of the solar system. *Astrophys. J. (Lett.)* **447**, L53–7.

Carlson, W. D. (1990). Mechanisms and kinetics of apatite fission-track annealing. *Amer. Miner.* **75**, 1120–39.

Carlson, R. W. and Irving, A. J. (1994). Depletion and enrichment history of subcontinental lithospheric mantle: an Os, Sr, Nd and Pb isotopic study of ultramafic xenoliths from the northwestern Wyoming Craton. *Earth Planet. Sci. Lett.* **126**, 457–72.

Casse, M., Lehoucq, R. and Vangioni-Flam, E. (1995). Production and evolution of light elements in active star-forming regions. *Nature* **373**, 318–19.

Chabaux, F. and Allegre, C. J. (1994). ^{238}U–^{230}Th–^{226}Ra disequilibria in volcanics: a new insight into melting conditions. *Earth Planet. Sci. Lett.* **126**, 61–74.

Chabaux, F., Cohen, A. S., O'Nions, R. K. and Hein, J. R. (1995). ^{238}U–^{234}U–^{230}Th chronometry of Fe–Mn crusts: growth processes and recovery of thorium isotopic ratios of seawater. *Geochim. Cosmochim. Acta* **59**, 633–8.

Chauvel, C., Goldstein, S. L. and Hofmann, A. W. (1995). Hydration and dehydration of oceanic crust controls Pb evolution in the mantle. *Chem. Geol.* **126**, 65–75.

Chen. J. H. and Wasserburg, G. J. (1981). The isotopic composition of uranium and lead in Allende inclusions and meteoritic phosphates. *Earth Planet. Sci. Lett.* **52**, 1–15.

Christensen, U. R. and Hofmann, A. W. (1994). Segregation of subducted oceanic crust in the convecting mantle. *J. Geophys. Res.* **99**, 867–84.

Clemens, S. C., Gromet, L. P. and Farrell, J. W. (1995). Artifacts in Sr isotope records. *Nature* **373**, 201.

Cochran, J. K., Livingston, H. D., Hirschberg, D. J. and Surprenant, L. D. (1987). Natural and anthropogenic radionuclide distributions in the northwest Atlantic Ocean. *Earth Planet. Sci. Lett.* **84**, 135–52.

Corrigan, J. (1991). Inversion of apatite fission track data for thermal history information. *J. Geophys. Res.* **96**, 10 347–60.

Corrigan, D., Culshaw, N. G. and Mortensen, J. K. (1994). Pre-Grenvillian evolution and Grenville overprinting of the Parautochthonous Belt in Key Harbour, Ontario: U–Pb and field constraints. *Can. J. Earth. Sci.* **31**, 583–96.

Crowley, K. D., Cameron, M. and Schaefer, R. L. (1991). Experimental studies of annealing of etched fission tracks in fluorapatite. *Geochim. Cosmochim. Acta* **55**, 1449–65.

Davis, R. and Schaeffer, O. A. (1955). Chlorine-36 in Nature. *Ann. N. Y. Acad. Sci.* **62**, 105–22.

Davis, W. J., Gariepy, C. and van Breemen, O. (1996). Pb isotopic composition of late Archean granites and the extent of recycling early Archean crust in the Slave Province, northwest Canada. *Chem. Geol.* **130**, 255–69.

Derry, L. A. and France-Lanord, C. (1996). Neogene Himalayan weathering history and river ^{87}Sr/^{86}Sr: impact on the marine Sr record. *Earth Planet. Sci. Lett.* **142**, 59–74.

DeWolf, C. P., Zeissler, C. J., Halliday, A. N., Mezger, K. and Essene, E. J. (1996). The role of inclusions in U–Pb and Sm–Nd garnet geochronology: stepwise dissolution experiments and trace uranium mapping by fission track analysis. *Geochim. Cosmochim. Acta* **60**, 121–34.

DeWolf, C. P. and Mezger, K. (1994). Lead isotope analysis of leached feldspars: constraints on the early crustal history of the Grenville Orogen. *Geochim. Cosmochim. Acta* **58**, 5537–50.

Dickin, A. P. (1996). HIMU and EM1: complementary reservoirs from the 'slab' and 'wedge'. *EOS* **77**(17), S282 (abs.).

Dodson, A., DePaolo, D. J. and Kennedy, B. M. (1996). Noble gases in the Columbia River basalt. *EOS* **77**(17), S288 (abs.).

Dong, H., Hall, C. M., Peacor, D. R. and Halliday, A. N. (1995). Mechanisms of argon retention in clays revealed by laser ^{40}Ar–^{39}Ar dating. *Science* **267**, 355–9.

Duddy, I. R., Green, P. F. and Laslett, G. M. (1988). Thermal annealing of fission tracks in apatite 3. Variable temperature behaviour. *Chem. Geol. (Isot. Geosci. Sect.)* **73**, 25–38.

Eiler, J. M., Farley, K. A., Valley, J. W., Stolper, E. M., Hauri, E. H. and Craig, H. (1995). Oxygen isotope evidence against bulk recycled sediment in the mantle source of Pitcairn Island lavas. *Nature* **377**, 138–41.

Ellam, R. M. and Cox, K. G. (1989). A Proterozoic lithospheric source for Karoo magmatism: evidence from the Nuanetsi picrites. *Earth Planet. Sci. Lett.* **92**, 207–18.

Ellam, R. M. and Cox, K. G. (1991). An interpretation of Karoo picrite basalts in terms of interaction between asthenospheric magmas and the mantle lithosphere. *Earth Planet. Sci. Lett.* **105**, 330–42.

Endress, M., Zinner, E. and Bischoff, A. (1996). Early aqueous activity on primitive meteorite parent bodies. *Nature* **379**, 701–3.

Esser, B. K. and Turekian, K. K. (1993a). The osmium isotopic composition of the continental crust. *Geochim. Cosmochim. Acta* **57**, 3093–104.

Esser, B. K. and Turekian, K. K. (1993b). Anthropogenic osmium in coastal deposits. *Envir. Sci. Tech.* **27**, 2719–24.

Farley, K. A. (1995a). Rapid cycling of subducted sediments into the Samoan mantle plume. *Geology* **23**, 531–4.

Farley, K. A. (1995b). Cenozoic variations in the flux of interplanetary dust recorded by ^{3}He in a deep-sea sediment. *Nature* **376**, 153–6.

Farley, K. A. and Patterson, D. B. (1995). A 100-kyr periodicity in the flux of extraterrestrial ^{3}He to the sea floor. *Nature* **378**, 600–3.

Farnetani, C. G. and Richards, M. A. (1995). Thermal entrainment and melting in mantle plumes. *Earth Planet. Sci. Lett.* **136**, 251–67.

Farrell, J. W., Clemens, S. C. and Gromet, L. P. (1995). Improved chronostratigraphic reference curve of late Neogene seawater ^{87}Sr/^{86}Sr. *Geology* **23**, 403–6.

Fitz Gerald, J. D. and Harrison, T. M. (1993). Argon diffusion domains in K-feldspar I: microstructures in MH-10. *Contrib. Mineral. Petrol.* **113**, 367–80.

Foster, J. G., Lambert, D. D., Frick, L. R. and Maas, R. (1996). Re–Os isotopic evidence for genesis of Archean nickel ores from uncontaminated komatiites. *Nature* **382**, 703–6.

Frank, M., Eisenhauer, A., Bonn, W. J., Walter, P., Grobe, H., Kubik, P. W., Dittrich-Hannan, B. and Mangini, A. (1995). Sediment redistribution versus paleoproductivity change: Weddell Sea margin sediment stratigraphy and biogenic particle flux of the last 250 000 years deduced from ^{230}Th$_{ex}$, ^{10}Be and biogenic barium profiles. *Earth Planet. Sci. Lett.* **136**, 559–73.

Galbraith, R. F. (1988). Graphical display of estimates having differing standard errors. *Tectonometrics* **30**, 271–81.

Galer, S. J. G. and Goldstein, S. L. (1996). Influence of accretion on lead in the Earth. In: Hart, S. and Basu, A. (Eds), *Earth Process: Reading the Isotopic Code. Geophys. Monograph 95*, Amer. Geophys. Union, pp. 75–98.

Gallagher, K. (1995). Evolving temperature histories from apatite fission-track data. *Earth Planet. Sci. Lett.* **136**, 421–35.

Getty, S. R. and DePaolo, D. J. (1995). Quaternary geochronology using the U–Th–Pb method. *Geochim. Cosmochim. Acta* **59**, 3267–72.

Gibson, S. A., Thompson, R. N., Dickin, A. P. and Leonardos, O. H. (1996). High-Ti and low-Ti mafic potassic magmas: key to plume-lithosphere interactions and continental flood-basalt genesis. *Earth Planet. Sci. Lett.* **141**, 325–41.

Gopel, C., Manhes, G. and Allegre, C. J. (1994). U–Pb systematics of phosphates from equilibrated ordinary chondrites. *Earth Planet. Sci. Lett.* **121**, 153–71.

Green, P. F., Duddy, I. R., Gleadow, A. J. W., Tingate, P. R. and Laslett, G. M. (1986). Thermal annealing of fission tracks in apatite. 1. A qualitative description. *Chem. Geol. (Isot. Geosci. Sect.)* **59**, 237–53.

Green, P. F., Laslett, G. M. and Duddy, I. R. (1993). Mechanisms and kinetics of apatite fission-track annealing – Discussion. *Amer. Miner.* **78**, 441–5.

Halliday, A. N., Rehkamper, M., Lee, D.-C. and Yi, W. (1996). Early evolution of the Earth and Moon: new constraints from Hf–W isotope geochemistry. *Earth Planet. Sci. Lett.* **142**, 75–89.

Hanan, B. B. and Graham, D. W. (1996). Lead and helium isotope evidence from oceanic basalts for a common deep source of mantle plumes. *Science* **272**, 991–5.

Harmon, R. S. and Hoefs, J. (1995). Oxygen isotope heterogeneity of the mantle deduced from global ^{18}O systematics of basalts from different geotectonic settings. *Contrib. Mineral. Petrol.* **120**, 95–114.

Harper, C. L., Volkening, J., Heumann, K. G., Shih, C.-Y. and Weismann, H. (1991). ^{182}Hf–^{182}W: new cosmochronometric constraints on terrestrial accretion, core formation, the astrophysical site of the r-process, and the origin of the solar system. *Lunar Planet. Sci.* **XXII**, 515–16.

Harper, C. L. (1996). Astrophysical site of the origin of the solar system inferred from extinct radionuclide abundances. *Astrophys. J.* **466**, 1026–38.

Harper, C. L. and Jacobsen, S. B. (1996). Evidence for ^{182}Hf in the early Solar System and constraints on the timescale of terrestrial accretion and core formation. *Geochim. Cosmochim. Acta* **60**, 1131–53.

Harris, N. (1995). Significance of weathering Himalayan metasedimentary rocks and leucogranites for the Sr isotope evolution of seawater during the early Miocene. *Geology* **23**, 795–8.

Harrison, T. M., Lovera, O. M. and Heizler, M. T. (1991). ^{40}Ar/^{39}Ar results for alkali feldspars containing diffusion domains with differing activation energy. *Geochim. Cosmochim. Acta* **55**, 1435–48.

Hart, S. R. (1993). Equilibration during mantle melting: a fractal tree model. *Proc. Natl. Acad. Sci. USA* **90**, 11914–18.

Hauri, E. H. (1996). Major-element variability in the Hawaiian mantle plume. *Nature* **382**, 415–19.

Hauri, E. H., Whitehead, J. A. and Hart, S. R. (1994). Fluid dynamic and geochemical aspects of entrainment in mantle plumes. *J. Geophys. Res.* **99**, 24275–300.

Hauri, E. H., Lassiter, J. C. and DePaolo, D. J. (1996). Osmium isotope systematics of drilled lavas from Mauna Loa, Hawaii. *J. Geophys. Res.* **101**, 11793–806.

Hawkesworth, C. J., Gallagher, K., Kelley, S., Mantovani, M. S. M., Peate, D. W., Regelous, M. and Rogers, N. W. (1992). Parana magmatism and the opening of the South Atlantic. *Geol. Soc. Lond. Spec. Pub.* **68**, 221–40.

Hejl, E. (1995). Evidence for unetchable gaps in apatite fission tracks. *Chem. Geol. (Isot. Geosci. Sect.)* **122**, 259–69.

Hemond, C., Hofmann, A. W., Heusser, G., Condomines, M., Raczek, I. and Rhodes, J. M. (1994). U–Th–Ra systematics in Kilauea and Mauna Loa basalts, Hawaii. *Chem. Geol.* **116**, 163–80.

Henderson, G. M. and O'Nions, R. K. (1995). ^{234}U/^{238}U ratios in Quaternary planktonic foraminifera. *Geochim. Cosmochim. Acta* **59**, 4685–94.

Henderson, G. M., Martel, D. J., O'Nions, R. K. and Shackleton, N. J. (1994). Evolution of seawater ^{87}Sr/^{86}Sr over the last 400 ka: the absence of glacial/interglacial cycles. *Earth Planet. Sci. Lett.* **128**, 643–51.

Hergt, J. M., Peate, D. W. and Hawkesworth, C. J. (1991). The petrogenesis of Mesozoic Gondwana low-Ti flood basalts. *Earth Planet. Sci. Lett.* **105**, 134–48.

Hilton, D. R., Hammerschmidt, K., Teufel, S. and Friedrichsen, H. (1993b), Helium isotope characteristics of Andean geothermal fluids and lavas. *Earth Planet. Sci. Lett.* **120**, 265–82.

Hilton, D. R., Barling, J. and Wheller, G. E. (1995). Effect of shallow-level contamination on the isotope systematics of ocean-island lavas. *Nature* **373**, 330–3.

Hilton, D. R., Hammerschmidt, K., Loock, G. and Friedrichsen, H. (1993a). Helium and argon isotope systematics of the central Lau Basin and Valu Fa Ridge: evidence of crust/mantle interactions in a back-arc basin. *Geochim. Cosmochim. Acta* **57**, 2819–41.

Hirschmann, M. M. and Stolper, E. M. (1996). A possible role for garnet pyroxenite in the origin of the 'garnet signature' in MORB. *Contrib. Mineral. Petrol.* **124**, 185–208.

Hodges, K. V., Hames, W. E. and Bowring, S. A. (1994). $^{40}Ar/^{39}Ar$ age gradients in micas from a high-temperature–low-pressure metamorphic terrane: evidence for very slow cooling and implications for the interpretation of age spectra. *Geology* **22**, 55–8.

Hoke, L., Hilton, D. R., Lamb, S. H., Hammerschmidt, K. and Friedrichsen, H. (1994). ^{3}He evidence for a wide zone of active mantle melting beneath the Central Andes. *Earth Planet. Sci. Lett.* **128**, 341–55.

Ireland, T. R. (1991). The abundance of ^{182}Hf in the early solar system. *Lunar Planet. Sci. Conf.* **XXII**, 609–10 (abstract).

Iwamori, H. (1994). $^{238}U–^{230}Th–^{226}Ra$ and $^{235}U–^{231}Pa$ disequilibria produced by mantle melting with porous and channel flows. *Earth Planet. Sci. Lett.* **125**, 1–16.

Jacobsen, S. B. and Harper, C. L. (1996). Accretion and early differentiation history of the Earth based on extinct radionuclides. In: Hart, S. and Basu, A. (Eds), *Earth Processes: Reading the Isotopic Code. Geophys. Monograph* **95**, Amer. Geophys. Union, pp. 47–74.

Jahn, B.-M. and Cuvellier, H. (1994). Pb–Pb and U–Pb geochronology of carbonate rocks: an assessment. *Chem. Geol. (Isot. Geosci. Sect.)* **115**, 125–51.

Jones, C. E., Halliday, A. N. and Lohmann, K. C. (1995). The impact of diagenesis on high-precision U–Pb dating of ancient carbonates: an example from the Late Permian of New Mexico. *Earth Planet. Sci. Lett.* **134**, 409–23.

Kelley, S. P. and Turner, G. (1991). Laser probe $^{40}Ar–^{39}Ar$ measurements of loss profiles within individual hornblende grains from the Giants Range granite, northern Minnesota, USA. *Earth Planet. Sci. Lett.* **107**, 634–48.

Kelley, S. P., Arnaud, N. O. and Turner, S. P. (1994). High spatial resolution $^{40}Ar/^{39}Ar$ investigations using an ultraviolet laser probe extraction technique. *Geochim. Cosmochim. Acta* **58**, 3519–25.

Kellogg, L. H. and Wasserburg, G. J. (1990). The role of plumes in mantle helium fluxes. *Earth Planet. Sci. Lett.* **99**, 276–89.

Kerr, A. C., Saunders, A. D., Tarney, J., Berry, N. H. and Hards, V. L. (1995). Depleted mantle-plume geochemical signatures: no paradox for plume theories. *Geology* **23**, 843–6.

Ku, T. L., Wang, L., Luo, S. and Southon, J. R. (1995). ^{26}Al in seawater and $^{26}Al/^{10}Be$ as paleo-flux tracer. *Geophys. Res. Lett.* **22**, 2163–6.

Ku, T. L., Kusakabe, M., Measures, C. I., Southon, J. R., Cusimano, G., Vogel, J. S., Nelson, D. E. and Nakaya, S. (1990). Beryllium isotope distribution in the western North Atlantic: a comparison to the Pacific. *Deep-Sea Res.* **37**, 795–808.

Lahaye, Y., Arndt, N., Byerly, G., Chauvel, C., Fourcade, S. and Gruau, G. (1995). The influence of alteration on the trace-element and Nd isotopic compositions of komatiites. *Chem. Geol.* **126**, 43–64.

Lambert, D. D., Walker, R. J., Morgan, J. W., Shirey, S. B., Carlson, R. W., Zientek, M. L., Lipin, B. R., Koski, M. S. and Cooper, R. L. (1994). Re–Os and Sm–Nd isotope geochemistry of the Stillwater Complex, Montana: implications for the petrogenesis of the J-M Reef. *J. Petrol.* **35**, 1717–53.

Lao, Y., Anderson, R. F., Broecker, W. S., Trumbore, S. E., Hofmann, H. J. and Wolfli, W. (1992). Increased production of cosmogenic ^{10}Be during the Last Glacial Maximum. *Nature* **357**, 576–8.

Laslett, G. M., Galbraith, R. F. and Green, P. F. (1994). The analysis of projected fission track lengths. *Rad. Meas.* **23**, 103–23.

Lee, J. (1995a). Rapid uplift and rotation of mylonitic rocks from beneath a detachment fault: insights from potassium feldspar $^{40}Ar/^{39}Ar$ thermochronology, northern Snake range, Nevada. *Tectonics* **14**, 54–77.

Lee, J. K. W. (1995b). Multipath diffusion in geochronology. *Contrib. Mineral. Petrol.* **120**, 60–82.

Lee, J. K. W. and Aldama, A. A. (1992). Multipath diffusion: a general numerical model. *Comput. Geosci.* **18**, 531–55.

Lee, D.-C. and Halliday, A. N. (1995). Hafnium–tungsten chronometry and the timing of terrestrial core formation. *Nature* **378**, 771–4.

Lee, D.-C. and Halliday, A. N. (1996). Hf–W isotopic evidence for rapid accretion and differentiation in the early solar system. *Science* **274**, 1876–9.

Liu, B., Phillips, F. M., Fabryka-Martin, J. T., Fowler, M. M. and Stone, W. D. (1994). Cosmogenic ^{36}Cl accumulation in unstable landforms 1. Effects of the thermal neutron distribution. *Water Resour. Res.* **30**, 3115–25.

Lovera, O. M. (1992). Computer programs to model $^{40}Ar/^{39}Ar$ diffusion data from multidomain samples. *Comput. Geosci.* **18**, 789–813.

Lovera, O. M., Harrison, T. M. and Grove, M. (1996). Comment on "Multipath Ar transport in K-feldspar deduced from isothermal heating experiments" by Igor Villa. *Earth Planet. Sci. Lett.* **140**, 281–3.

Lovera, O. M., Heizler, M. T. and Harrison, T. M. (1993). Argon diffusion domains in K-feldspar II: kinetic properties of MH-10. *Contrib. Mineral. Petrol.* **113**, 381–93.

Ludwig, K. R. and Titterington, D. M. (1994). Calculation of $^{230}Th/U$ isochrons, ages, and errors. *Geochim. Cosmochim. Acta* **58**, 5031–42.

Lundstrom, C. C., Gill, J. B., Williams, Q. and Perfit, M. R. (1995). Mantle melting and basalt extraction by equilibrium porous flow. *Science* **270**, 1958–61.

Lundstrom, C. C., Shaw, H. F., Ryerson, F. J., Phinney, D. L., Gill, J. B. and Williams, Q. (1994). Compositional controls on the partitioning of U, Th, Ba, Pb, Sr and Zr between clinopyroxene and haplobasaltic melts: implications for uranium series disequilibria in basalts. *Earth Planet. Sci. Lett.* **128**, 407–23.

Mahoney, J. J., White, W. M., Upton, B. G. J., Neal, C. R. and Scrutton, R. A. (1996). Beyond EM-1: lavas from Afanasy-Nikitin Rise and the Crozet Archipelago, Indian Ocean. *Geology* **24**, 615–18.

Manga, M. (1996). Mixing of heterogeneities in the mantle: effect of viscosity differences. *Geophys. Res. Lett.* **23**, 403–6.

Marcantonio, F., Zindler, A., Elliot, T. and Staudigel, H. (1995). Os isotope systematics of La Palma, Canary Islands: evidence for recycled crust in the mantle source of HIMU ocean islands. *Earth Planet. Sci. Lett.* **133**, 397–410.

Marcantonio, F., Kumar, N., Stute, M., Anderson, R. F., Seidl, M. A., Schlosser, P. and Mix, A. (1995). A comparative study of accumulation rates derived by He and Th isotope analysis of marine sediments. *Earth Planet. Sci. Lett.* **133**, 549–55.

Mazaud, A., Laj, C. and Bender, M. (1994). A geomagnetic chronology for Antarctic ice accumulation. *Geophys. Res. Lett.* **21**, 337–40.

McDougall, I. and Harrison, T. M. (1988). *Geochronology and Thermochronology by the $^{40}Ar/^{39}Ar$ Method*. Oxford Univ. Press, 212 p.

McKenzie, D. and Bickle, M. J. (1988). The volume and composition of melt generated by extension of the lithosphere. *J. Petrol.* **29**, 625–79.

Meisel, T., Walker, R. J. and Morgan, J. W. (1996). The osmium isotopic composition of the Earth's primitive upper mantle. *Nature* **383**, 517–20.

Meshik, A. P., Shukolyukov, Y. A. and JeBberger, E. K. (1995). Chemically fractionated fission xenon (CCF-Xe) on the Earth and in meteorites. In: Busso, M. *et al.* (Eds), *Nuclei in the Cosmos III, Amer. Inst. Phys. Conf. Proc.* **327**, 603–6.

Mezger, K., Essene, E. J. and Halliday, A. N. (1992). Closure temperatures of the Sm–Nd system in metamorphic garnets. *Earth Planet. Sci. Lett.* **113**, 397–409.

Mezger, K., Hanson, G. N. and Bohlen, S. R. (1989). U–Pb systematics in garnet: dating the growth of garnet in the Late Archean Pikwitonei granulite domain at Cauchon and Natawahunan Lakes, Manitoba, Canada. *Contrib. Mineral. Petrol.* **101**, 136–48.

Mezger, K., Rawnsley, C. M., Bohlen, S. R. and Hanson, G. N. (1991). U–Pb garnet, sphene, monazite, and rutile ages: implications for the duration of high-grade metamorphism and cooling histories, Adirondack Mts., New York. *J. Geol.* **99**, 415–28.

Moorbath, S. and Whitehouse, M. J. (1996). Sm–Nd isotopic data and Earth's evolution. *Science* **273**, 1878.

Muller, R. A. and Macdonald, G. J. (1995). Glacial cycles and orbital inclination. *Nature* **377**, 107–8.

Norman, E. B. and Schramm, D. N. (1993). ^{182}Hf chronometer for the early Solar System. *Nature* **304**, 515–17.

Normile, D. (1996). Flood of new isotopes offers keys to stellar evolution. *Science* **273**, 433.

O'Nions, R. K. and Tolstikhin, I. N. (1994). Behaviour and residence times of lithophile and rare gas tracers in the upper mantle. *Earth Planet. Sci. Lett.* **124**, 131–8.

O'Sullivan, P. B. and Parrish, R. R. (1995). The importance of apatite composition and single-grain ages when interpreting fission track data from plutonic rocks: a case study from the Coast Ranges, British Columbia. *Earth Planet. Sci. Lett.* **132**, 213–24.

Onstott, T. C., Miller, M. L., Ewing, R. C., Arnold, G. W. and Walsh, D. S. (1995). Recoil refinements: implications for the $^{40}Ar/^{39}Ar$ dating technique. *Geochim. Cosmochim. Acta* **59**, 1821–34.

Onstott, T. C., Phillips, D. and Pringle-Goodell, L. (1991). Laser microprobe measurement of chlorine and argon zonation in biotite. *Chem. Geol.* **90**, 145–68.

Pearson, D. G., Shirey, S. B., Carlson, R. W., Boyd, F. R., Pokhilenko, N. P. and Shimizu, N. (1995a). Re–Os, Sm–Nd, and Rb–Sr isotope evidence for thick Archean lithospheric mantle beneath the Siberian craton modified by multistage metasomatism. *Geochim. Cosmochim. Acta* **59**, 959–77.

Pearson, D. G., Snyder, G. A., Shirey, S. B., Taylor, L. A., Carlson, R. W. and Sobolev, N. V. (1995b). Archean Re–Os age for Siberian eclogites and constraints on Archean tectonics. *Nature* **374**, 711–13.

Pegram, W. J., Esser, B. K., Krishnaswami, S. and Turekian, K. K. (1994). The isotopic composition of leachable osmium from river sediments. *Earth Planet. Sci. Lett.* **128**, 591–9.

Pepin, R. O. (1992). Origin of noble gases in the terrestrial planets. *Ann. Rev. Earth. Planet. Sci.* **20**, 389–430.

Peucker-Ehrenbrink, B. and Ravizza, G. (1996). Continental runoff of osmium into the Baltic Sea. *Geology* **24**, 327–30.

Peucker-Ehrenbrink, B., Ravizza, G. and Hofmann, A. W. (1995). The marine $^{187}Os/^{186}Os$ record of the past 80 million years. *Earth Planet. Sci. Lett.* **130**, 155–67.

Phillips, F. M., Leavy, B. D., Jannik, N. O., Elmore, D. and Kubik, P. W. (1986). The accumulation of cosmogenic chlorine-36 in rocks: a method for surface exposure dating. *Science* **231**, 41–3.

Phillips, D. and Onstott, T. C. (1988). Argon isotopic zoning in mantle phlogopite. *Geology* **16**, 542–6.

Porcelli, D. and Wasserburg, G. J. (1995a). Mass transfer of xenon through a steady-state upper mantle. *Geochim. Cosmochim. Acta* **59**, 1991–2007.

Porcelli, D. and Wasserburg, G. J. (1995b). Mass transfer of helium, neon, argon, and xenon through a steady-state upper mantle. *Geochim. Cosmochim. Acta* **59**, 4921–37.

Raisbeck, G. M., Yiou, F. and Zhou, S. Z. (1994). Palaeointensity puzzle. *Nature* **371**, 207–8.

Raisbeck, G. M., Yiou, F., Bourles, D., Lorius, C., Jouzel, J. and Barkov, N. I. (1987). Evidence for two intervals of enhanced ^{10}Be deposition in Antarctic ice during the last glacial period. *Nature* **326**, 273–7.

Ravizza, G. (1993). Variations of the $^{187}Os/^{186}Os$ ratio of seawater over the past 28 million years as inferred from metalliferous carbonates. *Earth Planet. Sci. Lett.* **118**, 335–48.

Reid, M. R. (1995). Processes of mantle enrichment and magmatic differentiation in the eastern Snake River Plain: Th isotope evidence. *Earth Planet. Sci. Lett.* **131**, 239–54.

Reisberg, L. and Lorand, J-P. (1995). Longevity of subcontinental mantle lithosphere from osmium isotope systematics in orogenic peridotite massifs. *Nature* **376**, 159–62.

Richardson, C. and McKenzie, D. (1994). Radioactive disequilibria from 2D models of melt generation by plumes and ridges. *Earth Planet. Sci. Lett.* **128**, 425–37.

Richter, F. M., Lovera, O. M., Harrison, T. M. and Copeland, P. (1991). Tibetan tectonics from $^{40}Ar/^{39}Ar$ analysis of a single K-feldspar sample. *Earth Planet. Sci. Lett.* **105**, 266–78.

Robinson, C., Raisbeck, G. M., Yiou, F., Lehman, B. and Laj, C. (1995). The relationship between ^{10}Be and geomagnetic field strength records in central North Atlantic sediments during the last 80 ka. *Earth Planet. Sci. Lett.* **136**, 551–7.

Roy-Barman, M., Luck, J.-M. and Allegre, C. J. (1996). Os isotopes in orogenic lherzolite massifs and mantle heterogeneities. *Chem. Geol.* **130**, 55–64.

Roy-Barman, M. and Allegre, C. J. (1994). $^{187}Os/^{186}Os$ ratios of mid-ocean ridge basalts and abyssal peridotites. *Geochim. Cosmochim. Acta* **58**, 5043–54.

Russell, S. S., Srinivasan, G., Huss, G. R., Wasserburg, G. J. and MacPherson, G. J. (1996). Evidence for widespread ^{26}Al in the Solar Nebula and constraints for nebular time scales. *Science* **273**, 757–62.

Salters, V. J. M. (1996). The generation of mid-ocean ridge basalts from the Hf and Nd isotope perspective. *Earth Planet. Sci. Lett.* **141**, 109–23.

Salters, V. J. M. and Zindler, A. (1995). Extreme $^{176}Hf/^{177}Hf$ in the sub-oceanic mantle. *Earth Planet. Sci. Lett.* **129**, 13–30.

Sharma, M., Papanastassiou, D. A., Wasserburg, G. J. and Dymek, R. F. (1966). The issue of the terrestrial record of ^{146}Sm. *Geochim. Cosmochim. Acta* **60**, 2037–47.

Shen, J. J., Papanastassiou, D. A. and Wasserburg, G. J. (1996). Precise Re–Os determinations and systematics of iron meteorites. *Geochim. Cosmochim. Acta* **60**, 2887–2900.

Shepard, M. K., Arvidson, R. E., Caffee, M., Finkel, R. and Harris, L. (1995). Cosmogenic exposure ages of basalt flows: Lunar Crater volcanic field, Nevada. *Geology* **23**, 21–4.

Shirey, S. B. and Walker, R. J. (1995). Carius tube digestion for low-blank Re–Os analyses. *Anal. Chem.* **67**, 2136–41.

Sims, K. W. W., DePaolo, D. J., Murrell, M. T., Baldridge, W. S., Goldstein, S. J. and Clague, D. A. (1995). Mechanisms of magma generation beneath Hawaii and mid-ocean ridges: uranium/thorium and samarium/neodymium isotopic evidence. *Science* **267**, 508–12.

Smoliar, M. I., Walker, R. J. and Morgan, J. W. (1996). Re–Os ages of group IIA, IIIA, IVA, and IVB iron meteorites. *Science* **271**, 1099–102.

Spiegelman, M. and Kenyon, P. (1992). The requirements for chemical disequilibrium during magma migration. *Earth Planet. Sci. Lett.* **109**, 611–20.

Srinivasan, G., Sahijpal, S., Ulyanov, A. A. and Goswami, J. N. (1996). Ion microprobe studies of Efremovka CAIs: II. Potassium isotope composition and ^{41}Ca in the early Solar System. *Geochim. Cosmochim. Acta* **60**, 1823–35.

Srinivasan, G., Ulyanov, A. A. and Goswami, J. N. (1994). ^{41}Ca in the early Solar System. *Astrophys. J. (Lett.)* **431**, L67–70.

Stone, J. O., Allan, G. L., Fifield, L. K. and Cresswell, R. G. (1996). Cosmogenic chlorine-36 from calcium spallation. *Geochim. Cosmochim. Acta* **60**, 679–92.

Stuiver, M., Quay, P. D. and Ostlund, H. G. (1983). Abyssal water carbon-14 distribution and the age of the world oceans. *Science* **219**, 849–51.

Takayanagi, M. and Ozima, M. (1987). Temporal variation of $^3He/^4He$ ratio recorded in deep-sea sediment cores. *J. Geophys. Res.* **92**, 12 531–8.

Thirlwall, M. F., Upton, B. G. J. and Jenkins, C. (1994). Interaction between continental lithosphere and the Iceland plume – Sr–Nd–Pb isotope geochemistry of Tertiary basalts, NE Greenland. *J. Petrol.* **35**, 839–79.

Tilton, G. R. (1988). Age of the solar system. In: Kerridge, J. F. and Matthews, M. S. (Eds), *Meteorites and the Early Solar System*. Univ. Arizona Press, pp. 259–75.

Tolstikhin, I. N. and O'Nions, R. K. (1996). Some comments on isotopic structure of terrestrial xenon. *Chem. Geol.* **129**, 185–99.

Valbracht, P. J., Staudigel, H., Honda, M., McDougall, I. and Davies, G. R. (1996). Isotopic tracing of volcanic source regions from Hawaii: decoupling of gaseous from lithophile magma components. *Earth Planet. Sci. Lett.* **144**, 185–98.

Villa, I. M. (1994). Multipath Ar transport in K-feldspar deduced from isothermal heating experiments. *Earth Planet. Sci. Lett.* **122**, 393–401.

von Blanckenburg, F., O'Nions, R. K., Belshaw, N. S., Gibb, A. and Hein, J. R. (1996). Global distribution of beryllium isotopes in deep ocean water as derived from Fe–Mn crusts. *Earth Planet. Sci. Lett.* **141**, 213–26.

Walker, R. J., Hanski, E., Vuollo, J. and Liipo, J. (1996). The Os isotopic composition of Proterozoic upper mantle: evidence for chondritic upper mantle from the Outokumpu ophiolite, Finland. *Earth Planet. Sci. Lett.* **141**, 161–73.

Walker, R. J., Morgan, J. W. and Horan, M. F. (1995). Osmium-187 enrichment in some plumes: evidence for core-mantle interaction? *Science* **269**, 819–22.

Wang, L., Ku, T. L., Luo, S., Southon, J. R. and Kusakabe, M. (1996). ^{26}Al–^{10}Be systematics in deep-sea sediments. *Geochim. Cosmochim. Acta* **60**, 109–19.

Wartho, J-A. (1995). Apparent argon diffusive loss $^{40}Ar/^{39}Ar$ age spectra in amphiboles. *Earth Planet. Sci. Lett.* **134**, 393–407.

Wasserburg, G. J., Busso, M. and Gallino, R. (1996). Abundances of actinides and short-lived non-actinides in the interstellar medium: diverse supernova sources for the r-process. *Astrophys. J. (Lett.)* **431**, L109–13.

Widom, E. and Shirey, S. B. (1996). Os isotope systematics in the Azores: implications for mantle plume sources. *Earth Planet. Sci. Lett.* **142**, 451–65.

Yu, E.-F., Francois, R. and Bacon, M. P. (1996). Similar rates of modern and last-glacial ocean thermohaline circulation inferred from radiochemical data. *Nature* **379**, 689–94.

Zreda, M. G., Phillips, F. M., Elmore, D., Kubik, P. W., Sharma, P. and Dorn, R. I. (1991). Cosmogenic chlorine-36 production rates in terrestrial rocks. *Earth Planet. Sci. Lett.* **105**, 94–109.

Zreda, M. G. and Phillips, F. M. (1995). Surface exposure dating by cosmogenic chlorine-36 accumulation. In: Beck, C. (Ed.), *Dating in Exposed and Surface Contexts*. Univ. New Mexico Press, pp. 161–83.

Zreda, M. G., Phillips, F. M. and Elmore, D. (1994). Cosmogenic ^{36}Cl accumulation in unstable landforms 2. Simulations and measurements on eroding moraines. *Water Resour. Res.* **30**, 3127–36.

INDEX